Gray's Anatomy

Gray's Anatomy

Anatomy Descriptive and Surgical

by

Henry Gray, F.R.S.

Lecturer on Anatomy at St. George's Hospital

The Drawings

By H. V. Carter, M.D.

Late Demonstrator of Anatomy at St. George's Hospital

The Dissections

Jointly by the Author and Dr. Carter

Introduction by Professor R. Shane Tubbs, MS, PA-C, PhD

Canterbury Classics
San Diego

Canterbury Classics
An imprint of Printers Row Publishing Group
9717 Pacific Heights Blvd, San Diego, CA 92121
www.canterburyclassicsbooks.com • mail@canterburyclassicsbooks.com

Introduction copyright © 2024 Printers Row Publishing Group

All rights reserved. No part of this publication may be reproduced, distributed, or transmitted in any form or by any means, including photocopying, recording, or other electronic or mechanical methods, without the prior written permission of the publisher, except in the case of brief quotations embodied in critical reviews and certain other noncommercial uses permitted by copyright law.

Printers Row Publishing Group is a division of Readerlink Distribution Services, LLC.
Canterbury Classics is a registered trademark of Readerlink Distribution Services, LLC.

Correspondence regarding the content of this book should be sent to:
Canterbury Classics, Editorial Department, at the above address.

Publisher: Peter Norton • Associate Publisher: Ana Parker
Art Director: Charles McStravick
Editorial Director: April Graham
Editor: Dan Mansfield
Production Team: Beno Chan, Julie Greene

Cover design: Linda Lee Mauri

Library of Congress Control Number: 2023948415

ISBN: 978-1-6672-0473-4

Printed in Dongguan, China
First printing, June 2024. RRD/06/24
28 27 26 25 24 1 2 3 4 5

Editor's Note: This work has been published in its original form to preserve the author's intent and style. Illustrations are preserved in their original printed form, so some notations may be slightly illegible.

To
Sir Benjamin Collins Brodie, Bart.,
F.R.S., D.C.L.,

Serjeant-Surgeon to the Queen,
Corresponding Member of the Institute of France,

This work is dedicated,

in admiration of his great talents,

and

in remembrance of many acts of kindness

shown to the author,

from an early period of his professional career.

CONTENTS

Introduction xxxix
Preface xlvii

OSTEOLOGY

General Properties of Bone 1
Chemical Composition of Bone 1
Structure of Bone 2
Form of Bones 2
Vessels of Bone 3
Development of Bone 4
Growth of Bone 4
The Skeleton 5

The Spine
General Characters of the Vertebrae 6
Characters of the Cervical Vertebrae.... 6
    Atlas 7
    Axis 8
Vertebra Prominens 9
Characters of the Dorsal Vertebrae 9
Peculiar Dorsal Vertebrae 10
Characters of the Lumbar Vertebrae ... 11
Structure of the Vertebrae 12
Development of the Vertebrae 12
    Atlas 13
    Axis 13
    7th Cervical 13
    Lumbar Vertebrae 13
Progress of Ossification in the Spine ... 13
False Vertebrae 13
The Sacrum 14
The Coccyx 18
Development of the Coccyx 18
Of the Spine in general 19

The Skull
Bones of the Cranium 21
    Occipital Bone 21
    Parietal Bones 25
    Frontal Bone 27
    Temporal Bones 30
    Sphenoid Bone 36
    Sphenoidal Spongy Bones 40
    Ethmoid Bone 41
    Wormian Bones 43
Bones of the Face 43
    Nasal Bones 44

Superior Maxillary Bone 44
Lachrymal Bones 49
Malar Bones 50
Palate Bones 52
Inferior Turbinated Bones 54
Vomer 56
Inferior Maxillary Bone 56
Articulations of the Cranial Bones 60
Sutures of the Skull 61
Vertex of the Skull 62
Base of the Skull, Internal Surface 62
    Anterior Fossa 62
    Middle Fossa 62
    Posterior Fossa 64
Base of Skull, External Surface 65
Lateral Regions of the Skull 68
Temporal Fossae 69
Zygomatic Fossae 69
Spheno-maxillary Fossae 70
Anterior Region of Skull 70
Orbits 72
Nasal Fossae 73
Os Hyoides 76

The Thorax
The Sternum 77
Development of the Sternum 79
The Ribs 80
Peculiar Ribs 83
Costal Cartilages 84

The Pelvis
Os Innominatum 85
    Ilium 86
    Ischium 89
    Pubes 90
Development of the Os Innominatum . 92
Boundaries of Pelvis 93
Position of Pelvis 94
Axes of Pelvis 94
Differences between the Male and
    Female Pelvis 94

The Upper Extremities

The Clavicle 95
The Scapula 97
Development of the Scapula 102
The Humerus 102
Development of the Humerus 107
The Ulna 107
The Radius 112
The Hand 115
The Carpus 115
    Bones of Upper Row 115
    Bones of Lower Row 117
The Metacarpus 121
Peculiar Metacarpal Bones 121
Phalanges 122
Development of the Hand 123

The Lower Extremities

The Femur 124
Development of the Femur 129

The Leg

The Leg 130
Patella 130
Tibia 131
Development of Tibia 135
Fibula 135
Development of Fibula 137
The Foot 138
Tarsus 138
Os Calcis 138
Cuboid 140
Astragalus 141
Scaphoid 143
Internal Cuneiform 143
Middle Cuneiform 144
External Cuneiform 145
Metatarsal Bones 145
Peculiar Metatarsal Bones 146
Phalanges 147
Development of the Foot 147
Sesamoid Bones 148

THE ARTICULATIONS

General Anatomy of the Joints 150
    Cartilage 150
    Fibro-cartilage 151
    Ligament 151
    Synovial Membrane 151
Forms of Articulation 152
    Synarthrosis 152
    Amphiarthrosis 153
    Diarthrosis 154
Movements of Joints 156
    Gliding Movement 156
    Angular Movement 156
    Circumduction 156
    Rotation 156

Articulations of the Trunk

Articulations of the Vertebral Column . 156
    Atlas with the Axis 160
    Atlas with the Occipital Bone 163
    Axis with the Occipital Bone 164
Temporo-maxillary Articulation 165
Articulation of the Ribs with
    the Vertebrae 166
    Costo-vertebral 167
    Costo-transverse 168
Costo-sternal Articulations 169

Costo-xiphoid Ligaments 171
Intercostal Articulations 171
Ligaments of the Sternum 171
Articulation of the Pelvis with
    the Spine 171
    Sacrum and Ilium 172
    Sacrum and Ischium 174
    Sacrum and Coccyx 174
Inter-pubic 175

Articulations of the Upper Extremity

Sterno-clavicular 177
Scapulo-clavicular 178
Ligaments of the Scapula 180
Shoulder-joint 180
Elbow-joint 182
Radio-ulnar Articulation 184
Wrist-joint 187
Articulations of the Carpus 188
Carpo-metacarpal Articulations 190
Metacarpo-phalangeal Articulations 192
Articulation of the Phalanges 192

Articulations of the Lower Extremity

Hip-joint 193
Knee-joint 195

Articulations between the Tibia and Fibula	200
Ankle-joint	201
Articulations of the Tarsus	203
Tarso-metatarsal Articulations	207
Articulations of the Metatarsus	208
Metatarso-phalangeal Articulation	208
Articulations of the Phalanges	208

MUSCLES AND FASCIAE

General Anatomy of Muscles	209
of Fasciae	210

Muscles and Fasciae of the Head and Face

Subdivision into Groups	212

Epicranial Region

Dissection	213
Fascia of Head, Occipito-frontalis	214

Auricular Region

Dissection	215
Attollens Aurem, Attrahens Aurem	215
Retrahens Aurem, Actions	216

Palpebral Region

Dissection	216
Orbicularis Palpebrarum	216
Corrugator Supercilii	217
Tensor Tarsi, Actions	218

Orbital Region

Dissection	218
Levator Palpebrae	218
Rectus Superior, Inferior and External Recti	219
Superior Oblique	219
Inferior Oblique	219
Actions, Surgical Anatomy of	220

Nasal Region

Pyramidalis Nasi	221
Levator Labii Superioris	221
Alaeque Nasi	221
Dilator Naris, Anterior and Posterior	221
Compressor Nasi	221
Narium Minor	221
Depressor Alae Nasi	221
Actions	221

Superior Maxillary Region

Levator Labii Superioris Proprius	222
Levator Anguli Oris	222
Zygomatici, Actions	222

Inferior Maxillary Region

Dissection	223
Levator Labii Inferioris	223
Depressor Labii Inferioris	223
Depressor Anguli Oris	223

Intermaxillary Region

Dissection	223
Orbicularis Oris	224
Buccinator	224
Risorius	225
Actions	225

Temporo-Maxillary Region

Masseter	225
Temporal Fascia	226
Dissection of Temporal Muscle	226
Temporal	226

Pterygo-Maxillary Region

Dissection	227
Internal Pterygoid	227
External Pterygoid	227
Actions	228

Muscles and Fasciae of the Neck

Subdivision into Groups	228

Superficial Region

Dissection	229
Superficial Cervical Fascia	229
Platysma Myoides	229
Deep Cervical Fascia	230
Sterno-cleido-mastoid	231
Boundaries of the Triangles of the Neck	232
Actions	232

Infra-Hyoid Region

Dissection	233

Sterno-hyoid 233
Sterno-thyroid, Thyro-hyoid 234
Omo-hyoid 234
Actions 235

Supra-Hyoid Region
Dissection 235
Digastric 235
Stylo-hyoid, Mylo-hyoid 236
Genio-hyoid 236
Actions 237

Lingual Region
Dissection 237
Genio-hyo-glossus 237
Hyo-glossus, Lingualis 238
Stylo-glossus, Palato-glossus 239
Actions 239

Pharyngeal Region
Dissection 239
Inferior Constrictor 240
Middle Constrictor 240
Superior Constrictor 241
Stylo-pharyngeus, Actions 241

Palatal Region
Dissection 242
Levator Palati 242
Tensor Palati, Azygos Uvulae 243
Palato-glossus 243
Palato-pharyngeus 243
Actions, Surgical Anatomy 244

Vertebral Region (Anterior)
Rectus Capitis Anticus Major 245
Rectus Capitis Anticus Minor 245
Rectus Lateralis 246
Longus Colli 246

Vertebral Region (Lateral)
Scalenus Anticus 246
Scalenus Medius 247
Scalenus Posticus, Actions 247

MUSCLES AND FASCIAE OF THE TRUNK
Subdivision into Groups 247

Muscles of the Back
Subdivision into Layers 247

First Layer
Dissection 248
Trapezius 248
Ligamentum Nuchae 250
Latissimus Dorsi 250

Second Layer
Dissection 251
Levator Anguli Scapulae 251
Rhomboideus Minor 251
Rhomboideus Major, Actions 252

Third Layer
Dissection 253
Serratus Posticus Superior and Inferior ... 253
Vertebral Aponeurosis 253
Splenius Capitis and Colli 254
Actions 254

Fourth Layer
Dissection 254
Erector Spinae 254
Sacro-lumbalis 256
Musculus Accessorius ad
    Sacro-lumbalem 256
Cervicalis Ascendens 256
Longissimus Dorsi 256
Transversalis Colli 256
Trachelo-mastoid 257
Spinalis Dorsi, Spinalis Cervicis 257
Complexus 257
Biventer Cervicis 257

Fifth Layer
Dissection 258
Semi-spinalis Dorsi and Colli 258
Multifidus Spinae 258
Rotatores Spinae 259
Supra Spinales 259
Inter-spinales 259
Extensor Coccygis, Inter-transversales .. 259
Rectus Posticus Major and Minor 260
Obliquus Superior and Inferior 260
Actions 260

Muscles of the Abdomen
Dissection 261
Obliquus Externus 261
Obliquus Internus 264
Transversalis 265

Lumbar Fascia 266
Rectus 267
Pyramidalis, Quadratus Lumborum 268
Linea Alba 268
Lineae Semilunares 269
Lineae Transversae 269
Actions 269

Muscles and Fasciae of the Thorax
Intercostal Fasciae 269
Intercostales Interni et Externi 270
Infracostales, Triangularis Sterni 270
Levatores Costarum 271
Actions 271

Diaphragmatic Region
Diaphragm 271
Actions 274

Muscles and Fasciae of the Upper Extremity
Subdivision into Groups 274
Dissection of Pectoral Region
  and Axilla 275
Fasciae of the Thorax 275

Anterior Thoracic Region
Pectoralis Major 276
Costo-coracoid Membrane 278
Pectoralis Minor 278
Subclavius 278
Actions 279

Lateral Thoracic Region
Serratus Magnus, Actions 280

Acromial Region
Deltoid 281
Action 282

Anterior Scapular Region
Subscapular Aponeurosis 282
Subscapularis, Actions 282

Posterior Scapular Region
Supra-spinous Aponeurosis 283
Supra-spinatus 283
Infra-spinous Aponeurosis 283
Infra-spinatus 283
Teres Minor 284

Teres Major 284
Actions 285

Anterior Humeral Region
Deep Fascia of Arm 285
Coraco-brachialis, Biceps 286
Brachialis Anticus, Actions 287

Posterior Humeral Region
Triceps 287
Subanconeus 288
Actions 289

Muscles of Fore-arm
Deep Fascia of Fore-arm 289

*Anterior Brachial Region,
Superficial Layer*
Pronator Radii Teres 289
Flexor Carpi Radialis 290
Palmaris Longus 290
Flexor Carpi Ulnaris 291
Flexor Digitorum Sublimis 291

Anterior Brachial Region, Deep Layer
Flexor Profundus Digitorum 292
Flexor Longus Pollicis 293
Pronator Quadratus 293
Actions 294

Radial Region
Dissection 294
Supinator Longus 294
Extensor Carpi Radialis Longior 295
Extensor Carpi Radialis Brevior 295

*Posterior Brachial Region,
Superficial Layer*
Extensor Communis Digitorum 296
Extensor Minimi Digiti 297
Extensor Carpi Ulnaris 297
Anconeus 297

Posterior Brachial Region, Deep Layer
Supinator Brevis 297
Extensor Ossis Metacarpi Pollicis 298
Extensor Primi Internodii Pollicis 299
Extensor Secundii Internodii Pollicis ... 299
Extensor Indicis 299
Actions 299

Muscles and Fasciae of the Hand
Dissection 300
Anterior Annular Ligament 300
Posterior Annular Ligament................. 300
Palmar Fascia 300

Muscles of the Hand
Radial Group 301
Ulnar Group 303
Middle Palmar Group 304
Actions 305

Surgical Anatomy of the Muscles of the Upper Extremity
Fractures of the Clavicle 306
   Acromian Process 306
   Coracoid Process 307
   Humerus 307
   Ulna 308
   Olecranon 309
   Radius 309

MUSCLES AND FASCIAE OF THE LOWER EXTREMITY
Subdivision into groups 311

Iliac Region
Dissection 312
Iliac Fascia 312
Psoas Magnus 313
Psoas Parvus 314
Iliacus 314
Actions 314

Anterior Femoral Region
Dissection 315
Fasciae of the Thigh, Superficial Fascia .. 315
Deep Fascia (Fascia Lata) 315
Saphenous Opening 317
Iliac and Pubic Portions of Fascia Lata .. 317
Tensor Vaginae Femoris, Sartorius 317
Quadriceps Extensor Cruris 318
Rectus Femoris 318
Vastus Exteruus 319
Vastus Internus and Cruraeus 319
Sub-cruraeus 319
Actions 320

Internal Femoral Region
Dissection 320

Gracilis 320
Pectineus, Adductor Longus 321
Adductor Brevis, Adductor Magnus 322
Actions 323

Gluteal Region
Dissection 323
Gluteus Maximus 324
Gluteus Medius 325
Gluteus Minimus 325
Pyriformis, Obturator Internus 326
Gemelli 327
Quadratus Femoris 327
Obturator Externus 328
Actions 328

Posterior Femoral Region
Dissection 328
Biceps, Semi-tendinosus 329
Semi-membranosus 329
Actions 330
Surgical Anatomy of Hamstring Tendons 330

Muscles and Fasciae of Leg
Dissection of Front of Leg 330
Fascia of the Leg 330
Muscles of the Leg 331

Anterior Tibio-Fibular Region
Tibialis Anticus 331
Extensor Proprius Pollicis 332
Extensor Longus Digitorum 332
Peroneus Tertius 332
Actions 333

Posterior Tibio-Fibular Region, Superficial Layer
Dissection 333
Gastrocnemius 333
Soleus, Tendo Achillis 334
Plantaris 335
Actions 335

Posterior Tibio-Fibular Region, Deep Layer
Deep Fascia of Leg 335
Popliteus, Flexor Longus Pollicis 336

Flexor Longus Digitorum,
Tibialis Posticus 337
Actions 338

Fibular Region
Peroneus Longus 338
Peroneus Brevis 339
Actions 339
Surgical Anatomy of Tendons
around Ankle 339

Muscles and Fasciae of Foot
Anterior Annular Ligament 340
Internal Annular Ligament 340
External Annular Ligament 340
Plantar Fascia 340

Muscles of the Foot, Dorsal Region
Extensor Brevis Digitorum 341

Plantar Region
Subdivision into Groups 342
Subdivision into Layers 342
First Layer 342
Second Layer 344
Third Layer 345
Interossei 346

Surgical Anatomy of the Muscles of the Lower Extremity
Fracture of the Neck of the Femur 347
the Femur below Trochanter Minor ... 348
the Femur above the Condyles 348
the Patella 349
the Tibia 349
the Fibula, with Dislocation
of the Tibia 349

OF THE ARTERIES

General Anatomy
Subdivision into Pulmonary
and Systemic 351
Distribution of — Where found 351
Mode of Division — Anastomoses 351
Capillaries — Structure of Arteries 352
Sheath — Vasa Vasorum 353

Aorta
Arch of Aorta 353
Dissection 353
Ascending Part of Arch 354
Transverse Part of Arch 355
Descending Part of Arch 356
Peculiarities, Surgical Anatomy 357
Branches 358
Peculiarities of Branches 358

Coronary Arteries 359

Arteria Innominata
Relations 360
Peculiarities 360
Surgical Anatomy 361

Common Carotid Arteries
Course and Relations 362
Peculiarities 364

Surgical Anatomy 365

External Carotid Artery
Relations 366
Surgical Anatomy 367
Branches 367

Superior Thyroid Artery
Course and Relations 368
Surgical Anatomy 368

Lingual Artery
Course and Relations 368
Branches 369
Surgical Anatomy 369

Facial Artery
Course and Relations 370
Branches 370
Peculiarities 372
Surgical Anatomy 372

Occipital Artery
Course and Relations 373
Branches 373

Posterior Auricular Artery 374
Ascending Pharyngeal Artery 374

Temporal Artery
Course and Relations........................... 375
Branches, Surgical Anatomy 375

Internal Maxillary Artery
Course, Relations 376
Peculiarities 377
Branches from First Portion................. 377
    Second Portion.............................. 378
    Third Portion 378

SURGICAL ANATOMY OF THE
TRIANGLES OF THE NECK
Anterior Triangular Space
Inferior Carotid Triangle...................... 380
Superior Carotid Triangle 381
Submaxillary Triangle 381

Posterior Triangular Space
Occipital Triangle................................ 382
Subclavian Triangle.............................. 382

Internal Carotid Artery
Cervical Portion 384
Petrous Portion................................... 384
Cavernous Portion............................... 384
Cerebral Portion.................................. 384
Peculiarities 384
Surgical Anatomy 385
Branches.. 385

Ophthalmic Artery............................ 385

Cerebral Branches of Internal Carotid 385

Subclavian Arteries
First Part of Right Subclavian Artery ... 390
First Part of Left Subclavian Artery...... 391
Second Part of Subclavian Artery......... 391
Third Part of Subclavian Artery 392
Peculiarities 392
Surgical Anatomy 393
Branches.. 395
    Vertebral Artery............................. 395
    Spinal Branches of Vertebral 396
    Basilar Artery 397
    Cerebral Branches of Vertebral 397
    Cerebellar Branches of Vertebral...... 397
    Circle of Willis 398
    Thyroid Axis.................................. 398

Supra-scapular Artery 398
Transversalis Colli............................... 399
Internal Mammary 400
Superior Intercostal 401
Deep Cervical Artery....................... 401

SURGICAL ANATOMY OF THE AXILLA
Axillary Artery
First Portion 403
Second Portion.................................. 404
Third Portion 404
Peculiarities, Surgical Anatomy........... 405
Branches... 406

Brachial Artery
Relations 408
Bend of the Elbow............................. 408
Peculiarities of Brachial Artery............. 409
Surgical Anatomy 410
Branches... 411

Radial Artery
Relations 413
Deep Palmar Arch 413
Peculiarities, Surgical Anatomy........... 414
Branches... 414

Ulnar Artery
Relations 416
Superficial Palmar Arch 417
Peculiarities of Ulnar Artery 417
Surgical Anatomy 418
Branches... 418

DESCENDING AORTA 420

THORACIC AORTA
Relations 421
Surgical Anatomy 421
Branches... 421

ABDOMINAL AORTA
Relations 423
Surgical Anatomy 423
Branches... 425
    Coeliac Axis, Gastric Artery........... 425
    Hepatic Artery 425
    Branches....................................... 426
    Splenic Artery............................... 427
    Superior Mesenteric Artery............ 428

Inferior Mesenteric Artery 431
Supra-renal Arteries 431
Renal Arteries 431
Spermatic Arteries 432
Phrenic Arteries 432
Lumbar Arteries 433
Middle Sacral Artery 433

COMMON ILIAC ARTERIES
Course and Relations 433
Peculiarities, Surgical Anatomy 435

Internal Iliac Artery
Course and Relations 436
Peculiarities, Surgical Anatomy 437
Branches 437
    Vesical Arteries 437
    Haemorrhoidal Arteries 438
    Uterine and Vaginal Arteries 438
    Obturator Artery 438
    Internal Pudic Artery 439
    Sciatic Artery 441
    Gluteal, Ilio-lumbar, and Lateral
       Sacral Arteries 442

External Iliac Artery
Course and Relations 443
Surgical Anatomy 444
Epigastric Artery 444
Circumflex Iliac Artery 445

Femoral Artery
Course and Relations 445
Scarpa's Triangle 446
Peculiarities of Femoral Artery 447

Surgical Anatomy 447
Branches 448
Profunda Artery 449

Popliteal Space 451

Popliteal Artery
Course and Relations 452
Peculiarities, Surgical Anatomy 453
Branches 453

Anterior Tibial Artery
Course and Relations 455
Peculiarities 455
Surgical Anatomy 456
Branches 457

Dorsalis Pedis Artery
Course and Relations 457
Peculiarities, Surgical Anatomy 458
Branches 458

Posterior Tibial Artery
Course and Relations 459
Peculiarities, Surgical Anatomy 459
Branches 460

Peroneal Artery
Course and Relations 461
Peculiarities 461

Plantar Arteries 462

Pulmonary Artery 463

OF THE VEINS

General Anatomy
Subdivision into Pulmonary, Systemic,
   and Portal 465
Anastomoses of Veins 465
Superficial Veins, Deep Veins, or
   Venae Comites 465
Sinuses, their Structure 466
Structure of Veins 466
Coats of Veins 466
Valves of Veins 466
Vessels and Nerves of Veins 467

Veins of the Head and Neck
Facial Vein 468
Temporal Vein 468
Internal Maxillary Vein 469
Temporo-maxillary Vein 469
Posterior Auricular Vein,
   Occipital Vein 469

Veins of the Neck
External Jugular Vein 469
Posterior External Jugular Vein 470
Anterior Jugular Vein 470

Internal Jugular Vein 470
Lingual, Pharyngeal, and Thyroid
    Veins 471
Veins of the Diploë 471

Cerebral Veins
Superficial Cerebral Veins 472
Deep Cerebral Veins 473
Cerebellar Veins 473

Sinuses of the Dura Mater
Superior Longitudinal Sinus 473
Inferior Longitudinal and Straight
    Sinuses 474
Lateral and Occipital Sinuses 475
Cavernous Sinuses 475
Circular, Inferior Petrosal, and
    Transverse Sinuses 476
Superior Petrosal Sinus 476

Veins of the Upper Extremity
Superficial Veins 476
Deep Veins 477
Axillary Vein 478
Subclavian Vein 479
Vertebral Vein 479
Innominate Veins 480
Peculiarities of 480
Internal Mammary Vein 480
Inferior Thyroid Veins 481

Superior Intercostal Veins 481
Superior Vena Cava 481
Azygos Veins 481
Spinal Veins 482

Veins of the Lower Extremity
Internal Saphenous Vein 484
External Saphenous Vein 485
Popliteal Vein 486
Femoral Vein 486
External Iliac Vein 486
Internal Iliac Vein 486
Common Iliac Vein 487
Inferior Vena Cava 487
    Peculiarities 488
Lumbar and Spermatic Veins 488
Ovarian Veins 488
Renal and Supra-renal Veins 489
Phrenic Veins, Hepatic Veins 489

Portal System of Veins
Inferior and Superior Mesenteric
    Veins 489
Splenic and Gastric Veins 490
Portal Vein 491

Cardiac Veins
Coronary Sinus 491

Pulmonary Veins 492

OF THE LYMPHATICS

General Anatomy
Structure of, where found 493
Subdivision into Deep and Superficial . 493
Coats of Lymphatics 493
Valves of Lymphatics 494
Lymphatic or Conglobate Glands 494
Structure of Lymphatic Glands 494
Thoracic Duct 495
Right Lymphatic Duct 496

Lymphatics of Head, Face, and Neck
Superficial Lymphatic Glands
    of Head 496
Lymphatics of Head 496
    of the Face 497
Deep Lymphatics of the Face 497
    of the Cranium 497

Lymphatic Glands of the Neck 498
Superficial Cervical Glands 498
Deep Cervical Glands 498
Superficial and Deep Cervical
    Lymphatics 498

Lymphatics of the Upper Extremity
Superficial Lymphatic Glands 499
Deep Lymphatic Glands 499
Axillary Glands 499
Superficial Lymphatics of Upper
    Extremity 500
Deep Lymphatics of Upper Extremity . 501

Lymphatics of the Lower Extremity
Superficial Inguinal Glands 501
Deep Lymphatic Glands 501

Anterior Tibial Gland 501	Lymphatic Glands of Stomach 505
Deep Popliteal Glands 501	Lymphatics of Stomach 505
Deep Inguinal Glands 501	Lymphatic Glands of Spleen 506
Gluteal and Ischiatic Glands 501	Lymphatics of Spleen 506
Superficial Lymphatics of Lower	
Extremity 502	*Lymphatic System of the Intestines*
Internal Group 502	Lymphatic Glands of Small Intestines
External Group 502	(Mesenteric Glands) 506
Deep Lymphatics of Lower Extremity .. 502	Lymphatic Glands of Large Intestine ... 506
	Lymphatics of Small Intestine
Lymphatics of Pelvis and Abdomen	(Lacteals) 506
Deep Lymphatic Glands of Pelvis 502	of Great Intestine 506
External Iliac Glands 502	
Internal Iliac Glands 502	*Lymphatics of Thorax*
Sacral Glands 502	Deep Lymphatic Glands of Thorax 506
Lumbar Glands 503	Intercostal Glands 506
Lymphatics of Pelvis and Abdomen 504	Internal Mammary Glands 506
Superficial Lymphatics of Wall	Anterior Mediastinal Glands 507
of Abdomen 504	Posterior Mediastinal Glands 507
of Gluteal Region 504	Superficial Lymphatics on Front
of Scrotum and Perinaeum 504	of Thorax 507
of Penis 504	Deep Lymphatics of Thorax 507
of Labia, Nymphae, and Clitoris 504	Intercostal Lymphatics 507
Deep Lymphatics of Pelvis and	Internal Mammary Lymphatics 507
Abdomen 504	Lymphatics of Diaphragm 507
Lymphatics of Bladder 504	Bronchial Glands 507
of Rectum 504	Lymphatics of Lung 507
of Uterus 504	Cardiac Lymphatics 507
of Testicle 505	Thymic Lymphatics 508
Lymphatics of Kidney 505	Thyroid Lymphatics 508
of Liver 505	Lymphatics of Oesophagus 508

NERVOUS SYSTEM

General Anatomy	Sympathetic Nerve 512
Subdivision into Cerebro-spinal Axis,	
Ganglia, and Nerves 509	*The Spinal Cord and its Membranes*
	Dissection 513
Cerebro-Spinal Axis	Membranes of the Cord 513
Grey Substance 509	Dura Mater 513
White Substance 509	Arachnoid 514
Chemical Composition 509	Pia Mater 515
	Ligamentum Denticulatum 515
Ganglia	Spinal Cord 516
Where found, Structure 510	Fissures of Cord 516
	Columns of Cord 517
Nerves	Grey Matter of Cord 517
Subdivision into Afferent, Efferent,	Mode of Arrangement of Grey and
and Excito-motory 510	White Matter 518
Cerebro-spinal Nerves 511	White Matter of Cord 518

The Brain and its Membranes
Membranes of the Brain 518

Dura Mater
Structure 519
Arteries and Veins 519
Nerves 520
Glandulae Pacchioni 520
Processes of the Dura Mater 520
    Falx Cerebri 520
    Tentorium Cerebelli 520
    Falx Cerebelli 520

Arachnoid Membrane
Sub-arachnoid Space 521
Cerebro-spinal Fluid 521

Pia Mater 522

THE BRAIN
Subdivision into Cerebrum, Cerebellum,
    Pons Varolii, Medulla Oblongata 522
Weight of Brain 523

MEDULLA OBLONGATA
Anterior Pyramids 524
Lateral Tract, and Olivary Body 524
Restiform Bodies 524
Posterior Pyramids 524
Posterior Surface of Medulla
    Oblongata 525
Structure of Medulla Oblongata 525
    of Anterior Pyramid 525
    of Lateral Tract 525
    of Olivary Body 525
    of Restiform Body 526
Septum of Medulla Oblongata 526
Grey Matter of Medulla Oblongata 526

PONS VAROLII
Structure 527
Transverse Fibres 527
Longitudinal Fibres 527
Septum 527

CEREBRUM
Upper Surface of Cerebrum 528
Convolutions and Sulci 529
Base of the Brain 530
General Arrangement of the Parts
    composing the Cerebrum 530
Interior of the Cerebrum 534
    Corpus Callosum 534
    Lateral Ventricles 537
Boundaries of, and Parts forming the
    Lateral Ventricles 537
    Septum Lucidum 540
    Fornix 540
    Velum Interpositum 541
    Thalami Optici 541
    Third Ventricle 542
Anterior, Middle, and Posterior
    Commissures 543
Grey Matter of Third Ventricle 543
Pineal Gland 543
Corpora Quadrigemina 544
Valve of Vieussens 544
Corpora Geniculata 544
Structure of Cerebrum 545
    1. Diverging or Peduncular Fibres 545
    2. Transverse Commissural Fibres 545
    3. Longitudinal Commissural Fibres. 545

CEREBELLUM
Its Position, Size, Weight, etc. 545
Cerebellum, Upper Surface 546
    Under Surface 546
Lobes of the Cerebellum 547
Fourth Ventricle 548
Boundaries of Ventricle 548
Lining Membrane 548
Choroid Plexus of 549
Grey Matter of 549
Structure of the Cerebellum 549
    Its Laminae 549
    Corpus Dentatum 549
Peduncles of Cerebellum 549

CRANIAL NERVES
Subdivision into Groups 551
    Nerves of Special Sense 551
    of Motion 551
    Compound Nerves 551

Nerves of Special Sense
Olfactory Nerve 551
Optic Nerve 552
    Tracts 553

Commissure 553
Auditory Nerve 553

Motor Cranial Nerves
Third Nerve (Motor Oculi) 554
Fourth Nerve (Trochlearis) 555
Sixth Nerve (Abducens) 555
Relations of the Orbital Nerves 556
   in the Cavernous Sinus 556
   in the Sphenoidal Fissure 556
   in the Orbit 556
Facial Nerve 557
Branches of Facial Nerve 558
Ninth or Hypo-glossal Nerve 561

Compound Cranial Nerves
Fifth Nerve 562
Gaserian Ganglion 563
Ophthalmic Nerve 563
Lachrymal and Frontal Branches 563
Nasal Branch 564
Ophthalmic Ganglion 565
Superior Maxillary Nerve 566
Spheno-palatine Ganglion 567
Inferior Maxillary Nerve 570
Auriculo-temporal and Gustatory
   Branches 571
Inferior Dental Branch 572
Otic Ganglion 572
Submaxillary Ganglion 573
Eighth Pair 573
   Glosso-pharyngeal 574
   Spinal Accessory 576
   Pneumogastric (Vagus) 576

SPINAL NERVES
Roots of the Spinal Nerves 581
   Origin of Anterior Roots 581
      of Posterior Roots 581
Ganglia of the Spinal Nerves 582
Anterior Branches of the Spinal Nerves .. 582
Posterior Branches of the Spinal Nerves . 582

CERVICAL NERVES
Roots of the Cervical Nerves 582
Anterior Branches of the Cervical
   Nerves 583

CERVICAL PLEXUS
Superficial Branches of the Cervical
   Plexus 584
Deep Branches of the Cervical Plexus .. 585
Posterior Branches of the Cervical
   Nerves 586

BRACHIAL PLEXUS
Branches above the Clavicle
Posterior Thoracic, Supra Scapular 589

Branches below the Clavicle
Anterior Thoracic, and Subscapular
   Nerves 590
Circumflex Nerve 590
Musculo-cutaneous Nerve 591
Internal, and Lesser Internal
   Cutaneous Nerves 592
Median Nerve 592
Ulnar Nerve 595
Musculo-spiral Nerve 596
   Radial Nerve 597
   Posterior Interosseous Nerve 598

DORSAL NERVES
Roots of the Dorsal Nerves 598
Posterior Branches of the Dorsal
   Nerves 598
Intercostal Nerves 599
Upper Intercostal Nerves 599
Intercosto-humeral Nerve 599
Lower Intercostal Nerves 600
Peculiar Dorsal Nerves 600
   First Dorsal Nerve 600
   Last Dorsal Nerve 600

LUMBAR NERVES
Roots of Lumbar Nerves 600
Posterior Branches of Lumbar Nerves .. 601
Anterior Branches of Lumbar Nerves ... 601

LUMBAR PLEXUS
Branches of Lumbar Plexus 601
Ilio-hypogastric Nerve 602
Ilio-inguinal, and Genito-crural
   Nerves 603
External Cutaneous Nerve 603
Obturator Nerve 605
Accessory Obturator Nerve 605

Anterior Crural Nerve 605
Branches of Anterior Crural................. 606
Middle Cutaneous................................ 606
Internal Cutaneous............................... 606
Long Saphenous 607
Muscular and Articular Branches 607

Sacral and Coccygeal Nerves
Roots of, Origin of 608
Posterior Sacral Nerves 608
Anterior Sacral Nerves 608
Posterior Branch of Coccygeal Nerve... 608
Anterior Branch of Coccygeal Nerve.... 609

Sacral Plexus
Superior Gluteal Nerve......................... 610
Pudic Nerve... 610
Small Sciatic Nerve.............................. 612
Great Sciatic Nerve.............................. 612
Internal Popliteal Nerve....................... 613
Posterior Tibial Nerve.......................... 613
Plantar Nerves 613
External Popliteal or Peroneal Nerve.... 615
Anterior Tibial Nerve 615
Musculo-cutaneous Nerve 615

Sympathetic Nerve
Subdivision of, into Parts..................... 617
Branches of the Ganglia, General
 Description of 617

Cervical Portion of the Sympathetic
Superior Cervical Ganglion
Carotid Plexus..................................... 619
Cavernous Plexus 620

Middle Cervical Ganglion.................. 621
Inferior Cervical Ganglion 621

Cardiac Nerves
Superior Nerve 621
Middle and Inferior Cardiac Nerves 622
Deep Cardiac Plexus............................ 622
Superficial Cardiac Plexus.................... 622
Anterior and Posterior Coronary
 Plexus... 623

Thoracic Part of the Sympathetic
Great Splanchnic Nerve....................... 623
Lesser Splanchnic Nerve 623
Smallest Splanchnic Nerve................... 624
Epigastric or Solar Plexus..................... 624
Semilunar Ganglia............................... 624
Phrenic and Supra-renal Plexuses......... 624
Renal Plexus 625
Spermatic, Coeliac, and Gastric
 Plexuses 625
Hepatic, Splenic, and Superior
 Mesenteric Plexuses........................ 625
Aortic, and Inferior Mesenteric
 Plexuses 626

Lumbar Portion of Sympathetic
Pelvic Portion of Sympathetic.............. 626
Hypogastric Plexus 627
Inferior Hypogastric or Pelvic Plexus ... 627
Inferior Haemorrhoidal Plexus 627
Vesical Plexus...................................... 627
Prostatic Plexus................................... 627
Vaginal Plexus..................................... 627
Uterine Nerves 627

ORGANS OF SENSE

Skin
Derma, or True Skin............................ 629
 Corium 630
 Papillary Layer............................. 630
Epidermis or Cuticle............................ 631
Vessels and Nerves of the Skin............. 632

Appendages of the Skin
Nails.. 632
Hairs 633
Sebaceous and Sudoriferous Glands..... 634

Tongue
Papillae of, Structure of Papillae........... 636
Follicles, and Mucous Glands 637
Fibrous Septum of............................... 637
Muscular Fibres of 638
Arteries and Nerves of 638

Nose
Cartilages of, Muscles.......................... 639
Skin, Mucous Membrane 640
Arteries, Veins, and Nerves 640

Nasal Fossae

Mucous Membrane of 640
Peculiarities of, in Superior, Middle,
  and Inferior Meatuses 641
Arteries, Veins, and Nerves of
  Nasal Fossse 641

Eye

Situation, Form of 642
Tunics of, Sclerotic 642
Cornea, Structure of Cornea 644
Choroid, Structure of Choroid 646
Ciliary Processes 647
Iris 648
Membrana Pupillaris 648
Ciliary Ligament 649
Ciliary Muscle 649
Retina 649
Structure of Retina 650
  Jacob's Membrane 650
  Granular Layer 650
  Nervous Layer 650
  Radiating Fibres of the Retina 651
Arteria Centralis Retinae 651
Structure of Retina, at Yellow Spot 651

Humours of the Eye

Aqueous Humour 651
  Anterior Chamber 651
  Posterior Chamber 651
Vitreous Body 652
Crystalline Lens and its Capsule 652
Changes produced in the Lens
  by Age 652
Suspensory Ligament of Lens 653
Canal of Petit 653
Vessels of the Globe of the Eye 653
Arteries, Veins, and Nerves of Eyeball .. 654

Appendages of the Eye

Eyebrows 654
Eyelids 654
Structure of the Eyelids 654
Tarsal Cartilages 655
Meibomian Glands 655
Eyelashes 656
Conjunctiva, and Caruncula
  Lachrymalis 656

Lachrymal Apparatus

Lachrymal Gland 657
  Canals 657
  Sac 658
Nasal Duct 658

Ear

Pinna or Auricle 658
Structure of Auricle 659
Ligaments of the Pinna 659
Muscles of the Pinna 659
Arteries, Veins, and Nerves of the Pinna. 660
Auditory Canal 661

Middle Ear or Tympanum

Eustachian Tube 664
Membrana Tympani 665
  Structure of 665
Ossicles of the Tympanum 665
Ligaments of the Ossicula 666
Muscles of the Tympanum 666
Mucous Membrane of Tympanum 667
Arteries, Veins, and Nerves of
  Tympanum 667

Internal Ear or Labyrinth

Vestibule 668
Semicircular Canals 669
  Superior Semicircular Canal 669
  Posterior Semicircular Canal 670
  External Semicircular Canal 670
Cochlea 670
  Central Axis of, or Modiolus 670
  Spiral Canal of 671
  Lamina Spiralis of 671
Scala Tympani, Scala Vestibuli 671
Membranous Labyrinth 672
Utricle and Sacculus 672
Membranous Semicircular Canals 672
Vessels of the Labyrinth 673
Auditory Nerve, Vestibular Nerve,
  Cochlear Nerve 674

Viscera

Organs of Digestion and Their Appendages

Alimentary Canal 675
Its Subdivisions.................................... 675
The Mouth 675
The Lips 675
The Cheeks 676
The Gums 677

Teeth
General Characters of........................... 677
Permanent Teeth.................................. 678
Incisors, Canine, Bicuspid 678
Molars.. 679
Temporary or Milk Teeth..................... 679
Structure of the Teeth 679
Ivory or Dentine................................... 680
Chemical Composition 681
Enamel, Cortical Substance 681
Development of the Teeth 682
    of the Permanent Teeth 683
Growth of the Teeth 684
Eruption of the Teeth 684

Palate
Hard Palate.. 685
Soft Palate 685
Uvula, Pillars of the Soft Palate........... 686
Mucous Membrane, Aponeurosis,
    and Muscles of Soft Palate 686

Tonsils
Arteries, Veins, and Nerves of Tonsils... 686

Salivary Glands
Parotid Gland
Steno's Duct 688
Vessels and Nerves of Parotid Gland 688

Submaxillary Gland
Wharton's Duct 688
Vessels and Nerves of Submaxillary
    Gland... 688

Sublingual Gland
Vessels and Nerves of 689
Structure of Salivary Glands 689

Pharynx
Structure of 689
Oesophagus.. 690
Relations 690
Surgical Anatomy and Structure of...... 691

Abdomen
Boundaries 692
Apertures of, Regions........................... 692

Peritoneum
Reflections traced 694
Foramen of Winslow 696
Lesser Omentum 696
Great Omentum................................... 696
Gastro-splenic Omentum 697
Mesentery... 697
Meso-caecum, Meso-colon 697

Stomach
Situation 698
Splenic end, Pyloric end 698
Cardiac and Pyloric Orifices................ 698
Greater and Lesser curvatures 699
Surfaces 699
Ligaments of.. 699
Alterations in Position 699
Pylorus 699
Structure of Stomach........................... 700
Serous and Muscular Coats 700
Mucous Membrane 700
Gastric Follicles 701
Vessels and Nerves of Stomach............ 701

Small Intestines
Duodenum... 702
    Ascending portion........................... 702
    Descending portion......................... 702
    Transverse portion 702
Vessels and Nerves of Duodenum 703
Jejunum... 703
Ileum... 703
Structure of Small Intestines................ 703
Serous, Muscular and Cellular Coats ... 703
Mucous Membrane 703

Epithelium and Valvulae
    Conniventes 704
Villi—their Structure 704
Simple Follicles, Duodenal Glands 704
Solitary Glands, Aggregate Glands 705

Large Intestine

Caecum 706
    Appendix Caeci Vermiformis 706
    Ileo-caecal Valve 706
Colon 707
    Ascending 707
    Transverse 707
    Descending 708
    Sigmoid Flexure 708
Rectum 708
    Upper Portion 708
    Middle Portion 708
    Lower Portion 708
Structure of Large Intestine 709
Serous and Muscular Coats 709
Cellular and Mucous Coats 709
Epithelium, Simple Follicles and
    Solitary Glands of Large Intestine 710

Liver

Size, weight, position of 710
Its Surfaces and Borders 710
Changes of Position 711

Ligaments

Longitudinal, Lateral 711
Coronary 712
Round Ligament 712

Fissures

Longitudinal 712
Fissure of Ductus Venosus, Portal
    Fissure 713
Fissures for Gall Bladder and
    Vena Cava 713

Lobes

Right 713
Left 714
Quadratus, Spigelii, Caudatus 714
Vessels of Liver 714
Lymphatics, Nerves 714
Structure of Liver 714
    Serous and Fibrous Coats 715

Lobules 715
Hepatic Cells, Biliary Ducts,
    Portal Vein 716
Hepatic Artery 716
Hepatic Veins 717

Gall Bladder

Structure 717
Biliary Ducts 718
Hepatic, Cystic, and Common
    Choledic Ducts 718
Structure of Biliary Ducts 718

Pancreas

Dissection 719
Relations 719
Duct, Structure 720
Vessels and Nerves 720

Spleen

Relations 721
Size and Weight 721
Structure of Serous and Fibrous
    Coats 721
Proper Substance 722
Malpighian Corpuscles 723
Splenic Artery, distribution 724
Capillaries of Spleen 725
Veins of Spleen 725
Lymphatics and Nerves 725

Thorax

Boundaries of 726
Superior Opening, Base 726
Parts passing through Upper Opening . 726

Pericardium

Structure 726
Fibrous layer, Serous Layer 727

Heart

Position, Size 727
Subdivision into Four Cavities 728
Circulation of Blood in Adult 728
Auriculo-ventricular, and Ventricular
    Grooves 728

Right Auricle

Openings 729
Valves 730

Relics of Foetal Structure 730
Musculi Pectinati 730

Right Ventricle
Openings 731
Tricuspid and Semilunar Valves 731
Chordae Tendineae 731
Columnae Carneae 732

Left Auricle
Sinus 732
Appendix 733
Openings 733
Musculi Pectinati 734

Left Ventricle
Openings 734

Mitral and Semilunar Valves 734
Endocardium 735

Structure of Heart
Fibrous Rings 735
Muscular Structure 735
   of Auricles 735
   of Ventricles 736
Vessels and Nerves of Heart 736
Peculiarities in Vascular System
   of Foetus 737
Foramen Ovale, Eustachian Valve 737
Ductus Arteriosus 737
Umbilical or Hypogastric Arteries 737
Foetal Circulation 738
Changes in Vascular System at Birth 740

ORGANS OF VOICE AND RESPIRATION

The Larynx
Cartilages of the Larynx 741
   Thyroid Cartilage 741
   Cricoid Cartilage 742
   Arytenoid Cartilage 743
   Cartilages of Santorini, and Wrisberg. 743
   Epiglottis. Its structure 743
Ligaments of the Larynx 744
Ligaments connecting the Thyroid
   Cartilage with the Os Hyoides 744
Ligaments connecting the Thyroid
   Cartilage with the Cricoid 744
Ligaments connecting the Arytenoid
   Cartilages to the Cricoid 744
Ligaments of the Epiglottis 745
Interior of the Larynx 745
Cavity of the Larynx 745
Glottis 745
False Vocal Cords 746
True Vocal Cords 746
Ventricle of Larynx, Sacculus Laryngis . 746
Muscles of Larynx 747
   Crico-thyroid 747
   Crico-arytaenoideus posticus 747
                      lateralis 747
   Thyro-arytaenoideus 747
   Muscles of the Epiglottis 748
   Thyro-epiglottideus 748
   Arytaeno-epiglottideus, superior 748
                           inferior 748

Actions of Muscles of Larynx 748
Mucous Membrane of Larynx 749
Glands, Vessels and Nerves of 749

Trachea
Relations 750
Bronchi 751
Structure of Trachea 751
Surgical Anatomy of
   Laryngo-tracheal Region 752

The Pleurae
Reflections 754
Vessels and Nerves 755

Mediastinum
Anterior Mediastinum 755
Middle Mediastinum 756
Posterior Mediastinum 756

The Lungs
Surfaces, Lobes 756
Root of Lung 757
Weight, Colour, and Properties of
   Substance of Lung 758
Structure of Lung
Serous Coat, and Subserous
   Areolar Tissue 758
Parenchyma and Lobules of Lung 758

Bronchi, arrangement of Branches
  in Substance of Lung 758
Structure of smaller Bronchial Tubes ... 759
The Air Cells 759
Pulmonary Artery 759
Pulmonary Capillaries and Veins 759
Bronchial Arteries and Veins 759
Lymphatics and Nerves of Lung 760

Thyroid Gland
Structure 761
Vessels and Nerves 761
Chemical Composition 761

Thymus Gland
Structure 762
Vessels and Nerves 762
Chemical Composition 762

THE URINARY ORGANS

Kidneys
Relations 763
Dimensions, Weight 763
Cortical Substance 764
Medullary Substance 764
Minute Structure 764
Malpighian Bodies 765
Ureter, Pelvis, Infundibula 765
Renal Artery, Renal Veins 766
Lymphatics and Nerves 766

Ureters
Situation, Course, Relations 766
Structure 766

Supra-Renal Capsules
Relations 767

Structure 767
Vessels and Nerves 768

Pelvis
Boundaries and Contents 768

Bladder
Shape, Position, Relations 768
  Subdivisions 769
  Ligaments 770
  Structure 771
Interior of Bladder 771
Vessels and Nerves 772

Male Urethra
Structure 774

MALE GENERATIVE ORGANS

Prostate Gland 775
  Structure 775
  Vessels and Nerves 776
Prostatic Secretion 776
Cowper's Glands 776

Penis
Root 776
Glans Penis 776
Body 777
Corpora Cavernosa 777
  Structure 777
Corpus Spongiosum 778
The Bulb 778
Structure of Corpus Spongiosum 778
Erectile Tissue 778

Arteries of the Penis 779
Lymphatics of the Penis 779
Nerves of the Penis 779

The Testes and their Coverings
Scrotum 779
Other Coverings of the Testis 780
Vessels and Nerves of the Coverings
  of the Testis 780

Spermatic Cord
Its Composition 780
Relations of in Inguinal Canal 781
Arteries of the Cord 781
Veins of the Cord 781
Lymphatics and Nerves of the Cord 781

Testes
- Form and Situation 781
- Size and Weight 782
- Coverings 782
  - Tunica Vaginalis 782
  - Tunica Albuginea 782
  - Mediastinum Testis 783
  - Tunica Vasculosa 783
- Structure of Testis 783
- Lobules of the Testis 783
- Number, Size, Shape, Position 783
- Structure of the Lobuli Testes 783
- Tubuli Seminiferi 783
- Arrangement in the Lobuli 783
- in the Mediastinum 784
- in the Epididymis 784
- Vasculum Aberrans 784
- Vas Deferens, Course, Relations 784
  - Structure 784
- Vesiculae Seminales 785
  - Form and Size 785
  - Relations 785
  - Structure 785
- Ejaculatory Ducts 786
  - Structure of 786
- The Semen 786
- Descent of the Testes 786
- Gubernaculum Testis 786

FEMALE ORGANS OF GENERATION

- Mons Veneris, Labia Majora 788
- Labia Minora, Clitoris, Meatus Urinarius 789
- Hymen 789
- Glands of Bartholine 790
- *Bladder* 790
- *Urethra* 791
- *Rectum* 791

Vagina
- Relations 792
- Structure 792

Uterus
- Situation, Form, Dimensions 792
- Fundus and Body 792
- Cervix 793
- Ligaments 793
- Cavity of the Uterus 793
- Structure 793
- Vessels and Nerves 794
- Its Form, Size, and Situation 794
  - in the Foetus 794
  - at Puberty 794
  - during and after Menstruation 795
  - after Parturition 795
  - in Old Age 795

Appendages of the Uterus
- Fallopian Tubes 795
  - Structure 796
- Ovaries 796
  - Structure 796
  - Graafian Vesicles 796
- Ligament of the Ovary 797
- Round Ligaments 797
- Vessels and Nerves of Appendages 797

Mammary Glands
- Structure of Mamma 798
- Vessels and Nerves

SURGICAL ANATOMY OF INGUINAL HERNIA

- Dissection 800
- Superficial Fascia 800
- Superficial Vessels 800
- Superficial Nerves 801
- Deep Layer of Superficial Fascia 801
- Aponeurosis of External Obhque 802
- External Abdominal Ring 802
- Pillars of the Ring 802
- Intercolumnar Fibres 802
- Fascia 802
- Poupart's Ligament 803
- Gimbernat's Ligament 803
- Internal Oblique Muscle 803
- Triangular Ligament 803
- Cremaster 803
- Transversalis Muscle 805
- Spermatic Canal 805
- Fascia Transversalis 805

Internal Abdominal Ring 806
Subserous Areolar Tissue 806
Epigastric Artery 806
Peritoneum 806

Inguinal Hernia
Oblique Inguiual Hernia 807
Course and Coverings of Oblique
  Hernia 807
Seat of Stricture 807
Scrotal Hernia 808
Bubonocele 808

Congenital Hernia 808
Infantile Hernia 808

Direct Inguinal Hernia
Course and Coverings of the Hernia 808
Seat of Stricture 809
Incomplete Direct Hernia 809
Comparative Frequency of Oblique
  and Direct Hernia 809
Diagnosis of Oblique and Direct
  Hernia 809

SURGICAL ANATOMY OF FEMORAL HERNIA

Dissection 809
Superficial Fascia 809
Cutaneous Vessels 809
Internal Saphena Vein 810
Superficial Inguinal Glands 810
Cutaneous Nerves 811
Deep Layer of Superficial Fascia 811
Cribriform Fascia 811
Fascia Lata 811
  Iliac Portion 811
  Pubic Portion 811
Saphenous Opening 812

Crural Arch 813
Gimbernat's Ligament 813
Crural Sheath 813
Deep Crural Arch 814
Crural Canal 815
Femoral or Crural Ring 815
Position of Parts around the Ring 815
Septum Crurale 816
Descent of Femoral Hernia 817
Coverings of Femoral Hernia 817
Varieties of Femoral Hernia 817
Seat of Stricture 818

SURGICAL ANATOMY OF PERINAEUM AND ISCHIO-RECTAL REGION

Ischio-Rectal Region
Dissection 819
Superficial Fascia 820
External Sphincter 820
Internal Sphincter 821
Ischio-rectal Fossa 821
Position of Parts contained in 821

Perinaeum
Boundaries, and Extent 821
Superficial Layer of Superficial Fascia ... 822
Deep Layer of Superficial Fascia 822
Course taken by the Urine in Rupture
  of the Urethra 823

Muscles of the Perinaeum (Male)
Accelerator Urinae 823
Erector Penis 824

Transversus Perinaei 824
Superficial Perinaeal Vessels and
  Nerves 824
Transversus Perinaei Artery 824

Muscles of the Perinaeum (Female)
Sphincter Vaginae 825
Erector Clitoridis 825
Transversus Perinaei 825
Compressor Urethrae 825
Sphincter Ani 825
Levator Ani 825
Coccygeus 825
Deep Perinaeal Fascia 825
  Anterior Layer 826
  Posterior Layer 826
Parts between the two Layers 826
Compressor Urethrae 826

Cowper's Glands 826
Pudic Vessels and Nerves 826
Artery of the Bulb................................. 826
Levator Ani 827
    Relations, Actions.......................... 827
Coccygeus, Relations, Actions 827
Position of Viscera at Outlet
    of Pelvis.. 827
Parts concerned in the Operation
    of Lithotomy................................... 829

Parts divided in the Operation............. 830
Parts to be avoided in the Operation ... 830
Abnormal Course of Arteries in the
    Perinaeum 830
Pelvic Fascia... 831
    Obturator Fascia.............................. 831
    Recto-vesical Fascia 832

Surgical Anatomy of the Triangles of the Neck..............................	379–82
"	"	Axilla...	401
"	"	Bend of Elbow...	410
"	"	Scarpa's Triangle..	447
"	"	Popliteal Space...	453
"	"	Laryngo-Tracheal Region..............................	752

Index... 833

LIST OF ILLUSTRATIONS

☞ The Illustrations when copied from any other work, have the author's name affixed; when no such acknowledgment is made, the drawing is to be considered original.

OSTEOLOGY

1. A Cervical Vertebra 7
2. Atlas 7
3. Axis 8
4. A Dorsal Vertebra 9
5. Peculiar Dorsal Vertebrae 10
6. A Lumbar Vertebra 11
7 to 12. Development of a Vertebra...*Quain* 12
13. Sacrum, anterior surface 14
14. Sacrum, posterior surface 15
15. Development of Sacrum...*Quain* 17
16. Coccyx, anterior and posterior surfaces 18
17. Lateral View of Spine 19
18. Occipital Bone, outer surface 22
19. Occipital Bone, inner surface 23
20. Occipital Bone, development of...*Quain* 25
21. Parietal Bone, external surface 26
22. Parietal Bone, inner surface 27
23. Frontal Bone, outer surface 28
24. Frontal Bone, inner surface 29
25. Temporal Bone, outer surface 31
26. Temporal Bone, inner surface 32
27. Temporal Bone, Petrous portion 34
28. Temporal Bone, development of...*Quain* 35
29. Sphenoid Bone, superior surface 36
30. Sphenoid Bone, anterior surface 37
31. Sphenoid Bone, posterior surface 39
32. Plan of the Development of Sphenoid 40
33. Ethmoid Bone, outer surface 41
34. Perpendicular plate of Ethmoid, enlarged 41
35. Nasal Bone, outer surface 44
36. Nasal Bone, inner surface 44
37. Superior Maxillary Bone, outer surface 45
38. Superior Maxillary Bone, inner surface 46
39. Development of Superior Maxillary Bone...*Quain* 49
40. Lachrymal Bone, outer surface 49
41. Malar Bone, outer surface 50
42. Malar Bone, inner surface 51
43. Palate Bone, internal view, enlarged 52
44. Palate Bone, posterior view 53
45. Inferior Turbinated Bone, inner surface 55
46. Inferior Turbinated Bone, outer surface 55
47. Vomer 56

48. Lower Jaw, outer surface 57
49. Lower Jaw, inner surface 58
50. Base of Skull, inner surface 63
51. Base of Skull, external surface 66
52. Anterior Region of Skull 71
53. Nasal Fossae, outer wall 74
54. Nasal Fossae, inner wall or septum 75
55. Hyoid Bone, anterior surface 76
56. Sternum and Costal Cartilages, anterior surface 78
57. Sternum, posterior surface 78
58 to 61. Development of Sternum...*Quain*..................................80
62. A Rib 81
63. Vertebral Extremity of a Rib 82
64 to 68. Peculiar Ribs 83
69. Os Innominatum, external surface 86
70. Os Innominatum, internal surface 88
71. Plan of Development of Os Innominatum 91
72. Left Clavicle, superior surface 95
73. Left Clavicle, inferior surface 95
74. Left Scapula, anterior surface, or venter 97
75. Left Scapula, posterior surface, or dorsum 98
76. Plan of the Development of the Scapula 101
77. Left Humerus, anterior view 103
78. Left Humerus, posterior surface 105
79. Plan of the Development of the Humerus 106
80. Bones of the Left Fore-arm, anterior surface 108
81. Bones of the Left Fore-arm, posterior surface 111
82. Plan of the Development of the Ulna 112
83. Plan of the Development of the Radius 114
84. Bones of the Left Hand, dorsal surface 118
85. Bones of the Left Hand, palmar surface 120
86. Plan of the Development of the Hand 124
87. Right Femur, anterior surface 125
88. Right Femur, posterior surface 127
89. Plan of the Development of the Femur 130
90. Right Patella, anterior surface 131
91. Right Patella, posterior surface 131
92. Tibia and Fibula, anterior surface..132
93. Tibia and Fibula, posterior surface 134
94. Plan of the Development of the Tibia 135
95. Plan of the Development of the Fibula 137
96. Bones of the Right Foot, dorsal surface 139
97. Bones of the Right Foot, plantar surface 142
98. Plan of the Development of the Foot 148

Articulations

99. Vertical-Section of Two Vertebrae and their Ligaments, front view.....158
100. Occipito-Atloid and Alto-axoid ligaments, front view......................161
101. Occipito-Atloid and Alto-axoid ligaments, posterior view...............161

102. Articulation between Odontoid Process and Atlas... *Arnold* 162
103. Occipito-Axoid, and Alto-axoid Ligaments 162
104. Temporo-Maxillary Articulation, external view 164
105. Temporo-Maxillary Articulation, internal view 165
106. Temporo-Maxillary Articulation, vertical section 166
107. Costo-Vertebral and Costo-Transverse Articulations,
   anterior view 167
108. Costo-Transverse Articulation... *Arnold* 168
109. Costo-Sternal, Costo-Xiphoid and Intercostal Articulations,
   front view 170
110. Articulations of Pelvis and Hip, front view 172
111. Articulations of Pelvis and Hip, back view 173
112. Vertical Section of Symphysis pubis 176
113. Sterno-Clavicular Articulation 177
114. Shoulder Joint, Scapulo-Clavicular Articulation and proper
   Ligaments of Scapula 179
115. Left Elbow Joint, shewing anterior and internal Ligaments 182
116. Left Elbow Joint, shewing posterior and external Ligaments 183
117. Ligaments of Wrist and Hand, anterior view... *Arnold* 185
118. Ligaments of Wrist and Hand, posterior view... *do* 186
119. Vertical Section of Wrist, shewing the Synovial Membranes 190
120. Articulations of the Phalanges 192
121. Left Hip Joint, laid open 193
122. Right Knee Joint, anterior view 195
123. Right Knee Joint, posterior view 196
124. Right Knee Joint, shewing internal Ligaments 197
125. Head of Tibia, with semi-lunar Cartilages, seen from above 198
126. Ankle Joint, Tarsal, and Tarso-Metatarsal Articulations,
   internal view 202
127. Ankle Joint, Tarsal, and Tarso-Metatarsal Articulations,
   external view 203
128. Ligaments of Plantar surface of the Foot 204
129. Synovial Membranes of the Tarsus and Metatarsus... *Arnold* 206

Muscles and Fasciae

130. Plan of Dissection of Head, Face and Neck 213
131. Muscles of the Head, Face and Neck 214
132. Muscles of the right Orbit 217
133. The relative position and attachment of the Muscles of the
   left Eyeball 218
134. The Temporal Muscle 225
135. The Pterygoid Muscles 227
136. Muscles of the Neck and boundaries of the Triangles 231
137. Muscles of the Neck, anterior view... *Quain* 233
138. Muscles of the Tongue, left side 238
139. Muscles of the Pharynx, external view 240
140. Muscles of the Soft Palate 242
141. The Pre-vertebral Muscles... *Quain* 245
142. Plan of Dissection of the Muscles of the Back 248

143. Muscles of the Back — first, second, and part of the third layers...*Quain*..................249
144. Muscles of the Back — deep layers 255
145. Plan of Dissection of Abdomen..................262
146. The External Oblique Muscle..................263
147. The Internal Oblique Muscle 264
148. The Transversalis, Rectus and Pyramidalis..................266
149. Transverse Section of Abdomen in Lumbar Region...*Quain*..................267
150. The Diaphragm, under surface..................272
151. Plan of Dissection of Upper Extremity..................276
152. Muscles of the Chest and Front of the Arm, superficial view..................277
153. Muscles of the Chest and Front of the Arm, with the boundaries of the Axilla..................279
154. Muscles on the Dorsum of the Scapula and the Triceps..................284
155. Front of the left Fore-arm, superficial muscles..................290
156. Front of left Fore-arm, deep muscles..................293
157. Posterior surface of Fore-arm, superficial muscles..................295
158. Posterior surface of the Fore-arm, deep muscles..................298
159. Muscles of the left Hand, palmar surface..................302
160. Dorsal Interossei of the left Hand..................304
161. Palmar Interossei of Left Hand..................305
162. Fracture of the Middle of the Clavicle...*Hind*..................306
163. Fracture of the Surgical Neck of the Humerus...*do*..................307
164. Fracture of the Humerus above the Condyles...*do*..................308
165. Fracture of the Olecranon...*do*..................309
166. Fracture of the Shaft of the Radius...*do*..................309
167. Fracture of the lower end of the Radius...*do*..................311
168. Plan of Dissection of Lower Extremity, front view..................315
169. Muscles of Iliac and Anterior Femoral Regions..................316
170. Muscles of the Internal Femoral Region...*Quain*..................321
171. Plan of Dissection of Lower Extremity, posterior view..................323
172. Muscles of the Gluteal and Posterior Femoral Regions..................324
173. Muscles of the front of the Leg..................331
174. Muscles of the back of Leg, superficial layer..................333
175. Muscles of the back of the Leg, deep layer..................336
176. Muscles of the sole of the Foot, first layer..................343
177. Muscles of the sole of the Foot, second layer..................344
178. Muscles of the sole of the Foot, third layer..................345
179. The Dorsal Interossei..................346
180. The Plantar Interossei..................346
181. Fracture of the Neck of the Femur within the Capsular Ligament...*Hind*..................347
182. Fracture of the Femur below the Trochanter Minor...*do*..................348
183. Fracture of the Femur above the Condyles...*do*..................348
184. Fracture of the Patella...*do*..................349
185. Oblique fracture of the shaft of the Tibia...*do*..................349
186. Fracture of the Fibula, with dislocation of the Tibia inwards...*do*..................350

Arteries

187. The Arch of the Aorta and its branches 354
188. Plan of the branches of the Arch of the Aorta 354
189. Surgical anatomy of the Arteries of the Neck 363
190. Plan of the branches of the External Carotid 363
191. The Arteries of the Face and Scalp 371
192. The Internal Maxillary Artery, and its branches 376
193. Plan of the branches of the Internal Maxillary Artery 376
194. The Internal Carotid and Vertebral Arteries 383
195. The Ophthalmic Artery and its branches 386
196. The Arteries at the base of the Brain 388
197. Plan of the branches of the Right Subclavian Artery 395
198. The Scapular and Circumflex Arteries 399
199. The Axillary Artery and its branches 402
200. The Surgical Anatomy of the Brachial Artery 407
201. The Surgical Anatomy of the Radial and Ulnar Arteries 412
202. Ulnar and Radial Arteries, deep view 415
203. Arteries of the back of the Fore-Arm and Hand 419
204. The Abdominal Aorta and its branches 424
205. The Coeliac Axis and its branches 426
206. The Coeliac Axis and its branches, the Stomach having been raised, and the Transverse Meso-colon removed 427
207. The Superior Mesenteric Artery and its branches 429
208. The Inferior Mesenteric Artery and its branches 430
209. Arteries of the Pelvis 434
210. The Arteries of the Gluteal and Posterior Femoral Regions 441
211. Surgical Anatomy of the Femoral Artery 445
212. The Popliteal, Posterior Tibial, and Peroneal Arteries 452
213. Surgical Anatomy of the Anterior Tibial and Dorsalis Pedis Arteries 456
214. The Plantar Arteries, superficial view 462
215. The Plantar Arteries, deep view 462

Veins

216. Veins of the Head and Neck 467
217. Veins of the Diploë, as displayed by the removal of the outer table of the Skull 471
218. Vertical Section of the Skull, showing the Sinuses of the Dura Mater 472
219. The Sinuses at the Base of the Skull 474
220. The Superficial Veins of the Upper Extremity 476
221. The Venae Cavae and Azygos Veins, with their Formative Branches 479
222. Transverse Section of a Dorsal Vertebra, showing the Spinal Veins...*Breschet* 483
223. Vertical Section of two Dorsal Vertebrae, showing the Spinal Veins...*Breschet* 483
224. The Internal Saphenous Vein and its Branches 484

225. The External, or short Saphenous Vein 485
226. The Portal Vein and its Branches...*Quain* 490

Lymphatics

227. The Thoracic and Right Lymphatic Ducts 495
228. The Superficial Lymphatics and Glands of the Head, Face, and Neck...*Mascagni* 497
229. The Deep Lymphatics and Glands of the Neck and Thorax...*do.* 498
230. The Superficial Lymphatics and Glands of the Upper Extremity...*do.* 499
231. The Superficial Lymphatics and Glands of the Lower Extremity...*do.* 500
232. The Deep Lymphatic Vessels and Glands of the Abdomen and Pelvis...*do.* 503

Nervous System

233. The Spinal Cord and its Membranes 514
234. Transverse Section of the Spinal Cord and its Membranes...*Arnold* 515
235. Spinal Cord, side view. Plan of the Fissures and Columns...*Quain* ... 516
236. Transverse Sections of the Cord...*Arnold* 518
237. Medulla Oblongata and Pons Varolii, anterior surface 523
238. Medulla Oblongata and Pons Varolii, posterior surface 524
239. Transverse Section of Medulla Oblongata...*Arnold* 525
240. The Columns of the Medulla Oblongata, and their Connection with the Cerebrum and Cerebellum...*Altered from Arnold* 527
241. Upper Surface of the Brain, the Pia Mater having been removed 528
242. Base of the Brain 531
243. Section of the Brain, made on a level with the Corpus Callosum 535
244. The Lateral Ventricles of the Brain 536
245. The Fornix, Velum Interpositum, and Middle or Descending Horn of the Lateral Ventricle 539
246. The Third and Fourth Ventricles 542
247. The Cerebellum, upper surface 546
248. The Cerebellum, under surface 547
249. The Cerebellum, vertical section...*Arnold* 548

Cranial Nerves

250. The Optic Nerves and Optic Tracts 552
251. Course of the Fibres in the Optic Commissure...*Bowman* 553
252. Nerves of the Orbit, seen from above...*After Arnold* 554
253. Nerves of the Orbit and Ophthalmic Ganglion, side view...*After Arnold* 556
254. The Course and Connections of the Facial Nerve in the Temporal Bone...*After Bidder* 557

255. The Nerves of the Scalp, Face, and Side of the Neck 560
256. The Hypoglossal Nerve, Cervical Plexus, and their Branches 562
257. Distribution of the Second and Third Divisions of the Fifth
     Nerve and Sub-Maxillary Ganglion...*After Arnold*...........................*566*
258. The Spheno-Palatine Ganglion and its Branches 568
259. The Otic Ganglion and its Branches...*After Arnold*........................*573*
260. Origin of the Eighth Pair, their Ganglion and
     Communications...*Bendz* 574
261. Course and Distribution of the Eighth Pair of Nerves......................575

Spinal Nerves

262. Plan of the Brachial Plexus...588
263. Cutaneous Nerves of Right Upper Extremity, anterior view.............590
264. Cutaneous Nerves of Right Upper Extremity, posterior view 591
265. Nerves of the Left Upper Extremity, front view................................593
266. The Supra-Scapular, Circumflex, and Musculo-Spiral Nerves..........596
267. The Lumbar Plexus and its Branches...*Altered from Quinn*..............602
268. The Cutaneous Nerves of Lower Extremity, front view 604
269. Nerves of the Lower Extremity, front view.......................................604
270. Cutaneous Nerves of Lower Extremity, posterior view 611
271. Nerves of the Lower Extremity, posterior view 611
272. The Plantar Nerves 614
273. The Sympathetic Nerve...618

Organs of Sense

274. A Sectional View of the Skin, magnified 629
275. Upper Surface of the Tongue 636
276. The three kinds of Papillae of the Tongue, magnified...*Bowman*.......636
277. Cartilages of the Nose...*Arnold*...638
278. Bones and Cartilages of the Septum of the Nose...*do.* 639
279. Nerves of Septum of Nose...*do.*...641
280. A Vertical Section of the Eyeball, enlarged 643
281. The Choroid and Iris, enlarged...*Altered from Zinn*..........................645
282. The Veins of the Choroid, enlarged...*Arnold*....................................646
283. The Arteries of the Choroid and Iris, the Sclerotic has been
     mostly removed, enlarged...*do.* 647
284. The Arteria Centralis Retinae, Yellow Spot, &c., the anterior
     half of the Eyeball being removed, enlarged 649
285. The Crystalline Lens, hardened and divided, enlarged...*Arnold*........653
286. The Meibomian Glands, &c., seen from the Inner Surface
     of the Eyelids...*do.* 655
287. The Lachrymal Apparatus, right side..657
288. The Pinna, or Auricle, outer surface 658
289. The Muscles of the Pinna...*Arnold*..660
290. A Front View of the Organ of Hearing, right side...*Scarpa*..............661
291. View of Inner Wall of Tympanum, enlarged 663
292. The Small Bones of the Ear, seen from the outside,
     enlarged...*Arnold*..666

293. The Osseous Labyrinth, laid open, enlarged...*Soemmering*................669
294. The Cochlea laid open, enlarged...*Arnold* 670
295. The Membranous Labyrinth detached, enlarged...*Breschet*.............673

ORGANS OF DIGESTION AND THEIR APPENDAGES

296. Sectional View of the Nose, Mouth, Pharynx, &c...................676
297. The Permanent Teeth, external view................................677
298. The Temporary, or Milk Teeth, external view 680
299. Vertical Section of a Molar Tooth 680
300. Vertical Section of a Bicuspid Tooth, magnified...*After Retzius*.........681
301. to 306. Development of the Teeth...*Goodsir*.......................683
307. The Salivary Glands...687
308. The Regions of the Abdomen and their contents (edge of Costal Cartilages in dotted Outline)...693
309. Diagram shewing the Reflections of the Peritoneum, as seen in a Vertical Section of the Abdomen...*Altered from Quain*..............694
310. The Mucous Membrane of the Stomach and Duodenum, with the Bile Ducts 698
311. The Muscular Coat of the Stomach (the innermost Layer is not seen) 700
312. Minute Anatomy of Mucous Membrane of Stomach...*Dr. Sprott Boyd*..701
313. Two Villi, magnified 704
314. Patch of Peyer's Glands from the lower part of the Ileum 705
315. A portion of Peyer's Glands magnified...*Boehm*...................705
316. The Caecum and Colon laid open, to show the Ilio-caecal Valve......706
317. Minute structure of Large Intestine...*Boehm* 710
318. The Liver, upper surface...712
319. The Liver, under surface...713
320. Longitudinal section of an Hepatic Vein...*Kiernan*................715
321. Longitudinal section of a small Portal Vein and Canal...*do*...............716
322. A transverse section of a small Portal Canal and its vessels...*do*.........717
323. The Pancreas and its relations 719
324. Transverse section of the Spleen, showing the Trabecular Tissue, the Splenic Vein, and its branches..722
325. The Malpighian Corpuscles, and their relation with the Splenic Artery and its branches..723
326. One of the Splenic Corpuscles, showing its relations with the blood-vessels...724
327. Transverse section of the Human Spleen, showing the distribution of the Splenic Artery and its branches 723

ORGANS OF CIRCULATION

328. The right Auricle and Ventricle laid open, the anterior walls of both being removed 728
329. The left Auricle and Ventricle laid open, the anterior walls of both being removed 733
330. Plan of the Foetal Circulation...738

Organs of Voice and Respiration

331. Side view of Thyroid and Cricoid Cartilages 741
332. The Cartilages of the Larynx, posterior view 742
333. Interior of the Larynx, seen from above, enlarged... *Willis*................745
334. Muscles of Larynx, side view, right ala of Thyroid Cartilage removed 748
335. Front view of Cartilages of Larynx: the Trachea and Bronchi 750
336. Surgical anatomy of the Laryngo-tracheal Region 752
337. A transverse section of the Thorax, showing the relative position of the Viscera, and the reflections of the Pleura 754
338. Front view of the Heart and Lungs 756

The Urinary and Generative Organs

339. Vertical section of the Kidney 764
340. A Plan to shew the minute structure of the Kidney... *Bowman*..........764
341. Vertical section of Bladder, Penis, and Urethra..............................769
342. The Bladder and Urethra laid open, seen from above 772
343. The Testis in situ, the Tunica Vaginalis having been laid open 782
344. Plan of a vertical section of the Testicle, to shew the arrangement of the ducts 783
345. Base of the Bladder, with the Vasa Deferentia and Vesiculae Seminales... *Haller*..785
346. The Vulva, External Female Organs of Generation.........................788
347. Section of Female Pelvis, shewing Position of Viscera.....................790
348. The Uterus and its Appendages, anterior view... *Wilson*..................796

349. Inguinal Hernia, Superficial Dissection..801
350. Inguinal Hernia, showing the Internal Oblique, Cremaster, and Spermatic Canal 804
351. Inguinal Hernia, shewing the Transversalis Muscle, the Transversalis Fascia, and the Internal Abdominal Ring 806
352. Femoral Hernia, Superficial Dissection 810
353. Femoral Hernia, shewing Fascia Lata and Saphenous Opening........812
354. Femoral Hernia, Iliac Portion of Fascia Lata removed, and Sheath of Femoral Vessels and Femoral Canal exposed 814
355. Hernia; the Eolations of the Femoral and Internal Abdominal Rings, seen from within the Abdomen, right side 816
356. Variations in Origin and Course of Obturator Artery 816
357. Plan of Dissection of Perinaeum and Ischio-Rectal Region 820
358. The Perinaeum; the Integument and Superficial Layer of Superficial Fascia reflected 822
359. The Superficial Muscles and Vessels of the Perinaeum 823
360. Deep Perinseal Fascia; on the Left Side the Anterior Layer has been removed 825
361. A View of the Position of the Viscera at the Outlet of the Pelvis 828
362. A Transverse Section of the Pelvis, shewing the Pelvic Fascia... *After Wilson* 829
363. Side View of the Pelvic Viscera of the Male Subject, shewing the Pelvic and Perinaeal Fasciae 831

COLOR ILLUSTRATIONS

187. The Arch of the Aorta and its branches A1
189. Surgical anatomy of the Arteries of the Neck A2
197. Plan of the branches of the Right Subclavian Artery.........................A3
199. The Axillary Artery and its branches..A3
200. The Surgical Anatomy of the Brachial Artery A3
204. The Abdominal Aorta and its branches A4
205. The Coeliac Axis and its branches...A5
206. The Coeliac Axis and its branches, the Stomach having been raised, and the Transverse Meso-colon removed...............................A5
207. The Superior Mesenteric Artery and its branches A6
209. Arteries of the Pelvis A6
211. Surgical Anatomy of the Femoral Artery..A7
212. The Popliteal, Posterior Tibial, and Peroneal Arteries A7
222. Transverse Section of a Dorsal Vertebra, showing the Spinal Veins...*Breschet*..A8
223. Vertical Section of two Dorsal Vertebrae, showing the Spinal Veins...*Breschet*..A8
227. The Thoracic and Right Lymphatic Ducts A8

Introduction

Few works have left such an indelible mark on the field of anatomy as the seminal masterpiece by the British surgeon and anatomist Henry Gray. His book *Anatomy: Descriptive and Surgical*, now known simply as *Gray's Anatomy*, was first published in 1858, and it immediately took its place as a cornerstone of medical education. Countless generations of physicians have found it indispensable during their training and careers, and no doctor's library today would be complete without a copy. *Gray's Anatomy* is generally regarded as synonymous with medicine, even by the lay public. It is often called "the doctor's Bible."

More than 160 years since its first publication, the book remains a foundational resource for aspiring and practicing physicians. Still considered the gold standard reference for human anatomy, it remains relevant to the modern medical curriculum. Its current edition is the forty-second; a forty-third, currently in preparation, is due to be released in 2025.

ANATOMY IN GENERAL

The study of the human form is almost as old as medicine itself. Ancient physicians realized that a thorough knowledge of human anatomy was necessary to ensure the best care for their patients. Many of those physicians, such as Galen during the second century CE, were therefore among the first anatomists. The term itself comes from the Greek word *anatomía* (ανατομία), meaning "to cut up"; in other words, "to dissect." Thus, human dissection has long been integral to the study of anatomy, and it remains so today. Over time, the word came to denote the study of morphology or form, and as in the days of Henry Gray, the best way to learn this discipline is via cadaveric dissection. Along with studies of function and disease, anatomy has been at the core of medical education for millennia. Almost a century ago, David Cheever (1933) summed up this intimate connection by referring to anatomy as the oldest child of "Mother Medicine." He said of the relationship:

> These elder daughters (anatomy, physiology, and pathology) were long the chief support of Mother Medicine's family, but she has been prolific, and many other offspring have been born; the end is not yet, and these newcomers are lusty youngsters who, like most self-confident adolescents, are not conspicuously aware of the rights of others, and are apt to crowd

and jostle their elders as they seek their rightful places in the sun. The two younger of the elder trio, pathology and physiology, are still growing, the latter indeed with no evident slackening of pace; the eldest, however, is quite matured and though just as useful as ever in the family economy, is thought by some too old-fashioned and out-of-date and is in danger of being neglected. She has children of her own, by the way—first, histology and embryology, and then a flock of later arrivals such as comparative morphology, cytology and experimental embryology, all promising youngsters who are adding much to the family reputation, but too are necessarily crowding the old home and claiming all of the family income they can get. And unfortunately, the eldest sister, anatomy, cannot be dispossessed, because though neglected, she is perfectly certain to live just as long as Mother Medicine herself.

Creation of "the Book"

Henry Gray first envisaged a textbook of anatomy that would serve the needs of both medical student and practitioner; specifically, a surgical anatomy book for the learner and the surgeon. Although other anatomical textbooks were available at the time, Gray believed that a new, practical, fully illustrated, and student-friendly publication would be well received. To that end, he discussed the proposed book with his junior medical colleague at St. George's Hospital in London, Henry Vandyke Carter, and then enlisted Carter's services to create the now-famous drawings that accompanied the early editions. It is not clear how or where Gray and Carter first met, but it most likely happened at St. George's Hospital.

Nor is it clear how much writing Carter contributed to the project, but it is known that both he and Gray performed the cadaveric dissections used for most of the anatomical drawings in the book. The synergy between the two authors resulted in a publication that would stand the test of time. The book was prepared between December 1855 and July 1857. For two men to complete such a huge project in just twenty months, with a final length of over 700 pages with 363 figures, was a monumental task.

Gray's Early Life

Henry Gray was born on February 23, 1827, and grew up in Belgravia in London. However, a letter from his great-nephew, F. Lawrence Gray, to the publisher of the American edition of *Gray's Anatomy* states his year of birth as 1825.

Matriculating at the age of eighteen, Gray entered St. George's Hospital on May 6, 1845. Five years later, he became a house surgeon at St. George's, curator of the museum, and a lecturer on anatomy. Little is known about his life, but he was described by those who knew him as "a most painstaking and methodical worker, and one who learned his anatomy by the slow but invaluable method of making dissections for himself." The *Philosophical Transactions of the Royal Society of London* described him as an accomplished and lucid teacher of anatomy.

ANATOMY SCHOLARSHIP OF HENRY GRAY BEFORE *GRAY'S ANATOMY*

Before his book was published, Gray had shown himself to be an avid research anatomist. Interestingly, in the records of St. George's Hospital, he signs himself as a "perpetual student" in the students' book. His anatomical studies included works on the nerves of the orbit, the adrenal glands, and the spleen. His study of the eye earned him a prize from the Royal College of Surgeons, and his study of the spleen won him the triennial Astley Cooper Prize in 1853; that paper was republished as a book in 1854. Thanks to these and other accomplishments, Gray was elected Fellow of the Royal Society on June 3, 1852, at the age of twenty-five. When asked for names of potential anatomy demonstrators, British surgeon and pathologist Sir James Paget referred to him as one of London's "best men."

THE DRAWINGS

Most of the figures in the first edition of *Gray's Anatomy* were original. Those that were borrowed from other anatomy texts of the day by such authors as Friedrich Arnold, Gilbert Breschet, and Jones Quain were duly acknowledged. Up to a third of those borrowed works were from Arnold, who worked at the University of Heidelberg. Carter's renderings have been deemed "superb drawings that captured the human anatomy with a pristine clarity and simplicity of expression that are worth continued regard." Carter labeled his drawings directly with their anatomical names, as the anatomy textbook author Luther Holden had done before him. This obviated the use of leader lines and complex arrays of abbreviations, which had complicated other anatomical figures of the time with detailed figure legends. Also, the illustrations in other student-oriented anatomy books of the day were much smaller than those in *Gray's Anatomy*.

Henry Vandyke Carter was born in Hull, England, on May 22, 1831. In 1848 he entered medical school at St. George's Hospital. He spent thirty years in India before returning to England in 1888. He died at home of tuberculosis on May 4, 1897, at the age of sixty-five.

Contemporary Acclaim

Anatomy: Descriptive and Surgical made its debut in 1858. It garnered widespread acclaim from the medical community and became a standard work within just a few years. Contemporary reviews praised the textbook for its academic rigor and artistic beauty. For example, the *Lancet* called it "a work of no ordinary labour" and commented that it:

> ... demanded the highest accomplishments both as anatomist and surgeon for its successful completion ... we may say with truth, that there is not a treatise in any language, in which the relations of anatomy and surgery are so clearly and fully shown.

The *British Medical Journal* hailed it as:

> ... far superior to all other treatises on anatomy ... a book which must take its place as *the* manual of Anatomy Descriptive and Surgical ... it is by far the best book which we have on the subject, and there is nothing else which can compare with it for clearness of style and thoroughness of detail ... a monument of industry and accuracy.

Controversies

Carter made significant contributions to the book and would certainly have been considered an equal coauthor in most works. However, the preface states: "The Author gratefully acknowledges the great services he has derived, in the execution of this work, from the assistance of his friend, Dr. H. V. Carter, late Demonstrator of Anatomy at St. George's Hospital. He executed all the drawings from which the engravings were made. In the majority of cases, they have been copied from, or corrected by, recent dissections, made jointly by the Author and Dr. Carter." These comments, although appreciative, place Carter in a lesser role than "the Author." Moreover, Gray's edits to the title page prior to publication indicated that Carter's placement on the page should be located below his and in smaller type. He also demoted Carter by striking through his title of professor of anatomy at his new assignment at Grant College in Bombay (Mumbai),

India. It is unclear what Carter thought of this arrangement when he first received his copy of the book, but Gray's slight did not pass unnoticed contemporaneously or subsequently. In fact, Carter's contributions were better emphasized in later editions to ameliorate the perceived injustice.

Additionally, although the vast majority of evaluations of the book were glowing, some reviews of the day were pointed. For example, a review in the *Medical Times* of 1859 accused Gray of plagiarism from earlier texts such as that by Quain and Sharpey. However, such criticisms were refuted. For example, in his short biography of Gray, Clinton Dent, a surgeon in St. George's Hospital, said of such accusations:

> Undeniably many passages in Gray's work could be cited in which the phraseology and description closely resemble that of Quain and Sharpey's "Anatomy"; but it is somewhat absurd in ordinary anatomical descriptions to accuse one writer of paraphrasing passages to be found in the work of another; forgetfulness of the source from which we are borrowing is an extremely common form of originality.

Lastly, the cadavers used by Carter and Gray for their dissections would have been acquired under the terms of the 1832 Anatomy Act. This law was designed to suppress the grave-robbing that had previously supplied many medical schools, but it also decreed that if a patient died and no one claimed them within forty-eight hours, the body could be used for anatomical dissection.

Gray's Later Life

Henry Gray's life was tragically cut short on June 13, 1861, at the age of thirty-four. He had contracted smallpox from his nephew, whom he had been treating. Only three years had passed since the publication of his book, so he never had the opportunity to witness its immense success and lasting impact. He had just finished preparing the second edition of the text. Sir Benjamin Brodie, president of the Royal Society, surgeon at St. George's Hospital, and surgeon to Queen Victoria, wrote:

> I am much grieved about poor Gray. His death, just as he was on the point of obtaining the reward of his labors, is a sad event indeed . . . Gray is a great loss to the hospital and school. Who is there to take his place?

Interestingly, Gray had dedicated his book on anatomy to Brodie. The *Medical Times and Gazette* summed up the short professional life of Henry Gray and its impact:

... the feeling of general regret expressed at the great loss the Profession sustains by the death of Mr. Gray. We do not remember another instance in which so young a surgeon has attained such varied honours ... the works he leaves behind him [serve] as encouragements to others to follow the path by which he gained the respect of his contemporaries, and an enduring place in the Roll of Medical Worthies.

POSTHUMOUS SUCCESS OF *GRAY'S ANATOMY*

Despite Gray's death, his masterpiece on anatomy was to become one of the world's most widely used and respected books. Some have called it the "leading medical textbook of the modern world." Retrospectively, it differed from other books of its day by its lucid and logical arrangement of a mass of detailed descriptions, its clear drawings based on dissections by the surgeon-author and an artist, and its treatment of the surgical anatomy of variously defined areas. Ruth Richardson, in her book *The Making of Mr. Gray's Anatomy* (2008), extolled the key to *Gray's Anatomy*'s success: "... to make every word tell, to keep it well organized, in small manageable chunks, well signposted, and to give variety to the student's interest. With a diversity of accessible data."

LEGACY AND ENDURING INFLUENCE

The publishing house Lea & Blanchard purchased the American rights to the book and released it in 1859, one year after J. W. Parker and Son published the first edition in London. Dr. R. J. Dunglision was the editor of the first several editions published in Philadelphia. Interestingly, Dunglision's father, R. T. Dunglision, was personal physician to Thomas Jefferson. The last "American" edition of *Gray's* was published in 1984 under the editorship of Dr. C. D. Clemente.

In 1938 the twenty-seventh edition officially renamed the book *Gray's Anatomy*. It is said to be the oldest continuously revised textbook currently used by medical professionals. Charles Goss, the editor of the twenty-fifth to twenty-ninth American editions, stated that with the centennial edition, published in Philadelphia in 1959, the total number of copies printed was almost 600,000, which would form a stack twenty-four miles high and weighing 2,000 tons! Goss dedicated the centennial edition to Gray and Carter "in admiration of their imagination and skill in originating this book and to the many thousands of students and practitioners of medicine and surgery who have used it." Sixty-one years earlier, as an earlier illustration of

Henry Gray's impact, Thomas Cooke and his son F. G. Hamilton Cooke (1898) had dedicated their anatomy book to the memory of Henry Gray and elevated him to the status of three other influential anatomists of the time, George Viner Ellis, Christopher Heath, and Luther Holden.

Because the landscape of medical discoveries and treatment paradigms is constantly changing, medical textbooks are usually short-lived. *Gray's Anatomy* stands out as an anomaly, having never gone out of print. Obviously, this is a testament to Gray and Carter's vision and collaboration. However, the book's enduring impact on medicine is also due, as the British librarian and medical historian Frederick Poyntor said, to:

> . . . the intelligence and skill with which it has been treated, at critical times, by its medical attendants, in this case, a succession of distinguished editors and gifted draughtsman. They have amputated, excised, grated, and injected to such effect that Gray himself would regard with amazement the prodigy which has grown from his own modest offspring, although a closer inspection would enable him to recognize its original features. He was once an author of a book, but he might more truly be regarded as the founder of an institution.

Over recent years, this institution, *Gray's Anatomy*, has given rise to an extensive progeny. For example, the anatomy student can now access titles such as *Gray's Anatomy Review*, *Gray's Surgical Anatomy*, *Gray's Anatomy for Students*, *Gray's Clinical Photographic Dissector of the Human Body*, *Gray's Atlas of Anatomy*, *Gray's Surface Anatomy and Ultrasound*, and *Gray's Clinical Neuroanatomy*. As these diverse titles show, the name "Gray" has passed the test of time and continues to carry authority in the medical literature. I believe Henry Gray would be proud of the continued use of his surname until modern times and the sustained usage and appreciation of his 1858 classic thesis on human anatomy. As a clinical anatomist, I frequently go back to this original treatise to see how anatomical structures were first described by Henry Gray and illustrated by Henry Vandyke Carter, and I consider it a jewel of the medical literature.

Conclusion

The original edition of *Gray's Anatomy* is such a masterpiece that the exactness, practicality, and precision of its illustrations continue to impress generation after generation of readers. The value of the first edition, with its pivotal role in scholarship and its place in medical history, cannot be exaggerated. This anatomy book, first published in 1858, gives the name "Gray" its continuing stature in modern medicine.

Today, the first edition offers the reader several things. First, it exemplifies a shift in anatomical drawings from other anatomical books of its day and is thus of historical significance. Secondly, its many unique and clear depictions that are quintessentially Gray's are still of use as a learning tool to the modern student and physician. Lastly, the clear and succinct writing style found in the book is as useful today as it was in 1858, as anatomy does not change—only our methods of presenting it.

Professor R. Shane Tubbs
Co-Editor-in-Chief
Gray's Anatomy, 43rd edition
Tulane University School of Medicine
New Orleans, Louisiana

PREFACE

This Work is intended to furnish the Student and Practitioner with an accurate view of the Anatomy of the Human Body, and more especially the application of this science to Practical Surgery.

One of the chief objects of the Author has been, to induce the Student to apply his anatomical knowledge to the more practical points in Surgery, by introducing, in small type, under each subdivision of the work, such observations as shew the necessity of an accurate knowledge of the part under examination.

Osteology. Much time and care have been devoted to this part of the work, the basis of anatomical knowledge. It contains a concise description of the anatomy of the bones, illustrated by numerous accurately-lettered engravings, shewing the various markings and processes on each bone. The attachments of each muscle are shewn in dotted lines (after the plan recently adopted by Mr. Holden), copied from recent dissections. The articulations of each bone are shewn on a new plan; and a method has been adopted, by which the hitherto complicated account of the development of the bones is made more simple.

The Articulations. In this section, the various structures forming the joints are described; a classification of the joints is given; and the anatomy of each carefully described: abundantly illustrated by engravings, all of which are taken from, or corrected by, recent dissections.

The Muscles and Fasciae. In this section, the muscles are described in groups, as in ordinary anatomical works. A series of illustrations, shewing the lines of incision necessary in the dissection of the muscles in each region, are introduced, and the muscles are shewn in fifty-two engravings. The Surgical Anatomy of the muscles in connection with fractures, of the tendons or muscles divided in operations, is also described and illustrated.

The Arteries. The course, relations, and Surgical Anatomy of each artery are described in this section, together with the anatomy of the regions containing the arteries more especially involved in surgical operations. This part of the work is illustrated by twenty-seven engravings.

The Veins are described as in ordinary anatomical works; and illustrated by a series of engravings, shewing those in each region. The veins of the spine are described and illustrated from the well-known work of Breschet.

The Lymphatics are described, and figured in a series of illustrations copied from the elaborate work of Mascagni.

The Nervous System and Organs of Sense. A concise and accurate description of this important part of anatomy has been given, illustrated by seventy-two engravings, shewing the spinal cord and its membranes; the anatomy of the brain, in a series of sectional views; the origin, course, and distribution of the cranial, spinal, and sympathetic nerves; and the anatomy of the organs of sense.

The Viscera. A detailed description of this essential part of anatomy has been given, illustrated by fifty large, accurately-lettered engravings.

Regional Anatomy. The anatomy of the perinaeum, of the ischio-rectal region, and of femoral and inguinal herniae, is described at the end of the work; the region of the neck, the axilla, the bend of the elbow, Scarpa's triangle, and the popliteal space, in the section on the arteries; the laryngo-tracheal region, with the anatomy of the trachea and larynx. The regions are illustrated by many engravings.

Microscopical Anatomy. A brief account of the microscopical anatomy of some of the tissues, and of the various organs, has also been introduced.

The Author gratefully acknowledges the great services he has derived, in the execution of this work, from the assistance of his friend, Dr. H. V. Carter, late Demonstrator of Anatomy at St. George's Hospital. All the drawings from which the engravings were made, were executed by him. In the majority of cases, they have been copied from, or corrected by, recent dissections, made jointly by the Author and Dr. Carter.

The Author has also to thank his friend, Mr. T. Holmes, for the able assistance afforded him in correcting the proof-sheets in their passage through the press.

The engravings have been executed by Messrs. Butterworth and Heath; and the Author cannot omit thanking these gentlemen for the great care and fidelity displayed in their execution.

WILTON-STREET, BELGRAVE-SQUARE,
August, 1858.

Descriptive and Surgical Anatomy

Descriptive anatomy comprises a detailed account of the numerous organs of which the body is formed, especially with reference to their outward form, their internal structure, the mutual relations they bear to each other, and the successive conditions they present during their development.

Surgical Anatomy is, to the student of medicine and surgery, the most essential branch of anatomical science, having reference more especially to an accurate knowledge of the more important regions, and consisting in the application of anatomy generally to the practice of surgery.

The Study of Anatomy is commonly divided into several distinct branches, according to the mutual resemblance of the organs; and these branches have corresponding denominations.

OSTEOLOGY

In the construction of the human body, it would appear essential, in the first place, to provide some dense and solid texture capable of giving support and attachment to the softer parts of the frame, and at the same time to protect in closed cavities the more important vital organs; and such a structure we find provided in the various bones, which form what is called the Skeleton.

Bone is one of the hardest structures of the animal body; it possesses also a certain degree of toughness and elasticity. Its colour, in a fresh state, is of a pinkish white externally, and deep red within. Chemical analysis resolves bone into an *organic*, or animal, and an *inorganic*, or earthy material, intimately combined together; the animal matter giving to bone its elasticity and toughness, the earthy part its hardness and solidity. The animal constituent may be separated from the earthy, by steeping bone in a dilute solution of nitric or muriatic acid: by this process the earthy constituents are gradually dissolved out, leaving a tough semi-transparent substance which retains, in every respect, the original form of the bone. This is often called cartilage, but differs from it in being softer, more flexible, and, when boiled under a high pressure, it is almost entirely resolved into gelatine. The earthy constituent may be obtained by subjecting a bone to strong heat in an open fire with free access of air. By these means, the animal matter is entirely consumed, the earthy part remaining as a white brittle substance still preserving the original shape of the bone.

The organic or animal constituent of bone, forms about *one-third*, or 33.3 percent; the inorganic or earthy matter, *two-thirds*, or 66.7 percent: as is seen in the subjoined analysis by Berzelius:—

Animal Matter	Gelatine and Blood-vessels	33.30
Inorganic or Earthy Matter	Phosphate of Lime	51.04
	Carbonate of Lime	11.30
	Fluoride of Calcium	2.00
	Phosphate of Magnesia	1.16
	Soda and Chloride of Sodium	1.20
		100.00

The proportion between these two constituents varies at different periods of life, as is seen in the following table from Schreger:—

	Child	Adult	Old Age
Animal matter	47.20	20.18	12.2
Earthy matter	48.48	74.84	84.1

There are facts of some practical interest, bearing upon the difference here seen in the amount of the two constituents of bone, at different periods of life. Thus, in the child, where the animal matter forms nearly one-half of the weight of the bone, it is not uncommon to find, after an injury happening to the bones, that they become bent, or only partially broken, from the large amount of flexible animal matter which they contain. Again, also in aged people, where the bones contain a large proportion of earthy matter, the animal matter at the same time being deficient in quantity and quality, the bones are more brittle, their elasticity is destroyed; and, hence, fracture takes place more readily. Some of the diseases, also, to which bones are liable, mainly depend on the disproportion between the two constituents of bone. Thus, in the disease called rickets, so common in the children of scrofulous parents, the bones become bent and curved, either from the superincumbent weight of the body, or under the action of certain muscles. This depends upon some deficiency of the nutritive system, by which bone becomes minus its normal proportion of earthy matter, whilst the animal matter is of unhealthy quality. In the vertebra of a rickety subject, Dr. Bostock found in 100 parts 79.75 animal, and 20.25 earthy matter.

The relative proportions of the two constituents of bone are found to differ in different bones of the skeleton. Thus the petrous portion of the temporal bone contains a large proportion of earthy matter, the bones of the limbs contain more earthy matter than those of the trunk, and those of the upper extremity, a larger proportion than those of the lower.

On examining a section of any bone, it is seen to be composed of two kinds of tissue, one of which is dense and *compact* in texture like ivory; the other open, reticular, spongy, enclosing cancelli or spaces, and hence called *spongy* or *cancellated* tissue. The compact tissue is always placed on the exterior of a bone; the cancellous tissue is always internal. The relative quantity of these two kinds of tissue varies in different bones, and in different parts of the same bone, as strength or lightness is requisite.

Form of Bones. The various mechanical purposes for which bones are employed in the animal economy require them to be of very different forms. All the scientific principles of Architecture and Dynamics are more or less exemplified in the construction of this part of the human body. The power of the arch in resisting superincumbent pressure is well exhibited in various parts of the skeleton, such as the human foot, and more especially in the vaulted roof of the cranium.

Bones are divisible into four classes: *Long, Short, Flat,* and *Irregular.*

The long bones are found chiefly in the limbs, where they form a system of levers, which have to sustain the weight of the trunk, and to confer extensive powers of locomotion. A long bone consists of a lengthened cylinder or shaft, and two extremities. The *shaft* is a hollow cylinder, the walls consisting of dense compact tissue of great thickness in the middle, and becoming thinner towards the extremities; the spongy tissue is scanty, and the bone is hollowed out in its interior to form the *medullary canal*. The *extremities* are generally somewhat expanded for greater convenience of mutual connexion, and for the purposes of articulation. Here the bone is made up of spongy tissue with only a thin coating of compact substance. The long bones are the *clavicle, humerus, radius, ulna, femur, tibia, fibula, metacarpal,* and *metatarsal* bones and the *phalanges*.

Short Bones. Where a part is intended for strength and compactness, and the motion at the same time slight and limited, it is divided into a number of small pieces united together by ligaments, and the separate bones are short and compressed, such as the bones of the *carpus* and *tarsus*. These bones, in their structure, are spongy throughout, excepting at their surface, where there is a thin crust of compact substance.

Flat Bones. Where the principal requirement is either extensive protection, or the provision of broad surfaces for muscular attachment, we find the osseous structure remarkable for its slight thickness, becoming expanded into broad flat plates, as is seen in the bones of the skull and shoulder-blade. These bones are composed of two thin layers of compact tissue, enclosing a layer of cancellous tissue of variable thickness. In the cranial bones, these layers of compact tissue are familiarly known as the *tables* of the skull; the outer one is thick and tough, the inner one thinner, denser, and more brittle, and hence termed the *vitreous table*. The intervening cancellous tissue is called the *diploë*. The flat bones are the *occipital, parietal, frontal, nasal, lachrymal, vomer, scapulae,* and *ossa innominata*.

The *Irregular* or *Mixed* bones are such as, from their peculiar form, cannot be grouped under either of the preceding heads. Their structure is similar to that of other bones, consisting of an external layer of compact, and of a spongy cancellous substance within. The irregular bones are the *vertebrae, sacrum, coccyx, temporal, sphenoid, ethmoid, superior maxillary, inferior maxillary, palate, inferior turbinated,* and *hyoid*.

Vessels of Bone. The blood-vessels of bone are very numerous. Those of the compact tissue consist of a close and dense network of vessels, which ramify in a fibrous membrane termed the *periosteum*, which covers the entire surface of the bone in nearly every part. From this membrane, vessels pass through all parts of the compact tissue, running through the canals which traverse its substance. The cancellous tissue is supplied in a similar way, but by a less numerous set of larger vessels, which, perforating the outer compact tissue, are distributed to the cavities of the spongy portion of the bone. In the long bones, numerous apertures may be seen at the ends near the articular surfaces, some of which give passage to the arteries referred to; but the greater number, and these are the largest of them, are for the veins of the cancellous tissue which run separately from the arteries. The medullary canal is supplied by one large artery (or sometimes more), which enters the bone at the nutritious foramen (situated, in most cases, near the centre of the shaft), and perforates obliquely the compact substance. This vessel, usually accompanied by one or two veins, sends branches upwards and downwards, to supply the medullary membrane, which lines the central cavity and the adjoining canals. The ramifications of this vessel anastomose with the arteries both of the cancellous and compact tissues. The veins of bone are large, very numerous, and run in tortuous canals in the cancellous texture, the sides of which are constructed of a thin lamella of bone, perforated here and there for the passage of branches from the adjacent cancelli. The veins thus enclosed and supported by the hard structure, have exceedingly thin coats; and when the bony structure is divided, they remain patulous, and do not contract in the canals in which they are contained. Hence the constant occurrence of purulent absorption after amputation, in those cases where the stump becomes inflamed, and the cancellous tissue is infiltrated and bathed in pus. *Lymphatic* vessels have been traced into the substance of bone. *Nerves*, also, accompany the nutritious arteries into their interior.

Development of Bone. From the peculiar uses to which bone is applied, in forming a hard skeleton or framework for the softer materials of the body, and in enclosing and protecting some of the more important vital organs, we find its development takes place at a very early period. Hence the parts that appear soonest in the embryo, are the vertebral column and the skull, the great central column, to which the other parts of the skeleton are appended. At an early period of embryonic life, the parts destined to become bone consist of a congeries of cells, which constitutes the simplest form of cartilage. This *temporary cartilage*, as it is termed, is an exact miniature of the bone which in due course is to take its place; and as the process of ossification is slow, and not completed until adult life, it increases in bulk by an interstitial development of new cells. The next step in this process is the ossification of the intercellular substance, and of the cells composing the cartilage. Ossification commences in the interior of the cartilage at certain points, called *points* or *centres of ossification*, from which it extends into the surrounding substance. The period of ossification varies much in different bones. It commences first in the clavicle, in which the primitive point appears during the fifth week; next in the lower jaw. The ribs also, and the long bones of the limbs, appear soon after. The number of ossific centres varies in different bones. In most of the short bones, it commences by a single point in the centre, and proceeds towards the circumference. In the long bones, there is a central point of ossification for the shaft or diaphysis; and one for each extremity, the epiphyses. That for the shaft is the first to appear; those for the extremities appear later. For a long period after birth, a thin layer of unossified cartilage remains between the diaphysis and epiphyses, until their growth is finally completed. Processes such as the trochanters that have separate centres of ossification, are called *epiphyses* previous to their union.

Growth of Bone. Increase in the length of a bone, is provided for by the development of new bone from either end of the shaft (diaphysis); and in the thickness, by the deposition of new matter upon the surface: but when growth is at an end, the epiphyses become solidly united to the ends of the diaphysis, and the bone is completely formed. A knowledge of the exact periods when the epiphyses become joined to the shaft, aids the surgeon in the diagnosis of many of the injuries to which the joints are liable; for it not unfrequently happens, that on the application of severe force to a joint, the epiphyses become separated from the shaft, and such injuries may be mistaken for fracture.

The order in which the epiphyses become united to the shaft, follows a peculiar law, which appears to be regulated by the direction of the nutritious artery of the bone. Thus the arteries of the bones of the arm and forearm converge towards the elbow, and the epiphyses of the bones forming this joint become united to the shaft before those at the opposite extremity. In the lower extremities, on the contrary, the nutritious arteries pass in a direction from the knee; that is, upwards in the femur, downwards in the tibia and fibula; and in them it is observed, that the upper epiphysis of the femur, and the lower epiphyses of the tibia and fibula, become first united to the shaft.

A diseased condition of any joint makes considerable variation in the period of development of the several bones which enter into its formation. Thus, in chronic inflammation occurring in a joint at an early period of life, the epiphysal cartilages take on premature ossification; this process proceeding so rapidly, that it speedily becomes converted into bone, which becomes united to the shaft, and the bone ever after is considerably diminished in length: hence partial atrophy of the limb is the result.

The entire skeleton in an adult, consists of 206 distinct bones. These are—

Cranium	8
Ossicula auditus	6
Face	14
Vertebral column (sacrum and coccyx included)	26
Os hyoides, sternum, and ribs	26
Upper extremities	64
Lower extremities	62
	206

In this enumeration, the *sesamoid* and *Wormian* bones are excluded, as also are the teeth, which difier from bone both in structure, development, and mode of growth. The skeleton consists of a central column or *Spine*; of three great cavities, the *Skull*, *Thorax*, and *Pelvis*; and of the *Superior* and *Inferior Extremities*.

THE SPINE

The Spine is a flexuous column, formed of a series of bones called *Vertebrae*.

The Vertebrae are divided into *true* and *false*.

The true vertebrae are twenty-four in number, and have received the names *cervical*, *dorsal*, and *lumbar*, according to the position which they occupy; seven being found in the cervical region, twelve in the dorsal, and five in the lumbar.

The false vertebrae, nine in number, are firmly united, so as to form two bones—five entering into the formation of the upper bone or *sacrum*, and four into the terminal bone of the spine or *coccyx*.

True Vertebrae, 24 { 7 Cervical
12 Dorsal
5 Lumbar

False Vertebrae, 9 { 5 Sacrum
4 Coccyx

General Characters of the Vertebrae

Each vertebra consists of two parts, an anterior solid segment or body, forming the chief pillar of support; a posterior segment, the arch, forming part of a hollow cylinder for protection. The arch is formed of two pedicles and two laminae, supporting seven processes; viz. four articular, two transverse, and one spinous process.

The Body is the largest and most solid part of a vertebra, serving to support the weight of the cranium and trunk. Above and below it is slightly concave, presenting a rim around its circumference; and its surfaces are rough, for the attachment of the intervertebral fibro-cartilages. In front it is convex from side to side, concave from above downwards. Behind, flat from above downwards, and slightly concave from side to side. Its anterior surface is perforated by a few small apertures, for the passage of nutrient vessels; whilst on the posterior surface is a single irregular-shaped, or occasionally several large apertures, for the exit of veins from the body of the vertebra, the *venae basis vertebrae*.

The *Pedicles* project backwards, one on each side, from the upper part of the body of the vertebrae, at the line of junction of its posterior and lateral surfaces; they form the lateral parts of the arch, which is completed posteriorly by the two laminae. The concavities above and below the pedicles are the *intervertebral notches*; they are four in number, two on each side, the inferior ones being always the deeper.

The *Laminae* consist of two broad plates of bone, which complete the vertebral arch behind, enclosing a foramen which serves for the protection of the spinal cord; they are connected to the body through the intervention of the pedicles. Their upper and lower borders are rough, for the attachment of the *ligamenta subflava*.

The *Spinous Process* projects backwards from the junction of the two laminae, and serves for the attachment of muscles.

The *Transverse Processes*, two in number, project one at each side from the point where the articular processes join the pedicle. They also serve for the attachment of muscles.

The *Articular Processes* are four in number; two superior, the smooth surfaces of which are directed more or less backwards; and two inferior, the articular surfaces of which look more or less forwards.

Characters of the Cervical Vertebrae (fig. 1)

The Body is smaller than in any other region of the spine, thicker before than behind, and broader from side to side than from before backwards. Its upper surface is concave transversely, and presents a projecting lip on each side; its lower surface being convex from side to side, concave from before backwards, and presenting laterally a shallow concavity, which receives the corresponding projecting lip of the adjacent vertebra. The *pedicles* are directed obliquely outwards, and the superior intervertebral notches are slightly deeper, but narrower, than the inferior. The *laminae* are narrow, long, thinner above than below, and imbricated, *i.e.* overlapping each other; enclosing the foramen, which is very large, and of a triangular form. The *spinous processes* are short, bifid at the extremity, the two divisions being often of unequal size. They increase in length from the fourth to the seventh. The *transverse processes* are short, directed downwards, outwards, and forwards, bifid at their extremity, and marked by a groove along their upper surface, which runs downwards and outwards from the superior intervertebral notch, and serves for the transmission of one of the cervical nerves. The transverse processes are pierced at their base by a foramen, for the transmission of the vertebral artery, vein, and plexus of nerves. Each of these processes is formed by two roots: the anterior or smaller, which is attached to the side of the body, corresponds to the ribs in the dorsal region; the posterior is larger, springs from the pedicle, and corresponds to the true transverse processes. It is by the junction of these two processes, that the vertebral foramen is formed. The extremities of each of these roots form the *anterior* and *posterior tubercles* of the transverse processes. The *articular processes* are oblique: the superior are of an oval form, flattened, and directed upwards and backwards; the inferior downwards and forwards.

1.—A Cervical Vertebra.

The peculiar vertebrae in the cervical region are the first or *Atlas*; the second or *Axis*; and the seventh or *Vertebra prominens*.

2.—1st Cervical Vertebra, or Atlas.

The *Atlas* (fig. 2) (so named from supporting the globe of the head). The chief peculiarities of this bone are, that it has neither body, spinous process, nor pedicles. It consists of an anterior arch, a posterior arch, and two lateral masses. The *anterior* or lesser arch, which forms about one-fifth of the bone, represents the front part of the body of a vertebra; its anterior surface is convex, and presents about its centre a tubercle for the attachment of the Longus colli muscle; posteriorly it is concave, and marked by a smooth oval surface, for articulation with the odontoid process of the axis. The *posterior* or greater arch, which forms about two-fifths of the circumference of the bone, terminates behind in a tubercle, which is the rudiment of a spinous process, and gives origin to the Rectus capitis posticus minor. The posterior part of the arch presents above a rounded edge; whilst in front, immediately behind the superior articular processes, are two grooves, sometimes converted into foramina by delicate bony spiculae. These grooves represent the superior intervertebral notches, and are peculiar from being situated behind the articular processes, instead of before them, as in the other vertebrae. They serve for the transmission of the vertebral artery, which, ascending through the foramen in the transverse process, winds around the lateral mass in a direction backwards and inwards. They also transmit the sub-occipital nerves. On the under surface of the posterior arch, in the same situation, are two other grooves, placed behind the lateral masses, and representing the inferior intervertebral notches of other vertebrae. They are much less marked than the superior. The *lateral masses*, which are the most bulky and solid parts of the Atlas, present two articulating processes above, and two below. The two superior are of large size, oval, concave, and approach towards one another in front, but diverge behind; they are directed upwards, inwards, and a little backwards, forming a kind of cup for the condyles of the occipital bone, and are admirably adapted to the nodding movements of the head; whilst the inferior, which are circular in form, and flattened, are directed downwards, inwards, and a little backwards, articulating with the axis, and permitting the rotatory movements. Just below the inner margin of each superior articular surface, is a small tubercle, for the attachment of a ligament which, stretching across the ring of the Atlas, divides it into two unequal parts; the anterior or smaller segment receiving the odontoid process of the Axis, the posterior allowing the transmission of the spinal cord and its membranes. This ligament and the odontoid process are marked in the figure in dotted outline. The transverse processes are of large size, long, not bifid, perforated at their base by a canal for the vertebral artery, which is directed from below, upwards and backwards.

3.—2nd Cervical Vertebra, or Axis.

The *Axis* (fig. 3) (so named from forming the pivot upon which the head rotates). The most distinctive character of this bone is the existence of a strong prominent process, tooth-like in form (hence the name odontoid), which arises perpendicularly from the upper part of the body. The body is of a triangular form; its anterior surface deeper than the posterior, presents a median longitudinal ridge, separating two lateral depressed surfaces for the attachment of the Longi colli muscles. The odontoid process presents two oval articulating surfaces: one in front, for articulation with the Atlas; another behind, for the transverse ligament; the apex is pointed, and on either side of it is seen a rough impression for the attachment of the odontoid or cheek ligaments; whilst the base, where attached to the body, is constricted, so as to prevent displacement from the transverse ligament, which binds it in this situation to the anterior arch of the Atlas. On each side of this process are seen the superior articular surfaces; they are round, convex, directed upwards and outwards, and are peculiar in being supported on the body, pedicles, and transverse processes. The inferior articular surfaces, which are posterior and external to these, have the same direction as those of the other cervical vertebrae. The superior intervertebral notches are very shallow, and lie behind the articular processes; the inferior in front of them, as in the other cervical vertebrae. The transverse processes are very small, not bifid, and perforated by the vertebral foramen, which is directed obliquely upwards, and outwards. The laminae are thick and strong, and the spinous process is of large size, very strong, deeply channelled on its under surface, and presenting a bifid tubercular extremity for the attachment of muscles.

Seventh Cervical. The most distinctive character of this vertebra is the existence of a very large, long, and prominent spinous process; hence the name "Vertebra prominens." This process is thick, nearly horizontal in direction, not bifurcated, and has attached to it the ligamentum nuchae. The foramina in the transverse processes are small, often wanting, and when present do not give passage to the vertebral artery; their upper surface presents only a slight groove, and generally only a trace of bifurcation at their extremity.

Characters of the Dorsal Vertebrae

The *Dorsal Vertebrae* (fig. 4) are intermediate in size between the cervical and lumbar. The body is somewhat triangular in form, broader in the antero-posterior than in the lateral direction, more particularly in the middle of the dorsal region,

4.—A Dorsal Vertebra.

thicker behind than in front, flat above and below, deeply concave behind, and marked on each lateral surface, near the root of the pedicle, by two demi-facets, one above, the other below. These are covered with cartilage in the recent state; and, when articulated with the adjoining vertebrae, form oval surfaces for the reception of the heads of the corresponding ribs. The pedicles are strong, and the inferior intervertebral notches of large size. The laminae are broad and thick, and the spinal foramen small, and of a round or slightly oval form. The articular surfaces are flat, the superior being directed backwards and a little outwards and upwards, the inferior forwards and a little inwards and downwards. The transverse processes are thick, strong, and of great length, directed obliquely backwards and outwards, presenting a clubbed extremity, lipped on its anterior part by a small concave surface, for articulation with the tubercle of a rib. The spinous processes are long, directed obliquely downwards, and terminated by a tubercle.

The peculiar dorsal vertebrae are the *first*, *ninth*, *tenth*, *eleventh*, and *twelfth* (fig. 5).

5.—Peculiar Dorsal Vertebræ.

The *First Dorsal Vertebra* may be distinguished by the existence on each side of the body, of a single entire articular facet for the head of the first rib, and a half facet for the upper half of the second. The upper surface of the body is like that of a cervical vertebra, being broad transversely, concave, and lipped on each side. The *superior articular surfaces* are oblique, and the *spinous process* thick, long, and almost horizontal.

The *Ninth Dorsal* has no demi-facet below.

The *Tenth Dorsal* has an entire articular facet at each side above; no demi-facet below.

In the *Eleventh Dorsal*, the body approaches in its form to the lumbar; and has a single entire articular surface on each side. The *transverse* processes are very short, and have no articular surfaces at their extremities.

The *Twelfth Dorsal* has the same characters as the eleventh; but may be distinguished from it by the transverse processes being quite rudimentary, and the inferior articular surfaces being convex and turned outwards, like those of the lumbar vertebrae.

The smallest dorsal vertebra is the fourth. The vertebrae increase in size from that point downwards to the twelfth, and upwards to the first. The spinous processes also, from the eighth downwards, become shorter, and are directed more horizontally.

Characters of the Lumbar Vertebrae

The Lumbar Vertebrae (fig. 6) are the largest segments of the vertebral column. The Body is large, broad from side to side, flat above, and below, and thicker before than behind. The pedicles are very strong, directed backwards; and the inferior intervertebral notches are of large size. The laminae are short, but broad and strong;

6.—A Lumbar Vertebra.

and the foramen triangular, larger than in the dorsal, smaller than in the cervical region. The superior articular processes are concave, and directed backwards and inwards; the inferior, convex, and directed forwards and outwards. Projecting backwards from each of the superior articular processes is a tubercle, the representative of the transverse processes in the dorsal and cervical regions. The transverse processes are long, slender, directed a little backwards, and present, at the posterior part of their base, a small tubercle, which is directed downwards. The spinous processes are thick and broad, somewhat quadrilateral, horizontal in direction, and thicker below than above.

The *Fifth Lumbar* vertebra is peculiar from having the body much thicker in front than behind, which accounts for the prominence of the sacro-vertebral articulation.

Structure of the Vertebrae. The structure of a vertebra differs in different parts. The Body is composed almost entirely of light spongy cancellous tissue, having a thin coating of compact tissue on its external surface, permeated throughout its interior with large canals for the reception of veins, which converge towards a single large irregular or several small apertures at the posterior part of the body of each bone. The arch and processes projecting from it have, on the contrary, an exceedingly thick covering of compact tissue.

Development. Each vertebra is formed of three primary centres of ossification (fig. 7), one for each lamella and its processes, and one for the body. Those for the lamellae appear about the sixth week of foetal life, in the situation where the transverse processes afterwards project, the ossific granules shooting backwards to the spine, forwards to the body, and outwards into the transverse and articular processes. That for the body makes its appearance in the middle of the cartilage about the eighth week. At birth, these three pieces are perfectly separate. During the first year, the lateral portions become partly united behind, in the situation of the spinous process, and thus the arch is formed. About the third year, the body is joined to the arch on each side, in such a manner, that the body is formed from the three original centres of ossification. Before puberty, no other changes occur, excepting a gradual increase of growth of these primary centres, the upper and under surface of the bodies, and the ends of the transverse and spinous processes, being tipped with cartilage, in which ossific

7.—Development of a Vertebra.
By 3 primary centres
1 for Body (8th week)
1 for each Lamella (6th week)

8.
By 4 Secondary Centres
1 for each Trans. Proc. 16 yrs.
2 sometimes 1 for Spin. proc. (16 yrs.)

9.
By 2 additional plates
1 for upper surface of body
1 for under surface of body
21 yrs

10.—Axis.
By 3 centres
1 for anter. arch (1st yr.)
1 for each lateral mass } before birth

11.—Axis.
By 5 centres
2 for odontoid proc. (6th mo.)
1 for each lateral mass
1 for body (6th mo.)

12.—Lumbar Vertebra.
2 additional centres

for tubercles on Sup. Artic. Proc.

Exceptional Cases.

granules are not as yet deposited. At sixteen years (fig. 8), four secondary centres appear, one for the tip of each transverse process, and two (sometimes united into one) for the spinous process. At twenty-one years (fig. 9), two thin circular plates of bone are formed, one for the upper, and one for the under surface of the body. All these become joined, and the bone is completely formed about the thirtieth year of life.

Exceptions to this mode of development occur in the first, second, and seventh cervical, and in those of the lumbar region.

The *Atlas* (fig. 10) is developed by *three* centres. One (sometimes two) for the anterior arch, and one for each lateral mass. The ossific centres for each lateral mass commence before birth. At birth, the anterior arch is altogether cartilaginous, and the two lateral pieces are separated from one another behind. The nucleus for the anterior arch appears in the first year, between the second and third years the two lateral pieces unite, and join the anterior part at the age of five or six years. There is frequently a separate epiphysis for the rudimentary spine.

The *Axis* (fig. 11) is developed by *five* centres; three for its anterior part, and two for the posterior. The three anterior centres are, one for the lower part of the body, and two for the odontoid process and upper part of the body; the two posterior ones are, one for each lamella. At about the sixth month of foetal life, those for the body and odontoid process make their appearance, the two for the odontoid process joining before birth. At birth the bone consists of four pieces, two anterior and two lateral. At the fourth year the body and odontoid process are completely joined.

The Seventh Cervical. The anterior or costal part of the transverse process of the seventh cervical, is developed from a separate osseous centre at about the sixth month of foetal life, and joins the body and posterior division of the transverse process between the fifth and sixth years. Sometimes this process continues as a separate piece, and becoming lengthened outwards, constitutes what is known as a cervical rib.

The Lumbar Vertebrae (fig. 12) have *two additional centres* (besides those peculiar to the vertebrae generally), for the tubercles, which project from the back part of the superior articular processes. The transverse process of the first lumbar is sometimes developed as a separate piece, which may remain permanently unconnected with the remaining portion of the bone; thus forming a lumbar rib, a peculiarity which is sometimes, though rarely, met with.

PROGRESS OF OSSIFICATION IN THE SPINE GENERALLY. Ossification of the laminae of the vertebrae commences at the upper part of the spine, and proceeds gradually downwards; hence the frequent occurrence of spina bifida in the lower part of the spinal column. Ossification of the bodies, on the other hand, commences a little below the centre of the spinal column, and extends both upwards and downwards. Although, however, the ossific nuclei make their first appearance in the lower dorsal vertebrae (about the ninth), the lumbar and first sacral are those in which these nuclei are largest at birth.

THE FALSE VERTEBRAE

The False Vertebrae consist of nine pieces, which are united so as to form two bones, five entering into the formation of the sacrum, four the coccyx.

THE SACRUM (fig. 13) is a large triangular bone, situated at the lower part of the vertebral column, and at the upper and back part of the pelvic cavity, where it is inserted like a wedge between the two ossa innominata; its upper part, or base, articulating with the last lumbar vertebra, its apex with the coccyx. The sacrum is curved upon itself, and placed very obliquely, its upper extremity projecting forwards, forming, with the last lumbar vertebra, a very prominent angle, called the *promontory* or *sacro-vertebral angle*, whilst its central part is directed backwards, so as to give increased capacity to the pelvic cavity. It presents for examination an anterior and posterior surface, two lateral surfaces, a base, an apex, and a central canal.

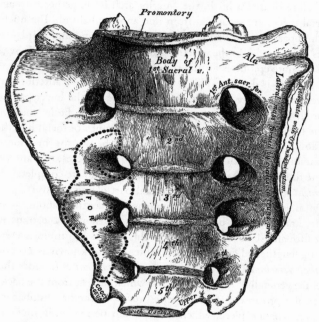

13.—Sacrum, Anterior Surface.

The Anterior Surface is deeply concave from above downwards, and slightly so from side to side. In the middle are seen four transverse lines, indicating the original division of the bone into five separate pieces. The portions of bone intervening between the lines correspond to the bodies of the vertebrae; they are slightly concave longitudinally, and diminish in size from above downwards. At the extremities of each of these lines, are seen the *anterior sacral foramina*, analogous to the intervertebral foramina, four in number on each side, somewhat rounded in form, diminishing in size from above downwards, and directed outwards and forwards; they transmit the anterior branches of the sacral nerves. External to these foramina, is the *lateral mass*, formed by the coalesced *transverse processes* of the sacral vertebrae, traversed by four broad shallow grooves, which lodge the anterior sacral nerves as they pass outwards, the grooves being separated by prominent ridges of bone, which give attachment to the slips of the Pyriformis muscle.

The *Posterior Surface* (fig. 14) is convex, and much narrower than the anterior. In the middle line, are three or four tubercles, sometimes connected together, which represent the rudimentary spinous processes. Of these tubercles, the first is usually very prominent, and perfectly separate from the rest; the second, third, and fourth,

14.—Sacrum Posterior Surface.

existing either separate, or united into a ridge, which diminishes in size as it descends; the fifth, and sometimes the fourth, remaining undeveloped, and exposing below, the lower end of the sacral canal. External to the spinous processes on each side, are the *laminae*, broad and well marked in the three first pieces; the lower part of the fourth, and the whole of the fifth, being undeveloped: in this situation the sacral canal is exposed. External to the laminae are a linear series of indistinct tubercles representing the *articular processes*; the upper pair are well developed; the second and third are small; the fourth and fifth (usually blended together) are situated on each side of the sacral canal: they are called the *sacral cornua*, and articulate with the cornua of the coccyx. External to the articular processes are the four *posterior sacral foramina*; they are smaller in size, and less regular in form than the anterior, and transmit the posterior branches of the sacral nerves. On the outer side of the posterior sacral foramina are a series of tubercles, representing the rudimentary *transverse processes*. The first pair of transverse tubercles are very distinct, and correspond with each superior angle of the bone; the second, small in size, enter into the formation of the sacro-iliac articulation;

the third give attachment to the oblique sacro-iliac ligaments; and the fourth and fifth to the great sacro-ischiatic ligaments. The interspace between the spinous and transverse processes of the sacrum, presents a wide shallow concavity, called the *sacral groove*; it is continuous above with the vertebral groove, and lodges the origin of the Erector Spinae.

The *Lateral Surface*, broad above, becomes narrowed into a thin edge below. Its upper half presents in front a broad ear-shaped surface for articulation with the ilium. This is called the *auricular* or *ear-shaped* surface, and in the fresh state is coated with cartilage. It is bounded posteriorly by deep and rough impressions, for the attachment of the sacro-iliac ligaments. The lower half is thin and sharp, and gives attachment to the greater and lesser sacro-ischiatic ligaments; below, it presents a deep notch, which is converted into a foramen by articulation with the transverse process of the upper piece of the coccyx, and transmits the anterior branch of the fifth sacral nerve.

The *Base* of the sacrum, which is broad and expanded, is directed upwards and forwards. In the middle is seen an oval articular surface, which corresponds with the under-surface of the body of the last lumbar vertebra, bounded behind by the large triangular orifice of the sacral canal. This orifice is formed behind by the spinous process and laminae of the first sacral vertebra, whilst projecting from it on each side are the superior articular processes; they are oval, concave, directed backwards and inwards, like the superior articular processes of a lumbar vertebra, in front of each articular process in an intervertebral notch, which forms the lower half of the last intervertebral foramen. Lastly, on each side of the articular surface is a broad and flat triangular surface of bone, called the *alae* of the sacrum; they extend outwards, and are continuous on each side with the iliac fossae.

The *Apex*, directed downwards and forwards, presents a small oval concave surface for articulation with the coccyx.

The *Sacral Canal* runs throughout the greater part of the bone; it is large and triangular in form above, small and flattened from before backwards below. In this situation, its posterior wall is incomplete, from the non-development of the laminae and spinous processes. It lodges the sacral nerves, and is perforated by the anterior and posterior sacral foramina, through which these pass out.

Structure. It consists of much loose spongy tissue within, invested externally by a thin layer of compact tissue.

DIFFERENCES IN THE SACRUM OF THE MALE AND FEMALE. The sacrum in the female is usually wider than in the male, and it is much less curved, the upper half of the bone being nearly straight, the lower half presenting the greatest amount of curvature. The bone is also directed more obliquely backwards; which increases the size of the pelvic cavity, and forms a more prominent sacro-vertebral angle. In the male the curvature is more evenly distributed over the whole length of the bone, and is altogether greater than in the female.

PECULIARITIES OF THE SACRUM. This bone, in some cases, consists of six instead of five pieces; occasionally the number is reduced to four. Sometimes the bodies of the first and second vertebrae are not joined, or the laminae and spinous processes have not coalesced with the rest of the bone. Occasionally the superior transverse tubercles are not joined to the rest of the bone on one or both sides; and, lastly, the sacral canal may be open for nearly the lower half of the bone, in consequence of the imperfect development of the laminae and spinous processes. The sacrum also varies considerably with respect to its degree of curvature. From the examination of a large number of skeletons, it would appear, that, in one set of cases, the anterior surface of this bone was nearly straight, the curvature, which was very slight, affecting only its lower end. In another set of cases, the bone was curved throughout its whole length, but especially towards its middle. In a third set, the degree of curvature was less marked, and affected especially the lower third of the bone.

Development of Sacrum (fig. 15). The sacrum, formed by the union of five vertebrae, has *thirty-five* centres of ossification. Each of the three first pieces is developed by seven centres; viz., three for the body, one for its central part, one for each epiphysal lamella on its upper and under surface, and one for each of the laminae: so far the first three sacral vertebrae, as well as the two last, are developed like the other pieces of the vertebral column. One of the characteristic points in the development of this bone, consists in the existence of two additional centres for each of the first three pieces, which appear one on each side, close to the anterior sacral foramina, and correspond to the transverse processes of the lumbar vertebrae.

Each of the two last pieces is developed by five centres: three for the body; viz., one for its central part, and one for each of the epiphysal lamellae; and one for each of the laminae.

15.—Development of Sacrum.
Formed by union of 5 Vertebræ.
2 characteristic points.

1st
2 Additional centres for the first 3 pieces *

at birth

2nd.
2 Epiphysal laminæ for each lateral surface. *

at 25th yr

A second characteristic point in the development of this bone consists in each lateral surface of the sacrum being developed by two epiphysal points, one for the auricular surface, and one for the thin lateral border of the bone.

Period of Development. At about the eighth or ninth week of foetal life, ossification of the central part of the bodies of the three first vertebrae commences, and, at a somewhat later period, that of the two last. Between the sixth and eighth months, ossification of the lamellae takes place; and at about the same period the characteristic osseous tubercles for the three first sacral vertebrae make their appearance. The lateral pieces join to form the arch, and are united to the bodies, first, in the lowest vertebrae. This occurs about the second year, the uppermost segment appearing as a single piece about the fifth or sixth year. About the sixteenth year the epiphysal lamellae for the bodies are formed; and between the eighteenth and twentieth years those for each lateral surface of the sacrum make their appearance. At about this period, the two last segments are joined to one another; and this process gradually extending upwards, all the pieces become united, and the bone completely formed from the twenty-fifth to the thirtieth year of life.

Articulations. With four bones; the last lumbar vertebra, coccyx, and the two ossa innominata.

Attachment of Muscles. The Pyriformis and Coccygeus on either side, behind the Gluteus maximus and Erector Spinae.

The Coccyx

The Coccyx (κόκκυξ, cuckoo), so called from resembling a cuckoo's beak, (fig. 16) is formed of four small segments of bone, the most rudimentary parts of the vertebral column. Of these, the first is the largest, and often exists as a separate piece, the three last diminishing in size from above downwards, are blended together so as to form a single bone. The gradual diminution in the size of the pieces gives this bone a triangular form, articulating by its base with the apex of the sacrum. It presents for examination an anterior and posterior surface, two borders, a base, and an apex. The *anterior surface* is slightly concave, and marked with three transverse grooves, indicating the points of junction of the different pieces. It has attached to it the anterior sacro-coccygeal ligament and levator ani muscle, and supports the lower end of the rectum. The *posterior surface* is convex, marked by grooves similar to those on the anterior surface, and presents on each side a linear row of tubercles, which represent the articular processes of the coccygeal vertebrae. Of these, the superior pair are very large; they are called the *cornua of the coccyx*, and projecting upwards, articulate with the cornua of the sacrum, the junction between these two bones completing the fifth sacral foramen for the transmission of the posterior branch of the fifth sacral nerve. The lateral borders are thin, and present a series of small eminences, which represent the transverse processes of the coccygeal vertebrae. Of these, the first on each side is of large size, flattened from before backwards, and often ascends upwards to join the lower part of the thin lateral edge of the sacrum, thus completing the fifth sacral foramen: the others diminish in size from above downwards, and are often wanting. The borders of the coccyx are narrow, and give attachment on each side to the sacro-sciatic ligaments and Coccygeus muscle. The *base* presents an oval surface for articulation with the sacrum. The *apex* is rounded, and has attached to it the tendon of the external Sphincter ani muscle. It is occasionally bifid, and sometimes deflected to one or other side.

16.—Coccyx. *Anterior Surface. Posterior Surface.*

Development. The coccyx is developed by *four* centres, one for each piece. Occasionally one of the three first pieces of this bone is developed by two centres, placed side by side. The periods when the ossific nuclei make their appearance is the following: in the first segment, at birth; in the second piece, at from five to ten years; in the third, from ten to fifteen years; in the fourth, from fifteen to twenty years. As age advances, these various segments become united in the following order: the two first pieces join, then the third and fourth; and, lastly the bone is completed by the union of the second and third. At a late period of life, especially in females, the coccyx becomes joined to the end of the sacrum.

Articulation. With the sacrum.

Attachment of Muscles. On either side, the Coccygeus; behind, the Gluteus maximus; at its apex, the Sphincter ani; and in front, the Levator ani.

OF THE SPINE IN GENERAL. — The spinal column, formed by the junction of the vertebrae, is situated in the median line, at the posterior part of the trunk: its average length is about two feet two or three inches; the lumbar region contributing seven parts, the dorsal eleven, and the cervical five.

Viewed in front, it presents two pyramids joined together at their bases, the upper one being formed by all the true vertebrae from the second cervical to the last lumbar; the lower one by the false vertebrae, the sacrum, and coccyx. Viewed somewhat more closely, the uppermost pyramid is seen to be formed of three smaller pyramids. Of these the most superior one consists of the six lower cervical vertebrae, its apex being formed by the axis or second cervical, its base by the first dorsal. The second pyramid, which is inverted, is formed by the four upper dorsal vertebrae, the base being at the first dorsal, the smaller end at the fourth. The third pyramid commences at the fourth dorsal, and gradually increases in size to the fifth lumbar.

Viewed laterally (fig. 17), the spinal column presents several curves. In the dorsal region, the seat of the principal curvature, the spine is concave anteriorly; whilst in the cervical and lumbar regions it is convex anteriorly, especially

17.—Lateral View of Spine.

in the latter. The spine has also a slight lateral curvature, the convexity of which is directed towards the right side. This is most probably produced, as Bichat first explained, from the effect of muscular action; most persons using the right arm in preference to the left, especially in making long-continued efforts, when the body is curved to the right side. In support of this explanation, it has been found by Beclard, that in one or two individuals who were left-handed, the lateral curvature was directed to the left side.

The spinal column presents for examination an anterior, a posterior, and two lateral surfaces, a base, summit, and vertebral canal.

The *anterior surface* presents the bodies of the vertebrae separated in the fresh state by the intervertebral discs. The bodies are broad in the cervical region, narrow in the upper part of the dorsal, and broadest in the lumbar region. The whole of this surface is convex transversely, concave from above downwards in the dorsal region, and convex in the same direction in the cervical and lumbar regions.

The *posterior surface* presents in the median line the spinous processes. These are short, horizontal, with bifid extremities in the cervical region. In the dorsal region, they are directed obliquely above, assume almost a vertical direction in the middle, and are horizontal, like the spines of the lumbar vertebrae, below. They are separated by considerable intervals in the loins, by narrower intervals in the neck, and are closely approximated in the middle of the dorsal region. On either side of the spinous processes; extending the whole length of the column, is the vertebral groove, formed by the laminae in the cervical and lumbar regions, where it is shallow, and by the laminae and transverse processes in the dorsal region, where it is deep and broad. In the recent state, these grooves lodge the deep muscles of the back. External to the vertebral grooves are the articular processes, and still more externally the transverse processes. In the dorsal region, these latter processes stand backwards, on a place considerably posterior to the same processes in the cervical and lumbar regions. The transverse processes in certain regions of the spine are formed of two different parts, or segments. In the cervical region, these two segments are distinct; the one arising from the side of the body, the other from the pedicle of the vertebra; and these uniting, enclose the vertebral foramen. In the dorsal region, the anterior segment is wanting; the posterior segment retaining the name of the transverse process. In the lumbar region, the anterior segments (which are largely developed) are called the transverse processes; but, in reality, they are lumbar ribs, the posterior segments or true transverse processes existing in a rudimentary state, and being developed from the superior articular processes, as in the cervical region. In the cervical region, the transverse processes are placed in front of the articular processes, and between the intervertebral foramina. In the lumbar, they are placed also in front of the articular processes, but behind the intervertebral foramina. In the dorsal region, they are posterior both to the articular processes and foramina.

The *lateral surfaces* are separated from the posterior by the articular processes in the cervical and lumbar regions, and by the transverse processes in the dorsal. These surfaces present in front the sides of the bodies of the vertebrae, marked in the dorsal region by the facets for articulation with the heads of the ribs. More posteriorly are the intervertebral foramina, formed by the juxtaposition of the intervertebral notches, oval in shape, smallest in the cervical and upper part of the dorsal regions, and gradually increasing in size to the last lumbar. They are situated between the transverse processes in the neck, and in front of them in the back and loins, and transmit the spinal nerves. The *base* of the vertebral column is formed by the under surface of the body of the fifth lumbar vertebra, and the summit by the upper surface of the atlas. The *vertebral canal* follows the different curves of the spine; it is largest in those regions in which the spine enjoys the greatest freedom of movement, as in the neck and loins, where it is wide and triangular; and narrow and rounded in the back, where motion is more limited.

THE SKULL

The Skull is divided into two parts, the Cranium and the Face. The Cranium is composed of *eight* bones; viz., the *occipital, two parietal, frontal, two temporal, sphenoid,* and *ethmoid.* The Face is composed of *fourteen* bones; viz., the *two nasal, two superior maxillary, two lachrymal, two malar, two palate, two inferior turbinated, vomer, inferior maxillary.* The *ossicula auditus,* the *teeth,* and *Wormian bones,* are not included in this enumeration.

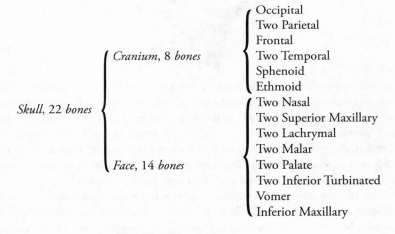

Skull, 22 *bones*
- *Cranium,* 8 *bones*
  - Occipital
  - Two Parietal
  - Frontal
  - Two Temporal
  - Sphenoid
  - Ethmoid
- *Face,* 14 *bones*
  - Two Nasal
  - Two Superior Maxillary
  - Two Lachrymal
  - Two Malar
  - Two Palate
  - Two Inferior Turbinated
  - Vomer
  - Inferior Maxillary

The Occipital Bone

The *Occipital Bone* (fig. 18) is situated at the posterior and inferior part of the cranium, is trapezoid in form, curved upon itself, and presents for examination two surfaces, four borders, and four angles.

External Surface. Midway between the summit of the bone and the posterior margin of the foramen magnum is a prominent tubercle, the external occipital protuberance, for the attachment of the ligamentum nuchae; and descending from it, as far as the foramen, a vertical ridge, the external occipital crest. Passing outwards from the occipital protuberance on each side are two semicircular ridges, the superior curved lines; and running parallel with these from the middle of the crest, are the two inferior curved lines. The surface of the bone above the superior curved lines presents on each side a smooth surface, which, in the recent state, is covered by

18.—Occipital Bone. Outer Surface.

the occipito-frontalis muscle, whilst the ridges, as well as the surfaces of the bone between them, serve for the attachment of numerous muscles. The superior curved line gives attachment internally to the Trapezius, externally to the Occipito-frontalis, and Sterno-cleido mastoideus; to the extent shewn in the figure. The depressions between the curved lines to the Complexus internally, the Splenius capitis and Obliquus superior externally. The inferior curved line, and the depressions below it, afford insertion to the Rectus capitis posticus, major and minor.

The *foramen magnum* is a large aperture, with rounded shelving margins, oblong in form, and wider behind than in front; it transmits the spinal cord and its membranes, the spinal accessory nerves, and the vertebral arteries. On each side of the foramen magnum are the occipital condyles, for articulation with the Atlas; they are convex articular surfaces, oval in form, and directed downwards and outwards; they approach each other anteriorly, and encroach more upon the anterior than the posterior segment of the foramen. On their inner surface is a rough tubercle, for the attachment of the check ligaments; whilst external to them is a rough tubercular prominence, the transverse or jugular process, channelled in front by a deep notch, which forms part of the jugular foramen. The under surface of this process affords attachment to the Rectus capitis lateralis muscle; its upper or cerebral surface presents a deeply curved groove, which lodges part of the lateral sinus, whilst its prominent extremity is marked by a quadrilateral rough surface, covered with cartilage in the fresh state, and articulating with a similar surface on the petrous portion of the temporal bone. On the outer side of each condyle is a depression, the anterior condyloid fossa, perforated at the bottom by the anterior condyloid foramen. This foramen (sometimes double) is directed downwards outwards, and forwards, and transmits the lingual nerve. Behind each condyle is seen an irregular fossa, also perforated at the bottom by a foramen, the posterior condyloid, for the transmission of a vein to the lateral sinus. This fossa and foramen are less regular in form and size than the anterior, and do not always exist. Sometimes they are found on one side only, and sometimes are altogether absent. In front of the foramen magnum is the basilar process, somewhat quadrilateral in form, wider behind than in front; its under surface, which is rough, presenting in the median line a tubercular ridge, the pharyngeal spine, for the attachment of the tendinous raphe and Superior constrictor of the pharynx; and on each side of it, rough depressions for the attachment of the Recti capitis antici, major and minor.

The *Internal or Cerebral Surface* (fig. 19) is deeply concave. The occipital part is divided by a crucial ridge into four fossae. The two superior, the smaller, receive the posterior lobes of the cerebrum, and present eminences and depressions corresponding to their convolutions. The two inferior, which receive the lateral lobes of the cerebellum, are larger than the former, and comparatively smooth; both are marked by slight grooves for the lodgment of arteries. At the point of meeting of the four divisions of the crucial ridge is an eminence, the internal occipital protuberance, which rarely corresponds to that on the outer surface. From this eminence, the

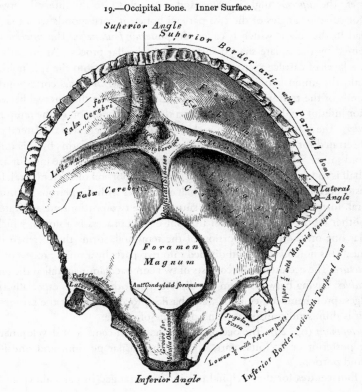

19.—Occipital Bone. Inner Surface.

superior division of the crucial ridge, called *sulcus longitudinalis*, runs upwards to the superior angle of the bone; it presents a deep groove for the superior longitudinal sinus, whilst its margins give attachment to the falx cerebri. The inferior division, the internal occipital crest, runs to the margin of the foramen magnum, on the edge of which it becomes gradually lost: this ridge, which is bifurcated below, serves for the attachment of the falx cerebelli, and is slightly grooved for the lodgment of the occipital sinuses. The transverse grooves (*sulci transversales*) pass outwards to the lateral angles; they are deeply grooved, for the lodgment of the lateral sinuses, their prominent margins affording attachment to the tentorium cerebelli. At the point of meeting of these four grooves is a depression for the torcular Herophili, placed a little to the right of the internal occipital protuberance. In the centre of the basilar portion of the bone is the foramen magnum, and above its margin, but nearer its anterior than its posterior part, the internal openings of the anterior condyloid foramina; the internal openings of the posterior condyloid foramina being a little external and posterior to them, and protected above by a small arch of bone. In front of the foramen magnum is the basilar process, presenting a shallow longitudinal depression, the basilar groove, for supporting the medulla oblongata; whilst on its lateral margins is observed a narrow channel on each side, which, when united with a similar channel on the petrous portion of the temporal bone, forms a groove, the inferior petrosal, which lodges the inferior petrosal sinus.

Angles. The *superior* angle is acute, and is received into the interval between the posterior superior angles of the two parietal bones: it corresponds with that part of the head in the foetus which is called the *posterior fontanelle*. The *inferior* angle is represented by the square-shaped surface of the basilar process. At an early period of life, a layer of cartilage separates this part of the bone from the sphenoid; but in the adult, the union between them is osseous. The *lateral angles* correspond to the outer ends of the transverse grooves, and are received into the interval between the posterior inferior angles of the parietal and the mastoid portion of the temporal.

Borders. The *superior* extends on each side from the superior to the lateral angle, is deeply serrated for articulation with the parietal bone, and forms by this union the lambdoid suture. The *inferior* border extends from the lateral to the inferior angle; its upper half is rough, and articulates with the mastoid portion of the temporal, forming the masto-occipital suture: the inferior half articulates with the petrous portion of the temporal, forming the petro-occipital suture: these two portions are separated from one another by the jugular process. In front of this process is a notch, which, with a similar one on the petrous portion of the temporal, forms the foramen lacerum posterius. This notch is often subdivided into two parts by a small process of bone.

Structure. The occipital bone consists of two compact laminae, called the *outer* and *inner tables*, having between them the diploic tissue; this bone is especially thick, at the ridges, protuberances, condyles, and basilar process; whilst at the bottom of the fossae it is thin, semi-transparent, and destitute of diploë.

Development (fig. 20). The occipital bone has *seven* centres of development; four for the posterior or occipital part, one for the basilar portion; and one for each condyloid portion.

The four centres for the occipital portion are arranged in pairs above and below the occipital protuberance, and appear about the tenth week of foetal life; the inferior pair make their appearance first, and join; the superior pair become also united: these two segments now join together, and form a single piece. The condyloid portions then ossify; and, lastly, the basilar portion. At birth, the bone consists of these four parts, separate from one another, the posterior being fissured in the direction of the original segments. At about the fourth year, the occipital and the two condyloid pieces join; and at about the sixth year the bone consists of a single piece. At a later period, between the eighteenth and twenty-fifth years, the occipital and sphenoid become united, forming a single bone.

20.—Development of Occipital Bone.

Articulations. With six bones; two parietal, two temporal, sphenoid, and Atlas.

Attachment of Muscles. To the superior curved line are attached the Occipito-frontalis, Trapezius, and Sterno-cleido-mastoid. To the space between the curved lines, the Complexus, Splenius capitis, and Obliquus superior; to the inferior curved line, and the space between it and the foramen magnum, the Rectus posticus major and minor; to the transverse process, the Rectus lateralis; and to the basilar process, the Recti antici majores and minores, and Superior Constrictor of the pharynx.

The Parietal Bones

The *Parietal Bones* form the sides and roof of the skull; they are of an irregular quadrilateral form, and present for examination two surfaces, four borders, and four angles.

Surfaces. The *External Surface* (fig. 21) is convex, smooth, and presents about its centre an eminence, called the parietal eminence, which indicates the point where ossification commenced. Crossing the centre of the bone in an arched direction is a curved ridge, the temporal ridge, for the attachment of the temporal fascia. Above this ridge, the surface of the bone is rough and porous, and covered by the aponeurosis of the Occipito-frontalis; below it the bone is smooth, and affords attachment to the Temporal muscle. At the back part of the superior border is a small foramen, the parietal foramen, which transmits a vein to the superior longitudinal sinus. Its existence is not constant, and its position varies considerably.

The *Internal Surface* (fig. 22), concave, presents numerous eminences and depressions for lodging the convolutions of the brain, and minute furrows for the ramifications of the meningeal arteries: these run upwards and backwards from deep grooves, which commence in the anterior inferior angle, and at the central and posterior part of the lower border of the bone. Along the upper margin is part of a shallow groove, which, when joined to the opposite parietal, forms a channel for the superior longitudinal sinus, the elevated edges of which afford attachment to the

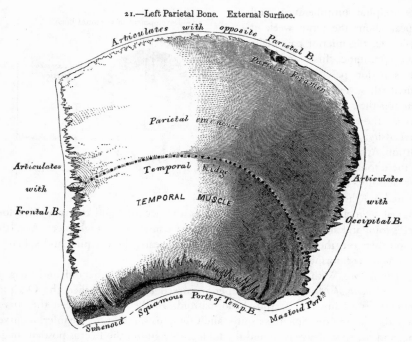

21.—Left Parietal Bone. External Surface.

falx cerebri. Near the groove are seen several depressions; they lodge the Pacchionian bodies. The internal opening of the parietal foramen is also seen when that aperture exists.

Borders. The *superior*, the longest, is dentated to articulate with its fellow of the opposite side, forming the sagittal suture. The *inferior* is divided into three parts; of these, the anterior is thin and pointed, bevelled at the expense of the outer surface, and overlapped by the tip of the great wing of the sphenoid; the middle portion is arched, bevelled at the expense of the outer surface, and overlapped by the squamous portion of the temporal; the posterior portion being thick and serrated for articulation with the mastoid portion of the temporal. The *anterior border*, deeply serrated, is bevelled at the expense of the outer surface above, and of the inner below; it articulates with the frontal bone, forming the coronal suture. The *posterior* border, deeply denticulated, articulates with the occipital, forming the lambdoid suture.

Angles. The *anterior superior*, thin and pointed, corresponds with that portion of the skull which in the foetus is membranous, and is called the *anterior fontanelle*. The *anterior inferior angle* is thin and lengthened, being received in the interval between the great wing of the sphenoid and the frontal. Its inner surface is marked by a deep groove, sometimes a canal, for the middle meningeal artery. The *posterior superior angle* corresponds with the junction of the sagittal and lambdoid sutures. In the foetus this part of the skull is membranous, and is called the *posterior fontanelle*. The *posterior inferior* articulates with the mastoid portion of the temporal bone, and presents on its inner surface a broad shallow groove for the lateral sinus.

Development. The parietal bone is developed by *one* centre, which corresponds

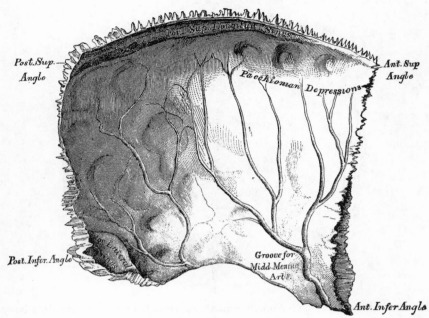
22.—Left Parietal Bone. Internal Surface.

with the parietal eminence, and makes its first appearance about the fifth or sixth week of foetal life.

Articulations. With five bones; the opposite parietal, the occipital, frontal, temporal, and sphenoid.

Attachment of Muscles. To one only, the Temporal.

The Frontal Bone

This bone, which resembles a cockle-shell in form, consists of two portions—a *vertical* or *frontal* portion, situated at the anterior part of the cranium, forming the forehead; and a *horizontal* or *orbito-nasal* portion, which enters into the formation of the roof of the orbits and nose.

Vertical Portion. External Surface (fig. 23). In the median line, traversing the bone from the upper to its lower part, is a slightly elevated ridge, and in young subjects a suture, which represents the point of union of its two lateral halves: in the adult, this suture usually disappears. On either side of this ridge, a little below the centre of the bone, is a rounded eminence, the frontal eminence, which indicates the point where ossification commenced. The whole surface of the bone above this part is smooth, and covered by the aponeurosis of the Occipito-frontalis muscle. Below the frontal protuberance, and separated from it by a slight groove, is the superciliary ridge, a curved eminence, broad internally where it is continuous with the nasal eminence, less distinct externally as it arches outwards. Beneath the superciliary ridge is the supra-orbital arch, a curved and prominent margin, which forms the upper boundary of the orbit, and separates the vertical from the horizontal portion of the bone. At

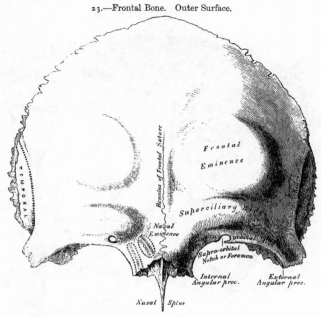

23.—Frontal Bone. Outer Surface.

the inner third of this arch is a notch, sometimes converted into a foramen by a bony process or ligament, and called the *supra-orbital notch* or *foramen*. It transmits the supra-orbital artery, veins, and nerve. The supra-orbital arch terminates externally in the external angular process, and internally in the internal angular process. The external angular is a strong prominent process, which articulates with the malar: running upwards and backwards from it is a sharp curved line, the temporal ridge, for the attachment of the temporal fascia; and beneath it a slight concavity, that forms part of the temporal fossa, and gives origin to the Temporal muscle. The internal angular processes are less marked than the external, and articulate with the lachrymal bones. Between the two is a rough, uneven interval, called the *nasal notch*, which articulates in the middle line with the nasal, and on either side with the nasal process of the superior maxillary bones.

Vertical Portion. Internal Surface (fig. 24). Along the middle line of this surface is a vertical groove, sulcus longitudinalis, the edges of which unite below to form a ridge, the frontal crest; the groove lodges the superior longitudinal sinus, whilst its edges afford attachment to the falx cerebri. The crest terminates below, at a small opening, the foramen caecum, which is generally completed behind by the ethmoid; it lodges a process of the falx cerebri, and occasionally transmits a small vein from the nose to the superior longitudinal sinus. On either side of the groove, the bone is deeply concave, presenting eminences and depressions for the convolutions of the brain, and numerous small furrows for lodging the ramifications of the anterior meningeal arteries. Several small, irregular fossae are also seen on either side of the groove, for the reception of the Pacchionian bodies.

Horizontal Portion. External Surface. This portion of the bone consists of two thin plates, which form the vault of the orbits, separated from one another by

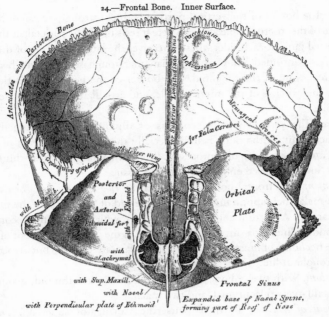

24.—Frontal Bone. Inner Surface.

the ethmoidal notch. Each orbital vault consists of a smooth, concave, triangular plate of bone, marked at its anterior and external part (immediately beneath the external angular process) by a shallow depression, the lachrymal fossa, for lodging the lachrymal gland; and at its anterior and internal part, by a depression, sometimes a tubercle, for the attachment of the fibrous pulley of the superior oblique muscle. The ethmoidal notch separates the two orbital plates: it is quadrilateral; and filled up, when the bones are united, by the cribriform plate of the ethmoid. The edges of this notch present several half-cells, which, when united with corresponding half-cells on the upper surface of the ethmoid, complete the ethmoidal cells: two grooves are also seen crossing these edges transversely; they are converted into canals by articulation with the ethmoid, and are called the *anterior* and *posterior* ethmoidal canals; they open on the inner wall of the orbit. In front of the ethmoidal notch is the nasal spine, a sharp eminence, which projects downwards and forwards, and the grooved base of which forms part of the roof of the nose. It articulates in front with the crest of the nasal bones, behind with the perpendicular plate of the ethmoid. On either side of this spine are the openings of the frontal sinuses. These are two irregular cavities, which extend upwards and outwards, a variable distance, between the two tables of the skull, and are separated from one another by a thin bony septum. They give rise to the prominences above the root of the nose, called the *nasal eminences*. In the child they are absent, and they become gradually developed as age advances. They are lined by mucous membrane, and communicate with the nose by the infundibulum.

The *Internal Surface* of the *Horizontal Portion* presents the convex upper surfaces of the orbital plates, separated from each other in the middle line by the ethmoidal notch, and marked by eminences and depressions for the convolutions of the anterior lobes of the brain.

Borders. The border of the vertical portion is thick, strongly serrated, bevelled at the expense of the internal table above, where it rests upon the parietal, at the expense of the external table at each side, where it receives the lateral pressure of those bones: this border is continued below, into a triangular rough surface, which articulates with the great wing of the sphenoid. The border of the horizontal portion is thin, bevelled at the expense of the internal table, and articulates with the lesser wing of the sphenoid.

Structure. The vertical portion consists of diploic tissue, contained between two compact laminae, the bone being especially thick in the situation of the nasal eminences and external angular processes. The horizontal portion is thinner, more translucent, and composed entirely of compact tissue.

Development. The frontal bone is developed by *two* centres, one for each lateral half, which make their appearance, at an early period of foetal life, in the situation of the orbital arches. At birth it consists of two pieces, which afterwards become united along the middle line, by a suture which runs from the vertex to the root of the nose. This suture becomes obliterated within a few years after birth; but it occasionally remains throughout life.

Articulations. With twelve bones; two parietal, sphenoid, ethmoid; two nasal, two superior maxillary, two lachrymal, and two malar.

Attachment of Muscles. To three pairs; the Corrugator supercilii, Orbicularis palpebrarum, and Temporal.

The Temporal Bones

The Temporal bones, situated at the side and base of the skull, present for examination a *squamous*, *mastoid*, and *petrous* portion.

The *Squamous Portion* (fig. 25), the most anterior and superior part of the bone, is flattened and scale-like in form, thin and translucent in texture. Its outer surface is smooth, convex, and grooved for the deep temporal arteries; it affords attachment to the fibres of the Temporal muscle, and forms part of the temporal fossa. At its back part may be seen a curved ridge—part of the temporal ridge; it serves for the attachment of the temporal fascia, limits the origin of the Temporal muscle, and marks the boundary between the squamous and mastoid portions of the bone. Projecting from the lower part of the squamous portion, is a long and arched process of bone, the zygomatic process. It is at first directed outwards, its two surfaces looking upwards and downwards; it then appears as if twisted upon itself, and takes a direction forwards, its surfaces now looking inwards and outwards. The superior border of this process is long, thin, and sharp, and serves for the attachment of the temporal fascia. The inferior, short, thick, and arched, has attached to it some of the fibres of the Masseter muscle. Its outer surface is convex and subcutaneous. Its inner, concave, also affords attachment to the Masseter. The extremity, broad and deeply serrated, articulates with the malar bone. This process is connected to the temporal bone by three divisions, called *the roots of the zygomatic process*, an anterior, middle, and posterior. The anterior, which is short, but broad and strong, runs transversely inwards into a rounded eminence, the eminentia articularis. This eminence forms the front boundary of the glenoid fossa, and in the recent state is covered with cartilage.

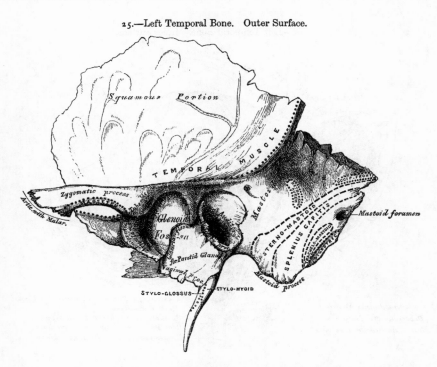

25.—Left Temporal Bone. Outer Surface.

The middle root runs obliquely inwards, and terminates at the edge of a well-marked fissure, the Glaserian fissure; whilst the posterior, which is strongly marked, runs from the upper border of the zygoma, in an arched direction, upwards and backwards, forming the posterior part of the temporal ridge. At the junction of the anterior and middle roots is a projection, called the *tubercle*, for the attachment of the external lateral ligament of the lower jaw; and between these roots is a large oval depression, forming part of the glenoid fossa, for the reception of the condyle of the lower jaw. This fossa is bounded in front by the eminentia articularis; behind, by the vaginal process; and externally by the auditory process; and is divided into two parts by a narrow slit, the Glaserian fissure: the anterior part, formed by the squamous portion of the bone is smooth, covered in the recent state with cartilage, and articulates with the condyle of the lower jaw; the posterior part, rough and uneven, and formed chiefly by the vaginal process of the petrous portion, lodges part of the parotid gland. The Glaserian fissure, dividing the two, leads into the tympanum; it lodges the processus gracilis of the malleus, and transmits the laxator tympani muscle and the anterior tympanic artery. The chorda tympani nerve passes through a separate canal parallel to the Glaserian fissure, on the outer side of the Eustachian tube, in the retiring angle between the squamous and petrous portions of the temporal bone.

The *internal surface* of the squamous portion (fig. 26) is concave, presents numerous eminences and depressions for the convolutions of the cerebrum, and two well marked grooves for the branches of the middle meningeal artery.

Borders. The superior border is thin, bevelled at the expense of the internal surface,

26.—Left Temporal Bone. Inner Surface.

so as to overlap the lower border of the parietal bone, forming the squamous suture. The anterior inferior border is thick, serrated, and bevelled alternately at the expense of the inner and outer surfaces, for articulation with the great wing of the sphenoid.

The *Mastoid Portion* is situated at the posterior part of the bone, its outer surface is rough, and perforated by numerous foramina, one of these, of large size, situated at the posterior border of the bone, is termed the *mastoid foramen*, it transmits a vein to the lateral sinus and a small artery. The position and size of this foramen are very variable, being sometimes situated in the occipital bone, or in the suture between it and the occipital. The mastoid portion is continued below into a conical projection, the mastoid process, the size and form of which varies considerably in different individuals. This process serves for the attachment of the Sterno-mastoid, Splenius capitis and Trachelo-mastoid (see fig. 25); on the inner side of the mastoid process is a deep groove, the digastric fossa, for the attachment of the Digastric muscle, and running parallel with it, but more internal, the occipital groove, which lodges the occipital artery. The internal surface presents a deeply curved groove which lodges the lateral sinus, and into it may be seen opening the mastoid foramen.

Borders. The superior border of the mastoid portion is rough and serrated for articulation with the posterior inferior angle of the parietal bone. The posterior border, also uneven and serrated, articulates with the inferior border of the occipital bone between its lateral angle and jugular process.

The *Petrous Portion*, so named from its extreme density and hardness, is a pyramidal process of bone, wedged in at the base of the skull between the sphenoid and occipital bones. Its direction from without is forwards, inwards, and a little downwards. It presents for examination a base, an apex, three surfaces, and three borders. The *base* is applied against the internal surface of the squamous and mastoid portions, its upper half being concealed, but its lower half is exposed by their divergence, which brings into view the oval expanded orifice of a canal leading into the tympanum, the meatus auditorius externus. This canal is situated between the mastoid process and the posterior and middle roots of the zygoma; its upper margin is smooth and rounded, its lower surrounded by a curved plate of bone, the auditory process, the free margin of which is rough for the attachment of the cartilage of the ear. The *apex* of the petrous portion, rough and uneven, is received into the angular interval between the spinous process of the sphenoid, and the basilar process of the occipital; it presents the anterior orifice of the carotid canal, and forms the posterior and external boundary of the foramen lacerum medium. The *anterior surface* (fig. 26) of the petrous portion, forms the posterior boundary of the middle fossa of the skull. This surface is continuous with the squamous portion, to which it is united by a suture, the temporal suture, the remains of which are distinct at a late period of life. Proceeding from the base to the apex, this surface presents five points for examination. 1. An eminence which indicates the situation of the superior semicircular canal. 2. A shallow groove, sometimes double, leading backwards to an oblique opening, the hiatus Fallopii, for the passage of the petrosal branch of the Vidian nerve. 3. A smaller opening immediately beneath and external to the latter for the passage of the smaller petrosal nerve. 4. Near the apex of the bone is seen the termination of the carotid canal, the wall of which in this situation is deficient in front. 5. Above the foramen is a shallow depression for the reception of the Gasserian ganglion.

The *posterior surface* forms the front boundary of the posterior fossa of the skull, and is continuous with the inner surface of the mastoid portion of the bone. It presents three points for examination. 1. About its centre is a large orifice, the meatus auditorius internus. This aperture varies considerably in size, its margins are smooth and rounded, and it leads into a short and oblique canal which is directed outwards and forwards. It transmits the auditory and facial nerves and auditory artery. 2. Behind the meatus auditorius is a small slit almost hidden by a thin plate of bone, and leading to a canal, the aquaeductus vestibuli; it transmits a small artery and vein, and lodges a process of the dura mater. 3. In the interval between these two openings, but above them, is an angular depression which lodges a process of the dura mater, and transmits a small vein into the cancellous tissue of the bone.

The *inferior* or *basilar Surface* (fig. 27) is rough and irregular, and forms part of the base of the skull. Passing from the apex to the base, this surface presents eleven points for examination; 1. A rough surface, quadrilateral in form, which serves partly for the attachment of the Levator palati, and Tensor tympani muscles. 2. The opening of the carotid canal, a large circular aperture, which ascends at first vertically upwards, and then making a bend, runs horizontally forwards and inwards. It transmits the internal carotid artery, and the carotid plexus. 3. The aquaeductus cochleae, a small triangular opening, lying on the inner side of the latter, close to the posterior border of the bone; it transmits a vein from the cochlea, which joins the internal jugular.

4. Behind these openings is a depression, the jugular fossa, which varies in depth and size in different skulls; it lodges the internal jugular vein, and with a similar depression on the margin of the occipital bone, forms the foramen lacerum posterius. 5. A small foramen for the passage of Jacobson's nerve (the tympanic branch of the glosso-pharyngeal). This is seen on the ridge of bone dividing the carotid canal from the jugular fossa. 6. The canal for Arnold's nerve, seen on the inner wall of the jugular fossa. 7. Behind the jugular fossa is a smooth square-shaped facet, the jugular surface, which articulates with the jugular process of the occipital bone. 8. The vaginal process, a very broad sheath-like plate of bone, which extends from the carotid canal to the mastoid process; it divides behind into two laminae, receiving between them the 9th point for examination, the styloid process; a long sharp spine, about an inch in length, continuous with the vaginal process, between the laminae of which it is received, and directed downwards, forwards, and inwards. It affords attachment to three muscles, the Stylo-pharyngeus, Stylo-glossus, and Stylo-hyoideus, and two ligaments, the stylo-hyoid and stylo-maxillary. 10. The stylo-mastoid foramen, a rather large orifice, placed between the styloid and mastoid processes; it is the termination of the aquaeductus Fallopii, and transmits the facial nerve and stylo-mastoid artery. 11. The auricular fissure, situated between the vaginal and mastoid processes, and transmitting the auricular branch of the pneumogastric nerve.

27.—Petrous Portion. Inferior Surface.

Borders. The *superior*, the longest, is grooved for the superior petrosal sinus, and has attached to it the tentorium cerebelli: at its inner extremity is a semilunar notch, upon which reclines the fifth nerve. The *posterior* border is intermediate in length between the superior and the anterior. Its inner half is marked by a groove, which, when completed by its articulation with the occipital, forms the channel for the inferior petrosal sinus. Its outer half presents a deep excavation for the jugular fossa, which, with a similar notch on the occipital, forms the foramen lacerum posterius. A projecting eminence of bone occasionally stands out from the centre of the notch, and divides the foramen into two parts. The *anterior* border is divided into two parts, an outer, joined to the squamous portion by a suture the remains of which are distinct; an inner, free, articulating with the spinous process of the sphenoid. At the angle of junction of these two parts, are seen two canals, separated from one another by a thin plate of bone, the processus cochleariformis; they both lead into the tympanum, the upper one transmitting the Tensor tympani muscle, the lower one the Eustachian tube.

Structure. The squamous portion is like that of the other cranial bones, the mastoid portion cellular, and the petrous portion dense and hard.

Development (fig. 28). The temporal bone is developed by *four* centres, exclusive of those for the internal ear and the ossicula, viz.;—one for the squamous portion including the zygoma, one for the petrous and mastoid parts, one for the styloid, and one for the auditory process (tympanic bone). The first traces of the development of this bone are found in the squamous portion, they appear about the time when osseous matter is deposited in the vertebrae; the auditory process succeeds next, it consists of an elliptical portion of bone, forming about three-fourths of a circle, the deficiency being above; it is grooved along its concave surface for the attachment of the membrana tympani, and becomes united

28.—Development of Temporal Bone.
By four Centres.

by its extremities to the squamous portion during the last months of intra-uterine life. The petrous and mastoid portions then become ossified, and lastly the styloid process, which remains separate a considerable period, and is occasionally never united to the rest of the bone. At birth the temporal bone, excluding the styloid process, is formed of three pieces, the squamous and zygomatic, the petrous and mastoid, and the auditory. The auditory process joins with the squamous at about the ninth month. The petrous and mastoid join with the squamous during the first year, and the styloid process becomes united between the second and third years. The subsequent changes in this bone are the extension outwards of the auditory process so as to form the meatus auditorius, the glenoid fossa becomes deeper, and the mastoid part enlarges from the development of numerous cellular cavities in its interior.

Articulations. With five bones, occipital, parietal, sphenoid, inferior maxillary and malar.

Attachment of Muscles. To the squamous portion, the Temporal; to the zygoma, the Masseter; to the mastoid portion, the Occipito-frontalis, Sterno-mastoid, Splenius capitis, Trachelo-mastoid, Digastricus and Retrahens aurem; to the styloid process, the Stylo-pharyngeus, Stylo-hyoideus and Stylo-glossus; and to the petrous portion, the Levator palati, Tensor tympani, and Stapedius.

The Sphenoid Bone

The Sphenoid (σφὴν, a "wedge"; εἶδος, "like") is situated at the anterior part of the base of the skull, articulating with all the other cranial bones, which it binds firmly and solidly together. In its form it somewhat resembles a bat, with its wings extended; and is divided into a central portion or body, two greater and two lesser wings extending outwards on each side of the body; and two processes, the pterygoid processes, which project from it below.

The *Body* presents for examination *four* surfaces—a superior, an inferior, an anterior, and a posterior.

The *superior surface* (fig. 29). From before, backwards, is seen a prominent spine, the ethmoidal spine, for articulation with the ethmoid; behind this a smooth surface, presenting in the median line a slight longitudinal eminence, with a depression on each side, for lodging the olfactory nerves. A narrow transverse groove, the optic groove, bounds the above-mentioned surface behind; it lodges the optic commissure, and terminates on either side in the optic foramen, for the passage of the optic nerve and ophthalmic artery. Behind the optic groove is a small eminence, olive-like in shape, the olivary process; and still more posteriorly, a deep depression, the pituitary fossa, or sella Turcica, which lodges the pituitary body. This fossa is perforated by numerous foramina, for the transmission of nutrient vessels to the substance of the bone. It is bounded in front by two small eminences, one on either side, called the middle clinoid processes, and behind by a square-shaped plate of bone, terminating at each superior angle in a tubercle, the posterior clinoid processes, the size and form of which vary considerably in different individuals. The sides of this plate of bone are notched below, for the passage of the sixth pair of nerves; and behind, it presents a shallow depression, which slopes obliquely backwards, and is continuous with the

29.—Sphenoid Bone, Superior Surface.

30.—Sphenoid Bone, Anterior Surface.

basilar groove of the occipital bone; it supports the medulla oblongata. On either side of the body may be seen a broad groove, curved somewhat like the italic letter *f*; it lodges the internal carotid artery and the cavernous sinus, and is called the *cavernous groove*. The *posterior surface*, quadrilateral in form, articulates with the basilar process of the occipital bone. During childhood, a separation between these bones exists by means of a layer of cartilage; but in after-life this becomes ossified, and the two bones are immoveably connected together. The *anterior surface* (fig. 30) presents, in the middle line, a vertical lamella of bone, which articulates in front with the perpendicular plate of the ethmoid. On either side of it are the irregular openings leading into the sphenoidal sinuses. These are two large, irregular cavities, hollowed out of the interior of the body of the sphenoid bone, and separated from one another by a more or less complete perpendicular septum; their form varies considerably, being often subdivided by irregular osseous laminae. These sinuses do not exist in children; but they increase in size as age advances. They are partially closed, in front and below, by two thin triangular plates of bone, the sphenoidal turbinated bones, leaving a round opening at their upper parts, by which they communicate with the upper and back part of the nose, and occasionally with the posterior ethmoidal cells. The lateral margins of this surface present a serrated edge, which articulates with the os planum of the ethmoid, completing the posterior ethmoidal cells; the lower margin, also rough and serrated, articulates with the orbital process of the palate bone; and the upper margin with the orbital plate of the frontal bone. The *inferior surface* presents, in the middle line, a triangular spine, the rostrum, which is continuous with the vertical plate on the anterior surface, and is received into a deep fissure between the alae of the vomer. On each side may be seen a projecting lamina of bone, which runs horizontally inwards from near the base of the pterygoid process: these plates, termed the vaginal processes, articulate with the edges of the vomer. Close to the root of the pterygoid process is a groove, formed into a complete canal when articulated with the sphenoidal process of the palate bone; it is called the pterygo-palatine canal, and transmits the pterygo-palatine vessels and pharyngeal nerve.

The *Greater Wings* are two strong processes of bone, which arise at the sides of the body, and are curved in a direction upwards, outwards, and backwards; being prolonged behind into a sharp-pointed extremity, called the *spinous process of the sphenoid*. Each wing presents three surfaces and a circumference. The *superior* or *cerebral* surface forms part of the middle fossa of the skull; it is deeply concave, and presents eminences and depressions for the convolutions of the brain. At its anterior and internal part is seen a circular aperture, the foramen rotundum, for the transmission of the second division of the fifth nerve. Behind and external to this, a large oval foramen, the foramen ovale, for the transmission of the third division of the fifth, the small meningeal artery, and the small petrosal nerve. At the inner side of the foramen ovale, a small aperture may occasionally be seen opposite the root of the pterygoid process; it is the foramen Vesalii, transmitting a small vein. Lastly, in the apex of the spine of the sphenoid is a short canal, sometimes double, the foramen spinosum; it transmits the middle meningeal artery. The *external* surface is convex, and divided by a transverse ridge, the pterygoid ridge, into two portions. The superior or larger, convex from above downwards, concave from before backwards, enters into the formation of the temporal fossa, and attaches part of the Temporal muscle. The inferior portion, smaller in size and concave, enters into the formation of the zygomatic fossa, and affords attachment to the External pterygoid muscle. It presents, at its posterior part, a sharp-pointed eminence of bone, the spinous process, to which is connected the internal lateral ligament of the lower jaw, and the Laxator tympani muscle. The pterygoid ridge, dividing the temporal and zygomatic portions, gives attachment to the upper origin of the External pterygoid muscle. At its inner extremity is a long triangular spine of bone, which serves to increase the extent of origin of this muscle. The *anterior* or *orbital* surface, smooth and quadrilateral in form, assists in forming the outer wall of the orbit. It is bounded above by a serrated edge, for articulation with the frontal bone; below, by a rounded border, which enters into the formation of the spheno-maxillary fissure; internally, it enters into the formation of the sphenoidal fissure; whilst externally it presents a serrated margin, for articulation with the malar bone. At the upper part of the inner border is a notch, for the transmission of a branch of the ophthalmic artery; and at its lower part a small pointed spine of bone, which serves for the attachment of part of the lower head of the external rectus. One or two small foramina may occasionally be seen, for the passage of arteries; they are called the *external orbitar foramina*. Circumference: from the body of the sphenoid to the spine (commencing from behind), the outer half of this margin is serrated, for articulation with the petrous portion of the temporal bone; whilst the inner half forms the anterior boundary of the foramen lacerum medium, and presents the posterior aperture of the Vidian canal. In front of the spine, the circumference of the great wing presents a serrated edge, bevelled at the expense of the inner table below, and of the external above, which articulates with the squamous portion of the temporal bone. At the tip of the great wing a triangular portion is seen, bevelled at the expense of the internal surface, for articulation with the anterior inferior angle of the parietal bone. Internal to this is a broad serrated edge, for articulation with the frontal bone: this surface is continuous internally with the sharp inner edge of the orbital plate, which assists in the formation of the sphenoidal fissure.

The *Lesser Wings* (processes of Ingrassias) are two thin triangular plates of bone, which arise from the upper and anterior part of the body of the sphenoid; and, projecting transversely outwards, terminate in a more or less acute point. The superior surface of each is smooth, flat, broader internally than externally, and supports the anterior lobe of the brain. The inferior surface forms the back part of the roof of the orbit, and the upper boundary of the sphenoidal fissure, or foramen lacerum anterius. This fissure is of a triangular form, and leads from the cavity of the cranium into the orbit; it is bounded internally by the body of the sphenoid; above, by the lesser wing; and below, by the orbital surface of the great wing; and is converted into a foramen by the articulation of this bone with the frontal. It transmits the third, fourth, ophthalmic division of the fifth and sixth nerves, and the ophthalmic vein. The anterior border of the lesser wing is serrated, for articulation with the frontal bone; the posterior, smooth and rounded, is received into the fissure of Sylvius of the brain. The inner extremity of this border forms the anterior clinoid process. The lesser wing is connected to the side of the body by two roots, the upper thin and flat, the lower thicker, obliquely directed, and presenting on its outer side a small tubercle, for the attachment of the common tendon of the muscles of the eye. Between the two roots is the optic foramen, for the transmission of the optic nerve and ophthalmic artery.

The *Pterygoid* processes (fig. 31), one on each side, descend perpendicularly from the point where the body and great wing unite. Each process consists of an external and an internal plate, separated behind by an intervening notch; but joined partially in front. The *external pterygoid plate* is broad and thin, turned a little outwards, and forms part of the inner wall of the zygomatic fossa. It gives attachment, by its outer surface, to the External pterygoid; its

31.—Sphenoid Bone. Posterior Surface.

inner surface forms part of the pterygoid fossa, and gives attachment to the Internal pterygoid. The *internal pterygoid plate* is much narrower and longer, curving outwards at its extremity, into a hook-like process of bone, the hamular process, around which turns the tendon of the Tensor-palati muscle. At the base of this plate is a small, oval, shallow depression, the scaphoid fossa, from which arises the Tensor-palati, and above which is seen the posterior orifice of the Vidian canal. The outer surface of this plate forms part of the pterygoid fossa, the inner surface forming the outer boundary of the posterior aperture of the nares. The two pterygoid plates are separated below by an angular notch, in which the pterygoid process, or tuberosity, of the palate bone is received. The anterior surface of the pterygoid process is very broad at its base, and supports Meckel's ganglion. It presents, above, the anterior orifice of the Vidian canal; and below, a rough margin, which articulates with the perpendicular plate of the palate bone.

Development. The sphenoid bone is developed by *ten* centres, six for the posterior sphenoidal division, and four for the anterior sphenoid. The six centres for the post-sphenoid are, one for each greater wing and external pterygoid plate; one for each internal pterygoid plate; two for the posterior part of the body. The four for the anterior sphenoid are, one for each lesser wing and anterior part of the body; and one for each sphenoidal turbinated bone. Ossification takes place in these pieces in the following order: the greater wing and external pterygoid plate are first formed, ossific granules being deposited close to the foramen rotundum on each side, at about the second month of foetal life; ossification spreading outwards into the great wing, and downwards into the external pterygoid process. Each internal pterygoid plate is then formed, and becomes united to the external about the middle of foetal life. The two centres for the posterior part of the body appear as separate nuclei, side by side, beneath the sella Turcica; they join about the middle of foetal life into a single piece, which remains ununited to the rest of the bone until after birth. Each lesser wing is formed by a separate centre, which appears on the outer side of the optic foramen, at about the third month; they become united and join with the body at about the eighth month of foetal life. At about the end of the third year, ossification has made its appearance in the sphenoidal spongy bones.

At birth, the sphenoid consists of three pieces; viz. the greater wing and pterygoid processes on each side; the lesser wings and body united. At the first year after birth, the greater wings and body are united. From the tenth to the twelfth year, the spongy bones commence their junction to the sphenoid, and become completely united by the twentieth year. Lastly, the sphenoid joins the occipital.

Articulations. The sphenoid articulates with *all* the bones of the cranium, and five of the face; the two malar, two palate, and vomer: the exact extent of articulation with each bone is shewn in the accompanying figures.

Attachment of Muscles. The Temporal, External pterygoid, Internal pterygoid, Superior constrictor, Tensor-palati, Laxator-tympani, Levator-palpebrae, Obliquus superior, Superior rectus, Internal rectus, Inferior rectus, External rectus. For the exact attachment of the muscles of the eye to the sphenoid bone, see fig. 133.

The Sphenoidal Spongy Bones

The *Sphenoidal Spongy Bones* are two thin, curved plates of bone, which exist as separate pieces up to the fifteenth year. They are situated at the anterior and inferior part of the body of the sphenoid, serving to close in the sphenoidal sinuses in this situation. They are irregular in form, thick, and tapering to a point behind, broader and thinner in front. Their inner surface, which looks towards the cavity of the sinus, is concave; their outer surface convex. Each bone articulates in front with the ethmoid, an aperture of variable size being left in their anterior wall, which communicates with the posterior ethmoidal cells: behind, its point is placed under the vomer, and is received between the root of the pterygoid process on the outer side, and the rostrum of the sphenoid on the inner: externally, it articulates with the palate.

The Ethmoid

The *Ethmoid* (ηθμος, a sieve), is an exceedingly light spongy bone, of a cubical form, situated at the anterior part of the base of the cranium, between the two orbits, at the root of the nose. It consists of three parts: a horizontal plate, which forms part of the base of the cranium; a perpendicular plate, which forms part of the septum nasi; and two lateral masses of cells.

33.—Ethmoid Bone. Outer Surface (enlarged).

The *Horizontal* or *Cribriform Plate* (fig. 33) forms part of the anterior fossa of the base of the skull, and is received into the ethmoid notch of the frontal bone between the two orbital plates. Projecting upwards from the middle line of this plate, at its fore part, is a thick smooth triangular process of bone, the crista galli, so called from its resemblance to a cock's-comb. Its base joins the cribriform plate. Its posterior border, long, thin, and slightly curved, serves for the attachment of the falx cerebri. Its anterior, short and thick, articulates with the frontal bone, and presents at its lower part two small projecting alae, which are received into corresponding depressions in the frontal, completing the foramen caecum behind. Its sides are smooth, and sometimes bulging, when it is found to enclose a small sinus. On each side of the crista galli, the cribriform plate is concave, to support the bulb of the olfactory nerves, and perforated by numerous foramina for the passage of its filaments. These foramina consist of three sets, corresponding to the three sets of olfactory nerves; an inner, which are lost in grooves on the upper part of the septum; an outer set, continued on to the surface of the upper spongy bones; whilst the middle set run simply through the bone, and transmit nerves distributed to the roof of the nose. At the front part of the cribriform plate, by the side of the crista galli, is a small fissure, which transmits the nasal branch of the ophthalmic nerve; and at its posterior part a triangular notch, which receives the ethmoidal spine of the sphenoid.

34.—Perpendicular Plate of Ethmoid (enlarged). Shewn by removing the Right Lateral Mass.

The *Perpendicular Plate* (fig. 34) is a thin central lamella of bone, which descends from the under surface of the cribriform plate, and assists in forming the septum of the nose. Its anterior border articulates with the frontal spine and crest of the nasal bones. Its posterior, divided into two parts, is connected by its upper half with the rostrum of the sphenoid; its lower half with the vomer. The inferior border serves for the attachment of the triangular cartilage of the nose. On each side of the perpendicular plate numerous grooves and canals are seen, leading from the foramina on the cribriform plate; they lodge the filaments of the olfactory nerves.

The *Lateral Masses* of the ethmoid are made up of a number of thin walled cellular cavities, called the *ethmoidal cells*. In the disarticulated bone, many of these appear to be broken; but when the bones are articulated, they are closed in in every part. The superior surface of each lateral mass presents a number of these apparently half-broken cellular spaces; these, however, are completely closed in when articulated with the edges of the ethmoidal fissure of the frontal bone. Crossing this surface are seen two grooves on each side, converted into canals by articulation with the frontal; they are the anterior and posterior ethmoidal foramina. They open on the inner wall of the orbit, and transmit; the anterior, the anterior ethmoidal vessels and nasal nerve; the posterior, the posterior ethmoidal artery and vein. The posterior surface also presents large irregular cellular cavities, which are closed in by articulation with the sphenoidal turbinated bones, and orbital process of the palate. The cells at the anterior surface are completed by the lachrymal bone and nasal process of the superior maxillary, and those below also by the superior maxillary. On the outer surface of each lateral mass is a thin smooth square plate of bone, called the *os planum*; it forms part of the inner wall of the orbit, and articulates above with the frontal; below, with the superior maxillary and orbital process of the palate; in front, with the lachrymal; and behind, with the sphenoid.

The cellular cavities of each lateral mass, thus walled in by the os planum on the outer side, and by its articulation with the other bones already mentioned, are divided by a thin transverse bony partition into two sets, which do not communicate with each other; they are termed the *anterior* and *posterior ethmoidal cells*; the former, the most numerous, communicate with the frontal sinuses above, and the middle meatus below, by means of a long flexuous cellular canal, the infundibulum; the posterior, the smallest and least numerous, open into the superior meatus, and communicate (occasionally) with the sphenoidal sinuses behind. If the inner wall of each lateral mass is now examined, it will be seen how these cellular cavities communicate with the nose. The internal surface of each lateral mass presents, at its upper and back part, a narrow horizontal fissure, the superior meatus of the nose, bounded above by a thin curved plate of bone, the superior turbinated bone of the ethmoid. By means of an orifice at the top part of this fissure, the posterior ethmoidal cells open into the nose. Below the superior meatus is seen the convex surface of another thin convoluted plate of bone, the middle turbinated bone. It extends along the whole length of the inner wall of each lateral mass; its lower margin is free and thick, and its concavity, directed outwards, assists in forming the middle meatus. It is by means of a large orifice at the upper and front part of this fissure, that the anterior ethmoid cells, and through them the frontal sinuses, by means of a funnel-shaped canal, the infundibulum, communicate with the nose. It will be remarked, that the whole of this surface is rough, and marked with numerous grooves and orifices, which run nearly vertically downwards from the cribriform plate; they lodge the branches of the olfactory nerve, which are distributed on the mucous membrane covering this surface. From the inferior part of each lateral mass, immediately beneath the os planum, there projects downwards and backwards an irregular lamina of bone, called the *unciform process*, from its hook-like form: it serves to close in the upper part of the orifice of the antrum, and articulates with the inferior turbinated bone.

Development. By *three* centres; one for the perpendicular lamella, and one for each lateral mass.

The lateral masses are first developed, ossific granules making their first appearance in the os planum between the fourth and fifth months of foetal life, and afterwards in the spongy bones. At birth, the bone consists of the two lateral masses, which are small and ill-developed; but when the perpendicular and horizontal plates begin to ossify, as they do about the first year after birth, the lateral masses become joined to the cribriform plate. The formation and increase in the ethmoidal cells, which complete the formation of the bone, take place about the fifth or sixth year.

Articulations. With fifteen bones; the sphenoid, two sphenoidal turbinated, the frontal, and eleven of the face—two nasal, two superior maxillary, two lachrymal, two palate, two inferior turbinated, and vomer.

The Wormian Bones

The Wormian* bones, called also, from their generally triangular form, *ossa triquetra*, are irregular plates of bone, presenting much variation in situation, number, and size. They are most commonly found in the course of the sutures, especially the lambdoid and sagittal, where they occasionally exist of large size; the superior angle of the occipital, and the anterior superior angle of the parietal, being occasionally replaced by large Wormian bones. They are not limited to the vertex, for they are occasionally found at the side of the skull, in the situation of the anterior inferior angle of the parietal bone, and in the squamous suture; and more rarely they have been found at the base, in the suture between the sphenoid and ethmoid bones. Their size varies, in some cases not being larger than a pin's head, and confined entirely to the outer table; in other cases so large, that a pair of these bones formed the whole of that portion of the occipital bone above the superior curved lines, as described by Beclard and others. Their number is most generally limited to two or three; but more than a hundred have been found in the skull of an adult hydrocephalic skeleton. It appears most probable that they are *separate accidental points of ossification*, which, during their development and growth, remain separate from the adjoining bones. In their development, structure, and mode of articulation, they resemble the other cranial bones.

Bones of the Face

The Facial Bones are fourteen in number, viz., the

Two Nasal,	Two Palate,
Two Superior Maxillary,	Two Inferior Turbinated,
Two Lachrymal,	Vomer,
Two Malar,	Inferior Maxillary.

---

* Wormius, a physician in Copenhagen.

Nasal Bones

The Nasal Bones (figs. 35, 36) are two small oblong bones, varying in size and form in different individuals; they are placed side by side at the middle and upper part of the face, forming by their junction the bridge of the nose. Each bone presents for examination two surfaces, and four borders. The *outer* surface is concave from above downwards, convex from side to side, it is covered by the Compressor nasi muscle, marked by numerous small arterial furrows, and perforated about its centre by a foramen, sometimes double, for the transmission of a small vein. Sometimes this foramen is absent on one or both sides, and occasionally the foramen caecum opens on this surface. The *inner* surface is concave from side to side, convex from above downwards; in which direction it is traversed by a well marked longitudinal groove, sometimes a canal, for the passage of a branch of the nasal nerve. The superior border is narrow, thick, and serrated for articulation with the nasal notch of the frontal bone. The inferior border is broad, thin, sharp, directed obliquely downwards, outwards and backwards; serving for the attachment of the lateral cartilage of the nose. This border presents about its centre a notch, which transmits the branch of the nasal nerve above referred to, and is prolonged at its inner extremity into a sharp spine, which, when articulated with the opposite bone, forms the nasal angle. The external border is serrated, bevelled at the expense of the internal surface above, and of the external below, to articulate with the nasal process of the superior maxillary. The internal border, thicker above than below, articulates with its fellow of the opposite side, and is prolonged behind into a vertical crest, which forms part of the septum of the nose; this crest articulates with the nasal spine of the frontal above, and the perpendicular plate of the ethmoid below.

35.—Right Nasal Bone. Outer Surface.

36.—Right Nasal Bone. Inner Surface.

Development. By *one* centre for each bone, which appears about the same period as in the vertebrae.

Articulations. With four bones; two of the cranium, the frontal and ethmoid, and two of the face, the opposite nasal and the superior maxillary.

No muscles are directly attached to this bone.

Superior Maxillary Bone

The Superior Maxillary is one of the most important bones of the face in a surgical point of view, on account of the number of diseases to which some of its parts are liable. Its minute examination becomes, therefore, a matter of considerable importance. It is the largest bone of the face, excepting the lower jaw, and forms, by its union with its fellow of the opposite side, the whole of the upper jaw. Each bone assists in the formation of three cavities, the roof of the mouth, the floor and outer wall of the nose, and the floor of the orbit; enters into the formation of two fossae, the zygomatic and spheno-maxillary, and two fissures, the spheno-maxillary, and pterygo-maxillary. Each bone presents for examination a body and four processes, malar, nasal, alveolar, and palatine.

The body is somewhat quadrilateral, and is hollowed out in its interior to form a large cavity, the antrum of Highmore. It presents for examination four surfaces, an external or facial, a posterior or zygomatic, a superior or orbital, and an internal.

The *external* or *facial surface* (fig. 37) is directed forwards and outwards. In the median line of the bone, just above the incisor teeth, is a depression, the incisive or myrtiform fossa, which gives origin to the Depressor labii superioris alaeque nasi. Above and a little external to it, the Compressor naris arises. More external and immediately beneath the orbit, is another depression, the canine fossa, larger and deeper than the incisive fossa, from which it is separated by a vertical ridge, the canine eminence, corresponding to the socket of the canine tooth. The canine fossa gives origin to the Levator anguli oris. Above the canine fossa is the infra orbital foramen, the termination of the infra-orbital canal; it transmits the infra-orbital nerve and artery. Above the infra-orbital foramen is the margin of the orbit, which affords partial attachment to the Levator labii superioris proprius muscle.

37.—Left Superior Maxillary Bone. Outer Surface.

The *posterior* or *zygomatic surface* is convex, directed backwards and outwards, and forms part of the zygomatic fossa. It presents about its centre two or three grooves leading to canals in the substance of the bone; they are termed the *posterior dental canals*, and transmit the posterior dental vessels and nerves. At the lower part of this surface is a rounded eminence, the maxillary tuberosity, especially prominent after the growth of the wisdom-tooth, rough on its inner side for articulation with the tuberosity of the palate bone. At the upper and inner part of this surface is the commencement of a groove, which, running down on the nasal surface of the bone, is converted into a canal by articulation with the palate bone, forming the posterior palatine canal.

The *superior* or *orbital surface* is thin, smooth, irregularly quadrilateral, and forms part of the floor of the orbit. It is bounded internally by an irregular margin which articulates with three bones; in front, with the lachrymal; in the middle, with the os planum of the ethmoid; and behind, with the orbital process of the palate bone; posteriorly, by a smooth rounded edge which enters into the formation of the spheno-maxillary fissure, and which sometimes articulates at its anterior extremity with the orbital plate of the sphenoid; bounded externally by the malar process, and in front by part of the circumference of the orbit continuous, on the inner side, with the nasal, on the outer side, with the malar process. Along the middle line of this surface is a deep groove, the infra-orbital, for the passage of the infra-orbital nerve and artery. This groove commences at the middle of the posterior border of the bone, and, passing forwards, terminates in a canal which subdivides into two branches; one of the canals, the infra-orbital, opens just below the margin of the orbit; the other, the smaller and most posterior one, runs in the substance of the anterior wall of the antrum; it is called the anterior dental, transmitting the anterior dental vessels and nerves to the front teeth of the upper jaw.

The *internal surface* (fig. 38) is unequally divided into two parts by a horizontal projection of bone, the palatine process; that portion above the palate-process forms part of the outer wall of the nose; the portion below it forms part of the cavity of the mouth. The superior division of this surface presents a large irregular shaped opening leading into the antrum of Highmore. At the upper border of this aperture are a number of broken cellular cavities, which, in the articulated skull, are closed in by the

38.—Left Superior Maxillary Bone. Inner Surface.

ethmoid and lachrymal bones. Below the aperture, is a smooth concavity which forms part of the inferior meatus of the nose, traversed by a fissure, the maxillary fissure, which runs from the lower part of the orifice of the antrum obliquely downwards and forwards, and receives the maxillary process of the palate. Behind it, is a rough surface which articulates with the perpendicular plate of the palate bone, traversed by a groove which, commencing near the middle of the posterior border, runs obliquely downwards and forwards, and forms, when completed by its articulation with the palate bone, the posterior palatine canal. In front of the opening in the antrum is a deep groove, converted into a canal by the lachrymal and inferior turbinated bones, and lodging the nasal duct. More anteriorly is a well marked rough ridge, the inferior turbinated crest, for articulation with the inferior turbinated bone. The concavity above this ridge forms part of the middle meatus of the nose, whilst that below it forms part of the inferior meatus. The inferior division of this surface is concave, rough and uneven, and perforated by numerous small foramina for the passage of nutrient vessels.

The *Antrum of Highmore*, or Maxillary Sinus, is a large triangular-shaped cavity, hollowed out of the body of the maxillary bone; its apex, directed outwards, is formed by the malar process; its base, by the outer wall of the nose. Its walls are everywhere exceedingly thin, its roof being formed by the orbital plate, its floor by the alveolar process, bounded in front by the facial surface, and behind by the zygomatic. Its inner wall, or base, presents, in the disarticulated bone, a large irregular aperture, which communicates with the nasal fossae. The margins of this aperture are thin and ragged, and the aperture itself is much contracted by its articulation with the ethmoid above, the inferior turbinated below, and the palate bone behind. In the articulated skull, this cavity communicates with the middle meatus of the nose generally by two small apertures left between the above-mentioned bones. In the recent state, usually only one small opening exists, near the upper part of the cavity, sufficiently large to admit the end of a probe, the rest being filled in by the lining membrane of the sinus.

Crossing the cavity of the antrum are often seen several projecting laminae of bone, similar to those seen in the sinuses of the cranium; and on its outer wall are the posterior dental canals, transmitting the posterior dental vessels and nerves to the teeth. Projecting into the floor are several conical processes, corresponding to the roots of the first and second molar teeth; in some cases, the floor is even perforated by the teeth in this situation. It is from the extreme thinness of the walls of this cavity, that we are enabled to explain how tumours, growing from the antrum, encroach upon the adjacent parts, pushing up the floor of the orbit and displacing the eyeball, projecting inward into the nose, protruding forwards on to the cheek, and making their way backwards into the zygomatic fossa, and downwards into the mouth.

The *Malar Process* is a rough triangular eminence, situated at the angle of separation of the facial from the zygomatic surface. In front, it is concave, forming part of the facial surface; behind, it is also concave, and forms part of the zygomatic fossa; superiorly, it is rough and serrated for articulation with the malar bone; whilst below, a prominent ridge, marks the division between the facial and zygomatic surfaces.

The *Nasal Process* is a thick triangular plate of bone, which projects upwards, inwards, and backwards, by the side of the nose, forming its lateral boundary. Its external surface is concave, smooth, perforated by numerous foramina, and gives attachment to the Levator labii superioris alaeque nasi, the Orbicularis palpebrarum, and Tendo oculi. Its internal surface forms part of the inner wall of the nares; it articulates above with the frontal, and presents a rough uneven surface which articulates with the ethmoid bone, closing in the anterior ethmoid cells; below this is a transverse ridge, the superior turbinated crest, for articulation with the middle turbinated bone of the ethmoid, bounded below by a smooth concavity, which forms part of the middle meatus; below this is the inferior turbinated crest (already described), for articulation with the inferior turbinated bone; and still more inferiorly, the concavity which forms part of the inferior meatus. The anterior border of the nasal process is thin, and serrated for articulation with the nasal bone: its posterior border thick, and hollowed into a groove for the nasal duct; of the two margins of this groove, the inner one articulates with the lachrymal bone, the outer one forming part of the circumference of the orbit. Just where this border joins the orbital surface is a small tubercle, the lachrymal tubercle. This serves as a guide to the surgeon in the performance of the operation for fistula lachrymalis. The lachrymal groove in the articulated skull is converted into a canal by the lachrymal bone, and lachrymal process of the inferior turbinated; it is directed downwards, and a little backwards and outwards, is about the diameter of a goose-quill, slightly narrower in the middle than at either extremity, and lodges the nasal duct.

The *Alveolar Process* is the thickest part of the bone, broader behind than in front, and excavated into deep cavities for the reception of the teeth. These cavities are eight in number, and vary in size and depth according to the teeth they contain: those for the canine teeth being the deepest; those for the molars being widest, and subdivided into minor cavities; those for the incisors being single, but deep and narrow.

The *Palate Process*, thick and strong, projects horizontally inwards from the inner surface of the bone. It is much thicker in front than behind, and forms the floor of the nares, and the roof of the mouth. Its upper surface is concave from side to side, smooth, and forms part of the floor of the nose. In front is seen the upper orifice of the anterior palatine (incisor) canal, which leads into a fossa formed by the junction of the two superior maxillary bones, and situated immediately behind the incisor teeth. It transmits the anterior palatine vessels, the naso-palatine nerves passing through the inter-maxillary suture. The inferior surface, also concave, is rough and uneven, and forms part of the roof of the mouth. This surface is perforated by numerous foramina for the passage of nutritious vessels, channelled at the back part of its alveolar border by a longitudinal groove, sometimes a canal, for the transmission of the posterior palatine vessels, and a large nerve, and presents little depressions for the lodgment of the palatine glands. This surface presents anteriorly the lower orifice of the anterior palatine fossa. The outer border is firmly united with the rest of the bone. The inner border is thicker in front than behind, raised above into a ridge, which, with the corresponding ridge in the opposite bone, forms a groove for the reception of the vomer. The anterior margin is bounded by the thin concave border of the opening of the nose, prolonged forwards internally into a sharp process, forming, with a similar process of the opposite bone, the anterior nasal spine. The posterior border is serrated for articulation with the horizontal plate of the palate bone.

Development (fig. 39). This bone is formed at such an early period, and ossification proceeds in it with such rapidity, that it has been found impracticable hitherto to determine with accuracy its number of centres. It appears, however, probable that it has *four* centres of development, viz., one for the nasal and facial portions, one for the orbital and malar, one for the incisive, and one for the palatal portion, including the entire palate except the incisive segment. The incisive portion is indicated in young bones by a fissure, which marks off a small segment of the palate, including the two incisor teeth. In some animals, this remains permanently as a separate piece, constituting the intermaxillary bone; and in the human subject, where the jaw is malformed, a detached piece is often found in this situation, most probably depending upon arrest of development of this centre. The maxillary sinus appears at an earlier period than any of the other sinuses, its development commencing about the fourth month of foetal life.

39.—Development of Superior Maxillary Bone. By Four Centres.

Articulations. With *nine* bones; two of the cranium—the frontal and ethmoid, and with seven of the face, viz., the nasal, malar, lachrymal, inferior turbinated, palate, vomer, and its fellow of the opposite side. Sometimes it articulates with the orbital plate of the sphenoid.

Attachment of Muscles. Orbicularis palpebrarum, Obliquus inferior oculi, Levator labii superioris alaeque nasi, Levator labii superioris proprius, Levator angulioris, Compressor naris, Depressor alae nasi, Masseter, Buccinator.

The Lachrymal Bones

The *Lachrymal* are the smallest and most fragile of all the bones of the face, situated at the front part of the inner wall of the orbit, and resemble somewhat in form, thinness, and size, a finger-nail; hence they are termed the *ossa unguis*. Each bone presents for examination, two surfaces and four borders. The external (fig. 40) or orbital surface is divided by a vertical ridge into two parts. The portion of bone in front of this ridge presents a smooth, concave, longitudinal groove, the free margin of which unites with the nasal process of the superior maxillary bone, completing the lachrymal groove. The upper part of this groove lodges the lachrymal sac; the lower part is continuous with the lachrymal canal, and lodges the nasal duct. The portion of bone behind the ridge is smooth, slightly concave, and forms part of the inner wall of the orbit. The ridge, and part of the orbital surface immediately

40.—Left Lachrymal Bone. External Surface.

(*Slightly enlarged*)

behind it, affords attachment to the Tensor tarsi: it terminates below in a small hook-like process, which articulates with the lachrymal tubercle of the superior maxillary bone, and completes the upper orifice of the lachrymal canal. It sometimes exists as a separate piece, which is then called the *lesser lachrymal bone*. The internal or nasal surface presents a depressed furrow, corresponding to the elevated ridge on its outer surface. The surface of bone in front of this forms part of the middle meatus; and that behind it articulates with the ethmoid bone, filling in the anterior ethmoidal cells. Of the *four borders*, the anterior is the longest, and articulates with the nasal process of the superior maxillary bone. The posterior, thin and uneven, articulates with the os planum of the ethmoid. The superior border, the shortest and thickest, articulates with the internal angular process of the frontal bone. The inferior is divided by the lower edge of the vertical crest into two parts, the posterior articulating with the orbital plate of the superior maxillary bone; the anterior portion being prolonged downwards into a pointed process, which articulates with the lachrymal process of the inferior turbinated bone, assisting in the formation of the lachrymal canal.

Development. By a single centre, which makes its appearance soon after ossification of the vertebrae has commenced.

Articulations. With four bones; two of the cranium, the frontal and ethmoid, and two of the face, the superior maxillary and the inferior turbinated.

Attachment of Muscles. The Tensor tarsi.

The Malar Bones

The *Malar* are two small quadrangular bones, situated at the upper and outer part of the face, forming the prominence of the cheek, part of the outer wall and floor of the orbit, and part of the temporal and zygomatic fossae. Each bone presents for examination an external and an internal surface; four processes, the frontal, orbital, maxillary, and zygomatic; and four borders. The external surface (fig. 41) is smooth, convex, perforated near its centre by one or two small apertures, the malar canals, for the passage of small nerves and vessels, covered by the Orbicularis palpebrarum muscle, and affords attachment to the Zygomaticus major and minor muscles.

41.—Left Malar Bone. Outer Surface.

The internal surface (fig. 42), directed backwards and inwards, is concave, presenting internally a rough triangular surface, for articulation with the superior maxillary bone; and externally, a smooth concave surface, which forms the anterior boundary of the temporal fossa above, wider below, where it forms part of the zygomatic fossa. This surface presents a little above its centre the aperture of one or two malar canals, and affords attachment to part of two muscles, the temporal above, and the masseter below. Of the four processes, the *frontal* is thick and serrated, and articulates with the external angular process of the frontal bone. The *orbital* process is a thick and strong plate, which projects backwards from the orbital margin of the bone. Its upper surface, smooth and concave, forms, by its junction with the great ala of the sphenoid, the outer wall of the orbit. Its under surface, smooth and convex, forms part of the temporal fossa. Its anterior margin is smooth and rounded, forming part of the circumference of the orbit. Its superior margin, rough, and directed horizontally, articulates with the frontal behind the external angular process.

42.—Left Malar Bone. Inner Surface.

Its posterior margin is rough and serrated, for articulation with the sphenoid; internally it is also serrated for articulation with the orbital process of the superior maxillary. At the angle of junction of the sphenoid and maxillary portions, a short rounded non-articular margin is sometimes seen; this forms the anterior boundary of the spheno-maxillary fissure: occasionally, no such non-articular surface exists, the fissure being completed by the direct junction of the maxillary and sphenoid bones, or by the interposition of a small Wormian bone in the angular interval between them.

On the upper surface of the orbital process are seen the orifices of one or two malar canals; one of these usually opens on the posterior surface, the other (occasionally two), on the facial surface: they transmit filaments of the orbital branch of the superior maxillary nerve. The *maxillary* process is a rough triangular surface, which articulates with the superior maxillary bone. The *zygomatic* process, long, narrow, and serrated, articulates with the zygomatic process of the temporal bone. *Of the four borders*, the superior, or orbital, is smooth, arched, and forms a considerable part of the circumference of the orbit. The inferior, or zygomatic, is continuous with the lower border of the zygomatic arch, affording attachment by its rough edge to the Masseter muscle. The anterior or maxillary border is rough, and bevelled at the expense of its inner table, to articulate with the superior maxillary bone; affording attachment by its outer margin to the levator labii superioris proprius, just at its point of junction with the superior maxillary. The posterior or temporal border, curved like an italic *f*, is continuous above with the commencement of the temporal ridge; below, with the upper border of the zygomatic arch; it affords attachment to the temporal fascia.

Development. By a single centre of ossification, which appears at about the same period when ossification of the vertebrae commences.

Articulations. With four bones: three of the cranium, frontal, sphenoid, and temporal; and one of the face, the superior maxillary.

Attachment of Muscles. Levator labii superioris proprius, Zygomaticus major and minor, Masseter, and Temporal.

The Palate Bones

The Palate Bones are situated at the posterior part of the nasal fossae, wedged in between the superior maxillary and the pterygoid process of the sphenoid. In form they are somewhat like the letter L. Each bone assists in the formation of three cavities; the floor and outer wall of the nose, the roof of the mouth, and the floor of the orbit; and enters into the formation of three fossae; the zygomatic, spheno-maxillary, and pterygoid. Each bone consists of two portions; an inferior or horizontal plate, a superior or vertical plate.

The *Horizontal Plate* is thick, of a quadrilateral form, and presents two surfaces and four borders. The superior surface, concave from side to side, forms the back part of the floor of the nares. The inferior surface, slightly concave and rough, forms the back part of the hard palate. At its posterior part may be seen a transverse ridge, more or less marked, for the attachment of the tendon of the Tensor palati muscle. At the outer extremity of this ridge is a deep groove, converted into a canal by its articulation with the tuberosity of the superior maxillary bone, and forming the posterior palatine canal. Near this groove, the orifices of one or two small canals, accessory posterior palatine, may frequently be seen. The anterior border is serrated, bevelled at the expense of its inferior surface, and articulates with the palate process of the superior maxillary bone. The posterior border is concave, free, and serves for the attachment of the soft palate. Its inner extremity is sharp and pointed, and when united with the opposite bone, forms a projecting process, the posterior nasal spine, for the attachment of the Azygos uvulae. The external border is united with the lower part of the perpendicular plate almost at right angles. The internal border, the thickest, is serrated for articulation with its fellow of the opposite side; its superior edge is raised into a ridge, which, united with the opposite bone, forms a groove, in which the vomer is received.

The *Vertical Plate* (fig. 43) is thin, of an oblong form, and directed upwards and a little inwards. It presents two surfaces, an external and an internal, and four borders.

43.—Left Palate Bone. Internal View (enlarged).

The *internal surface* presents at its lower part a broad shallow depression, which forms part of the lateral boundary of the inferior meatus. Immediately above this is a well marked horizontal ridge, the inferior turbinated crest, for the articulation of the inferior turbinated bone; above this, a second broad shallow depression may be seen, which forms part of the lateral boundary of the middle meatus, surmounted above by a horizontal ridge, less prominent than the inferior, the superior turbinated crest, for the articulation of the middle turbinated bone. Above the superior turbinated crest is a narrow horizontal groove, which forms part of the superior meatus. The external surface is rough and irregular throughout the greater part of its extent, for articulation with the inner surface of the superior maxillary bone, its upper and back part being smooth where it enters into the formation of the zygomatic fossa; it is also smooth in front, where it covers the orifice of the antrum. This surface presents towards its back part a deep groove, converted into a canal, the posterior palatine, by its articulation with the superior maxillary bone. It transmits the posterior palatine vessels and a large nerve. The anterior border is thin, irregular, and presents opposite the inferior turbinated crest, a pointed projecting lamina, the maxillary process of the palate bone, which is directed forwards, and closes in the lower and back part of the opening of the antrum, being received into a fissure that exists at the inferior part of this aperture. The posterior border (fig. 44) presents a deep groove, the edges of which are serrated for articulation with the pterygoid process of the sphenoid. At the lower part of this border is seen a pyramidal process of bone, the *pterygoid process* or tuberosity of the palate, which is received into the angular interval between the two pterygoid plates of the sphenoid at their inferior extremity. This process presents at its back part three grooves, a median and two lateral ones. The former is smooth, and forms part of the pterygoid fossa, affording attachment to the Internal pterygoid muscle; whilst the lateral grooves are rough and uneven, for articulation with the anterior border of each pterygoid plate. The base of this process, continuous with the horizontal portion of the bone, presents the apertures of the accessory descending palatine canals; whilst its outer surface is rough, for articulation with the inner surface of the body of the superior maxillary bone. The superior border of the vertical plate presents two well marked processes, separated by an intervening notch or foramen. The anterior, or larger, is called the *orbital process*; the posterior, the *sphenoidal*. The Orbital Process, directed upwards and outwards, is placed on a higher level than the sphenoidal. It presents five surfaces, which enclose a hollow cellular cavity, and is connected to the perpendicular plate by a narrow constricted neck. Of these five surfaces, three are articular, two non-articular, or free surfaces. The three articular are the anterior or maxillary surface, which is directed forwards, outwards, and downwards, is of an oblong form, and rough for articulation with the superior maxillary bone. The posterior or sphenoidal surface, is directed backwards, upwards, and inwards. It ordinarily presents a small half-cellular cavity which communicates with the sphenoidal sinus, and the margins of which

44.—Left Palate Bone. Posterior View (enlarged).

are serrated for articulation with the vertical part of the sphenoidal turbinated bone. The internal or ethmoidal surface is directed inwards, upwards and forwards, and articulates with the lateral mass of the ethmoid bone. In some cases, the cellular cavity above-mentioned opens on this surface of the bone, it then communicates with the posterior ethmoidal cells. More rarely it opens on both surfaces, and then communicates with the posterior ethmoidal cells, and the sphenoidal sinus. The non-articular or free surfaces of the orbital process are the superior or orbital, directed upwards and outwards, of a triangular form, concave, smooth, articulating with the superior maxillary bone, and forming the back part of the floor of the orbit. The external or zygomatic surface, directed outwards, backwards and downwards, is of an oblong form, smooth, and forms part of the zygomatic fossa. This surface is separated from the orbital by a smooth rounded border, which enters into the formation of the spheno-maxillary fissure.

The *Sphenoidal Process* of the palate bone is a thin compressed plate, much smaller than the orbital, and directed upwards and inwards. It presents three surfaces and two borders. The superior surface, the smallest of the three, articulates with the horizontal part of the sphenoidal turbinated bone; it presents a groove which contributes to the formation of the pterygo-palatine canal. The internal surface is concave, and forms part of the outer wall of the nasal fossa. The external surface is divided into two parts, an articular, and a non-articular portion; the non-articular portion is smooth and free, forming part of the zygomatic fossa, whilst behind is a rough surface for articulation with the inner surface of the pterygoid process of the sphenoid. The anterior border forms the posterior boundary of the spheno-palatine foramen. The posterior border, serrated at the expense of the outer table, articulates with the internal surface of the pterygoid process.

The orbital and sphenoidal processes are separated from one another by a deep notch, which is converted into a foramen, the spheno-palatine, by articulation with the sphenoidal turbinated bone. Sometimes the two processes are united above, and form between them a complete foramen, or the notch is crossed by one or more spiculas of bone, so as to form two or more foramina. In the articulated skull, this foramen opens into the back part of the outer wall of the superior meatus, and transmits the spheno-palatine vessels and nerves.

Development. From a single centre, which makes its appearance at the angle of junction of the two plates of the bone. From this point ossification spreads; inwards, to the horizontal plate; downwards, into the tuberosity; and upwards, into the vertical plate. In the foetus, the horizontal plate is much longer than the vertical; and even after it is fully ossified, the whole bone is remarkable for its shortness.

Articulations. With seven bones; the sphenoid, ethmoid, superior maxillary, inferior turbinated, vomer, opposite palate, and sphenoidal turbinated.

Attachment of Muscles. The Tensor palati, Azygos uvulae, Internal and External pterygoid.

The Inferior Turbinated Bones

The *Inferior Turbinated* bones are situated one on each side of the outer wall of the nasal fossae. Each bone consists of a layer of thin "spongy" bone, curled upon itself like a scroll, hence its name "turbinated"; and extending horizontally across the outer wall of the nasal fossa, immediately below the orifice of the antrum. Each bone presents two surfaces, two borders, and two extremities.

The *internal surface* (fig. 45) is convex, perforated by numerous apertures, and traversed by longitudinal grooves and canals for the lodgment of arteries and veins. In the recent state it is covered by the lining membrane of the nose. The *external surface* is concave (fig. 46), and forms part of the inferior meatus. Its upper border is thin, irregular, and connected to various bones along the outer wall of the nose. It may be divided into three portions; of these, the anterior articulates with the inferior turbinated crest of the superior maxillary bone; the posterior with the inferior turbinated crest of the palate bone; the middle portion of the superior border presents three well marked processes, which vary much in their size and form. Of these the anterior and smallest, is situated at the junction of the anterior fourth with the posterior three-fourths of the bone; it is small and pointed, and is called the *lachrymal process*, for it articulates with the anterior inferior angle of the lachrymal bone, and by its margins, with the groove on the back of the nasal process of the superior maxillary, and thus assists in forming the lachrymal canal. At the junction of the two middle fourths of the bone, but encroaching on the latter, a broad thin plate, the *ethmoidal process*, ascends to join the unciform process of the ethmoid; from the lower border of this process, a thin lamina of bone curves downwards and outwards, hooking over the lower edge of the orifice of the antrum, which it narrows below; it is called the *maxillary process*, and fixes the bone firmly on to the outer wall of the nasal fossa. The inferior border is free, thick and cellular in structure, more especially in the centre of the bone. Both extremities are more or less narrow and pointed. If the bone is held so that its outer concave surface is directed backwards (i.e., towards the holder), and its superior border, from which the lachrymal and ethmoidal processes project, upwards, the lachrymal process will be directed to the side to which the bone belongs.

45.—Right Inferior Turbinated Bone. Inner Surface.

46.—Right Inferior Turbinated Bone. Outer Surface.

Development. By a single centre which makes its appearance about the middle of foetal life.

Articulations. With four bones; one of the cranium, the ethmoid, and three of the face, the superior maxillary, lachrymal and palate.

No muscles are attached to this bone.

The Vomer

The *Vomer* (fig. 47) is a single bone, situated vertically at the back part of the nasal fossae, and forming part of the septum of the nose. It is thin, somewhat like a ploughshare in form, but it varies in different individuals, being frequently bent to one or the other side; it presents for examination two surfaces and four borders. The lateral surfaces are smooth, marked with small furrows for the lodgment of blood-vessels, and by a groove on each side, sometimes a canal, the naso-palatine, which runs obliquely downwards and forwards to the intermaxillary suture between the two anterior palatine canals; it transmits the naso-palatine nerve. The superior border, the thickest, presents a deep groove, bounded on each side by a horizontal projecting ala of bone; the groove receives the rostrum of the sphenoid, whilst the alae are overlapped and retained by laminae which project from the under surface of the body of the sphenoid at the base of the pterygoid processes. At the anterior part of the groove a fissure is left for the transmission of blood-vessels to the substance of the bone. The inferior border, the longest, is broad and uneven in front, where it articulates with the two superior maxillary bones; thin and sharp behind where it joins with the palate bones. The upper half of the anterior border usually presents two laminae of bone, which receive between them the perpendicular plate of the ethmoid, the lower half consisting of a single rough edge, also occasionally channelled, which is united to the triangular cartilage of the nose. The posterior border is free, concave, and separates the nasal fossae from one another behind. It is thick and bifid above, thin below.

47.—Vomer.

Development. The vomer at an early period consists of two laminae united below, but separated above by a very considerable interval. Ossification commences in it at about the same period as in the vertebrae.

Articulations. With six bones; two of the cranium, the sphenoid and ethmoid; and four of the face, the two superior maxillary, the two palate bones, and with the cartilage of the septum.

The vomer has no muscles attached to it.

The Inferior Maxillary Bone

The *Inferior Maxillary* Bone, the largest and strongest bone of the face, serves for the reception of the inferior teeth. It consists of a curved horizontal portion, the body, and of two perpendicular portions, the rami, which join the former nearly at right angles behind.

The *Horizontal* portion, or body (fig. 48), is convex in its general outline, and curved somewhat like a horse-shoe. It presents for examination two surfaces and two borders. The *External Surface* is convex from side to side, concave from above downwards. In the median line is a well marked vertical ridge, the symphysis; it

48.—Inferior Maxillary Bone. Outer Surface. Side View.

extends from the upper to the lower border of the bone, and indicates the point of junction of the two pieces of which the bone is composed at an early period of life. The lower part of the ridge terminates in a prominent triangular eminence, the mental process. On either side of the symphysis, just below the roots of the incisor teeth, is a depression, the incisive fossa, for the attachment of the Levator menti; and still more externally, a foramen, the mental foramen, for the passage of the mental nerve and artery. This foramen is placed just below the root of the second bicuspid tooth. Running outwards from the base of the mental process on each side, is a well marked ridge, the external oblique line. This ridge is at first nearly horizontal, but afterwards inclines upwards and backwards, and is continuous with the anterior border of the ramus; it affords attachment to the Depressor labii inferioris and Depressor anguli oris, below it, to the Platysma myoides.

The *Internal Surface* (fig. 49) is concave from side to side, convex from above downwards. In the middle line is an indistinct linear depression, corresponding to the symphysis externally; on either side of this depression, just below its centre, are four prominent tubercles, placed in pairs, two above and two below; they are called the *genial tubercles*, and afford attachment, the upper pair to the Genio-hyoglossi muscles, the lower pair to the Genio-hyodei muscles. Sometimes the tubercles on each side are blended into one, or they all unite into an irregular eminence of bone, or nothing but an irregularity may be seen on the surface of the bone at this part. On either side of the genial tubercles is an oval depression, the sublingual fossa, for lodging the sublingual gland; and beneath it a rough depression on each side, which gives attachment to the anterior belly of the Digastric muscle. At the back part of the sublingual fossa, the internal oblique line (mylo-hyoidean) commences; it is faintly marked at its commencement, but becomes more distinct as it passes upwards and outwards, and is especially prominent opposite the two last molar teeth; it divides the lateral surface of the bone into two portions, and affords attachment throughout its whole extent to the Mylo-hyoid muscle, the Superior constrictor being attached above its posterior extremity, nearer the alveolar margin. The portion of bone above

49.—Inferior Maxillary Bone. Inner Surface. Side View.

this ridge is smooth, and covered by the mucous membrane of the mouth; whilst that below it presents an oblong depression, wider behind than in front, the submaxillary fossa, for the lodgment of the submaxillary gland. The *superior* or *alveolar border* is wider, and its margins thicker behind than in front. It is hollowed into numerous cavities, for the reception of the teeth; these are sixteen in number, and vary in depth and size according to the teeth which they contain. At an early period of life, before the eruption of the teeth, the alveolar process is proportionally larger and deeper than in the adult, and the chief part of the body is above the oblique line. In adult life the base of the bone attains its maximum of development. In old age, on the contrary, after the loss of the teeth, the alveolar process becomes absorbed, and the chief part of the body is that which exists below the oblique line. At this period, the dental canal and mental foramen are situated close to the upper border of the bone. The *inferior border*, longer than the superior, and thicker in front than behind, is rounded; it presents a shallow groove, just where the body joins the ramus, over which the facial artery turns.

The *Perpendicular Portions*, or *Rami*, are of a quadrilateral form, and differ in their direction at various periods of life. In the foetus, they are almost parallel with the body; in youth they are oblique; in manhood they are nearly vertical, joining the body at almost a right angle. In old age, after the loss of the teeth, they again decline and assume an oblique direction. Each ramus presents for examination two surfaces, four borders, and two processes. The *external surface* is flat, marked with ridges, and gives attachment throughout nearly the whole of its extent to the Masseter muscle. The *internal surface* presents about its centre the oblique aperture of the inferior dental canal, for the passage of the inferior dental vessels and nerve. The margins of this opening are irregular, and present in front a prominent ridge, surmounted by a sharp spine, which gives attachment to the internal lateral ligament of the lower jaw; and at its lower and back part is seen a notch leading to a groove, which runs obliquely downwards to the posterior extremity of the submaxillary fossa; this groove is the mylo-hyoidean, and lodges the mylo-hyoid vessels and nerve; behind the groove is a rough surface, for the insertion of the Internal pterygoid muscle. The inferior dental canal descends obliquely downwards and forwards in the substance of the ramus, and then horizontally forwards in the body; it is here placed under the alveoli, with which it communicates by small openings. On arriving at the incisor teeth, it turns back to communicate with the mental foramen, giving off two small canals, which run forward, to be lost in the cancellous tissue of the bone beneath the incisor teeth. This canal, in the posterior two-thirds of the bone, runs nearest the internal surface of the jaw; and in the anterior third, nearer its external surface. Its walls are composed of compact tissue at either extremity, cancellous in the centre. It contains the inferior dental vessels and nerve, from which branches are distributed to the teeth through the small apertures at the bases of the alveoli. The *superior border* is thin, and presents two processes, separated by a deep concavity, the sigmoid notch. Of these processes, the anterior is the coronoid, the posterior the condyloid.

The *Coronoid Process* is a thin, flattened, triangular eminence of bone, which varies in length in different subjects. Its *external surface* is smooth, and affords attachment to the masseter and temporal muscles. Its *internal surface* gives attachment to the temporal muscle, and presents the commencement of a longitudinal ridge, which is continued to the posterior part of the alveolar process. In front of this ridge is a deep groove, continued below on to the outer side of the alveolar process; this ridge and part of the groove afford attachment above to the Temporal, below to the Buccinator muscle.

The *Condyloid Process*, shorter but thicker than the coronoid, consists of two portions; the condyle, and the constricted portion which supports the condyle, the neck. The condyle is of an oval form, its long axis being transverse, and placed in such a manner that its outer end is a little more forward and a little higher than its inner. It is convex from before backwards, and from side to side, the articular surface extending further on the posterior than on the anterior surface. The neck of the condyle is flattened from before backwards. Its posterior surface is convex; its anterior is hollowed out on its inner side by a depression (the pterygoid fossa), for the attachment of the External pterygoid. The *lower border* of the ramus is thick, straight, and continuous with the body of the bone. At its junction with the posterior border is the angle of the jaw, which is somewhat everted, rough on each side for the attachment of the masseter externally, and the internal pterygoid internally, and, between them, serving for the attachment of the stylo-maxillary ligament. The *anterior* border is thin above, thicker below, and continuous with the external oblique line. The *posterior* border is thick, smooth, and rounded, and covered by the parotid gland.

The *Sigmoid Notch*, separating the two processes, is a deep semilunar depression, crossed by the masseteric artery and nerve.

Development. This bone is formed at such an early period of life, before, indeed, any other bone excepting the clavicle, that it has been found impossible at present to determine its earliest condition. It appears probable, however, that it is developed by *two* centres, one for each lateral half, the two segments meeting at the symphysis, where they become united. Additional centres have also been described for the coronoid process, the condyle, the angle, and the thin plate of bone, which forms the inner side of the alveolus. At birth it consists of two lateral halves. These join at the symphysis at the end of the first year; but a trace of separation at their upper part is seen at the commencement of the second year.

Articulations. With the glenoid fossae of the two temporal bones.

Attachment of Muscles. By its external surface, commencing at the symphysis, and proceeding backwards; Levator menti, Depressor labii inferioris, Depressor anguli oris, Platysma myoides, Buccinator, Masseter. By its internal surface, commencing at the same point; Genio-hyo-glossus, Genio-hyoideus, Mylo-hyoideus, Digastric, Superior constrictor, Temporal, Internal pterygoid, External pterygoid.

Articulations of the Cranial Bones

The bones of the cranium and face, are connected to each other by means of sutures. The *Cranial Sutures* may be divided into three sets: 1. Those of the vertex of the skull. 2. Those at the side of the skull. 3. Those at the base.

The sutures at the vertex of the skull are three, the sagittal, coronal, and lambdoid.

The *Sagittal Suture* (*sagitta*, an arrow) is formed by the junction of the two parietal bones, and extends from the middle of the frontal bone, backwards to the superior angle of the occipital. In childhood, and occasionally in the adult, when the two halves of the frontal bone are not united, it is continued forwards to the root of the nose. This suture sometimes presents, near its posterior extremity, the parietal foramen on each side; and in front, where it joins the coronal suture, a space is occasionally left, which encloses a large Wormian bone.

The *Coronal Suture* extends transversely across the vertex of the skull, and connects the frontal with the parietal bones. It commences at the extremity of the great wing of the sphenoid on one side, and terminates at the same point on the opposite side. The dentations of this suture are more marked at the sides than at the summit, and are so constructed that the frontal rests on the parietal above, whilst laterally the parietal supports the frontal.

The *Lambdoid Suture*, so called from its resemblance to the Greek letter λ, connects the occipital with the parietal bones. It commences on each side at the angle of the mastoid portion of the temporal bone, and inclines upwards to the end of the sagittal suture. The dentations of this suture are very deep and distinct, and are often interrupted by several small Wormian bones.

The sutures at the side of the skull are also three in number; the spheno-parietal, squamo-parietal, and masto-parietal. They are subdivisions of a single suture, formed between the lower border of the parietal, and the temporal and sphenoid bones, and extending from the lower end of the lambdoid suture behind, to the lower end of the coronal suture in front.

The *Spheno-parietal* is very short, and formed by the tip of the great wing of the sphenoid, and the anterior inferior angle of the parietal bone.

The *Squamo-parietal*, or squamous suture, is arched. It is formed by the squamous portion of the temporal bone overlapping the middle division of the lower border of the parietal.

The *Masto-parietal* is a short suture, deeply dentated, formed by the posterior inferior angle of the parietal, and the superior border of the mastoid portion of the temporal.

The sutures at the base of the skull are the basilar in the centre, and on each side, the petro-occipital, the masto-occipital, the petro-sphenoidal, and the squamo-sphenoidal.

The *Basilar Suture* is formed by the junction of the basilar surface of the occipital bone with the posterior surface of the body of the sphenoid. At an early period of life a thin plate of cartilage exists between these bones, but in the adult they become inseparably united. Between the outer extremity of the basilar suture, and the termination of the lambdoid, an irregular suture exists which is subdivided into two portions. The inner portion, formed by the union of the petrous part of the temporal, with the occipital bone, is termed the *petro-occipital*. The outer portion, formed by the junction of the mastoid part of the temporal with the occipital, is called the *masto-occipital*. Between the bones forming the petro-occipital suture, a thin plate of cartilage exists; in the masto-occipital is occasionally found the opening of the mastoid foramen. Between the outer extremity of the basilar suture and the spheno-parietal, an irregular suture may be seen formed by the union of the sphenoid with the temporal bone. The inner and smaller portion of this suture is termed the *petro-sphenoidal*; it is formed between the petrous portion of the temporal, and the great wing of the sphenoid; the outer portion, of greater length, and arched, is formed between the squamous portion of the temporal and the great wing of the sphenoid, it is called the *squamo-sphenoidal*.

The cranial bones are connected with those of the face, and the facial with each other, by numerous sutures, which, though distinctly marked, have received no special names. The only remaining suture deserving especial consideration is the *transverse*. This extends across the upper part of the face, and is formed by the junction of the frontal with the facial bones; it extends from the external angular process of one side, to the same point on the opposite side, and connects the frontal with the malar, the sphenoid, the ethmoid, the lachrymal, the superior maxillary, and the nasal bones on each side.

The Skull

The Skull, formed by the union of the several cranial and facial bones already described, when considered as a whole, is divisible into five regions; a superior region or vertex, an inferior region or base, two lateral regions, and an anterior region, the face.

The Superior Region, or vertex, presents two surfaces, an external and an internal. The *External Surface* is bounded in front by the nasal eminences, and superciliary ridges; behind, by the occipital protuberance and superior curved lines of the occipital bone, laterally, by an imaginary line extending from the outer end of the superior curved line, along the temporal ridge, to the external angular process of the frontal. This surface includes the vertical portion of the frontal, the greater part of the parietal, and the superior third of the occipital bone; it is smooth, convex, of an elongated oval form, crossed transversely by the coronal suture, and from before backwards by the sagittal, which terminates behind in the lambdoid. From before backwards may be seen the frontal eminences and remains of the suture connecting the two lateral halves of the frontal bone; on each side of the sagittal suture is the parietal foramen and parietal eminence, and still more posteriorly the smooth convex surface of the occipital bone.

The *Internal Surface* of the vertex is concave, presents eminences and depressions for the convolutions of the brain, and numerous furrows for the lodgment of branches of the meningeal arteries. Along the middle line of this surface is a longitudinal groove, narrow in front, where it terminates in the frontal crest, broader behind; it lodges the superior longitudinal sinus, and its margins afford attachment to the falx cerebri. On either side of it are several depressions for the Pacchionian bodies, and at its back part, the internal openings of the parietal foramina. This surface is also crossed in front by the coronal suture; from before backwards, by the sagittal; behind, by the lambdoid.

Base of the Skull

The *Inferior Region*, or *base* of the skull presents two surfaces, an internal or cerebral, and an external or basilar.

The *Internal*, or *Cerebral Surface* (fig. 50), is divisible into three parts, or fossae, called the *anterior*, *middle*, and *posterior* fossae of the cranium.

The *Anterior Fossa* is formed by the orbital plate of the frontal, the cribriform plate of the ethmoid, the ethmoidal process and lesser wing of the sphenoid. It is the most elevated of the three fossae, convex on each side where it corresponds to the roof of the orbits, concave in the median line in the situation of the cribriform plate of the ethmoid. It is traversed by three sutures, the ethmoido-frontal, ethmo-sphenoidal, and fronto-sphenoidal, and lodges the anterior lobes of the cerebrum. It presents in the median line from before backwards, the commencement of the groove for the superior longitudinal sinus, and crest for the attachment of the falx cerebri; the foramen caecum, this aperture is formed by the frontal and crista galli of the ethmoid, and if pervious, transmits a small vein from the nose to the superior longitudinal sinus. Behind the foramen caecum is the crista galli, the posterior margin of which affords attachment to the falx cerebri. On either side of the crista galli is the olfactory groove, which supports the bulb of the olfactory nerves, perforated by the three sets of orifices which give passage to its filaments; and in front by a slit-like opening, which transmits the nasal branch of the ophthalmic nerve. On each side are the internal openings of the anterior and posterior ethmoidal foramina, the former, situated about the middle of its outer margin, transmitting the nasal nerve, which runs in a groove along its surface, to the slit-like opening above mentioned; whilst the latter, the posterior ethmoidal foramen, opens at the back part of this margin under cover of a projecting lamina of the sphenoid, it transmits the posterior ethmoidal artery and vein to the posterior ethmoidal cells. Further back in the middle line is the ethmoidal spine, bounded behind by an elevated ridge, separating a longitudinal groove on each side which support the olfactory nerves. The anterior fossa presents laterally eminences and depressions for the convolutions of the brain, and grooves for the lodgment of the anterior meningeal arteries.

The *Middle Fossa*, somewhat deeper than the preceding, is narrow in the middle, and becomes wider as it expands laterally. It is bounded in front by the posterior margin of the lesser wing of the sphenoid, the anterior clinoid process, and the anterior margin of the optic groove; behind, by the petrous portion of the temporal, and basilar suture; externally, by the squamous portion of the temporal, and anterior inferior angle of the parietal bone, and is divided into two lateral parts by the sella Turcica. It is traversed by four sutures, the squamous, spheno-parietal, spheno-temporal, and petro-sphenoidal.

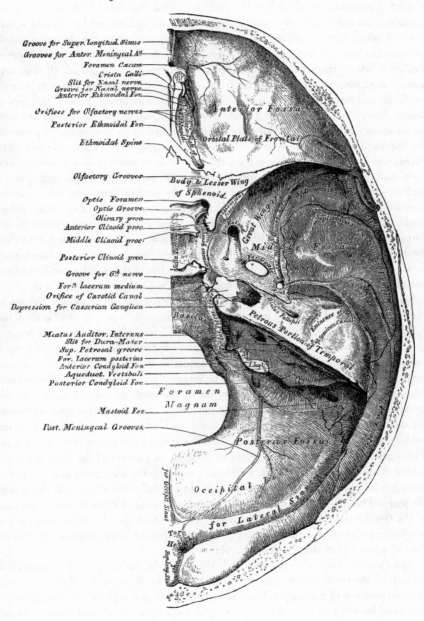

50.—Base of Skull. Inner or Cerebral Surface.

In the middle line, from before backwards, is the optic groove, which supports the optic commissure, terminating on each side in the optic foramen, for the passage of the optic nerve and ophthalmic artery, behind is seen the olivary process, and laterally the anterior clinoid processes, which afford attachment to the folds of the dura mater, which form the cavernous sinus. In the centre of the middle fossa is the sella Turcica, a deep depression, which lodges the pituitary gland, bounded in front by a small eminence on either side, the middle clinoid process, and behind by a broad square plate of bone, surmounted at each superior angle by a tubercle, the posterior clinoid process; beneath the latter process is a groove, for the lodgment of the sixth nerve. On each side of the sella Turcica is the cavernous groove; it is broad, shallow, and curved somewhat like the italic letter ∫; it commences behind at the foramen lacerum medium, and terminates on the inner side of the anterior clinoid process. This groove lodges the cavernous sinus, the internal carotid artery, and the orbital nerves. The sides of the middle fossa are of considerable depth; they present eminences and depressions for the middle lobes of the brain, and grooves for lodging the branches of the middle meningeal artery; these commence on the outer side of the foramen spinosum, and consist of two large branches, an anterior and a posterior; the former passing upwards and forwards to the anterior inferior angle of the parietal bone, the latter passing upwards and backwards. The following foramina may also be seen from before backwards. Most anteriorly is the foramen lacerum anterius, or sphenoidal fissure, formed above by the lesser wing of the sphenoid; below, by the greater wing; internally, by the body of the sphenoid; and completed externally by the orbital plate of the frontal bone. It transmits the third, fourth, the three branches of the ophthalmic division of the fifth, the sixth nerve, and the ophthalmic vein. Behind the inner extremity of the sphenoidal fissure is the foramen rotundum, for the passage of the second division of the fifth or superior maxillary nerve; still more posteriorly is seen a small orifice, the foramen Vesalii; this opening is situated between the foramen rotundum and ovale, a little internal to both; it varies in size in different individuals, and transmits a small vein. It opens below in the pterygoid fossa, just at the outer side of the scaphoid depression. Posteriorly and externally is the foramen ovale, which transmits the third division of the fifth or inferior maxillary nerve, the small meningeal artery, and the small petrosal nerve. On the outer side of the foramen ovale is the foramen spinosum, for the passage of the middle meningeal artery; and on the inner side of the foramen ovale, the foramen lacerum medium. This aperture in the recent state is filled up with cartilage. On the anterior surface of the petrous portion of the temporal bone is seen from without inwards, the eminence caused by the projection of the superior semicircular canal, the groove leading to the hiatus Fallopii, for the transmission of the petrosal branch of the Vidian nerve; beneath it, the smaller groove, for the passage of the smaller petrosal nerve; and near the apex of the bone, the depression for the semilunar ganglion, and the orifice of the carotid canal, for the passage of the internal carotid artery and carotid plexus of nerves.

The *Posterior Fossa*, deeply concave, is the largest of the three, and situated on a lower level than either of the preceding. It is formed by the occipital, the petrous and mastoid portions of the temporal, and the posterior inferior angle of the parietal bones; is crossed by three sutures, the petro-occipital, masto-occipital, and masto-parietal;

and lodges the cerebellum, pons varolii, and medulla oblongata. It is separated from the middle fossa in the median line by the basilar suture, and on each side by the superior border of the petrous portion of the temporal bone. This serves for the attachment of the tentorium cerebelli, is grooved externally for the superior petrosal sinus, and at its inner extremity presents a notch, upon which rests the fifth nerve. Its circumference is bounded posteriorly by the groove for the lateral sinus. In the centre of this fossa is the foramen magnum, bounded on either side by a rough tubercle, which gives attachment to the odontoid ligaments; and a little above these are seen the internal openings of the anterior condyloid foramina. In front of the foramen magnum is the basilar process, grooved for the support of the medulla oblongata and pons varolii, and articulating on each side with the petrous portion of the temporal bone, forming the petro-occipital suture, the anterior half of which is grooved for the inferior petrosal sinus, the posterior half being encroached upon by the foramen lacerum posterius, or jugular foramen. This foramen is partially subdivided into two parts; the posterior and larger division transmits the internal jugular vein, the anterior the eighth pair of nerves. Above the jugular foramen is the internal auditory foramen, for the auditory and facial nerves and auditory artery; behind and external to this is the slit-like opening leading into the aquaeductus vestibuli; whilst between these two latter, and near the superior border of the petrous portion, is a small triangular depression, which lodges a process of the dura mater, and occasionally transmits a small vein into the substance of the bone. Behind the foramen magnum are the inferior occipital fossae, which lodge the lateral lobes of the cerebellum, separated from one another by the internal occipital crest, which serves for the attachment of the falx cerebelli, and lodges the occipital sinuses. These fossae are surmounted, above, by the deep transverse grooves for the lodgment of the lateral sinuses, that on the right side being usually larger than the left. These channels, in their passage outwards, groove the occipital bone, the posterior inferior angle of the parietal, the mastoid portion of the temporal, and the occipital just behind the jugular foramen, at the back part of which they terminate. Where this sinus grooves the mastoid part of the temporal bone, the orifice of the mastoid foramen may be seen; and just previous to its termination it has opening into it the posterior condyloid foramen.

The *External Surface* of the base of the Skull (fig. 51) is extremely irregular. It is bounded in front by the incisor teeth in the upper jaws; behind, by the superior curved lines of the occipital bone; and laterally, by the lower border of the malar bone, the zygomatic arch, and an imaginary line, extending from the zygoma to the mastoid process and extremity of the superior curved line of the occiput. It is formed by the palate processes of the two superior maxillary and palate bones, the vomer, the pterygoid, under surface of the great wing, spinous process and part of the body of the sphenoid, the under surface of the squamous, mastoid, and petrous portions of the temporal, and occipital bones. The anterior part of the base of the skull is raised above the level of the rest of this surface (when the skull is turned over for the purpose of examination), surrounded by the alveolar process, which is thicker behind than in front, and excavated by sixteen depressions for lodging the teeth of the upper jaw; they vary in depth and size according to the teeth they contain. Immediately behind the incisor teeth is the anterior palatine fossa or canal. At the bottom of this fossa may usually be seen four apertures, two placed laterally, which open above, one in the floor of each nostril, and transmit the anterior palatine vessels, and two in the median line of the intermaxillary suture, one in front of the other, the most anterior one transmitting the left, and the posterior one (the larger) the right naso-palatine nerve.

51.—Base of Skull. External Surface.

These two latter canals are sometimes wanting, or they may join to form a single one, or one of them may open into one of the lateral canals above referred to. The palatine vault is concave, uneven, perforated by numerous nutritious foramina, marked by depressions for the palatal glands, and crossed by a crucial suture, which indicates the point of junction of the four bones of which it is composed. One or two small foramina, seen in the alveolar margin behind the incisor teeth, occasionally seen in the adult, almost constant in young subjects, are called the *incisive foramina*; they transmit nerves and vessels to the incisor teeth. At each posterior angle is the posterior palatine foramen, for the transmission of the posterior palatine vessels and anterior palatine nerve, and running forwards and inwards from it a groove, which lodges the same vessels and nerve. Behind the posterior palatine foramen is the tuberosity of the palate bone, perforated by one or more accessory posterior palatine canals, and marked by the commencement of a ridge, which runs transversely inwards, and serves for the attachment of the tendinous expansion of the tensor palati muscle. Projecting backwards from the centre of the posterior border of the hard palate is the posterior nasal spine, for the attachment of the Azygos uvulae. Behind and above the hard palate is the posterior aperture of the nares, divided into two parts by the vomer, bounded above by the body of the sphenoid, below by the horizontal plate of the palate bone, and laterally by the pterygoid processes of the sphenoid. Each aperture measures about an inch in the vertical, and half an inch in the transverse direction. At the base of the vomer may be seen the expanded alae of this bone, receiving between them the rostrum of the sphenoid. Near the lateral margins of the vomer, at the root of the pterygoid process, are the pterygo-palatine canals. The pterygoid process, which bounds the posterior nares on each side, presents near its base the pterygoid or Vidian canal, for the Vidian nerve and artery. Each process consists of two plates, which bifurcate at the extremity to receive the tuberosity of the palate bone, and are separated behind by the pterygoid fossa, which lodges the Internal pterygoid muscle. The internal plate is long and narrow, presenting on the outer side of its base the scaphoid fossa, for the origin of the Tensor palati muscle, and at its extremity the hamular process, around which the tendon of this muscle turns. The external pterygoid plate is broad, forms the inner boundary of the zygomatic fossa, and affords attachment to the External pterygoid muscle.

Behind the nasal fossa in the middle line is the basilar surface of the occipital bone, presenting in its centre the pharyngeal spine for the attachment of the Superior constrictor muscle of the pharynx, with depressions on each side for the insertion of the Rectus anticus major and minor. At the base of the external pterygoid plate is the foramen ovale; behind this, the foramen spinosum, and the prominent spinous process of the sphenoid, which gives attachment to the internal lateral ligament of the lower jaw and the Laxator tympani muscle. External to the spinous process is the glenoid fossa, divided into two parts by the Glaserian fissure, the anterior portion being concave, smooth, bounded in front by the eminentia articularis, and serving for the articulation of the condyle of the lower jaw; the posterior portion rough, bounded behind by the vaginal process, and serving for the reception of part of the parotid gland. Emerging from between the laminae of the vaginal process is the styloid process; and at the base of this process is the stylo-mastoid foramen, for the exit of the facial nerve, and entrance of the stylo-mastoid artery. External to the stylo-mastoid foramen is the auricular fissure for the auricular branch of the pneumogastric, bounded behind by the mastoid process. Upon the inner side of this process is a deep groove, the digastric fossa; and a little more internally, the occipital groove, for the occipital artery. At the base of the internal pterygoid plate is a large and somewhat triangular aperture, the foramen lacerum medium, bounded in front by the great wing of the sphenoid, behind by the apex of the petrous portion of the temporal bone, and internally by the body of the sphenoid and basilar process of the occipital bone; it presents in front the posterior orifice of the Vidian canal, behind the aperture of the carotid canal. This opening is filled up in the recent subject by a fibro-cartilaginous substance; across its upper or cerebral aspect passes the internal carotid artery and Vidian nerve. External to this aperture, the petro-sphenoidal suture is observed, at the outer termination of which is seen the orifice of the canal for the Eustachian tube, and that for the Tensor tympani muscle. Behind this suture is seen the under surface of the petrous portion of the temporal bone, presenting, from within outwards, the quadrilateral rough surface, part of which affords attachment to the Levator palati and Tensor tympani muscles; behind this surface are the orifices of the carotid canal and the aquaeductus cochleae, the former transmitting the internal carotid artery and the ascending branches of the superior cervical ganglion of the sympathetic, the latter serving for the passage of a small artery and vein to the cochlea. Behind the carotid canal is a very large irregular aperture, the jugular fossa, formed in front by the petrous portion of the temporal, and behind by the occipital; it is generally larger on the right than on the left side, and is perforated at the bottom by an irregular aperture; it is divided into two parts by a ridge of bone, which projects usually from the temporal; the anterior, or smaller portion, transmitting the three divisions of the eighth pair of nerves; the posterior, transmitting the internal jugular vein and the two ascending meningeal vessels, from the occipital and ascending pharyngeal arteries. On the ridge of bone dividing the carotid canal from the jugular fossa, is the small foramen for the transmission of the tympanic nerve; and on the outer wall of the jugular foramen, near the root of the styloid process, is the small aperture for the transmission of Arnold's nerve. Behind the basilar surface of the occipital bone is the foramen magnum, bounded on each side by the condyles, rough internally for the attachment of the alar ligaments, and presenting externally a rough surface, the jugular process, which serves for the attachment of the Rectus lateralis. On either side of each condyle anteriorly is the anterior condyloid fossa, perforated by the anterior condyloid foramen, for the passage of the lingual nerve. Behind each condyle are the posterior condyloid fossae, perforated on one or both sides by the posterior condyloid foramina, for the transmission of a vein to the lateral sinus. Behind the foramen magnum is the external occipital crest, terminating above at the external occipital tuberosity, whilst on each side are seen the superior and inferior curved lines; these, as well as the surfaces of the bone between them, being rough for the attachment of numerous muscles.

Lateral Regions of the Skull

The *Lateral Regions* of the Skull are somewhat of a triangular form, their base being formed by a line extending from the external angular process of the frontal bone along the temporal ridge backwards to the outer extremity of the superior curved line of the occiput; and the sides being formed by two lines, the one drawn downwards and backwards from the external angular process of the frontal bone to the angle of the lower jaw, the other from the angle of the jaw upwards and backwards to the extremity of the superior curved line. This region is divisible into three portions, temporal, mastoid, and zygomatic.

THE TEMPORAL FOSSAE

The *Temporal Portion*, or fossa, is bounded above and behind by the temporal ridge, which extends from the external angular process of the frontal upwards and backwards across the frontal and parietal bones, curving downwards behind to terminate at the root of the zygomatic process. In front, it is bounded by the frontal, malar, and great wing of the sphenoid: externally, by the zygomatic arch, formed conjointly by the malar and temporal bones; below, it is separated from the zygomatic fossa by the pterygoid ridge, seen on the under surface of the great wing of the sphenoid. This fossa is formed by five bones, part of the frontal, great wing of the sphenoid, parietal, squamous portion of the temporal, and malar bones, and is traversed by five sutures, the transverse facial, coronal, spheno-parietal, squamo-parietal, and squamo-sphenoidal. It is deeply concave in front, convex behind, traversed by numerous grooves for lodging the branches of the deep temporal arteries, and filled by the temporal muscle.

The *Mastoid Portion* is bounded in front by the anterior horizontal root of the zygoma; above, by a line which corresponds with the posterior root of the zygoma and the masto-parietal suture; behind and inferiorly, by the masto-occipital suture. It is formed by the mastoid and part of the squamous portion of the temporal bone; its surface is convex and rough for the attachment of muscles, and presents, from behind forwards, the mastoid foramen, below the mastoid process. In front of the mastoid process is the external auditory meatus, surrounded by the auditory process. Anterior to the meatus is the Glenoid fossa, bounded in front by the tubercle of the zygoma, behind by the auditory process, and above by the middle root of the zygoma, which terminates at the Glaserian fissure.

THE ZYGOMATIC FOSSAE

The *Zygomatic* fossae, are two irregular-shaped cavities, situated one on each side of the head, below, and on the inner side of the zygoma; bounded in front by the tuberosity of the superior maxillary bone and the ridge which descends from its malar process; behind, by the posterior border of the pterygoid process; above, by the pterygoid ridge on the under surface of the great wing of the sphenoid and squamous portion of the temporal; below, by the alveolar border of the superior maxilla; internally, by the external pterygoid plate; and externally, by the zygomatic arch and ramus of the jaw. It contains the lower part of the Temporal, the External, and Internal pterygoid muscles, the internal maxillary artery, the inferior maxillary nerve, and their branches. At its upper and inner part may be observed two fissures, the spheno-maxillary and pterygo-maxillary.

The *Spheno-maxillary* fissure, horizontal in direction, opens into the outer and back part of the orbit. It is formed above by the lower border of the orbital surface of the great wing of the sphenoid; below, by the posterior rounded border of the superior maxilla and a small part of the palate bone; externally, by a small part of the malar bone; internally, it joins at right angles with the pterygo-maxillary fissure. This fissure opens a communication from the orbit into three fossae, the temporal, zygomatic, and spheno-maxillary; it transmits the superior maxillary nerve, infra-orbital artery, and ascending branches from Meckel's ganglion.

The *Pterygo-maxillary* fissure is vertical, and descends at right angles from the inner extremity of the preceding; it is a triangular interval, formed by the divergence of the superior maxillary bone from the pterygoid process of the sphenoid. It serves to connect the spheno-maxillary fossa with the zygomatic, and transmits branches of the internal maxillary artery.

The Spheno-maxillary Fossa

The Spheno-maxillary fossa is a small triangular space situated at the angle of junction of the spheno-maxillary and pterygo-maxillary fissures, and placed beneath the apex of the orbit. It is formed above by a small part of the under surface of the body of the sphenoid; in front, by the superior maxillary bone; behind, by the pterygoid process of the sphenoid; internally by the vertical plate of the palate; externally, it communicates with the spheno-maxillary fissure. This fossa has three fissures terminating in it, the sphenoidal, spheno-maxillary, and pterygo-maxillary; it communicates with three fossae, the orbital, nasal, and zygomatic, and with the cavity of the cranium, and has opening into it five foramina. Of these there are three on the posterior wall, the foramen rotundum above, the Vidian below and internal, and still more inferior and internal, the pterygo-palatine. On the inner wall is the spheno-palatine foramen by which it communicates with the nasal fossa, and below, the superior orifice of the posterior palatine canal, besides occasionally the orifices of two or three accessory posterior palatine canals.

Anterior Region of the Skull (Fig. 52)

The Anterior Region of the Skull, which forms the face, is of an oval form, presents an irregular surface, and is excavated for the reception of the two principal organs of sense, the eye and the nose. It is bounded above by the nasal eminences and margins of the orbit; below, by the prominence of the chin; on each side, by the malar bone, and anterior margin of the ramus of the jaw. In the median line are seen from above downwards, the nasal eminences, which indicate the situation of the frontal sinuses; diverging outwards from them, the superciliary ridges which support the eyebrows. Beneath the nasal eminences is the arch of the nose, formed by the nasal bones, and the nasal process of the superior maxillary. The nasal arch is convex from side to side, concave from above downwards, presenting in the median line the inter-nasal suture, formed between the nasal bones, laterally the naso-maxillary suture, formed between the nasal and the nasal process of the superior maxillary bones, both these sutures terminating above in that part of the transverse suture which connects the nasal bones and nasal processes of the superior maxillary with the frontal. Below the nose is seen the heart-shaped opening of the anterior nares, the narrow end

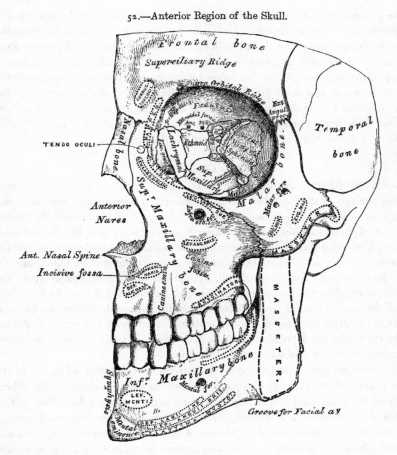

52.—Anterior Region of the Skull.

upwards, and broad below; it presents laterally the thin sharp margins which serve for the attachment of the lateral cartilages of the nose, and in the middle line below, a prominent process, the anterior nasal spine, bounded by two deep notches. Below this is the intermaxillary suture, and on each side of it the incisive fossa. Beneath this fossa is the alveolar process of the upper and lower jaw, containing the incisive teeth, and at the lower part of the median line, the symphysis of the chin, the mental eminence, and the incisive fossa of the lower jaw.

Proceeding from above downwards, on each side are the supra orbital ridges, terminating externally in the external angular process at its junction with the malar, and internally in the internal angular process; towards the inner third of this ridge is the supra orbital notch or foramen, for the passage of the supra-orbital vessels and nerve, and at its inner side a slight depression for the attachment of the cartilaginous pulley of the superior oblique muscle. Beneath the supra-orbital ridges are the openings of the orbits, bounded externally by the orbital ridge of the malar bone; below, by the orbital ridge formed by the malar, superior maxillary, and lachrymal bones; internally, by the nasal process of the superior maxillary, and the internal angular process of the frontal bone. On the outer side of the orbit, is the quadrilateral anterior surface of the malar bone, perforated by one or two small malar foramina. Below the inferior margin of the orbit, is the infra-orbital foramen, the termination of the infra-orbital canal, and beneath this, the canine fossa, which gives attachment to the Levator anguli oris; bounded below by the alveolar processes, containing the teeth of the upper and lower jaw. Beneath the alveolar arch of the lower jaw is the mental foramen for the passage of the mental nerve and artery, the external oblique line, and at the lower border of the bone, at the point of junction of the body with the ramus, a shallow groove for the passage of the facial artery.

The Orbits

The Orbits (fig. 52) are two quadrilateral hollow cones, situated at the upper and anterior part of the face, their bases being directed forwards and outwards, and their apices backwards and inwards. Each orbit is formed of *seven* bones, the frontal, sphenoid, ethmoid, superior maxillary, malar, lachrymal and palate; but three of these, the frontal, ethmoid and sphenoid, enter into the formation of *both* orbits, so that the two cavities are formed of *eleven* bones only. Each cavity presents for examination, a roof, a floor, an inner and an outer wall, a circumference or base, and an apex. The *Roof* is concave, directed downwards and forwards, and formed in front by the orbital plate of the frontal; behind, by the lesser wing of the sphenoid. This surface presents internally the depression for the fibro-cartilaginous pulley of the superior oblique muscle; externally, the depression for the lachrymal gland, and posteriorly, the suture connecting the frontal and lesser wing of the sphenoid.

The *Floor* is nearly flat, and of less extent than the roof; it is formed chiefly by the orbital process of the superior maxillary; in front, to a small extent, by the orbital process of the malar, and behind, by the orbital surface of the palate. This surface presents at its anterior and internal part, just external to the lachrymal canal, a depression for the attachment of the tendon of origin of the inferior oblique muscle; externally, the suture between the malar and superior maxillary bones; near its middle, the infra-orbital groove; and posteriorly, the suture between the maxillary and palate bones.

The *Inner Wall* is flattened, and formed from before backwards by the nasal process of the superior maxillary, the lachrymal, os planum of the ethmoid, and a small part of the body of the sphenoid. This surface presents the lachrymal groove, and crest of the lachrymal bone, and the sutures connecting the ethmoid, in front, with the lachrymal, behind, with the sphenoid.

The *Outer Wall* is formed in front by the orbital process of the malar bone; behind, by the orbital plate of the sphenoid. On it are seen the orifices of one or two malar canals, and the suture connecting the sphenoid and malar bones.

Angles. The *superior external angle* is formed by the junction of the upper and outer walls; it presents from before backwards, the sutures connecting the frontal with the malar in front, and with the orbital plate of the sphenoid behind; quite posteriorly is the foramen lacerum anterius, or sphenoidal fissure, which transmits the third, fourth, ophthalmic division of the fifth, and sixth nerves, and the ophthalmic vein. The *superior internal angle* is formed by the junction of the upper and inner wall, and presents the suture connecting the frontal with the lachrymal in front, and with the ethmoid behind. This suture is perforated by two foramina, the anterior and posterior ethmoidal, the former transmitting the anterior ethmoidal artery and nasal nerve, the latter the posterior ethmoidal artery and vein. The *inferior external angle*, formed by the junction of the outer wall and floor, presents the spheno-maxillary fissure, which transmits the infra-orbital vessels and nerve, and the ascending branches from the spheno-palatine ganglion. The *inferior internal angle* is formed by the union of the lachrymal and os planum of the ethmoid, with the superior maxillary and palate bones. The *circumference*, or base, of the orbit, quadrilateral in form, is bounded above by the supra-orbital arch; below, by the anterior border of the orbital plate of the malar, superior maxillary, and lachrymal bones; externally, by the external angular process of the frontal and malar bone; internally, by the internal angular process of the frontal and nasal process of the superior maxillary. The circumference is marked by three sutures, the fronto-maxillary internally, the fronto-malar externally, and the malo-maxillary below; it contributes to the formation of the lachrymal groove, and presents above, the supra-orbital notch (or foramen), for the passage of the supra-orbital artery, veins, and nerve. The *apex*, situated at the back of the orbit, corresponds to the optic foramen, a short circular canal, which transmits the optic nerve and ophthalmic artery. It will thus be seen that there are nine openings communicating with each orbit, viz., the optic, foramen lacerum anterius, spheno-maxillary fissure, supra-orbital foramen, infra-orbital canal, anterior and posterior ethmoidal foramina, malar foramina, and lachrymal canal.

The Nasal Fossae

The *Nasal Fossae* are two large irregular cavities, situated in the middle line of the face, separated from each other by a thin vertical septum, and extending from the base of the cranium to the roof of the mouth. They communicate by two large apertures, the anterior nares, with the front of the face, and with the pharynx behind by the two posterior nares. These fossae are much narrower above than below, and in the middle than at the anterior or posterior openings: their depth, which is considerable, is much greater in the middle than at either extremity. Each nasal fossa communicates with four sinuses, the frontal in front, the sphenoidal behind, and the maxillary and ethmoidal on either side. Each fossa also communicates with four cavities: with the orbit by the lachrymal canal, with the mouth by the anterior palatine canal, with the cranium by the olfactory foramina, and with the spheno-maxillary fossa by the spheno-palatine foramen; and they occasionally communicate with each other by an aperture in the septum. The bones entering into their formation are fourteen in number: three of the cranium, the frontal, sphenoid, and ethmoid, and all the bones of the face excepting the malar and lower jaw. Each cavity has four walls, a roof, a floor, an inner, and an outer wall.

The *upper wall*, or roof (fig. 53), is long, narrow, and concave from before backwards; it is formed in front by the nasal bones and nasal spine of the frontal, which are directed downwards and forwards; in the middle, by the cribriform lamella of the ethmoid, which is horizontal; and behind, by the under surface of the body of the sphenoid, and sphenoidal turbinated bones, which are directed downwards and backwards. This surface presents, from before backwards, the internal aspect of the nasal bones; on their outer side, the suture formed between the nasal, with the nasal process of the superior maxillary; on their inner side, the elevated crest which receives the nasal spine of the frontal and the perpendicular plate of the ethmoid, and articulates with its fellow of the opposite side; whilst the surface of the bones is perforated by a few small vascular apertures, and presents the longitudinal groove for the nasal nerve: further back is the transverse suture, connecting the frontal with the nasal in front, and the ethmoid behind, the olfactory foramina on the under surface of the cribriform plate, and the suture between it and the sphenoid behind: quite posteriorly are seen the sphenoidal turbinated bones, the orifice of the sphenoidal sinuses, and the articulation of the alae of the vomer with the under surface of the body of the sphenoid.

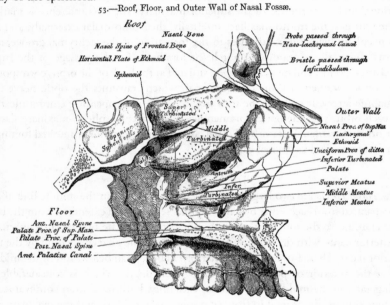

53.—Roof, Floor, and Outer Wall of Nasal Fossæ.

The *floor* is flattened from before backwards, concave from side to side, and wider in the middle than at either extremity. It is formed in front by the palate process of the superior maxillary; behind, by the palate process of the palate bone. This surface presents, from before backwards, the anterior nasal spine; behind this, the upper orifice of the anterior palatine canal; internally, the elevated crest which articulates with the vomer; and behind, the suture between the palate and superior maxillary bones, and the posterior nasal spine.

The *inner wall*, or septum (fig. 54), is a thin vertical septum, which separates the nasal fossae from one another; it is occasionally perforated, so that they communicate, and is frequently deflected considerably to one side. It is formed, in front, by the crest of the nasal bones and nasal spine of the frontal; in the middle, by the perpendicular lamella of the ethmoid; behind, by the vomer and rostrum of the sphenoid; below, by the crest of the superior maxillary and palate bones. It presents, in front, a large triangular notch, which receives the triangular cartilage of the nose; above, the lower orifices of the olfactory canals; and behind, the guttural edge of the vomer. Its surface is marked by numerous vascular and nervous canals, and traversed by sutures connecting the bones of which it is formed.

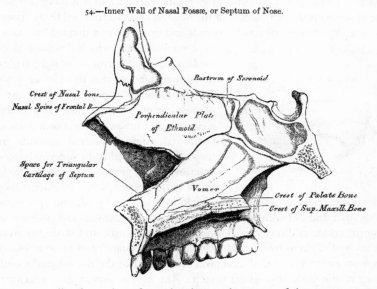

54.—Inner Wall of Nasal Fossæ, or Septum of Nose.

The *outer wall* is formed, in front, by the nasal process of the superior maxillary and lachrymal bones; in the middle, by the ethmoid and inner surface of the superior maxillary and inferior turbinated bones; behind, by the vertical plate of the palate bone. This surface presents three irregular longitudinal passages, or meatuses, formed between three horizontal plates of bone that spring from it; they are termed the superior, middle, and inferior meatuses of the nose. The *superior meatus*, the smallest of the three, is situated at the upper and back part of each nasal fossa, occupying the posterior third of the outer wall. It is situated between the superior and middle turbinated bones, and has opening into it two foramina, the spheno-palatine, at the back part of its outer wall, the posterior ethmoidal cells, at the front part of the upper wall. The opening of the sphenoidal sinuses is usually at the upper and back part of the nasal fossae, immediately behind the superior turbinated bone. The *middle meatus*, situated between the middle and inferior turbinated bones, occupies the posterior two-thirds of the outer wall of each nasal fossa. It presents two apertures. In front is the orifice of the infundibulum, by which the middle meatus communicates with the anterior ethmoidal cells, and through these with the frontal sinuses. At the centre of the outer wall is the orifice of the antrum, which varies somewhat as to its exact position in different skulls. The *inferior meatus*, the largest of the three, is the space between the inferior turbinated bone and the floor of the nasal fossa. It extends along the entire length of the outer wall of the nose, is broader in front than behind, and presents anteriorly the lower orifice of the lachrymal canal.

Os Hyoides

The Hyoid bone is named from its resemblance to the Greek Upsilon; it is also called the *lingual bone*, from supporting the tongue, and giving attachment to its numerous muscles. It is a bony arch, shaped like a horse-shoe, and consisting of five segments, a central portion or body, two greater cornua, and two lesser cornua.

The *Body* forms the central part of the bone, is of a quadrilateral form, its *anterior surface* (fig. 55) convex, directed forwards and upwards, and divided into two parts by a vertical ridge, which descends along the median line, and is crossed at right angles by a horizontal ridge, so that this surface is divided into four muscular depressions. At the point of meeting of these two lines is a prominent elevation, the tubercle. The portion above the horizontal ridge is directed upwards, and is sometimes described as the superior border. The anterior surface gives attachment to the Genio-hyoid in the greater part of its extent; above, to the Genio-hyo-glossus; below, to the Mylo-hyoid, Stylo-hyoid, and aponeurosis of the Digastric; and between these to part of the Hyo-glossus. The *posterior surface* is smooth, concave, directed backwards and downwards, and separated from the epiglottis by the thyro-hyoid membrane, and by a quantity of loose areolar tissue. The *superior border* is rounded, and gives attachment to the thyro-hyoid membrane, and part of the Genio-hyo-glossi muscles. The *inferior border* gives attachment in front to the Sterno-hyoid, behind to part of the Thyro-hyoid, and to the Omo-hyoid at its junction with the great cornu. The *lateral surfaces* are small, oval, convex facets, covered with cartilage for articulation with the greater cornua.

55.—Hyoid Bone. Anterior Surface.

The *Greater Cornua* project backwards from the lateral surfaces of the body, they are flattened from above downwards, diminish in size from before backwards, and terminate posteriorly in a tubercle for the attachment of the thyro-hyoid ligament. Their outer surface gives attachment to the Hyo-glossus; their upper border, to the Middle constrictor of the pharynx; their lower border, to part of the Thyro-hyoid muscle.

The *Lesser Cornua* are two small conical shaped eminences, attached by their bases to the angles of junction between the body and greater cornua, and giving attachment by their apices to the stylo-hyoid ligaments. In youth the cornua are connected to the body by cartilaginous surfaces and held together by ligaments; in middle life, the body and greater cornua usually become joined; and in old age all the segments are united together, forming a single bone.

Development. By *five* centres; one for the body and one for each cornu. Ossification commences in the body and greater cornua towards the end of foetal life, those for the cornua first appearing. Ossification of the lesser cornua commences some months after birth.

Attachment of Muscles. Sterno-hyoid, Thyro-hyoid, Omo-hyoid, aponeurosis of the Digastricus, Stylo-hyoid, Mylo-hyoid, Genio-hyoid, Genio-hyo-glossus, Hyo-glossus, Middle constrictor of the pharynx, and occasionally a few fibres of the

Lingualis. It also gives attachment to the thyro-hyoidean membrane, and the stylo-hyoid, thyro-hyoid, and hyo-epiglottic ligaments.

THE THORAX

The Thorax or chest is an osseo-cartilaginous cage, intended to contain and protect the principal organs of respiration and circulation. It is the largest of the three cavities connected with the spine, and is formed by the sternum and costal cartilages in front, the twelve ribs on each side, and the bodies of the dorsal vertebrae behind.

The Sternum

The Sternum (figs. 56, 57) is a flat narrow bone, situated in the median line of the front of the chest, and consisting in the adult of three portions. Its form resembles an ancient sword: the upper piece representing the handle, is termed the *manubrium*, the middle and largest piece which represents the chief part of the blade, is termed the *gladiolus*, and the inferior piece like the point of the sword, is termed the *ensiform* or *xiphoid appendix*. In its natural position, its direction is oblique, its anterior surface looking upwards and forwards, its posterior downwards and backwards. It is flattened in front, concave behind, broad above, becoming narrowed at the point where the first and second pieces are connected, after which it again widens a little, and is pointed at its extremity.

The *First Piece* of the sternum or *Manubrium*, is of a somewhat triangular form, broad and thick above, narrowed below at its junction with the middle piece. Its *anterior surface* convex from side to side, concave from above downwards, is smooth and affords attachment on each side to the Pectoralis major and sternal origin of the Sterno-cleido-mastoid muscle. In well marked bones, ridges limiting the attachment of these muscles are very distinct. Its *posterior surface*, concave and smooth, affords attachment on each side to the Sterno-hyoid and Sterno-thyroid muscles. The *superior border*, the thickest, presents at its centre the interclavicular notch, and on each side an oval articular surface, directed upwards backwards and outwards, for articulation with the sternal end of the clavicle. The *inferior border* presents an oval rough surface, covered in the recent state with a thin layer of cartilage, for articulation with the second portion of the bone. The *lateral borders* are marked superiorly by an articular depression for the first costal cartilage, and below by a half facet, which, with a similar facet on the upper angle of the middle portion of the bone, forms a notch for the reception of the costal cartilage of the second rib. These articular surfaces are separated by a curved edge which slopes from above downwards and inwards.

The *Second Piece* of the sternum, or *gladiolus*, considerably longer, narrower, and thinner than the superior, is broader below than above. Its *anterior surface* is nearly flat, directed upwards and forwards, and marked by three transverse lines which cross the bone opposite the third, fourth, and fifth articular depressions. These lines indicate the point of union of the four separate pieces of which this part of the bone consists at an early period of life. At the junction of the third and fourth pieces, is occasionally seen an orifice, the sternal foramen; it varies in size and form in different individuals, and pierces the bone from before backwards. This surface affords attachment on each side to the sternal origin of the Pectoralis major. The *posterior surface*, slightly

56.—Sternum and Costal Cartilages. Anterior Surface.

57.—Posterior Surface of Sternum.

concave, is also marked by three transverse lines; but they are less distinct than those on the anterior surface; this surface affords attachment below, on each side, to the Triangularis sterni muscle, and occasionally presents the posterior opening of the sternal foramen. The *superior border* is marked by an oval surface for articulation with the manubrium. The *inferior border* is narrow and articulates with the ensiform appendix. Each *lateral border* presents five articular depressions; the first, at each superior angle, is a half facet for the lower half of the cartilage of the second rib, the three succeeding depressions receive the cartilages of the third, fourth, and fifth ribs, whilst each inferior angle presents a half facet for the upper half of the cartilage of the seventh rib. These depressions are separated by a series of curved inter-articular notches, which diminish in length from above downwards.

The *Third Piece* of the sternum, the *ensiform* or *xiphoid appendix*, is the smallest of the three; it is thin and elongated in form, cartilaginous in structure in youth, but more or less ossified at its upper part in the adult. Its *anterior surface* affords attachment to the costo-xiphoid ligaments. Its *posterior surface*, to some of the fibres of the Diaphragm and Triangularis sterni muscles. Its *lateral borders*, to the aponeurosis of the abdominal muscles. Above, it is continuous with the lower end of the gladiolus; below, by its pointed extremity, it gives attachment to the linea alba, and at each superior angle presents a half facet for the lower half of the cartilage of the seventh rib. This portion of the sternum is very various in appearance, being sometimes pointed, broad and thin, sometimes bifid, or perforated by a round hole, occasionally curved, or deflected considerably to one or the other side.

Structure. This bone is composed of a considerable amount of loose spongy tissue within, covered externally with a very thin layer of compact tissue.

Development. The sternum, including the ensiform appendix, is developed by *six* centres. One for the first piece or manubrium, four for the second piece or gladiolus, and one for the ensiform appendix. The sternum is entirely cartilaginous up to the middle of foetal life, and when ossification takes place, the ossific granules are deposited in the middle of the intervals between the articular depressions for the costal cartilages, in the following order (fig. 58). In the first piece, between the fifth and sixth months; in the second and third, between the sixth and seventh; in the fourth piece, at the ninth month; in the fifth, within the first year, or between the first and second years after birth; and in the ensiform appendix, between the second and the seventeenth or eighteenth years, by a single centre which makes its appearance at the upper part, and proceeds gradually downwards. To these may be added the occasional existence, as described by Breschet, of two small episternal centres, which make their appearance one on each side of the interclavicular notch. It occasionally happens that some of these divisions are formed from more than one centre, the number and position of which vary (fig. 60). Thus the first piece may have two, three, or even six centres; the second piece has seldom more than one; the third, fourth, and fifth pieces, are often formed from two centres placed laterally, the irregular union of which will serve to explain the occasional occurrence of the sternal foramen (fig. 61), or of the vertical fissure which occasionally intersects this part of the bone. Union of these various parts commences from below, and proceeds upwards, taking place in the following order (fig. 59). The fifth piece is joined to the fourth soon after puberty; the

58.—Development of Sternum, by Six Centres.

59.

60.—Peculiarities.

61.

fourth to the third, between the twentieth and twenty-fifth years; the third to the second, between the thirty-fifth and fortieth years; the second is rarely joined to the first except in very advanced age.

Articulations. With the clavicles, and seven costal cartilages on each side.

Attachment of Muscles. The Pectoralis major, Sterno-cleido-mastoid, Sterno-hyoid, Sterno-thyroid, Triangularis sterni, aponeurosis of the Obliquus externus, Obliquus internus, and Transversalis muscles, Rectus and Diaphragm.

The Ribs

The Ribs are elastic arches of bone, which form the chief part of the thoracic walls. They are twelve in number on each side; but this number may be increased by the development of a cervical or lumbar rib, or may be diminished to eleven. The first seven are connected behind with the spine, and in front with the sternum, through the intervention of the costal cartilages, they are called *vertebro-sternal*, or true ribs. The remaining five are false ribs; of these the first three, being connected behind with the spine, and in front with the costal cartilages, are called the *vertebro-costal ribs*; the last two are connected with the vertebrae only, being free at their anterior extremities, they are termed *vertebral* or *floating ribs*. The ribs vary in their direction, the upper ones being placed nearly at right angles with the spine; the lower ones are placed obliquely, so that the anterior extremity is lower than the posterior. The extent of obliquity reaches its maximum at the ninth rib, gradually decreasing from that point towards the twelfth. The ribs are situated one beneath the other in such a manner that spaces are left between them; these are called *intercostal spaces*. Their length corresponds to the length of the ribs, their breadth is more considerable in front than behind, and between the upper than between the lower ribs. The ribs increase in length from the first to the eighth, when they again diminish to the twelfth. In breadth they decrease from above downwards; in each rib the greatest breadth is at the sternal extremity.

Common characters of the Ribs (fig. 62). Take a rib from the middle of the series in order to study its common characters. Each rib presents two extremities, a posterior or vertebral, an anterior or sternal, and an intervening portion, the body or shaft. The *posterior* or *vertebral extremity*, presents for examination a head, neck, and tuberosity.

The *head* (fig. 63) is marked by a kidney-shaped articular surface, divided by a horizontal ridge into two facets for articulation with the costal cavity formed by the junction of the bodies of two contiguous dorsal vertebrae; the upper facet is small, the inferior one of large size; the ridge separating them, serves for the attachment of the inter-articular ligament.

The *neck* is that flattened portion of the rib which extends outwards from the head; it is about an inch long, and rests upon the transverse process of the inferior of the two vertebrae with which the head articulates. Its *anterior surface* is flat and smooth, its *posterior* rough, for the attachment of the middle costo-transverse ligament. Of its two borders, the *superior* presents a rough crest for the attachment of the anterior costo-transverse ligament; its *inferior border* is rounded. On the outer surface of the neck, just where it joins the shaft, is an eminence, the tuberosity; it consists of two portions, an articular and a non-articular. The *articular portion*, the

62.—A Central Rib of Right Side. Inner Surface.

most internal and inferior of the two, presents a small oval surface directed downwards, backwards and inwards, for articulation with the extremity of the transverse process of the vertebra below it. The *non-articular portion* is a rough elevation, which affords attachment to the posterior costo-transverse ligament.

63.—Vertebral Extremity of a Rib. External Surface.

The *shaft* presents two surfaces, an external and an internal; and two borders, a superior and an inferior. The *external surface* is convex, and marked for the attachment of muscles. At its posterior part, a little in front of the tuberosity, is seen a prominent line, directed obliquely from above, downwards and outwards; this gives attachment to a tendon of the Sacro-lumbalis muscle, and is called the *angle*. At this point, the rib is bent upon itself in two directions. If the rib is laid upon its lower border, it will be seen that the anterior portion of the shaft, as far as the angle, rests upon this surface, while the vertebral end of the bone, beyond the angle, is bent inwards and at the same time tilted upwards. The distance between the angle and the tuberosity increases gradually from the second to the tenth rib. This portion of bone is rounded, rough, and irregular, and serves for the attachment of the Longissimus dorsi. The portion of bone between the angle and sternal extremity is also slightly twisted upon its own axis, the external surface looking downwards behind the angle, a little upwards in front of it. This surface presents, towards its sternal extremity, an oblique line, the anterior angle. The *internal surface* is concave, smooth, and presents the orifices of two or three nutrient foramina, the course of which is directly backwards towards the vertebral extremity. This surface looks a little upwards, behind the angle; a little downwards, in front of it. The *superior border*, thick and rounded, is marked by an external and an internal lip, more distinct behind than in front; they serve for the attachment of the External and Internal intercostal muscles. The *inferior border*, thin and sharp, has attached the External intercostal muscle. This border is marked on its inner side by a deep groove which commences at the tuberosity and gradually becomes lost at the junction of the anterior with the middle third of the bone. At the back part of the bone, this groove belongs to the inferior border; but just in front of the angle, where it is deepest and broadest, it corresponds to the internal surface; it lodges the intercostal vessels and nerve. Its superior edge is rounded and continued back as far as the vertebral extremity; it serves for the attachment of the Internal intercostal muscle. Its inferior edge corresponds to the lower margin of the rib, and gives attachment to the External intercostal. The anterior or sternal extremity, is flattened, and presents a porous oval concave depression, into which the costal cartilage is received.

Peculiar Ribs

The peculiar ribs which require especial consideration, are five in number, viz., the first, second, tenth, eleventh and twelfth.

The first rib (fig. 64) is one of the shortest and the most curved of all the ribs; it is broad, flat, and placed horizontally at the upper part of the thorax, its surfaces looking upwards and downwards; and its borders, inwards and outwards. The *head* is of small size, rounded, and presents only a single articular facet for articulation with

Peculiar Ribs.

the body of the first dorsal vertebra. The *neck* is narrow and rounded. The *tuberosity*, thick and prominent, rests on the outer border. There is no angle, and it is not twisted on its axis. The upper surface of the shaft is marked towards its anterior part by two shallow depressions, separated from one another by a ridge, which becomes more prominent towards the internal border, where it terminates in a tubercle; this tubercle and ridge serve for the attachment of the Scalenus anticus muscle, the groove in front of it transmitting the subclavian vein; that behind it, the subclavian artery. The *inferior surface* is smooth, and destitute of the groove observed on the other ribs. The *outer border* is convex, thick, and rounded. The *inner*, concave, thin, and sharp, and marked about its centre by the tubercle before mentioned. The *anterior extremity* is larger and thicker than any of the other ribs.

The *second rib* (fig. 65) is much longer than the first, but bears a very considerable resemblance to it in the direction of its curvature. The non-articular portion of the tuberosity is occasionally only slightly marked. The *angle* is slight, and situated close to the tuberosity, and the rib is not twisted, so that both ends touch any plane surface upon which it may be laid. The shaft is not horizontal, like that of the first rib; its *external surface*, which is convex, looking upwards and a little outwards; it presents near the middle a rough eminence for the attachment of part of the first, and the second serration of the serratus magnus. The *inner surface* smooth and concave, is directed downwards and a little inwards; it presents a short groove towards its posterior part.

The *tenth rib* (fig. 66) has only a single articular surface on its head.

The *eleventh* and *twelfth ribs* (figs. 67 and 68) have each a single articular surface on the head, which is of rather large size; they have no neck or tuberosity, and are pointed at the extremity. The eleventh has a slight angle and a shallow groove on the lower border. The twelfth has neither, and is much shorter than the eleventh.

Structure. The ribs consist of a quantity of cancellous tissue, enclosed in a thin compact layer.

Development. Each rib, with the exception of the last two, is developed by *three* centres, one for the shaft, one for the head, and one for the tubercle. The last two have only *two* centres, that for the tubercle being wanting. Ossification commences in the body of the ribs at a very early period, before its appearance in the vertebrae. The epiphysis of the head, which is of a slightly angular shape, and that for the tubercle, of a lenticular form, make their appearance between the sixteenth and twentieth years, and do not become united to the rest of the bone until about the twenty-fifth year.

Attachment of Muscles. The Intercostals, Scalenus anticus, Scalenus medius, Scalenus posticus, Pectoralis minor, Serratus magnus, Obliquus externus, Transversalis, Quadratus lumborum, Diaphragm, Latissimus dorsi, Serratus posticus superior, Serratus posticus inferior, Sacro-lumbalis, Musculus accessorius ad sacro-lumbalem, Longissimus dorsi, Cervicalis ascendens, Levatores costarum.

The Costal Cartilages

The *Costal Cartilages* (fig. 56) are white elastic structures, which serve to prolong the ribs forward to the front of the chest, and contribute very materially to the elasticity of this cavity. The seven first are connected with the sternum, the three next with the lower border of the cartilage of the preceding rib. The cartilages of the two last ribs, which have pointed extremities, float freely in the parietes of the abdomen. Like the ribs, the costal cartilages vary in their length, breadth, and direction. They increase in length from the first to the seventh, and gradually diminish to the last. They diminish in breadth, as well as the intervals between them, from the first to the last. They are broad at their attachment to the ribs, and taper towards their sternal extremities, excepting the two first, which are of the same breadth throughout, and the sixth, seventh, and eighth, which are enlarged where their margins are in contact. In direction they also vary; the first descends a little, the second is horizontal, the third ascends slightly, whilst all the rest follow the course of the ribs for a short extent, and then ascend to the sternum or preceding cartilage. Each costal cartilage presents two surfaces, two borders, and two extremities. The *anterior surface* is convex, and looks forwards and upwards; that of the first gives attachment to the costo-clavicular ligament; that of the first, second, third, fourth, fifth, and sixth at their sternal ends to the Pectoralis major. The others are covered, and give partial attachment to some of the great flat muscles of the abdomen. The *posterior surface* is concave, and directed backwards and downwards, the six or seven inferior ones affording attachment to the Transversalis and Diaphragm muscles. Of the two borders, the superior is concave, the inferior, convex; they afford attachment to the Intercostal muscles, the upper border of the sixth giving attachment to the Pectoralis major muscle. The contiguous borders also of the sixth, seventh, and eighth, and sometimes the ninth and tenth costal cartilages present smooth oblong surfaces at the points where they articulate. Of the two extremities, the outer one is continuous with the osseous tissue of the rib to which it belongs. The inner extremity of the first is continuous with the sternum; the six next have rounded extremities, which are received into shallow concavities on the lateral margins of the sternum. The inner extremities of the eighth, ninth and tenth costal cartilages are pointed, and lie in contact with the cartilage above. Those of the eleventh and twelfth are free and pointed.

In the male, the first costal cartilage becomes more or less ossified in the adult, and is often connected to the sternum by bone. Ossification of the remaining cartilages also occurs to a variable extent after the middle of life, those of the true ribs being first ossified. In the female, the process of ossification does not take place until old age. The costal cartilages are most elastic in youth, those of the false ribs being more so than the true. In old age they become of a deep yellow colour.

Attachment of Muscles. The Subclavius, Sterno-thyroid, Pectoralis major, Internal oblique, Transversalis, Rectus, Diaphragm, Triangularis sterni, Internal and External intercostals.

THE PELVIS

The *Pelvis* is composed of four bones. The two Ossa Innominata, which bound it in front and at the sides, and the Sacrum and Coccyx, which complete it behind.

The Os Innominatum

The *Os Innominatum*, so called from bearing no resemblance to any known object, is a large irregular-shaped bone, which, with its fellow of the opposite side, forms the sides and anterior wall of the pelvic cavity. In young subjects, it consists of three separate parts, which meet and form the large cup-like cavity, situated near the middle of the outer side of the bone; and, although in the adult these have become united, it is usual to describe the bone as divisible into three portions, the ilium, the ischium, and the pubes.

The *ilium* is the superior broad and expanded portion which runs upwards from the upper and back part of the acetabulum, and forms the prominence of the hip.

The *ischium* is the inferior and strongest portion of the bone; it proceeds downwards from the acetabulum, expands into a large tuberosity, and then curving upwards, forms with the descending ramus of the pubes a large aperture, the obturator foramen.

The *pubes* is that portion which runs horizontally inwards from the inner side of the acetabulum for about two inches, then makes a sudden bend, and descends to the same extent: it forms the front of the pelvis, and supports the external organs of generation.

The *Ilium* presents for examination two surfaces, an external and an internal, a crest, and two borders, an anterior and a posterior.

69.—Right Os Innominatum. External Surface.

External Surface or *Dorsum of the Ilium* (fig. 69). The back part of this surface is directed backwards, downwards, and outwards; its front part forwards, downwards and outwards. It is smooth, convex in front, deeply concave behind; bounded above by the crest, below by the upper border of the acetabulum, in front and behind by the anterior and posterior borders. This surface is crossed in an arched direction by three semicircular lines, the superior, middle, and inferior curved lines. The superior curved line, the shortest of the three, commences at the crest, about two inches in front of its posterior extremity; it is at first distinctly marked, but as it passes downwards and outwards to the back part of the great sacro-sciatic notch, where it terminates, it becomes less marked, and is often altogether lost. The rough surface included between this line and the crest, affords attachment to part of the Gluteus maximus above, a few fibres of the Pyriformis below. The middle curved line, the longest of the three, commences at the crest, about an inch behind its anterior extremity, and, taking a curved direction downwards and backwards, terminates at the upper part of the great sacro-sciatic notch. The space between the middle, the superior curved lines, and the crest, is concave, and affords attachment to the Gluteus medius muscle. Near the central part of this line may often be observed the orifice of a nutritious foramen. The inferior curved line, the least distinct of the three, commences in front at the upper part of the anterior inferior spinous process, and taking a curved direction backwards and downwards, terminates at the anterior part of the great sacro-sciatic notch. The surface of bone included between the middle and inferior curved lines, is concave from above downwards, convex from before backwards, and affords attachment to the Gluteus minimus muscle. Beneath the inferior curved line, and corresponding to the upper part of the acetabulum, is a smooth eminence (sometimes a depression), to which is attached the reflected tendon of the Rectus femoris muscle.

The *Internal Surface* (fig. 70) of the ilium is bounded above by the crest, below by a prominent line, the linea-ileo pectinea, and before and behind by the anterior and posterior borders. It presents anteriorly a large smooth concave surface called the *internal iliac fossa*, or *venter of the ilium*; it lodges the Iliacus muscle, and presents at its lower part, the orifice of a nutritious canal. Behind the iliac fossa is a rough surface, divided into two portions, a superior and an inferior. The inferior, or auricular portion, so called from its resemblance to the external ear, is coated with cartilage in the recent state, and articulates with a similar shaped surface on the side of the sacrum. The superior portion is concave and rough for the attachment of the posterior sacro-iliac ligaments.

The crest of the ilium is convex in its general outline and sinuously curved, being bent inwards anteriorly, outwards posteriorly. It is longer in the female than in the male, very thick behind, and thinner at the centre than at the extremities. It terminates at either end in a prominent eminence, the anterior superior, and posterior superior spinous process. The surface of the crest is broad, and divided into an external lip, an internal lip, and an intermediate space. To the external lip is attached the Tensor vaginae femoris, Obliquus externus abdominis, and Latissimus dorsi, and by its whole length the fascia lata; to the interspace between the lips, the Internal oblique; to the internal lip, the Transversalis, Quadratus lumborum, and Erector spinae.

The anterior border of the ilium is concave. It presents two projections separated

70.—Right Os Innominatum. Internal Surface.

by a notch. Of these, the uppermost, formed by the junction of the crest and anterior border, is called the anterior superior spinous process of the ilium, the outer border of which gives attachment to the fascia lata, and the origin of the Tensor vaginae femoris, its inner border, to the Iliacus internus, whilst its extremity affords attachment to Poupart's ligament and the origin of the Sartorius. Beneath this eminence, is a notch which gives attachment to the Sartorius muscle, and across which passes the external cutaneous nerve. Below the notch is the anterior inferior spinous process, which terminates in the upper lip of the acetabulum; it gives attachment to the straight tendon of the Rectus femoris muscle. On the inner side of the anterior inferior spinous process, is a broad shallow groove, over which passes the Iliacus muscle. The posterior border, shorter than the anterior, also presents two projections separated by a notch, the posterior superior, and the posterior inferior spinous processes. The former corresponds with that portion of the posterior surface of the ilium, which serves for the attachment of the sacro-iliac ligaments, the latter, to the auricular portion which articulates with the sacrum. Below the posterior inferior spinous process, is a deep notch, the great sacro-sciatic.

The *Ischium* forms the inferior and posterior part of the os innominatum. It is divisible into a thick and solid portion, the body, and a thin ascending part, the ramus. The body, somewhat triangular in form, presents three surfaces, an external, internal, and posterior, and three borders. The external surface corresponds to that portion of the acetabulum formed by the ischium; it is smooth and concave above, and forms a little more than two-fifths of that cavity; its outer margin is bounded by a prominent rim or lip, to which the cotyloid-fibro-cartilage is attached. Below the acetabulum, between it and the tuberosity, is a deep groove, along which the tendon of the Obturator externus muscle runs, as it passes outwards to be inserted into the digital fossa of the femur. The internal surface is smooth, concave, and forms the lateral boundary of the true pelvic cavity; it is broad above, and separated from the venter of the ilium by the linea-ileo-pectinea, narrow below, its posterior border being encroached upon a little below its centre, by the spine of the ischium, above and below which are the greater and lesser sacro-sciatic notches; in front it presents a sharp margin, which forms the outer boundary of the obturator foramen. This surface is perforated by two or three large vascular foramina, and affords attachment to part of the Obturator internus muscle.

The *posterior surface* is quadrilateral in form, broad and smooth above, narrow below where it becomes continuous with the tuberosity; it is limited in front by the margin of the acetabulum, behind by the front part of the great sacro-sciatic notch. This surface supports the Pyriformis, the two Gemelli, and the Obturator internus muscles, in their passage outwards to the great trochanter. The body of the ischium presents three borders, posterior, inferior, and internal. The posterior border presents, a little below the centre, a thin and pointed triangular eminence, the spine of the ischium, more or less elongated in different subjects. Its external surface gives attachment to the Gemellus superior, its internal surface to the Coccygeus and Levator ani, whilst to the pointed extremity is connected the lesser sacro-sciatic ligament. Above the spine is a notch of large size, the great sacro-sciatic, converted into a foramen by the lesser sacro-sciatic ligament; it transmits the Pyriformis muscle, the gluteal vessels and nerve passing out above this muscle, the sciatic, and internal pudic vessels and nerve, and a small nerve to the Obturator internus muscle below it. Below the spine is a smaller notch, the lesser sacro-sciatic; it is smooth, coated with cartilage in the recent state, the surface of which presents numerous markings corresponding to the subdivisions of the tendon of the Obturator internus which winds over it. It is converted into a foramen by the sacro-sciatic ligaments, and transmits the tendon of the Obturator internus, the nerve which supplies this muscle, and the pudic vessels and nerve. The inferior border is thick and broad: at its point of junction with the posterior, is a large rough eminence upon which the body rests in sitting; it is called the tuberosity of the ischium. The internal border is thin, and forms the outer circumference of the obturator foramen.

The *tuberosity*, situated at the junction of the posterior and inferior borders, presents for examination an external lip, an internal lip, and an intermediate space. The external lip gives attachment to the Quadratus femoris and part of the Adductor magnus muscles. The inner lip is bounded by a sharp ridge for the attachment of a falciform prolongation of the great sacro-sciatic ligament, presents a groove on the inner side of this for the lodgment of the internal pudic vessels and nerve, and more anteriorly has attached the Transversus perinei, Erector penis, and Compressor urethrae muscles. The intermediate surface presents four distinct impressions. Two of these seen at the front part of the tuberosity are rough, elongated, and separated from each other by a prominent ridge; the outer one gives attachment to the Adductor magnus, the inner one to the great sacro-sciatic ligament. Two situated at the back part are smooth, larger in size and separated by an oblique ridge: from the upper and outer arises the Semi-membranosus; from the lower and inner, the Biceps and Semi-tendinosus. The most superior part of the tuberosity gives attachment to the Gemellus inferior.

The *ramus* is the thin flattened part of the ischium, which ascends from the tuberosity upwards and inwards, and joins the ramus of the pubes, their point of junction being indicated in the adult by a rough eminence. Its outer surface is rough for the attachment of the Obturator externus muscle. Its inner surface forms part of the anterior wall of the pelvis. Its inner border is thick, rough, slightly everted, forms part of the outlet of the pelvis, and serves for the attachment of the crus penis. Its outer border is thin and sharp, and forms part of the inner margin of the obturator foramen.

The *Pubes* forms the anterior part of the os innominatum; it is divisible into a horizontal ramus or body, and a perpendicular ramus.

The *body*, or horizontal ramus, presents for examination two extremities, an outer and an inner, and four surfaces. The *outer extremity*, the thickest part of the bone, forms one-fifth of the cavity of the acetabulum: it presents above, a rough eminence, the ilio-pectineal, which serves to indicate the point of junction of the ilium and pubes. The *inner extremity* of the body of the bone is the symphysis; it is oval, covered by eight or nine transverse ridges, or a series of nipple-like processes arranged in rows, separated by grooves; they serve for the attachment of the interarticular fibro-cartilage, placed between it and the opposite bone. Its *upper surface*, triangular in form, wider externally than internally, is bounded behind by a sharp ridge, the pectineal line, or linea-ilio-pectinea, which, running outwards, marks the brim of the true pelvis. The surface of bone in front of the pubic portion of the linea-ilio-pectinea, serves for the attachment of the Pectineus muscle. This ridge terminates internally at a tubercle, which projects forwards, and is called the *spine* of the pubes. The portion of bone included between the spine and inner extremity of the pubes is called the *crest*; it serves for the attachment of the Rectus, Pyramidalis, and conjoined tendon of the Internal oblique and Transversalis. The point of junction of the crest with the symphysis is called the *angle of the pubes*. The *inferior surface* presents externally a broad and deep oblique groove, for the passage of the obturator vessels and nerve; and internally a sharp margin, which forms part of the circumference of the obturator foramen. Its *external surface* is flat and compressed, and serves for the attachment of muscles. Its *internal surface*, convex from above downwards, concave from side to side, is smooth, and forms part of the anterior wall of the pelvis. The descending ramus of the pubes passes outwards and downwards, becoming thinner and narrower as it descends, and joins with the ramus of the ischium. Its *external surface* is rough, for the attachment of muscles. Its *inner surface* is smooth. Its *inner border* is thick, rough, and everted, especially in females. In the male it serves for the attachment of the crus penis. Its *outer border* forms part of the circumference of the obturator foramen.

The *cotyloid cavity* or *acetabulum*, is a deep cup-shaped hemispherical depression; formed internally by the pubes, above by the ilium, behind and below by the ischium, a little less than two-fifths being formed by the ilium, a little more than two-fifths by the ischium, and the remaining fifth by the pubes. It is bounded by a prominent uneven rim, which is thick and strong above, and serves for the attachment of a fibro-cartilaginous structure, which contracts its orifice and deepens the surface for articulation. It presents on its inner side a deep notch, the cotyloid notch, which transmits the nutrient vessels into the interior of the joint, and is continuous with a deep circular depression at the bottom of the cavity: this depression is perforated by numerous apertures, lodges a mass of fat, and its margins serve for the attachment of the ligamentum teres. The notch is converted, in the natural state, into a foramen by a dense ligamentous band which passes across it. Through this foramen the nutrient vessels and nerves enter the joint.

The *obturator* or *thyroid foramen* is a large aperture, situated between the ischium and pubes. In the male it is large, of an oval form, its longest diameter being obliquely from above downwards; in the female smaller, and more triangular. It is bounded by a thin uneven margin, to which a strong membrane is attached; and presents at its upper and outer part a deep groove, which runs from the pelvis obliquely forwards, inwards, and downwards. It is converted into a foramen by the obturator membrane, and transmits the obturator vessels and nerve.

71.—Plan of the Development of the Os Innominatum.

The 3 Primary centres unite through Y Shaped piece, about puberty
Epiphyses appear about puberty, & unite about 25th year

Structure. This bone consists of much cancellous tissue, especially where it is thick, enclosed between two layers of thick and dense compact tissue. In the thinner parts of the bone, as at the bottom of the acetabulum, and centre of the iliac fossa, it is usually semi-transparent, and composed entirely of compact tissue.

Development (fig. 71). By *eight* centres: three primary—one for the ilium, one for the ischium, and one for the pubes; and *five* secondary—one for the crest of the ilium its whole length, one for the anterior inferior spinous process (said to occur more frequently in the male than the female), one for the tuberosity of the ischium, one for the symphysis pubis (more frequent in the female than the male), and one for the Y-shaped piece at the bottom of the acetabulum. These various centres appear in the following order: First, in the ilium, at the lower part of the bone, immediately above the sciatic notch, at about the same period that the development of the vertebrae commences. Secondly, in the body of the ischium, at about the third month of foetal life. Thirdly, in the body of the pubes, between the fourth and fifth months. At birth, these centres are quite separate; the crest, the bottom of the acetabulum, and the rami of the ischium and pubes, being still cartilaginous. At about the sixth year, the rami of the pubes and ischium are almost completely ossified. About the thirteenth or fourteenth year, the three divisions of the bone have extended their growth into the bottom of the acetabulum, being separated from each other by a Y-shaped portion of cartilage, which now presents traces of ossification. The ilium and ischium then become joined, and lastly the pubes, through the intervention of the portion above mentioned. At about the age of puberty, ossific matter appears in each of the remaining portions, and they become joined to the rest of the bone about the twenty-fifth year.

Articulations. With its fellow of the opposite side, the sacrum and femur.

Attachment of Muscles. Ilium. To the outer lip of the crest, the Tensor vaginae femoris, Obliquus externus abdominis, and Latissimus dorsi; to the internal lip, the Transversalis, Quadratus lumborum, and Erector spinae; to the interspace between the lips, the Obliquus internus. To the outer surface of the ilium, the Gluteus maximus, Gluteus medius, Gluteus minimus, reflected tendon of Rectus, portion of Pyriformis; to the internal surface, the Iliacus; to the anterior border, the Sartorius and straight tendon of the Rectus. To the ischium. Its outer surface, the Obturator externus; internal surface, Obturator internus and Levator ani. To the spine. The Gemellus superior, Levator ani, and Coccygeus. To the tuberosity, the Biceps, Semi-tendinosus, Semi-membranosus, Quadratus femoris, Adductor magnus, Gemellus inferior, Transversus perinaei, Erector penis. To the pubis, the Obliquus externus, Obliquus internus, Transversalis, Rectus, Pyramidalis, Psoas parvus, Pectineus, Adductor longus, Adductor brevis, Gracilis, Obturator externus and internus, Levator ani, Compressor urethrae, and occasionally a few fibres of the Accelerator urinae.

THE PELVIS

The pelvis is stronger and more massively constructed than either of the other osseous cavities already considered; it is connected to the lower end of the spine, which it supports, and transmits its weight to the lower extremities, upon which it rests. It is composed of four bones—the two ossa innominata, which bound it on either side and in front; and the sacrum and coccyx, which complete it behind.

The pelvis is divided by a prominent line, the linea ileo pectinea, into a false and true pelvis.

The *false pelvis* is all that expanded portion of the pelvic cavity which is situated above the linea ileo pectinea. It is bounded on each side by the ossa ilii; in front it is incomplete, presenting a wide interval between the anterior superior spinous processes of the ilia on either side, filled up in the recent state by the parietes of the abdomen; behind, in the middle line, is a deep notch. This broad shallow cavity is admirably adapted to support the intestines, and to transmit part of their weight to the anterior wall of the abdomen.

The *true pelvis* is all that part of the pelvic cavity which is situated beneath the linea ileo pectinea. It is smaller than the false pelvis, but its walls are more perfect. For convenience of description, it may be divided into a superior circumference or inlet, an inferior circumference or outlet, and a cavity.

The *superior circumference* forms the margin or brim of the pelvis, the included space being called the *inlet*. It is formed by the linea ileo pectinea, completed in front by the spine and crest of the pubes, and behind by the anterior margin of the base of the sacrum and sacro-vertebral angle.

The *inlet* of the pelvis is somewhat cordate in form, obtusely pointed in front, diverging on either side, and encroached upon behind by the projection forwards of the promontory of the sacrum. It has three principal diameters: antero-posterior (sacro-pubic), transverse, and oblique. The antero-posterior extends from the sacro-vertebral angle to the symphysis pubis; its average measurement is four inches. The transverse extends across the greatest width of the pelvis, from the middle of the brim on one side, to the same point on the opposite; its average measurement is five inches. The oblique extends from the margin of the pelvis corresponding to the ileo pectineal eminence on one side, to the sacro-iliac symphysis on the opposite side; its average measurement is also five inches.

The *cavity* of the true pelvis is bounded in front by the symphysis pubis; behind, by the concavity of the sacrum and coccyx, which, curving forwards above and below, contracts the inlet and outlet of the canal; and laterally it is bounded by a broad, smooth, quadrangular plate of bone, corresponding to the inner surface of the body of the ischium. This cavity is shallow in front, measuring at the symphysis an inch and a half in depth, three inches and a half in the middle, and four inches and a half posteriorly. From this description, it will be seen that the cavity of the pelvis is a short, curved canal, considerably deeper on its posterior than on its anterior wall, and broader in the middle than at either extremity, from the projection forwards of the sacro-coccygeal column above and below. This cavity contains, in the recent subject, the rectum, bladder, and part of the organs of generation. The rectum is placed at the back of the pelvis, and corresponds to the curve of the sacro-coccygeal column, the bladder in front, behind the symphysis pubis. In the female, the uterus and vagina occupy the interval between these parts.

The *lower circumference* of the pelvis is very irregular, and forms what is called the *outlet*. It is bounded by three prominent eminences: one posterior, formed by the point of the coccyx; and one on each side, the tuberosities of the ischia. These eminences are separated by three notches; one in front, the *pubic arch*, formed by the convergence of the rami of the ischia and pubes on each side. The other notches, one on each side, are formed by the sacrum and coccyx behind, the ischium in front, and the ilium above: these are called the *greater* and *lesser sacro-sciatic notches*; in the natural state they are converted into foramina by the lesser and greater sacro-sciatic ligaments.

The diameters of the outlet of the pelvis are two, antero-posterior and transverse. The *antero-posterior* extends from the tip of the coccyx to the lower part of the symphysis pubis; and the *transverse* from the posterior part of one ischiatic tuberosity, to the same point on the opposite side: the average measurement of both is four inches. The antero-posterior diameter varies with the length of the coccyx, and is capable of increase or diminution, on account of the mobility of this bone.

Position of the Pelvis. In the erect posture, the pelvis is placed obliquely with regard to the trunk of the body; the pelvic surface of the symphysis pubis looking upwards and backwards, the concavity of the sacrum and coccyx looking downwards and forwards. The base of the sacrum, in well-formed female bodies, being nearly four inches above the upper border of the symphysis pubis, and the apex of the coccyx a little more than half an inch above its lower border. This obliquity is much greater in the foetus, and at an early period of life, than in the adult.

Axes of the Pelvis. The plane of the inlet of the true pelvis will be represented by a line drawn from the base of the sacrum to the upper margin of the symphysis pubis. A line carried at right angles with this at its middle, would correspond at one extremity with the umbilicus, and by the other with the middle of the coccyx; the axis of the inlet is therefore directed downwards and backwards. The axis of the outlet produced upwards, would touch the base of the sacrum; and is therefore directed downwards and forwards. The axis of the cavity is curved like the cavity itself: this curve corresponds to the concavity of the sacrum and coccyx, the extremities being indicated by the central points of the inlet and outlet.

Differences between the Male and Female Pelvis. In the *male*, the bones are heavier, stronger, and more solid, and the muscular impressions and eminences on their surfaces more strongly marked. It is altogether more massive in its general form; its cavity is deeper and narrower, and its apertures small. In the *female*, the bones are lighter and more delicate, the muscular impressions on its surface only slightly marked, and the pelvis generally is less massive in structure. The iliac fossae are large, and the ilia widely expanded; hence the great prominence of the hips. The cavity is shallow, but capacious, being very broad both in the antero-posterior and transverse diameters; the inlet and outlet are also large. The obturator foramen is triangular; the tuberosities of the ischia are widely separated; the sacrum is wider and less curved; the symphysis pubis not so deep; and the arch of the pubis is greater, and its edges more everted.

In the *foetus*, and for several years after birth, the cavity of the pelvis is small; the viscera peculiar to this cavity in the adult, being situated in the lower part of the abdomen.

THE UPPER EXTREMITY

The Upper Extremity consists of four parts—the shoulder, the arm, the forearm, and the hand. The *shoulder* consists of two bones, the clavicle and the scapula.

The Clavicle

The *Clavicle* (*clavis*, a "key"), or collar-bone, is a long bone, curved somewhat like the italic letter ∫, and placed horizontally at the upper and lateral part of the thorax, immediately above the first rib. It articulates internally with the upper border of the sternum, and with the acromion process of the scapula by its outer extremity; serving to sustain the upper extremity in the various positions which it assumes, whilst at the same time it allows it great latitude of motion. The horizontal plane of the clavicle is nearly straight; but in the vertical plane it presents a double curvature, the convexity being in front at the sternal end, and behind at the scapular end. Its inner two-thirds are of a triangular prismatic form, and extend, in the natural position of the bone, from the sternum to the coracoid process of the scapula; the outer fourth being flattened from above downwards, and extending from the coracoid process to the acromion. It presents for examination two surfaces, two borders, and two extremities.

The *superior surface* (fig. 72), for the inner three-fourths of its extent, is narrow, smooth, of equal diameter throughout, and presents near the sternal end impressions for the attachment of the Sterno-mastoid muscle behind, the Pectoralis major in front. Its outer fourth is broad, flat, uneven, perforated by numerous foramina, and covered by the fibres of the Deltoid and Trapezius muscles, which encroach upon it considerably in front and behind.

The *inferior surface* (fig. 73) is also narrow for the inner three-fourths of its extent, broader and more flattened externally. Commencing at the sternal extremity, may be seen a small facet for articulation with the cartilage of the first rib, continuous with the articular surface at the sternal end of the bone. External to this a rough

72.—Left Clavicle. Superior Surface.

73.—Left Clavicle. Inferior Surface.

impression, the rhomboid, for the attachment of the costo-clavicular (rhomboid) ligament. The middle third of this surface is occupied by a longitudinal groove, the subclavian groove, broader externally than internally; it gives attachment to the Subclavius muscle, and by its anterior margin to the strong aponeurosis which encloses it; internally is a rough surface, the limit of the attachment of the Pectoralis major below. At the junction of the prismatic with the flattened portion of the bone, at its posterior border, may be seen a rough eminence, the *tubercle* of the clavicle. This, in the natural position of the bone, surmounts the coracoid process of the scapula, and affords attachment to the conoid ligament. From this tubercle an oblique line passes forwards and outwards on the under surface of the acromial extremity. It is called the *oblique line* of the clavicle, and affords attachment to the trapezoid ligament.

The *anterior border* is broad and convex for its sternal half, and presents a rough impression for the attachment of the Pectoralis major muscle. Its outer half is a narrow, concave margin, serving for the attachment of the Deltoid: a small interval is usually left between the attachments of these muscles, where this border is smooth, receiving no muscular fibres. This is the narrowest part of the clavicle, and hence the most common seat of fracture.

The *posterior border* is broad and deeply concave for the inner two-thirds of its extent, affording attachment internally by a small extent of surface, to the Sterno-hyoid; convex, narrow, and irregular in its outer third, for the attachment of the Trapezius. This border corresponds to the subclavian vessels and brachial plexus of nerves, and presents, towards its centre, the foramen for the nutritious artery of the bone.

The *internal* or *sternal* end of the clavicle is directed inwards, and a little downwards and forwards; it presents a large triangular facet, concave from before backwards, convex from above downwards, which articulates with the sternum through the intervention of an inter-articular fibro-cartilage; the circumference of the articular surface is rough, for the attachment of numerous ligaments.

The *outer extremity*, directed forwards and outwards, presents a small oval facet, for articulation with the acromion process of the scapula.

Peculiarities of this Bone in the Sexes and in Individuals. In the female, the clavicle is less curved, smoother, longer, and more slender than in the male. In those persons who perform considerable manual labour, which brings into constant action the muscles connected with this bone, it acquires considerable bulk, becomes shorter, more curved, its ridges for muscular attachment become prominently marked, and its sternal end of a prismatic or quadrangular form.

Structure. The shaft as well as the extremities consists of cancellous tissue, invested in a compact layer much thicker in the centre than at either end. The clavicle is highly elastic, by reason of its curves. From the experiments of Mr. Ward, it has been shewn that it possesses sufficient longitudinal elastic force to raise its own weight nearly two feet on a level surface; and sufficient transverse elastic force, opposite the centre of its anterior convexity, to raise its own weight about a foot. This extent of elastic power must serve to moderate very considerably the effect of concussions received upon the point of the shoulder.

Development. By *two* centres: one for the shaft, and one for the sternal end of the bone. The centre for the shaft appears very early, before any other bone; the second centre makes its appearance about the eighteenth or twentieth year, and unites with the rest of the bone a few years after.

Articulations. With the sternum, scapula, and cartilage of the first rib.

Attachment of Muscles. The Sterno cleido-mastoid, Trapezius, Pectoralis major, Deltoid, Subclavius, and Sterno-hyoid.

The Scapula

The *Scapula* is a large flat bone, triangular in shape, which forms the back part of the shoulder. It is situated at the posterior aspect and side of the thorax, between the first and seventh ribs, and presents for examination two surfaces, three borders, and three angles.

The *anterior surface*, or *venter* (fig. 74), presents a broad concavity, the subscapular fossa. It is marked, in the posterior two thirds, by several oblique ridges, which pass from behind obliquely forwards and upwards, the anterior third being smooth. The oblique ridges above-mentioned, give attachment to the tendinous intersections, and the surfaces between them, to the fleshy fibres of the Subscapularis muscle.

74.—Left Scapula. Anterior Surface, or Venter.

The anterior third of the fossa, which is smooth, is covered by, but does not afford attachment to, the fibres of this muscle. This surface is separated from the posterior border, by a smooth triangular margin at the superior and inferior angles, and in the interval between these, by a narrow margin which is often deficient. This marginal surface affords attachment throughout its entire extent to the Serratus magnus muscle. The subscapular fossa presents a transverse depression at its upper part, called the *subscapular angle*; it is in this situation that the fossa is deepest, and consequently the thickest part of the Subscapularis muscle lies in a line parallel with the glenoid cavity, and must consequently operate most effectively on the humerus which is contained in it.

The *posterior surface*, or *dorsum* (fig. 75) is convex from above downwards, alternately convex and concave from side to side. It is subdivided unequally into two parts by the spine; that portion above the spine is called the supra spinous fossa, and that below it, the infra spinous fossa.

The *supra spinous fossa*, the smaller of the two, is concave, smooth, and broader towards the vertebral than at the humeral extremity. It affords attachment by its inner two-thirds to the fibres of the Supra spinatus muscle.

75.—Left Scapula. Posterior Surface, or Dorsum.

The *infra spinous fossa* is much larger than the preceding; towards its inner side a shallow concavity is seen at its upper part; its centre presents a prominent convexity, whilst towards the axillary border is a deep groove, which runs from the upper towards the lower part. The inner three-fourths of this surface affords attachment to the Infra-spinatus muscle; the outer fourth is only covered by it, without giving origin to its fibres. This surface is separated from the axillary border by an elevated ridge, which runs from the lower part of the glenoid cavity, downwards and backwards to the posterior border, about an inch above the inferior angle. This ridge serves for the attachment of a strong aponeurosis, which separates the Infra-spinatus from the two Teres muscles. The surface of bone between this line and the axillary border is narrow for the upper two-thirds of its extent, and traversed near its centre by a groove for the passage of the dorsalis scapulae artery; it affords attachment to the Teres minor. Its lower third presents a broader, somewhat triangular surface, which gives origin to the Teres major, and over which glides the Latissimus dorsi muscle; sometimes this muscle takes origin by a few fibres from this part. The broad and narrow portions of bone above alluded to are separated by an oblique line, which runs from the axillary border, downwards and backwards; to it is attached the aponeurosis separating the two Teres muscles from each other.

The *Spine* is a prominent plate of bone, which crosses obliquely the inner three-fourths of the dorsum of the scapula at its upper part, and separates the supra from the infra spinous fossa: it commences at the vertebral border by a smooth triangular surface, over which the trapezius glides, separated by a bursa; and, gradually becoming more elevated as it passes forwards, terminates in the acromion process which overhangs the shoulder joint. The spine is triangular and flattened from above downwards, its apex corresponding to the posterior border, its base, which is directed outwards, to the neck of the scapula. It presents two surfaces and three borders. Its *superior surface* is concave, assists in forming the supra-spinous fossa, and affords attachment to part of the Supra-spinatus muscle. Its *inferior surface* forms part of the infra-spinous fossa, gives origin to part of the Infra-spinatus muscle, and presents near its centre the orifice of a nutritious canal. Of the three borders, the *anterior* is attached to the dorsum of the bone; the *posterior*, or *crest* of the spine, is broad, and presents two lips and an intervening rough interval. To the superior lip is attached the Trapezius, to the extent shown in the figure. A very rough prominence is generally seen occupying that portion of the spine which receives the insertion of the middle and inferior fibres of this muscle. To the inferior lip, its whole length, is attached the Deltoid. The interval between them is also partly covered by the fibres of these muscles. The *external border*, the shortest of the three, is slightly concave, its edges thick and round, continuous above with the under surface of the acromion process, below with the neck of the scapula. The narrow portion of bone external to this border, serves to connect the supra and infra spinous fossae.

The *Acromion process*, so called from forming the summit of the shoulder joint (ακρον, a summit; ωμος, the shoulder), is a large, and somewhat triangular process, flattened from behind forwards, directed at first a little outwards, and then curving forwards and upwards, so as to overhang the glenoid cavity. Its *upper surface* directed upwards, backwards, and outwards, is convex, rough, and partly covered by some of the fibres of origin of the Deltoid. Its *under surface* is smooth and concave. Its *outer border*, which is thick and irregular, affords attachment to the Deltoid muscle. Its *inner margin*, shorter than the outer, is concave, gives attachment to a portion of the Trapezius muscle, and presents about its centre a small oval surface, for articulation with the scapular end of the clavicle. Its *apex*, formed at the point of meeting of these two borders in front, is thin, and has attached to it the coraco-acromion ligament.

Of the three borders or costae of the scapula, the *superior* is the shortest and thinnest; it is concave, terminating at its inner extremity at the superior angle, at its outer extremity at the coracoid process. At its outer part is a deep semicircular notch, formed partly by the base of the coracoid process. This notch is converted into a foramen by the transverse ligament, and serves for the passage of the supra scapular nerve. The adjacent margin of the superior border affords attachment to the Omohyoid muscle. The *external*, or *axillary border*, is the thickest of the three. It commences above at the lower margin of the glenoid cavity, and inclines obliquely downwards and backwards to the inferior angle. Immediately below the glenoid cavity, is a rough depression about an inch in length, which affords attachment to the long head of the Triceps muscle; to this succeeds a longitudinal groove which extends as far as its lower third, and affords origin to part of the Subscapularis muscle. The inferior third of this border which is thin and sharp, serves for the attachment of a few fibres of the Teres major behind, and of the Subscapularis in front. The *internal*, or *vertebral border*, also named the base, is the longest of the three, and extends from the superior to the inferior angle of the bone. It is convex, intermediate in thickness between the superior and the external, and that portion of it above the spine bent considerably outwards, so as to form an obtuse angle with the lower part. This border presents an anterior lip, a posterior lip, and an intermediate space. The *anterior lip* affords attachment to the Serratus magnus; the *posterior lip*, to the Supra-spinatus above the spine, the Infra-spinatus below; the interval between the two lips, to the Levator anguli scapulae above the triangular surface at the commencement of the spine, the Rhomboideus minor, to the edge of that surface; the Rhomboideus major being attached by means of a fibrous arch, connected above to the lower part of the triangular surface at the base of the spine, and below to the lower part of the posterior border.

Of the three angles, the *superior*, formed by the junction of the superior and internal borders, is thin, smooth, rounded, somewhat inclined outwards, and gives attachment to a few fibres of the Levator anguli scapulae muscle. The *inferior* angle thick and rough, is formed by the union of the vertebral and axillary borders, its outer surface affording attachment to the Teres major, and occasionally a few fibres of the Latissimus dorsi. The *anterior* angle is the thickest part of the bone, and forms what is called the *head* of the scapula. The head presents a shallow, pyriform, articular surface, the *glenoid cavity* (γληνη, a superficial cavity; ειδος, like); its longest diameter is from above downwards, and its direction outwards and forwards. It is broader below than above; at its apex is attached the long tendon of the Biceps muscle. It is covered with cartilage in the recent state; and its margins, slightly raised, give attachment to a fibro-cartilaginous structure, the glenoid ligament, by which its cavity is deepened. The neck of the scapula is the slightly depressed surface which surrounds the head, it is more distinct on the posterior than on the anterior surface, and below than above.

In the latter situation, it has, arising from it, a thick prominence, the coracoid process.

The *Coracoid process*, so called from its fancied resemblance to a crow's beak (κοραξ, a crow; ειδος, like), is a thick curved process of bone, which arises by a broad base from the upper part of the neck of the scapula; it ascends at first upwards and inwards, then becoming smaller, it changes its direction and passes forwards and outwards. The ascending portion, flattened from before backwards, presents in front a smooth concave surface, over which passes the sub-scapularis muscle. The horizontal portion is flattened from above downwards; its upper surface is convex and irregular; its under surface is smooth; its anterior border is rough, and gives attachment to the Pectoralis minor, its posterior also rough to the coraco-acromion ligament, while the apex is embraced by the conjoined tendon of origin of the short head of the Biceps and Coraco-brachialis muscles. At the inner side of the root of the coracoid process is a rough depression for the attachment of the conoid ligament, and running from it obliquely forwards and outwards on the upper surface of the horizontal portion, an elevated ridge for the attachment of the trapezoid ligament.

Structure. In the head, processes, and all the thickened parts of the bone, it is cellular in structure, of a dense compact tissue in the rest of its extent. The centre and upper part of the dorsum, but especially the former, is usually so thin as to be semi-transparent; occasionally the bone is found wanting in this situation, and the adjacent muscles come into contact.

76.—Plan of the Development of the Scapula. By Seven Centres.

Epiphyses (except one for the Coracoid process) appear at fifteen to seventeen years, and unite at twenty-two to twenty-five years.

Development (fig. 76). By *seven* centres; one for the body, two for the coracoid process, two for the acromion, one for the posterior border, and one for the inferior angle.

That for the body makes its first appearance at about the same period that osseous matter is deposited in the vertebrae, and forms the chief part of the bone. At birth, all the other centres are cartilaginous. About the first year after birth, osseous deposition occurs in the middle of the coracoid process; which usually becomes joined with the rest of the bone at the time when the other centres make their appearance. Between the fifteenth and seventeenth years, osseous matter is deposited in the remaining centres in quick succession, and in the following order: first, near the base of the acromion, and in the upper part of the coracoid process, the latter appearing in the form of a broad scale; secondly, in the inferior angle and contiguous part of the posterior border; thirdly, near the extremity of the acromion; fourthly, in the posterior border. The acromion process, besides being formed of two separate nuclei, has its base formed by an extension into it of the centre of ossification which belongs to the spine, the extent of which varies in different cases. The two separate nuclei unite, and then join with the extension carried in from the spine. These various epiphyses become united to the bone between the ages of twenty-two and twenty-five years.

Articulations. With the humerus and clavicle.

Attachment of Muscles. To the anterior surface, the Subscapularis; posterior surface, Supra-spinatus, Infra-spinatus; spine, Trapezius, Deltoid; superior border, Omo-hyoid; vertebral border, Serratus magnus, Levator anguli scapulae, Rhomboideus minor and Major; axillary border, Triceps, Teres minor, Teres major, glenoid cavity; long head of the Biceps, coracoid process; short head of Biceps, Coraco-brachialis, Pectoralis minor; and to the inferior angle occasionally a few fibres of the Latissimus dorsi.

The Humerus

The *Humerus* is the longest and largest bone of the upper extremity; it presents for examination a shaft and two extremities.

The *Superior Extremity* is the largest part of the bone; it presents a rounded head, a constriction around the base of the head, the neck, and two other eminences, the greater and lesser tuberosities (fig. 77).

The *head*, nearly hemispherical in form, is directed inwards, upwards, and a little backwards; its surface is smooth, coated with cartilage in the recent state, and articulates with the glenoid cavity of the scapula. The circumference of its articular surface is slightly constricted, and is termed the *anatomical neck*, in contradistinction to the constriction which exists below the tuberosities, and is called the *surgical neck*, from its being the seat of the accident called by surgeons, "fracture of the neck of the humerus."

The *neck*, which is obliquely directed, forming an obtuse angle with the shaft, is more distinctly marked in the lower half of its circumference, than in the upper half, where it presents a narrow groove, separating the head from the tuberosities. Its circumference affords attachment to the capsular ligament, and is perforated by numerous vascular foramina.

77.—Left Humerus. Anterior View.

103

The *greater tuberosity* is situated on the outer side of the head and lesser tuberosity. Its superior surface is rounded and marked by three flat facets, separated by two slight ridges, the most anterior giving attachment to the tendon of the Supra-spinatus; the middle, to the Infra-spinatus; the posterior, to the Teres minor. The external surface of the great tuberosity is convex, rough, and continuous with the outer side of the shaft.

The *lesser tuberosity* is more prominent, although smaller than the greater; it is situated in front of the head, and is directed inwards and forwards. Its summit presents a prominent facet for the insertion of the tendon of the Subscapularis muscle. These two tuberosities are separated from one another by a deep groove, the *bicipital groove*, so called from its lodging the long tendon of the Biceps muscle. It commences above between the two tuberosities, passes obliquely downwards and a little inwards, and terminates at the junction of the upper with the middle third of the bone. It is deep and narrow at its commencement, and becomes shallow and a little broader as it descends. In the recent state it is covered with a thin layer of cartilage, lined by a prolongation of the synovial membrane of the shoulder joint, and receives part of the tendon of insertion of the Latissimus dorsi about its centre.

The *Shaft* of the humerus is almost cylindrical in the upper half of its extent; prismatic and flattened below, it presents three borders and three surfaces for examination.

The *external border* runs from the back part of the greater tuberosity to the external condyle, and separates the external from the posterior surface. It is rounded and indistinctly marked in its upper half, and serves for the attachment of the external head of the triceps muscle; its centre is traversed by a broad but shallow oblique depression, the musculo-spiral groove; its lower part is marked by a prominent rough margin, a little curved from behind forwards, which presents an anterior lip for the attachment of the Supinator longus above, the Extensor carpi radialis longior below, a posterior lip for the Triceps, and an interstice for the attachment of the external intermuscular aponeurosis.

The *internal border* extends from the lesser tuberosity above to the internal condyle below. Its upper third is marked by a prominent ridge, forming the inner lip of the bicipital groove, and giving attachment from above downwards to the tendons of the Latissimus dorsi, Teres major, and part of the origin of the inner head of the Triceps. About its centre is a rough ridge for the attachment of the Coraco-brachialis, and just below this is seen the entrance of the nutritious canal directed downwards. Its inferior third is raised into a slight ridge, which becomes very prominent below; it presents an anterior lip for the attachment of the Brachialis anticus, a posterior lip for the internal head of the Triceps, and an intermediate space for the internal intermuscular aponeurosis.

The *anterior border* runs from the front of the great tuberosity above, to the coronoid depression below, separating the internal from the external surface. Its upper part is very prominent and rough, forms the outer lip of the bicipital groove, and serves for the attachment of the tendon of the Pectoralis major. About its centre is seen the rough deltoid impression; below, it is smooth and rounded, affording attachment to the Brachialis anticus.

The *external surface* is directed outwards above, where it is smooth, rounded, and covered by the Deltoid muscle; forwards below, where it is slightly concave from above downwards, and gives origin to part of the Brachialis anticus muscle. About the middle of this surface, is seen a rough triangular impression for the insertion of the Deltoid muscle, and below it the musculo-spiral groove, directed obliquely from behind, forwards and downwards; it transmits the musculo-spiral nerve and superior profunda artery.

The *internal surface*, less extensive than the external, is directed forwards above, inwards and forwards below: at its upper part it is narrow, and forms the bicipital groove. The middle part of this surface is slightly rough for the attachment of the Coraco-brachialis; its lower part is smooth, concave, and gives attachment to the Brachialis anticus muscle.

The *posterior surface* (fig. 78) appears somewhat twisted, so that its superior part is directed a little inwards, its inferior part backwards and a little outwards. Nearly the whole of this surface is covered by the external and internal heads of the triceps, the former being attached to its upper and outer part, the latter to its inner and back part, their origin being separated by the musculo-spiral groove.

The *Lower Extremity* is flattened from before backwards, and curved slightly forwards; it terminates below in a broad articular surface, which is divided into two parts by a shallow groove; on either side of the articulate surface are the external and internal condyles. The articular surface extends a little lower than the condyles, and is curved slightly forwards, so as to occupy the more anterior part of the bone; its greatest breadth is in the transverse diameter, and it is obliquely directed, so that its inner

78.—Left Humerus. Posterior surface.

extremity occupies a lower level than the outer. The outer portion of this articular surface presents a smooth rounded eminence, which has received the name of the *lesser* or *radial head* of the humerus; it articulates with the cup-shaped depression on the head of the radius, is limited to the front and lower part of the bone, and does not extend as far back as the other portion of the articular surface. On the inner side of this eminence is a shallow groove, in which is received the inner margin of the cup-like cavity of the head of the radius. The inner or trochlear portion of the articular surface presents a deep depression between two well-marked borders. This surface is curved from before backwards, concave from side to side, and occupies the anterior lower and posterior part of the bone. The external border, less prominent than the internal, corresponds to the interval between the radius and ulnar. The internal border is thicker, more prominent, and, consequently, of greater length than the external. The grooved portion of the articular surface fits accurately within the greater sigmoid cavity of the ulna; it is broader and deeper on the posterior than on the anterior aspect of the bone, and is directed obliquely from behind forwards, and from without inwards. Immediately above the back part of the trochlear surface, is a deep triangular depression, the olecranon depression, in which is received the summit of the olecranon process in extension of the fore-arm. Above the front part of the trochlear surface, is seen a smaller depression, the coronoid depression; it receives the coronoid process of the ulna during flexion of the fore-arm. These fossae are separated from one another by a thin lamina of bone, which is sometimes perforated; their margins afford attachment to the anterior and posterior ligaments of the elbow joint, and they are lined in the recent state by the synovial membrane of this articulation. Above the front part of the radial tuberosity, is seen a slight depression which receives the anterior border of the head of the radius when the fore-arm is strongly flexed. The external condyle is a small tubercular eminence, less prominent than the internal, curved a little forwards, and giving attachment to the external lateral ligament of the elbow joint, and to a tendon common to the origin of some of the extensor and supinator muscles. The internal condyle, larger and more prominent than the external, is directed a little backwards, it gives attachment to the internal lateral ligament, and to a tendon common to the origin of some of the flexor muscles of the fore-arm. These eminences are directly continuous above with the external and internal borders.

79.—Plan of the Development of the Humerus. By 7 centres.

Structure. The extremities consist of cancellous tissue, covered with a thin compact layer; the shaft is composed of a cylinder of compact tissue, thicker at the centre than at the extremities, and hollowed out by a large medullary canal.

Development. By *seven* centres (fig. 79); one for the shaft, one for the head, one for the greater tuberosity, one for the radial, and one for the trochlear portion of the articular surface, and one for each condyle. The centre for the shaft appears very early, soon after ossification has commenced in the clavicle, and soon extends towards the extremities. At birth, it is ossified nearly in its whole length, the extremities remaining cartilaginous. Between the first and second years, ossification commences in the head of the bone, and between the second and third years the centre for the tuberosities marks its appearance usually by a single ossific point, but sometimes, according to Beclard, by one for each tuberosity, that for the lesser being small, and not appearing until after the fourth year. By the fifth year, the centres for the head and tuberosities have enlarged and become joined, so as to form a single large epiphysis.

The lower end of the humerus is developed in the following manner: At the end of the second year, ossification commences in the radial portion of the articular surface, and from this point extends inwards, so as to form the chief part of the articular end of the bone, the centre for the inner part of the articular surface not appearing until about the age of twelve. Ossification commences in the internal condyle about the fifth year, and in the external one not until between the age of thirteen or fourteen. About sixteen or seventeen years, the outer condyle and both portions of the articulating surface (having already joined) unite with the shaft; at eighteen years, the inner condyle becomes joined, whilst the upper epiphysis, although the first formed, is not joined until about the twentieth year.

Articulations. With the glenoid cavity of the scapula, and with the ulna and radius.

Attachment of Muscles. To the greater tuberosity, the Supra-spinatus, Infra-spinatus, and Teres minor; to the lesser tuberosity, the Subscapularis; to the anterior bicipital ridge, the Pectoralis major; to the posterior bicipital ridge and groove, the Latissimus dorsi and Teres major; to the shaft, the Deltoid, Coraco-brachialis, Brachialis anticus, External and Internal heads of the Triceps; to the internal condyle, the Pronator radii teres, and common tendon of the Flexor carpi radialis, Palmaris longus, Flexor digitorum sublimis, and Flexor carpi ulnaris; to the external condyloid ridge, the Supinator longus, and Extensor carpi radialis longior; to the external condyle, the common tendon of the Extensor carpi radialis brevior, Extensor communis digitorum, Extensor minimi digiti, and Extensor carpi ulnaris, the Anconeus, and Supinator brevis.

The *Fore-arm* is that portion of the upper extremity, situated between the elbow and wrist. It is composed of two bones, the Ulna and Radius.

The Ulna

The *Ulna* (fig. 80, 81) is a long bone, prismatic in form, placed at the inner side of the fore-arm, parallel with the radius, being the largest and longest of the two. Its upper extremity, of great thickness and strength, forms a large part of the articulation of the elbow joint; it gradually tapers as it descends, its inferior extremity being very small, and excluded from the wrist joint by the interposition of an interarticular fibro-cartilage. It is divisible into a shaft and two extremities.

The *Upper Extremity*, the strongest part of the bone, presents for examination two large curved processes, the Olecranon process and the Coronoid process, and two concave articular cavities, the greater and lesser Sigmoid cavities.

80.—Bones of the Left Fore-Arm. Anterior Surface.

The *Olecranon Process* is a large thick curved eminence, situated at the upper and back part of the ulna. It rises somewhat higher than the coronoid, is contracted where it joins the shaft, and curved forwards at the summit so as to present a prominent tip. Its posterior surface, directed backwards, is of a triangular form, smooth, subcutaneous, and covered by a bursa. Its superior surface, directed upwards, is of a quadrilateral form, marked behind by a rough surface for the attachment of the Triceps muscle, and in front, near the margin, by a slight transverse groove for the attachment of part of the posterior ligament of the elbow joint. Its anterior surface is smooth, concave, covered with cartilage in the recent state, and forms the upper and back part of the great sigmoid cavity. The lateral borders present a continuation of the same groove that was seen on the margin of the superior surface, they serve for the attachment of ligaments, viz., the back part of the internal lateral ligament internally; the posterior ligament externally. The Olecranon process, in its structure as well as in its position and use, resembles the Patella in the lower limb, and, like it, sometimes exists as a separate piece, not united to the rest of the bone.

The *Coronoid Process* (κορωνη, a crow's beak; ειδος) is a rough triangular eminence of bone which projects horizontally forwards from the upper and front part of the ulna, forming the lower part of the great sigmoid cavity. Its base is continuous with the shaft. Its apex, pointed, slightly curved upwards, is received into the coronoid depression of the humerus in flexion of the fore-arm. Its superior surface is smooth, concave, and forms the lower part of the great sigmoid cavity. The inferior surface is concave, directed downwards and forwards and marked internally by a rough impression for the insertion of the Brachialis anticus. At the junction of this surface with the body, is a rough eminence, the tubercle of the ulna, for the attachment of the oblique ligament. Its *outer surface* presents a narrow, oblong, articular depression, the lesser sigmoid cavity. The *inner surface*, by its prominent free margin, serves for the attachment of the front part of the internal lateral ligament. At the front part of this surface is a small rounded eminence for the attachment of one head of the Flexor digitorum sublimis. Behind the eminence, a depression for part of the origin of the Flexor profundus digitorum, and descending from it a ridge, lost below on the inner border of the shaft, which gives attachment to one head of the Pronator radii teres.

The *Greater Sigmoid Cavity* (σιγμα, ειδος, form), so called from its resemblance to the Greek letter Σ, is a semi-lunar depression of large size, situated between the olecranon and coronoid processes, and serving for articulation with the trochlear surface of the humerus. About the middle of either lateral border of this cavity is a notch, which contracts it somewhat, and serves to indicate the junction of the two processes of which it is formed. The cavity is concave from above downwards, and divided into two lateral parts by a smooth elevated ridge, which runs from the summit of the olecranon to the tip of the coronoid process. Of these two portions, the internal is the largest, and slightly concave transversely; the external the smallest, being nearly plane from side to side.

The *Lesser Sigmoid Cavity* is a narrow, oblong, articular depression, placed on the outer side of the coronoid process, and serving for articulation with the head of the radius. It is concave from before backwards; and its extremities, which are prominent, serve for the attachment of the orbicular ligament.

The *Shaft* is prismatic in form at its upper part, and curved from behind forwards, and from within outwards, so as to be convex behind and externally; its central part is quite straight; its lower part rounded, smooth, and bent a little outwards; it tapers gradually from above downwards, and presents for examination three borders and three surfaces.

The *anterior border* commences above at the prominent inner angle of the coronoid process, and terminates below in front of the styloid process. It is well marked above, smooth and rounded in the middle of its extent, and affords attachment to the Flexor profundus digitorum, sharp and prominent in its lower fourth for the attachment of the Pronator quadratus. It separates the anterior from the internal surface.

The *posterior border* commences above at the apex of the triangular surface at the back part of the olecranon, and terminates below at the back part of the styloid process; it is well marked in the upper three-fourths, and gives attachment to an aponeurosis common to the Flexor carpi ulnaris, the Extensor carpi ulnaris, and the Flexor profundus digitorum muscles; its lower fourth is smooth and rounded. This border separates the internal from the posterior surface.

The *external border* commences above by two lines, which converge one from each extremity of the lesser sigmoid cavity, enclosing between them a triangular space for the attachment of part of the Supinator brevis, and terminates below at the middle of the articular surface for the radius. Its two middle-fourths are very prominent, and serve for the attachment of the interosseous membrane; its lower fourth is smooth and rounded. This border separates the anterior from the posterior surface.

The *anterior surface*, much broader above than below, is concave in the upper three-fourths of its extent, and affords attachment to the Flexor profundus digitorum. Its lower fourth, also concave, to the Pronator quadratus. The lower fourth is separated from the remaining portion of the bone by a prominent ridge, directed obliquely from above downwards and inwards; this ridge marks the extent of attachment of the Pronator above. At the junction of the upper with the middle third of the bone is the nutritious canal, directed obliquely upwards and inwards.

The *posterior surface*, directed backwards and outwards, is broad and concave above, somewhat narrower and convex in the middle of its course, narrow, smooth, and rounded below. It presents above an oblique ridge, which runs from the posterior extremity of the lesser sigmoid cavity, downwards to the posterior border, marking off a small triangular surface above it for the insertion of the Anconeus muscle, whilst the ridge itself affords attachment to the Supinator brevis. The surface of bone below this is subdivided by a longitudinal ridge into two parts, the internal part is smooth, concave, and gives origin (occasionally is merely covered by) the Extensor carpi ulnaris. The external portion, wider and rougher, gives attachment from above downwards to part of the Supinator brevis, the Extensor ossis metacarpi pollicis, Extensor secundii internodii pollicis, and Extensor indicis muscles.

The *internal surface* is broad and concave above, narrow and convex below. It gives attachment by its upper three-fourths to the Flexor profundus digitorum muscle; its lower fourth is subcutaneous.

The *Lower Extremity* of the ulna is of small size, and excluded from the articulation of the wrist joint. It presents for examination two eminences; the outer and larger is

GRAY'S ANATOMY

81.—Bones of the Left Forearm. Posterior Surface.

a rounded articular eminence, termed the head of the ulna. The inner, narrower and more projecting, is a non-articular eminence, the styloid process. The *head* presents an articular facet, part of which, of an oval form, is directed downwards, and plays on the surface of the triangular fibro-cartilage, which separates this bone from the wrist joint; the remaining portion, directed outwards, is narrow, convex, and received into the sigmoid cavity of the radius. The *styloid process* projects from the inner and back part of the bone, and descends a little lower than the head, terminating in a rounded summit, which affords attachment to the internal lateral ligament of the wrist. The head is separated from the styloid process below and in front, by a depression for the attachment of the triangular inter-articular fibro-cartilage; behind, by a shallow groove for the passage of the tendon of the Extensor carpi ulnaris.

82.—Plan of the Development of the Ulna. By Three Centres.

Structure. Similar to that of the other long bones.

Development. By *three* centres; one for the shaft, one for the inferior extremity, and one for the olecranon (fig. 82). The centre for the shaft appears a short time after the radius, and soon extends through the greater part of the bone. At birth, the ends are cartilaginous. About the fourth year, a separate osseous nucleus appears in the middle of the head, which soon extends into the styloid process. About the tenth year, ossific matter appears in the upper cartilaginous end of the bone near its extremity, the chief part of the olecranon being formed from an extension of the shaft of the bone into it. At about the sixteenth year, the upper epiphysis becomes joined, and at about the twentieth the inferior one.

Articulations. With the humerus and radius.

Attachment of Muscles. To the olecranon; the Triceps, Anconeus, and one head of the Flexor carpi ulnaris. To the coronoid process; the Brachialis anticus, Pronator radii teres, Flexor sublimis digitorum, and Flexor profundus digitorum. To the shaft; the Flexor profundus digitorum, Pronator quadratus, Flexor carpi ulnaris, Extensor carpi ulnaris, Anconeus, Supinator brevis, Extensor ossis metacarpi pollicis, Extensor secundi internodii pollicis, and Extensor indicis.

The Radius

The *Radius* is situated on the outer side of the fore-arm, lying parallel with the ulna, which exceeds it in length and size. Its upper end is small, and forms only a small part of the elbow-joint; but its lower end is large, and forms the chief part of the wrist. It is one of the long bones, having a prismatic form, slightly curved longitudinally, and presenting for examination a shaft and two extremities.

The *Upper Extremity* presents a head, neck, and tuberosity. The *head* is of a cylindrical form, depressed on its upper surface into a shallow cup, which receives the radial or lesser head of the humerus. Around the circumference of the head is a smooth articular surface, coated with cartilage in the recent state, broad internally where it articulates with the lesser sigmoid cavity of the ulna, narrow in the rest of its circumference, to play in the orbicular ligament. The head is supported on a round, smooth, and constricted portion of bone, called the *neck*, which presents, behind, a slight ridge, for the attachment of part of the Supinator brevis. Beneath the neck, at the inner and front aspect of the bone, is a rough eminence, the *tuberosity*. Its surface is divided into two parts by a vertical line—a posterior rough portion, for the insertion of the tendon of the Biceps muscle; and an anterior smooth portion, on which a bursa is interposed between the tendon and the bone.

The *Shaft* of the bone is prismoid in form, narrower above than below, and slightly curved, so as to be convex outwards. It presents three surfaces, separated by three borders.

The *anterior border* extends from the lower part of the tuberosity above, to the anterior part of the base of the styloid process below. It separates the anterior from the external surface. Its upper third is very prominent; and, from its oblique direction downwards and outwards, has received the name of the *oblique line of the radius*. It affords attachment, externally, to the Supinator brevis; internally, to the Flexor longus pollicis, and between these to the Flexor digitorum sublimis. The middle third of the anterior border is indistinct and rounded. Its lower fourth is sharp, prominent, affords attachment to the Pronator quadratus, and terminates in a small tubercle, into which is inserted the tendon of the Supinator longus.

The *posterior border* commences above, at the back part of the neck of the radius, and terminates below, at the posterior part of the base of the styloid process; it separates the posterior from the external surface. It is indistinct above and below, but well marked in the middle third of the bone.

The *internal* or *interosseous border* commences above, at the back part of the tuberosity, where it is rounded and indistinct, becomes sharp and prominent as it descends, and at its lower part bifurcates into two ridges, which descend to the anterior and posterior margins of the sigmoid cavity. This border separates the anterior from the posterior surface, and has the interosseous membrane attached to it throughout the greater part of its extent.

The *anterior surface* is narrow and concave for its upper two-thirds, and gives attachment to the Flexor longus pollicis muscle; below, it is broad and flat, its lower fourth giving attachment to the Pronator quadratus. At the junction of the upper and middle thirds of this surface is the nutritious foramen, which is directed obliquely upwards.

The *posterior surface* is rounded, convex, and smooth in the upper third of its extent, and covered by the Supinator brevis muscle. Its middle third is broad, slightly concave, and gives attachment to the Extensor ossis metacarpi pollicis above, the Extensor primi internodii pollicis below. Its lower third is broad, convex, and covered by the tendons of the muscles which subsequently run in the grooves on the lower end of the bone.

The *external surface* is rounded and convex throughout its entire extent. Its upper third gives attachment to the Supinator brevis muscle. About its centre is seen a rough ridge, for the insertion of the Pronator radii teres muscle. Its lower part is narrow, and covered by the tendons of the Extensor ossis metacarpi, and Extensor primi internodii pollicis muscles.

The *Lower Extremity* of the radius is large, of quadrilateral form, and provided with two articular surfaces, one at the extremity, and one at the inner side of the bone; it presents, also, three borders, an anterior, posterior, and external. The articular surface at the extremity of the bone is of triangular form, concave, smooth, and divided by a slight ridge into two parts. Of these, the external is large, of a triangular form, and articulates with the scaphoid bone; the inner, smaller and quadrilateral, articulates with the semi-lunar. The articular surface at the inner side of the bone is called the *sigmoid cavity* of the radius; it is narrow, concave, smooth, and articulates with the head of the ulna.

Its *anterior border*, rough and irregular, affords attachment to the anterior ligament of the wrist-joint. Its *external border* is prolonged obliquely downwards upon the margin of a strong conical projection, the styloid process, which gives attachment by its base to the tendon of the Supinator longus, and by its apex to the external lateral ligament of the wrist-joint. The outer surface of this process is marked by two grooves, which run obliquely downwards and forwards, and are separated from one another by an elevated ridge. The most anterior one gives passage to the tendon of the Extensor ossis metacarpi pollicis, the posterior one to the tendon of the Extensor primi internodii pollicis. Its *posterior border* is convex, affords attachment to the posterior ligament of the wrist, and is marked by three grooves. The most external is broad, but shallow, and subdivided into two by a slightly elevated ridge. The external groove transmits the tendon of the Extensor carpi radialis longior, the inner one the tendon of the Extensor carpi radialis brevior. Near the centre of the bone is a deep, but narrow, groove, directed obliquely from above downwards and outwards; it transmits the tendon of the Extensor secundi internodii pollicis. Internally is a broad groove, for the passage of the tendons of the Extensor communis digitorum, and that of the Extensor indicis; the tendon of the Extensor minimi digiti passing through the groove at its point of articulation with the ulna.

Development (fig. 83). By *three* centres: one for the shaft, and one for each extremity. That for the shaft, makes its appearance near the centre of the

83.—Plan of the Development of the Radius. By Three Centres.

bone, soon after the development of the humerus commences. At birth, the shaft is ossified; but the ends of the bone are cartilaginous. About the end of the second year, ossification commences in the lower epiphysis; and about the fifth year, in the upper one. At the age of puberty, the upper epiphysis becomes joined to the shaft; the lower epiphysis becoming united about the twentieth year.

Articulations. With four bones; the humerus, ulna, scaphoid, and semi-lunar.

Attachment of Muscles. To the tuberosity, the Biceps; to the oblique ridge, the Supinator brevis, Flexor digitorum sublimis, and Flexor longus pollicis; to the shaft (its anterior surface), the Flexor longus pollicis and Pronator quadratus, (its posterior surface) the Extensor ossis metacarpi pollicis, and Extensor primi internodii pollicis; to the outer surface, the Pronator radii teres; and to the styloid process, the Supinator longus.

THE HAND

The Hand is subdivided into three segments, the Carpus or wrist, the Metacarpus or palm, and the Phalanges or fingers.

Carpus

The bones of the Carpus, eight in number, are arranged in two horizontal rows. Those of the upper row, enumerated from the radial to the ulnar side, are the scaphoid, semi-lunar, cuneiform, and pisiform; those of the lower row, enumerated in the same order, are the trapezium, trapezoid, magnum, and unciform.

Common Characters of the Carpal Bones

Each bone (excepting the pisiform) presents six surfaces. Of these, the *anterior* or *palmar*, and the *posterior* or *dorsal*, are rough, for ligamentous attachment, the dorsal surface being generally the broadest of the two. The *superior* and *inferior* are articular, the superior generally convex, the inferior concave; and the *internal* and *external* are also articular when in contact with contiguous bones, otherwise rough and tubercular. Their structure in all is similar, consisting within of a loose cancellous tissue enclosed in a thin layer of compact tissue. Each bone is also developed from a single centre of ossification.

Bones of the Upper Row (Figs. 84, 85)

The *Scaphoid* is the largest bone of the first row. It has received its name from its fancied resemblance to a boat, being broad at one end, and narrowed like a prow at the opposite. It is situated at the superior and external part of the carpus, its direction being from above downwards, outwards, and forwards. Its *superior surface* is convex, smooth, of triangular shape, and articulates with the lower end of the radius. Its *inferior surface*, directed downwards, outwards, and backwards, is smooth, convex, also triangular, and divided by a slight ridge into two parts, the external of which articulates with the trapezium, the inner with the trapezoid. Its *posterior* or *dorsal surface* presents a narrow, rough groove, which runs the entire breadth of the bone, and serves for the attachment of ligaments. The *anterior* or *palmar surface* is concave above, and elevated at its lower and outer part into a prominent rounded tubercle, which projects forwards from the front of the carpus, and gives attachment to the anterior annular ligament of the wrist. The *external surface* is rough and narrow, and gives attachment to the external lateral ligament of the wrist. The *internal surface* presents two articular facets: of these, the superior or smaller one is flattened, of semi-lunar form, and articulates with the semi-lunar; the inferior or larger is concave, forming, with the semi-lunar bone, a concavity for the head of the os magnum.

To ascertain to which hand this bone belongs, hold the convex radial articular surface upwards, and the dorsal surface backwards; the prominent tubercle will be directed to the side to which the bone belongs.

Articulations. With five bones; the radius above, trapezium and trapezoid below, os magnum and semi-lunar internally.

The *Semi-lunar* bone may be distinguished by its deep concavity and crescentic outline. It is situated in the centre of the upper range of the carpus, between the scaphoid and cuneiform. Its *superior surface*, convex, smooth, and quadrilateral in form, articulates with the radius. Its *inferior surface* is deeply concave, and of greater extent from before backwards, than transversely; it articulates with the head of the os magnum, and by a long narrow facet (separated by a ridge from the general surface) with the unciform bone. Its *anterior* or *palmar* and *posterior* or *dorsal surfaces* are rough, for the attachment of ligaments, the former being the broader, and of somewhat rounded form. The *external surface* presents a narrow, flattened, semi-lunar facet, for articulation with the scaphoid. The *internal surface* is marked by a smooth, quadrilateral facet, for articulation with the cuneiform.

To ascertain to which hand this bone belongs, hold it with the dorsal surface upwards, and the convex articular surface backwards; the quadrilateral articular facet will then point to the side to which the bone belongs.

Articulations. With five bones: the radius above, os magnum and unciform below, scaphoid and cuneiform on either side.

The *Cuneiform* (*l' Os Pyramidal*), may be distinguished by its pyramidal shape, and from having an oval-shaped, isolated facet, for articulation with the pisiform bone. It is situated at the upper and inner side of the carpus. The *superior surface* presents an internal, rough, non-articular portion; and an external or articular portion, which is convex, smooth, and separated from the lower end of the ulna by the inter-articular fibro-cartilage of the wrist. The *inferior surface*, directed outwards, is concave, sinuously curved, and smooth, for articulation with the unciform. Its *posterior* or *dorsal surface* is rough, for the attachment of ligaments. Its *anterior* or *palmar surface* presents, at its inner side, an oval-shaped facet, for articulation with the pisiform; and is rough externally, for ligamentous attachment. Its *external surface*, the base of the pyramid, is marked by a flat, quadrilateral, smooth facet, for articulation with the semi-lunar. The *internal surface*, the summit of the pyramid, is pointed and roughened, for the attachment of the internal lateral ligament of the wrist.

To ascertain to which hand this bone belongs, hold it so that the base is directed backwards, and the articular facet for the pisiform bone upwards; the concave articular facet will point to the side to which the bone belongs.

Articulations. With three bones: the semi-lunar externally, the pisiform in front, the unciform below, and with the triangular inter-articular fibro-cartilage which separates it from the lower end of the ulna.

The *Pisiform* bone may be known by its small size, and from its presenting a single articular facet. It is situated at the anterior and inner side of the carpus, is nearly circular in form, and presents on its *posterior surface* a smooth, oval facet, for articulation with the cuneiform bone. This facet approaches the superior, but not the inferior, border of the bone. Its *anterior* or *palmar surface* is rounded and rough, and gives attachment to the anterior annular ligament. The *outer* and *inner surfaces* are also rough, the former being convex, the other usually concave.

To ascertain to which hand it belongs, hold the bone with its posterior or articular facet downwards, and the non-articular portion of the same surface backwards; the inner concave surface will then point to the side to which the bone belongs.

Articulations. With one bone, the cuneiform.

Attachment of Muscles. To two: the Flexor carpi ulnaris, and Abductor minimi digiti; and to the anterior annular ligament.

Bones of the Lower Row (Figs. 84, 85)

The *Trapezium* is of very irregular form. It may be distinguished by a deep groove, for the tendon of the Flexor carpi radialis muscle. It is situated at the external and inferior part of the carpus, between the scaphoid and first metacarpal bone. The *superior surface*, concave and smooth, is directed upwards and inwards, and articulates with the scaphoid. Its *inferior surface*, directed downwards and outwards, is oval, concave from side to side, convex from before backwards, so as to form a saddle-shaped surface, for articulation with the base of the first metacarpal bone. The *anterior* or *palmar surface* is narrow and rough. At its upper part is a deep groove, running from above obliquely downwards and inwards; it transmits the tendon of the Flexor carpi radialis, and is bounded externally by a prominent ridge, the oblique ridge of the trapezium. This surface gives attachment to the Abductor pollicis, Flexor ossis metacarpi, and Flexor brevis pollicis muscles; and the anterior annular ligament. The *posterior* or *dorsal surface* is rough, and the *external surface* also broad and rough, for the attachment of ligaments. The *internal surface* presents two articular facets; the upper one, large and concave, articulates with the trapezoid; the lower one, narrow and flattened, with the base of the second metacarpal bone.

To ascertain to which hand it belongs, hold the bone with the grooved palmar surface upwards, and the external, broad, non-articular surface backwards; the saddle-shaped surface will then be directed to the side to which the bone belongs.

Articulations. With four bones: the scaphoid above, the trapezoid and second metacarpal bones internally, the first metacarpal below.

Attachment of Muscles. Abductor pollicis, Flexor ossis metacarpi, part of the Flexor brevis pollicis, and the anterior annular ligament.

The *Trapezoid* is the smallest bone in the second row. It may be known by its wedge-shaped form; its broad end occupying the dorsal, its narrow end the palmar surface of the hand. Its *superior surface*, quadrilateral in form, smooth and slightly concave, articulates with the scaphoid. The *inferior surface* articulates with the upper end of the second metacarpal bone; it is convex from side to side, concave from before backwards, and subdivided, by an elevated ridge, into two unequal lateral facets. The *posterior* or *dorsal*, and *anterior* or *palmar surfaces* are rough, for the attachment of ligaments; the former being the larger of the two. The *external surface*, convex and smooth, articulates with the trapezium. The *internal surface* is concave and smooth below, for articulation with the os magnum, rough above, for the attachment of an interosseous ligament.

To ascertain to which side this bone belongs, let the broad dorsal surface be held upwards, and its inferior concavo-convex surface forwards; the internal concave surface will then point to the side to which the bone belongs.

84.—Bones of the Left Hand. Dorsal Surface.

Articulations. With four bones; the scaphoid above, second metacarpal bone below, trapezium externally, os magnum internally.

Attachment of Muscles. Part of the Flexor brevis pollicis.

The *Os Magnum* is the largest bone of the carpus, and occupies the centre of the wrist. It presents, above, a rounded portion or head, which is received into the concavity formed by the scaphoid and semi-lunar bones; a constricted portion or neck; and, below, the body. Its *superior surface* is rounded, smooth, and articulates with the semi-lunar. Its *inferior surface* is divided, by two ridges, into three facets, for articulation with the second, third, and fourth metacarpal bones; that for the third (the middle facet) being the largest of the three. The *posterior* or *dorsal surface* is broad and rough, and the *anterior* or *palmar* narrow, rounded, but also rough, for the attachment of ligaments. The *external surface* articulates with the trapezoid by a small facet at its anterior inferior angle, behind which is a rough depression, for the attachment of an interosseous ligament. Above this is a deep and rough groove, which forms part of the neck, and serves for the attachment of ligaments, bounded superiorly by a smooth, convex surface, for articulation with the scaphoid. The *internal surface* articulates with the unciform by a smooth, concave, oblong facet, which occupies its posterior and superior parts; rough in front, for the attachment of an interosseous ligament.

To ascertain to which hand this bone belongs, the rounded head should be held upwards, and the broad dorsal surface forwards; the internal concave articular surface will point to its appropriate side.

Articulations. With seven bones: the scaphoid and semi-lunar above; the second, third, and fourth metacarpal below; the trapezoid on the radial side; and the unciform on the ulnar side.

Attachment of Muscles. Part of the Flexor brevis pollicis.

The *Unciform* bone may be readily distinguished by its wedge-shaped form, and the hook-like process that projects from its palmar surface. It is situated at the inner and lower angle of the carpus, with its base downwards, resting on the two inner metacarpal bones, and its apex directed upwards and outwards. Its *superior surface*, the apex of the wedge, is narrow, convex, smooth, and articulates with the semi-lunar. Its *inferior surface* articulates with the fourth and fifth metacarpal bones, the concave surface for each being separated by a ridge, which runs from before backwards. The *posterior* or *dorsal surface* is triangular and rough, for ligamentous attachment. The *anterior* or *palmar surface* presents at its lower and inner side a curved, hook-like process of bone, the unciform process, directed from the palmar surface forwards and outwards. It gives attachment, by its apex, to the annular ligament; by its inner surface, to the Flexor brevis minimi digiti, and the Flexor ossis metacarpi minimi digiti; and is grooved on its outer side, for the passage of the Flexor tendons into the palm of the hand. This is one of the four eminences on the front of the carpus, to which the anterior annular ligament is attached; the others being the pisiform internally, the oblique ridge of the trapezium and the tuberosity of the scaphoid externally. The *internal surface* articulates with the cuneiform by an oblong surface, cut obliquely from above downwards and inwards. Its *external surface* articulates with the os magnum by its upper and posterior part, the remaining portion being rough, for the attachment of ligaments.

To ascertain to which hand it belongs, hold the apex of the bone upwards, and the broad dorsal surface backwards; the concavity of the unciform process will be directed to the side to which the bone belongs.

Articulations. With five bones: the semi-lunar above, the fourth and fifth metacarpal below, the cuneiform internally, the os magnum externally.

85.—Bones of the Left Hand. Palmar Surface.

Attachment of Muscles. To two: the Flexor brevis minimi digiti, the Flexor ossis metacarpi minimi digiti; and to the anterior annular ligament.

The Metacarpus

The Metacarpal bones are five in number; they are long cylindrical bones, presenting for examination a shaft and two extremities.

Common Characters of the Metacarpal Bones

The *shaft* is prismoid in form, and curved longitudinally, so as to be convex in the longitudinal direction behind, concave in front. It presents three surfaces; two lateral, and one posterior. The *lateral surfaces* are concave, for the attachment of the Interossei muscles, and separated from one another by a prominent line. The *posterior* or *dorsal surface* is triangular, smooth, and flattened below, and covered, in the recent state, by the tendons of the Extensor muscles. In its upper half, it is divided by a ridge into two narrow lateral depressions, for the attachment of the Dorsal interossei muscles. This ridge bifurcates a little above the centre of the bone, and its branches run to the small tubercles on each side of the digital extremity.

The *carpal extremity*, or *base*, is of a cuboidal form, and broader behind than in front: it articulates, above, with the carpus; and on each side with the adjoining metacarpal bones; its *dorsal* and *palmar surfaces* being rough, for the attachment of tendons and ligaments.

The *digital extremity*, or *head*, presents an oblong surface, flattened at each side, for articulation with the first phalanx; it is broader and extends farther forwards in front than behind; and longer in the antero-posterior, than in the transverse diameter. On either side of the head is a deep depression, surmounted by a tubercle, for the attachment of the lateral ligament of the metacarpo-phalangeal joint. The *posterior surface*, broad and flat, supports the Extensor tendons; and the *anterior surface* presents a median groove, bounded on each side by a tubercle, for the passage of the Flexor tendons.

Peculiar Metacarpal Bones

The *metacarpal bone of the thumb* is shorter and wider than the rest, diverges to a greater degree from the carpus, and its *palmar surface* is directed inwards towards the palm. The *shaft* is flattened and broad on its dorsal aspect, and does not present the bifurcated ridge peculiar to the other metacarpal bones; concave from before backwards on its palmar surface. The *carpal extremity*, or *base*, presents a concavo-convex surface, for articulation with the trapezium, and has no lateral facets. The *digital extremity* is less convex than that of the other metacarpal bones, broader from side to side than from before backwards, and terminates anteriorly in a small articular eminence on each side, over which play two sesamoid bones.

The *metacarpal bone of the index finger* is the longest, and its base the largest of the other four. Its *carpal extremity* is prolonged upwards and inwards; and its *dorsal* and *palmar surfaces* are rough, for the attachment of tendons and ligaments. It presents four articular facets: one at the end of the bone, which has an angular depression for articulation with the trapezoid; on the radial side, a flat quadrilateral facet, for articulation with the trapezium; its ulnar side being prolonged upwards and inwards, to articulate above with the os magnum, internally with the third metacarpal bone.

The *metacarpal bone of the middle finger* is a little less in size than the preceding; it presents a pyramidal eminence on the radial side of its base (dorsal aspect), which extends upwards behind the os magnum. The carpal-articular facet is concave behind, flat and horizontal in front, and corresponds to the os magnum. On the radial side is a smooth, concave facet, for articulation with the second metacarpal bone; and on the ulnar side two small oval facets, for articulation with the third metacarpal.

The *metacarpal bone of the ring-finger* is shorter and smaller than the preceding, and its base small and quadrilateral; its carpal surface presenting two facets, for articulation with the unciform and os magnum. On the radial side are two oval facets, for articulation with the third metacarpal bone; and on the ulnar side a single concave facet, for the fifth metacarpal.

The *metacarpal bone of the little finger* may be distinguished by the concavo-convex form of its carpal surface, for articulation with the unciform, and from having only one lateral articular facet, which corresponds with the fourth metacarpal bone. On its ulnar side, is a prominent tubercle for the insertion of the tendon of the Extensor carpi ulnaris. The dorsal surface of the shaft is marked by an oblique ridge, which extends from near the inner side of the upper extremity, to the outer side of the lower. The outer division of this surface serves for the attachment of the fourth Dorsal interosseous muscle; the inner division is smooth, and covered by the Extensor tendons of the little finger.

Articulations. The first, with the trapezium; the second, with the trapezium, trapezoides, os magnum, and third metacarpal bones; the third, with the os magnum, and second and fourth metacarpal bones; the fourth, with the os magnum, unciform, and third and fifth metacarpal bones; and the fifth, with the unciform and fourth metacarpal.

Attachment of Muscles. To the metacarpal bone of the thumb, three: the Flexor ossis metacarpi pollicis, Extensor ossis metacarpi pollicis, and first Dorsal interosseous. To the second metacarpal bone, five: the Flexor carpi radialis, Extensor carpi radialis longior, first and second Dorsal interosseous, and first Palmar interosseous. To the third, five: the Extensor carpi radialis brevior, Flexor brevis pollicis, Adductor pollicis, and second and third Dorsal interosseous. To the fourth, three: the third and fourth Dorsal interosseous and second Palmar. To the fifth, four: the Extensor carpi ulnaris, Flexor carpi ulnaris, Flexor ossis metacarpi minimi digiti, and third Dorsal interosseous.

Phalanges

The Phalanges are the bones of the fingers; they are fourteen in number, three for each finger and two for the thumb. They are long bones, and present for examination a shaft, and two extremities. The *shaft* tapers from above downwards, is convex posteriorly, concave in front from above downwards, flat from side to side, and marked laterally by rough ridges, which give attachment to the fibrous sheaths of the Flexor tendons. The *metacarpal extremity* or *base*, in the first row, presents an oval concave articular surface, broader from side to side, than from before backwards;

and the same extremity in the other two rows, a double concavity separated by a longitudinal median ridge, extending from before backwards. The *digital extremities* are smaller than the others, and terminate, in the first and second row, in two small lateral condyles, separated by a slight groove, the articular surface being prolonged farther forwards on the palmar, than on the dorsal surface, especially in the first row.

The *Ungual phalanges* are convex on their dorsal, flat on their palmar surfaces, they are recognised by their small size, and from their ungual extremity presenting, on its palmar aspect, a roughened elevated surface of a horse-shoe form, which serves to support the sensitive pulp of the finger.

Articulations. The first row with the metacarpal bones, and the second row of phalanges; the second row, with the first and third; the third, with the second row.

Attachment of Muscles. To the base of the first phalanx of the thumb, four muscles: the Extensor primi internodii pollicis, Flexor brevis pollicis, Abductor pollicis, Adductor pollicis. To the second phalanx, two: the Flexor longus pollicis, and the Extensor secundi internodii. To the base of the first phalanx of the index finger, the first Dorsal and the first Palmar interosseous; to that of the middle finger, the second and third Dorsal interosseous; to the ring finger, the fourth Dorsal and the second Palmar interosseous; and to that of the little finger, the third Palmar interosseous, the Flexor brevis minimi digiti, and Abductor minimi digiti. To the second phalanges, the Flexor sublimis digitorum, Extensor communis digitorum; and, in addition, the Extensor indicis, to the index finger; the Extensor minimi digiti, to the little finger. To the third phalanges, the Flexor profundus digitorum and Extensor communis digitorum.

Development of the Hand

The *Carpal bones* are each developed by a *single* centre; at birth they are all cartilaginous. Ossification proceeds in the following order (fig. 86); in the os magnum and unciform an ossific point appears during the first year, the former preceding the latter; in the cuneiform, at the third year; in the trapezium and semi-lunar, at the fifth year, the former preceding the latter; in the scaphoid, at the sixth year; in the trapezoid, during the eighth year; and in the pisiform, about the twelfth year.

The *Metacarpal bones* are developed each by *two* centres: one for the shaft, and one for the digital extremity, for the four inner metacarpal bones; one for the shaft and one for the base, for the metacarpal bone of the thumb, which, in this respect, resembles the phalanges. Ossification commences in the centre of the shaft about the sixth week, and gradually proceeds to either end of the bone; about the third year the digital extremity of the four inner metacarpal bones and the base of the first metacarpal, commence to ossify, and they unite about the twentieth year.

The *Phalanges* are each developed by *two* centres: one for the shaft and one for the base. Ossification commences in the shaft, in all three rows, at about the sixth week, and gradually involves the whole of the bone excepting the upper extremity. Ossification of the base commences in the first row between the third and fourth years, and a year later in those of the second and third row. The two centres become united between the eighteenth and twentieth years.

86.—Plan of the Development of the Hand.

THE LOWER EXTREMITY

The Lower Extremities, two in number, are connected with the inferior part of the trunk. They are divided into three parts, the thigh, the leg, and the foot, which correspond to the arm, the forearm, and hand in the upper extremity.

The thigh is formed of a single bone, the femur.

THE FEMUR

The Femur is the longest, largest, and heaviest bone in the skeleton, and almost perfectly cylindrical in the greater part of its extent. In the erect position of the body it is not vertical, but presents a general curvature in the longitudinal direction, which renders the bone convex in front and slightly concave behind; it also gradually inclines from above downwards and inwards, approaching its fellow towards its lower part, but separated from it above by a very considerable interval which corresponds to the entire breadth of the pelvis. The degree of this inclination varies in different persons, and is greater in the female than in the male. The femur, like other long bones, is divisible into a shaft and two extremities.

The Upper Extremity presents for examination a head, neck, and the greater and lesser trochanters.

The *head*, which is globular, and forms rather more than a hemisphere, is directed upwards, inwards, and a little forwards, the greater part of its convexity being above and in front. Its surface is smooth, coated with cartilage in the recent state, and presents a little behind and below its centre a rough depression, for the attachment of the ligamentum teres. The *neck* is a flattened pyramidal process of bone, which connects the head with the shaft. It varies in length and obliquity at various periods of life, and under different circumstances. In the adult male, it forms an obtuse angle with the shaft, being directed upwards, inwards, and a little forwards. In the female, it approaches more nearly a right angle. Occasionally, in very old subjects, and more especially in those greatly debilitated, its direction becomes horizontal, so that the head sinks below the level of the trochanter, and its length diminishes to such a degree, that the head becomes almost contiguous with the shaft. The neck, is flattened from before backwards, contracted in the middle, and broader at its outer extremity, where it is connected with the shaft, than at its

87.—Right Femur. Anterior Surface.

summit, where it is attached to the head. It is much broader in the vertical than in the anterior posterior diameter, on account of the greater amount of resistance required in sustaining the weight of the trunk. Its *anterior surface*, narrower than the posterior, is perforated by numerous vascular foramina. Its *posterior surface* is smooth, broader, and more concave than the anterior, and receives towards its outer side the attachment of the capsular ligament of the hip. Its *superior border* is short and thick, bounded externally by the great trochanter, and its surface perforated by large foramina. Its *inferior border*, long and narrow, curves a little backwards, to terminate at the lesser trochanter.

The *Greater Trochanter* is a large irregular quadrilateral eminence, situated at the outer side of the neck, at its junction with the upper part of the shaft. It is directed a little outwards and backwards, and rises less high than the head. It presents for examination two surfaces and four borders.

Its *external surface*, quadrilateral in form, is broad, rough, convex, and marked by a prominent diagonal line, which extends from the posterior superior to the anterior inferior angle: this line serves for the attachment of the tendon of the Gluteus medius. Above the line is a triangular surface, sometimes rough for part of the tendon of the same muscle, sometimes smooth for the interposition of a bursa between that tendon and the bone. Below and behind the diagonal line is a smooth triangular surface, over which the tendon of the Gluteus maximus muscle plays, a bursa being interposed. The *internal surface* is of much less extent than the external, and presents at its base a deep depression, the digital or trochanteric fossa, for the attachment of the tendon of the Obturator externus muscle.

The *superior border* is free; it is thick and irregular, and marked by impressions for the attachment of the Pyriformis behind, the Obturator internus and Gemelli in front. The *inferior border* is placed at the point of junction of the trochanter with the outer surface of the shaft; it is rough, prominent, slightly curved, and gives attachment to the upper part of the Vastus externus muscle. The *anterior border* is prominent, somewhat irregular, as well as the surface of bone immediately below it; it affords attachment by its outer part to the Gluteus minimus. The *posterior border* is very prominent, and appears as a free rounded edge, which forms the back part of the digital fossa.

The *Lesser Trochanter* is a conical eminence, which varies in size in different subjects; it is situated at the lower and back part of the base of the neck. Its base is triangular, and connected with the adjacent parts of the bone by three well-marked borders: of these the *superior* is continuous with the lower border of the neck; the *posterior*, with the posterior intertrochanteric line; and the *inferior* with the middle bifurcation of the linea aspera. Its summit, which is directed inwards and backwards, is rough, and gives insertion to the tendon of the Psoas magnus. The Iliacus is inserted into the shaft below the lesser trochanter, between the Vastus internus in front, and the Pectineus behind. A well marked prominence, but of variable size, situated at the upper and front part of the neck, at its junction with the great trochanter, is called the *tubercle of the femur*; it is the point of meeting of three muscles, the Gluteus minimus externally, the Vastus externus below, and the tendon of the Obturator internus and Gemelli above. Running obliquely downwards and inwards from the tubercle is the spiral line of the femur, or anterior intertrochanteric line; it winds around the inner side of the shaft, below the lesser trochanter, and terminates in the linea aspera, about two inches below this eminence. Its upper half is rough, and affords attachment to the capsular ligament of the hip joint; its lower half is less prominent and gives attachment to the upper part of the Vastus internus. The posterior intertrochanteric line is very prominent, and runs from the summit of the great trochanter downwards and inwards to the upper and back part of the lesser trochanter. Its upper half forms the posterior border of the great trochanter. A well-marked eminence commences about the centre of the posterior intertrochanteric line, and passes vertically downwards for about two inches along the back part of the shaft: it is called the *linea quadrati*, and gives attachment to the Quadratus femoris, and a few fibres of the Adductor magnus muscles.

The *Shaft*, almost perfectly cylindrical in form, is a little broader above than in the centre, and somewhat flattened from before backwards below. It is curved from before backwards, smooth and convex in front, and strengthened behind by a prominent longitudinal ridge, the linea aspera. It presents for examination three borders separating three surfaces. Of the three borders, one, the linea aspera, is posterior, the other two are placed laterally.

The *linea aspera* (fig. 88) is a prominent longitudinal ridge or crest, presenting on the middle third of the bone an external lip, an internal lip, and a rough intermediate space. A little above the centre of the shaft, this crest divides into three lines; the most external one becomes very rough,

88.—Right Femur. Posterior Surface.

and is continued almost vertically upwards to the base of the great trochanter; the middle one, the least distinct, is continued to the base of the trochanter minor; and the internal one is lost above in the spiral line of the femur. Below, the linea aspera divides into two bifurcations, which enclose between them a triangular space (the popliteal space), upon which rests the popliteal artery. Of these two bifurcations, the outer branch is the most prominent, and descends to the summit of the outer condyle. The inner branch is less marked, presents a broad and shallow groove for the passage of the femoral artery, and terminates at a small tubercle at the summit of the internal condyle.

To the inner lip of the linea aspera, its whole length, is attached the Vastus internus; and to the whole length of the outer lip the Vastus externus. The Adductor magnus is also attached to the whole length of the linea aspera, being connected with the outer lip above, and the inner lip below. Between the Vastus externus and the Adductor magnus are attached two muscles, viz., the Gluteus maximus above, and the short head of the Biceps below. Between the Adductor magnus and the Vastus internus four muscles are attached: the Iliacus and Pectineus above (the latter to the middle division of the upper bifurcation); below these, the Adductor brevis and Adductor longus. The linea aspera is perforated a little below its centre by the nutritious canal, which is directed obliquely from below upwards.

The two lateral borders of the femur are only very slightly marked, the *external* extending from the anterior inferior angle of the great trochanter to the anterior extremity of the external condyle; the *internal* passes from the spiral line, at a point opposite the trochanter minor, to the anterior extremity of the internal condyle. The internal border marks the limit of attachment of the Cruraeus muscle internally.

The *anterior surface* includes that portion of the shaft which is situated between the two lateral borders. It is smooth, convex, broader above and below than in the centre, slightly twisted, so that its upper part is directed forwards and a little outwards, its lower part forwards and a little inwards. The upper three-fourths of this surface serve for the attachment of the Cruraeus; the lower fourth is separated from this muscle by the intervention of the synovial membrane of the knee-joint, and affords attachment to the Sub-cruraeus to a small extent. The *external surface* includes the portion of bone between the external border and the outer lip of the linea aspera; it is continuous above with the outer surface of the great trochanter, below with the outer surface of the external condyle: to its upper three-fourths is attached the outer portion of the Cruraeus muscle. The *internal surface* includes the portion of bone between the internal border and the inner lip of the linea aspera; it is continuous above with the lower border of the neck, below with the inner side of the internal condyle: it is covered by the Vastus internus muscle.

The *Lower Extremity*, larger than the upper, is of a cuboid form, flattened from before backwards, and divided by an interval presenting a smooth depression in front, and a notch of considerable size behind, into two large eminences, the condyles. The interval is called the *inter-condyloid notch*. The *external condyle* is the most prominent anteriorly, and is the broadest both in the antero-posterior and transverse diameters. The *internal condyle* is the narrowest, longest, and most prominent internally. This difference in the length of the two condyles depends upon the obliquity of the thigh-bones, in consequence of their separation above at the articulation with the pelvis. If the femur is held in this oblique position, the surfaces of the two condyles will be seen to be nearly horizontal. The two condyles are joined together anteriorly, and form a smooth trochlear surface, the external border of which is more prominent, and ascends higher than the internal one. This surface articulates with the patella. It presents a median groove, which extends downwards and backwards to the inter-condyloid notch; and two lateral convexities, of which the external is the broader, more prominent, and prolonged farther upwards upon the front of the outer condyle. The inter-condyloid notch lodges the crucial ligaments; it is bounded laterally by the opposed surfaces of the two condyles, and in front by the lower end of the shaft.

Outer Condyle. The *outer surface* of the external condyle presents, a little behind its centre, an eminence, the *outer tuberosity*; it is less convex and prominent than the inner tuberosity, and gives attachment to the external lateral ligament of the knee. Immediately beneath it is a groove, which commences at a depression a little behind the centre of the lower border of this surface: the depression is for the tendon of origin of the Popliteus muscle; the groove in which this tendon is contained is smooth, covered with cartilage in the recent state, and runs upwards and backwards to the posterior extremity of the condyle. The *inner surface* of the outer condyle forms one of the lateral boundaries of the inter-condyloid notch, and gives attachment, by its posterior part, to the anterior crucial ligament. The *inferior surface* is convex, smooth, and broader than that of the internal condyle. The posterior extremity is convex and smooth: just above the articular surface is a depression, for the tendon of the outer head of the Gastrocnemius.

Inner Condyle. The *inner surface* of the inner condyle presents a convex eminence, the *inner tuberosity*, rough, for the attachment of the internal lateral ligament. Above this tuberosity, at the termination of the inner bifurcation of the linea aspera, is a tubercle, for the insertion of the tendon of the Adductor magnus; and behind and beneath the tubercle a depression, for the tendon of the inner head of the Gastrocnemius. The *outer side* of the inner condyle forms one of the lateral boundaries of the inter-condyloid notch, and gives attachment, by its anterior part, to the posterior crucial ligament. Its *inferior* or *articular surface* is convex, and presents a less extensive surface than the external condyle.

Structure. Like that of the other cylindrical bones, the linea aspera is composed of a very dense, ivory-like, compact tissue.

Articulations. With three bones; the os innominatum, tibia, and patella.

Development (fig. 89). The femur is developed by *five* centres; one for the shaft, one for each extremity, and one for each trochanter. Of all the long bones, it is the first to show traces of ossification: this first commences in the shaft, at about the fifth week of foetal life, the centres of ossification in the epiphyses appearing in the following order. First, in the lower end of the bone, at the ninth month of foetal life; from this the condyles and tuberosities are formed; in the head, at the end of the first year after birth; in the great trochanter, during the fourth year; and in the lesser trochanter, between the thirteenth and fourteenth. The order in which the epiphyses are joined to the shaft, is the direct reverse of their appearance; their junction does not commence until after puberty, the lesser trochanter being first joined, then the greater, then the

89.—Plan of the Development of the Femur, by Five Centres.

head, and, lastly, the inferior extremity (the first in which ossification commenced), which is not united until the twentieth year.

Attachment of Muscles. To the great trochanter, the Gluteus medius, Gluteus minimus, Pyriformis, Obturator internus, Obturator externus, Gemellus superior, Gemellus inferior, and Quadratus femoris. To the lesser trochanter, the Psoas magnus, and the Iliacus below it. To the shaft, its posterior surface, the Vastus externus, Gluteus maximus, short head of the Biceps, Vastus internus, Adductor magnus, Pectineus, Adductor brevis, and Adductor longus; to its anterior surface, the Cruraeus and Sub-cruraeus. To the condyles, the Gastrocnemius, Plantaris, and Popliteus.

THE LEG

The Leg consists of three bones: the Patella, a large sesamoid bone, placed in front of the knee, analogous to the olecranon process of the ulna; and the Tibia and Fibula.

THE PATELLA (Figs. 90, 91)

The *Patella* is a small, flat, triangular bone, situated at the anterior part of the knee-joint. It resembles the sesamoid bones, from being developed in the tendon of the Quadriceps extensor; but, in relation with the tibia, it may be regarded as analogous to the olecranon process of the ulna, which occasionally exists as a separate piece, connected to the shaft of that bone by a continuation of the tendon of the Triceps muscle. It presents an anterior and posterior surface, three borders, a base, and an apex.

The *anterior surface* is convex, perforated by small apertures, for the passage of nutrient vessels, and marked by numerous rough, longitudinal striae. This surface is covered, in the recent state, by an expansion from the tendon of the Quadriceps extensor, separated from the integument by a synovial bursa, and gives attachment below to the ligamentum patellae. The *posterior surface* presents a smooth, oval-shaped, articular surface, covered with cartilage in the recent state, and divided into two facets by a vertical ridge, which descends from the superior towards the inferior angle of the bone. The ridge corresponds to the groove on the trochlear surface of the femur, and the two facets to the articular surfaces of the two condyles;

the outer facet, for articulation with the outer condyle, being the broader and deeper, serves to indicate the leg to which the bone belongs. This surface presents, inferiorly, a rough, convex, non-articular depression, the lower half of which gives attachment to the ligamentum patellae; the upper half being separated from the head of the tibia by adipose tissue.

90.—Right Patella, Anterior Surface.

Its *superior* and *lateral borders* give attachment to the tendon of the Quadriceps extensor; to the *superior border*, that portion of the tendon which is derived from the Rectus and Cruraeus muscles; and to the *lateral borders*, the portion derived from the external and internal Vasti muscles.

91.—Posterior Surface.

The *base*, or *superior border*, is thick, directed upwards, and cut obliquely at the expense of its outer surface; it receives the attachment, as already mentioned, of part of the Quadriceps extensor tendon.

The *apex* is pointed, and gives attachment to the ligamentum patellae.

Structure. It consists of loose cancellous tissue, covered by a thin compact lamina.

Development. By a single centre, which makes its appearance, according to Beclard, about the third year. In two instances, I have seen this bone cartilaginous throughout, at a much later period (six years). More rarely, the bone is developed by two centres, placed side by side.

Articulations. With the two condyles of the femur.

Attachment of Muscles. Four muscles are attached to the patella, viz., the Rectus, Cruraeus, Vastus internus, and Vastus externus. The tendons of these muscles joined at their insertion, constitute the Quadriceps extensor cruris.

The Tibia

The Tibia (figs. 92, 93) is situated at the anterior and inner side of the leg, and, excepting the femur, is the longest and largest bone in the skeleton. It is prismoid in form, expanded above, where it enters into formation with the knee joint, and more slightly below. In the male, its direction is vertical, and parallel with the bone of the opposite side, but in the female it has a slight oblique direction downwards and outwards, to compensate for the oblique direction of the femur inwards. It presents for examination a shaft and two extremities.

The *Superior Extremity*, or head, is large and expanded on each side into two lateral eminences, the tuberosities. Superiorly, the tuberosities present two smooth concave surfaces, which articulate with the condyles of the femur; the internal articular surface is longer than the external, oval from before backwards, to articulate with the internal condyle; the external one being broader, flatter, and more circular, to articulate with the external condyle. Between the two articular surfaces, and nearer the posterior than the anterior aspect of the bone, is an eminence, the spinous process of the tibia,

92.—Bones of the Right Leg. Anterior Surface.

surmounted by a prominent tubercle on each side, which give attachment to the extremities of the semilunar fibro-cartilages; and in front and behind the spinous process, a rough depression for the attachment of the anterior and posterior crucial ligaments and the semilunar cartilages. Anteriorly the tuberosities are continuous with one another, presenting a large and somewhat flattened triangular surface, broad above, and perforated by large vascular foramina, narrow below, where it terminates in a prominent oblong elevation of large size, the tubercle of the tibia; the lower half of this tubercle is rough, for the attachment of the ligamentum patellae; the upper half is a smooth facet corresponding, in the recent state, with a bursa which separates this ligament from the bone. Posteriorly, the tuberosities are separated from each other above by a shallow depression, the popliteal notch, which gives attachment to the posterior crucial ligament. The posterior surface of the inner tuberosity presents a deep transverse groove, for the insertion of the tendon of the Semi-membranosus; and the posterior surface of the outer one, a flat articular facet, nearly circular in form, directed downwards, backwards, and outwards, for articulation with the fibula. The lateral surfaces are convex and rough, the internal one, the most prominent, gives attachment below to the internal lateral ligament.

The *Shaft* of the tibia is of a triangular prismoid form, broad above, gradually decreasing in size to the commencement of its lower fourth, its most slender part, and then enlarges again towards its lower extremity. It presents for examination three surfaces and three borders.

The *anterior border*, the most prominent of the three, is called the *crest of the tibia*, or in popular language, the *shin*; it commences above at the tubercle, and terminates below at the anterior margin of the inner malleolus. This border is very prominent in the upper two-thirds of its extent, smooth and rounded below. It presents a very flexuous course, being curved outwards above, and inwards below; it gives attachment to the deep fascia of the leg.

The *internal border* is smooth and rounded above and below, but more prominent in the centre; it commences at the back part of the inner tuberosity, and terminates at the posterior border of the internal malleolus; its upper third gives attachment to the internal lateral ligament of the knee, and to some fibres of the Popliteus muscle; its middle third, to some fibres of the Soleus and Flexor longus digitorum muscles.

The *external border* is thin and prominent, especially its central part, and gives attachment to the interosseous membrane; it commences above in front of the fibular articular facet, and bifurcates below, forming the boundaries of a triangular rough surface, for the attachment of the inferior interosseous ligament, connecting the tibia and fibula.

The *internal surface* is smooth, convex, and broader above than below; its upper third, directed forwards and inwards, is covered by the aponeurosis derived from the tendon of the Sartorius, and by the tendons of the Gracilis and Semi-tendinosus, all of which are inserted nearly as far forwards, as the anterior border; in the rest of its extent it is sub-cutaneous.

The *external surface* is narrower than the internal, its upper two-thirds present a shallow groove for the attachment of the Tibialis anticus muscle; its lower third is smooth, convex, curves gradually forwards to the anterior part of the bone, and is covered from within outwards by the tendons of the following muscles: Tibialis anticus, Extensor proprius pollicis, Extensor longus digitorum, Peroneus tertius.

The *posterior surface* (fig. 93) presents at its upper part a prominent ridge, the oblique line of the tibia, which extends from the back part of the articular facet for the fibula, obliquely downwards, to the internal border, at the junction of its upper and middle thirds. It marks the limit for the insertion of the Popliteus muscle, and serves for the attachment of the popliteal fascia, and part of the Soleus, Flexor longus digitorum, and Tibialis posticus muscles; the triangular concave surface, above, and to the inner side of this line, gives attachment to the Popliteus muscle. The middle third of the posterior surface is divided by a vertical ridge into two lateral halves; the ridge is well marked at its commencement at the oblique line, but becomes gradually indistinct below; the inner and broadest half gives attachment to the Flexor longus digitorum, the outer and narrowest, to part of the Tibialis posticus. The remaining part of the bone is covered by the Tibialis posticus, Flexor longus digitorum, and Flexor longus pollicis muscles. Immediately below the oblique line is the medullary foramen, which is directed obliquely downwards.

The *Lower Extremity*, much smaller than the upper, is somewhat quadrilateral in

93.—Bones of the Right Leg. Posterior Surface.

form, and prolonged downwards, on its inner side, into a strong process, the internal malleolus. The *inferior surface* of the bone presents a quadrilateral smooth surface, for articulation with the astragalus; narrow internally, where it becomes continuous with the articular surface of the inner malleolus, broader externally, and traversed from before backwards by a slight elevation, separating two lateral depressions. The *anterior surface* is smooth and rounded above, and covered by the tendons of the Extensor muscles of the toes; its lower margin presents a rough transverse depression, for the attachment of the anterior ligament of the ankle joint. The *posterior surface* presents a superficial groove directed obliquely downwards and inwards, continuous with a similar groove on the posterior extremity of the astragalus, it serves for the passage of the tendon of the Flexor longus pollicis. The *external surface* presents a triangular rough depression, the lower part of which, in some bones, is smooth, covered with cartilage in the recent state and articulates with the fibula, the remaining part is rough for the attachment of the inferior interosseous ligament, which connects it with the fibula. This surface is bounded by two prominent ridges, continuous above with the interosseous ridge; they afford attachment to the anterior and posterior tibio-fibular ligaments. The *internal surface* is prolonged downwards to form a strong pyramidal-shaped process, flattened from without inwards, the inner malleolus; its *inner surface* is convex and subcutaneous. Its *outer surface*, smooth and slightly concave, deepens the articular surface for the astragalus. Its *anterior border* is rough, for the attachment of ligamentous fibres. Its *posterior border* presents a broad and deep groove, directed obliquely downwards and inwards; it is occasionally double, and transmits the tendons of the Tibialis posticus and Flexor longus digitorum muscles. Its *summit* is marked by a rough depression behind, for the attachment of the internal lateral ligament of the ankle joint.

Structure. Like that of the other long bones.

Development. By three centres (fig. 94): one for the shaft, and one for each extremity. Ossification commences in the centre of the shaft about the same time as in the femur, the fifth week, and gradually extends towards either extremity. The centre for the upper epiphysis appears at birth; it is flattened in form, and has a thin tongue-shaped process in front, which forms the tubercle. That for the lower epiphysis appears in the second year. The lower epiphysis joins the shaft at about the twentieth year, and the upper one about the twenty-fifth year. Two additional centres occasionally exist, one for the tongue-shaped process of the upper epiphysis, the tubercle, and one for the inner malleolus.

94.—Plan of the Development of the Tibia. By Three Centres.

Articulations. With three bones: the femur, fibula, and astragalus.

Attachment of Muscles. To the inner tuberosity, the Semi-membranosus. To the outer tuberosity, the Tibialis anticus and Extensor longus digitorum: to the shaft; its internal surface, the Sartorius, Gracilis, and Semi-tendinosus: to its external surface, the Tibialis anticus: to its posterior surface, the Popliteus, Soleus, Flexor longus digitorum, and Tibialis posticus: to the tubercle, the ligamentum patellae.

THE FIBULA

The Fibula (fig. 92, 93) is situated at the outer side of the leg. It is the smaller of the two bones, and, in proportion to its length, the most slender of all the long bones; it is placed nearly parallel with the tibia, its lower extremity inclining a little forwards, so as to be on a plane anterior to that of the upper end. It presents for examination a shaft and two extremities.

The *Superior Extremity* or *Head*, is of an irregular rounded form, presenting above a flattened articular facet, directed upwards and inwards, for articulation with a corresponding facet on the external tuberosity of the tibia. On the outer side is a thick and rough prominence, continued behind into a pointed eminence, the styloid process, which projects upwards from the posterior part of the head. The prominence above mentioned gives attachment to the tendon of the Biceps muscle, and to the long external lateral ligament of the knee, the ligament dividing this tendon into two parts. The summit of the styloid process gives attachment to the short external lateral ligament. The remaining part of the circumference of the head is rough, for the attachment, in front, of the anterior superior tibio-fibular ligament, and the upper and anterior part of the Peroneus longus; and behind, to the posterior superior tibio-fibular ligament, and the upper fibres of the outer head of the Soleus muscle.

The *Lower Extremity*, called the *malleolus externus*, is of a pyramidal form, somewhat flattened from without inwards, and is longer, and descends lower than the internal malleolus. Its *external surface* is convex, sub-cutaneous, and continuous with a triangular (also sub-cutaneous) surface on the outer side of the shaft. The *internal surface* presents in front a smooth triangular facet, broader above than below, convex from above downwards, which articulates with a corresponding surface on the outer side of the astragalus. Behind and beneath the articular surface is a rough depression, which gives attachment to the posterior fasciculus of the external lateral ligament of the ankle. Its *anterior border* is thick and rough, and marked below by a depression for the attachment of the anterior fasciculus of the external lateral ligament. The *posterior border* is broad and marked by a shallow groove, for the passage of the tendons of the Peroneus longus and brevis muscles. Its *summit* is rounded, and gives attachment to the middle fasciculus of the external lateral ligament.

The *Shaft* presents three surfaces, and three borders. The *anterior border* commences above in front of the head, runs vertically downwards to a little below the middle of the bone, and then curving a little outwards, bifurcates below into two lines, which bound the triangular sub-cutaneous surface immediately above the outer side of the malleolus externus. It gives attachment to an inter-muscular septum, which separates the muscles on the anterior surface from those on the external.

The *internal border* or *interosseous ridge*, is situated close to the inner side of the preceding, it runs nearly parallel with it in the upper third of its extent, but diverges from it so as to include a broader space in the lower two-thirds. It commences above just beneath the head of the bone (sometimes it is quite indistinct for about an inch below the head), and terminates below at the apex of a rough triangular surface immediately above the articular facet of the external malleolus. It serves for the attachment of the interosseous membrane, and separates the extensor muscles in front, from the flexor muscles behind. The portion of bone included between the anterior and interosseous lines, forms the anterior surface.

The *posterior border* is sharp and prominent; it commences above at the base of the styloid process, and terminates below in the posterior border of the outer malleolus. It is directed outwards above, backwards in the middle of its course, backwards and a little inwards below, and gives attachment to an aponeurosis which separates the muscles on the outer from those on the inner surface of the shaft. The portion of bone included between this line and the interosseous ridge, forms the internal surface. Its upper three-fourths are subdivided into two parts, an anterior and a posterior, by a very prominent ridge, the oblique line of the tibia, which commences above at the inner side of the head, and terminates by being continuous with the interosseous ridge at the lower fourth of the bone. It attaches an aponeurosis which separates the Tibialis posticus from the Soleus above, and the Flexor longus pollicis below. This ridge sometimes ceases just before approaching the interosseous ridge.

The *anterior surface* is the interval between the anterior and interosseous lines. It is extremely narrow and flat in the upper third of its extent; broader and grooved longitudinally in its lower third; it serves for the attachment of three muscles, the Extensor longus digitorum, Peroneus tertius, and Extensor longus pollicis.

The *external surface*, much broader than the preceding, is directed outwards in the upper two-thirds of its course, backwards in the lower third, where it is continuous with the posterior border of the external malleolus. This surface is completely occupied by the Peroneus longus and brevis muscles.

The *internal surface* is the interval between the interosseous ridge and the posterior border, and occupies nearly two-thirds of the circumference of the bone. Its upper three-fourths are divided into an anterior and a posterior portion by a very prominent ridge already mentioned, the oblique line of the fibula. The anterior portion is directed inwards, and is grooved for the attachment of the Tibialis posticus muscle. The posterior portion is continuous below with the rough triangular surface above the articular facet of the outer malleolus; it is directed backwards above, backwards and inwards at its middle, directly inwards below. Its upper fourth is rough, for the attachment of the Soleus muscle; its lower part presents a triangular rough surface, connected to the tibia by a strong interosseous ligament, and between these two points, the entire surface is covered by the fibres of origin of the Flexor longus pollicis muscle. At about the middle of this surface is the nutritious foramen, which is directed downwards.

In order to distinguish the side to which the bone belongs, hold it with the lower extremity downwards, and the broad groove for the Peronei tendons backwards, towards the holder, the triangular sub-cutaneous surface will then be directed to the side to which the bone belongs.

Articulations. With two bones; the tibia and astragalus.

Development. By three centres (fig. 95); one for the shaft, and one for each extremity. Ossification commences in the shaft about the sixth week of foetal life, a little later than in the tibia, and extends gradually towards the extremities. At birth both ends are cartilaginous. Ossification commences in the lower end in the second year, and in the upper one about the fourth year. The lower epiphysis, the first in which ossification commences, becomes united to the shaft about the twentieth year, contrary to the law which appears to prevail with regard to the junction of the epiphyses with the shaft; the upper one is joined about the twenty-fifth year.

95.—Plan of the Development of the Fibula. By Three Centres.

Attachment of Muscles. To the head, the Biceps, Soleus, and Peroneus longus: to the shaft, its anterior surface, the Extensor longus digitorum, Peroneus tertius, and Extensor longus pollicis: to the internal surface, the Soleus, Tibialis posticus, and Flexor longus pollicis: to the external surface, the Peroneus longus and brevis.

THE FOOT

The Foot (fig. 96, 97) is the terminal part of the inferior extremity; it serves to support the body in the erect posture, and as an important instrument of locomotion. It consists of three divisions: the Tarsus, Metatarsus, and Phalanges.

The Tarsus

The bones of the Tarsus are seven in number; viz., the calcaneum, or os calcis, astragalus, cuboid, scaphoid, internal, middle, and external, cuneiform bones. These bones may be conveniently arranged into two lateral rows. The outer row, remarkable for its great solidity and strength, forms the basis of support to the foot; it consists of two bones, the os calcis and cuboid. The inner row, which contributes chiefly to its elasticity, is formed by the astragalus, scaphoid, and three cuneiform bones.

The Calcaneum

The *Calcaneum*, or *Os Calcis*, is the largest bone of the tarsus. It is irregularly cuboidal in form, and situated at the lower and back part of the foot. It presents for examination six surfaces; superior, inferior, external, internal, anterior, and posterior.

The *superior surface* is formed behind, of the upper edge of that process of the os calcis which projects backwards to form the heel. This process varies in length in different individuals; it is convex from side to side, concave from before backwards, and corresponds above to a mass of adipose substance placed in front of the tendo Achillis. In the middle of this surface are two (sometimes three) articular facets, separated by a broad shallow groove, directed obliquely forwards and outwards, and rough for the attachment of the interosseous ligament connecting the astragalus and os calcis. Of these two articular surfaces, the *external* is the larger, and situated upon the body of the bone; it is of an oblong form, broader behind than in front, and convex from before backwards. The *internal articular surface* is supported on a projecting process of bone, called the *lesser process* of the calcaneum (sustentaculum tali); it is of an oblong form, concave longitudinally, and sometimes subdivided into two, which differ in size and shape. More anteriorly is seen the upper surface of the *greater process*, marked by a rough depression for the attachment of numerous ligaments, and the tendon of origin of the Extensor brevis digitorum muscle.

The *inferior surface* is narrow, rough, uneven, broader behind than in front, and convex from side to side; it is bounded posteriorly by two tuberosities, separated by a rough depression: the *external*, small, prominent, and rounded, gives attachment to part of the Abductor minimi digiti; the *internal*, broader and larger, for the support of the heel, gives attachment, by its prominent inner margin, to the Abductor pollicis, and in front to the Flexor brevis digitorum muscles, and the depression between the tubercles to the Abductor minimi digiti, and plantar fascia. The rough surface in front of these tubercles gives attachment to the long plantar ligament; and to a

96.—Bones of the Right Foot. Dorsal Surface.

prominent tubercle nearer the anterior part of the bone, as well as to the transverse groove in front, is attached the short plantar ligament.

The *external surface* is subcutaneous, and presents near its centre a tubercle, for the attachment of the middle fasciculus of the external lateral ligament. Behind the tubercle is a broad smooth surface, giving attachment, at its upper and anterior part, to the external astragalo-calcanean ligament; and in front a narrow surface marked by two oblique grooves, separated by an elevated ridge: the *superior groove* transmits the tendon of the Peroneus brevis; the *inferior*, the tendon of the Peroneus longus; the intervening ridge gives attachment to a prolongation from the external annular ligament.

The *internal surface* presents a deep concavity, directed obliquely downwards and forwards, for the transmission of the plantar vessels and nerves and Flexor tendons into the sole of the foot; it affords attachment to part of the Flexor accessorius muscle. This surface presents in front an eminence of bone, the *lesser process*, which projects horizontally inwards from the upper and front part of this surface. This process is concave above, and supports the anterior articular surface of the astragalus; below, it is convex, and grooved for the tendon of the Flexor longus pollicis. Its free margin is rough, for the attachment of ligaments.

The *anterior surface*, of a somewhat triangular form, is smooth, concavo-convex, and articulates with the cuboid. It is surmounted, on its outer side, by a rough prominence, which forms an important guide to the surgeon in the performance of Chopart's operation.

The *posterior surface* is rough, prominent, convex, and wider below than above. Its lower part is rough, for the attachment of the tendo Achillis; its upper part smooth, coated with cartilage, and corresponds to a bursa which separates this tendon from the bone.

Articulations. With two bones: the astragalus and cuboid.

Attachment of Muscles. Part of the Tibialis posticus, the tendo Achillis, Plantaris, Abductor pollicis, Abductor minimi digiti, Flexor brevis digitorum, Flexor accessorius, and Extensor brevis digitorum.

The Cuboid

The *Cuboid* bone is placed on the outer side of the foot, immediately in front of the os calcis. It is of a pyramidal shape, its base being directed upwards and inwards, its apex downwards and outwards. It may always be known from all the other tarsal bones, by the existence of a deep groove on its under surface, for the tendon of the Peroneus longus muscle. It presents for examination six surfaces; three articular, and three non-articular: the non-articular surfaces are the superior, inferior, and external.

The *superior* or *dorsal surface*, directed upwards and outwards, is rough, for the attachment of numerous ligaments. The *inferior* or *plantar surface* presents in front a deep groove, which runs obliquely from without, forwards and inwards; it lodges the tendon of the Peroneus longus, and is bounded behind by a prominent ridge, terminating externally in an eminence, the tuberosity of the cuboid, the surface of which presents a convex facet, for articulation with the sesamoid bone of the tendon contained in the groove. The ridge and surface of bone behind it are rough, for the attachment of the long and short plantar ligaments. The *external surface*, the smallest and narrowest of the three, presents a deep notch, formed by the commencement of the peroneal groove.

The articular surfaces are the posterior, anterior, and internal. The *posterior* is a smooth, triangular, concavo-convex surface, for articulation with the anterior surface of the os calcis. The *anterior*, of smaller size, but also irregularly triangular, is divided by a vertical ridge into two facets; the inner quadrilateral in form, to articulate with the fourth metatarsal bone; the outer larger and more triangular, for articulation with the fifth metatarsal. The *internal surface* is broad, rough, irregularly quadrilateral, presenting at its middle and upper part a small oval facet, for articulation with the external cuneiform bone; and behind this (occasionally) a smaller facet, for articulation with the scaphoid; it is rough in the rest of its extent, for the attachment of strong interosseous ligaments.

To ascertain to which foot it belongs, hold the bone so that its under surface, marked by the peroneal groove, looks downwards, and the large concavo-convex articular surface backwards, towards the holder; the small non-articular surface marked by the commencement of the peroneal groove, will point to the side to which the bone belongs.

Articulations. With four bones: the os calcis, external cuneiform, and the fourth and fifth metatarsal bones, occasionally with the scaphoid.

Attachment of Muscles. Part of the Flexor brevis pollicis.

The Astragalus

The *Astragalus* (fig. 96), next to the os calcis, is the largest of the tarsal bones. It is placed at the middle and upper part of the tarsus, supporting the tibia above, articulating with the malleoli on either side, resting below upon the os calcis, and joined in front to the scaphoid. This bone may easily be recognised by its large rounded head, the broad articular facet on its upper convex surface, and by the two articular facets separated by a deep groove on its under concave surface. It presents six surfaces for examination.

The *superior surface* presents, behind, a broad smooth trochlear surface, for articulation with the tibia; it is broader in front than behind, convex from before backwards, slightly concave from side to side. In front of the trochlea is the upper surface of the neck of the astragalus, rough for the attachment of ligaments. The *inferior surface* presents two articular facets separated by a deep groove. The groove runs obliquely forwards and outwards, becoming gradually broader and deeper in front: it corresponds with a similar groove upon the upper surface of the os calcis, and forms, when articulated with that bone, a canal, filled up in the recent state by the calcaneo-astragaloid interosseous ligament. Of the two articular facets, the posterior is the larger, of an oblong form, and deeply concave from side to side; the anterior, although nearly of equal length, is narrower, of an elongated oval form, convex from side to side, and often subdivided into two by an elevated ridge; the posterior articulates with the lesser process of the os calcis; the anterior, with the upper surface of the calcaneo-scaphoid ligament. The *internal surface* presents at its upper part a pear-shaped articular facet for the inner malleolus, continuous above with the trochlear surface; below the articular surface is a rough depression, for the

97.—Bones of the Right Foot. Plantar Surface.

attachment of the deep portion of the internal lateral ligament. The *external surface* presents a large triangular facet, concave from above downwards, for articulation with the external malleolus; it is continuous above with the trochlear surface: in front is a deep rough margin, for the attachment of the anterior fasciculus of the external lateral ligament. The *anterior surface*, convex and rounded, forms the head of the astragalus; it is smooth, of an oval form, and directed obliquely inwards and downwards; it is continuous below with that part of the anterior facet on the under surface which rests upon the calcaneo-scaphoid ligament. The head is surrounded by a constricted portion, the neck of the astragalus. The *posterior surface* is narrow, and traversed by a groove, which runs obliquely downwards and inwards, and transmits the tendon of the Flexor longus pollicis.

To ascertain to which foot it belongs, hold the bone with the broad articular surface upwards, and the rounded head forwards; the lateral triangular articular surface for the external malleolus will then point to the side to which the bone belongs.

Articulations. With four bones; tibia, fibula, os calcis, and scaphoid.

The Scaphoid

The *Scaphoid* or *Navicular* bone, so called from its fancied resemblance to a boat, is situated at the inner side of the tarsus, between the astragalus behind and the three cuneiform bones in front. This bone may be distinguished by its boat-like form, being concave behind, convex, and subdivided into three facets in front.

The *anterior surface*, of an oblong form, is convex from side to side, and subdivided by two ridges into three facets, for articulation with the three cuneiform bones. The *posterior surface* is oval, concave, broader externally than internally, and articulates with the rounded head of the astragalus. The *superior surface* is convex from side to side, and rough for the attachment of ligaments. The *inferior*, somewhat concave, irregular, and also rough for the attachment of ligaments. The *internal surface* presents a rounded tubercular eminence, the tuberosity of the scaphoid, which gives attachment to part of the tendon of the Tibialis posticus. The *external surface* is broad, rough, and irregular, for the attachment of ligamentous fibres, and occasionally presents a small facet for articulation with the cuboid bone.

To ascertain to which foot it belongs, hold the bone with the concave articular surface backwards, and the broad dorsal surface upwards; the broad external surface will point to the side to which the bone belongs.

Articulations. With four bones; astragalus and three cuneiform; occasionally also with the cuboid.

Attachment of Muscles. Part of the Tibialis posticus.

The Cuneiform Bones have received their name from their wedge-like form. They form the most anterior row of the inner division of the tarsus, being placed between the scaphoid behind, the three innermost metatarsal bones in front, and the cuboid externally. They are called the *first*, *second*, and *third*, counting from the inner to the outer side of the foot, and from their position, *internal*, *middle*, and *external*.

The Internal Cuneiform

The *Internal Cuneiform* is the largest of the three. It is situated at the inner side of the foot, between the scaphoid behind and the base of the first metatarsal in front. It may be distinguished by its large size, as compared with the other two, and from its more irregular wedge-like form. It presents for examination six surfaces.

The *internal surface* is subcutaneous, and forms part of the inner border of the foot; it is broad, quadrilateral, and presents at its anterior inferior angle a smooth oval facet, over which the tendon of the Tibialis anticus muscle glides; rough in the rest of its extent, for the attachment of ligaments. The *external surface* is concave, presenting, along its superior and posterior borders, a narrow surface for articulation with the middle cuneiform behind, and second metatarsal bone in front; in the rest of its extent, it is rough for the attachment of ligaments, and prominent below, where it forms part of the tuberosity. The *anterior surface*, reniform in shape, articulates with the metatarsal bone of the great toe. The *posterior surface* is triangular, concave, and articulates with the innermost and largest of the three facets on the anterior surface of the scaphoid. The *inferior* or *plantar surface* is rough, and presents a prominent tuberosity at its back part for the attachment of part of the tendon of the Tibialis posticus. It also gives attachment in front of this to part of the tendon of the Tibialis anticus. The *superior surface* is the narrow pointed end of the wedge, which is directed upwards and outwards; it is rough for the attachment of ligaments.

To ascertain to which side it belongs, hold the bone so that its superior narrow edge looks upwards, and the long articular surface forwards; the external surface marked by its vertical and horizontal articular facets will point to the side to which it belongs.

Articulations. With four bones; scaphoid, middle cuneiform, and first and second metatarsal bones.

Attachment of Muscles. The Tibialis anticus and posticus.

The Middle Cuneiform

The *Middle Cuneiform*, the smallest of the three, is of very regular wedge-like form; the broad extremity being placed upwards, the narrow end downwards. It is situated between the other two bones of the same name, and corresponds to the scaphoid behind, and the second metatarsal in front.

The *anterior surface*, triangular in form, and narrower than the posterior, articulates with the base of the second metatarsal bone. The *posterior surface*, also triangular, articulates with the scaphoid. The *internal surface* presents an articular facet, running along the superior and posterior borders, for articulation with the internal cuneiform, and is rough below for the attachment of ligaments. The *external surface* presents posteriorly a smooth facet for articulation with the external cuneiform bone. The *superior surface* forms the base of the wedge; it is quadrilateral, broader behind than in front, and rough for the attachment of ligaments. The *inferior surface*, pointed and tubercular, is also rough for ligamentous attachment.

To ascertain to which foot the bone belongs, hold its superior or dorsal surface upwards, the broadest edge being towards the holder, and the smooth facet (limited to the posterior border) will point to the side to which it belongs.

Articulations. With four bones; scaphoid, internal and external cuneiform, and second metatarsal bone.

The External Cuneiform

The *External Cuneiform*, intermediate in size between the two preceding, is of a very regular wedge-like form, the broad extremity being placed upwards, the narrow end downwards. It occupies the centre of the front row of the tarsus between the middle cuneiform internally, the cuboid externally, the scaphoid behind, and the third metatarsal in front. It has six surfaces for examination.

The *anterior surface* triangular in form, articulates with the third metatarsal bone. The *posterior surface* articulates with the most external facet of the scaphoid, and is rough below for the attachment of ligamentous fibres. The *internal surface* presents two articular facets separated by a rough depression; the anterior one, situated at the superior angle of the bone, articulates with the outer side of the base of the second metatarsal bone; the posterior one skirts the posterior border, and articulates with the middle cuneiform; the rough depression between the two gives attachment to an interosseous ligament. The *external surface* also presents two articular facets, separated by a rough non-articular surface; the anterior facet, situated at the superior corner of the bone, is small, and articulates with the inner side of the base of the fourth metatarsal; the posterior, and larger one, articulates with the cuboid; the rough non-articular surface serves for the attachment of an interosseous ligament. The three facets for articulation with the three metatarsal bones are continuous with one another, and covered by a prolongation of the same cartilage; the facets for articulation with the middle cuneiform and scaphoid are also continuous, but that for articulation with the cuboid is usually separate and independent. The *superior* or *dorsal surface*, of an oblong form, is rough for the attachment of ligaments. The *inferior* or *plantar surface* is an obtuse rounded margin, and serves for the attachment of part of the tendon of the Tibialis posticus, some of the fibres of origin of the Flexor brevis pollicis, and ligaments.

To ascertain to which side it belongs, hold the bone with the broad dorsal surface upwards, the prolonged edge backwards; the separate articular facet for the cuboid will point to the proper side.

Articulations. With six bones: the scaphoid, middle cuneiform, cuboid, and second, third, and fourth metatarsal bones.

Attachment of Muscles. Part of Tibialis posticus, and Flexor brevis pollicis.

The Metatarsal Bones

The Metatarsal bones are five in number; they are long bones, and subdivided into a shaft, and two extremities.

The *Shaft* is prismoid in form, tapers gradually from the tarsal to the phalangeal extremity, and is slightly curved longitudinally, so as to be concave below, slightly convex above.

The *Posterior Extremity*, or *Base*, is wedge-shaped, articulating by its terminal surface with the tarsal bones, and by its lateral surfaces with the contiguous bones; its dorsal and plantar surfaces being rough, for the attachment of ligaments.

The *Anterior Extremity*, or *Head*, presents a terminal rounded articular surface, oblong from above downwards, and extending further backwards below than above. Its sides are flattened, and present a depression, surmounted by a tubercle, for ligamentous attachment. Its under surface is grooved in the middle line, for the passage of the Flexor tendon, and marked on each side by an articular eminence continuous with the terminal articular surface.

Peculiar Metatarsal Bones

The *First* is remarkable for its great size, but is the shortest of all the metatarsal bones. The *shaft* is strong, and of well-marked prismoid form. The *posterior extremity* presents no lateral articular facets; its terminal articular surface is of large size, of semi-lunar form, and its circumference grooved for the tarso-metatarsal ligaments; its inferior angle presents a rough oval prominence, for the insertion of the tendon of the Peroneus longus. The *head* is of large size; on its plantar surface are two grooved facets, over which glide sesamoid bones, separated by a smooth elevated ridge.

The *Second* is the longest and largest of the remaining metatarsal bones; its *posterior extremity* being prolonged backwards, into the recess formed between the three cuneiform bones. Its *tarsal extremity* is broad above, narrow and rough below. It presents four articular surfaces: one behind, of a triangular form, for articulation with the middle cuneiform; one at the upper part of its internal lateral surface, for articulation with the internal cuneiform; and two on its external lateral surface, a superior and an inferior, separated by a rough depression. Each articular surface is divided by a vertical ridge into two parts; the anterior segment of each facet articulates with the third metatarsal; the two posterior (sometimes continuous) with the external cuneiform.

The *Third* articulates behind, by means of a triangular smooth surface, with the external cuneiform; on its inner side, by two facets, with the second metatarsal; and on its outer side, by a single facet, with the third metatarsal. This facet is of circular form, and situated at the upper angle of the base.

The *Fourth* is smaller in size than the preceding; its *tarsal extremity* presents a terminal quadrilateral surface, for articulation with the cuboid; a smooth facet on the inner side, divided by a ridge into an anterior portion for articulation with the third metatarsal, and a posterior portion for articulation with the external cuneiform; on the outer side a single facet, for articulation with the fifth metatarsal.

The *Fifth* is recognised by the tubercular eminence on the outer side of its base; it articulates behind, by a triangular surface cut obliquely from without inwards, with the cuboid, and internally with the fourth metatarsal.

Articulations. Each bone articulates with the tarsal bones by one extremity, and by the other with the first row of phalanges. The number of tarsal bones with which each metatarsal articulates, is one for the first, three for the second, one for the third, two for the fourth, and one for the fifth.

Attachment of Muscles. To the first metatarsal bone, three: part of the Tibialis anticus, Peroneus longus, and First dorsal interosseous. To the second, three: the Adductor pollicis, and First and Second dorsal interosseous. To the third, four: the Adductor pollicis, Second and Third dorsal interosseous, and First plantar. To the fourth, four: the Adductor pollicis, Third and Fourth dorsal, and Second plantar interosseous. To the fifth, five: the Peroneus brevis, Peroneus tertius, Flexor brevis minimi digiti, Fourth dorsal, and Third plantar interosseous.

Phalanges

The *Phalanges* of the foot, both in number and general arrangement, resemble those in the hand; there being two in the great toe, and three in each of the other toes.

The phalanges of the *first row* resemble closely those of the hand. The *shaft* is compressed from side to side, convex above, concave below. The *posterior extremity* is concave; and the *anterior extremity* presents a trochlear-articular surface, for articulation with the second phalanges.

The phalanges of the *second row* are remarkably small and short, but rather broader than those of the first row.

The *ungual* phalanges in form resemble those of the fingers; but they are smaller, flattened from above downwards, presenting a broad base for articulation with the second row, and an expanded extremity for the support of the nail and end of the toe.

Articulations. The first row with the metatarsal bones, and second phalanges; the second of the great toe with the first phalanx, and of the other toes with the first and third phalanges; the third with the second row.

Attachment of Muscles. To the first phalanges, great toe: innermost tendon of Extensor brevis digitorum, Abductor pollicis, Adductor pollicis, Flexor brevis pollicis, Transversus pedis. Second toe: First and Second dorsal interosseae. Third toe: Third dorsal and First plantar interosseae. Fourth toe: Fourth dorsal and Second plantar interosseae. Fifth toe: Flexor brevis minimi digiti, Abductor minimi digiti, and Third plantar interosseous.—Second phalanges, great toe: Extensor longus pollicis, Flexor longus pollicis. Other toes: Flexor brevis digitorum, one slip from the Extensor brevis digitorum and Extensor longus digitorum.—Third phalanges: two slips from the common tendon of the Extensor longus and Extensor brevis digitorum, and the Flexor longus digitorum.

Development of the Foot (Fig. 98)

The Tarsal bones are each developed by a single centre, excepting the os calcis, which has an epiphysis for its posterior extremity. The centres make their appearance in the following order: in the os calcis, at the sixth month of foetal life; in the astragalus, about the seventh month; in the cuboid, at the ninth month; external cuneiform, during the first year; internal cuneiform, in the third year; middle cuneiform, in the fourth year. The epiphysis for the posterior tuberosity of the os calcis appears at the tenth year, and unites with the rest of the bone soon after puberty.

The Metatarsal bones are each developed by *two* centres: one for the shaft, and one for the digital extremity in the four outer metatarsal; one for the shaft, and one for the base in the metatarsal bone of the great toe. Ossification commences in the centre of the shaft about the seventh week, and extends towards either extremity, and in the digital epiphyses about the third year; they become joined between the eighteenth and twentieth years.

The Phalanges are developed by *two* centres for each bone: one for the shaft, and one for the metatarsal extremity.

98.—Plan of the Development of the Foot.

Sesamoid Bones

These are small rounded masses, cartilaginous in early life, osseous in the adult, which are developed in those tendons which exert a certain amount of pressure upon the parts over which they glide. It is said that they are more commonly found in the male than in the female, and in persons of an active muscular habit than in those that are weak and debilitated. They are invested throughout their whole surface by the fibrous tissue of the tendon in which they are found, excepting upon that side which lies in contact with the part over which they play, where they present a free articular facet. They may be divided into two kinds: those which glide over the articular surfaces of joints; those which play over the cartilaginous facets found on the surfaces of certain bones.

The sesamoid bones of the joints are, in the lower extremity, the patella (already described), which is developed in the tendon of the Quadriceps extensor. Two small sesamoid bones are found opposite the metatarso-phalangeal joint of the great toe in each foot, in the tendons of the Flexor brevis pollicis, and occasionally one in the metatarso-phalangeal joints of the second toe, the little toe, and, still more rarely, in the third and fourth toes.

In the upper extremity, there are two on the palmar surface, opposite the metacarpo-phalangeal joint in the thumb, developed in the tendons of the Flexor brevis pollicis. Occasionally one or two opposite the metacarpo-phalangeal articulations of the fore and little fingers, and, still more rarely, one opposite the same joints of the third and fourth fingers.

Those found in tendons which glide over certain bones occupy the following situations. One in the tendon of the Peroneus longus, where it glides through the groove in the cuboid bone. One appears late in life in the tendon of the Tibialis anticus, opposite the smooth facet on the internal cuneiform bone. One in the tendon of the Tibialis posticus, opposite the inner side of the astragalus. One in the outer head of the Gastrocnemius, behind the outer condyle of the femur; and one in the Psoas and Iliacus, where they glide over the body of the pubes. Occasionally in the tendon of the Biceps, opposite the tuberosity of the radius; in the tendon of the Gluteus maximus, as it passes over the great trochanter; and in the tendons which wind around the inner and outer malleoli.

THE ARTICULATIONS

The various bones of which the Skeleton consists are connected together at different parts of their surfaces, and such connection is designated by the name of *Joint* or *Articulation*. If the joint is *immoveable*, as between the cranial bones, their adjacent margins are applied in almost close contact, a thin layer of fibrous membrane only being interposed; but in the *moveable* joints, the bones forming the articulation are generally expanded for greater convenience of mutual connexion, covered by an elastic structure, called *cartilage*, held together by strong bands, or capsules, of fibrous tissue, called *ligament*, and lined throughout by a membrane, the *synovial membrane*, which secretes a fluid which lubricates the various parts of which the joint is formed, so that the structures which enter into the formation of a joint are bone, cartilage, fibro-cartilage, ligament, and synovial membrane.

Bone constitutes the fundamental element of all the joints. In the *long* bones the extremities are the parts which form the articulations; they are generally somewhat enlarged and expanded, consisting of loose spongy cancellous tissue, with a thin coating of compact substance, which forms their articular surface, and is called the *articular lamella*. In the flat bones, the articulations usually take place at the edges; and in the short bones, by various parts of their surface.

Cartilage is a firm, opaque, blueish-white substance, highly elastic, extremely flexible, and possessed of considerable cohesive power. That form of cartilage which enters into the formation of the joints is called *articular cartilage*; it forms a thin incrustation upon the articular extremities, or surfaces, of bones, and is admirably adapted, by its elastic property, to break the force of concussions, and afford perfect ease and freedom of movement between the bones. Where it covers the rounded ends of bones, as the extremities of the femur and humerus, it is thick at the centre, and becomes gradually thinner towards the circumference: an opposite arrangement exists where it lines the corresponding cavities. On the articular surfaces of the short bones, as the carpus and tarsus, the cartilage is disposed in a layer of uniform thickness throughout. The attached surface of articular cartilage is closely adapted to the articular lamella; the free surface is smooth, polished, and covered in the foetus by an extremely thin prolongation of synovial membrane, which, however, at a later period of life, cannot be demonstrated. Articular cartilage in the adult does not contain blood-vessels; its nutrition being derived from the vessels of the synovial membrane which skirt the circumference of the cartilage, and from those of the adjacent bone, which are, however, separated from direct contact with the cartilage by means of the articular lamella. Mr. Toynbee has shown, that the minute vessels as they approach the articular lamella dilate, and forming arches, return into the cancellous tissue of the bone. The vessels of the synovial membrane advance forwards with it upon the circumference of the cartilage for a very short distance, and then return in loops; they are only found on the parts not subjected to pressure. In the foetus, and also in certain diseased conditions of the joints, the vessels advance for some distance upon the cartilage. Lymphatic vessels and nerves have not, as yet, been traced in its substance.

Fibro-cartilage is also employed in the construction of the joints, contributing to their strength and elasticity. This consists of a mixture of white fibrous and cartilaginous tissues in various proportions; it is to the first of these two constituents that its strength and toughness is chiefly owing and to the latter its elasticity. The fibro-cartilages admit of arrangement into four groups, inter-articular, inter-osseous, circumferential, and stratiform.

The *Inter-articular fibro-cartilages* (*menisci*) are flattened fibro-cartilaginous plates, of a round, oval, or sickle-like form, interposed between the articular surfaces of certain joints. They are free on both surfaces, thinner toward their centre than at their circumference, and held in position by their extremities being connected to the surrounding ligaments. The synovial membrane of the joint is prolonged over them a short distance from their attached margin. They are found in the temporo-maxillary, sterno-clavicular, acromio-clavicular, wrist and knee joints.

The *Inter-osseous fibro-cartilages* are interposed between the bony surfaces of those joints which admit of only slight mobility, as between the bodies of the vertebrae and the symphysis of the pubes; they exist in the form of discs, intimately adherent to the opposed surfaces, being composed of concentric rings of fibrous tissue, with cartilaginous laminae interposed, the former tissue predominating towards the circumference, the latter towards the centre.

The *Circumferential fibro-cartilages* consist of a rim of fibro-cartilage, which surrounds the margins of some of the articular cavities, as the cotyloid cavity of the hip, and the glenoid cavity of the shoulder; they serve to deepen the articular surface and protect the edges of the bone.

The *Stratiform fibro-cartilages* are those which form a thin layer in the osseous grooves, through which the tendons of certain muscles glide.

Ligaments are found in nearly all the moveable articulations; they consist of bands of various forms, serving to connect together the articular extremities of bones, and composed mainly of bundles of white fibrous tissue, placed parallel with, or closely interlaced with, one another, and presenting a white, shining, silvery aspect. Ligament is pliant and flexible, so as to allow of the most perfect freedom of movement, but strong, tough, and inextensile, so as not readily to yield under the most severely applied force; it is, consequently, admirably adapted to serve as the connecting medium between the bones. There are some ligaments which consist entirely of yellow elastic tissue, as the ligamenta subflava, which connect together the adjacent arches of the vertebrae.

Synovial Membrane is a thin, delicate membrane, which invests the articular extremities of the bones, and is then reflected on the inner surface of the various ligaments which connect the articulating surfaces. It resembles the serous membrane in being a shut sac, but differs in the nature of its secretion, which is thick, viscid, and glairy, like the white of egg; and hence termed *synovia*. The synovial membranes found in the body admit of subdivision into three kinds, articular, bursal, and vaginal.

The *Articular Synovial Membranes* are found in all the freely moveable (diarthrodial) joints. In the foetus, this membrane may be traced over the entire surface of the cartilages; but in the adult it is wanting, excepting at their circumference, upon which it encroaches for a short distance: it then invests the inner surface of the capsular or other ligaments enclosing the joint, and is reflected over the surface of any tendons passing through its cavity, as the tendon of the Popliteus in the knee, and the tendon of the Biceps in the shoulder. In some joints, the synovial membrane is thrown into numerous folds, which project forward into the cavity. These folds consist of a reduplication of the synovial membrane, some of them containing fat, and, more rarely, isolated cartilage-cells; the free borders of the longer processes being subdivided into vascular fringe-like processes, the vessels of which have a convoluted arrangement. They are especially distinct in the knee, where they are known as the mucous and alar ligaments, and were described by Clopton Havers as *mucilaginous glands*, and as the source of the synovial secretion, a view lately revived by Mr. Rainey, who finds them in the bursal and vaginal, as well as in the articular synovial membranes.

The *Bursal Synovial Membranes* (*Bursae mucosae*) are found interposed between surfaces which move upon each other, producing friction, as in the gliding of a tendon, or of the integument over projecting bony surfaces. They are small shut sacs, connected by areolar tissue with the adjacent parts, and secreting a fluid in their interior analogous to synovia. The bursae admit of a subdivision into two kinds, subcutaneous and deep-seated. The *subcutaneous* are found in various situations, as between the integument and front of the patella, over the olecranon, the malleoli, and other prominent parts. The *deep seated* are more numerous, and usually found interposed between muscles or their tendons as they play over projecting bony surfaces, as between the Glutei muscles and surface of the great trochanter. Where one of these exists in the neighbourhood of a joint, it usually communicates with the cavity of the synovial membrane, as is generally the case with the bursa between the tendon of the Psoas and Iliacus, and the capsular ligament of the hip, or the one interposed between the under surface of the Subscapularis and the neck of the scapula.

The *Vaginal Synovial Membranes* (synovial sheaths) serve to facilitate the gliding of tendons in the osseo-fibrous canals through which they pass. The membrane is here arranged in the form of a sheath, one layer of which adheres to the wall of the canal, and the other is reflected upon the outer surface of the contained tendon; the space between the two free surfaces of the membrane, being partially filled with synovia. These sheaths are chiefly found surrounding the tendons of the Flexor and Extensor muscles of the fingers and toes, as they pass through the osseo-fibrous canals in the hand or foot.

Synovia is a transparent, yellowish-white, or slightly reddish fluid, viscid like the white of egg, having an alkaline reaction, and slightly saline taste. It consists, according to Frerichs, in the ox, of 94.85 water, 0.56 mucus and epithelium, 0.07 fat, 3.51 albumen and extractive matter, and 0.99 salts.

The Articulations are divided into three classes: *Synarthrosis*, or immoveable; *Amphiarthrosis*, or mixed; and *Diarthrosis*, or moveable.

1. Synarthrosis: Immoveable Articulations

Synarthrosis (συν, with, αρθρον, a joint), or *Immoveable Joints*, include all those articulations in which the surfaces of the bones are in almost direct contact, not separated by an intervening synovial cavity, and immoveably connected with each other, as between the bones of the cranium and face, excepting the lower jaw. The varieties of synarthrosis are three in number: Sutura, Schindylesis, Gomphosis.

Sutura (a seam). Where the articulating surfaces are connected by a series of processes and indentations interlocked together, it is termed *sutura vera*; of which there are three varieties: sutura dentata, serrata, and limbosa. The surfaces of the bones are not in direct contact, being separated by a layer of membrane continuous externally with the pericranium, internally with the dura mater. The *sutura dentata* (*dens*, a tooth) is so called from the tooth-like form of the projecting articular processes, as in the suture between the parietal bones. In the *sutura serrata* (*serra*, a saw), the edges of the two bones forming the articulation are serrated like the teeth of a fine saw, as between the two portions of the frontal bone. In the *sutura limbosa* (*limbus*, a selvage), besides the dentated processes, there is a certain degree of bevelling of the articular surfaces, so that the bones overlap one another, as in the suture between the parietal and occipital bones. Where the articulation is formed by roughened surfaces placed in apposition with one another, it is termed the *false suture, sutura notha*, of which there are two kinds: the *sutura squamosa* (*squama*, a scale), formed by the overlapping of two contiguous bones by broad bevelled margins, as in the temporo-parietal suture; and the *sutura harmonia* (ἀρεῖν, to adapt), where there is simple apposition of two contiguous rough bony surfaces, as in the articulation between the two superior maxillary bones, or of the palate processes of the palate bones with each other. The sutures present a great tendency to obliteration as age advances, the intervening fibrous-tissue becoming ossified. The frontal suture seldom exists after puberty; and it rarely happens that all the others are distinct in a skull beyond the age of fifty.

Schindylesis (σχινδυλησις, a fissure) is that form of articulation in which a thin plate of bone is received into a cleft or fissure formed by the separation of two laminae of another, as in the articulation of the rostrum of the sphenoid, and descending plate of the ethmoid with the vomer, or in the reception of the latter in the fissure between the superior maxillary and palate bones.

Gomphosis (γομφος, a nail) is an articulation formed by the insertion of a conical process into a socket, as a nail is driven into a board; and is illustrated in the articulation of the teeth in the alveoli of the maxillary bones.

2. Amphiarthrosis: Mixed Articulations

Amphiarthrosis (αμφι "on all sides," αρθρον a "joint"), or *Mixed Articulation*. In this form of articulation, the contiguous osseous surfaces are connected together by broad flattened discs of fibro-cartilage, which adhere to the ends of both bones, as in the articulation between the bodies of the vertebrae, and first two pieces of the sternum; or the articulating surfaces are covered with fibro-cartilage, lined by a partial synovial membrane, and connected together by external ligaments, as in the sacro-iliac and pubic symphyses; both these forms being capable of limited motion in every direction. The former resemble the synarthrodial joints in the continuity of their surfaces, and absence of synovial sac; the latter, the diarthrodial. These joints occasionally become obliterated in old age: this is frequently the case in the inter-pubic articulation, and occasionally in the intervertebral and sacro-iliac.

3. Diarthrosis: Moveable Articulations

Diarthrosis (δια "through," αρθρον "a joint"). This form of articulation includes the greater number of the joints in the body, mobility being their distinguishing character. They are formed by the approximation of two contiguous bony surfaces, covered with cartilage, connected by ligaments, and having a synovial sac interposed. The varieties of joints in this class, have been determined by the kind of motion permitted in each; they are four in number: Arthrodia, Enarthrosis, Ginglymus, Diarthrosis Rotatorius.

Arthrodia is that form of joint which admits of a gliding movement; it is formed by the approximation of plane surfaces, or one slightly concave, the other slightly convex; the amount of motion between them being limited by the ligaments, or osseous processes, surrounding the articulation; as in the articular processes of the vertebrae, temporo-maxillary, sterno and acromio-clavicular, inferior radio-ulnar, carpal, carpo-metacarpal, superior tibio-fibular, tarsal, and tarso-metatarsal articulations.

Enarthrosis is that form of joint which is capable of motion in all directions. It is formed by the reception of a globular head into a deep cup-like cavity (hence the name "ball and socket"), the parts being kept in apposition by a capsular ligament strengthened by accessory ligamentous bands, and the contiguous cartilaginous surfaces having a synovial sac interposed. Examples of this form of articulation are found in the hip and shoulder-joints.

Ginglymus, Hinge-joint (γιγγλυμος, a hinge). In this form of joint, the articular surfaces are moulded to each other in such a manner, as to permit motion only in one direction, forwards and backwards, the extent of motion at the same time being considerable. The articular surfaces are connected together by strong lateral ligaments, which form their chief bond of union. The most perfect forms of ginglymi are the elbow and ankle; the knee is less perfect, as it allows a slight degree of rotation in certain positions of the limb: there are also the metatarso-phalangeal and phalangeal joints in the lower extremity, metacarpo-phalangeal and phalangeal joints in the upper extremity.

Diarthrosis rotatorius (Lateral Ginglymus). Where the mobility is limited to rotation, the joint is formed by a pivot-like process turning within a ring, or the ring on the pivot, the ring being formed partly of bone, partly of ligament. In the articulation of the odontoid process of the axis with the atlas, the ring is formed in front by the anterior arch of the atlas; behind, by the transverse ligament; here the ring rotates around the odontoid process. In the superior radio-ulnar articulation, the ring is formed partly by the lesser sigmoid cavity of the ulna; in the rest of its extent, by the orbicular ligament; here, the neck of the radius rotates within the ring.

Subjoined, in a tabular form, are the names, distinctive characters, and examples of the different kinds of articulations.

Synarthrosis, or immoveable joint. Surfaces separated by fibrous membrane, no intervening synovial cavity, and immoveably connected with each other.
Example: bones of cranium and face (except lower jaw).

- *Sutura*. Articulation by processes and indentations interlocked together.
  - *Sutura vera* (true) articulate by indented borders.
    - *Dentata*, having tooth-like processes.
      Inter-parietal suture.
    - *Serrata*, having serrated edges, like the teeth of a saw.
      Inter-frontal suture.
    - *Limbosa*, having bevelled margins, and dentated processes.
      Occipito-parietal suture.
  - *Sutura notha* (false) articulate by rough surfaces
    - *Squamosa*, formed by thin bevelled margins overlapping each other.
      Temporo-parietal suture.
    - *Harmonia*, formed by the apposition of contiguous rough surfaces.
      Inter-maxillary suture.
- *Schindylesis*. Articulation formed by the reception of a thin plate of bone into a fissure of another.
  Rostrum of sphenoid with vomer.
- *Gomphosis*. An articulation formed by the insertion of a conical process into a socket.
  Tooth in socket.

Amphiarthrosis, Mixed Articulation.

1. Surfaces connected by fibro-cartilage, not separated by synovial membrane, and having limited motion. Bodies of vertebrae.
2. Surfaces connected by fibro-cartilage, lined by a partial synovial membrane. Sacro-iliac and pubic symphyses.

Diarthrosis, Moveable Joint.

- *Arthrodia*. Gliding joint; articulation by plane surfaces, which glide upon each other. As in sterno and acromio-clavicular articulations.
- *Enarthrosis*. Ball and socket joint; capable of motion in all directions. Articulation by a globular head received into a cup-like cavity. As in hip and shoulder joints.
- *Ginglymus*. Hinge joint; motion limited to one direction, forwards and backwards. Articular surfaces fitted together so as to permit of movement in one plane. As in the elbow, ankle, and knee.
- *Diarthrosis rotatorius*. Articulation by a pivot process turning within a ring, or ring around a pivot. As in superior radio-ulnar articulation, and atlo-axoid joint.

The Kinds of Movement admitted in Joints

The movements admissible in joints may be divided into four kinds, gliding, angular movement, circumduction, and rotation.

Gliding movement is the most simple kind of motion that can take place in a joint, one surface gliding over another. This kind of movement is common to all moveable joints; but in some, as in the articulations of the carpus and tarsus, is the only motion permitted. This motion is not confined to plane surfaces, but may exist between any two contiguous surfaces, of whatever form, limited by the ligaments which enclose the articulation.

Angular movement occurs only between the long bones, and may take place in four directions, forwards or backwards, constituting flexion and extension, or inwards and outwards, which constitutes abduction and adduction. Flexion and extension is confined to the strictly ginglymoid or hinge joints. Abduction and adduction, combined with flexion and extension, are met with only in the most moveable joints; as in the hip, shoulder, and thumb, and partially in the wrist and ankle.

Circumduction is that limited degree of motion which takes place between the head of a bone and its articular cavity, whilst the extremity and sides of a limb are made to circumscribe a conical space, the base of which corresponds with the inferior extremity of the limb, the apex to the articular cavity; and is best seen in the shoulder and hip joints.

Rotation is the movement of a bone upon its own axis, the bone retaining the same relative situation with respect to the adjacent parts; as in the articulation between the atlas and axis, where the odontoid process serves as a pivot around which the atlas turns; or in the rotation of the radius against the humerus, and also in the hip and shoulder.

The articulations may be arranged into those of the trunk, those of the upper extremity, and those of the lower extremity.

ARTICULATIONS OF THE TRUNK

These may be divided into the following groups viz.—

1. Of the vertebral column.
2. Of the atlas with the axis.
3. Of the atlas with the occipital bone.
4. Of the axis with the occipital bone.
5. Of the lower jaw.
6. Of the ribs with the vertebrae.
7. Of the cartilages of the ribs with the sternum, and with each other.
8. Of the sternum.
9. Of the vertebral column with the pelvis.
10. Of the Pelvis.

1. Articulations of the Vertebral Column

The different segments of the vertebral column are connected together by ligaments, which admit of the same arrangement as the vertebrae themselves. They may be divided into five sets. 1. Those connecting the *bodies* of the vertebrae. 2. Those connecting the *laminae*. 3. Those connecting the *articular processes*. 4. The ligaments connecting the *spinous processes*. 5. Those of the *transverse processes*.

The articulation of the *bodies* of the vertebrae with each other, form a series of amphiarthrodial joints; whilst those between the *articular processes* form a series of arthrodial joints.

1. The Ligaments of the Bodies are
Anterior Common Ligament.   Posterior Common Ligament.
Intervertebral Substance.

The *Anterior Common Ligament* (fig. 107) is a broad and strong band of ligamentous fibres, which extends along the front surface of the bodies of the vertebrae, from the axis to the sacrum. It is broader below than above, and thicker in the dorsal than in the cervical or lumbar regions. It is attached, above, to the body of the axis by a pointed process, which is connected with the tendon of origin of the Longus colli muscle; and, as it descends, is somewhat broader opposite the centre of the body of each vertebra, than opposite the intervertebral substance. It consists of dense longitudinal fibres, which are intimately adherent to the intervertebral substance and prominent margins of the vertebrae; but less closely with the centre of the bodies. In this situation the fibres are exceedingly thick, and serve to fill up the concavities on their front surface, and to make the anterior surface of the spine more even. This ligament is composed of several layers of fibres, which vary in their length, but are closely interlaced with each other. The most superficial or longest fibres extend between four or five vertebrae. A second subjacent set extend between two or three vertebrae; whilst a third set, the shortest and deepest, extend from one vertebra to the next. At the sides of the bodies, this ligament consists of a few short fibres, which pass from one vertebra to the next, separated from the median portion by large oval apertures, for the passage of vessels.

The *Posterior Common Ligament* is situated within the spinal canal, and extends along the posterior surface of the bodies of the vertebrae, from the body of the axis above, where it is continuous with the occipito-axoid ligament, to the sacrum below. It is broader at the upper than at the lower part of the spine, and thicker in the dorsal than in the cervical or lumbar regions. In the situation of the intervertebral substance and contiguous margins of the vertebrae, where the ligament is more intimately adherent, it is broad, and presents a series of dentations with intervening concave margins; but it is narrow and thick over the centre of the bodies, from which it is separated by the *venae basis vertebrae*. This ligament is composed of smooth, shining, longitudinal fibres, denser and more compact than the anterior ligament, and composed of a superficial layer occupying the interval between three or four vertebrae, and of a deeper layer, which extends between one vertebra and the next adjacent to it. It is separated from the dura mater of the spinal cord by some loose filamentous tissue, very liable to serous infiltration.

The *Intervertebral Substance* (fig. 99) is a lenticular disc of fibro-cartilage, interposed between the adjacent surfaces of the bodies of the vertebrae, from the axis to the base of the sacrum. These discs vary in shape, size, and thickness, in different parts of the spine. In *shape* they accurately correspond with the surfaces of the bodies between which they are placed, being oval in the cervical and lumbar regions, circular in the dorsal. Their *size* is greatest in the lumbar region. In *thickness* they vary not only in the different regions of the spine, but in different parts of the same region: thus, they are uniformly thick in the lumbar region; thickest in front in the cervical and lumbar regions which are convex forwards, and behind, to a slight extent, in the dorsal region.

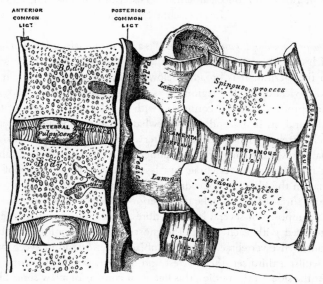

99.—Vertical Section of two Vertebræ and their Ligaments, from the Lumbar Region.

They thus contribute, in a great measure, to the curvatures of the spine in the neck and loins; whilst the concavity of the dorsal region is chiefly due to the shape of the bodies of the vertebrae. The intervertebral discs form about one-fourth of the spinal column, exclusive of the first two vertebrae; they are not equally distributed, however, between the various bones; the dorsal portion of the spine having, in proportion to its length, a much smaller quantity than in the cervical and lumbar regions, which necessarily gives to the latter parts greater pliancy and freedom of movement. The intervertebral discs are adherent, by their surfaces, to the adjacent parts of the bodies of the vertebrae; and by their circumference are closely connected in front to the anterior, and behind to the posterior common ligament; whilst in the dorsal region they are connected laterally to the heads of those ribs which articulate with two vertebrae, by means of the inter-articular ligament; they consequently form part of the articular cavities in which the heads of these bones are received.

The intervertebral substance is composed, at its circumference, of laminae of fibrous tissue and fibro-cartilage; and at its centre of a soft, elastic, pulpy matter. The laminae are arranged concentrically one within the other, with their edges turned towards the corresponding surfaces of the vertebrae, and consist of alternate plates of fibrous tissue and fibro-cartilage. These plates are not quite vertical in their direction, those near the circumference being curved outwards and closely approximated; whilst those nearest the centre curve in the opposite direction, and are somewhat more widely separated. The fibres of which each plate is composed, are directed, for the most part, obliquely from above downwards; the fibres of an adjacent plate have an exactly opposite arrangement, varying in their direction in every layer; whilst in some few they are horizontal. This laminar arrangement belongs to about the outer half of each disc, the central part being occupied by a soft, pulpy, highly elastic substance, of a yellowish colour, which rises up considerably above the surrounding level, when the disc is divided horizontally. This substance presents no concentric arrangement, and consists of white fibrous tissue, having interspersed cells of variable shape and size. The pulpy matter is separated from immediate contact with the vertebrae, by the interposition of thin plates of cartilage.

2. Ligaments connecting the Laminae
Ligamenta Subflava.

The *Ligamenta Subflava* are interposed between the laminae of the vertebrae, from the axis to the sacrum. They are most distinct when seen from the inner surface of the spine; when viewed from the outer surface, they appear short, being overlapped by the laminae. Each ligament consists of two lateral portions, which commence on each side at the root of either articular process, and pass backwards to the point where the laminae converge to form the spinous process, where their margins are thickest, and separated by a slight interval, filled up with areolar tissue. These ligaments consist of yellow elastic tissue, the fibres of which, almost perpendicular in direction, are attached to the anterior surface of the margin of the lamina above, and to the posterior surface, as well as to the margin of the lamina below. In the cervical region, they are thin in texture, but very broad and long; they become thicker in the dorsal region; and in the lumbar acquire very considerable thickness. Their highly elastic property serves to preserve the upright posture, and to counteract the efforts of the Flexor muscles of the spine. These ligaments do not exist between the occiput and atlas, or between the atlas and axis.

3. Ligaments connecting the Articular Processes
Capsular.   Synovial Membranes.

The *Capsular Ligaments* are thin and loose bags of ligamentous fibre attached to the contiguous margins of the articulating processes of each vertebra, through the greater part of their circumference, and completed internally by the ligamenta subflava. They are longer and more loose in the cervical than in the dorsal or lumbar regions. The capsular ligaments are lined on their inner surface by a delicate synovial membrane.

4. Ligaments connecting the Spinous Processes
Inter-spinous.   Supra-spinous.

The *Inter-spinous Ligaments*, thin and membranous, are interposed between the spinous processes in the dorsal and lumbar regions. Each ligament extends from the root to near the summit of each spinous process, and connects together their adjacent margins. They are narrow and elongated in the dorsal region, broader, quadrilateral in form, and thicker in the lumbar region.

The *Supra-spinous Ligament* is a strong fibrous cord, which connects together the apices of the spinous processes from the seventh cervical to the spine of the sacrum. It is thicker and broader in the lumbar than in the dorsal region, and intimately blended, in both situations, with the neighbouring aponeuroses. The most superficial fibres of this ligament connect three or four vertebrae; those deeper seated pass between two or three vertebrae; whilst the deepest connect the contiguous extremities of neighbouring vertebrae.

5. Ligaments connecting the Transverse Processes
Inter-transverse.

The *Inter-transverse Ligaments* consist of a few thin scattered fibres, interposed between the transverse processes. They are generally wanting in the cervical region; in the dorsal, they are rounded cords; in the lumbar region, thin and membranous.

The two upper vertebrae, the Atlas and Axis, are connected together by ligaments distinct from those by which the rest are united.

2. Articulation of the Atlas with the Axis

The articulation of the anterior arch of the atlas with the odontoid process forms a lateral ginglymoid joint, whilst that between the articulating processes of the two bones forms a double arthrodia. The ligaments of this articulation are the

Two Anterior Atlo-Axoid.   Transverse.
Posterior Atlo-Axoid.   Two Capsular.
Four Synovial Membranes.

Of the *Two Anterior Atlo-Axoid Ligaments* (fig. 100), the most superficial is a rounded cord, situated in the middle line, attached, above, to the tubercle on the anterior arch of the atlas; below, to the base of the odontoid process and body of the axis. The deeper ligament is a membranous layer, attached, above, to the lower border of the anterior arch of the atlas; below, to the base of the odontoid process and body of the axis. These ligaments are in relation, in front, with the Recti antici majores.

The *Posterior Atlo-Axoid Ligament* (fig. 101) is a broad and thin membranous layer, attached, above, to the lower border of the posterior arch of the atlas; below, to the upper edge of the laminae of the axis. This ligament supplies the place of the ligamenta subflava, and is in relation, behind, with the Inferior oblique muscles.

The *Transverse Ligament* (figs. 102, 103) is a thick and strong ligamentous band, which arches across the ring of the atlas, and serves to retain the odontoid process in firm connection with its anterior arch. This ligament is flattened from before backwards, broader and thicker in the middle than at either extremity, and firmly attached on each side of the atlas to a small tubercle on the inner surface of each of its lateral masses. As it crosses the odontoid process, a small fasciculus is derived from its upper and lower borders; the former, passing upwards to be inserted into the basilar process of the occipital bone; the latter, downwards, to be attached to the root of the odontoid process: hence this ligament has received the name of *cruciform*. The transverse ligament divides the ring of the atlas into two unequal parts: of these, the posterior and larger serves for the transmission of the cord and its membranes; the anterior and smaller serving to

100.—Occipito-Atloid and Atlo-Axoid Ligaments. Front View.

101.—Occipito-Atloid and Atlo-Axoid Ligaments. Posterior View.

102.—Articulation between Odontoid Process and Atlas.

retain the odontoid process in its position. The lower border of the space formed between the atlas and transverse ligament being smaller than the upper (on account of the transverse ligament embracing firmly the narrow neck of the odontoid process), while the central part of the odontoid process is larger than its base; this process is still retained in firm connection with the anterior arch when all the other ligaments have been divided.

The *Capsular Ligaments* are two thin and loose capsules, connecting the articular surfaces of the atlas and axis, the fibres being strongest on the anterior and external part of the articulation.

There are *four Synovial Membranes* in this articulation. One lining the inner surface of each of the capsular ligaments: one between the anterior surface of the odontoid process and anterior arch of the atlas: and one between the posterior surface of the odontoid process and the transverse ligament. This synovial membrane often communicates with those between the condyles of the occipital bone and the articular surfaces of the atlas.

103.—Occipito-Axoid and Atlo-Axoid Ligaments. Posterior View.

Actions. This joint is capable of great mobility, and allows the rotation of the atlas, and, with it, of the cranium upon the axis, the extent of rotation being limited by means of the odontoid ligaments.

Articulation of the Spine with the Cranium

The ligaments connecting the spine with the cranium may be divided into two sets: Those connecting the occipital bone with the atlas; Those connecting the occipital bone with the axis.

3. Articulation of the Atlas with the Occipital Bone

This articulation is a double arthrodia. Its ligaments are the
>Two Anterior Occipito-Atloid.
>Posterior Occipito-Atloid.
>Two Lateral Occipito-Atloid.
>Two Capsular and Synovial Membranes.

Of the *Two Anterior Ligaments* (fig. 100), the most superficial is a strong, narrow, rounded cord, attached, above, to the basilar process of the occiput; below, to the tubercle on the anterior arch of the atlas: the deeper ligament is a broad and thin membranous layer, which passes between the anterior margin of the foramen magnum above, and the whole length of the upper border of the anterior arch of the atlas below. This ligament is in relation, in front, with the Recti antici minores; behind, with the odontoid ligaments.

The *Posterior Occipito-Atloid Ligament* (fig. 101) is a very broad but thin membranous lamina, intimately blended with the dura mater. It is connected, above, to the posterior margin of the foramen magnum; below, to the central part of the upper border of the posterior arch of the atlas. This ligament is incomplete at each side, and forms, with the superior intervertebral notches, an opening for the passage of the vertebral artery and sub-occipital nerve. It is in relation, behind, with the Recti postici minores and Obliqui superiores; in front, with the dura mater of the spinal canal, to which it is intimately adherent.

The *Lateral Ligaments* are strong bands of fibres, directed obliquely upwards and inwards, attached, above, to the jugular process of the occipital bone; below, to the base of the transverse process of the atlas.

The *Capsular Ligaments* surround the condyles of the occipital bone, and connect them with the articular surfaces of the atlas; they consist of thin and loose capsules, which enclose the synovial membrane of this articulation. The synovial membranes between the occipital bone and atlas communicate occasionally with that between the posterior surface of the odontoid process and transverse ligament.

Actions. The movements permitted in this joint are flexion and extension, which give rise to the ordinary forward or backward nodding of the head, besides slight lateral motion to one or the other side. When either of these actions is carried beyond a slight extent, the whole of the cervical portion of the spine assists in its production.

4. Articulation of the Axis with the Occipital Bone
Occipito-Axoid. Three Odontoid.

To expose these ligaments, the spinal canal should be laid open by removing the posterior arch of the atlas, the laminae and spinous process of the axis, and that portion of the occipital bone behind the foramen magnum, as seen in fig. 103.

The *Occipito-Axoid Ligament* (Apparatus ligamentosus colli) is situated at the upper part of the front surface of the spinal canal. It is an exceedingly broad and strong ligamentous band, which covers the odontoid process and its ligaments, and appears to be a prolongation upwards of the posterior common ligament of the spine. It is attached, below, to the posterior surface of the body of the axis, and becoming broader and expanded as it ascends, is inserted into the basilar groove of the occipital bone, in front of the foramen magnum.

Relations. By its anterior surface, it is intimately connected with the transverse ligament; by its posterior surface, with the dura mater. By dividing this ligament transversely across, and turning its ends aside, the transverse and odontoid ligaments are exposed.

The *Odontoid* or *Check Ligaments* are strong rounded fibrous cords, which arise one on either side of the apex of the odontoid process, and passing obliquely upwards and outwards, are inserted into the rough depressions on the inner side of the condyles of the occipital bone. In the triangular interval left between these ligaments and the margin of the foramen magnum, a third strong ligamentous band (ligamentum suspensorium) may be seen, which passes almost perpendicularly from the apex of the odontoid process to the anterior margin of the foramen, being intimately blended with the anterior occipito-atloid ligament.

Actions. The odontoid ligaments serve to limit the extent to which rotation of the cranium may be carried; hence they have received the name of *check ligaments*.

104.—Temporo-Maxillary Articulation. External View.

5. Temporo-Maxillary Articulation

This articulation is a double arthrodia. The parts entering into its formation are, on each side, the anterior part of the glenoid cavity of the temporal bone and the eminentia articularis above; with the condyle of the lower jaw below. The ligaments are the following.

External Lateral.	Capsular.
Internal Lateral.	Inter-articular Fibro-cartilage.
Stylo-maxillary.	Two Synovial Membranes.

The *External Lateral Ligament* (fig. 104) is a short, thin, and narrow fasciculus, attached above to the outer surface of the zygoma and to the rough tubercle on its lower border; below, to the outer surface and posterior border of the neck of the lower jaw. This ligament is a little broader above than below; its fibres are placed parallel with one another, and directed obliquely downwards and backwards. Externally, it is covered by the parotid gland and by the integument. Internally, it is in relation with the inter-articular fibro-cartilage and the synovial membranes.

105.—Temporo-Maxillary Articulation. Internal View.

The *Internal Lateral Ligament* (fig. 105) is a long, thin, and loose band, attached above by its narrow extremity to the spinous process of the sphenoid bone, and becoming broader as it descends, is inserted into the inner margin of the dental foramen. Its outer surface is in relation above with the External pterygoid muscle; lower down it is separated from the neck of the condyle by the internal maxillary artery; and still more inferiorly the inferior dental vessels and nerve separate it from the ramus of the jaw. Internally it is in relation with the Internal pterygoid.

The *Stylo-maxillary Ligament* is a thin aponeurotic cord, which extends from near the apex of the styloid process of the temporal bone, to the angle and posterior border of the ramus of the lower jaw, between the Masseter and Internal pterygoid muscles. This ligament separates the parotid from the sub-maxillary gland, and has attached to its inner side part of the fibres of origin of the Stylo-glossus muscle. Although usually classed among the ligaments of the jaw, it can only be considered as an accessory in the articulation.

The *Capsular Ligament* consists of a thin and loose ligamentous capsule, attached above to the circumference of the glenoid cavity and the articular surface immediately in front; below, to the neck of the condyle of the lower jaw. It consists of a few thin scattered fibres, and can hardly be considered as a distinct ligament; it is thickest at the back part of the articulation.

The *Inter-articular Fibro-cartilage* (fig. 106) is a thin plate of a transversely oval form, placed horizontally between the condyle of the jaw and the glenoid cavity. Its upper surface is concave from before backwards, and a little convex transversely, to accommodate itself to the form of the glenoid cavity. Its under surface, where it is in contact with the condyle, is concave. Its circumference is connected externally to the external lateral ligament; internally, to the capsular ligament; and in front to the tendon of the External pterygoid muscle. It is thicker at its circumference, especially behind, than at its centre, where it is sometimes perforated. The fibres of which it is composed have a concentric arrangement, more apparent at the circumference than at the centre. Its surfaces are smooth, and divide the joint into two cavities, each of which is furnished with a separate synovial membrane. When the fibro-cartilage is perforated, the synovial membranes are continuous with one another.

106.—Vertical Section of Temporo-Maxillary Articulation.

The *Synovial Membranes*, two in number, are placed one above, and the other below the fibro-cartilage. The upper one, the larger and looser of the two, after lining the cartilage covering the glenoid cavity and eminentia articularis, is continued over the upper surface of the fibro-cartilage. The lower one is interposed between the under surface of the fibro-cartilage and the condyle of the jaw, being prolonged downwards a little further behind than in front.

Actions. The movements permitted in this articulation are very extensive. Thus the jaw may be depressed or elevated, or it may be carried forwards or backwards, or from side to side. It is by the alternation of these movements performed in succession, that a kind of rotatory movement of the lower jaw upon the upper takes place, which materially assists in the mastication of the food.

If the movement of depression is carried only to a slight extent, the condyles remain in the glenoid cavities, their anterior part descending only to a slight extent, but if depression is considerable, the condyles glide from the glenoid fossae on to the eminentia articularis, carrying with them the inter-articular fibro-cartilages. When the jaw is elevated, the condyles and fibro-cartilages are carried backwards into their original position. When the jaw is carried forwards or backwards, a horizontal gliding movement of the fibro-cartilages and condyles upon the glenoid cavities takes place in the antero-posterior direction; whilst in the movement from side to side, this occurs in the lateral direction.

6. Articulation of the Ribs with the Vertebrae

The articulation of the ribs with the vertebral column, may be divided into two sets. 1. Those which connect the heads of the ribs with the bodies of the vertebrae 2. Those which connect the neck and tubercle of the ribs with the transverse processes.

1. Articulation between the Heads of the Ribs and the Bodies of the Vertebrae

These form a series of angular ginglymoid joints, connected together by the following ligaments:—

Anterior Costo-vertebral or Stellate.
Capsular.
Inter-articular.
Two Synovial Membranes.

107.—Costo-vertebral and Costo-transverse Ariculations. Anterior View.

The *Anterior Costo-vertebral* or *Stellate Ligament* (fig. 107) connects the anterior part of the head of each rib, with the sides of the bodies of the vertebrae, and the intervening intervertebral disc. It consists of three flat bundles of ligamentous fibres, which radiate from the anterior part of the head of the rib. The superior fasciculus passes upwards to be connected with the body of the vertebra above; the inferior one descends to the body of the vertebra below; and the middle one, the smallest and least distinct, passes horizontally inwards to be attached to the intervertebral substance.

Relations. In front with the thoracic ganglia of the sympathetic, the pleura, and on the right side, the vena azygos major; behind, with the inter-articular ligament and synovial membranes.

In the first rib, which articulates with a single vertebra only, this ligament does not present a distinct division into three fasciculi; its superior fibres, however pass to be attached to the body of the last cervical vertebra, as well as to that of the vertebra with which the rib articulates. In the eleventh and twelfth ribs, which also articulate with a single vertebra, the same division does not exist, but the superior fibres of the ligament, in each case, are connected with the vertebra above, as well as that with which the ribs articulate.

The *Capsular Ligament* is a thin and loose ligamentous bag, which surrounds the joint between the head of the rib and the articular cavity formed by the junction of the vertebrae. It is very thin, firmly connected with the anterior ligament, and most distinct at the upper and lower parts of the articulation.

The *Inter-articular Ligament* is situated in the interior of the articulation. It consists of a short band of fibres, flattened from above downwards, attached by one extremity to the sharp crest on the head of the rib, and by the other to the intervertebral disc. It divides the joint into two cavities, which have no communication with one another, and are each lined by a separate synovial membrane. In the first, eleventh, and twelfth ribs, the inter-articular ligament does not exist, consequently there is but one synovial membrane.

Actions. The movements permitted in these articulations are limited to elevation, depression, and slightly forwards and backwards. This movement varies however, very much in its extent in different ribs. The first rib is almost entirely immovable, excepting in strong and violent inspirations. The movement of the second rib is also not very extensive. In the other ribs, their mobility increases successively to the two last, which are very moveable. The ribs are generally more moveable in the female than in the male.

2. Articulation between the Neck and Tubercle of the Ribs with the Transverse Processes

The ligaments connecting these parts are—
Anterior Costo-Transverse.
Middle Costo-Transverse (Interosseous).
Posterior Costo-Transverse.
Capsular and Synovial Membrane.

108.—Costo-Transverse Articulation. Seen from above.

The *Anterior Costo-Transverse Ligament* (fig. 108) is a broad and strong band of fibres, attached below to the sharp crest on the upper border of the neck of each rib, and passing obliquely upwards and outwards, to the lower border of the transverse process immediately above. It is broader below than above, broader and thinner between the lower ribs than between the upper, and more distinct in front than behind. This ligament is in relation, in front, with the intercostal vessels and nerves; behind, with the Longissimus dorsi. Its *internal border* completes an aperture formed between it and the articular processes, through which pass the posterior branches of the intercostal vessels and nerve. Its *external border* is continuous with a thin aponeurosis, which covers the External intercostal muscle.

The *first* and *last ribs* have no anterior costo-transverse ligament.

The *Middle Costo-Transverse* or *Interosseous Ligament* consists of short, but strong, fibres, which pass between the rough surface on the posterior part of the neck of each rib, and the anterior surface of the adjacent transverse process. In order fully to expose this ligament, a horizontal section should be made across the transverse process and corresponding part of the rib; or the rib may be forcibly separated from the transverse process, and its fibres torn asunder.

In the *eleventh* and *twelfth ribs*, this ligament is quite rudimentary.

The *Posterior Costo-Transverse Ligament* is a short, but thick and strong, fasciculus, which passes obliquely from the summit of the transverse process to the rough non-articular portion of the tubercle of the rib. This ligament is shorter and more oblique in the upper, than in the lower ribs. Those corresponding to the superior ribs ascend, and those of the inferior ones slightly descend.

In the *eleventh* and *twelfth ribs*, this ligament is wanting.

The articular portion of the tubercle of the rib, and adjacent transverse process, form an arthrodial joint, provided with a thin *capsular ligament* attached to the circumference of the articulating surfaces, and enclosing a small *synovial membrane*.

In the *eleventh* and *twelfth ribs*, this articulation is wanting.

Actions. The movement permitted in these joints, is limited to a slight gliding motion of the articular surfaces one upon the other.

7. Articulation of the Cartilages of the Ribs with the Sternum

The articulation of the cartilages of the true ribs with the sternum are arthrodial joints. The ligaments connecting them are—

    Anterior Costo-Sternal.
    Posterior Costo-Sternal.
    Capsular.
    Synovial Membranes.

The *Anterior Costo-Sternal Ligament* (fig. 109) is a broad and thin membranous band that radiates from the inner extremity of the cartilages of the true ribs, to the anterior surface of the sternum. It is composed of fasciculi, which pass in different directions. The *superior* fasciculi ascend obliquely, the *inferior* pass obliquely downwards, and the *middle* fasciculi horizontally. The superficial fibres of this ligament are the longest; they intermingle with the fibres of the ligaments above and beneath them, with those of the opposite side, and with the tendinous fibres of origin of the Pectoralis major; forming a thick fibrous membrane, which covers the surface of the sternum, but is more distinct at the lower than at the upper part.

The *Posterior Costo-Sternal Ligament*, less thick and distinct than the anterior, is composed of fibres which radiate from the posterior surface of the sternal end of the cartilages of the true ribs, to the posterior surface of the sternum, becoming blended with the periosteum.

The *Capsular Ligament* surrounds the joints formed between the cartilages of the true ribs and the sternum. It is very thin, intimately blended with the anterior and posterior ligaments, and strengthened at the upper and lower part of the articulation by a few fibres, which pass from the cartilage to the side of the sternum. These ligaments protect the synovial membranes.

Synovial Membranes. The cartilage of the *first rib* is directly continuous with

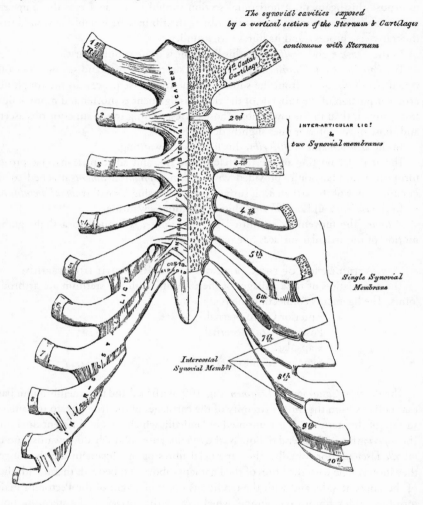

109.— Costo-Sternal, Costo-Xiphoid, and Inter-costal Articulations. Anterior View.

the sternum, the synovial membrane being absent. The cartilage of the *second rib* articulates with the sternum by means of an inter-articular ligament, attached by one extremity to the ridge which separates the two articular facets of the cartilage of the second rib, and by the other extremity to the cartilage which unites the first and second pieces of the sternum. This articulation is provided with two synovial membranes. That of the third rib has also two synovial membranes; and that of the fourth, fifth, sixth, and seventh, each a single synovial membrane. These synovial membranes may be demonstrated by removing a thin section from the anterior surface of the sternum and cartilages, as seen in the figure. After middle life, the articular surfaces lose their polish, become roughened, and the synovial membranes appear to be wanting. In old age, the articulations do not exist, the cartilages of most of the ribs becoming firmly united to the sternum. The cartilage of the *seventh rib*, and occasionally also that of the *sixth*, is connected to the anterior surface of the ensiform appendix, by a band of ligamentous fibres, which varies in length and breadth in different subjects. It is called the *costo-xiphoid ligament*.

Actions. The movements which are permitted in the costo-sternal articulations, are limited to elevation and depression; and these only to a slight extent.

Articulation of the Cartilages of the Ribs with each other

The cartilages of the sixth, seventh, and eighth ribs articulate, by their lower borders, with the corresponding margin of the adjoining cartilages, by means of a small, smooth, oblong-shaped facet. Each articulation has a perfect *synovial membrane* enclosed in a thin *capsular ligament*, strengthened externally and internally by some ligamentous fibres (intercostal ligaments), which pass from one cartilage to the other, and which are intimately united to the perichondrium. Sometimes the cartilage of the fifth rib, more rarely that of the ninth, articulate, by their lower borders, with the corresponding cartilages by small oval facets; more frequently they are connected together by a few ligamentous fibres. Occasionally, the articular surfaces above mentioned are found wanting.

Articulation of the Ribs with their Cartilages

The outer extremity of each costal cartilage is received into a depression in the sternal end of the ribs, and held together by the periosteum.

8. Ligaments of the Sternum

The first and second pieces of the Sternum are united by a layer of cartilage which rarely ossifies, except at an advanced period of life. These two segments are connected by an anterior and posterior ligament.

The *anterior sternal ligament* consists of a layer of fibres, having a longitudinal direction; they blend with the fibres of the anterior costo-sternal ligaments on both sides, and with the aponeurosis of origin of the Pectoralis major. This ligament is rough, irregular, and much thicker at the lower than at the upper part of this bone.

The *posterior sternal ligament* is disposed in a somewhat similar manner on the posterior surface of the articulation.

9. Articulation of the Pelvis with the Spine

The ligaments connecting the last lumbar vertebra with the sacrum are similar to those which connect the segments of the spine with each other, viz. 1. The continuation downwards of the anterior and posterior common ligaments. 2. The inter-vertebral substance connecting the flattened oval surfaces of the two bones, thus forming an amphiarthrodial joint. 3. Ligamenta subflava, connecting the arch of the last lumbar vertebra with the posterior border of the sacral canal. 4. Capsular ligaments and synovial membranes connecting the articulating processes and forming a double arthrodia. 5. Inter- and supra-spinous ligaments.

The two proper ligaments connecting the pelvis with the spine are the lumbo-sacral amd lumbo-iliac.

The *Lumbo-sacral Ligament* (fig. 110) is a short, thick, triangular fasciculus, connected above to the lower and front part of the transverse process of the last lumbar vertebra, and passing obliquely outwards, is attached below to the lateral surface of the base of the sacrum; becoming blended with the anterior sacro-iliac ligament. This ligament is in relation anteriorly with the Psoas muscle.

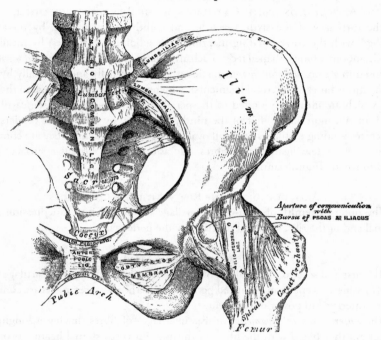

110.—Articulations of Pelvis and Hip. Anterior View.

The *Lumbo-iliac Ligament* (fig. 110) passes horizontally outwards from the apex of the transverse process of the last lumbar vertebra, to that portion of the crest of the ilium immediately in front of the sacro-iliac articulation. It is of a triangular form, thick and narrow internally, broad and thinner externally; and is in relation, in front, with the Psoas muscle; behind, with the muscles occupying the vertebral groove; above, with the Quadratus lumborum.

10. Articulations of the Pelvis

The Ligaments connecting the bones of the pelvis with each other may be divided into four groups. 1. Those connecting the sacrum and ilium. 2. Those passing between the sacrum and ischium. 3. Those connecting the sacrum and coccyx. 4. Those between the two pubic bones.

1. Articulation of the Sacrum and Ilium

The sacro-iliac articulation is an amphiarthrodial joint, formed between the lateral surfaces of the sacrum and ilium. The anterior or auricular portion of the articular surfaces is covered with a thin plate of cartilage, thicker on the sacrum than on the ilium. The surfaces of these cartilages in the adult are rough and irregular, and separated from one another by a soft yellow pulpy substance. At an early period of life, occasionally in the adult, and in the female during pregnancy, they are smooth and lined by a delicate synovial membrane. The ligaments connecting these surfaces are the anterior and posterior sacro-iliac.

The *Anterior Sacro-iliac Ligament* consists of numerous thin ligamentous bands, which connect the anterior surfaces of the sacrum and ilium.

The *Posterior Sacro-Iliac* (fig. 111) is a strong interosseous ligament, situated in the deep depression between the sacrum and ilium behind, and forming the chief bond of connexion between these bones. It consists of numerous strong fasciculi, which pass between the bones in various directions. Three of these are of large size; the *two superior*, nearly horizontal in direction, arise from the first and second transverse tubercles on the posterior surface of the sacrum, and are inserted into the rough uneven surface at the posterior part of the inner surface of the ilium. The third fasciculus, oblique in direction, is attached by one extremity to the third or fourth transverse tubercle on the posterior surface of the sacrum, and by the other to the posterior superior spine of the ilium; it is sometimes called the *oblique sacro-iliac ligament*.

111.—Articulations of Pelvis and Hip. Posterior View.

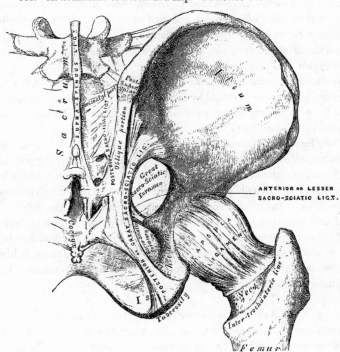

2. Articulation of the Sacrum and Ischium
The Great Sacro-Sciatic (Posterior).
The Lesser Sacro-Sciatic (Anterior).

The *Great* or *Posterior Sacro-Sciatic Ligament* is situated at the posterior and inferior part of the pelvis. It is thin, flat, and triangular in form; narrower in the middle than at the extremities; attached by its broad base to the posterior inferior spine of the ilium, to the third and fourth transverse tubercles on the sacrum, and to the lower part of the lateral margin of that bone and the coccyx; passing obliquely downwards, outwards, and forwards, it becomes narrow and thick; and at its insertion into the inner margin of the tuberosity, it increases in breadth, and is prolonged forwards along the inner margin of the ramus forming the falciform ligament. The free concave edge of this ligament has attached to it the obturator fascia, with which it forms a kind of groove, protecting the internal pudic vessels and nerve. One of its surfaces is turned towards the perinaeum, the other towards the Obturator internus muscle.

The *posterior surface* of this ligament gives origin, by its whole extent, to fibres of the Gluteus maximus. Its *anterior surface* is united to the lesser sacro-sciatic ligament. Its *superior border* forms the lower boundary of the lesser sacro-sciatic foramen. Its *lower border* forms part of the boundary of the perinaeum. This ligament is pierced by the coccygeal branch of the sciatic artery.

The *Lesser* or *Anterior Sacro-Sciatic Ligament*, much shorter and smaller than the preceding, is thin, triangular in form, attached by its apex to the spine of the ischium, and internally, by its broad base, to the lateral margins of the sacrum and coccyx, anterior to the attachment of the great sacro-sciatic ligament, with which its fibres are intermingled.

It is in relation, *anteriorly*, with the Coccygeus muscle; *posteriorly*, it is covered by the posterior ligament, and crossed by the pudic vessels and nerves. Its *superior border* forms the lower boundary of the great sacro-sciatic foramen. Its *inferior border*, part of the lesser sacro-sciatic foramen.

These two ligaments convert the sacro-sciatic notches into foramina. The *superior* or *larger* sacro-sciatic foramen is bounded, in front and above, by the posterior border of the os innominatum; behind, by the great sacro-sciatic ligament; and below, by the lesser ligament. It is partially filled up, in the recent state, by the Pyriformis muscle. Above this muscle, the gluteal vessels and nerve emerge from the pelvis; and below it, the ischiatic vessels and nerves, the internal pudic vessels and nerve, and the nerve to the Obturator internus. The *inferior* or *smaller* sacro-sciatic foramen is bounded, in front, by the tuber ischii; above, by the spine and lesser ligament; behind, by the greater ligament. It transmits the tendon of the Obturator internus muscle, its nerve, and the pudic vessels and nerve.

3. Articulation of the Sacrum and Coccyx

This articulation is an amphiarthrodial joint, formed between the oval surface on the summit of the sacrum, and the base of the coccyx. It is analogous to the joints between the bodies of the vertebrae, and is connected by similar ligaments. They are the

Anterior Sacro-Coccygeal.
Posterior Sacro-Coccygeal.
Inter-articular Fibro-Cartilage.

The *Anterior Sacro-Coccygeal Ligament* consists of a few irregular fibres, which descend from the anterior surface of the sacrum to the front of the coccyx, becoming blended with the periosteum.

The *Posterior Sacro-Coccygeal Ligament* is a flat band of ligamentous fibres, of a pearly tint, which arises from the margin of the lower orifice of the sacral canal, and descends to be inserted into the posterior surface of the coccyx. This ligament completes the lower and back part of the sacral canal. Its superficial fibres are much longer than the deep-seated; the latter extend from the apex of the sacrum to the upper cornua of the coccyx. Anteriorly, it is in relation with the arachnoid membrane of the sacral canal, a portion of the sacrum, and almost the whole of the posterior surface of the coccyx; posteriorly, with some aponeurotic fibres from the Gluteus maximus.

An *Inter-articular Fibro-Cartilage* is interposed between the contiguous surfaces of the sacrum and coccyx; it differs from that interposed between the bodies of the vertebrae, in being thinner, and its central part more firm in texture. It is somewhat thicker in front and behind, than at the sides. Occasionally a synovial membrane is found where the coccyx is freely movable, which is more especially the case during pregnancy.

The different segments of the coccyx are connected together by an extension downwards of the anterior and posterior sacro-coccygeal ligaments, a thin annular disc of fibro-cartilage being interposed between each of the bones. In the adult male, all the pieces become ossified; but in the female, this does not commonly occur until a later period of life. The separate segments of the coccyx are first united, and at a more advanced age the joint between the sacrum and the coccyx.

Actions. The movements which take place between the sacrum and coccyx, and between the different pieces of the latter bone, are slightly forwards and backwards; they are very limited. Their mobility increases during pregnancy.

4. Articulation of the Pubes

The articulation between the ossa pubis is an amphiarthrodial joint, formed by the junction of the two oval surfaces which have received the name of the *symphysis*. The ligaments of this articulation are the

Anterior Pubic.   Posterior Pubic.
Superior Pubic.   Sub-Pubic.
Inter-articular Fibro-Cartilage.

The *Anterior Pubic Ligament* consists of several superimposed layers, which pass across the anterior surface of the articulation. The superficial fibres pass obliquely from one bone to the other, decussating and forming an interlacement with the fibres of the aponeurosis of the External oblique muscle. The deep fibres pass transversely across the symphysis, and are blended with the interarticular fibro-cartilage.

The *Posterior Pubic Ligament* consists of a few thin, scattered fibres, which unite the two pubic bones posteriorly.

The *Superior Pubic Ligament* is a band of fibres, which connects together the two pubic bones superiorly.

The *Sub-Pubic Ligament* is a thick, triangular arch of ligamentous fibres, connecting together the two pubic bones below, and forming the upper boundary of the pubic arch. Above, it is blended with the inter-articular fibro-cartilage; laterally, with the rami of the pubes. Its fibres are of a yellowish colour, closely connected, and have an arched direction.

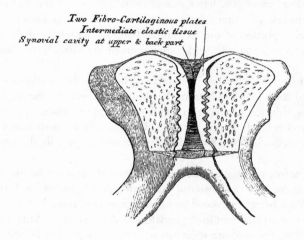

112.—Vertical Section of the Symphysis Pubis. Made near its Posterior Surface.

Two Fibro-Cartilaginous plates
Intermediate elastic tissue
Synovial cavity at upper & back part

The *Inter-articular Fibro-Cartilage* consists of two oval-shaped plates, one covering the surface of each symphysis pubis. They vary in thickness in different subjects, and project somewhat beyond the level of the bones, especially behind. The outer surface of each is firmly connected to the bone by a series of nipple-like processes, which accurately fit within corresponding depressions on the osseous surface. Their opposed surfaces are connected, in the greater part of their extent, by an intermediate fibrous elastic-tissue; and by their circumference to the various ligaments surrounding the joint. An interspace is left between the two plates at the upper and back part of the articulation, where the fibrous-tissue is deficient, and the surface of the fibro-cartilage lined by epithelium. This space is found at all periods of life, both in the male and female; but it is larger in the latter, especially during pregnancy, and after parturition. It is most frequently limited to the upper and back part of the joint; but it occasionally reaches to the front, and may extend the entire length of the cartilages. This structure may be easily demonstrated, by making a vertical section of the symphysis pubis near its posterior surface.

The *Obturator Ligament* is a dense membranous layer, consisting of fibres which interlace in various directions. It is attached to the circumference of the obturator foramen, which it closes completely, except at its upper and outer part, where a small oval canal is left for the passage of the obturator vessels and nerve. It is in relation, in front, with the Obturator externus; behind, with the Obturator internus; both of which muscles are in part attached to it.

113.—Sterno-Clavicular Articulation. Anterior View.

ARTICULATIONS OF THE UPPER EXTREMITY

The articulations of the Upper Extremity may be arranged into the following groups:—1. Sterno-clavicular articulation. 2. Scapulo-clavicular articulation. 3. Ligaments of the Scapula. 4. Shoulder-joint. 5. Elbow-joint. 6. Radio-ulnar articulation. 7. Wrist-joint. 8. Articulation of the Carpal bones. 9. Carpo-metacarpal articulation. 10. Metacarpo-phalangeal articulation. 11. Articulation of the Phalanges.

1. STERNO-CLAVICULAR ARTICULATION

The *Sterno-Clavicular* is an arthrodial joint. The parts entering into its formation are the sternal end of the clavicle, the upper and lateral part of the first piece of the sternum, and the cartilage of the first rib. The articular surface of the clavicle is much longer than that of the sternum, and invested with a layer of cartilage, which is considerably thicker than that on the latter bone. The ligaments of this joint are the

Anterior Sterno-Clavicular.	Costo-Clavicular (rhomboid).
Posterior Sterno-Clavicular.	Inter-Articular Fibro-Cartilage.
Inter-Clavicular.	Two Synovial Membranes.

The *Anterior Sterno-Clavicular Ligament* is a broad band of ligamentous fibres, which covers the anterior surface of the articulation, being attached, above, to the upper and front part of the inner extremity of the clavicle; and, passing obliquely downwards and inwards, is attached, below, to the front and upper part of the first piece of the sternum. This ligament is covered anteriorly by the sternal portion of the Sterno-cleido-mastoid and the integument; behind, it is in relation with the inter-articular fibro-cartilage and the two synovial membranes.

The *Posterior Sterno-Clavicular Ligament* is a broad band of fibres, which covers the posterior surface of the articulation, being attached, above, to the posterior part of the inner extremity of the clavicle; and, passing obliquely downwards and inwards, to be connected, below, to the posterior and upper part of the sternum. It is in relation, in front, with the inter-articular fibro-cartilage and synovial membranes; behind, with the Sterno-hyoid and Sterno-thyroid muscles.

The *Inter-Clavicular Ligament* is a flattened ligamentous band, which varies considerably in form and size in different individuals; it passes from the superior part of the inner extremity of one clavicle to the other, and is closely attached to the upper margin of the sternum. It is in relation, in front, with the integument; behind, with the Sterno-thyroid muscles.

The *Costo-Clavicular Ligament* (*rhomboid*) is a short, flat, and strong band of ligamentous fibres of a rhomboid form, attached, below, to the upper and inner part of the cartilage of the first rib; and, ascending obliquely backwards and outwards, to be attached, above, to the rhomboid depression on the under surface of the inner extremity of the clavicle. It is in relation, in front, with the tendon of origin of the Subclavius; behind, with the subclavian vein.

The *Inter-articular Fibro-Cartilage* is a flat and nearly circular disc, interposed between the articulating surfaces of the sternum and clavicle. It is attached above, to the upper and posterior border of the clavicle; below, to the cartilage of the first rib, at its junction with the sternum; and by its circumference to the anterior and posterior sterno-clavicular ligaments. It is thicker at the circumference, especially its upper and back part, than at its centre, or below. It divides the joint into two cavities, each of which is furnished with a separate synovial membrane; when the fibro-cartilage is perforated, which not unfrequently occurs, the synovial membranes communicate.

Of the *two Synovial Membranes* found in this articulation, one is reflected over the sternal end of the clavicle, the adjacent surface of the fibro-cartilage, and cartilage of the first rib; the other is placed between the articular surface of the sternum and adjacent surface of the fibro-cartilage; the latter is the more loose of the two. They seldom contain much synovia.

Actions. This articulation is the centre of the movements of the shoulder, and admits of motion in nearly every direction—upwards, downwards, backwards, forwards, as well as circumduction; the sternal end of the clavicle and the inter-articular cartilage gliding on the articular surface of the sternum.

2. Scapulo-Clavicular Articulation

The *Scapulo-Clavicular* is an arthrodial joint, formed between the outer extremity of the clavicle, and the upper edge of the acromion process of the scapula. Its ligaments are the

Superior Acromio-Clavicular.
Inferior Acromio-Clavicular.
Coraco-Clavicular { Trapezoid and Conoid.
Inter-articular Fibro-Cartilage.
Two Synovial Membranes.

The *Superior Acromio-Clavicular Ligament* is a broad band of fibres, of a quadrilateral form, which covers the superior part of the articulation, extending between the upper part of the outer end of the clavicle, and the superior part of the acromion. It is composed of parallel fibres, which interlace, above, with the aponeurosis of the Trapezius and Deltoid muscles; below, it is in contact with the inter-articular fibro-cartilage and synovial membranes.

The *Inferior Acromio-Clavicular Ligament*, somewhat thinner than the preceding, covers the inferior part of the articulation, and is attached to the adjoining surfaces of the two bones. It is in relation, above, with the inter-articular fibro-cartilage (when it exists) and the synovial membranes; below, with the tendon of the Supra-spinatus. These two ligaments are continuous with each other in front and behind, and form a complete capsule around the joint.

The *Coraco-Clavicular Ligament* serves to connect the clavicle with the coracoid process of the scapula. It consists of two distinct fasciculi, which have received separate names.

The *Trapezoid Ligament*, the anterior and external fasciculus, is a broad, thin, quadrilateral-shaped band of fibres, placed obliquely between the acromion process and the clavicle. It is attached, below, to a rough line at the inner and back part of the upper surface of the coracoid process; above, to the oblique line on the under surface of the clavicle. Its anterior border is free; its posterior is joined with the conoid ligament, forming by their junction a projecting angle.

114.—The Left Shoulder-Joint, Scapulo-Clavicular Articulations, and Proper Ligaments of Scapula.

The *Conoid Ligament*, the posterior and internal fasciculus, is a dense band of fibres, conical in form, the base being turned upwards, the summit downwards. It is attached by its apex to a rough depression at the anterior and inner side of the base of the coracoid process, internal to the preceding; above, by its expanded base, to the rough tubercle on the under surface of the clavicle. These ligaments are in relation, in front, with the Subclavius; behind, with the Trapezius: they serve to limit rotation of the scapula forwards and backwards.

The *Inter-articular Fibro-Cartilage* is most frequently absent in this articulation. When it exists, it generally only partially separates the articular surfaces, and occupies the upper part of the articulation. More rarely, it completely separates this joint into two cavities.

There are *two Synovial Membranes* where a complete inter-articular cartilage exists; more frequently there is only one synovial membrane.

Actions. The movements of this articulation are of two kinds. 1. A gliding motion of the articular end of the clavicle on the acromion. 2. Rotation of the scapula forwards and backwards upon the clavicle, the extent of this rotation being limited by the two portions of the coraco-clavicular ligament.

3. Proper Ligaments of the Scapula

The proper ligaments of the scapula are the

Coraco-acromial.   Transverse (Coracoid).

The *Coraco-acromial Ligament* is a broad, thin, and flat band, of a triangular shape, extended transversely across the upper part of the shoulder-joint, between the coracoid process and the acromion. It is attached by its apex to the summit of the acromion just in front of the articular surface for the clavicle, and by its broad base to the whole length of the outer border of the coracoid process. Its posterior fibres are directed obliquely backwards and outwards, its anterior fibres transversely. This ligament completes the vault formed by the acromion and coracoid processes for the protection of the head of the humerus. It is in relation, above, with the clavicle and under surface of the deltoid; below, with the tendon of the Supra-spinatus muscle, a bursa being interposed. Its anterior border is continuous with a dense cellular lamina that passes beneath the deltoid upon the tendons of the Supra- and Infra-spinati muscles.

The *Transverse* or *Coracoid Ligament*, is a thin and flat fasciculus, narrower at the middle than at the extremities, attached by one end to the base of the coracoid process, and by the other, to the inner extremity of the scapular notch, which it converts into a foramen. The supra-scapular nerve passes through this foramen, its accompanying vessels above it.

4. Shoulder Joint

The Shoulder is an enarthrodial or ball and socket joint. The bones entering into its formation are the large globular head of the humerus, which is received into the shallow glenoid cavity of the scapula, an arrangement which permits of very considerable movement, whilst the joint itself is protected against displacement by the strong ligaments and tendons which surround it, and above by an arched vault, formed by the under surface of the coracoid and acromion processes, and the coraco-acromion ligament. The two articular surfaces are covered by a layer of cartilage, which on the head of the humerus is thicker at the centre than at the circumference, the reverse being observed in the glenoid cavity. Its ligaments are the

        Capsular.   Glenoid.
        Coraco-humeral.   Synovial Membrane.

The *Capsular Ligament* completely encircles this articulation; being attached, above, to the circumference of the glenoid cavity beyond the glenoid ligament; below, to the margin of the neck of the humerus, approaching nearer to the articular cartilage above, than in the rest of its extent. It is thicker above than below, remarkably loose and lax, and much larger and longer than is necessary to keep the bones in contact, allowing them to be separated from each other more than an inch, an evident provision for that extreme freedom of movement which is peculiar to this articulation. Its external surface is strengthened above by the Supra spinatus; above and internally by the coraco-humeral ligament; below, where it is thin and weak, the long tendon of the Triceps is separated from it by a little loose areolar tissue; externally the tendons of the Infra-spinatus and Teres minor are firmly attached to it; and internally, the tendon of the Sub-scapularis. The capsular ligament usually presents three openings; one at its inner side, partially filled up by the tendon of the Sub-scapularis; it establishes a communication between the synovial membrane of the joint, and a bursa beneath the tendon of that muscle; a second, not constant, at its external part, where a communication exists between the joint and a bursal sac belonging to the Infra-spinatus muscle. The third is seen in the lower border of the ligament, between the two tuberosities, for the passage of the tendon of the Biceps muscle.

The *Coraco-humeral* or *Accessory Ligament*, is a broad band which strengthens the upper and inner part of the capsular ligament. It arises from the outer border of the coracoid process, and descends obliquely downwards and outwards to the anterior part of the great tuberosity of the humerus, being blended with the tendon of the Supra-spinatus muscle. This ligament is intimately united to the capsular in the greater part of its extent.

The *Glenoid Ligament* is a fibro-cartilaginous band attached around the margin of the glenoid cavity. It is triangular on section, the thickest portion being fixed to the circumference of the cavity, the free edge being thin and sharp. It appears to be mainly formed of the fibres of the long tendon of the Biceps muscle, bifurcating at the upper part of the glenoid cavity into two fasciculi, which encircle its margin and unite at its lower part. This ligament deepens the cavity for articulation, and protects the edges of the bone. It is lined by the synovial membrane.

The *Synovial Membrane* lines the glenoid cavity and the fibro-cartilaginous rim surrounding it; it is then reflected over the internal surface of the capsular ligament, lines the lower part and sides of the neck of the humerus, and is continued over the cartilage covering the head of this bone. The long tendon of the Biceps muscle which passes through the joint, is enclosed in a tubular sheath of synovial membrane, which is reflected upon it at the point where it perforates the capsule, and is continued around it as far as the summit of the glenoid cavity, where it is continuous with that portion of the membrane which covers its surface. The tendon of the Biceps is thus enabled to traverse the articulation, but is not contained in the interior of the synovial cavity. The synovial membrane communicates with a large bursal sac beneath the tendon of the Sub-scapularis, by an opening at the inner side of the capsular ligament; it also occasionally communicates with another bursal sac, beneath the tendon of the Infra-spinatus, through an orifice at its upper part. A third bursal sac, which does not communicate with the joint is placed between the under surface of the deltoid and the outer surface of the capsule.

115.—Left Elbow-Joint, showing Anterior and Internal Ligaments.

The Muscles in relation with this joint are, above, the Supra-spinatus; below, the long tendon of the Triceps; internally, the Sub-scapularis; externally, the Infra-spinatus and Teres minor; within, the long tendon of the Biceps. The Deltoid is placed most externally, and covers the articulation on its outer side, and in front and behind.

The Arteries supplying this joint are articular branches of the anterior and posterior circumflex, and supra-scapular.

The Nerves are derived from the circumflex and supra-scapular.

Actions. The shoulder joint is capable of movement in almost any direction, forwards, backwards, abduction, adduction, circumduction, and rotation.

5. Elbow Joint

The *Elbow* is a *ginglymoid* or hinge joint. The bones entering into its formation are the trochlear surface of the humerus, which is received in the greater sigmoid cavity of the ulna, and admits of the movements peculiar to this joint, those of flexion and extension, whilst the cup-shaped depression of the head of the radius articulates with the radial tuberosity of the humerus, its circumference with the lesser sigmoid cavity of the ulna, allowing of the movement of rotation of the radius on the ulna, the chief action of the superior radio-ulnar articulation. These various articular surfaces are covered with a thin layer of cartilage, and connected together by the following ligaments.

Anterior Ligament.
Posterior Ligament.
Internal Lateral.
External Lateral.
Synovial Membrane.

The *Anterior Ligament* (fig. 115) is a broad and thin membranous layer, which covers the anterior surface of the joint. It is attached to the humerus immediately above the coronoid fossa; below, to the anterior surface of the coronoid process of the ulna and orbicular ligament, being continuous on each side with the lateral ligaments. Its superficial or oblique fibres pass from the internal tuberosity of the humerus outwards to the orbicular ligament. The middle fibres, vertical in direction, pass from the upper part of the coronoid depression, and become blended with the preceding. A third, or transverse set, intersect these at right angles. This ligament is in relation, in front, with the Brachialis anticus; behind, with the synovial membrane.

The *Posterior Ligament* is a thin and loose membranous fold, attached, above, to the lower end of the humerus, immediately above the olecranon depression; below, to the margin of the olecranon. The superficial or transverse fibres pass between the adjacent margins of the olecranon fossa. The deeper portion consists of vertical fibres, which pass from the upper part of the olecranon fossa to the margin of the olecranon. This ligament is in relation, behind, with the tendon of the Triceps and Anconeus; in front, with the synovial membrane.

116.—Left Elbow-Joint, shewing Posterior and External Ligaments.

The *Internal Lateral Ligament* is a thick triangular band of ligamentous fibres, consisting of two distinct portions, an anterior and posterior. The *anterior portion*, directed obliquely forwards, is attached, above, by its apex, to the front part of the internal condyle of the humerus; and, below, by its broad base to the inner margin of the coronoid process. The *posterior portion*, also of triangular form, is attached, above, by its apex to the lower and back part of the internal condyle; below, to the inner margin of the olecranon. This ligament is in relation, internally, with the Triceps and Flexor carpi ulnaris muscles and the ulnar nerve.

The *External Lateral Ligament* (fig. 116) is a short and narrow fibrous fasciculus, less distinct than the internal, attached, above, to the external condyle of the humerus; below, to the orbicular ligament, some of its most posterior fibres passing over that ligament to be inserted into the outer margin of the greater sigmoid cavity. This ligament is intimately blended with the tendon of origin of the Supinator brevis muscle.

The *Synovial Membrane* is very extensive. It covers the articular surface of the humerus, and lines the coronoid and olecranon depressions on that bone; from these points, it is reflected over the anterior, posterior and lateral ligaments; lines the greater sigmoid cavity, the concave depression on the head of the radius; and forms a pouch between the lesser sigmoid cavity, the internal surface of the annular ligament, and the circumference of the radius.

The *Muscles* in relation with this joint are, in front, the Brachialis anticus; behind, the Triceps and Anconeus; externally, the Supinator brevis, and the common tendon of origin of the Extensor muscles; internally, the common tendon of origin of the Flexor muscles, the Flexor carpi ulnaris, and ulnar nerve.

The *Arteries* supplying this joint are derived from the communicating branches between the superior profunda, inferior profunda, and anastomatic branches of the Brachial, with the anterior, posterior and interosseous recurrent branches of the Ulnar, and the recurrent branch of the Radial. These vessels form a complete chain of inosculation around this joint.

The *Nerves* are derived from the ulnar, as it passes between the internal condyle and the olecranon.

Actions. The elbow is one of the most perfect hinge-joints in the body; its movements are consequently limited to flexion and extension, the exact apposition of the articular surfaces preventing the least lateral motion. The movement of flexion is limited by the coronoid process, and that of extension by the olecranon process.

6. Radio-Ulnar Articulations

The articulation of the radius with the ulna is effected by ligaments, which connect together both extremities as well as the centre of these bones. They may, consequently, be subdivided into three sets: 1, the superior radio-ulnar; 2, the middle radio-ulnar; and, 3, the inferior radio-ulnar articulations.

1. Superior Radio-Ulnar Articulation

This articulation is a lateral ginglymoid joint. The bones entering into its formation are the inner side of the circumference of the head of the radius, which is received into the lesser sigmoid cavity of the ulna. These surfaces are covered with cartilage, and invested with a duplicature of synovial membrane, continuous with that which lines the elbow-joint. Its only ligament is

The Annular or Orbicular

The *Orbicular Ligament* (fig. 116) is a strong flat band of ligamentous fibres, which surrounds the head of the radius, and retains it in firm connection with the lesser sigmoid cavity of the ulna. It forms about three-fourths of a fibrous ring, attached by each end to the extremities of this cavity, and is broader at the upper part of its circumference than below, which serves to hold the head of the radius more securely in its position. Its *outer surface* is strengthened by the external lateral ligament, and affords partial origin to the Supinator brevis muscle. Its *internal surface* is smooth, and lined by the synovial membrane of the elbow-joint.

Actions. The movement which takes place in this articulation is limited to rotation of the inner part of the head of the radius within the orbicular ligament, and upon the lesser sigmoid cavity of the ulna; rotation forwards being called *pronation*; rotation backward, *supination*.

2. Middle Radio-Ulnar Articulation

The interval between the radius and ulna in the middle of the forearm is occupied by two ligaments.

Oblique. Interosseous.

The *Oblique* or *Round Ligament* (fig. 115) is a small round fibrous cord, which extends obliquely downwards and outwards, from the tubercle of the ulna at the base of the coronoid process, to the radius a little below the bicipital tuberosity. Its fibres run in the opposite direction to those of the interosseous ligament; and it appears to be placed as a substitute for it in the upper part of the interosseous interval.

The *Interosseous Ligament* is a broad and thin plane of aponeurotic fibres, descending obliquely downwards and inwards, from the interosseous ridge on the radius to that on the ulna. It is deficient above, commencing about an inch beneath the tubercle of the radius; broader in the middle than at either extremity; and presents an oval aperture just above its lower margin for the passage of the anterior interosseous vessels to the back of the forearm. This ligament serves to connect the bones, and to increase the extent of surface for the attachment of the deep muscles. Between its upper border and the oblique ligament an interval exists, through which the posterior interosseous vessels pass. Two or three fibrous bands are occasionally found on the posterior surface of this membrane, which descend obliquely from the ulna towards the radius, and which have consequently a direction contrary to that of the other fibres. It is in relation, in *front*, by its upper three-fourths (radial margin) with the Flexor longus pollicis (ulnar margin), with the Flexor profundus digitorum (lying upon the interval between which are the anterior interosseous vessels and nerve), by its lower fourth with the Pronator quadratus; *behind*, with the Supinator

117.—Ligaments of Wrist and Hand. Anterior View.

118.—Ligaments of Wrist and Hand. Posterior View.

brevis, Extensor ossis metacarpi pollicis, Extensor primi internodii pollicis, Extensor secundi internodii pollicis, Extensor indicis; and, near the wrist, with the anterior interosseous artery and posterior interosseous nerve.

3. INFERIOR RADIO-ULNAR ARTICULATION

This is a lateral ginglymoid joint, formed by the head of the ulna being received into the sigmoid cavity at the inner side of the lower end of the radius. The articular surfaces are invested by a thin layer of cartilage, and connected together by the following ligaments.

> Anterior radio-ulnar.
> Posterior radio-ulnar.
> Triangular Inter-articular Fibro-cartilage.
> Synovial Membrane.

The *Anterior Radio-ulnar Ligament* (fig. 117) is a narrow band of fibres, extending from the anterior margin of the sigmoid cavity of the radius to the anterior surface of the head of the ulna.

The *Posterior Radio-ulnar Ligament* (fig. 118) extends between the same points on the posterior surface of the articulation.

The *Inter-articular Fibro-cartilage* (fig. 119) is a thick fibro-cartilaginous lamella, of a triangular form, placed transversely, completing the wrist-joint, and binding the lower ends of the radius and ulna firmly together. Its circumference is more dense than its centre, which is thin and occasionally perforated; and it is thinner and broader externally than internally. It is attached by its apex to a depression which separates the styloid process of the ulna from the head of that bone; by its base, which is thin, to the prominent edge of the radius, which separates the sigmoid cavity from the carpal articulating surface, and by its anterior and posterior margins to the ligaments of the radio-carpal articulation. Its *upper surface*, smooth and concave, is contiguous with the head of the ulna; its *under surface*, also concave and smooth, with the cuneiform bone. Both surfaces are lined by a synovial membrane: the superior surface, by one peculiar to the radio-ulnar articulation; the inferior surface, by the synovial membrane of the wrist.

The *Synovial Membrane* of this articulation has been called, from its extreme looseness, the *membrana sacciformis*; it covers the articular surface of the head of the ulna, and where reflected from this bone on to the radius, forms a very loose *cul-de-sac*; from the radius it is continued over the upper surface of the fibro-cartilage. The quantity of synovia which it contains is usually considerable. When the fibro-cartilage is perforated, this synovial membrane is continuous with that which lines the wrist-joint.

Actions. The movement which occurs in the inferior radio-ulnar articulation is just the inverse of that which takes place between the two bones above; it is limited to rotation of the radius around the head of the ulna; rotation forwards being termed *pronation*, rotation backwards *supination*. In pronation, the sigmoid cavity glides forward on the articular edge of the ulna; in supination, it rolls in the opposite direction, the extent of these movements being limited by the anterior and posterior ligaments.

7. Wrist Joint

The *Wrist* presents most of the characters of an enarthrodial joint. The parts entering into its formation are the lower end of the radius, and under surface of the triangular interarticular fibro-cartilage, above; and the scaphoid, semilunar, and cuneiform bones below. The articular surfaces of the radius and interarticular fibro-cartilage form a transversely elliptical concave surface. The radius is subdivided into two parts by a line extending from before backwards; and these, together with the interarticular cartilage, form three facets, one for each carpal bone. The three carpal bones are connected together, and form a rounded convex surface, which is received into the cavity above mentioned. All the bony surfaces of this articulation are covered with cartilage, and connected together by the following ligaments.

   External Lateral.   Anterior.
   Internal Lateral.   Posterior.
        Synovial Membrane.

The *External Lateral Ligament* extends from the summit of the styloid process of the radius to the outer side of the scaphoid, some of its fibres being prolonged to the trapezium and annular ligament.

The *Internal Lateral Ligament* is a rounded cord, attached, above, to the extremity of the styloid process of the ulna; below, it divides into two fasciculi, which are attached, one to the inner side of the cuneiform bone, the other to the pisiform bone and annular ligament.

The *Anterior Ligament* is a broad membranous band, consisting of three fasciculi, attached, above, to the anterior margin of the lower end of the radius, its styloid process, and the ulna; its fibres pass downwards and inwards, to be inserted into the anterior surface of the scaphoid, semilunar, and cuneiform bones. This ligament is perforated by numerous apertures for the passage of vessels, and is in relation, in front, with the tendons of the Flexor profundus digitorum and Flexor longus pollicis; behind, with the synovial membrane of the wrist-joint.

The *Posterior Ligament*, less thick and strong than the anterior, is attached, above, to the posterior border of the lower end of the radius; its fibres descend obliquely downwards and inwards to be attached to the posterior surface of the scaphoid, semilunar, and cuneiform bones, its fibres being continuous with those of the dorsal carpal ligaments. This ligament is in relation, behind, with the extensor tendons of the fingers; in front, with the synovial membrane of the wrist.

The *Synovial Membrane* lines the lower end of the radius and under surface of the triangular inter-articular fibro-cartilage above; and being reflected on the inner surface of the ligaments above mentioned, covers the convex surface of the scaphoid, semilunar, and cuneiform bones below.

Relations. The wrist-joint is covered in front by the flexor, and behind by the extensor tendons; it is also in relation with the radial and ulnar arteries.

The Arteries supplying this joint are the anterior and posterior carpal branches of the Radial and Ulnar, the anterior and posterior interosseous, and some ascending branches from the deep palmar arch.

The Nerves are derived from the posterior interosseous.

Actions. The movements permitted in this joint are flexion, extension, abduction, adduction, and circumduction. It is totally incapable of rotation, one of the characteristic movements in true enarthrodial joints.

8. Articulations of the Carpus

These articulations may be subdivided into three sets.

    1. The articulation of the first row of carpal bones.
    2. The articulation of the second row of carpal bones.
    3. The articulation of the two rows with each other.

1. Articulation of the First Row of Carpal Bones

These are arthrodial joints. The articular surfaces are covered with cartilage, and connected together by the following ligaments.

    Two Dorsal.   Two Palmar.
    Two Interosseous.

The *Dorsal Ligaments*, two in number, are placed transversely behind the bones of the first row; they connect the scaphoid and semilunar, and the semilunar and cuneiform.

The *Palmar Ligaments*, also two in number, connect the scaphoid and semilunar, and the semilunar and cuneiform bones; they are less strong than the dorsal, and placed very deep under the anterior ligament of the wrist.

The *Interosseous Ligaments* (fig. 119) are two narrow bundles of dense fibrous tissue, connecting the semilunar bone, on one side with the scaphoid, on the other with the cuneiform bone. They close the upper part of the interspaces between the scaphoid, semilunar, and cuneiform bones, their upper surfaces being smooth, and lined by the synovial membrane of the wrist-joint.

The articulation of the pisiform with the cuneiform is provided with a separate synovial membrane, protected by a thin capsular ligament. There are also two strong fibrous fasciculi, which connect this bone to the unciform, and base of the fifth metacarpal bone.

2. Articulation of the Second Row of Carpal Bones

These are also arthrodial joints, the articular surfaces being covered with cartilage, and connected by the following ligaments.

Three Dorsal.   Three Palmar.
Two Interosseous.

The *three Dorsal Ligaments* extend transversely from one bone to another on the dorsal surface, connecting the trapezium with the trapezoid, the trapezoid with the os magnum, and the os magnum with the unciform.

The *three Palmar Ligaments* have a similar arrangement on the palmar surface.

The *two Interosseous Ligaments*, much thicker than those of the first row, are placed one on each side of the os magnum, connecting it with the trapezoid externally, and the unciform internally. The former is less distinct than the latter.

3. Articulation of the Two Rows of Carpal Bones with each other

The articulation between the two rows of the carpus consists of an enarthrodial joint in the middle, formed by the reception of the os magnum into a cavity formed by the scaphoid and semilunar bones, and of an arthrodial joint on each side, the outer one formed by the articulation of the scaphoid with the trapezium and trapezoid, the internal one by the articulation of the cuneiform and unciform. The articular surfaces are covered by a thin layer of cartilage, and connected by the following ligaments.

Anterior or Palmar.   External Lateral.
Posterior or Dorsal.   Internal Lateral.
Synovial Membranes.

The *Anterior* or *Palmar Ligaments* consist of short fibres, which pass obliquely between the bones of the first and second row on the palmar surface.

The *Posterior* or *Dorsal Ligaments* have a similar arrangement on the dorsal surface of the carpus.

The *Lateral Ligaments* are very short; they are placed, one on the radial, the other on the ulnar side of the carpus; the former, the stronger and more distinct, connecting the scaphoid and trapezium bones, the latter the cuneiform and unciform: they are continuous with the lateral ligaments of the wrist-joint.

There are *two Synovial Membranes* found in the articulation of the carpal bones with each other. The first of these, the more extensive, lines the under surface of the scaphoid, semilunar, and cuneiform bones, sending upwards two prolongations between their contiguous surfaces; it is then reflected over the bones of the second row, and sends down three prolongations between them, which line their contiguous surfaces, and invest the carpal extremities of the four outer metacarpal bones. The second is the synovial membrane between the pisiform and cuneiform bones.

Actions. The partial movement which takes place between the bones of each row is very inconsiderable; the movement between the two rows is more marked, but limited chiefly to flexion and extension.

9. Carpo Metacarpal Articulations
Articulation of the First Metacarpal Bone with the Trapezium

This is an enarthrodial joint. Its ligaments are a capsular and synovial membrane. The *capsular ligament* is a thick but loose capsule, which passes from the circumference of the upper extremity of the metacarpal bone, to the rough edge bounding the articular surface of the trapezium; it is thickest externally and behind, and lined by a separate *synovial membrane.*

Articulation of the Four inner Metacarpal Bones with the Carpus

The joints formed between the carpus and four inner metacarpal bones, are connected together by dorsal, palmar, and interosseous ligaments.

The *Dorsal Ligaments*, the strongest and most distinct, connect the carpal and metacarpal bones on their dorsal surface. The second metacarpal bone receives two fasciculi, one from the trapezium, the other from the trapezoid; the third metacarpal receives one from the os magnum; the fourth two, one from the os magnum, and one from the unciform; the fifth receives a single fasciculus from the unciform bone.

119.—Vertical Section through the Articulations at the Wrist, showing the five Synovial Membranes.

The *Palmar Ligaments* have a somewhat similar arrangement on the palmar surface, with the exception of the third metacarpal, which has three ligaments, an external one from the trapezium, situated above the sheath of the tendon of the Flexor carpi radialis; a middle one, from the os magnum; and an internal one, from the unciform.

The *Interosseous Ligaments* consist of short thick fibres, which are limited to one part of the carpo-metacarpal articulation; they connect the inferior angles of the os magnum and unciform, with the adjacent surfaces of the third and fourth metacarpal bones.

The *Synovial Membrane* is a continuation of that between the two rows of carpal bones. Occasionally the unciform has a separate synovial membrane, lining it and the fourth and fifth metacarpal bones.

The Synovial Membranes of the wrist (fig. 119) are thus seen to be five in number. The *first*, the membrana sacciformis, lining the lower end of the ulna, the sigmoid cavity of the radius, and upper surface of the triangular inter-articular fibro-cartilage. The *second* lines the lower end of the radius and inter-articular fibro-cartilage above, and the scaphoid, semilunar, and cuneiform bones below. The *third*, the most extensive, covers the contiguous surfaces of the two rows of carpal bones, and passing between the bones of the second range, lines the carpal extremities of the four inner metacarpal bones. The *fourth* lines the adjacent surfaces of the trapezium and metacarpal bone of the thumb. And the *fifth* the adjacent surfaces of the cuneiform and pisiform bones.

Actions. The movement permitted in the carpo-metacarpal articulations is limited to a slight gliding of the articular surfaces upon each other, the extent of which varies in the different joints. Thus the articulation of the metacarpal bone of the thumb with the trapezium is most moveable, then the fifth metacarpal, and then the fourth. The second and third are almost immoveable. In the articulation of the metacarpal bone of the thumb with the trapezium, the movements permitted are flexion, extension, adduction, abduction, and circumduction.

Articulation of the Metacarpal Bones with each other

The carpal extremities of the metacarpal bones of the fingers, articulate with one another at each side by small surfaces covered with cartilage, and connected together by dorsal, palmar, and interosseous ligaments.

The *Dorsal* or *Palmar Ligaments* pass transversely from one bone to another on the dorsal and palmar surfaces. The *Interosseous Ligaments* passing between their contiguous surfaces, just beneath their lateral articular facets.

The *Synovial Membrane* lining the lateral facets, is a reflection of that between the two rows of carpal bones.

The digital extremities of the metacarpal bones of the fingers, are connected together by the transverse ligament, a narrow fibrous band, passing transversely across their under surfaces, and blended with the ligaments of the metacarpo-phalangeal articulations. Its *anterior surface* presents four grooves for the passage of the flexor tendons, and its sides are continuous with their sheaths. Its *posterior surface* blends with the ligaments of the metacarpo-phalangeal articulation.

10. Metacarpo-phalangeal Articulations (fig. 120)

These articulations are of the ginglymoid kind, formed by the reception of each of the rounded heads of the metacarpal bones of the four fingers, into a superficial cavity in the extremity of the first phalanges. They are connected by the following ligaments,

Anterior.
Two Lateral.
Synovial Membrane.

120.—Articulations of the Phalanges.

The *Anterior Ligaments* are very thick and dense, they are placed on the palmar surface of the joint in the interval between the lateral ligaments, to which they are connected; they are loosely united to the metacarpal bone, but very firmly to the base of the first phalanges. Their palmar surface is intimately united to the transverse ligament, each ligament forming with it a groove for the passage of the flexor tendons, the sheath surrounding which is connected to it at each side. By their internal surface they form part of the articular surface for the head of the metacarpal bone, and are lined by a synovial membrane.

The *Lateral Ligaments* are thick and strong rounded cords, placed one on each side of the joint, attached by one extremity to the sides of the head of the metacarpal bones, and by the other, to the contiguous extremity of the phalanges.

The Posterior Ligament is supplied by the extensor tendon of the fingers placed over the back of each joint.

Actions. The movements which occur in these joints are flexion, extension, adduction, abduction, and circumduction; the lateral movements are very limited.

11. Articulations of the Phalanges

These are ginglymoid joints, connected by the following ligaments;

Anterior.   Two Lateral.
Synovial Membrane.

The arrangement of these ligaments is similar to those in the preceding articulations; the extensor tendon supplies the place of a posterior ligament.

Actions. The only movements permitted in the phalangeal joints are flexion and extension; these movements are more extensive between the first and second phalanges than between the second and third. The movement of flexion is very extensive, but extension is limited by the anterior and lateral ligaments.

ARTICULATIONS OF THE LOWER EXTREMITY

The articulations of the lower extremity comprise the following groups. 1. The hip joint. 2. The knee joint. 3. The articulations between the tibia and fibula. 4. The ankle joint. 5. The articulations of the tarsus. 6. The tarso-metatarsal articulations. 7. The metatarso phalangeal articulations. 8. The articulation of the phalanges.

1. Hip Joint (fig. 121)

This articulation is an enarthrodial, or ball and socket joint, formed by the reception of the globular head of the femur into the cup-shaped cavity of the acetabulum. These two articulating surfaces are covered with cartilage, that on the head of the femur being thicker at the centre than at the circumference, and covering the entire surface with the exception of a depression just below its centre for the ligamentum teres; that covering the acetabulum is much thinner at the centre than at the circumference, and is deficient in the situation of the circular depression at the bottom of this cavity. The ligaments of this joint are the

Capsular.	Cotyloid.
Ilio-femoral.	Transverse.
Teres.	Synovial Membrane.

121.—Left Hip Joint laid open.

The *Capsular Ligament* is a strong, dense, ligamentous capsule, embracing the margin of the acetabulum above, and surrounding the neck of the femur below. Its *upper circumference* is attached to the acetabulum two or three lines external to the cotyloid ligament; but opposite the notch where the margin of this cavity is deficient, it is connected with the transverse ligament, and by a few fibres to the edge of the obturator foramen. Its *lower circumference* surrounds the neck of the femur, being attached, in front, to the spiral or anterior inter-trochanteric line; above, to the base of the neck; behind, to the middle of the neck of the bone, about three quarters of an inch from the posterior inter-trochanteric line. It is much thicker at the upper and anterior part of the joint where the greatest amount of resistance is required, than below, where it is thin, loose, and longer than in any other situation. Its external surface is rough, covered by numerous muscles, and separated in front from the Psoas and Iliacus by a synovial bursa, which not unfrequently communicates by a circular aperture with the cavity of the joint. It differs from the capsular ligament of the shoulder, in being much less loose and lax, and in not being perforated for the passage of a tendon.

The *Ilio-femoral Ligament* (fig. 110) is an accessory band of fibres, extending obliquely across the front of the joint: it is intimately connected with the capsular ligament, and serves to strengthen it in this situation. It is attached above to the anterior inferior spine of the ilium, below, to the anterior inter-trochanteric line.

The *Ligamentum Teres* is a flat triangular band of fibres, implanted by its apex into the depression just below the middle of the head of the femur, and by its broad base, which consists of two bundles of fibres, into the margins of the notch at the bottom of the acetabulum, becoming blended with the transverse ligament. It is formed of a bundle of fibres, the thickness and strength of which is very variable, surrounded by a tubular sheath of synovial membrane. Sometimes the synovial fold only exists, or the ligament may be altogether absent.

The *Cotyloid Ligament* is a fibro-cartilaginous rim attached to the margin of the acetabulum, the cavity of which it deepens, at the same time it protects the edges of the bone, and fills up the inequalities on its surface. It is prismoid in form, its base being attached to the margin of the acetabulum, its opposite edge being free and sharp; whilst its two surfaces are invested by synovial membrane, the external one being in contact with the capsular ligament, the internal being inclined inwards so as to narrow the acetabulum and embrace the cartilaginous surface of the head of the femur. It is much thicker above and behind than below and in front, and consists of close, compact fibres, which arise from different points of the circumference of the acetabulum, and interlace with each other at very acute angles.

The *Transverse Ligament* is a strong flattened band of fibres, which crosses the notch at the lower part of the acetabulum, and converts it into a foramen. It is continuous at each side with the cotyloid ligament, and consists of fibres which arise from each side of the notch, and pass across each other. An interval is left beneath this ligament for the passage of nutrient vessels to the joint.

The *Synovial Membrane* is very extensive. It invests the cartilaginous surface of the head of the femur, and all that portion of the neck which is contained within the joint; from this point it is reflected on the internal surface of the capsular ligament, covers both surfaces of the cotyloid ligament, and lines the cavity of the acetabulum, covers the mass of fat contained in the fossa at the bottom of this cavity, and is prolonged in the form of a tubular sheath around the ligamentum teres on to the head of the femur.

The Muscles in relation with this joint are, in front, the Psoas and Iliacus, separated from the capsular ligament by a synovial bursa; above, the short head of the Rectus and Gluteus minimus, the latter being closely adherent to it; internally, the Obturator externus and Pectineus; behind, the Pyriformis, Gemellus superior, Obturator internus, Gemellus inferior, Obturator externus, and Quadratus femoris.

The Arteries supplying it are derived from the obturator, sciatic, internal circumflex, and gluteal.

The Nerves are articular branches from the sacral plexus, great sciatic, obturator, and accessory obturator nerves.

Actions. The movements of the hip, like all enarthrodial joints, are very extensive; they are flexion, extension, adduction, abduction, circumduction, and rotation.

2. THE KNEE JOINT

The knee is a ginglymoid, or hinge joint; the bones entering into its formation are the condyles of the femur above, the head of the tibia below, and the patella in front. The articular surfaces are covered with cartilage, lined by synovial membrane, and connected together by ligaments, some of which are placed on the exterior of the joint, whilst others occupy its interior.

122.—Right Knee Joint. Anterior View.

External Ligaments

Anterior, or Ligamentum Patellae.   Internal Lateral.
Posterior, or Ligamentum Posticum Winslowii.   Two External Lateral.
Capsular.

Internal Ligaments

Anterior, or External Crucial.   Two Semilunar Fibro-cartilages.
Posterior, or Internal Crucial.   Transverse.
Coronary.
Synovial Membrane.   { Ligamentum mucosum.
    Ligamenta alaria.

The *Anterior Ligament,* or *Ligamentum Patellae* (fig. 122), is that portion of the common tendon of the extensor muscles of the thigh which is continued from

123.—Right-Knee Joint. Posterior View.

the patella to the tubercle of the tibia, supplying the place of an anterior ligament. It is a strong, flat, ligamentous band, attached, above, to the apex of the patella and the rough depression on its posterior surface; below, to the lower part of the tuberosity of the tibia; its superficial fibres being continuous across the front of the patella with those of the tendon of the Rectus femoris. Two synovial bursae are connected with this ligament and the patella; one is interposed between the patella and the skin covering its anterior surface; the other, of small size, between the ligamentum patellae and the upper part of the tuberosity of the tibia. The posterior surface of this ligament is separated above from the knee joint by a large mass of adipose tissue, its lateral margins are continuous with the aponeuroses derived from the Vasti muscles.

The *Posterior Ligament, Ligamentum Posticum Winslowii* (fig. 123), is a broad, flat, fibrous band, which covers over the whole of the back part of the joint. It consists of two lateral portions, formed chiefly of vertical fibres, which arise above from the condyles of the femur, and connected below with the back part of the head of the tibia, being closely united with the tendons of the Gastrocnemii, Plantaris, and Popliteus muscles; the central portion is formed of fasciculi obliquely directed and separated from one another by apertures for the passage of vessels. The strongest of these fasciculi is derived from the tendon of the Semi-membranosus, it passes from the back part of the inner tuberosity of the tibia, obliquely upwards and outwards to the back part of the outer condyle of the femur. The posterior ligament forms part of the floor of the popliteal space, and upon it rests the popliteal artery.

The *Internal Lateral Ligament* is a broad, flat, membranous band, thicker behind than in front, and situated nearer to the back than the front of the joint. It is attached, above, to the inner tuberosity of the femur; below, to the inner tuberosity and inner surface of the shaft of the tibia, to the extent of about two inches. It is crossed, at its lower part, by the aponeurosis of the Sartorius, and the tendons of the Gracilis and Semi-tendinosus muscles, a synovial bursa being interposed. Its *deep surface* covers the anterior portion of the tendon of the Semi-membranosus, the synovial membrane of the joint, and the inferior internal articular artery; it is intimately adherent to the internal semi-lunar fibro-cartilage.

The *Long External Lateral Ligament* is a strong, rounded, fibrous cord, situated nearer the posterior part of the articulation than the anterior. It is attached, above, to the outer tuberosity of the femur; below, to the outer part of the head of the fibula. Its *outer surface* is covered by the tendon of the Biceps, which divides into two parts, separated by this ligament, at its insertion. It has, passing beneath it, the tendon of the Popliteus muscle, and the inferior external articular artery.

124.—Right Knee-Joint. Shewing Internal Ligaments.

The *Short External Lateral Ligament* is an accessory bundle of fibres, placed behind and parallel with the preceding; attached, above, to the lower part of the outer tuberosity of the femur; below, to the summit of the styloid process of the fibula. This ligament is intimately connected with the capsular ligament, and has passing beneath it the tendon of the Popliteus muscle.

The *Capsular Ligament* consists of an exceedingly thin, but strong, fibrous membrane, which surrounds the joint in the intervals left by the preceding ligaments, being attached to the femur immediately above its articular surface; below, to the upper border and sides of the patella, the margins of the head of the tibia and inter-articular cartilages, and being continuous behind with the posterior ligament. This membrane

125.—Head of Tibia, with Semi-lunar Cartilages, etc. Seen from above. Right Side.

is strengthened by fibrous expansions, derived from the fascia lata and Vasti muscles, at their insertion into the sides of the patella.

The *Crucial* are two interosseous ligaments of very considerable strength, situated in the interior of the joint, nearer its posterior than its anterior part. They are called *crucial*, because they cross each other, somewhat like the lines of the letter X; and have received the names *anterior* and *posterior*, from the position of their attachment to the tibia.

The *Anterior* or *External Crucial Ligament* (fig. 124), smaller than the posterior, arises from the inner side of the depression in front of the spine of the tibia, being blended with the anterior extremity of the external semi-lunar fibro-cartilage, and passing obliquely upwards, backwards, and outwards, is inserted into the inner and back part of the outer condyle of the femur.

The *Posterior* or *Internal Crucial Ligament* is larger in size, but less oblique in its direction than the anterior. It arises from the back part of the depression behind the spine of the tibia, and from the posterior extremity of the external semi-lunar fibro-cartilage; passing upwards, forwards, and inwards, it is inserted into the outer and front part of the inner condyle of the femur. As it crosses the anterior crucial ligament, a fasciculus is given off from it, which blends with its posterior part. It is in relation, in front, with the anterior ligament; behind, with the ligamentum posticum Winslowii.

The *Semi-lunar Fibro-Cartilages* (fig. 125) are two crescentic lamellae attached to the margins of the head of the tibia, serving to deepen its surface for articulation with the condyles of the femur. The circumference of each cartilage is thick and convex; the inner free border, thin and concave. Their upper surfaces are concave, and in relation with the condyles of the femur; their lower surfaces are flat, and rest upon the head of the tibia. Each cartilage covers nearly the outer two-thirds of the corresponding articular surface of the tibia, the inner third being uncovered; both surfaces are smooth, and invested by synovial membrane.

The *Internal Semi-lunar Fibro-Cartilage* is nearly semicircular in form, a little elongated from before backwards, and broader behind than in front; its convex border is united to the internal lateral ligament, and to the head of the tibia, by means of the coronary ligaments; its anterior extremity, thin and pointed, is firmly implanted into the depression in front of the spine of the tibia; its posterior extremity to the depression behind the spine.

The *External Semi-lunar Fibro-Cartilage* forms nearly an entire circle, covering a larger portion of the articular surface than the internal one. It is grooved on its outer side, for the tendon of the Popliteus muscle. Its circumference is held in connexion with the head of the tibia, by means of the coronary ligaments; and by its two extremities is firmly implanted in the depressions in front and behind the spine of the tibia. These extremities, at their insertion, are interposed between the attachments of the internal cartilage. The external semi-lunar fibro-cartilage gives off from its *anterior border* a fasciculus, which forms the transverse ligament. By its *anterior extremity*, it is continuous with the anterior crucial ligament. Its *posterior extremity* divides into three slips; one, a strong cord, passes upwards and forwards, and is inserted into the outer side of the inner condyle, in front of the posterior crucial ligament; another fasciculus is inserted into the outer side of the inner condyle, behind the posterior crucial ligament; a third fasciculus is inserted into the back part of the anterior crucial ligament.

The *Transverse Ligament* is a band of fibres, which passes transversely between the anterior convex margin of the external cartilage, to the anterior extremity of the internal cartilage; its thickness varies considerably in different subjects.

The *Coronary Ligaments* consist of numerous short fibrous bands, which connect the convex border of the semi-lunar cartilages with the circumference of the head of the tibia, and with the other ligaments surrounding the joint.

The *Synovial Membrane* of the knee-joint is the largest and most extensive in the body. Commencing at the upper border of the patella, it forms a large *cul-de-sac* beneath the Extensor tendon of the thigh: this is sometimes replaced by a synovial bursa interposed between this tendon and the femur, which in some subjects communicates with the synovial membrane of the knee-joint, by an orifice of variable size. On each side of the patella, the synovial membrane extends beneath the aponeuroses of the Vasti muscles, and more especially beneath that of the Vastus internus; it covers the surface of the patella itself, and, beneath it, is separated from the anterior ligament by a considerable quantity of adipose tissue. In this situation, it sends off a triangular-shaped prolongation, containing a few ligamentous fibres, which extends from the anterior part of the joint below the patella, to the front of the inter-condyloid notch. This fold has been termed the *ligamentum mucosum*. The *ligamenta alaria* consist of two fringe-like folds, which extend from the sides of the ligamentum mucosum, upwards and outwards, to the sides of the patella. The synovial membrane covers both surfaces of the semi-lunar fibro-cartilages, and on the back part of the external one forms a *cul-de-sac* between the groove on its surface and the tendon of the Popliteus; it covers the articular surface of the tibia; surrounds the crucial ligaments, and inner surface of the ligaments which enclose the joint; lastly, it covers the entire surface of the condyles of the femur, and from them is continued on to the lower part of the front surface of the shaft. The pouch of synovial membrane between the Extensor tendons and front of the femur is supported, during the movements of the knee, by a small muscle, the Sub-cruraeus, which is inserted into it.

The Arteries supplying this joint are derived from the anastomotic branch of the Femoral, articular branches of the Popliteal, and recurrent branch of the Anterior Tibial.

The Nerves are derived from the obturator and external and internal popliteal.

Actions. The chief movements of this joint are flexion and extension; but it is also capable of performing some slight rotatory movement. During flexion, the articular surfaces of the tibia, covered by their inter-articular cartilages, glide backwards upon the condyles of the femur, the lateral posterior and crucial ligaments are relaxed, the ligamentum patellae is put upon the stretch, the patella filling up the vacuity in the front of the joint between the femur and tibia. In extension, the tibia and inter-articular cartilages glide forwards upon the femur; all the ligaments are stretched, with the exception of the ligamentum patellae, which is relaxed, and admits of considerable lateral movement. The movement of rotation is permitted when the knee is semi-flexed, rotation outwards being most extensive.

3. Articulations between the Tibia and Fibula

The articulations between the tibia and fibula are effected by ligaments which connect both extremities, as well as the centre of these bones. They may, consequently, be subdivided into three sets. 1. The Superior Tibio-Fibular articulation. 2. The Middle Tibio-Fibular articulation. 3. The Inferior Tibio-Fibular articulation.

1. Superior Tibio-Fibular Articulation

This articulation is an arthrodial joint. The contiguous surfaces of the bones present two flat oval surfaces covered with cartilage, and connected together by the following ligaments.

Anterior Superior Tibio-Fibular.
Posterior Superior Tibio-Fibular.
Synovial Membrane.

The *Anterior Superior Ligament* (fig. 124) consists of two or three broad and flat bands, which pass obliquely upwards and inwards, from the head of the fibula to the outer tuberosity of the tibia.

The *Posterior Superior Ligament* is a single thick and broad band, which passes from the back part of the head of the fibula to the back part of the outer tuberosity of the tibia. It is covered in by the tendon of the Popliteus muscle.

There is a distinct *Synovial Membrane* in this articulation. Occasionally the synovial membrane of the knee-joint is continuous with it at its upper and back part.

2. Middle Tibio-Fibular Articulation

The interval between the tibia and fibula is filled up by an interosseous membrane, which extends between the contiguous margins of the two bones. It consists of a thin aponeurotic lamina composed of oblique fibres, which pass between the interosseous ridges on the tibia and fibula. It is broader above than below, and presents at its upper part a large oval aperture for the passage of the anterior tibial artery forwards to the anterior aspect of the leg; and at its lower third, another opening, for the passage of the anterior peroneal vessels. It is continuous below with the inferior interosseous ligament; and is perforated in numerous parts for the passage of small vessels. By its anterior surface it is in relation with the Tibialis anticus, Extensor longus digitorum, Extensor proprius pollicis, Peroneus tertius, and the anterior tibial vessels and nerve; behind, with the Tibialis posticus and Flexor longus pollicis muscles.

3. Inferior Tibio-Fibular Articulation

This articulation, continuous with that of the ankle-joint, is formed by the convex surface at the lower end of the inner side of the fibula, being received into a concave surface on the outer side of the tibia. These surfaces, below, to the extent of about two lines, are smooth and covered with cartilage, which is continuous with that of the ankle-joint. Its ligaments are—

      Inferior Interosseous.   Posterior Inferior Tibio-fibular.
      Anterior Inferior Tibio-fibular.   Transverse.

The *Inferior Interosseous Ligament* consists of numerous short, strong fibrous bands, which pass between the contiguous rough surfaces of the tibia and fibula, constituting the chief bond of union between these bones. It is continuous, above, with the interosseous membrane.

The *Anterior Inferior Ligament* (fig. 127) is a flat triangular band of fibres, broader below than above, which extends obliquely downwards and outwards between the adjacent margins of the tibia and fibula on the front aspect of the articulation. It is in relation, in front, with the Peroneus tertius, the aponeurosis of the leg, and the integument; behind, with the inferior interosseous ligament, and lies in contact with the cartilage covering the astragalus.

The *Posterior Inferior Ligament*, smaller than the preceding, is disposed in a similar manner on the posterior surface of the articulation.

The *Transverse Ligament* is a long narrow band of ligamentous fibres, continuous with the preceding, passing transversely across the back of the joint, from the external malleolus to the tibia, a short distance from its malleolar process. The three preceding ligaments project somewhat below the margins of the bones, and form part of the articulating surface for the ankle-joint.

The *Synovial Membrane* lining the articular surfaces is derived from that of the ankle-joint.

Actions. The movement permitted in these articulations is limited to a very slight gliding of the articular surfaces upon one another.

4. Ankle Joint

The *Ankle* is a ginglymoid or hinge joint. The bones entering into its formation are the lower extremity of the tibia and its malleolus, and the malleolus of the fibula, above, which, united, form an arch, in which is received the upper convex surface of the astragalus and its two lateral facets. These surfaces are covered with cartilage, lined by synovial membrane, and connected together by the following ligaments:

      Anterior.   Internal Lateral.
              External Lateral.

The *Anterior Ligament* (fig. 126) is a broad, thin, membranous layer, attached, above, to the margin of the articular surface of the tibia; below, to the margin of the astragalus, in front of its articular surface. It is in relation, in front, with the extensor tendons of the toes, the tendons of the Tibialis anticus and Peroneus tertius, and the anterior tibial vessels and nerve; posteriorly, it lies in contact with the synovial membrane.

The *Internal Lateral* or *Deltoid Ligament* consists of two layers, superficial and deep. The *superficial layer* is a strong, flat, triangular band, attached, above, to the apex and

126.—Ankle-Joint: Tarsal and Tarso-Metatarsal Articulations. Internal View. Right Side.

anterior and posterior borders of the inner malleolus. The most anterior fibres pass forwards to be inserted into the scaphoid; the middle descend almost perpendicularly to be inserted into the os calcis; and the posterior fibres pass backwards and outwards to be attached to the inner side of the astragalus. The *deeper layer* consists of a short, thick, and strong fasciculus, which passes from the apex of the malleolus to the inner surface of the astragalus, below the articular surface. This ligament is covered in by the tendons of the Tibialis posticus and Flexor longus digitorum muscles.

The *External Lateral Ligament* (fig. 127) consists of three fasciculi, taking different directions, and separated by distinct intervals.

The *anterior fasciculus*, the shortest of the three, passes from the anterior margin of the summit of the external malleolus, downwards and forwards, to the astragalus, in front of its external articular facet.

The *posterior fasciculus*, the most deeply seated, passes from the depression at the inner and back part of the external malleolus to the astragalus, behind its external malleolar facet. Its fibres are directed obliquely downwards and inwards.

The *middle fasciculus*, the longest of the three, is a narrow rounded cord, passing from the apex of the external malleolus downwards and slightly backwards to the middle of the outer side of the os calcis. It is covered by the tendons of the Peroneus longus and brevis. There is no posterior ligament, its place being supplied by the transverse ligament of the tibia and fibula.

The *Synovial Membrane* invests the cartilaginous surfaces of the tibia and fibula, and sends a duplicature upwards between their lower extremities; it is then reflected on the inner surface of the ligaments surrounding the joint, and covers the upper surface of the astragalus and its two lateral facets below.

Relations. The tendons, vessels, and nerves in connection with this joint are, in front, from within outwards, the Tibialis anticus, Extensor proprius pollicis, anterior

127.—Ankle-Joint: Tarsal and Tarso-Metatarsal Articulations. External View. Right Side.

tibial vessels, anterior tibial nerve, Extensor communis digitorum, and Peroneus tertius; behind, from within outwards, Tibialis posticus, Flexor longus digitorum, posterior tibial vessels, posterior tibial nerve, Flexor longus pollicis, and, in the groove behind the external malleolus, the tendons of the Peroneus longus and brevis.

The *Arteries* supplying the joint are derived from the malleolar branches of the anterior tibial and peroneal.

The *Nerves* are derived from the anterior tibial.

Actions. The movements of this joint are limited to flexion and extension. There is no lateral motion.

5. ARTICULATIONS OF THE TARSUS

These articulations may be subdivided into three sets: 1. The articulation of the first row of tarsal bones. 2. The articulation of the second row of tarsal bones. 3. The articulation of the two rows with each other.

1. ARTICULATION OF THE FIRST ROW OF TARSAL BONES

The articulation between the astragalus and os calcis is an arthrodial joint, connected together by three ligaments.

   External Calcaneo-Astragaloid.   Interosseous.
   Posterior Calcaneo-Astragaloid.   Two Synovial Membranes.

The *External Calcaneo-Astragaloid Ligament* (fig. 127) is a short, strong fasciculus, passing from the outer surface of the astragalus, immediately beneath its external malleolar facet, to the outer edge of the os calcis. It is placed in front of the middle fasciculus of the external lateral ligament of the ankle-joint, with the fibres of which it is parallel.

128.—Ligaments of Plantar Surface of the Foot.

The *Posterior Calcaneo-Astragaloid Ligament* (fig. 126) connects the posterior extremity of the astragalus with the upper contiguous surface of the os calcis; it is a short narrow band, the fibres of which are directed obliquely backwards and inwards.

The *Interosseous Ligament* forms the chief bond of union between these bones. It consists of numerous vertical and oblique fibres, attached by one extremity to the groove between the articulating surfaces of the astragalus, by the other, to a corresponding depression on the upper surface of the os calcis. It is very thick and strong, being at least an inch in breadth from side to side, and serves to unite the os calcis and astragalus solidly together.

The *Synovial Membranes* (fig. 129) are two in number; one for the posterior calcaneo-astragaloid articulation, a second for the anterior calcaneo-astragaloid joint. The latter synovial membrane is continued forwards between the contiguous surfaces of the astragalus and scaphoid bones.

2. ARTICULATIONS OF THE SECOND ROW OF TARSAL BONES

The articulations between the scaphoid, cuboid, and three cuneiform are effected by the following ligaments.

Dorsal.   Plantar.
Interosseous.

The *Dorsal Ligaments* are small bands of parallel fibres, which pass from each bone to the neighbouring bones with which it articulates.

The *Plantar Ligaments* have the same arrangement on the plantar surface.

The *Interosseous Ligaments* are four in number. They consist of strong transverse fibres, which pass between the rough non-articular surfaces of adjoining bones. There is one between the sides of the scaphoid and cuboid, a second between the internal and middle cuneiform bones, a third between the middle and external cuneiform, and a fourth between the external cuneiform and cuboid. The scaphoid and cuboid, when in contact, present each a small articulating facet, covered with cartilage, and lined either by a separate synovial membrane, or by an offset from the common tarsal synovial membrane.

3. Articulations of the Two Rows of the Tarsus with each other

These articulations consist of ligaments that may be conveniently divided into three sets. 1. The articulation of the os calcis with the cuboid. 2. The os calcis with the scaphoid. 3. The astragalus with the scaphoid.

1. The ligaments connecting the os calcis with the cuboid are four in number.

Dorsal.
{ Superior Calcaneo-Cuboid.
Internal Calcaneo-Cuboid (Interosseous).

Plantar.
{ Long Calcaneo-Cuboid.
Short Calcaneo-Cuboid.

Synovial Membrane.

The *Superior Calcaneo-Cuboid Ligament* (fig. 127) is a thin and narrow fasciculus, which passes between the contiguous surfaces of the os calcis and cuboid, on the dorsal surface of the joint.

The *Internal Calcaneo-Cuboid (Interosseous) Ligament* (fig. 127) is a short, but thick and strong, band of fibres, arising from the os calcis, in the deep groove which intervenes between it and the astragalus; being closely blended, at its origin, with the superior calcaneo-scaphoid ligament. It is inserted into the inner side of the cuboid bone. This ligament forms one of the chief bonds of union between the first and second row of the tarsus.

The *Long Calcaneo-Cuboid* (fig. 128), the most superficial of the two plantar ligaments, is the longest of all the ligaments of the tarsus, being attached posteriorly to the under surface of the os calcis, as far forwards as the anterior tubercle, and passing horizontally forwards to the tuberosity on the under surface of the cuboid bone, the more superficial fibres being continued forwards to the bases of the second, third, and fourth metatarsal bones. This ligament crosses the groove on the under surface of the cuboid bone, converting it into a canal for the passage of the tendon of the Peroneus longus.

The *Short Calcaneo-Cuboid* lies nearer to the bones than the preceding, from which it is separated by a little areolar adipose tissue. It is exceedingly broad, and about an inch in length; passing from the tuberosity at the fore part of the under surface of the os calcis, to the inferior surface of the cuboid bone behind the peroneal groove. A synovial membrane lines the contiguous surfaces of the bones, and is reflected upon the ligaments connecting them.

2. The ligaments connecting the os calcis with the scaphoid are two in number.
Superior Calcaneo-Scaphoid.
Inferior Calcaneo-Scaphoid.
Synovial Membrane.

The *Superior Calcaneo-Scaphoid* arises (fig. 127), as already mentioned, with the internal calcaneo-cuboid in the deep groove between the astragalus and os calcis, it passes forward from the inner side of the anterior extremity of the os calcis to the outer side of the scaphoid bone. These two ligaments resemble the letter Y, being blended together behind, but separated in front.

The *Inferior Calcaneo-Scaphoid* (fig. 128) is by far the largest and strongest of the two ligaments of this articulation; it is a broad and thick band of ligamentous fibres, which passes forwards and inwards from the anterior and inner extremity of the os calcis, to the under surface of the scaphoid bone. This ligament not only serves to connect the os calcis and scaphoid, but supports the head of the astragalus, forming part of the articular cavity in which it is received. Its *upper surface* is lined by the synovial membrane continued from the anterior calcaneo-astragaloid articulation. Its *under surface* is in contact with the tendon of the Tibialis posticus muscle.

3. The articulation between the astragalus and scaphoid is an enarthrodial joint; the rounded head of the astragalus being received into the concavity formed by the posterior surface of the scaphoid, the anterior articulating surface of the calcaneum, and the upper surface of the calcaneo-scaphoid ligament, which fills up the triangular interval between these bones. The only ligament of this joint is the superior astragalo-scaphoid, a broad band of ligamentous fibres, which passes obliquely forwards from the neck of the astragalus, to the superior surface of the scaphoid bone. It is thin and weak in texture, and covered by the Extensor tendons. The inferior calcaneo-scaphoid supplies the place of an inferior ligament.

The *Synovial Membrane* which lines this joint is continued forwards from the anterior calcaneo-astragaloid articulation. This articulation permits of considerable mobility; but its feebleness is such as to occasionally allow of dislocation of the astragalus.

The *Synovial Membranes* (fig. 129) found in the articulations of the tarsus are four in number: *one* for the posterior calcaneo-astragaloid articulation; a *second* for the anterior calcaneo-astragaloid and astragalo-scaphoid articulations; a *third* for the calcaneo-cuboid articulation; and a *fourth* for the articulations between the scaphoid and the three cuneiform, the three cuneiform with each other, the external cuneiform with the cuboid, and the middle and external cuneiform with the bases of the second and third metatarsal bones. The prolongation which lines the metatarsal bones, passes forwards between the external and middle cuneiform bones. A small synovial membrane is sometimes found between the contiguous surfaces of the scaphoid and cuboid bones.

129.—Oblique Section of the Articulations of the Tarsus and Metatarsus. Shewing the Six Synovial Membranes.

Actions. The movements permitted between the bones of the first row, the astragalus, and os calcis, are limited to a gliding upon each other from before backwards, and from side to side. The gliding movement which takes place between the bones of the second row is very slight, the articulation between the scaphoid and cuneiform bones being more moveable than those of the cuneiform with each other and with the cuboid. The movement which takes place between the two rows is more extensive, and consists in a sort of rotation, by means of which the sole of the foot may be slightly flexed, and extended, or carried inwards and outwards.

6. Tarso-Metatarsal Articulations

These are arthrodial joints. The bones entering into their formation are the internal, middle, external cuneiform, and cuboid, which articulate with the metatarsal bones of the five toes. The metatarsal bone of the first toe articulates with the internal cuneiform; that of the second is deeply wedged in between the internal and external cuneiform, resting against the middle cuneiform, and being the most strongly articulated of all the metatarsal bones; the third metatarsal articulates with the extremity of the external cuneiform; the fourth with the cuboid and external cuneiform; and the fifth with the cuboid. These various articular surfaces are covered with cartilage, lined by synovial membrane, and connected together by the following ligaments.

Dorsal.   Plantar.
Interosseous.

The *Dorsal Ligaments* consist of strong, flat, fibrous bands, which connect the tarsal with the metatarsal bones. The first metatarsal is connected to the internal cuneiform by a single broad, thin, fibrous band; the second has three dorsal ligaments, one from each cuneiform bone; the third has one from the external cuneiform; and the fourth and fifth have one each from the cuboid.

The *Plantar Ligaments* consist of strong fibrous bands connecting the tarsal and metatarsal bones, but disposed with less regularity than on the dorsal surface. Those for the first and second metatarsal are the most strongly marked; the second and third receive strong fibrous bands, which pass obliquely across from the internal cuneiform; the plantar ligaments of the fourth and fifth consist of a few scanty fibres derived from the cuboid.

The *Interosseous Ligaments* are three in number: internal, middle, and external. The *internal* one passes from the outer extremity of the internal cuneiform, to the adjacent angle of the second metatarsal. The *middle* one, less strong than the preceding, connects the external cuneiform with the adjacent angle of the second metatarsal. The *external* interosseous ligament connects the outer angle of the external cuneiform with the adjacent side of the third metatarsal.

The *Synovial Membranes* of these articulations are three in number: one for the metatarsal bone of the great toe, with the internal cuneiform: one for the second and third metatarsal bones, with the middle and external cuneiform; this is continuous with the great tarsal synovial membrane: and one for the fourth and fifth metatarsal bones with the cuboid. The synovial membranes of the tarsus and metatarsus are thus seen to be six in number (fig. 129).

Articulations of the Metatarsal Bones with each other

At their tarsal extremities, the metatarsal bones are connected together by dorsal, plantar, and interosseous ligaments. The *dorsal* and *plantar ligaments* pass from one metatarsal bone to another. The *interosseous ligaments* lie deeply between the rough non-articular portions of their lateral surfaces. The articular surfaces are covered by synovial membrane, continued forwards from their respective tarsal joints. At their digital extremities, they are connected to each other by the transverse metatarsal ligament, which holds them loosely together. This ligament, which is analogous to the same structure in the hand, connects the great toe with the rest of the metatarsal bones, which in this respect differs from the same structure in the hand.

Actions. The movement permitted in the tarsal ends of the metatarsal bones is limited to a slight gliding of the articular surfaces upon one another; considerable motion, however, takes place in their digital extremities.

Metatarso-Phalangeal Articulations

The heads of the metatarsal bones are connected with the concave articular surfaces of the first phalanges by the following ligaments:

Anterior or Plantar.　　　Two Lateral.
Synovial Membrane.

They are arranged precisely similar to the corresponding parts in the hand. The expansion of the extensor tendon supplies the place of a posterior ligament.

Actions. The movements permitted in the metatarso-phalangeal articulations are flexion, extension, abduction, and adduction.

Articulation of the Phalanges

The ligaments of these articulations are similar to those found in the hand; each pair of phalanges being connected by an anterior or plantar and two lateral ligaments, and their articular surfaces lined by synovial membrane. Their actions are also similar.

THE MUSCLES AND FASCIAE

The Muscles and Fasciae are described conjointly, in order that the student may consider the arrangement of the latter in his dissection of the former. It is rare for the student of anatomy in this country to have the opportunity of dissecting the fasciae separately; and it is from this reason, as well as from the close connexion that exists between the muscles and their investing aponeuroses, that they are considered together. Some general observations are first made on the anatomy of the muscles and fasciae, the special description being given in connexion with the different regions.

The Muscles are the active organs of locomotion. They are formed of bundles of reddish fibres, consisting chemically of fibrine, and endowed with the property of contractility.

Muscle is of a deep red colour, the intensity of which varies considerably with the age and health of the individual. It is composed of bundles of parallel fibres, placed side by side, and connected together by a delicate web of areolar tissue. Each fasciculus consists of numerous smaller bundles, and these of single fibres, which, from their minute size and comparatively isolated appearance, have been called *ultimate fibres*. Two kinds of ultimate muscular fibre are found in the animal body, viz., that of *voluntary* or animal life, and that of *involuntary* or organic life. The ultimate fibre of animal life is capable of being either excited or controlled by the efforts of the will, and is characterised, on microscopic examination, by its size, its uniform calibre, and the presence of minute transverse bars, which are situated at short and irregular distances throughout its whole extent. Of such is composed the muscular tissue of the trunk and limbs; the fibres of the heart, and some of those of the oesophagus: the muscles of the internal ear, and those of the urethra, present a similar structure, although they are not capable of being acted upon by the will. Involuntary muscular fibre is entirely withdrawn from the influence of volition, and is characterised, on microscopic examination, by the ultimate fibrils being homogeneous in structure, of smaller size than those of animal life, flattened, and unstriped; of such the muscles of the digestive canal, the bladder, and uterus are composed.

Each muscle is invested externally by a thin cellular layer, forming what is called its *sheath*, which not only covers its outer surface, but penetrates into its interior in the intervals between the fasciculi, surrounding these, and serving as a bond of connection between them.

The *voluntary* muscular fibres terminate at either extremity in fibrous tissue, the separate fibrillae of which being, in some cases, aggregated together, form a rounded or flattened fibrous cord or tendon; in the flat muscles, the separate fibres are arranged in flattened membranous laminae, termed *aponeuroses*; and it is in one or other of these forms, that nearly every muscle is attached to the part which it is destined to move.

The *involuntary* muscular fibres, on the contrary, form a dense interlacement, crossing each other at various angles, forming a layer of variable thickness, which usually circumscribes the wall of some cavity, which, by its contraction, it constricts.

Muscles vary considerably in their form. In the limbs, they are of considerable length, especially the more superficial ones, the deep ones being generally broad; they surround the bones, and form an important protection to the various joints. In the trunk, they are broad, flattened, and expanded, forming the parietes of the cavities which they enclose; hence the reason of the terms, *long, broad, short*, etc., used in the description of a muscle.

There is considerable variation in the arrangement of the fibres of certain muscles, in relation to the tendon to which they are attached. In some, the fibres are arranged longitudinally, and terminate at either end in a narrow tendon, so that the muscle is broad at the centre, and narrowed at either extremity: such a muscle is said to he *fusiform* in shape, as the Rectus femoris. If the fibres converge, like the plumes of a pen, to one side of a tendon, which runs the entire length of the muscle, it is said to be *penniform*, as the Peronei; or, if they converge to both sides of a tendon, they are called *bipenniform*, as the Rectus femoris; if they converge from a broad surface to a narrow tendinous point, they are then said to be *radiated*, as the Temporal and Glutei muscles.

Their size presents considerable variation: the Gastrocnemius forms the chief bulk of the back of the leg, and the fibres of the Sartorius are nearly two feet in length, whilst the Stapedius, a small muscle of the internal ear, weighs about a grain, and its fibres are not more than two lines in length. In each case, however, they are admirably adapted to execute the various movements they are required to perform.

The names applied to the various muscles have been derived: 1, from their situation, as the Tibialis, Radialis, Ulnaris, Peroneus; 2, from their direction, as the Rectus abdominis, Obliqui capitis, Transversalis; 3, from their uses, as Flexors, Extensors, Abductors, etc.; 4, from their shape, as the Deltoid, Trapezius, Rhomboideus; 5, from the number of their divisions, as the Biceps (from having two heads), the Triceps (from having three heads); 6, from their points of attachment, as the Sterno-cleido-mastoid, Sterno-hyoid, Sterno-thyroid.

In the description of a muscle, the term *origin* is meant to imply its more fixed or central attachment; and the term *insertion*, the moveable point upon which the force of the muscle is directed: this holds true, however, for only a very small number of muscles, such as those of the face, which are attached by one extremity to the bone, and by the other to the moveable integument; in the greater number, the muscle can be made to act from either extremity.

In the dissection of the muscles, the student should pay especial attention to the exact *origin, insertion*, and *actions* of each, and its more important *relations* with surrounding parts. An accurate knowledge of the points of attachment of the muscles is of great importance in the determination of their action. By a knowledge of the action of the muscles, the surgeon is able at once to explain the causes of displacement in the various forms of fracture, or the causes which produce distortion in various forms of deformities, and, consequently, to adopt appropriate treatment in each case. The relations, also, of some of the muscles especially those in immediate apposition with the larger blood-vessels; and the surface-markings they produce should be especially remembered, as they form most useful guides to the surgeon in the application of a ligature to these vessels.

The Fasciae (*fascia*, a bandage) are fibro-areolar or aponeurotic laminae, of variable thickness and strength, found in all regions of the body, investing the softer and more delicate organs. The fasciae have been subdivided, from the structure which they present, into two groups, fibro-areolar or superficial fasciae, and aponeurotic or deep fasciae.

The *fibro-areolar fascia* is found immediately beneath the integument over almost the entire surface of the body, and is generally known as the *superficial fascia*. It connects the skin with the deep or aponeurotic fascia, and consists of fibro-areolar tissue, containing in its meshes pellicles of fat in varying quantity. In the eyelids and scrotum, where adipose tissue is never deposited, this tissue is very liable to serous infiltration. This fascia varies in thickness in different parts of the body: in the groin it is so thick as to be capable of being subdivided into several laminae, but in the palms of the hands it is of extreme thinness, and intimately adherent to the integument. The superficial fascia is capable of separation into two or more layers, between which are found the superficial vessels and nerves, and superficial lymphatic glands; as the superficial epigastric vessels in the abdominal region, the radial and ulnar veins in the forearm, the saphenous veins in the leg and thigh, as well as in certain situations cutaneous muscles, as the Platysma myoides in the neck, Orbicularis palpebrarum around the eyelids. It is most distinct at the lower part of the abdomen, the scrotum, perinaeum, and in the extremities; is very thin in those regions where muscular fibres are inserted into the integument, as on the side of the neck, the face, and around the margin of the anus, and almost entirely wanting in the palms of the hands and soles of the feet, where the integument is adherent to the subjacent aponeurosis. The superficial fascia connects the skin to the subjacent parts, serves as a soft nidus, for the passage of vessels and nerves to the integuments, and retains the warmth of the body from the adipose tissue contained in its areolae, being a bad conductor of caloric.

The *aponeurotic* or *deep fascia* is a dense inelastic and unyielding fibrous membrane, forming sheaths for the muscles, and affording them broad surfaces for attachment, it consists of shining tendinous fibres, placed parallel with one another, and connected together by other fibres disposed in a reticular manner. It is usually exposed on the removal of the superficial fascia, forming a strong investment, which not only binds down collectively the muscles in each region, but gives a separate sheath to each, as well as to the vessels and nerves. The fasciae are thick in unprotected situations, as on the outer side of a limb, and thinner on the inner side. By Bichat, aponeurotic fasciae were divided into two classes, aponeuroses of insertion, and aponeuroses of investment.

The *aponeuroses of insertion* serve for the insertion of muscles. Some of these are formed by the expansion of a tendon into an aponeurosis, as, for instance, the tendon of the Sartorius; others do not originate in tendons, as the aponeuroses of the abdominal muscles.

The *aponeuroses of investment* form a sheath for the entire limb, as well as for each individual muscle. Many aponeuroses, however, serve both for investment and insertion. Thus the deep fascia on the front of the leg gives attachment to the muscles in this region; and the aponeurosis of insertion given off from the tendon of the Biceps is continuous with the investing fascia of the forearm, and gives origin to the muscles in this region. The deep fasciae assist the muscles in their action, by the degree of tension and pressure they make upon their surface; and in certain situations this is increased and regulated by muscular action, as, for instance, by the Tensor vaginae femoris and Gluteus maximus in the thigh, by the Biceps in the leg, and Palmaris longus in the hand. In the limbs, the fasciae not only invest the entire limb, but give off septa, which separate the various muscles, and are attached beneath to the periosteum; these prolongations of fasciae are usually spoken of as intermuscular septa.

The Muscles and Fasciae may be arranged, according to the general division of the body, into, 1. Those of the head, face, and neck. 2. Those of the trunk. 3. Those of the upper extremity. 4. Those of the lower extremity.

MUSCLES AND FASCIAE OF THE HEAD AND FACE

The Muscles of the Head and Face consist of ten groups, arranged according to the region in which they are situated.

1. Cranial Region.
2. Auricular Region.
3. Palpebral Region.
4. Orbital Region.
5. Nasal Region.
6. Superior Maxillary Region.
7. Inferior Maxillary Region.
8. Inter-Maxillary Region.
9. Temporo-Maxillary Region.
10. Pterygo-Maxillary Region.

The Muscles contained in each of these groups are the following.

1. *Epicranial Region.*
Occipito-frontalis.

2. *Auricular Region.*
Attollens aurem.
Attrahens aurem.
Retrahens aurem.

3. *Palpebral Region.*
Orbicularis palpebrarum.
Corrugator supercilii.
Tensor tarsi.

4. *Orbital Region.*
Levator palpebrae.
Rectus superior.
Rectus inferior.
Rectus internus.
Rectus externus.
Obliquus superior.
Obliquus inferior.

5. *Nasal Region.*
Pyramidalis nasi.
Levator labii superioris alaeque nasi.
Levator proprius alae nasi posterior.
Levator proprius alae nasi anterior.
Compressor nasi.
Compressor narium minor.
Depressor alae nasi.

6. *Superior Maxillary Region.*
Levator labii superioris proprius.
Levator anguli oris.
Zygomaticus major.
Zygomaticus minor.

7. *Inferior Maxillary Region.*
Levator labii inferioris.
Depressor labii inferioris.
Depressor anguli oris.

130.—Dissection of the Head, Face, and Neck.

1 Dissection of SCALP

2.3. of AURICULAR REGION

4.5.6. of FACE

7.8. of NECK

8. *Inter-Maxillary Region.*
Buccinator.
Risorius.
Orbicularis oris.

9. *Temporo-Maxillary Region.*
Masseter.
Temporal.

10. *Pterygo-Maxillary Region.*
Pterygoideus externus.
Pterygoideus internus.

1. Epicranial Region—Occipito-Frontalis

Dissection (fig. 130). The head being shaved, and a block placed beneath the back of the neck, make a vertical incision through the skin from before backwards, commencing at the root of the nose in front, and terminating behind at the occipital protuberance; make a second incision in a horizontal direction along the forehead and around the side of the head, from the anterior to the posterior extremity of the preceding. Raise the skin in front from the subjacent muscle from below upwards; this must be done with extreme care, on account of their intimate union. The tendon of this muscle is best avoided by removing the integument from the outer surface of the vessels and nerves which lie between the two.

The *superficial fascia* in the epicranial region is a firm, dense layer, intimately adherent to the integument, and to the Occipito-frontalis and its tendinous aponeurosis; it is continuous, behind, with the superficial fascia at the back part of the neck; and, laterally, is continued over the temporal aponeurosis: it contains between its layers the small muscles of the auricle, and the superficial temporal vessels and nerves.

The *Occipito-frontalis* (fig. 131) is a broad musculo-fibrous layer, which covers over the whole of one side of the vertex of the skull, from the occiput to the eyebrow. It consists of two muscular bellies, separated by an intervening tendinous aponeurosis.

131.—Muscles of the Head, Face, and Neck.

The *occipital portion*, thin, quadrilateral in form, and about an inch and a half in length, arises from the outer two-thirds of the superior curved line of the occipital bone, and from the mastoid portion of the temporal. Its fibres of origin are tendinous, but they soon become muscular, and ascend in a parallel direction to terminate in the tendinous aponeurosis. The *frontal portion* is thin, of a quadrilateral form, and intimately adherent to the skin. It is broader, its fibres are longer, and their structure more pale than the occipital portion. Its internal fibres are continuous with those of the Pyramidalis nasi. Its middle fibres become blended with the Corrugator supercilii and Orbicularis: and the outer fibres are also blended with the latter muscle over the external angular process. The inner margins of the two frontal portions of the muscle are joined together for some distance above the root of the nose; but between the occipital portions there is a considerable but variable interval.

The *aponeurosis* covers over the whole of the vertex of the skull without any separation into two lateral parts, and is connected with the occipital and frontal portions of the muscle. Behind, it is attached, in the interval between the occipital origins, to the occipital protuberance and superior curved lines above the attachment of the trapezius; in front, it forms a short angular prolongation between the frontal portions; and on each side, it has connected with it the Attollens and Attrahens aurem muscles: in this situation it loses its aponeurotic character, and is continued over the temporal fascia to the zygoma by a layer of laminated areolar tissue. This aponeurosis is closely connected to the integument by a dense fibro-cellular tissue, which contains much granular fat, and in which ramify the numerous vessels and nerves of the integument; it is loosely connected with the pericranium by a quantity of loose cellular tissue, which allows of a considerable degree of movement of the integument.

Nerves. The Occipito-frontalis is supplied (frontal portion) by the supra-orbital and facial nerves; (occipital portion) by the posterior auricular branch of the facial and the small occipital.

Actions. This muscle raises the eyebrows and the skin over the root of the nose; at the same time it throws the integument of the forehead into transverse wrinkles, a predominant expression in the emotions of delight. It also moves the scalp from before backwards, by bringing alternately into action the occipital and frontal portions.

Auricular Region (fig. 131).
Attollens Aurem.   Attrahens Aurem.
Retrahens Aurem.

These three small muscles are placed immediately beneath the skin around the external ear. In man, in whom the external ear is almost immoveable, they are rudimentary. They are the analogues of large and important muscles in some of the mammalia.

Dissection. This requires considerable care, and should be performed in the following manner. To expose the Attollens aurem; draw the pinna or broad part of the ear downwards, when a tense band will be felt beneath the skin, passing from the side of the head to the upper part of the concha; by dividing the skin over the tendon, in a direction from below upwards, and then reflecting it on each side, the muscle is exposed. To bring into view the Attrahens aurem, draw the helix backwards by means of a hook, when the muscle will be made tense, and may be exposed in a similar manner to the preceding. To expose the Retrahens aurem, draw the pinna forwards, when the muscle being made tense may be felt beneath the skin, at its insertion into the back part of the concha, and may be exposed in the same manner as the other muscles.

The *Attollens Aurem* (superior auriculae), the largest of the three, is thin, and of a radiated form; it arises from the aponeurosis of the Occipito-frontalis, and is inserted by a thin, flattened tendon into the upper and anterior part of the concha.

Relations. Externally, with the integument; internally, with the Temporal aponeurosis.

The *Attrahens Aurem* (anterior auriculae), the smallest of the three, is of a triangular form, very thin in texture, and its fibres pale and indistinct. It arises from the lateral edge of the aponeurosis of the Occipito-frontalis; its fibres converge to be inserted into the front of the helix.

Relations. Externally, with the skin; *internally*, with the temporal fascia, which separates it from the temporal artery and vein.

The *Retrahens Aurem* (posterior auriculae) consists of two or three fleshy fasciculi, which arise from the mastoid portion of the temporal bone by short aponeurotic fibres. They are inserted into the back part of the concha.

Relations. Externally, with the integument; *internally*, with the mastoid portion of the temporal bone.

Nerves. The Attollens aurem is supplied by the small occipital; the Attrahens aurem, by the facial and auriculo-temporal branch of the inferior maxillary, and the Retrahens aurem, by the posterior auricular branch of the facial.

Actions. In man these muscles possess very little action; their use is sufficiently expressed in their names.

PALPEBRAL REGION (fig. 131).
Orbicularis Palpebrarum.   Levator Palpebrae.
Corrugator Supercilii.   Tensor Tarsi.

Dissection (fig. 130–4). In order to expose the muscles of the face, continue the longitudinal incision made in the dissection of the Occipito-frontalis, down the median line of the face to the tip of the nose, and from this point onwards to the upper lip; another incision should be carried along the margin of the lip to the angle of the mouth, and transversely across the face to the angle of the jaw. The integument should also be divided by an incision made in front of the external ear, from the angle of the jaw, upwards, to the transverse incision made in exposing the Occipito-frontalis. These incisions include a square-shaped flap which should be carefully removed in the direction marked in the figure, as the muscles at some points are intimately adherent to the integument.

The *Orbicularis Palpebrarum* is a sphincter muscle which surrounds the whole circumference of the orbit and eyelids. It arises from the internal angular process of the frontal bone, from the nasal process of the superior maxillary in front of the lachrymal groove, and from the anterior surface and borders of a short tendon, the Tendo palpebrarum, placed at the inner angle of the orbit. The muscle, thus arising, forms a broad, thin, and flat plane of elliptical fibres, which cover the eyelids, surround the circumference of the orbit, and spread out over the temple, and downwards on the cheek, becoming blended with the Occipito-frontalis and Corrugator supercilii. The palpebral portion (ciliaris) of the Orbicularis is thin and pale; it arises from the bifurcation of the Tendo palpebrarum, and forms a series of concentric curves, which are united on the outer side of the eyelids at an acute angle by a cellular raphe, some being inserted into the external tarsal ligament and malar bone. The orbicular portion (orbicularis latus) is thicker, of a reddish colour, its fibres well developed, forming a complete ellipse.

The *tendo palpebrarum* (oculi) is a short tendon, about two lines in length and one in breadth, attached to the nasal process of the superior maxillary bone anterior to the lachrymal groove. Crossing the lachrymal sac, it divides into two parts, each division being attached to the inner extremity of the corresponding tarsal cartilage. As the tendon crosses the lachrymal sac, a strong aponeurotic lamina is given off from its posterior surface, which expands over the sac, and is attached to the ridge on the lachrymal bone. This is the reflected aponeurosis of the Tendo palpebrarum.

Relations. By its *superficial surface*, the orbicular portion is closely adherent to the integument, more especially over the upper segment of the muscle; the palpebral portion being separated from the skin by loose areolar tissue. By its *deep surface*, above, with the Occipito-frontalis and Corrugator supercilii, with which it is intimately blended, and with the supra-orbital vessels and nerve; below, it covers the lachrymal sac and the origin of the Levator labii superioris, Levator labii superioris alaeque nasi, and the Zygomaticus major and minor muscles. *Internally*, it is occasionally blended with the Pyramidalis nasi. *Externally*, it lies on the temporal fascia. On the eyelids, it is separated from the conjunctiva by a fibrous membrane and the tarsal cartilages.

The *Corrugator Supercilii* is a small, narrow, pyramidal muscle, placed at the inner extremity of the eyebrow beneath the Occipito-frontalis and Orbicularis palpebrarum muscles. It arises from the inner extremity of the superciliary ridge; its fibres pass upwards and outwards, to be inserted into the under surface of the orbicularis, opposite the middle of the orbital arch.

Relations. By its *anterior surface*, with the Occipito-frontalis and Orbicularis palpebrarum muscles. By its *posterior surface*, with the frontal bone and supra-orbital vessels and nerve.

The *Levator Palpebrae* will be described with the muscles of the orbital region.

The *Tensor Tarsi* is a small thin muscle, about three lines in breadth and six in length, situated at the inner side of the orbit, beneath the Tendo oculi. It arises from the crest and adjacent part of the orbital surface of the lachrymal bone, and passing across the lachrymal sac, divides into two slips, which cover the lachrymal canals, and are inserted into the tarsal cartilages near the Puncta lachrymalia. Its fibres appear to be continuous with those of the palpebral portion of the Orbicularis; it is occasionally very indistinct.

Nerves. The Orbicularis palpebrarum and Corrugator supercilii are supplied by the facial and supra-orbital nerves; the Tensor tarsi by the facial.

132.—Muscles of the Right Orbit.

Actions. The Orbicularis palpebrarum is the sphincter muscle of the eyelids. The palpebral portion acts involuntarily in closing the lids, and independently of the orbicular portion, which is subject to the will. When the entire muscle is brought into action, the integuments of the forehead, temple, and cheek are drawn inwards towards the inner angle of the eye, and the eyelids are firmly closed. The Levator palpebrae is the direct antagonist of this muscle; it raises the upper eyelid, and exposes the globe. The Corrugator supercilii draws the eyebrow downwards and inwards, producing the vertical wrinkles of the forehead. This muscle may be regarded as the principal agent in the expression of grief. The Tensor tarsi draws the eyelids and the extremities of the lachrymal canals inwards, and compresses them against the surface of the globe of the eye; thus placing them in the most favourable situation for receiving the tears. It serves, also, to compress the lachrymal sac.

ORBITAL REGION (fig. 132)

Levator Palpebrae.   Rectus Internus.
Rectus Superior.   Rectus Externus.
Rectus Inferior.   Obliquus Superior.
           Obliquus Inferior.

133.—The relative Position and Attachment of the Muscles of the Left Eyeball.

Dissection. To open the cavity of the orbit, the skull-cap and brain should be first removed; then saw through the frontal bone at the inner extremity of the supra-orbital ridge, and externally at its junction with the malar. The thin roof of the orbit should then be comminuted by a few slight blows with the hammer, and the superciliary portion of the frontal bone driven forwards by a smart stroke; but must not be removed. The several fragments may then be detached, when the periosteum of the orbit will be exposed: this being removed, together with the fat which fills the cavity of the orbit, the several muscles of this region can be examined. To facilitate their dissection, the globe of the eye should be distended; this may be effected by puncturing the optic nerve near the eyeball, with a curved needle, and pushing it onwards into the globe. Through this aperture the point of a blow-pipe should be inserted, and a little air forced into the cavity of the eyeball; then apply a ligature around the nerve, so as to prevent the air escaping. The globe should now be drawn forwards, when the muscles will be put upon the stretch.

The *Levator Palpebrae* is a thin, flat, triangular muscle. It arises from the under surface of the lesser wing of the sphenoid, immediately above the optic foramen; and is inserted, by a broad aponeurosis, into the upper border of the superior tarsal cartilage. At its origin it is narrow and tendinous, but soon becomes broad and fleshy, and finally terminates in a broad aponeurosis.

Relations. By its *upper surface*, with the frontal nerve and artery, the periosteum of the orbit; and in front with the inner surface of the broad tarsal ligament. By its *under surface*, with the Superior rectus; and, in the lid, with the conjunctiva.

The *Rectus Superior* (*Attollens*), the thinnest and narrowest of the four Recti, arises from the upper margin of the optic foramen, beneath the Levator palpebrae and Superior oblique, and from the fibrous sheath of the optic nerve; and is inserted, by a tendinous expansion, into the sclerotic coat of the eyeball, about three or four lines from the margin of the cornea.

Relations. By its *upper surface*, with the Levator palpebrae. By its *under surface*, with the optic nerve, the ophthalmic artery, and nasal nerve; and in front with the tendon of the Superior oblique and the globe of the eye.

The *Inferior* and *Internal Recti* arise by a common tendon (the ligament of Zinn), which is attached around the circumference of the optic foramen, except at its upper and outer part. The External rectus has two heads: the upper one arises from the outer margin of the optic foramen, immediately beneath the Superior rectus; the lower head, partly from the ligament of Zinn, and partly from a small pointed process of bone on the lower margin of the sphenoidal fissure. Each muscle passes forward in the position implied by its name, to be inserted, by a tendinous expansion, into the sclerotic coat of the eyeball, about three or four lines from the margin of the cornea. Between the two heads of the External rectus is a narrow interval, through which pass the third, nasal branch of the fifth, and sixth nerves, and the ophthalmic vein. Although nearly all these muscles present a common origin, and are inserted in a similar manner into the sclerotic coat, there are certain differences to be observed in them, as regards their length and breadth. The Internal rectus is the broadest, the External the longest, and the Superior the thinnest and narrowest.

The *Superior Oblique* is a fusiform muscle, placed at the upper and inner side of the orbit, internal to the Levator palpebrae. It arises about a line above the inner margin of the optic foramen, and, passing forwards to the front and inner side of the orbit, terminates in a rounded tendon, which passes through a fibro-cartilaginous pulley attached to a depression beneath the internal angular process of the frontal bone, the contiguous surfaces of the tendon and pulley being lined by a delicate synovial membrane, and enclosed in a thin fibrous investment. The tendon is then reflected backwards and outwards beneath the Superior rectus to the outer and posterior part of the globe of the eye, and inserted into the sclerotic coat between the Superior and External recti muscles, midway between the cornea and entrance of the optic nerve.

Relations. By its *upper surface*, with the periosteum covering the roof of the orbit, and the fourth nerve. By its *under surface*, with the nasal nerve, and the upper border of the Internal rectus muscle.

The *Inferior Oblique* is a thin, narrow muscle, which arises from a depression in the orbital plate of the superior maxillary bone, immediately external to the lachrymal groove. Passing outwards and backwards beneath the Inferior rectus, it terminates in a tendinous expansion, which is inserted into the outer and posterior part of the sclerotic coat of the eyeball.

Relations. By its *superior surface*, with the globe of the eye, and with the Inferior rectus. By its *under surface*, with the periosteum covering the floor of the orbit, and with the External rectus.

Nerves. The Levator palpebrae, Inferior oblique, and all the recti excepting the External, are supplied by the third nerve; the Superior oblique by the fourth; the External rectus by the sixth.

Actions. The Levator palpebrae raises the upper eyelid, and is the direct antagonist of the Orbicularis palpebrarum. The four Recti muscles are attached in such a manner to the globe of the eye, that, acting singly, they will turn it either upwards, downwards, inwards, or outwards, as expressed by their names. If any two Recti act together, they carry the globe of the eye in the diagonal of these directions, viz. upwards and inwards, upwards and outwards, downwards and inwards, or downwards and outwards. By some anatomists, these muscles have been considered the chief agent in adjusting the sight at different distances, by compressing the globe, and so lengthening its antero-posterior diameter. The Oblique are the "rotatory muscles" of the eyeball. The Superior oblique acting alone, would rotate the globe, so as to carry the pupil outwards and downwards to the lower and outer side of the orbit; the Inferior oblique rotating the globe in such a direction, as to carry the pupil upwards and outwards to the upper and outer angle of the eye.

Surgical Anatomy. The position and exact point of insertion of the tendons of the Internal and External recti muscles into the globe, should be carefully examined from the front of the eyeball, as the surgeon is often required to divide one or the other muscle for the cure of strabismus. In convergent strabismus, which is the most common form of the disease, the eye is turned inwards, requiring the division of the Internal rectus. In the divergent form, which is more rare, the eye is turned outwards, the External rectus being especially implicated. The deformity produced in either case is considerable, and is easily remedied by division of one or the other muscle. This operation is readily effected by having the lids well separated by retractors held by an assistant, and the eyeball being drawn outwards by a blunt hook; the conjunctiva should be raised by a pair of forceps, and divided immediately beneath the lower border of the tendon of the Internal rectus, a little behind its insertion into the sclerotic; the submucous areolar tissue is then divided, and into the small aperture thus made a blunt hook is passed upwards between the muscle and the globe, and the tendon of the muscle and conjunctiva covering it divided by a pair of blunt-pointed scissors. Or the tendon may be divided by a subconjunctival incision, one blade of the scissors being passed upwards between the tendon and the conjunctiva, and the other between the tendon and sclerotic. The student, when dissecting these muscles, should remove on one side of the subject the conjunctiva from the front of the eye, in order to see more accurately the position of these tendons, and on the opposite side the operation may be performed.

NASAL REGION (fig. 131)

Pyramidalis Nasi.
Levator Labii Superioris Alaeque Nasi.
Levator Proprius Alae Nasi Posterior.
Levator Proprius Alae Nasi Anterior.
Compressor Nasi.
Compressor Narium Minor.
Depressor Alae Nasi.

The *Pyramidalis Nasi* is a small pyramidal slip of muscular fibre, prolonged downwards from the Occipito-frontalis upon the bridge of the nose, where it becomes tendinous, and blends with the Compressor nasi. As the two muscles descend, they diverge, leaving an angular interval between them, which is filled up by cellular tissue.

Relations. By its *upper surface*, with the skin. By its *under surface*, with the frontal and nasal bones. By its *outer border*, it is connected with the fleshy fibres of the Orbicularis palpebrarum.

The *Levator Labii Superioris Alaeque Nasi* is a thin triangular muscle, situated along the side of the nose, and extending between the inner margin of the orbit and upper lip. It arises by a pointed extremity from the upper part of the nasal process of the superior maxillary bone, and passing obliquely downwards and outwards, divides into two slips, one of which is inserted into the cartilage of the ala of the nose; the other is prolonged into the upper lip, becoming blended with the Orbicularis and Levator labii proprius.

Relations. In front, with the integument; and with a small part of the Orbicularis palpebrarum above.

Lying upon the superior maxillary bone, beneath this muscle, is a longitudinal muscular fasciculus about an inch in length. It is attached by one end near the origin of the Compressor naris, and by the other to the nasal process about an inch above it; it was described by Albinus as the Musculus "anomalus," and by Santorini, as the "Rhomboideus."

The *Levator Proprius Alae Nasi Posterior* (*dilator naris posterior*) is a small muscle, which is placed partly beneath the proper elevator of the nose and lip. It arises from the margin of the nasal notch of the superior maxilla, and from the sesamoid cartilages, and is inserted into the skin near the margin of the nostril.

The *Levator Proprius Alae Nasi Anterior* (*dilator naris anterior*) is a thin, delicate fasciculus, passing from the cartilage of the ala of the nose to the integument near its margin. This muscle is situated in front of the preceding.

The *Compressor Nasi* is a small, thin, triangular muscle, arising by its apex from the superior maxillary bone, above and a little external to the incisive fossa; its fibres proceed upwards and inwards, expanding into a thin aponeurosis which is attached to the fibro-cartilage of the nose, and is continuous on the bridge of the nose with that of the muscle of the opposite side, and with the aponeurosis of the pyramidalis nasi.

The *Compressor Narium Minor* is a small muscle, attached by one end to the alar cartilage, and by the other to the integument at the end of the nose.

The *Depressor Alae Nasi* (*myrtiformis*) is a short, radiated muscle, arising from the incisive fossa of the superior maxilla; its fibres diverge upwards and outwards, the upper, or ascending set, being inserted into the septum, and back part of the ala of the nose; the lower, or descending, into the back part of the upper segment of the orbicularis.

Nerves. All the muscles of this group are supplied by the facial nerve.

Actions. The Pyramidalis nasi draws down the inner angle of the eyebrow; by some anatomists it is also considered as an elevator of the ala, and, consequently, a dilator of the nose. The Levator labii superioris alaeque nasi draws upwards the upper lip and ala of the nose; its most important action is upon the nose, which it dilates to a considerable extent. The action of this muscle produces a marked influence over the countenance, and is the principal agent in the expression of contempt. The two Levatores alae nasi are the dilators of the pinna of the nose, and the Compressores nasi appear to act as a dilator of the nose rather than as a constrictor. The Depressor alae nasi is a direct antagonist of the preceding muscles, drawing the upper lip and ala of the nose downwards, and thereby constricting the aperture of the nares.

Superior Maxillary Region (fig. 131)

Levator Labii Superioris Proprius.   Zygomaticus major.
Levator Anguli Oris.   Zygomaticus minor.

The *Levator Labii Superioris Proprius* is a thin muscle of a quadrilateral form. It arises from the lower margin of the orbit immediately above the infra-orbital foramen, some of its fibres being attached to the superior maxilla, some to the malar bone; its fibres converge downwards and inwards to be inserted into the muscular substance of the upper lip.

Relations. By its *superficial surface*, with the lower segment of the Orbicularis palpebrarum; below, it is sub-cutaneous. By its *deep surface*, it conceals the origin of the Compressor nasi and Levator anguli oris muscles, and the infra-orbital vessels and nerves, as they escape from the infra-orbital foramen.

The *Levator Anguli Oris* (*musculus caninus*) arises by a broad attachment from the canine fossa, immediately below the infra-orbital foramen; its fibres incline downwards and a little outwards, to be inserted into the angle of the mouth, intermingling its fibres with those of the Zygomatici, the Depressor anguli oris, and the Orbicularis.

Relations. Its *superficial surface*, is covered above by the Levator labii superioris proprius and the infra-orbital vessels and nerves; below, by the integument. By its *deep surface*, it is in relation with the superior maxilla, the Buccinator, and the mucous membrane.

The *Zygomaticus major* is a slender cylindrical fasciculus, which arises from the malar bone, in front of the zygomatic suture, and, descending obliquely downwards and inwards, is inserted into the angle of the mouth, where it blends with the fibres of the Orbicularis and Depressor anguli oris.

Relations. By its *superficial surface*, occasionally with the Orbicularis palpebrarum, above; and below, with the sub-cutaneous adipose tissue. By its *deep surface*, with the malar bone, the Masseter and Buccinator muscles.

The *Zygomaticus Minor* arises from the malar bone, in front of the Zygomaticus major, immediately behind the maxillary suture, and, passing downwards and inwards, is continuous with the outer margin of the Levator labii superioris proprius.

Relations. By its *superficial surface*, with the integument and the Orbicularis palpebrarum above. By its *deep surface*, with the Levator anguli oris.

Nerves. This group of muscles is supplied by the facial nerve.

Actions. The Levator labii superioris proprius is the proper elevator of the upper lip, carrying it at the same time a little outwards. The Levator anguli oris raises the angle of the mouth and draws it inwards; whilst the Zygomatici raise the upper lip, and draw it somewhat outwards, as in laughing.

Inferior Maxillary Region (fig. 131)
Levator Labii Inferioris.
Depressor Labii Inferioris (Quadratus menti).
Depressor Anguli Oris (Triangularis menti).

Dissection. The Muscles in this region may be dissected by making a vertical incision through the integument from the margin of the lower lip to the chin: a second incision should then be carried along the margin of the lower jaw as far as the angle, and the integument carefully removed in the direction shewn in fig. 130.

The *Levator Labii Inferioris* (*Levator menti*) is to be dissected by everting the lower lip and raising the mucous membrane. It is a small conical muscular fasciculus, which arises from the incisive fossa, external to the symphysis of the lower jaw; its fibres expand downwards and forwards, to be inserted into the integument of the chin.

Relations. On its *inner surface*, with the buccal mucous membrane; in the *median line*, it is blended with the muscle of the opposite side; and on its *outer side*, with the Depressor labii inferioris.

The *Depressor Labii Inferioris* (*Quadratus menti*) is a small quadrilateral muscle, situated at the outer side of the preceding. It arises from the external oblique line of the lower jaw, between the symphysis and mental foramen, and passes obliquely upwards and inwards, to be inserted into the integument of the lower lip, its fibres blending with the Orbicularis, and with those of its fellow of the opposite side. It is continuous with the fibres of the Platysma at its origin.

Relations. By its *superficial surface*, with part of the Depressor anguli oris, and with the integument, to which it is closely connected. By its *deep surface*, with the mental vessels and nerves, the mucous membrane of the lower lip, the labial glands and the Levator labii inferioris, with which it is intimately united.

The *Depressor Anguli Oris* is a triangular muscle, arising, by its broad base, from the external oblique line of the lower jaw; its fibres pass upwards, to be inserted, by a thick and narrow fasciculus, into the angle of the mouth, being continuous with the Orbicularis, Levator anguli oris, and Zygomaticus major.

Relations. By its *superficial surface*, with the integument. By its *deep surface*, with the Depressor labii inferioris, the Platysma, and Buccinator.

Nerves. This group of muscles is supplied by the facial nerve.

Actions. The Levator labii inferioris raises the lower lip, and protrudes it forwards; at the same time it wrinkles the integument of the chin. The Depressor labii inferioris draws the lower lip directly downwards and a little outwards. The Depressor anguli oris depresses the angle of the mouth, being the great antagonist to the Levator anguli oris and Zygomaticus major: acting with those muscles, it will draw the angle of the mouth directly backwards.

Inter-Maxillary Region

Orbicularis Oris.   Buccinator.   Risorius.

Dissection. The dissection of these muscles may be considerably facilitated by filling the cavity of the mouth with tow, so as to distend the cheeks and lips; the mouth should then be closed by a few stitches, and the integument carefully removed from the surface.

The *Orbicularis Oris* is a sphincter muscle, elliptic in form, composed of concentric fibres, which surround the orifice of the mouth. It consists of two thick semicircular planes of muscular fibre, which surround the oral aperture, and interlace on either side with those of the Buccinator and other muscles inserted into this part. On the free margin of the lips the muscular fibres are continued uninterruptedly from one side to the other, forming a roundish fasciculus of fine pale fibres closely approximated. To the outer part of each segment some special fibres are added, by which the lips are connected directly with the maxillary bones and septum of the nose. The additional fibres for the upper segment consist of four bands, two of which (Accessorii orbicularis superioris) arise from the alveolar border of the superior maxilla, opposite the incisor teeth, and arching outwards on each side, are continuous at the angles of the mouth with the other muscles inserted into this part.

The two remaining muscular slips, called the Naso-labialis, connect the upper lip to the septum of the nose: as they descend from the septum, an interval is left between them, which corresponds to that left by the divergence of the accessory portions of the Orbicularis above described. It is this interval which forms the depression seen on the surface of the skin beneath the septum of the nose. Those for the lower segment (Accessorii orbicularis inferior) arise from the inferior maxilla, external to the Levator labii inferioris, near the root of the canine teeth, being separated from each other by a considerable interval; arching outwards to the angles of the mouth, they join the Buccinator and the other muscles attached to this part.

Relations. By its *superficial surface*, with the integument, to which it is closely connected. By its *deep surface*, with the mucous membrane, the labial glands, and coronary vessels. By its *outer circumference*, it is blended with the numerous muscles, which converge to the mouth from various parts of the face. Its *inner circumference* is free, and covered by mucous membrane.

The *Buccinator* is a broad, thin muscle, quadrilateral in form, occupying the interval between the jaws at the side of the face. It arises, above, from the external surface of the alveolar process of the upper jaw, between the first molar tooth and the tuberosity; below, from the external surface of the alveolar process of the lower jaw, corresponding to the three last molar teeth; and, behind, from the anterior border of the pterygo-maxillary ligament. The fibres of this muscle converge towards the angle of the mouth, where those occupying its centre intersect each other, the inferior fibres being continuous with the upper segment of the Orbicularis oris; the superior fibres, with the inferior segment; but the upper and lower fibres continue forward uninterruptedly into the corresponding segment of the lip.

Relations. By its *superficial surface*, behind, with a large mass of fat, which separates it from the ramus of the lower jaw, the Masseter, and a small portion of the Temporal muscle; anteriorly, with the Zygomaticus, Risorius, Levator anguli oris, Depressor anguli oris, and Stenon's duct, which pierces it opposite the second molar tooth of the upper jaw; the transverse facial artery and vein lie parallel with its fibres, and the facial artery and vein cross it from below upwards; it is also crossed by the branches of the facial and buccal nerves. By its *internal surface*, with the buccal glands and mucous membrane of the mouth.

The Pterygo-maxillary ligament separates the Buccinator muscle from the Superior constrictor of the pharynx. It is a tendinous band, attached by one extremity to the apex of the internal pterygoid plate, and by the other, to the posterior extremity of the internal oblique line of the lower jaw. Its *inner surface* corresponds to the cavity of the mouth, and is lined by mucous membrane. Its *outer surface* is separated from the ramus of the jaw by a quantity of adipose tissue. Its *posterior border* gives attachment to the Superior constrictor of the pharynx; its *anterior border*, to the fibres of the Buccinator.

The *Risorius (Santorini)* consists of a delicate bundle of muscular fibres, which arises in the fascia over the Masseter muscle, and passing horizontally forwards, is inserted into the angle of the mouth, joining with the fibres of the Depressor anguli oris. It is placed superficial to the Platysma, and is broadest at its outer extremity. This muscle varies much in its size and form.

Nerves. The Orbicularis oris is supplied by the facial, the Buccinator by the facial and buccal branch of the inferior maxillary nerve.

Actions. The Orbicularis oris is the direct antagonist of all those muscles which converge to the lips from the various parts of the face, its action producing the direct closure of the lips; and its forcible action throwing the integument into wrinkles, on account of the firm connection between the latter and the surface of the muscle. The Buccinators contract and compress the cheeks, so that, during the process of mastication, the food is kept under the immediate pressure of the teeth.

Temporo-Maxillary Region (fig. 134)
Masseter.   Temporal.

The Masseter muscle has been already exposed by the removal of the integument from the side of the face (fig. 131).

The *Masseter* is a short thick muscle, somewhat quadrilateral in form, consisting of two portions, superficial and deep. The *superficial portion*, the largest part of the

134.—The Temporal Muscle, the Zygoma and Masseter having been removed.

muscle, arises by a thick tendinous aponeurosis from the malar process of the superior maxilla, and from the anterior two-thirds of the lower border of the zygomatic arch: its fibres pass downwards and backwards, to be inserted into the lower half of the ramus and angle of the lower jaw. The *deep portion* is much smaller, more muscular in texture, and the direction of its fibres is forwards; it arises from the posterior third of the lower border and whole of the inner surface of the zygomatic arch and is inserted into the upper half of the ramus and coronoid process of the jaw. The deep portion of the muscle is partly concealed, in front, by the superficial portion; behind, it is covered by the parotid gland. The fibres of the two portions are united at their insertion.

Relations. By its *superficial surface*, with the integument; above, with the Orbicularis palpebrarum and Zygomaticus major; and has passing across it transversely, Stenon's duct, the branches of the facial nerve, and the transverse facial artery. By its *deep surface*, with the ramus of the jaw, the Temporal muscle, and the Buccinator, from which it is separated by a mass of fat. Its *posterior margin* is covered by the parotid gland. Its *anterior margin* is in relation, below, with the facial artery.

At this stage of the dissection, the temporal fascia is seen covering in the Temporal muscle. It is a strong aponeurotic investment, affording attachment, by its inner surface, to the superficial fibres of this muscle. Above, it is a single uniform layer, attached to the entire extent of the temporal ridge; but below, where it is attached to the zygoma, it consists of two layers, one of which is inserted into the outer, and the other to the inner border of the zygomatic arch. A small quantity of fat, and the orbital branch of the temporal artery, are contained between these. It is covered, on its outer surface, by the aponeurosis of the Occipito frontalis, the Orbicularis palpebrarum, and Attollens and Attrahens aurem muscles; the temporal artery and vein, and ascending branches of the temporal nerves, cross it from below upwards.

Dissection. In order to expose the Temporal muscle, this fascia should be removed: this may be effected by separating it at its attachment along the upper border of the zygoma, and dissecting it upwards from the surface of the muscle. The zygomatic arch should then be divided in front at its junction with the malar bone, and, behind, near the external auditory meatus, and drawn downwards with the masseter, which should be detached from its insertion into the ramus and angle of the jaw. The whole extent of the Temporal muscle is then exposed.

The *Temporal* is a broad radiating muscle, situated at the side of the head, and occupying the entire extent of the temporal fossa. It arises from the whole of the temporal fossa, which extends from the external angular process of the frontal in front, to the mastoid portion of the temporal behind, and from the curved line on the frontal and parietal bones above, to the pterygoid ridge on the great wing of the sphenoid below. It is also attached to the inner surface of the temporal fascia. Its fibres converge as they descend, the anterior passing obliquely backwards, the posterior obliquely forwards, and the middle fibres descend vertically, and terminate in an aponeurosis, the fibres of which, radiated at its commencement, converge into a thick and flat tendon, which is inserted into the inner surface, apex, and anterior border of the coronoid process of the lower jaw.

Relations. By its *superficial surface*, with the integument, the temporal fascia, aponeurosis of the Occipito-frontalis, the Attollens and Attrahens aurem muscles, the temporal vessels and nerves, the zygoma and Masseter. By its *deep surface*, with the temporal fossa, the External pterygoid and part of the Buccinator muscles, the internal maxillary artery, and its deep temporal branches.

Nerves. Both muscles are supplied by the inferior maxillary nerve.

10. PTERYGO-MAXILLARY REGION
Internal Pterygoid.   External Pterygoid.

Dissection. The Temporal muscle having been examined, the muscles in the pterygo-maxillary region may be exposed by sawing through the base of the coronoid process, and drawing it upwards, together with the Temporal muscle, which should be detached from the surface of the temporal fossa. Divide the ramus of the jaw just below the condyle, and also, by a transverse incision extending across the commencement of its lower third, just above the dental foramen, remove the fragment, and the Pterygoid muscles will be exposed.

The *Internal Pterygoid* is a thick quadrilateral muscle, and resembles the Masseter in form, structure, and in the direction of its fibres. It arises from the pterygoid fossa, its fibres being attached to the inner surface of the external pterygoid plate of the sphenoid, and to the grooved surface of the tuberosity of the palate bone; its fibres descend downwards, outwards, and backwards, to be inserted, by strong tendinous laminae, into the lower and back part of the inner side of the ramus and angle of the lower jaw.

Relations. By its *external surface*, with the ramus of the lower jaw, from which it is separated at its upper part by the External Pterygoid, the internal lateral ligament, the internal maxillary artery, and the superior dental vessels and nerves. By its *internal surface*, with the Tensor palati, being separated from the Superior constrictor of the pharynx by a cellular interval.

The *External Pterygoid* is a short thick muscle, somewhat conical in form, being broader at its origin than at its insertion. The two extremities of the muscle are

135.—The Pterygoid Muscles, the Zygomatic Arch and a portion of the Ramus of the Jaw having been removed.

tendinous, the intervening portion being fleshy. It arises by two heads, separated by a cellular interval. The upper head is attached to the pterygoid ridge on the great ala of the sphenoid, and the portion of bone included between it and the base of the external pterygoid plate; the other, the larger fasciculus, from the outer surface of the external pterygoid plate, and part of the tuberosity of the palate bone. From this origin, its fibres proceed horizontally backwards and outwards, to be inserted into a depression on the anterior part of the neck of the condyle of the lower jaw, and into the corresponding part of the interarticular fibro-cartilage.

Relations. By its *external surface*, with the ramus of the lower jaw, the internal maxillary artery, which crosses it, the tendon of the Temporal muscle, and the Masseter. By its *internal surface*, it rests against the upper part of the Internal pterygoid, the internal lateral ligament, the middle meningeal artery, and inferior maxillary nerve; by its *upper border* it is in relation with the temporal and masseteric branches of the inferior maxillary nerve.

Nerves. These muscles are supplied by the inferior maxillary nerve.

Actions. The Temporal, Masseter, and Internal pterygoid raise the lower jaw against the upper with great force. The two latter muscles, from the obliquity in the direction of their fibres, assist the External pterygoid in drawing the lower jaw forwards upon the upper, the jaw being drawn back again by the deep fibres of the Masseter, and posterior fibres of the Temporal. The External pterygoid muscles are the direct agents in the trituration of the food, drawing the lower jaw directly forwards, so as to make the lower teeth project beyond the upper. If the muscle of one side acts, the corresponding side of the jaw is drawn forwards, and the other condyle remaining fixed, the symphysis deviates to the opposite side. The alternation of these movements on the two sides, produces trituration.

MUSCLES AND FASCIAE OF THE NECK

The muscles of the Neck may be arranged into groups, corresponding with the region in which they are situated.

These groups are nine in number.

1. Superficial Region.
2. Depressors of the Os Hyoides and Larynx.
3. Elevators of the Os Hyoides and Larynx.
4. Muscles of the Tongue.
5. Muscles of the Pharynx.
6. Muscles of the Soft Palate.
7. Muscles of the Anterior Vertebral Region.
8. Muscles of the Lateral Vertebral Region.
9. Muscles of the Larynx.

1. *Superficial Region.*
Platysma myoides.
Sterno-cleido-mastoideus.

Infra-hyoid Region.
2. *Depressors of the Os Hyoides and Larynx.*
Sterno-hyoid.
Sterno-thyroid.
Thyro-hyoid.
Omo-hyoid.

Supra-hyoid Region.
3. *Elevators of the Os Hyoides and Larynx.*
Digastric.
Stylo-hyoid.
Mylo-hyoid.
Genio-hyoid.

Lingual Region.
4. *Muscles of the Tongue.*
Genio-hyo-glossus.
Hyo-glossus.
Lingualis.
Stylo-glossus.
Palato-glossus.

5. *Muscles of the Pharynx.*
Constrictor inferior.
Constrictor medius.
Constrictor superior.
Stylo-pharyngeus.
Palato-pharyngeus.

6. *Muscles of the Soft Palate.*
Levator palati.
Tensor palati.
Azygos uvulae.
Palato-glossus.
Palato-pharyngeus.

7. *Muscles of the Anterior Vertebral Region.*
Rectus capitis anticus major.
Rectus capitis anticus minor.
Rectus lateralis.
Longus colli.

8. *Muscles of the Lateral Vertebral Region.*
Scalenus anticus.
Scalenus medius.
Scalenus posticus.

9. *Muscles of the Larynx.*
Included in the description of the Larynx.

SUPERFICIAL REGION (fig. 131)
Platysma Myoides.   Sterno-Cleido-Mastoid.

Dissection. A block having been placed at the back of the neck, and the face turned to the side opposite to that to be dissected, so as to place the parts upon the stretch, two transverse incisions are to be made: one from the chin, along the margin of the lower jaw, to the mastoid process; and the other along the upper border of the clavicle. These are to be connected by an oblique incision made in the course of the Sterno-mastoid muscle, from the mastoid process to the sternum; the two flaps of integument having been removed in the direction shewn in fig. 130, the superficial fascia will be exposed.

The *Superficial Cervical Fascia* is exposed on the removal of the integument from the side of the neck; it is an extremely thin aponeurotic lamina, which is hardly demonstrable as a separate membrane. Beneath it is found the Platysma myoides muscle, the external jugular vein, and some superficial branches of the cervical plexus of nerves.

The *Platysma Myoides* is a broad thin flat plane of muscular fibres, of an irregular quadrilateral form, placed immediately beneath the skin on each side of the neck. It arises from the clavicle and acromion, and from the fascia covering the upper part of the Pectoral, Deltoid, and Trapezius muscles; its fibres proceed obliquely upwards and inwards along the side of the neck, to be inserted into the lower jaw beneath the external oblique line, some fibres passing forwards to the angle of the mouth, and others becoming lost in the cellular tissue of the face. The most anterior fibres interlace, in front of the jaw, with the fibres of the muscle of the opposite side; those next in order become blended with the Depressor labii inferioris and the Depressor anguli oris; others are prolonged upon the side of the cheek, and interlace, near the angle of the mouth, with the muscles in this situation, and may occasionally be traced to the Zygomatic muscles, or to the margin of the Orbicularis palpebrarum. The most posterior fibres, which are lost in the skin at the side of the face, are the rudiments of a remarkable accessory fasciculus, the *Risorius Santorini*, already described. Beneath this muscle, the external jugular vein may be seen descending from the angle of the jaw to the clavicle. It is essential to remember the direction of the fibres of the Platysma, in connection with the operation of bleeding from this vessel; for if the point of the lancet is introduced in the direction of the fibres of this muscle, the orifice made will be filled up by its contraction, and blood will not flow; but if the incision is made in a direction opposite to that of the course of the fibres, they will retract, and expose the orifice in the vein, and so facilitate the flow of blood. This operation is now, however, very rarely performed.

Relations. By its *external surface*, with the integument, to which it is united closely below, but more loosely above. By its *internal surface*, below the clavicle which it covers, with the Pectoralis major, Deltoid, and Trapezius. In the *neck*, with the external and anterior jugular veins, the deep cervical fascia, the superficial cervical plexus, the Sterno-mastoid, Sterno-hyoid, Omo-hyoid, and Digastric muscles. In front of the Sterno-mastoid, it covers the sheath of the carotid vessels; and behind it, the Scaleni muscles and the nerves of the brachial plexus. On the *face*, it is in relation with the parotid gland, the facial artery and vein, and the Masseter and Buccinator muscles.

The *Deep Cervical Fascia* is exposed on the removal of the Platysma myoides. It is a strong fibrous layer, which invests the muscles of the neck, and encloses the vessels and nerves. It commences, as an extremely thin layer, at the back part of the neck, where it is attached to the spinous processes of the cervical vertebrae, and to the ligamentum nuchae; and, passing forwards to the posterior border of the Sterno-mastoid muscle, divides into two layers, one of which passes in front, and the other behind it. These join again at its anterior border; and, being continued forwards to the front of the neck, blend with the fascia of the opposite side. The superficial layer of the deep cervical fascia (that which passes in front of the Sterno-mastoid), if traced upwards, is found to pass across the parotid gland and Masseter muscle, forming the parotid and masseteric fasciae, and is attached to the lower border of the Zygoma, and more anteriorly to the lower border of the body of the jaw; if the same layer is traced downwards, it is seen to pass to the upper border of the clavicle and sternum, being pierced just above the former bone for the external jugular vein. In the middle line of the neck, this layer is thin above, and connected to the hyoid bone; but it becomes thicker below, and divides, just below the thyroid gland, into two layers, the more superficial of which is attached to the upper border of the sternum and interclavicular ligament; the deeper and stronger layer is connected to the posterior border of that bone, covering in the Sterno-hyoid and Sterno-thyroid muscles. Between these two layers is a little areolar tissue and fat, and occasionally a small lymphatic gland. The deep layer of the cervical fascia (that which lies behind the posterior surface of the Sterno-mastoid) sends numerous prolongations, which invest the muscles and vessels of the neck; if traced upwards, a process of this fascia, of extreme density, passes behind and to the inner side of the parotid gland, and is attached to the base of the styloid process and angle of the lower jaw, forming the stylo-maxillary ligament; if traced downwards and outwards, it will be found to enclose the posterior belly of the Omo-hyoid muscle, binding it down by a distinct process, which descends to be inserted into the clavicle and cartilage of the first rib. The deep layer of the cervical fascia also assists in forming the sheath which encloses the common carotid artery, internal jugular vein, and pneumogastric nerve. There are fibrous septa intervening between each of these parts, which, however, are included together in one common investment. More internally, a thin layer is continued across the trachea and thyroid gland, beneath the Sterno-thyroid muscles; and at the root of the neck this may be traced, over the large vessels, to be continuous with the fibrous layer of the pericardium.

The *Sterno-Cleido-Mastoid* (fig. 136) is a large thick muscle which passes obliquely across the side of the neck, being enclosed between the two layers of the deep cervical

136.—Muscles of the Neck, and Boundaries of the Triangles.

fascia. It is thick and narrow at its central part, but is broader and thinner at each extremity. It arises, by two distinct heads, from the sternum and clavicle. The *sternal portion* arises by a rounded fasciculus, tendinous in front, fleshy behind, from the upper and anterior part of the first piece of the sternum, and is directed upwards and backwards. The *clavicular portion* arises from the inner third of the upper surface of the clavicle, being composed of fleshy and aponeurotic fibres; it is directed perpendicularly upwards. These two portions are separated from one another, at their origin, by a triangular cellular interval; but become gradually blended, below the middle of the neck, into a thick rounded muscle, which is inserted, by a strong aponeurosis, into the outer surface of the mastoid process, from the apex to its superior border, and into the outer two-thirds of the superior curved line of the occipital bone. This muscle varies much in its extent of attachment to the clavicle: in one case it may be as narrow as the sternal portion; in another, as much as three inches in breadth. When the clavicular origin is broad, it is occasionally subdivided into numerous slips, separated by narrow intervals. More rarely the corresponding margins of the Sterno-mastoid and Trapezius have been found in contact. In the application of a ligature to the third part of the subclavian artery, it will be necessary, where the muscles have an arrangement similar to that above mentioned, to divide a portion of one or of both, in order to facilitate the operation.

This muscle serves to divide the large quadrilateral space at the side of the neck into two large triangles, an anterior and a posterior. The boundaries of the great *anterior* triangle being, in front, the median line of the neck; above, the lower border of the body of the jaw, and an imaginary line drawn from the angle of the jaw to the mastoid process; behind, the anterior border of the Sterno-mastoid muscle. The boundaries of the great *posterior* triangle are, in front, the posterior border of the Sterno-mastoid; below, the upper border of the clavicle; behind, the anterior margin of the Trapezius.

The *anterior edge* of this muscle forms a very prominent ridge beneath the skin, which forms a guide to the surgeon in making the incisions for ligature of the common carotid artery, and for oesophagotomy.

Relations. By its *superficial surface*, with the integument and Platysma, from which it is separated by the external jugular vein, the superficial branches of the cervical plexus, and the anterior layer of the deep cervical fascia. By its *deep surface*, it rests on the deep layer of the cervical fascia, the sterno-clavicular articulation, the Sterno-hyoid, Sterno-thyroid, and Omo-hyoid muscles, the posterior belly of the Digastric, Levator anguli scapula, the Splenius and Scaleni muscles. Below, with the lower part of the common carotid artery, internal jugular vein, pneumogastric, descendens noni, and communicans noni nerves, and with the deep lymphatic glands; with the spinal accessory nerve, which pierces its upper third, the cervical plexus, the sympathetic nerve, and the parotid gland.

Nerves. The Platysma-myoides is supplied by the facial and superficial cervical nerves. The Sterno-cleido-mastoid by the spinal accessory and deep branches of the cervical plexus.

Actions. The Platysma-myoides produces a slight wrinkling of the surface of the skin of the neck, in a vertical direction, when the entire muscle is brought into action. Its anterior portion, the thickest part of the muscle, depresses the lower jaw; it also serves to draw down the lower lip and angle of the mouth on each side, being one of the chief agents in the expression of melancholy. The accessory transverse fibres draw the angle of the lips upwards and outwards, as in laughing. The Sterno-mastoid muscles, when both are brought into action, serve to depress the head upon the neck, and the neck upon the chest. Either muscle, acting singly, flexes the head, and (combined with the Splenius) draws it towards the shoulder of the same side, and rotates it so as to carry the face towards the opposite side.

Infra-Hyoid Region (figs. 136, 137)
Depressors of the Os Hyoides and Larynx

Sterno-Hyoid.　　　　　Thyro-Hyoid.
Sterno-Thyroid.　　　　Omo-Hyoid.

Dissection. The muscles in this region may be exposed by removing the deep fascia from the front of the neck. In order to see the entire extent of the Omo-hyoid, it is necessary to divide the Sterno-mastoid at its centre, and turn its ends aside, and to detach the Trapezius from the clavicle and scapula, if this muscle has been previously dissected; but not otherwise.

The *Sterno-Hyoid* is a thin, narrow, ribbon-like muscle, which arises from the inner extremity of the clavicle, and the upper and posterior part of the first piece of the sternum; and, passing upwards and inwards, is inserted, by short tendinous fibres, into the lower border of the body of the os hyoides. This muscle is separated, below, from its fellow by a considerable interval; they approach one another in the middle

137.—Muscles of the Neck. Anterior View.

of their course, and again diverge as they ascend. It often presents, immediately above its origin, a transverse tendinous intersection, analogous to those in the Rectus abdominis.

Variations in Origin. The origin of this muscle presents many variations. Thus, it may be found to arise from the inner extremity of the clavicle, and the posterior sterno-clavicular ligament; or from the sternum and this ligament; from either bone alone, or from all these parts; and occasionally has a fasciculus connected with the cartilage of the first rib.

Relations. By its *superficial surface*, below, with the sternum, sternal end of the clavicle, and the Sterno-mastoid; and, above, with the Platysma and deep cervical fascia. By its *deep surface*, with the Sterno-thyroid, Crico-thyroid, and Thyro-hyoid muscles, the thyroid gland, the superior thyroid artery, the crico-thyroid and thyro-hyoid membranes.

The *Sterno-Thyroid* is situated immediately beneath the preceding muscle, but is shorter and broader than it. It arises from the posterior surface of the first bone of the sternum, beneath the origin of the Sterno-hyoid, and occasionally from the edge of the cartilage of the first rib; and is inserted into the oblique line on the side of the ala of the thyroid cartilage. These muscles are in close contact at the lower part of the neck by their inner margins; and are frequently traversed by transverse or oblique tendinous intersections, analogous to those in the Rectus abdominis.

Variations. This muscle, at its insertion, is liable to some variations. A lateral prolongation is sometimes continued as far as the os hyoides; and it is sometimes continuous with the Thyro-hyoideus and Inferior constrictor of the pharynx.

Relations. By its *anterior surface*, with the Sterno-hyoid, Omo-hyoid, and Sterno-mastoid. By its *posterior surface*, from below upwards, with the trachea, vena innominata, common carotid (and on the right side the arteria innominata), the thyroid gland and its vessels, and the lower part of the larynx. The middle thyroid vein lies along its inner border, this should be remembered in the operation of tracheotomy.

The *Thyro-Hyoid* is a small quadrilateral muscle, appearing like a continuation of the Sterno-thyroid. It arises from the oblique line on the side of the thyroid cartilage, and passes vertically upwards to be inserted into the lower border of the body, and greater cornu of the hyoid bone.

Relations. By its *external surface*, with the Sterno-hyoid and Omo-hyoid muscles. By its *internal surface*, with the thyroid cartilage, and thyro-hyoid membrane. Interposed between this muscle and the membrane, is the superior laryngeal nerve and artery.

The *Omo-hyoid* passes across the side of the neck, from the scapula to the hyoid bone. It consists of two fleshy bellies, united by a central tendon. It arises from the upper border of the scapula, and occasionally from the transverse ligament which crosses the supra-scapular notch; its extent of attachment to the scapula varying from a few lines to an inch. From this origin, the posterior belly forms a flat, narrow fasciculus, which inclines forwards across the lower part of the neck; behind the Sterno-mastoid muscle, where it becomes tendinous, it changes its direction, forming an obtuse angle, and ascends almost vertically upwards, close to the outer border of the Sterno-hyoid, to be inserted into the lower border of the body of the os hyoides, just external to the insertion of the Sterno-hyoid. The tendon of this muscle, which much varies in its length and form in different subjects, is held in its position between two lamella of the deep cervical fascia, which include it in a sheath, and are prolonged down to be attached to the cartilage of the first rib. It is by this means that the angular form of the muscle is maintained.

This muscle subdivides each of the two large triangles at the side of the neck, formed by the Sterno-mastoid, into two smaller triangles. The two posterior ones being the *posterior superior* or *sub-occipital*, and the *posterior inferior* or *subclavian*; the two anterior, the *anterior superior* or *superior carotid*, and the *anterior inferior* or *inferior carotid* triangle.

Relations. By its *superficial surface*, with the Trapezius, Subclavius, the clavicle, the Sterno-mastoid, deep cervical fascia, Platysma, and integument. By its *deep surface*, with the Scaleni, brachial plexus, sheath of the common carotid artery, and internal jugular vein, the descendens noni nerve, Sterno-thyroid and Thyro-hyoid muscles.

Nerves. All the muscles of this group, excepting the Thyro-hyoid, which is supplied by the hypo-glossal, receive their nerves from the loop of communication between the descendens and communicans noni.

Actions. These muscles serve to depress the larynx and hyoid bone, after these parts have been drawn up with the pharynx in the act of deglutition. The Omo-hyoid muscles not only depress the hyoid bone, but carry it backwards, and to one or the other side. These muscles also are tensors of the cervical fascia. The Thyro-hyoid may act as an elevator of the thyroid cartilage, when the hyoid bone ascends, drawing upwards the thyroid cartilage behind the os hyoides.

Supra-Hyoid Region (figs. 136, 137)
Elevators of the Os Hyoides—Depressors of the Lower Jaw

Digastricus.   Mylo-Hyoid.
Stylo-Hyoid.   Genio-Hyoid.

Dissection. To dissect these muscles, a block should be placed beneath the back of the neck, and the head drawn backwards, and retained in that position. On the removal of the deep fascia, the muscles are at once exposed.

The *Digastric*, so called from its consisting of two fleshy bellies united by an intermediate rounded tendon, is a small muscle, situated immediately beneath the side of the body of the lower jaw, and extending, in a curved form, from the side of the head to the symphysis of the jaw. The *posterior belly*, longer than the anterior, arises from the digastric groove on the inner side of the mastoid process of the temporal bone, and passes downwards, forwards, and inwards. The *anterior belly*, being reflected upwards and forwards, is inserted into a depression on the inner side of the lower border of the jaw, close to the symphysis. The tendon of this muscle perforates the Stylo-hyoid, and is held in connection with the side of the body of the hyoid bone by an aponeurotic loop, lined by a synovial membrane. A broad aponeurotic layer is given off from the tendon of the digastric on each side, which is attached to the hyoid bone: this is termed the *supra-hyoid aponeurosis*. It forms a strong layer of fascia between the anterior portion of the two muscles, and forms a firm investment for the other muscles of the supra-hyoid region, which lie beneath it.

The Digastric muscle divides the anterior superior triangle of the neck into two smaller triangles; the upper, or sub-maxillary, being bounded above by the lower jaw; below, by the two bellies of the Digastric muscle: the lower, or superior carotid triangle, being bounded above by the posterior belly of the Digastric; behind, by the Sterno-mastoid; below, by the Omo-hyoid. (fig. 136).

Relations. By its *superficial surface*, with the Platysma, Sterno-mastoid, part of the Stylo-hyoid muscle, and the parotid and sub-maxillary glands. By its *deep surface*, its anterior belly lies on the Mylo-hyoid, the posterior belly lies on the Stylo-glossus, Stylo-pharyngeus, and Hyo-glossus muscles, the external carotid and its lingual and facial branches, the internal carotid, internal jugular vein, and hypoglossal nerve.

The *Stylo-Hyoid* is a small, slender muscle, lying in front of, and above, the posterior belly of the Digastric. It arises from the middle of the outer surface of the styloid process; and, passing downwards and forwards, is inserted into the body of the hyoid bone, just at its junction with the greater cornu, and immediately above the Omo-hyoid.

This muscle is perforated near its insertion by the tendon of the Digastric muscle.

Relations. By its *superficial surface*, with the Sterno-mastoid and Digastric muscles, the parotid and submaxillary glands. Its *deep surface* has the same relations as the posterior belly of the Digastric.

Dissection. The Digastric and Stylo-hyoid muscles should be removed, in order to expose the next muscle.

The *Mylo-Hyoid* is a flat triangular plane of muscular fibre, situated immediately beneath the anterior belly of the Digastric, and forming, with its fellow of the opposite side, a muscular floor for the cavity of the mouth. It arises from the whole length of the mylo-hyoid ridge, from the symphysis in front, to the last molar tooth behind. The posterior fibres pass obliquely forwards, to be inserted into the body of the os hyoides. The middle and anterior fibres are inserted into the median fibrous raphe, where they join at an angle with the fibres of the opposite muscle. This median raphe is sometimes wanting; the muscular fibres of the two sides are then directly continuous with one another.

Relations. By its *superficial* or *inferior surface* with the Platysma, the anterior belly of the Digastric, the supra-hyoid fascia, the submaxillary gland, and submental vessels. By its *deep* or *superior surface*, with the Genio-hyoid, part of the Hyo-glossus, and Stylo-glossus muscles, the lingual and gustatory nerves, the sublingual gland, and the buccal mucous membrane. Wharton's duct curves around its posterior border in its passage to the mouth.

Dissection. The Mylo-hyoid should now be removed, in order to expose the muscles which lie beneath; this is effected by detaching it from its attachments to the hyoid bone and jaw, and separating it by a vertical incision from its fellow of the opposite side.

The *Genio-Hyoid* is a narrow slender muscle, situated immediately beneath the inner border of the preceding. It arises from the inferior genial tubercle on the inner side of the symphysis of the lower jaw, and descends downwards and backwards, to be inserted into the anterior surface of the body of the os hyoides. This muscle lies in close contact with its fellow of the opposite side, and increases slightly in breadth as it descends.

Relations. It lies between the Mylo-hyoid and the Genio-hyo-glossus muscles.

Nerves. The Digastric is supplied, its anterior belly, by the mylo-hyoid branch of the inferior dental; its posterior belly, by the facial and glosso-pharyngeal; the Stylo-hyoid, by the facial and glosso-pharyngeal; the Mylo-hyoid, by the mylo-hyoid branch of the inferior dental; the Genio-hyoid, by the lingual.

Actions. This group of muscles performs two very important actions. They raise the hyoid bone, and with it the base of the tongue, during the act of deglutition; or, when the hyoid bone is fixed by its depressors and those of the larynx, they depress the lower jaw. During the first act of deglutition, when the mass is being driven from the mouth into the pharynx, the hyoid bone, and with it the tongue, is carried upwards and forwards by the anterior belly of the Digastric, the Mylo-hyoid, and Genio-hyoid muscles. In the second act, when the mass is passing, the direct elevation of the hyoid bone takes place by the combined action of all the muscles; and after the food has passed, the hyoid bone is carried upwards and backwards by the posterior belly of the Digastric and Stylo-hyoid muscles, which assists in preventing the return of the morsel into the cavity of the mouth.

Lingual Region

Genio-Hyo-Glossus. Lingualis.
Hyo-Glossus. Stylo-Glossus.
Palato-Glossus.

Dissection. After completing the dissection of the preceding muscles, saw through the lower jaw just external to the symphysis. The tongue should then be drawn forwards with a hook, and its muscles, which are thus put on the stretch, may be examined.

The *Genio-Hyo-Glossus* has received its name from its triple attachment to the chin, hyoid bone, and tongue; it is a thin, flat, triangular muscle, placed vertically in the middle line, its apex corresponding with its point of attachment to the lower jaw, its base with its insertion into the tongue and hyoid bone. It arises by a short tendon from the superior genial tubercle on the inner side of the symphysis of the chin, immediately above the Genio-hyoid; from this point the muscle spreads out in a fan-like form, the inferior fibres passing downwards, to be inserted into the upper part of the body of the hyoid bone, a few being continued into the side of the pharynx; the middle fibres passing backwards, and the anterior ones upwards and forwards, to be attached to the whole length of the under surface of the tongue, from the base to the apex.

Relations. By its *internal surface*, it is in contact with its fellow of the opposite side, from which it is separated, at the back part of the tongue, by a fibro-cellular structure, which extends forwards through the middle of the organ. By its *external surface*, with the Lingualis, Hyo-glossus, and Stylo-glossus, the lingual artery and hypoglossal nerve, the gustatory nerve, and the sublingual gland. By its *upper border*, with the mucous membrane of the floor of the mouth. By its *lower border*, with the Genio-hyoid.

138.—Muscles of the Tongue. Left Side.

The *Hyo-Glossus* is a thin, flat, quadrilateral plane of muscular fibres, arising from the body, the lesser cornu, and whole length of the greater cornu of the hyoid bone, and passing almost vertically upwards, is inserted into the side of the tongue, between the Stylo-glossus and Lingualis. Those fibres of this muscle which arise from the body are directed upwards and backwards, overlapping those from the greater cornu, which are directed obliquely forwards. Those from the lesser cornu extend forwards and outwards along the side of the tongue, under cover of the portion arising from the body.

The difference in the direction of the fibres of this muscle, and their separate origin from different segments of the hyoid bone, led Albinus and other anatomists to describe it as three muscles, under the names of the Basio-glossus, the Cerato-glossus, and the Chondro-glossus.

Relations. By its *external surface*, with the Digastric, the Stylo-hyoid, Stylo-glossus, and Mylo-hyoid muscles, the gustatory and hypoglossal nerves, Wharton's duct, and the sublingual gland. By its *deep surface*, with the Genio-hyo-glossus, Lingualis, and the origin of the middle Constrictor muscle of the pharynx, the lingual artery, and the glosso-pharyngeal nerve.

The *Lingualis* is a longitudinal band of muscular fibres, situated on the under surface of the tongue, lying in the interval between the Hyo-glossus and the Genio-hyo-glossus, and extending from the base to the apex of that organ. Posteriorly, some of its fibres are lost in the base of the tongue, and others are attached to the hyoid bone. It blends with the fibres of the Stylo-glossus, in front of the Hyo-glossus, and is continued forwards as far as the apex of the tongue. It is in relation, by its under surface, with the ranine artery.

The *Stylo-Glossus*, the shortest and smallest of the three styloid muscles, arises from the anterior and outer side of the styloid process, near its centre, and from the stylo-maxillary ligament, to which its fibres in most cases are attached by a thin aponeurosis. Passing downwards and forwards, so as to become nearly horizontal in its direction, it divides upon the side of the tongue into two portions; one longitudinal, which is inserted along the side of the tongue, blending with the fibres of the Lingualis, in front of the Hyo-glossus; the other oblique, which overlaps the Hyo-glossus muscle, and decussates with its fibres.

Relations. By its *external surface*, from above downwards, with the parotid gland, the Internal pterygoid muscle, the sublingual gland, the gustatory nerve, and the mucous membrane of the mouth. By its *internal surface*, with the tonsil, the Superior constrictor muscle of the pharynx, and the Hyo-glossus muscle.

The *Palato-Glossus*, or *Constrictor Isthmi Faucium*, although one of the muscles of the tongue, serving to draw its base upwards during the act of deglutition, is more nearly associated with the soft palate, both in its situation and function; it will, consequently, be described with that group of muscles.

Nerves. The muscles of the tongue are supplied by the hypoglossal nerve, excepting the Palato-glossus, which receives its nerves from the palatine branches of Meckel's ganglion.

Actions. The movements of the tongue, although numerous and complicated, may easily be explained by carefully considering the direction of the fibres of the muscles of this organ. The *Genio-hyo-glossi*, by means of their posterior and inferior fibres, draw upwards the hyoid bone, bringing it and the base of the tongue forwards, so as to protrude the apex from the mouth. The anterior fibres will restore it to its original position by retracting the organ within the mouth. The whole length of these two muscles acting along the middle line of the tongue will draw it downwards, so as to make it concave from before backwards, forming a channel along which fluids may pass towards the pharynx, as in sucking. The *Hyo-glossi* muscles draw down the sides of the tongue, so as to render it convex from side to side. The *Linguales*, by drawing downwards the centre and apex of the tongue, render it convex from before backwards. The Palato-glossi draw the base of the tongue upwards, and the Stylo-glossi upwards and backwards.

Pharyngeal Region

Constrictor Inferior.   Constrictor Superior.
Constrictor Medius.   Stylo-pharyngeus.
Palato-pharyngeus.

Dissection (fig. 139). In order to examine the muscles of the pharynx, cut through the trachea and oesophagus just above the sternum, and draw them upwards by dividing the loose areolar tissue connecting the pharynx with the front of the vertebral column. The parts being drawn well forwards, the edge of the saw should be applied immediately behind the styloid processes, and the base of the skull sawn

139.—Muscles of the Pharynx. External View.

through from below upwards. The pharynx and mouth should then be stuffed with tow, in order to distend its cavity and render the muscles tense and easier of dissection.

The *Inferior Constrictor*, the most superficial and thickest of the three, arises from the side of the cricoid and thyroid cartilages. To the cricoid cartilage it is attached in the interval between the crico-thyroid, in front, and the articular facet for the thyroid cartilage behind. To the thyroid cartilage, it is attached to the oblique line on the side of the great ala, the cartilaginous surface behind it, nearly as far as its posterior border, and to the inferior cornu. From these attachments, the fibres spread backwards and inwards, to be inserted into the fibrous raphe in the posterior median line of the pharynx. The inferior fibres are horizontal, and overlap the commencement of the oesophagus; the rest ascend, increasing in obliquity, and overlap the Middle constrictor. The superior laryngeal nerve passes near the upper border, and the inferior, or recurrent laryngeal, beneath the lower border of this muscle, previous to their entering the larynx.

Relations. It is covered by a dense cellular membrane which surrounds the entire pharynx. *Behind*, it lies on the vertebral column and the Longus colli. *Laterally*, it is in relation with the thyroid gland, the common carotid artery, and the Sterno-thyroid muscle. By its *internal surface*, with the Middle constrictor, the Stylo-pharyngeus, Palato-pharyngeus, and the mucous membrane of the pharynx.

The *Middle Constrictor* is a flattened, fan-shaped muscle, smaller than the preceding, and situated on a plane anterior to it. It arises from the whole length of the upper border of the greater cornu of the hyoid bone, from the apex of this cornu by a tendinous origin, from the lesser cornu, and from the stylo-hyoidean ligament. The fibres diverge from their origin in various directions; the lower ones descending and being overlapped by the inferior constrictor, the middle fibres passing transversely, and the upper fibres ascending to cover in the Superior constrictor. It is inserted into the posterior median fibrous raphe, blending in the middle line with the fibres of the opposite muscle.

Relations. This muscle is separated from the Superior constrictor by the glosso-pharyngeal nerve and the Stylo-pharyngeus muscle; and from the inferior constrictor, by the superior laryngeal nerve. *Behind*, it lies on the vertebral column, the Longus colli, and the Rectus anticus major. *On each side* it is in relation with the carotid vessels, the pharyngeal plexus, and some lymphatic glands. Near its origin, it is covered by the hyo-glossus, from which it is separated by the lingual artery. It covers in the Superior constrictor, the Stylo-pharyngeus, the Palato-pharyngeus, and the mucous membrane.

The *Superior Constrictor* is a quadrilateral plane of muscular fibres, thinner and paler than those of the other Constrictors, situated at the upper part of the pharynx. It arises from the lower third of the margin of the internal pterygoid plate and its hamular process, from the contiguous portion of the palate bone and the reflected tendon of the Tensor palati muscle, from the pterygo-maxillary ligament, from the alveolar process above the posterior extremity of the mylo-hyoid ridge, and by a few fibres from the side of the tongue in connexion with the Genio-hyo-glossus. From these points, the fibres curve backwards, to be inserted into the median raphe, being also prolonged by means of a fibrous aponeurosis to the pharyngeal spine on the basilar process of the occipital bone. Its superior fibres arch beneath the Levator palati and the Eustachian tube, the interval between this border of the muscle and the basilar process being deficient in muscular fibres, and closed by fibrous membrane.

Relations. By its *outer surface*, behind, with the vertebral column. *On each side*, with the carotid vessels, the internal jugular vein, the three divisions of the eighth and the ninth nerves, the Middle constrictor which overlaps it, and the Stylo-pharyngeus. *Internally*, it covers the Palato-pharyngeus and the tonsil and is lined by mucous membrane.

The *Stylo-pharyngeus* is a long, slender muscle, round above, broad and thin below. It arises from the inner side of the base of the styloid process, passes downwards and inwards to the side of the pharynx between the Superior and Middle constrictors, and spreading out beneath the mucous membrane, some of its fibres are lost in the Constrictor muscles, and others joining with the Palato-pharyngeus, are inserted into the posterior border of the thyroid cartilage. The glosso-pharyngeal nerve runs on the outer side of this muscle, and crosses over it in passing forward to the tongue.

Relations. Externally, with the Stylo-glossus muscle, the external carotid artery, the parotid gland, and the Middle constrictor. *Internally*, with the internal carotid, the internal jugular vein, the Superior constrictor, Palato-pharyngeus and mucous membrane.

Nerves. The muscles of this group are supplied by branches from the pharyngeal plexus and glosso-pharyngeal nerve; and the Inferior constrictor, by an additional branch from the external laryngeal nerve.

Actions. When deglutition is about to be performed, the pharynx is drawn upwards and dilated in different directions, to receive the morsel propelled into it from the mouth. The Stylo-pharyngei, which are much farther removed from one another at their origin than at their insertion, draw upwards and outwards the sides of this cavity, the breadth of the pharynx in the antero-posterior direction being increased, by the larynx and tongue being carried forwards in their ascent. As soon as the morsel is received in the pharynx, the elevator muscles relax, the bag descends, and the Constrictors contract upon the morsel, and convey it gradually downwards into the oesophagus. The pharynx also exerts an important influence in the modulation of the voice, especially in the production of the higher tones.

Palatal Region

Levator Palati.   Azygos Uvulae.
Tensor Palati.   Palato-glossus.
           Palato-pharyngeus.

Dissection (fig. 140). Lay open the pharynx from behind, by a vertical incision extending from its upper to its lower part, and the posterior surface of the soft palate is exposed. Having fixed the uvula so as to make it tense, the mucous membrane and glands should be carefully removed from the posterior surface of the soft palate and the muscles of this part are at once exposed.

The *Levator Palati* is a long, thin muscle, placed on the outer side of the posterior aperture of each nasal fossa. It arises from the apex of the basilar surface of the petrous portion of the temporal bone and from the adjoining cartilaginous portion of the

140.—Muscles of the Soft Palate. The Pharynx being laid open from behind.

Eustachian tube; after passing into the interior of the pharynx, above the upper concave margin of the Superior constrictor, it descends obliquely downwards and inwards, its fibres spreading out in the posterior surface of the soft palate as far as the middle line, where they blend with those of the opposite side.

Relations. Externally, with the Tensor palati and Superior constrictor. *Internally*, it is lined by the mucous membrane of the pharynx. *Posteriorly*, with the mucous lining of the soft palate. This muscle must be removed and the pterygoid attachment of the Superior constrictor dissected away, in order to expose the next muscle.

The Circumflexus or *Tensor Palati* is a broad, thin, flat muscle, placed on the outer side of the preceding, and consisting of two distinct portions, a vertical and horizontal. The vertical portion arises by a broad, thin, and flat lamella from the scaphoid fossa at the base of the internal pterygoid plate, its fibres of origin extending as far back as the spine of the sphenoid; it also arises from the anterior aspect of the cartilaginous portion of the Eustachian tube, descending vertically downwards between the internal pterygoid plate and the inner surface of the Internal pterygoid muscle; it terminates in a tendon which winds around the hamular process, being retained in this situation by a tendon of origin of the Internal pterygoid muscle, and lubricated by a synovial membrane. The tendon or horizontal portion then passes horizontally inwards, and expands into a broad aponeurosis on the anterior surface of the soft palate, which unites in the median line with the aponeurosis of the opposite muscle, the fibres of which are attached anteriorly to the transverse ridge on the posterior border of the horizontal portion of the palate bone.

Relations. Externally, with the Internal pterygoid. *Internally*, with the Levator palati, from which it is separated by the Superior constrictor, and the internal pterygoid plate. In the soft palate its aponeurotic expansion is anterior to that of the Levator palati, being covered by mucous membrane.

The *Azygos Uvulae* is not a single muscle as implied by its name, but a pair of small cylindrical fleshy fasciculi, placed side by side in the median line of the soft palate. Each muscle arises from the posterior nasal spine of the palate bone, and from the contiguous tendinous aponeurosis of the soft palate, and descending vertically downwards, is inserted into the uvula.

Relations. Anteriorly, with the tendinous expansion of the Levatores palati; *behind*, with the mucous membrane.

The two next muscles are exposed by removing the mucous membrane which covers the pillars of the soft palate on each side throughout their whole extent.

The *Palato-Glossus* (or, *Constrictor Isthmi Faucium*) is a small fleshy fasciculus, narrower in the middle than at either extremity, forming, with the mucous membrane covering its surface, the anterior pillar of the soft palate. It arises from the soft palate on each side of the uvula, and passing forwards and outwards in front of the tonsil, is inserted into the side and upper surface of the tongue, where it blends with the fibres of the Stylo-glossus muscle. In the soft palate, the fibres of origin of this muscle are continuous with those of the opposite side, and with the Palato-pharyngeus.

The *Palato-Pharyngeus* is a long fleshy fasciculus, narrower in the middle than at either extremity, forming, with the mucous membrane covering its surface, the posterior pillar of the soft palate. It is separated from the preceding by an angular interval, in which the tonsil is lodged. It arises from the soft palate by an expanded fasciculus, its fibres being divided into two unequal parts by the Levator palati, and being continuous partly with the muscle of the opposite side, and partly with the fibrous aponeurosis of the palate. Passing outwards and downwards behind the tonsil, it joins the Stylo-pharyngeus, and is inserted with it into the posterior border of the thyroid cartilage, some of its fibres being lost on the side of the pharynx.

Relations. In the soft palate, its *anterior* and *posterior surfaces* are covered by mucous membrane, from which it is separated by a layer of palatine glands. By its *superior border*, it is in relation with the Levator palati. Where it forms the posterior pillar of the fauces, it is covered by mucous membrane, excepting on its outer surface. In the *pharynx*, it lies between the mucous membrane and the constrictor muscles.

Nerves. The Tensor palati is supplied by a branch from the otic ganglion; the other muscles by the palatine branches of Meckel's ganglion.

Actions. When the morsel of food has been driven backwards into the fauces by the pressure of the tongue against the hard palate, the Palato-glossi muscles, the constrictors of the fauces, contract behind it, the soft palate is slightly raised (by the Levator palati), and made tense (by the Tensor palati), and the Palato-pharyngaei contract, and come nearly together, the Uvula filling up the slight interval between them. By these means, the food is prevented passing into the upper part of the pharynx or the posterior nares; at the same time the latter muscles form an inclined plane, directed obliquely downwards and backwards, along which the morsel descends into the pharynx.

Surgical Anatomy. The muscles of the soft palate should be carefully dissected, the relations they bear to the surrounding parts especially examined, and their action attentively studied upon the dead subject, as the surgeon is required to divide one or more of these muscles in the operation of staphyloraphy. Mr. Ferguson has shewn, that in the congenital deficiency, called *cleft palate*, the edges of the fissure are forcibly separated by the action of the Levatores palati and Palato-pharyngaei muscles, producing very considerable impediment to the healing process after the performance of the operation for uniting their margins by adhesion; he has, consequently, recommended the division of these muscles as one of the most important steps in the operation: by these means, the flaps are relaxed, lie perfectly loose and pendulous, and are easily brought and retained in apposition. The Palato-pharyngaei may be divided by cutting across the posterior pillar of the soft palate, just below the tonsil, with a pair of blunt-pointed curved scissors, and the anterior pillar may be divided also. To divide the Levator palati, the plan recommended by Mr. Pollock is to be greatly preferred. The flap being put upon the stretch, a double-edged knife is passed through the soft palate just on the inner side of the hamular process, and above the line of the Levator palati. The handle being now alternately raised and depressed, a sweeping cut is made along the posterior surface of the soft palate, and the knife withdrawn, leaving but a small opening in the mucous membrane on the anterior surface. If this operation is performed on the dead body, and the parts afterwards dissected, the Levator palati will be found completely divided.

Vertebral Region (Anterior)

Rectus Capitis Anticus Major. Rectus Lateralis.
Rectus Capitis Anticus Minor. Longus Colli.

The *Rectus Capitis Anticus Major* (fig. 141), broad and thick above, narrow below, appears like a continuation upwards of the Scalenus anticus. It arises by four tendons from the anterior tubercles of the transverse processes of the third, fourth, fifth, and sixth cervical vertebrae, and ascends, converging towards its fellow of the opposite side, to be inserted into the basilar process of the occipital bone.

Relations. By its *anterior surface*, with the pharynx, the sympathetic nerve, and the sheath enclosing the carotid artery, internal jugular vein, and pneumogastric nerve. By its *posterior surface*, with the Longus colli, the Rectus anticus minor, and the upper cervical vertebrae.

The *Rectus Capitis Anticus Minor* is a short muscle, situated immediately beneath the upper part of the preceding. It arises from the anterior surface of the lateral mass of the atlas, and from the root of its transverse process; passing obliquely upwards

141.—The Pre-Vertebral Muscles.

and inwards, it is inserted into the basilar process immediately behind the preceding muscle.

Relations. By its *anterior surface*, with the Rectus anticus major. By its *posterior surface*, with the anterior part of the occipito-atlantal articulation. *Externally*, with the superior cervical ganglion of the sympathetic.

The *Rectus Lateralis* is a short, flat muscle, situated between the transverse process of the atlas, and the jugular process of the occipital bone. It arises from the upper surface of the transverse process of the atlas, and is inserted into the under surface of the jugular process of the occipital bone.

Relations. By its *anterior surface*, with the internal jugular vein. By its *posterior surface*, with the vertebral artery.

The *Longus Colli* is a long, flat muscle, situated on the anterior surface of the spine, between the atlas and the third dorsal vertebra, being broad in the middle, narrow and pointed at each extremity. It consists of three portions, a superior oblique, an inferior oblique, and a vertical portion.

The *superior oblique portion* arises by a narrow tendon from the tubercle on the anterior arch of the atlas, and descending obliquely outwards, is inserted into the anterior tubercles of the transverse processes of the third, fourth, and fifth cervical vertebrae.

The *inferior oblique portion*, the smallest part of the muscle, arises tendinous from the transverse processes of the fifth and sixth cervical vertebrae, and passing obliquely inwards, is inserted into the bodies of the first two or three dorsal vertebrae.

The *vertical portion* lies directly on the front of the spine, and is extended between the bodies of the second, third, and fourth cervical vertebrae above, and the bodies of the three lower cervical and the three upper dorsal below.

Relations. By its *anterior surface*, with the pharynx, the oesophagus, sympathetic nerve, the sheath of the carotid artery, internal jugular vein, and pneumogastric nerve, inferior thyroid artery, and recurrent laryngeal nerve. By its *posterior surface*, with the cervical and dorsal portions of the spine.

VERTEBRAL REGION (LATERAL)

Scalenus Anticus.   Scalenus Medius.

Scalenus Posticus.

The *Scalenus Anticus* is a triangular muscle, situated deeply at the side of the neck, behind the Sterno-mastoid. It arises by a narrow, flat tendon from the tubercle on the inner border and upper surface of the first rib, and ascending vertically upwards, is inserted into the anterior tubercles of the transverse processes of the third, fourth, fifth, and sixth cervical vertebrae. The lower part of this muscle separates the subclavian artery and vein; the latter being in front, and the former, with the brachial plexus, behind.

Relations. By its *anterior surface*, with the Sterno-mastoid and Omo-hyoid muscles, the transversalis Colli, and descending cervical arteries, and the phrenic nerve. By its *posterior surface*, with the subclavian artery, and brachial plexus of nerves. It is separated from the Longus colli on the inner side by the subclavian artery.

The *Scalenus Medius*, the largest and longest of the three Scaleni, arises, by a broad origin, from the upper surface of the first rib, behind the groove for the subclavian artery, as far back as the tubercle, and ascending along the side of the vertebral column, is inserted, by separate tendinous slips, into the posterior tubercles of the transverse processes of the six lower cervical vertebrae. It is separated from the Scalenus anticus by the subclavian artery below, and the cervical nerves above.

Relations. By its *external surface*, with the Sterno-mastoid; it is crossed by the clavicle and Omo-hyoid muscle. To its *outer side*, is the Levator anguli scapulae and the Scalenus posticus muscle.

The *Scalenus Posticus*, the smallest of the three Scaleni, arises by a thin tendon from the outer surface of the second rib, behind the attachment of the Serratus magnus, and enlarging as it ascends, is inserted, by two or three separate tendons, into the posterior tubercles of the transverse processes of the two or three lower cervical vertebrae. This is the most deeply-placed of the three Scaleni, and is occasionally blended with the Scalenus medius.

Nerves. The Rectus capitis anticus major and minor are supplied by the suboccipital and deep branches of the cervical plexus; the Rectus lateralis by the sub-occipital; and the Longus colli and Scaleni by branches from the lower cervical nerves.

Actions. The Rectus anticus major and minor are the direct antagonists of those placed at the back of the neck, serving to restore the head to its natural position when drawn backwards by the posterior muscles. These muscles also serve to bow the head forwards. The Longus colli will flex and slightly rotate the cervical portion of the spine. The Scaleni muscles, taking their fixed point from below, draw down the transverse processes of the cervical vertebrae, flexing the spinal column to one or the other side. If the muscles of both sides act, the spine will be kept erect. When taking their fixed point from above, they elevate the first and second ribs, and are, therefore, inspiratory muscles.

MUSCLES AND FASCIAE OF THE TRUNK

The muscles of the Trunk may be subdivided into four groups.

1. Muscles of the Back.
2. Muscles of the Abdomen.
3. Muscles of the Thorax.
4. Muscles of the Perinaeum.

The Muscles of the Back are very numerous, and may be subdivided into five layers.

First Layer.
Trapezius.
Latissimus dorsi.

Second Layer.
Levator anguli scapulae.
Rhomboideus minor.
Rhomboideus major.

Third Layer.
Serratus posticus superior.
Serratus posticus inferior.
Splenius capitis.
Splenius colli.

Fourth Layer.
Sacral and Lumbar Regions.
Erector Spinae.
Dorsal Region.
Sacro-lumbalis.
Musculus accessorius ad sacro-lumbalem.
Longissimus dorsi.
Spinalis dorsi.
Cervical Region.
Cervicalis ascendens.
Transversalis cervicis.

FOURTH LAYER (continued).
Trachelo-mastoid.
Complexus.
Biventer cervicis.
Spinalis cervicis

FIFTH LAYER.
Semi-spinalis dorsi.
Semi-spinalis colli.
Multifidus spinae.

Rotatores spinae.
Supra-spinales.
Inter-spinales.
Extensor coccygis.
Inter-transversales.
Rectus posticus major.
Rectus posticus minor.
Obliquus superior
Obliquus inferior.

FIRST LAYER
Trapezius.   Latissimus Dorsi.

Dissection (fig. 142). The body should be placed in the prone position, with the arms extended over the sides of the table, and the chest and abdomen supported by several blocks, so as to render the muscles tense. An incision should then be made along the middle line of the back, from the occipital protuberance of the coccyx. From the upper end of this, a transverse incision should extend to the mastoid process; and from the lower end a third incision should be made along the crest of the ilium to about its middle. This large intervening space, for convenience of dissection, should be subdivided by a fourth incision, extending obliquely from the spinous process of the last dorsal vertebra, upwards and outwards, to the acromion process. This incision corresponds with the lower border of the Trapezius muscle. The flaps of integument should then be removed in the direction shewn in the accompanying figure.

142.—Dissection of the Muscles of the Back.

The *Trapezius* is a broad, flat, triangular muscle, placed immediately beneath the skin, and covering the upper and back part of the neck and shoulders. It arises from the occipital protuberance and inner third of the superior curved line of the occipital bone; from the ligamentum nuchae, the spinous processes of the seventh cervical, and all the dorsal vertebrae, and from the corresponding portion of the supra-spinous ligament. From these points the muscular fibres proceed, the superior ones downwards and outwards, the inferior ones upwards and outwards, and the middle fibres horizontally, and are inserted, the

143.—Muscles of the Back. On the Left Side is exposed the First Layer; on the Right Side, the Second Layer and part of the Third.

superior ones curving forwards into the outer third of the posterior border of the clavicle, the middle fibres into the upper margin of the acromion process, and into the whole length of the upper border of the spine of the scapula; the inferior fibres converge near the scapula, and are attached to a triangular aponeurosis, which glides over a small triangular surface at the inner extremity of the spine, and is inserted into a small tubercle in immediate connection with its outer part. The Trapezius is fleshy in the greater part of its extent, but tendinous at its origin and insertion. At its occipital origin, it is connected to the bone by a thin fibrous lamina, firmly adherent to the skin, and wanting the lustrous, shining appearance of aponeurosis. At its origin from the spines of the vertebrae, it is connected by means of a broad semi-elliptical aponeurosis, which occupies the space between the sixth cervical and the third dorsal vertebrae, and forms, with the muscle of the opposite side, a tendinous ellipse. The remaining part of the origin is effected by numerous short tendinous fibres. If the Trapezius is dissected on both sides, the two muscles resemble a trapezium, or diamond-shaped quadrangle; two angles, corresponding to the shoulders; a third, to the occipital protuberance; and the fourth, to the spinous process of the last dorsal vertebra.

The clavicular insertion of this muscle varies as to the extent of its attachment; it sometimes advances as far as the middle of the clavicle, and may even become blended with the posterior edge of the Sterno-mastoid, or overlap its margin. This should be borne in mind in the operation for tying the subclavian artery.

Relations. By its *superficial surface*, with the integument to which it is closely adherent above, but separated below by an aponeurotic lamina. By its *deep surface*, in the neck, with the Complexus, Splenius, Levator anguli scapulae, and Rhomboideus minor; in the back, with the Rhomboideus major, Supra-spinatus, Infra-spinatus, a small portion of the Serratus posticus superior, the intervertebral aponeurosis which separates it from the Erector spinae, and with the Latissimus dorsi. The spinal accessory nerve passes beneath the anterior border of this muscle, near the clavicle. The outer margin of its cervical portion forms the posterior boundary of the large posterior triangle of the neck, the other boundaries being the Sterno-mastoid in front, and the clavicle below.

The *Ligamentum Nuchae* (fig. 143) is a thin band of condensed cellulo-fibrous membrane, placed in the line of union between the two Trapezii in the neck. It extends from the external occipital protuberance to the spinous process of the seventh cervical vertebra, where it is continuous with the supra-spinous ligament. From its anterior surface a fibrous slip is given off to the spinous processes of each of the cervical vertebra, excepting the atlas, so as to form a septum between the muscles on each side of the neck. In the human subject, it is merely the rudiment of an important elastic ligament, which serves to sustain the weight of the head in some of the lower animals.

The *Latissimus Dorsi* is a broad flat muscle, which covers the lumbar and lower half of the dorsal regions, and is gradually contracted into a narrow fasciculus at its insertion into the humerus. It arises by tendinous fibres from the spinous processes of the six inferior dorsal, from those of the lumbar and sacral vertebrae, and from the supra-spinous ligament. Over the sacrum, the aponeurosis of this muscle blends with the tendon of the Erector spinae. It also arises from the external lip of the crest of the ilium, behind the origin of the External oblique, and by fleshy digitations from the three or four lower ribs, being interposed between similar processes of the External oblique muscle. From this extensive origin the fibres pass in different directions, the upper ones horizontally, the middle ones obliquely upwards, and the lower ones vertically upwards, so as to converge and form a thick fasciculus, which crosses the inferior angle of the scapula, and occasionally receives a few fibres from it. The muscle then curves around the lower border of the Teres major, and is twisted upon itself, so that the superior fibres become at first posterior and then inferior, and the vertical fibres at first anterior and then superior. It then terminates in a short quadrilateral tendon, about three inches in length, which, passing in front of the tendon of the Teres major, is inserted into the bottom of the bicipital groove of the humerus, above the insertion of the tendon of the Pectoralis major. The lower border of the tendon of this muscle is united with that of the Teres major, the surfaces of the two being separated by a synovial bursa; a second synovial bursa is interposed between the muscle and the inferior angle of the scapula.

The origin of this muscle from the spine and ilium is effected by an aponeurosis, which assists in forming the sheath for the Erector spinae. Its costal attachment takes place by means of three or four fleshy slips, which inter-digitate with the External oblique muscle of the abdomen.

Relations. Its *superficial surface* is subcutaneous, excepting at its upper part, where it is covered by the Trapezius. By its *deep surface*, it is in relation with the Erector spinae, the Serratus posticus inferior, Intercostal muscles and ribs, the Serratus magnus, inferior angle of the scapula, Rhomboideus major, Infra-spinatus, and Teres major. Its external margin is separated below, from the external oblique, by a small triangular interval; and another triangular interval exists between its superior border and the margin of the Trapezius, in which the Intercostal and Rhomboideus major muscles are exposed.

Nerves. The Trapezius is supplied by the spinal accessory and cervical plexus; the Latissimus dorsi, by the subscapular nerves.

Second Layer
Levator Anguli Scapulae.   Rhomboideus Minor.
Rhomboideus Major.

Dissection. The Trapezius must be removed in order to expose the next layer; to effect this, the muscle must be detached from its attachment to the clavicle and spine of the scapula, and turned back towards the spine.

The *Levator Anguli Scapulae* is a long, thick, and somewhat flattened muscle, situated at the posterior part and side of the neck. It arises by four tendons from the posterior tubercles of the transverse processes of the three or four upper cervical vertebrae, these becoming fleshy are united so as to form a flat muscle, which, passing downwards and backwards, is inserted into the posterior border of the scapula, between the superior angle and the triangular smooth surface at the root of the spine.

Relations. By its *superficial surface*, with the integument, Trapezius, and Sternomastoid. By its *deep surface*, with the Splenius colli, Transversalis colli, Cervicalis ascendens, and Serratus posticus superior, and with the transverse cervical and posterior scapular arteries.

The *Rhomboideus Minor* arises from the ligamentum nuchae, and spinous processes of the seventh cervical and first dorsal vertebrae, its fibres of origin being intimately united with those of the Trapezius. Passing downwards and outwards, it is inserted into the margin of the triangular smooth surface at the root of the spine of the scapula. This small muscle is usually separated from the Rhomboideus major by a slight cellular interval.

The *Rhomboideus Major* is situated immediately below the preceding, the adjacent margins of the two being occasionally united. It arises by tendinous fibres from the spinous processes of the four or five upper dorsal vertebrae and their inter-spinous ligaments, and is inserted into the posterior border of the scapula, between the triangular surface at the base of the spine and the inferior angle. The insertion of this muscle takes place by means of a narrow, tendinous arch, attached above, to the triangular surface near the spine; below, to the inferior angle, the arch being connected to the border of the scapula by a thin membrane. When the arch extends, as it occasionally does, but a short distance, the muscular fibres are inserted into the scapula itself.

Relations. By their *superficial surface*, with the integument, and Trapezius, the Rhomboideus major, with the Latissimus dorsi. By their *deep surface*, with the Serratus posticus superior, posterior scapular artery, part of the Erector spinae, the Intercostal muscles and ribs.

Nerves. These muscles are supplied by branches from the fifth cervical nerve, and additional filaments from the deep branches of the cervical plexus are distributed to the Levator anguli scapulae.

Actions. The movements effected by the preceding muscles are numerous, as may be conceived from their extensive attachment. If the head is fixed, the upper part of the Trapezius will elevate the point of the shoulder, as in supporting weights; when the middle and lower fibres are brought into action, partial rotation of the scapula upon the side of the chest is produced. If the shoulders are fixed both Trapezii acting together will draw the head directly backwards, or if only one acts, the head is drawn to the corresponding side.

The *Latissimus Dorsi*, when it acts upon the humerus, draws it backwards and downwards, and at the same time rotates it inwards. If the arm is fixed, the muscle may act in various ways upon the trunk; thus, it may raise the lower ribs and assist in forcible inspiration, or if both arms are fixed, the two muscles may conspire with the Abdominal and great Pectoral muscles in drawing the whole trunk forwards, as in climbing or walking on crutches.

The *Levator Anguli Scapulae* raises the superior angle of the scapula after it has been depressed by the Trapezius, whilst the Rhomboid muscles carry the inferior angle backwards and upwards, thus producing a slight rotation of the scapula upon the side of the chest. If the shoulder be fixed, the Levator scapulae may incline the neck to the corresponding side. The Rhomboid muscles acting together with the middle and inferior fibres of the Trapezius, will draw the scapula directly backwards towards the spine.

Third Layer

Serratus Posticus Superior. Serratus Posticus Inferior.

Splenius { Splenius Capitis.
          Splenius Colli.

Dissection. The third layer of muscles is brought into view by the entire removal of the preceding, together with the Latissimus dorsi. To effect this, the Levator anguli scapulae and Rhomboid muscles should be detached near their insertion, and reflected upwards, thus exposing the Serratus posticus superior; the Latissimus dorsi should then be divided in the middle by a vertical incision carried from its upper to its lower part, and the two halves of the muscle reflected.

The *Serratus Posticus Superior* is a thin, flat muscle, irregularly quadrilateral in form, and situated at the upper and back part of the thorax. It arises by a thin and broad aponeurosis, from the ligamentum nuchae and from the spinous processes of the last cervical and two or three upper dorsal vertebrae. Inclining downwards and outwards, it becomes muscular, and is inserted by four fleshy digitations, into the upper borders of the second, third, fourth, and fifth ribs, a little beyond their angles.

Relations. By its *superficial surface*, with the Trapezius, Rhomboidei, and Serratus magnus. By its *deep surface*, with the Splenius, upper part of the Erector spinae, Intercostal muscles and ribs.

The *Serratus Posticus Inferior* is situated at the lower part of the dorsal and upper part of the lumbar regions: it is of an irregularly quadrilateral form, broader than the preceding, and separated from it by a considerable interval. It arises by a thin aponeurosis from the spinous processes of the two lower dorsal and two or three upper lumbar vertebrae, and from the inter-spinous ligaments. Passing obliquely upwards and outwards, it becomes fleshy, and divides into four flat digitations, which are inserted into the lower borders of the four lower ribs, a little beyond their angles.

Relations. By its *superficial surface*, it is covered by the Latissimus dorsi, with the aponeurosis of which its own aponeurotic origin is inseparably blended. By its *deep surface*, with the posterior aponeurosis of the Transversalis, the Erector spinae, ribs and Intercostal muscles. Its upper margin is continuous with the vertebral aponeurosis.

The *Vertebral Aponeurosis* is a thin aponeurotic lamina, extending along the whole length of the posterior part of the thoracic region, serving to bind down the Erector spinae, and separating it from those muscles which connect the spine to the upper extremity. It consists of longitudinal and transverse fibres blended together, forming a thin lamella, which is attached in the median line to the spinous processes of the dorsal vertebrae; externally, to the angles of the ribs; and below, to the upper border of the Inferior serratus and tendon of the Latissimus dorsi; above, it passes beneath the Serratus posticus superior, and blends with the deep fascia of the neck.

The Serratus posticus superior should now be detached from its origin and turned outwards, when the Splenius muscle will be brought into view.

The *Splenius* is a broad muscle, situated at the posterior part of the neck and upper part of the dorsal region. At its origin, it is a single muscle, narrow and pointed in form; but it soon becomes broader, and divides into two portions, which have separate insertions. It arises, by tendinous fibres, from the lower half of the Ligamentum nuchae, from the spinous processes of the last cervical and of the six upper dorsal vertebrae, and from the supra-spinous ligament. From this origin, the fleshy fibres proceed obliquely upwards and outwards, forming a broad flat muscle, which divides as it ascends into two portions, the Splenius capitis and Splenius colli.

The *Splenius capitis* is inserted into the mastoid process of the temporal bone, and into the rough surface on the occipital bone beneath the superior curved line.

The *Splenius colli* is inserted, by tendinous fasciculi, into the posterior tubercles of the transverse processes of the three or four upper cervical vertebrae.

The Splenius is separated from its fellow of the opposite side by a triangular interval, in which is seen the Complexus.

Relations. By its *superficial surface*, with the Trapezius, from which it is separated below by the Rhomboidei and the Serratus posticus superior. It is also covered by the Sterno-mastoid and Levator anguli scapulae. By its *deep surface*, with the Spinalis dorsi, Longissimus dorsi, Semi-spinalis colli, Complexus, Trachelo-mastoid, and Transversalis colli.

Nerves. The Splenius and Superior serratus are supplied from the external posterior branches of the cervical nerves; the Inferior serratus, from the external branches of the dorsal nerves.

Actions. The Serrati are respiratory muscles acting in antagonism to each other. The Serratus posticus superior elevates the ribs; it is, therefore, an inspiratory muscle; while the Serratus inferior draws the lower ribs downwards, and is a muscle of expiration. This muscle is also probably a tensor of the vertebral aponeurosis. The Splenii muscles of the two sides, acting together, draw the head directly backwards, assisting the Trapezius and Complexus; acting separately, they draw the head to one or the other side, and slightly rotate it, turning the face to the same side. They also assist in supporting the head in the erect position.

Fourth Layer

Sacral and Lumbar Regions.	Cervical Region.
Erector Spinae.	Cervicalis ascendens.
Dorsal Region.	Transversalis cervicis.
Sacro-lumbalis.	Trachelo-mastoid.
Musculus accessorius ad sacro-lumbalem.	Complexus.
Longissimus dorsi.	Biventer cervicis.
Spinalis dorsi.	Spinalis cervicis.

Dissection. To expose the muscles of the fourth layer, the Serrati and vertebral aponeurosis should be entirely removed. The Splenius may then be detached by separating its attachments to the spinous processes, and reflecting it outwards.

The *Erector Spinae* (fig. 142), and its prolongations in the dorsal and cervical regions, fill up the vertebral groove on each side of the spine. They are covered in the lumbar region by the lumbar aponeurosis; in the dorsal region, by the Serrati muscles and the vertebral aponeurosis; and in the cervical region, by a layer of cervical fascia continued beneath the Trapezius. This large muscular and tendinous mass varies in size and structure at different parts of the spine. In the sacral region, the Erector spinae is narrow and pointed, and its origin chiefly tendinous in structure. In the lumbar region, it becomes enlarged, and forms a large fleshy mass. In the dorsal region, it subdivides into two parts, which gradually diminish in size as they ascend to be inserted into the vertebrae and ribs, and are gradually lost in the cervical region,

144.—Muscles of the Back. Deep Layers

where a number of special muscles are superadded, which are continued upwards to the head, which they support upon the spine.

The Erector spinae arises from the sacro-iliac groove, and from the anterior surface of a very broad and thick tendon, which is attached, internally, to the spines of the sacrum, to the spinous processes of the lumbar and three lower dorsal vertebra, and the supra-spinous ligament; externally, to the back part of the inner lip of the crest of the ilium, and to the series of eminences on the posterior part of the sacrum, representing the transverse processes, where it blends with the great sacro-sciatic ligament. The muscular fibres thus arising form a single large muscular mass, bounded in front by the transverse processes of the lumbar vertebrae, and by the middle lamella of the fascia of the Transversalis muscle. Opposite the last rib, this mass divides into two parts, one external, the Sacro-lumbalis, the other internal and larger, the Longissimus dorsi.

The *Sacro-Lumbalis*, the external and smaller portion of the Erector spinae, is inserted, by a series of separate tendons, into the angles of the six lower ribs. If this muscle is reflected outwards, it will be seen to be reinforced by a series of muscular slips, which arise from the angles of the ribs; by means of these the Sacro-lumbalis is continued upwards, to be connected with the upper ribs, and with the cervical portion of the spine, forming two additional muscles, the Musculus accessorius and the Cervicalis ascendens.

The *Musculus Accessorius ad Sacro-Lumbalem* arises by separate flattened tendons, from the upper margins of the angles of the six lower ribs; these become muscular, and are finally inserted, by separate tendons, into the angles of the six upper ribs.

The *Cervicalis Ascendens* is the continuation of the Sacro-lumbalis upwards into the neck: it is situated on the inner side of the tendons of the Accessorius, arising from the angles of the four or five upper ribs, and is inserted, by a series of slender tendons, into the posterior tubercles of the transverse processes of the fourth, fifth, and sixth cervical vertebrae.

Longissimus Dorsi. The inner portion of the Erector spinae, the larger and longer of the two, has received the name "Longissimus dorsi." It arises, with the Sacro-lumbalis, from the common origin already mentioned. In the lumbar region, where it is as yet blended with the Sacro-lumbalis, some of the fibres are directed forwards to be inserted into the posterior surface of the transverse processes of the lumbar vertebrae their whole length, into the tubercles at the back of the articular processes, and into the layer of lumbar fascia connected with the apices of the transverse processes. In the dorsal region, the Longissimus dorsi is inserted, by long and thin tendons, into the extremities of the transverse processes of all the dorsal vertebra, and into from seven to eleven ribs between their tubercles and angles.

This muscle is continued upwards to the cranium and cervical portion of the spine, by means of two additional slender fasciculi, the Transversalis colli, and Trachelo-mastoid.

The *Transversalis Colli*, placed on the inner side of the Longissimus dorsi, arises, by long thin tendons, from the summit of the transverse processes of the third, fourth, fifth, and sixth dorsal vertebrae, and is inserted, by similar tendons, into the posterior tubercles of the transverse processes of the five lower cervical.

The *Trachelo-Mastoid* lies on the inner side of the preceding, between it and the Complexus muscle, and may be regarded as the continuation of the Longissimus dorsi upwards to the head. It arises, by four tendons, from the transverse processes of the third, fourth, fifth, and sixth dorsal vertebrae, and from the articular processes of the three or four lower cervical; these joining form a small muscle, which ascends to be inserted into the posterior margin of the mastoid process, beneath the Splenius and Sterno-mastoid muscles. This small muscle is almost always crossed by a tendinous intersection near its insertion into the mastoid process.

The *spinous processes* of the upper lumbar and the dorsal vertebrae are connected together by a series of muscular and tendinous slips, which are intimately connected with the Longissimus dorsi, forming, in fact, part of this muscle; it is called the *Spinalis dorsi*.

The *Spinalis Dorsi* is situated at the inner side of the Longissimus dorsi. It arises, by three or four tendons, from the spinous processes of the two upper lumbar and the two lower dorsal vertebrae: these uniting, form a small muscle, which is inserted, by separate tendons, into the spinous processes of all the upper dorsal vertebrae, the number varying from four to eight. It is intimately united with the Semi-spinalis dorsi, which lies beneath it.

The *Spinalis Cervicis* is a small muscle, connecting together the spinous processes of the cervical vertebrae, and analogous to the Spinalis dorsi in the dorsal region. This muscle varies considerably in its size, and in its extent of attachment to the vertebrae, not only in different bodies, but on the two sides of the same body. It usually arises by fleshy or tendinous fibres, varying from two to four in number, from the spinous processes of the fifth and sixth cervical vertebrae, and occasionally from the first and second dorsal, and is inserted into the spinous process of the axis, and occasionally into the spinous processes of the two vertebrae below it. This muscle has been found absent in five cases out of twenty-four.

The *Complexus* is a broad thick muscle, situated at the upper and back part of the neck, lying beneath the Splenius, the direction of which it crosses obliquely from without inwards. It arises, by a series of tendons, about seven in number, from the posterior and upper part of the transverse processes of the three upper dorsal and seventh cervical, and from the articular processes of the three cervical above this. The tendons uniting form a broad muscle, which is directed obliquely upwards and inwards, and is inserted into the innermost depression between the two curved lines of the occipital bone. This muscle, about its middle, is traversed by a transverse tendinous intersection.

The *Biventer Cervicis*, is a small fasciculus, situated on the inner side of the preceding muscle, and in the majority of cases blended with it; it has received its name from presenting a tendon of considerable length with two fleshy bellies. It is sometimes described as a separate muscle, arising, by from two to four tendinous slips, from the transverse processes of as many upper dorsal vertebrae, and is inserted, on the inner side of the Complexus, into the superior curved line of the occipital bone.

Relations. By their *superficial surface*, with the Trapezius and Splenius. By their *deep surface*, with the Semi-spinalis dorsi and colli and the Recti and Obliqui. The Biventer cervicis is separated from its fellow of the opposite side by the ligamentum nuchae, and the Complexus from the Semi-spinalis colli by the profunda cervicis artery, the princeps cervicis branch of the occipital, and by the posterior cervical plexus of nerves.

Nerves. The Erector spinae and its subdivisions in the dorsal region are supplied by the external posterior branches of the lumbar and dorsal nerves. The Cervicalis ascendens, Transversalis colli, Trachelo-mastoid, and Spinalis cervicis, by the external posterior branches of the cervical nerves; the Complexus, by the internal posterior branches of the cervical nerves, the sub-occipital and great occipital.

Fifth Layer

Semi-spinalis Dorsi.	Extensor Coccygis.
Semi-spinalis Colli.	Inter-transversales.
Multifidus Spinae.	Rectus Capitis Posticus Major.
Rotatores Spinae.	Rectus Capitis Posticus Minor.
Supra-spinales.	Obliquus Superior.
Inter-spinales.	Obliquus Inferior.

Dissection. The muscles of the preceding layer must be removed by dividing and turning aside the Complexus, then detach the Spinalis and Longissimus dorsi from their attachments, and divide the Erector spinae at its connection below to the sacral and lumbar spines, and turn it outwards. The muscles filling up the interval between the spinous and transverse processes are then exposed.

The *Semi-spinales Muscles* connect together the transverse and spinous processes of the vertebrae, extending from the lower part of the dorsal region to the upper part of the cervical.

The *Semi-spinalis Dorsi* consists of a thin, narrow, fleshy fasciculus, interposed between tendons of considerable length. It arises by a series of small tendons from the transverse processes of the lower dorsal vertebrae, from the tenth or eleventh to the fifth or sixth; these uniting form a small muscular fasciculus, which subdividing into five or six tendons, is inserted into the spinous processes of the four upper dorsal and two lower cervical.

The *Semi-spinalis Colli*, thicker than the preceding, arises by a series of tendinous and fleshy points from the transverse processes of the four upper dorsal vertebrae, and is inserted into the spinous processes of the four upper cervical vertebrae, from the axis to the fifth cervical. The fasciculus connected with the axis is the largest, and chiefly muscular in structure.

Relations. By their *superficial surface*, from below upwards with the Longissimus dorsi, Spinalis dorsi, Splenius, Complexus, the profunda cervicis and princeps cervicis arteries, and the posterior cervical plexus of nerves. By their *deep surface*, with the Multifidus spinae.

The *Multifidus Spinae* consists of a number of fleshy and tendinous fasciculi, which fill up the groove on either side of the spinous processes of the vertebrae from the sacrum to the axis. In the sacral region, these fasciculi arise from the sacral groove, as low down as the fourth sacral foramen, being connected with the aponeurosis of origin of the Erector spinae. In the iliac region, from the inner surface of the posterior superior spine, and posterior sacro-iliac ligaments. In the lumbar and cervical regions they arise from the articular processes, and in the dorsal region, from the transverse processes. Each fasciculus, ascending obliquely upwards and inwards, is inserted into the lamina and whole length of the spinous process of the vertebra above. These fasciculi vary in length; the most superficial, the longest, pass from one vertebra to the third or fourth above; those next in order pass from one vertebra to the second or third above; whilst the deepest connect two contiguous vertebrae.

Relations. By its *superficial surface*, with the Longissimus dorsi, Spinalis dorsi, Semi-spinalis dorsi, and Semi-spinalis colli. By its *deep surface*, with the laminae and spinous processes of the vertebrae, and with the Rotatores spinae in the dorsal region.

The *Rotatores Spinae* are found only in the dorsal region of the spine, beneath the Multifidus spinae, they are eleven in number on each side. Each muscle, which is small and somewhat quadrilateral in form, arises from the upper and back part of the transverse process, and is inserted into the lower border and outer surface of the lamina of the vertebra above, the fibres extending as far inwards as the root of the spinous process. The first is found between the first and second dorsal, the last, between the eleventh and twelfth. Sometimes the number of these muscles is diminished by the absence of one or more from the upper or lower end.

The *Supra Spinales* consist of a series of fleshy bands, which lie on the spinous processes in the cervical region of the spine.

The *Inter-Spinales* are short muscular fasciculi, placed in pairs between the spinous processes of the contiguous vertebrae. In the cervical region they are most distinct, and consist of six pairs, the first being situated between the axis and third vertebra, and the last between the last cervical and the first dorsal. In the dorsal region they are found above, between the first and second vertebrae, and occasionally between the second and third; and below, between the eleventh and twelfth. In the lumbar region there are four pairs of these muscles in the intervals between the five lumbar vertebrae. There is also occasionally one in the interspinous space, between the last dorsal and first lumbar, and between the fifth lumbar and the sacrum.

The *Extensor Coccygis* is a slender muscular fasciculus, occasionally present, which extends over the lower part of the posterior surface of the sacrum and coccyx. It arises by tendinous fibres from the last bone of the sacrum, or first piece of the coccyx, and passes downwards to be inserted into the lower part of the coccyx. It is a rudiment of the Extensor muscle of the caudal vertebrae present in some animals.

The *Inter-Transversales* are small muscles placed between the transverse processes of the vertebrae. In the cervical region they are most developed, consisting of two rounded muscular and tendinous fasciculi, which pass between the anterior and posterior tubercles of the transverse processes of two contiguous vertebrae, being separated from one another by the anterior branch of a cervical nerve, which lies in the groove between them, and by the vertebral artery and vein. In this region there are seven pairs of these muscles, the first being between the atlas and axis, and the last between the seventh cervical and first dorsal vertebrae. In the dorsal region the Inter-transversales are least developed, consisting chiefly of rounded tendinous cords in the inter-transverse spaces of the upper dorsal vertebrae; but between the transverse processes of the three lower dorsal vertebrae and the first lumbar, they are muscular in structure. In the lumbar region they are four in number, and consist of a single muscular layer, which occupies the entire interspace between the transverse processes of the lowest lumbar vertebrae, whilst those between the transverse processes of the upper lumbar, are not attached to more than half the breadth of the process.

The *Rectus Capitis Posticus Major*, the larger of the two Recti, arises by a pointed tendinous origin from the spinous process of the axis, and becoming broader as it ascends, is inserted into the inferior curved line of the occipital bone and the surface of bone immediately beneath it. As the muscles of the two sides ascend upwards and outwards, they leave between them a triangular space, in which are seen the Recti capitis postici minores muscles.

Relations. By its *superficial surface*, with the Complexus, and at its insertion, with the Superior oblique. By its *deep surface*, with the posterior arch of the atlas, the posterior occipito-atloid ligament, and part of the occipital bone.

The *Rectus Capitis Posticus Minor*, the smallest of the four muscles in this region, is of a triangular shape, it arises by a narrow, pointed tendon from the tubercle on the posterior arch of the atlas, and becoming broader as it ascends, is inserted into the rough surface beneath the inferior curved line, nearly as far as the foramen magnum, nearer to the middle line than the preceding.

Relations. By its *superficial surface*, with the Complexus. By its *deep surface*, with the posterior occipito-atloid ligament.

The *Obliquus Inferior*, the largest of the two oblique muscles, arises from the apex of the spinous process of the axis, and forms a thick cylindrical muscle, which passes almost horizontally outwards, to be inserted into the apex of the transverse process of the atlas.

Relations. By its *superficial surface*, with the Complexus, and is crossed by the posterior branch of the second cervical nerve. By its *deep surface*, with the vertebral artery and posterior occipito-atloid ligament.

The *Obliquus Superior*, narrow below, wide and expanded above, arises by tendinous fibres from the upper part of the extremity of the transverse process of the atlas, joining with the insertion of the Inferior oblique muscle, and passing obliquely upwards and inwards, is inserted into the occipital bone, between the two curved lines, external to the Complexus. Between the two oblique muscles and the Rectus posticus major, a triangular interval exists, in which is seen the vertebral artery and the posterior branch of the sub-occipital nerve.

Relations. By its *superficial surface*, with the Complexus and Trachelo-mastoid. By its *deep surface*, with the posterior occipito-atloid ligament.

Nerves. The Semi-spinalis dorsi and Rotatores spinae are supplied by the internal posterior branches of the dorsal nerves. The Semi-spinalis colli, Supra-spinales, and Inter-spinales, by the internal posterior branches of the cervical nerves. The Inter-transversales, by the internal posterior branches of the cervical, dorsal, and lumbar nerves. And the Multifidus spinae, by the same, with the addition of the internal posterior branches of the sacral nerves. The Recti and Obliqui muscles are all supplied by the sub-occipital and great occipital nerves.

Actions. The Erector spinae, comprising the Sacro-lumbalis, with its accessory muscle, the Longissimus dorsi and Spinalis dorsi, serves, as its name implies, to maintain the spine in the erect posture; it also serves to bend the trunk backwards, when it is required to counter-balance the influence of any weight at the front of the body, as, for instance, when a heavy weight is suspended from the neck, or when there is any great abdominal development, as in pregnant women or in abdominal dropsy; the peculiar gait under such circumstances depends upon the spine being drawn backwards, by the counter-balancing action of the Erector spinae muscles. The continuation of these muscles upwards to the neck and head, steady and preserve the upright position of these several parts. If the Sacro-lumbalis and Longissimus dorsi of one side act, they serve to draw down the chest and spine to the corresponding side. The Musculus accessorius, taking its fixed point from the cervical vertebrae, elevates those ribs to which it is attached. The Multifidus spinae act successively upon the different segments of the spine; thus the lateral parts of the sacrum furnish a fixed point from which the fasciculi of this muscle act upon the lumbar region; these then become the fixed points for the fasciculi moving the dorsal region, and so on throughout the entire length of the spine; it is by the successive contraction and relaxation of the separate fasciculi of this and other muscles, that the spine preserves the erect posture without the fatigue that would necessarily have existed had this movement been accomplished by the action of a single muscle. The Multifidus spinae, besides preserving the erect position of the spine, serves to rotate it, so that the front of the trunk is turned to the side opposite to that from which the muscle acts, this muscle being assisted in its action by the Obliquus externus abdominis. The Complexi, the analogues of the Multifidus spinae in the neck, draw the head directly backwards; if one muscle acts, it draws the head to one side, and rotates it so that the face is turned to the opposite side. The Rectus capitis posticus minor and the Superior oblique draw the head backwards, and the latter from the obliquity in the direction of its fibres, may turn the face to the opposite side. The Rectus capitis posticus major and the Obliquus inferior, rotate the atlas, and with it the cranium around the odontoid process, and turn the face to the same side.

Muscles of the Abdomen

The muscles in this region are, the

- Obliquus Externus.
- Obliquus Internus.
- Transversalis.
- Rectus.
- Pyramidalis.
- Quadratus Lumborum.

Dissection (fig. 145). To dissect the abdominal muscles, a vertical incision should be made from the ensiform cartilage to the pubes; a second oblique incision should extend from the umbilicus upwards and outwards to the outer surface of the chest, as high as the lower border of the fifth or sixth rib; and a third, commencing midway between the umbilicus and pubes, should pass transversely outwards to the anterior superior iliac spine, and along the crest of the ilium as far as its posterior third. The three flaps included between these incisions should then be reflected from within outwards, in the direction indicated in the figure.

The *External Oblique Muscle* (Obliquus descendens) (fig. 146), so called from the direction of its fibres, is situated on the lateral and anterior aspects of the abdomen; being the largest and the most superficial of the three flat muscles in this region. It is

145.—Dissection of Abdomen.

broad, thin, irregularly quadrilateral in form, its muscular portion occupying the sides, its aponeurosis the anterior wall of that cavity. It arises, by eight fleshy digitations, from the external surface and lower borders of the eight inferior ribs; these digitations are arranged in an oblique line running downwards and backwards; the upper ones being attached close to the cartilages of the corresponding ribs; the lowest, to the apex of the cartilage of the last rib; the intermediate ones, to the ribs at some distance from their cartilages. The five superior serrations increase in size from above downwards, and are received between corresponding processes of the Serratus magnus; the three lower ones diminish in size from above downwards, receiving between them corresponding processes from the Latissimus dorsi. From these attachments the fleshy fibres proceed in various directions. Those from the lowest ribs pass nearly vertically downwards, to be inserted into the anterior half of the outer lip of the crest of the ilium; the middle and upper fibres, directed downwards and forwards, terminate in tendinous fibres, which spread out into a broad aponeurosis. This aponeurosis, joined with that of the opposite muscle along the median line, covers the whole of the front of the abdomen: above, it is connected with the lower border of the Pectoralis major; below, its fibres are closely aggregated together, and extend obliquely across from the anterior superior spine of the ilium to the spine of the os pubis and the pectineal line. In the median line, it interlaces with the aponeurosis of the opposite muscle, forming the linea alba, and extends from the ensiform cartilage to the symphysis pubis.

That portion of the aponeurosis which extends between the anterior superior spine of the ilium and the spine of the os pubis is a broad band, folded inwards, and continuous below with the fascia lata; it is called *Poupart's ligament*. The portion which is reflected from Poupart's ligament backwards and inwards into the pectineal line, is called *Gimbernat's ligament*.

In the aponeurosis of the External oblique, immediately above the crest of the os pubis, is a triangular opening, the *external abdominal ring*, formed by a splitting of the fibres of the aponeurosis in this situation; it serves for the transmission of the spermatic cord in the male, and the round ligament in the female. This opening is directed obliquely upwards and outwards, and corresponds with the course of the fibres of the aponeurosis. It is bounded below by the crest of the os pubis; above, by some curved fibres, which pass across the aponeurosis at the upper angle of the ring so as to increase its strength; and on either side, by the margins of the aponeurosis, which are called the *pillars of the ring*. Of these, the external, which is, at the same

146.—The External Oblique Muscle.

time inferior, from the obliquity of its direction, is inserted into the spine of the os pubis. The internal, or superior pillar, being attached to the front of the symphysis pubis, interlaces with the corresponding fibres of the opposite muscle. To the margins of the pillars of the external abdominal ring is attached an exceedingly thin and delicate fascia, which is prolonged down over the external surface of the cord and testis. This has received the name of *inter-columnar fascia* from its attachment to the pillars of the ring. It has also received the name of *external spermatic fascia*, from being the most external of the fasciae which cover the spermatic cord.

Relations. By its *external surface*, with the superficial fascia, superficial epigastric and circumflexa ilii vessels, and some cutaneous nerves. By its *internal surface*, with the Internal oblique, the lower part of the eight inferior ribs and Intercostal muscles, the cremaster, the spermatic cord in the male, and round ligament in the female. Its *posterior border* is occasionally overlapped by the Latissimus dorsi; sometimes an interval exists between the two muscles, in which is seen a portion of the Internal oblique.

Dissection. The External oblique should now be detached by dividing it across, just in front of its attachment to the ribs, as far as its posterior border, and by separating it below from the crest of the ilium as far as the spine; the muscle should then be carefully separated from the Internal oblique, which lies beneath, and turned towards the opposite side.

The *Internal Oblique Muscle* (fig. 147) (Obliquus ascendens), thinner and smaller than the preceding, beneath which it lies, is of an irregularly quadrilateral form, and situated at the anterior lateral and posterior parts of the abdomen. It arises, by fleshy fibres, from the outer half of Poupart's ligament, being attached to the groove on its upper surface; from the anterior two-thirds of the middle lip of the crest of the ilium, and from the lumbar fascia. From this origin, the fibres diverge in different directions. Those from Poupart's ligament, few in number and paler in colour than the rest, arch downwards and inwards across the spermatic cord, to be inserted, conjointly with those of the Transversalis, into the crest of the os pubis and pectineal line, to the extent of half an inch, forming the conjoined tendon of the Internal oblique

147.—The Internal Oblique Muscle.

and Transversalis; those from the anterior superior iliac spine are horizontal in their direction; whilst those which arise from the front part of the crest of the ilium pass obliquely upwards and inwards, and terminate in an aponeurosis, which is continued forwards to the linea alba; the most posterior fibres ascend almost vertically upwards, to be inserted into the lower borders of the cartilages of the four lower ribs, being continuous with the internal intercostal muscles.

The conjoined tendon of the Internal oblique and Transversalis is inserted into the crest of the os pubis and pectineal line immediately behind the external abdominal ring, serving to protect what would otherwise be a weak point in the abdomen. Sometimes this tendon is insufficient to resist the pressure from within, and is carried forward in front of the protrusion through the external ring, forming one of the coverings of direct inguinal hernia.

The aponeurosis of the Internal oblique is continued forward to the middle line of the abdomen, where it joins with the aponeurosis of the opposite muscle at the linea alba, and extends from the margin of the thorax to the pubes. At the outer margin of the sheath of the Rectus muscle, for the upper three-fourths of its extent, this aponeurosis divides into two lamellas, which pass, one in front and the other behind it, enclosing it in a kind of sheath, and reuniting on its inner border at the linea alba: the anterior layer is blended with the aponeurosis of the External oblique muscle; the posterior layer with that of the Transversalis. Along the lower fourth, the aponeurosis passes altogether in front of the Rectus without any separation.

Relations. By its *external surface*, with the External oblique, Latissimus dorsi, spermatic cord, and external ring. By its *internal surface*, with the Transversalis muscle, fascia transversalis, internal ring, and spermatic cord. Its lower border forms the upper boundary of the spermatic canal.

Dissection. The Internal oblique should now be detached in order to expose the Transversalis muscle beneath. This may be effected by dividing the muscle, above, at its attachment to the ribs; below, at its connexion with Poupart's ligament and the crest of the ilium; and behind, by a vertical incision extending from the last rib to the crest of the ilium. The muscle should previously be made tense by drawing upon it with the lingers of the left hand, and if its division is carefully effected, the cellular interval between it and the Transversalis, as well as the direction of the fibres of the latter muscle, will afford a clear guide to their separation; along the crest of the ilium the circumflex ilii vessels are interposed between them, and form an important aid in separating them. The muscle should then be thrown forwards towards the linea alba.

The *Transversalis muscle* (fig. 148), so called from the direction of its fibres, is the most internal flat muscle of the abdomen, being placed immediately beneath the Internal Oblique. It arises by fleshy fibres from the outer third of Poupart's ligament, from the inner lip of the crest of the ilium, its anterior two-thirds, from the inner surface of the cartilages of the six lower ribs, interdigitating with the Diaphragm, and by a broad aponeurosis from the spinous and transverse processes of the lumbar vertebrae. The lower fibres curve downwards, and are inserted together with those of the Internal oblique, into the crest of the os pubis and pectineal line, forming what was before mentioned as the conjoined tendon of these muscles. Throughout the rest

148.—The Transversalis, Rectus, and Pyramidalis Muscles.

of its extent the fibres pass horizontally inwards, and near the outer margin of the Rectus, terminate in an aponeurosis, which is inserted into the linea alba; its upper three-fourths passing behind the Rectus muscle, blending with the posterior lamella of the Internal oblique; its lower fourth passing in front of the Rectus.

Relations. By its *external surface*, with the Internal oblique, the inner surfaces of the lower ribs, and Internal intercostal muscles. By its *internal surface*, it is lined by the fascia transversalis, which separates it from the peritoneum. Its lower border forms the upper boundary of the spermatic canal.

Lumbar Fascia (fig. 149). The vertebral aponeurosis of the Transversalis divides into three layers, an anterior, very thin, which is attached to the front part of the apices of the transverse processes of the lumbar vertebrae, and, above, to the lower margin of the last rib, forming the ligamentum arcuatum externum; a middle layer, much stronger, which is attached to the apices of the transverse processes; and a posterior layer, attached to the apices of the spinous processes. Between the anterior and middle layers is situated the Quadratus lumborum, between the middle and posterior, the Erector spinae. The posterior lamella of this aponeurosis receives the attachment of the Internal oblique; it is also blended with the aponeurosis of the Serratus posticus inferior and with that of the Latissimus dorsi, forming the Lumbar fascia; the two anterior layers are connected solely with the Transversalis.

Dissection. To expose the Rectus muscle, its sheath should be opened by a vertical incision extending from the margin of the thorax to the pubes, the two portions should then be reflected from the surface of the muscle, which is easily effected, excepting at the lineae transversae, where so close an adhesion exists, that the greatest care is requisite in separating them. The outer edge of the muscle should now be raised, when the posterior layer of the sheath will be seen. By dividing the muscle in the centre, and turning its lower part downwards, the point where the posterior wall of the sheath terminates in a thin curved margin will be seen.

The *Rectus Abdominis* is a long, flat muscle, which extends along the whole length of the anterior wall of the abdomen, being separated from its fellow of the opposite side by the linea alba. It is much broader above than below, and arises by two tendons, the external or larger being attached to the crest of the os pubis; the internal, smaller portion, interlacing with its fellow of the opposite side, and being connected with the ligaments covering the symphysis pubis. The fibres ascend vertically upwards, and the muscle becoming broader and thinner at its upper part, is inserted by three portions of unequal size into the cartilages of the fifth, sixth, and seventh ribs. Some fibres are also occasionally connected with the costo-xiphoid ligaments, and side of the ensiform cartilage.

The Rectus muscle is traversed by a series of tendinous intersections, which vary from two to five in number, and have received the name lineae transversae. One of these is usually situated opposite the umbilicus, and two above that point; of these, one corresponds to the ensiform cartilage, and the other, to the interval between the ensiform cartilage and the umbilicus; there is occasionally one below the umbilicus.

149.—A Transverse Section of the Abdomen in the Lumbar Region.

These intersections pass transversely or obliquely across the muscle in a zigzag course; they rarely extend completely through its substance, sometimes pass only half way across it, and are intimately adherent to the sheath in which the muscle is enclosed.

The Rectus is enclosed in a sheath (fig. 149) formed by the aponeuroses of the Oblique and Transversalis muscles, which are arranged in the following manner. When the aponeurosis of the Internal oblique arrives at the margin of the Rectus, it divides into two lamellae, one of which passes in front of the Rectus, blending with the aponeurosis of the External oblique; the other, behind it, blending with the aponeurosis of the Transversalis; and these, joining again at its inner border, are inserted into the linea alba. This arrangement of the fasciae exists along the upper three-fourths of this muscle; at the commencement of the lower fourth, the posterior wall of the sheath terminates in a thin curved margin, the concavity of which looks downwards towards the pubes; the aponeuroses of all three muscles passing in front of the Rectus without any separation. The Rectus muscle in the situation where its sheath is deficient, is separated from the peritoneum by the transversalis fascia.

The *Pyramidalis* is a small muscle, triangular in form, situated at the lower part of the abdomen, one on each side of the linea alba. It arises by tendinous fibres from the front of the os pubis and anterior pubic ligament; the fleshy portion of the muscle passes upwards, diminishing in size as it ascends, and terminates by a pointed extremity, which is inserted into the linea alba, midway between the umbilicus and the os pubis. It rests against the lower part of the front of the Rectus, and is contained in the same sheath with that muscle. This muscle is sometimes found wanting on one or both sides; the lower end of the Rectus then becomes proportionally increased in size. Occasionally it has been found double on one side, or the muscles of the two sides are of unequal size. Sometimes its length exceeds that stated above.

The *Quadratus Lumborum* is situated in the lumbar region of the spine, it is irregularly quadrilateral in shape, broader below than above, and consists of two portions. One portion arises by aponeurotic fibres from the ilio-lumbar ligament, and the adjacent portion of the crest of the ilium for about two inches, and is inserted into the lower border of the last rib, about half its length, and by four small tendons, into the apices of the transverse processes of the third, fourth, and fifth lumbar vertebrae. The other portion of the muscle, situated anterior to the preceding, arises from the upper borders of the transverse processes of the third, fourth, and fifth lumbar vertebrae, and is inserted into the lower margin of the last rib. The Quadratus lumborum is contained in a sheath formed by the anterior and middle lamellae of the vertebral aponeurosis of the Transversalis.

Nerves. All the abdominal muscles are supplied by the lower intercostal, ilio-hypogastric, and ilio-inguinal nerves, excepting the Quadratus lumborum, which receives filaments from the anterior primary branches of the lumbar nerves.

In the description of the abdominal muscles, mention has frequently been made of the linea alba, lineae semilunares, lineae transversae; when the dissection of these muscles is completed, these structures should be examined.

The *Linea Alba* is a tendinous raphe or cord seen along the middle line of the abdomen, extending from the ensiform cartilage to the symphysis pubis. It is placed between the inner borders of the Recti muscles, and formed by the blending of the anterior aponeuroses of the Oblique and Transversalis muscles. It is narrow below, corresponding to the narrow interval existing between the Recti, but broader above, as these muscles diverge from one another in their ascent, becoming of considerable breadth after great distension of the abdomen from pregnancy or ascites. It presents numerous apertures for the passage of vessels and nerves; the largest of these is the umbilicus, which in the foetus transmits the umbilical vessels, but in the adult is obliterated, the cicatrix being stronger than the neighbouring parts; hence the occurrence of umbilical hernia in the adult above the umbilicus, whilst in the foetus it occurs at the umbilicus. The linea alba is in relation, in front, with the integument to which it is adherent, especially at the umbilicus; behind, it is separated from the peritoneum by the transversalis fascia; and below, by the urachus, and the bladder, when that organ is distended.

The *Lineae Semilunares* are two curved tendinous lines, placed one on each side of, and a little external to the linea alba. Each extends from the cartilage of the eighth rib to the pubes, and corresponds with the outer border of the Rectus muscle. They are formed by the aponeurosis of the Internal oblique at its point of division to enclose the Rectus.

The *Lineae Transversae* are three or four narrow transverse lines which intersect the Rectus muscle as already mentioned, they connect the lineae semilunares with the linea alba.

Actions. The abdominal muscles perform a three-fold action.

When the pelvis and thorax are fixed, they can compress the abdominal viscera, by constricting the cavity of the abdomen, in which action they are materially assisted by the descent of the diaphragm. By these means, the foetus is expelled from the uterus, the faeces from the rectum, the urine from the bladder, and the ingesta from the stomach in vomiting.

If the spine be fixed, these muscles compress the lower part of the thorax, materially assisting in the process of expiration. If the spine be not fixed, the thorax is bent directly forward, if the muscles of both sides act, or to either side if they act alternately, rotation of the trunk at the same time taking place to the opposite side.

If the thorax be fixed, these muscles act upon the pelvis, as in climbing, when the pelvis is drawn directly upwards, or to one or the other side. The Recti muscles may draw the pelvis forwards, and flex it upon the vertebral column. The Pyramidales are tensors of the linea alba.

Muscles and Fasciae of the Thorax

The muscles exclusively connected with the bones in this region are few in number. They are the

Intercostales Externi. Infra-Costales.
Intercostales Interni. Triangularis Sterni.
Levatores Costarum.

Intercostal Fasciae. A thin but firm layer of fascia covers the outer surface of the External intercostal and the inner surface of the Internal intercostal muscles; and a third layer, more delicate, is interposed between these two planes of muscular fibres.

These are the intercostal fasciae; they are best marked in those situations where the muscular fibres are deficient, as between the External intercostal muscles and sternum, in front; and between the Internal intercostals and spine, behind.

The *Intercostal Muscles* are two thin planes of muscular and tendinous structure, placed one over the other, filling up the intercostal spaces, and being directed obliquely between the margins of the adjacent ribs. These two planes have received the name "external" and "internal," from the position they bear to one another.

The *External Intercostals* are eleven in number on each side, being attached to the adjacent margins of each pair of ribs, and extending from the tubercles of the ribs, behind, to the commencement of the cartilages of the ribs, in front, where they terminate in a thin membranous aponeurosis, which is continued forwards to the sternum. They arise from the outer lip of the groove on the lower border of each rib, and are inserted into the upper border of the rib below. In the two lowest spaces they extend to the end of the ribs. Their fibres are directed obliquely downwards and forwards, in a similar direction with those of the External oblique muscle of the abdomen. They are thicker than the Internal intercostals.

Relations. The External intercostals, by their *outer surface*, are covered by the muscles which immediately invest the chest, viz., the Pectoralis major and minor, Serratus magnus, Rhomboideus major, Serratus posticus superior and inferior, Scalenus posticus, Sacro-lumbalis and Longissimus dorsi, Cervicalis ascendens, Transversalis colli, Levatores costarum, and the Obliquus externus abdominis. By their *internal surface*, they are in relation with a thin layer of fascia, which separates them from the intercostal vessels and nerve, the Internal intercostal muscles, and, behind, from the pleura.

The *Internal Intercostals*, also eleven in number on each side, are placed on the inner surface of the preceding, commencing anteriorly at the sternum, in the interspaces between the cartilages of the true ribs, and from the anterior extremities of the cartilages of the false ribs; and extend backwards as far as the angles of the ribs, where they are continued to the vertebral column by a thin aponeurosis. They arise from the inner lip of the groove on the lower border of each rib, as well as from the corresponding costal cartilage, and are inserted into the upper border of the rib below. Their fibres are directed obliquely downwards and backwards, decussating with the fibres of the preceding.

Relations. By their *external surface*, with the External intercostals, and the intercostal vessels and nerves. By their *internal surface*, with the pleura costalis, Triangularis sterni, and Diaphragm.

The Intercostal muscles consist of muscular and tendinous fibres, the latter being long and more numerous than the former; hence these spaces present very considerable strength, to which their crossing materially contributes.

The *Infra-Costales* consist of muscular and aponeurotic fasciculi, which vary in number and length; they arise from the inner surface of one rib, and are inserted into the inner surface of the first, second, or third rib below. Their direction is most usually oblique, like the Internal intercostals. They are most frequent between the lower ribs.

The *Triangularis Sterni* is a thin plane of muscular and tendinous fibres, situated upon the inner wall of the front of the chest. It arises from the lower part of the side of the sternum, from the inner surface of the ensiform cartilage, and from the sternal ends of the costal cartilages of the three or four lower true ribs. Its fibres diverge upwards and outwards, to be inserted by fleshy digitations into the lower border and inner surfaces of the costal cartilages of the second, third, fourth, and fifth ribs. The lowest fibres of this muscle are horizontal in their direction, and continuous with those of the Transversalis; those which succeed are oblique, whilst the superior fibres are almost vertical. This muscle varies much in its attachment, not only in different bodies, but on opposite sides of the same body.

Relations. In front with the sternum, ensiform cartilage, the costal cartilages, the Internal intercostal muscles, and internal mammary vessels. *Behind*, with the pleura, pericardium, and anterior mediastinum.

The *Levatores Costarum*, twelve in number on each side, are small tendinous and fleshy bundles, which arise from the extremities of the transverse processes of the dorsal vertebrae, and passing obliquely downwards and outwards, are inserted into the upper rough surface of the rib below them, between the tubercle and the angle. That for the first rib arises from the transverse process of the last cervical vertebra, and that for the last from the eleventh dorsal. The Inferior levatores divide into two parts, one being inserted as above described, the other fasciculus passing downwards to the second rib below their origin; thus each of the lower ribs receives fibres from the transverse processes of two vertebrae.

Nerves. The muscles of this group are supplied by the intercostal nerves.

Actions. The Intercostals are the chief agents in the movement of the ribs in ordinary respiration. The External intercostals raise the ribs, especially their fore part, and so increase the capacity of the chest from before backwards; at the same time they evert their lower borders, and so enlarge the thoracic cavity transversely. The Internal intercostals, at the side of the thorax, depress the ribs, and invert their lower borders, and so diminish the thoracic cavity; but at the fore part of the chest these muscles assist the External intercostals in raising the cartilages. The Levatores Costarum assist the external intercostals in raising the ribs. The Triangularis sterni draws down the costal cartilages; it is therefore an expiratory muscle.

Diaphragmatic Region
Diaphragm.

The *Diaphragm* (Διαφράσσω, to separate two parts) (fig. 150) is a thin musculo-fibrous septum, placed obliquely at the junction of the upper with the lower two-thirds of the trunk, and separating the thorax from the abdomen, forming the floor of the former cavity and the roof of the latter. It is elliptical, its longest diameter being from side to side, somewhat fan-shaped, the broad elliptical portion being horizontal, the narrow part, which represents the handle, being vertical, and joined at right angles with the former. It is from this circumstance that some anatomists describe it as consisting of two portions, the upper or great muscle of the diaphragm, and the lower or lesser muscle. This muscle arises from the whole of the internal circumference of the thorax, being attached, in front, by fleshy fibres to the ensiform cartilage; on either side, to the inner surface of the cartilages and bony portions of the six or seven inferior ribs, interdigitating with the Transversalis; and behind, to the ligamentum

150.—The Diaphragm. Under Surface.

arcuatum externum and internum. The fibres from these sources vary in length; those arising from the ensiform appendix are very short and occasionally aponeurotic; but those from the ligamenta arcuata, and more especially those from the ribs at the side of the chest, are the longest, describe well marked curves as they ascend, forming an arch on each side with the concavity downwards, this concavity being deeper on the right than on the left side. These fibres converge, to be inserted into the circumference of the central tendon. Between the sides of the muscular slip from the ensiform appendix and the cartilage of the adjoining rib, the fibres of the diaphragm are deficient, the interval being filled by areolar tissue, covered on the thoracic side by the pleurae, on the abdominal by the peritoneum. This is, consequently, a weak point, and a portion of the contents of the abdomen may protrude into the chest, forming phrenic or diaphragmatic hernia, or a collection of pus in the mediastinum may descend through it so as to point at the epigastrium.

The *Ligamentum Arcuatum Internum* is a tendinous arch, thrown across the upper part of the Psoas magnus muscle, on each side of the spine. It arises from the outer side of the body of the first, and occasionally from the second lumbar vertebra, being continuous with the outer side of the tendon of the corresponding crus, and, arching across the Psoas muscle, is attached to the front of the transverse process of the second lumbar vertebra.

The *Ligamentum Arcuatum Externum* is the thickened upper margin of the anterior lamella of the transversalis fascia; it arches across the upper part of the Quadratus lumborum, being attached by one extremity to the front of the transverse process of the second lumbar vertebra, and by the other to the apex and lower margin of the last rib.

To the spine the Diaphragm is connected by two crura, which are situated on the bodies of the lumbar vertebrae, one on each side of the aorta. The crura at their origin are tendinous in structure; the right crus, larger and longer than the left, arising from the anterior surface of the bodies and intervertebral substances of the second, third, and fourth lumbar vertebrae; the left from the second and third; and both blending with the anterior common ligament of the spine. A tendinous arch is thrown across the front of the vertebral column, from the tendon of one crus to that of the other, beneath which passes the aorta, vena azygos major, and thoracic duct. The tendons terminate in two large fleshy bellies, which, with the tendinous portions above alluded to, are called the *crura*, or *pillars of the diaphragm*. The outer fasciculi of the two crura are directed upwards and outwards to the central tendon, but the inner fasciculi decussate in front of the aorta, and then diverge, so as to surround the oesophagus before ending in the tendinous centre. The most anterior and larger of these fasciculi is formed by the right crus.

The *Central* or *Cordiform Tendon* of the Diaphragm is a thin tendinous aponeurosis, situated at the centre of the vault of this muscle, immediately beneath the pericardium, with which its circumference is blended in adults. It is shaped somewhat like a trefoil leaf, consisting of three divisions, or leaflets, separated from one another by slight indentations. The right leaflet is the largest; the middle one, directed towards the ensiform cartilage, the next in size; and the left the smallest. In structure, it is composed of several planes of fibres, which intersect one another at various angles, and unite into straight or curved bundles, an arrangement which affords additional strength to the tendon.

The *Openings* connected with the Diaphragm are three large and several smaller apertures. The former are the aortic, oesophageal, and the opening for the vena cava.

The *Aortic Opening* is the lowest and the most posterior of the three large apertures connected with this muscle. It is situated in the middle line, immediately in front of the bodies of the vertebrae. It is an osseo-aponeurotic aperture, formed by a tendinous arch thrown across the front of the bodies of the vertebrae, from the crus on one side to that on the other, and transmits the aorta, vena azygos major, thoracic duct, and occasionally the left sympathetic nerve.

The *Oesophageal Opening*, elliptical in form, muscular in structure, and formed by the two crura, is placed higher, and, at the same time, anterior, and a little to the left of the preceding. It transmits the oesophagus and pneumogastric nerves. The anterior margin of this aperture is occasionally tendinous, being formed by the margin of the central tendon.

The *Opening for the Vena Cava* is situated the highest; it is quadrilateral in form, tendinous in structure, and placed at the junction of the right and middle leaflets of the central tendon, its margins being bounded by four bundles of tendinous fibres, which meet at right angles.

The *Right Crus* transmits the sympathetic and the greater and lesser splanchnic nerves of the right side; the *left crus*, the greater and lesser splanchnic nerves of the left side, and the vena azygos minor.

The *Serous Membranes* in relation with the Diaphragm are four in number; three lining its upper or thoracic surface, one its abdominal. The three serous membranes on its upper surface are the pleura on either side, and the serous layer of the pericardium, which covers the upper surface of the tendinous centre. The serous membrane covering its under surface is a portion of the general peritoneal membrane of the abdominal cavity.

Peculiarities. The portion of the muscle described as arising from the last rib is occasionally aponeurotic in structure. The sternal attachment of the muscle is sometimes partially or entirely deficient.

Relations. Its *upper* or *thoracic surface* is convex on each side, and corresponds with the pleura and lungs, more flattened at the centre where it supports the heart. The convexity of this surface is greater on the right than on the left side, reaching in the former situation as high as the junction of the fifth rib with the sternum, and in the latter as high as the sixth rib. It reaches much higher in the foetus than in the adult.

Its *under* or *abdominal surface* is concave, more so on the right side, where it is in relation with the convex surface of the liver, than on the left, where it corresponds to the spleen and great end of the stomach behind; it is also in relation with the kidneys, suprarenal capsules, transverse portion of the duodenum, pancreas, and the solar plexus.

Nerves. The Diaphragm is supplied by the phrenic nerves.

Actions. The Diaphragm is the most important inspiratory muscle, being the only one brought into action in tranquil respiration. *During inspiration*, when the fibres of the Diaphragm contract, the muscle descends, forming an inclined plane, which extends from the ensiform cartilage to the tenth rib. During this action, the cavity of the thorax is enlarged considerably from above downwards, and the abdominal viscera are pushed into the lower and fore part of the abdomen, which is much diminished in size. If the abdominal muscles and Diaphragm act together, the viscera are compressed and forced to the lower part of the abdominal cavity, as in most expulsory efforts, which are usually accompanied by a deep inspiration. *During expiration*, when the Diaphragm is relaxed, the muscle is convex, encroaching considerably on the cavity of the chest, particularly at the sides, its upper border, in a forced expiration, being on a level with the lower border of the fourth rib on the right side, and with the fifth on the left. During the action of the Diaphragm the oesophagus is compressed, the aperture through which it passes being chiefly muscular; the apertures for the vena cava and aorta are also compressed, but only to a very trifling extent, as the openings for the passage of these vessels are completely tendinous. Hiccough and sobbing are the result of spasmodic contraction of this muscle; and laughing and crying are produced by its rapid alternation of contraction and relaxation, combined with laryngeal and facial movements.

MUSCLES AND FASCIAE OF THE UPPER EXTREMITY

The Muscles of the Upper Extremity are divisible into groups, corresponding with the different regions of the limb.

Anterior Thoracic Region.
Pectoralis major.
Pectoralis minor.
Subclavius.

Lateral Thoracic Region.
Serratus magnus.
Acromial Region.
Deltoid.

Anterior Scapular Region.
Subscapularis.
Posterior Scapular Region.
Supra-spinatus.
Infra-spinatus.
Teres minor.
Teres major.
Anterior Humeral Region.
Coraco-brachialis.
Biceps.
Brachialis anticus.
Posterior Humeral Region.
Triceps.
Sub-anconeus.
Anterior Brachial Region.

Superficial Layer
{ Pronator radii teres.
Flexor carpi radialis.
Palmaris longus.
Flexor carpi ulnaris.
Flexor sublimis digitorum. }

Deep Layer
{ Flexor profundus digitorum.
Flexor longus pollicis.
Pronator quadratus. }

Radial Region.
Supinator longus.
Extensor carpi radialis longior.
Extensor carpi radialis brevior.

Posterior Brachial Region.

Superficial Layer
{ Extensor communis digitorum.
Extensor minimi digiti.
Extensor carpi ulnaris.
Anconeus. }

Deep Layer
{ Supinator brevis.
Extensor ossis metacarpi pollicis.
Extensor primi internodii pollicis.
Extensor secundi internodii pollicis.
Extensor indicis. }

MUSCLES OF THE HAND.
Radial Region.
Abductor pollicis.
Flexor ossis metacarpi pollicis (opponens).
Flexor brevis pollicis.
Adductor pollicis.
Ulnar Region.
Palmaris brevis.
Abductor minimi digiti.
Flexor brevis minimi digiti.
Flexor ossis metacarpi minimi digiti.
Palmar Region.
Lumbricales.
Interossei palmares.
Interossei dorsales.

Dissection of Pectoral Region and Axilla (fig. 151). The arm being drawn away from the side nearly at right angles with the trunk, and rotated outwards, a vertical incision should be made through the integument in the median line of the chest, from the upper to the lower part of the sternum; a second incision should be carried along the lower border of the Pectoral muscle, from the ensiform cartilage to the outer side of the axilla; a third, from the sternum along the clavicle, as far as its centre; and a fourth, from the middle of the clavicle obliquely downwards, along the interspace between the Pectoral and Deltoid muscles, as low as the fold of the armpit. The flap of integument may then be dissected off in the direction indicated in the figure, but not entirely removed, as it should be replaced on completing the dissection. If a transverse incision is now made from the lower end of the sternum to the side of the chest, as far as the posterior fold of the armpit, and the integument reflected outwards, the axillary space will be more completely exposed.

FASCIAE OF THE THORAX

The *Superficial Fascia* of the thoracic region is a loose cellulo-fibrous layer, continuous with the superficial fascia of the neck and upper extremity above, and

151.—Dissection of Upper Extremity.

of the abdomen below; opposite the mamma it subdivides into two layers, one of which passes in front, and the other behind this gland; and from both of these layers numerous septa pass into its substance, supporting its various lobes: from the anterior layer, fibrous processes pass forward to the integument and nipple, enclosing in their areolae masses of fat. These processes were called by Sir A. Cooper, the *ligamenta suspensoria*, from the support they afford to the gland in this situation. On removing the superficial fascia, the *deep fascia* of the thoracic region is exposed: it is a thin aponeurotic lamina, covering in the outer surface of the great Pectoral muscle, and sending numerous prolongations between its fasciculi: it is attached, in the middle line, to the front of the sternum, and above to the clavicle: it is very thin over the upper part of the muscle, somewhat thicker in the interval between the Pectoralis major and Latissimus dorsi, where it closes in the axillary space, and divides at the margin of the latter muscle into two layers, one of which passes in front and the other behind it; these proceed as far as the spinous processes of the dorsal vertebrae, to which they are attached. At the lower part of the thoracic region this fascia is well developed, and is continuous with the fibrous sheath of the Recti muscles.

Anterior Thoracic Region
Pectoralis Major.   Pectoralis Minor.
Subclavius.

The *Pectoralis Major* (fig. 152) is a broad, thick, triangular muscle, situated at the upper and anterior part of the chest, in front of the axilla. It arises, by short tendinous fibres, from the entire breadth of the anterior border of the clavicle, its sternal half or two-thirds, from one half the breadth of the anterior surface of the sternum, as low down as the attachment of the cartilage of the sixth or seventh rib, its origin consisting of aponeurotic fibres, which intersect with those of the opposite muscle: it also arises from the cartilages of all the true ribs, and from the aponeurosis of the External oblique muscle of the abdomen. The fibres from this extensive origin converge towards its insertion, giving to the muscle a radiated appearance. Those fibres which arise from the clavicle pass obliquely downwards and outwards, and are usually separated from the rest by a cellular interval, those from the lower part of the

152.—Muscles of the Chest and Front of the Arm. Superficial View.

sternum and the cartilages of the lower true ribs pass upwards and outwards; whilst the middle fibres pass horizontally. As these three sets of fibres converge, they are so disposed that the upper overlap the middle, and the middle the lower portion, the fibres of the lower portion being folded backwards upon themselves; so that those fibres which are lowest in front, become highest at their point of insertion. They all terminate in a flat tendon, about two inches broad, which is inserted into the anterior lip of the bicipital groove of the humerus. This tendon consists of two laminae, placed one in front of the other, and usually blended together below. The anterior, the thicker, receives the clavicular and upper half of the sternal portion of the muscle; the posterior layer receiving the attachment of the lower half of the sternal portion. A peculiarity resulting from this arrangement is, that the fibres of the upper and middle portions of the muscle are inserted into the lower part of the bicipital ridge, those of the lower portion into the upper part. The tendon of the Pectoralis major, at its insertion, is connected with that of the Deltoid, and from its borders an expansion is given off above to the head of the humerus below to the fascia of the arm.

Relations. By its *anterior surface*, with the Platysma myoides, the mammary gland, the superficial fascia, and integument. By its *posterior surface:* its *thoracic portion*, with the sternum, the ribs and costal cartilages, the Subclavius, Pectoralis minor, Serratus magnus, and the Intercostals; by its *axillary portion*, it forms the anterior wall of the axillary space, and is in relation with the axillary vessels and nerves. By its *outer border*, it lies parallel with the Deltoid, from which it is separated by the cephalic vein and descending branch of the thoracico-acromialis artery. Its *lower border* forms the anterior margin of the axilla, being at first serrarated from the Latissimus dorsi by a considerable interval; but both muscles gradually converge towards the outer part of this space.

Peculiarities. In well developed muscular subjects, the sternal origins of the two Pectoral muscles are separated only by a very narrow interval; but this interval is enlarged in those cases where these muscles are ill developed. Very rarely, the whole of the sternal portion is deficient. Occasionally, one or two additional muscular slips arise from the aponeurosis of the External oblique, and become united to the lower margin of the Pectoralis major.

Dissection. The Pectoralis major should now be detached by dividing the muscle along its attachment to the clavicle, and by making a vertical incision through its substance a little external to its line of attachment to the sternum and costal cartilages. The muscle should then be reflected outwards, and its tendon carefully examined.

The Pectoralis minor is now exposed, and immediately above it, in the interval between its upper border and the clavicle, a strong fascia, the costo-coracoid membrane. This fascia, which protects the axillary vessels and nerves, is very thick and dense externally, where it is attached to the coracoid process, and is continuous with the fascia of the arm; more internally, it is connected with the lower border of the clavicle, as far as the inner extremity of the first rib: traced downwards, it passes behind the Pectoralis minor, surrounding, in a more or less complete sheath, the axillary vessels and nerves; and above, it sends a prolongation behind the Subclavius, which is attached to the lower border of the clavicle, and so encloses this muscle in a kind of sheath. The costo-coracoid membrane is pierced by the cephalic vein, the thoracico-acromialis artery and vein, superior thoracic artery, and anterior thoracic nerve.

The *Pectoralis Minor* (fig. 153) is a thin, flat, triangular muscle, situated at the upper part of the thorax, immediately beneath the Pectoralis major. It arises, by three delicate tendinous digitations, from the upper margin and external surface of the third, fourth, and fifth ribs, near their cartilages, and from the aponeurosis covering the Intercostal muscles: the fleshy fibres succeeding to these unite, and passing upwards and outwards, converge to form a flat tendon, which is inserted into the anterior and upper margin of the coracoid process of the scapula.

Relations. By its *anterior surface*, with the Pectoralis major, and the superior thoracic vessels and nerves. By its *posterior surface*, with the ribs, Intercostal muscles, Serratus magnus, the axillary space, and the axillary vessels and nerves. Its superior border is separated from the clavicle by a triangular interval, broad internally, narrow externally, bounded in front by the costo-coracoid membrane, and internally by the ribs. In this space are seen the axillary vessels and nerves.

The costo-coracoid membrane should now be removed, when the Subclavius muscle will be seen.

The *Subclavius* is a long, thin, spindle-shaped muscle, placed immediately beneath the clavicle, in the interval between it and the first rib. It arises by a short and thick

153.—Muscles of the Chest and Front of the Arm, with the boundaries of the Axilla.

tendon from the cartilage of the first rib, immediately in front of the rhomboid ligament; the fleshy fibres proceed outwards to be inserted by short tendinous fibres into a deep groove on the under surface of the middle third of the clavicle.

Relations. By its *upper surface*, with the clavicle. By its *under surface*, it is separated from the first rib by the axillary vessels and nerves. Its *anterior surface* is separated from the Pectoralis major by a strong aponeurosis, which with the clavicle, forms an osteo-fibrous sheath in which the muscle is enclosed. If the costal attachment of the Pectoralis minor is divided across, and the muscle reflected outwards, the axillary vessels and nerves are brought fully into view, and should be examined.

Nerves. The Pectoral muscles are supplied by the anterior thoracic nerves; the Subclavius, by a filament from the cord formed by the union of the fifth and sixth cervical nerves.

Actions. If the arm has been raised by the Deltoid, the Pectoralis major will, conjointly with the Latissimus dorsi and Teres major, depress it to the side of the chest; and, if acting singly, it will draw the arm across the front of the chest. The Pectoralis minor depresses the point of the shoulder, drawing the scapula downwards and inwards to the thorax. The Subclavius depresses the shoulder, drawing the clavicle downwards and forwards. When the arms are fixed, all three muscles act upon the ribs, drawing them upwards and expanding the chest, thus becoming very important agents in forced inspiration. Asthmatic patients always assume this attitude, fixing the shoulders, so that all these muscles may be brought into action to assist in dilating the cavity of the chest.

Lateral Thoracic Region
Serratus Magnus.

The *Serratus Magnus* is a broad, thin, and irregularly quadrilateral muscle, situated at the upper part and side of the chest. It arises by eight fleshy digitations from the external surface and upper borders of the eight upper ribs, and from the aponeurosis covering the upper intercostal spaces, and is inserted into the whole length of the inner margin of the posterior border of the scapula. This muscle has been divided into three portions, a superior, middle, and inferior, on account of the difference in the direction, and in the extent of attachment of each part. The superior portion, separated from the rest by a cellular interval, is a narrow, but thick fasciculus, consisting of the first digitation, which arises by a double origin from the first and second ribs, and from the aponeurotic arch between them (called by some authors, first and second serrations); its fibres proceed upwards, outwards and backwards, to be inserted into the triangular smooth surface on the inner side of the superior angle of the scapula. The middle portion of the muscle, the broadest and thinnest of the three, consists of the second, third, and fourth digitations, the fibres from which form a thin and broad muscular layer, which proceeds horizontally backwards, to be inserted by short tendinous fibres into the posterior border of the scapula, between the superior and inferior angles. The largest portion of this division of the muscle is formed by the third digitation. The inferior portion of the muscle consists of four digitations, in the intervals between which are received corresponding processes of the External oblique; the muscular fibres from these converging, pass upwards, outwards, and backwards, to be inserted into the inner surface of the inferior angle of the scapula, by an attachment partly muscular, partly tendinous.

Relations. This muscle is covered, in front, by the Pectoral muscles; behind, by the Subscapularis; above, by the axillary vessels and nerves. Its *deep surface* rests upon the ribs and intercostal spaces.

Nerves. The Serratus magnus is supplied by the posterior thoracic nerve.

Actions. The Serratus magnus is the most important external inspiratory muscle. When the shoulders are fixed, it elevates the ribs, and so dilates the cavity of the chest, assisting the Pectoral and Subclavius muscles. This muscle, especially its middle and lower segments, draws the base and inferior angle of the scapula forwards, and so raises the point of the shoulder by causing a rotation of the bone on the side of the chest; assisting the Trapezius muscle in supporting weights upon the shoulder, the thorax being at the same time fixed by preventing the escape of the included air.

Dissection. After completing the dissection of the axilla, if the muscles of the back have been dissected, the upper extremity should be separated from the trunk. Saw through the clavicle at its centre, and then cut through the muscles which connect the scapula and arm with the trunk, viz., the Pectoralis minor, in front, Serratus magnus, at the side, and behind, the Levator anguli scapulae, the Rhomboids, Trapezius, and

Latissimus dorsi. These muscles should be cleaned and traced to their respective insertions. An incision should then be made through the integument, commencing at the outer third of the clavicle, and extending along the margin of this bone, the acromion process, and spine of the scapula; the integument should be dissected from above downwards and outwards, when the fascia covering the Deltoid is exposed.

The *Superficial Fascia* of the upper extremity, is a thin cellulo-fibrous lamina, containing between its layers the superficial veins and lymphatics, and the cutaneous nerves. It is most distinct in front of the elbow, and contains between its laminae in this situation the large superficial cutaneous veins and nerves; in the hand it is hardly demonstrable, the integument being closely adherent to the deep fascia by dense fibrous bands. Small subcutaneous bursae are found in this fascia, over the acromion, the olecranon, and the knuckles. The deep fascia of the upper extremity comprises the aponeurosis of the shoulder, arm, and fore-arm, the anterior and posterior annular ligaments of the carpus, and the palmar fascia. These will be considered in the description of the muscles of these several regions.

Acromial Region
Deltoid.

The *Deep Fascia* covering the Deltoid (deltoid aponeurosis) is a thick and strong fibrous layer, which covers the outer surface of the muscle, and sends down numerous prolongations between its fasciculi; it is continuous internally with that covering the great Pectoral muscle; behind, with the aponeurosis covering the Infra-spinatus and back of the arm; above, it is attached to the clavicle, the acromion, and spine of the scapula.

The *Deltoid* is a large thick triangular muscle, which forms the convexity of the shoulder, and has received its name from its resemblance to the Greek letter Δ reversed. It surrounds the shoulder-joint in the greater part of its extent, covering it on its outer side, and in front and behind. It arises, by tendinous fibres, from the outer third of the anterior border and upper surface of the clavicle; from the external margin and upper surface of the acromion process; and from the whole length of the inferior border of the spine of the scapula, as far back as the triangular surface which terminates it. From this extensive origin, the muscular fibres proceed downwards, and converge towards their insertion, the middle passing vertically, the anterior obliquely backwards, the posterior obliquely forwards; they unite to form a thick tendon, which is inserted into a rough prominence on the middle of the outer side of the shaft of the humerus. This muscle is remarkably coarse in its texture, and intersected by three or four tendinous laminae, attached at intervals to the clavicle and acromion; these extend into the substance of the muscle, and give origin to a number of fleshy fibres. The largest of these laminae extends from the summit of the acromion.

Relations. By its *superficial surface*, with the Platysma, supra-acromial nerves, the superficial fascia, and integument. By its *deep surface*, it is separated from the Scapular muscles covering the head of the humerus by a large sacculated synovial bursa, and covers the coracoid process, coraco-acromial ligament, Pectoralis minor, Coraco-brachialis, both heads of the Biceps, tendon of the Pectoralis major, Teres major,

Scapular, and external heads of the Triceps, the circumflex vessels and nerve, and the humerus. Its *anterior border* is separated from the Pectoralis major by a cellular interspace, which lodges the cephalic vein and descending branch of the thoracico-acromialis artery. Its *posterior border* is thin above, thicker below, and bound down by the aponeurotic covering of the Infra-spinatus.

Nerves. The Deltoid is supplied by the circumflex nerve.

Actions. The Deltoid serves to raise the arm directly from the side, and to bring it at right angles with the trunk. Its anterior fibres, assisted by the Pectoralis major, draw the arm forwards; and its posterior fibres, aided by the Teres major and Latissimus dorsi, will draw it backwards.

Dissection. Divide the Deltoid across, near its upper part, by an incision carried along the margin of the clavicle, the acromion process, and spine of the scapula, and reflect it downwards; the bursa will be seen on its under surface, as well as the circumflex vessels and nerves, and External rotator muscle. The insertion of the muscle should be carefully examined.

Anterior Scapular Region
Subscapularis.

The *Subscapular Aponeurosis* is a thin membrane, attached to the entire circumference of the subscapular fossa, and affording attachment by its inner surface to some of the fibres of the Subscapularis muscle: when this is removed the Subscapularis muscle is exposed.

The *Subscapularis* is a large triangular muscle, which fills up the whole of the subscapular fossa, arising from its internal two-thirds, with the exception of a narrow margin along the posterior border, and the small triangular portions of bone on the inner side of the superior and inferior angles, which afford attachment to the Serratus magnus. Some of the fibres arise from tendinous laminae, which intersect the muscle, and are attached to ridges on the bone; and others from an aponeurosis attached to the anterior margin of the axillary border of the scapula, which separates this muscle from the Teres major and the long head of the Triceps. From this origin, the fibres pass outwards, and gradually converging, the muscle becomes narrow and thick, and terminates in a tendon, which is inserted into the lesser tuberosity of the humerus. Some of the muscular fibres which arise from the axillary border of the scapula are inserted into the neck of the bone to the extent of an inch below the tuberosity. The tendon of this muscle is in close contact with the capsular ligament of the shoulder-joint, and glides over a large bursa, which separates it from the base of the coracoid process. This bursa communicates with the cavity of the joint by an aperture in the capsular ligament.

Relations. By its *anterior surface*, with the Serratus magnus, some loose areolar tissue being interposed, the Coraco-brachialis, and Biceps, and the axillary vessels and nerves. By its *posterior surface*, with the scapula, the subscapular vessels and nerves, and the capsular ligament of the shoulder-joint.

Nerves. It is supplied by the subscapular nerves.

Actions. The Subscapularis rotates the head of the humerus inwards; when the arm is raised it draws the humerus downwards. It is a powerful defence to the front of the shoulder-joint, preventing displacement of the head of the bone forwards.

Posterior Scapular Region

Supra-spinatus. Teres Minor.
Infra-spinatus. Teres Major.

Dissection. To expose these muscles, and to examine their mode of insertion into the humerus, detach the Deltoid and Trapezius from their attachment to the spine of the scapula and acromion process. Remove the clavicle by dividing the ligaments connecting it with the coracoid process, and separate it at its articulation with the scapula: divide the acromion process near its root with a saw, and the fragment being removed, the tendons of the posterior Scapular muscles will be fully exposed, and can be examined. A block should be placed beneath the shoulder-joint, so as to make the muscles tense.

The *Supra-spinous Aponeurosis* is a thick and dense membranous layer, attached to the entire circumference of the supra-spinous fossa, and completing the osteofibrous case in which the Supra-spinatus muscle is contained: by its inner surface it affords attachment to some of the fibres of this muscle. It is very thick internally, but thinner externally under the coraco-acromion ligament. When this fascia is removed, the Supra-spinatus muscle is exposed.

The *Supra-spinatus* is a thick triangular muscle, which occupies the whole of the supra-spinous fossa, arising from its internal two-thirds, and from a strong fascia which covers the muscle and completes the osteo-fibrous sheath in which it is enclosed. From these points, the muscular fibres converge to a tendon, which passes across the capsular ligament of the shoulder-joint, to which it is intimately adherent, and is inserted into the highest of the three facets on the great tuberosity of the humerus.

Relations. By its *upper surface*, with the Trapezius, the clavicle, the acromion, the coraco-acromion ligament, and the Deltoid. By its *under surface*, with the scapula, the supra-scapular vessels and nerve, and upper part of the shoulder-joint.

The *Infra-spinous Aponeurosis* is a dense fibrous membrane, covering in the Infra-spinatus muscle, and attached to the entire circumference of the infra-spinous fossa; it affords attachment by its inner surface to some fibres of this muscle, is continuous externally with the fascia of the arm, and gives off from its under surface intermuscular septa, which separate it from the Teres minor, and the latter from the Teres major.

The *Infra-spinatus* is a thick triangular muscle, which occupies the chief part of the infra-spinous fossa, arising by fleshy fibres, from its internal two-thirds; and by tendinous fibres, from the ridges on its surface: it also arises from a strong fascia which covers it externally, and separates it from the Teres major and minor. The fibres converge to a tendon, which glides over the concave border of the spine of the scapula, and passing across the capsular ligament of the shoulder-joint, is inserted into the middle facet on the great tuberosity of the humerus. The tendon of this muscle is occasionally separated from the spine of the scapula by a synovial bursa, which communicates with the synovial membrane of the shoulder-joint.

Relations. By its *posterior surface*, with the Deltoid, the Trapezius, Latissimus dorsi,

154.—Muscles on the Dorsum of the Scapula and the Triceps.

and the integument. By its *anterior surface*, with the scapula, from which it is separated by the superior and dorsalis scapulae vessels, and with the capsular ligament of the shoulder-joint. Its *lower border* is in contact with the Teres minor, and occasionally united with it, and with the Teres major.

The *Teres Minor* is a narrow elongated muscle, which lies along the inferior border of the scapula. It arises from the dorsal surface of the axillary border of the scapula for the upper two-thirds of its extent, and from two aponeurotic laminae, which separate this muscle, one from the Infra-spinatus, the other from the Teres major; its fibres pass obliquely upwards and outwards, and terminate in a thick tendon, which is inserted below the Infra-spinatus into the lowest of the three facets on the great tuberosity of the humerus, and, by fleshy fibres, into the humerus immediately below it. The tendon of this muscle, passes across the capsular ligament of the shoulder-joint.

Relations. By its *posterior surface*, with the Deltoid, Latissimus dorsi, and integument. By its *anterior surface*, with the scapula, the dorsal branch of the subscapular artery, the long head of the Triceps, and the shoulder-joint. By its *upper border*, with the Infra-spinatus. By its *lower border*, with the Teres major, from which it separated anteriorly by the long head of the Triceps.

The *Teres Major* is a broad and somewhat flattened muscle, which arises from the triangular surface on the dorsal aspect of the inferior angle of the scapula, and from the fibrous septa interposed between it and the Teres minor and Infra-spinatus; the fibres are directed upwards and outwards, and terminate in a flat tendon, about two inches in length, which is inserted into the posterior border of the bicipital groove of the humerus. The tendon of this muscle lies immediately behind that of the Latissimus dorsi, from which it is separated by a synovial bursa; it is also placed a little below that muscle at its insertion into the humerus.

Relations. By its *posterior surface*, with the integument, from which it is separated internally by the Latissimus dorsi, and externally by the long head of the Triceps. By its *anterior surface*, with the Subscapularis, Latissimus dorsi, Coraco-brachialis, short head of the Biceps, the axillary vessels, and brachial plexus of nerves. Its *upper border*, is at first in relation with the Teres minor, from which it is afterwards separated by the long head of the Triceps. Its *lower border* forms, in conjunction with the Latissimus dorsi, part of the posterior boundary of the axilla.

Nerves. The Supra and Infra-spinati muscles are supplied by the supra-scapular nerve; the Teres minor, by the circumflex; and the Teres major by the subscapular.

Actions. The Supra-spinatus assists the Deltoid in raising the arm from the side; its action must, however, be very feeble, from the very disadvantageous manner in which the force is applied. The Infra-spinatus and Teres minor rotate the head of the humerus outwards; when the arm is raised, they assist in retaining it in that position, and carrying it backwards. One of the most important uses of these three muscles, is the great protection they afford to the shoulder joint, the Supra-spinatus supporting it above, and preventing displacement of the head of the humerus upwards, whilst the Infra-spinatus and Teres minor protect it behind, and prevent dislocation backwards. The Teres major assists the Latissimus dorsi in drawing the humerus downwards and backwards when previously raised, and rotating it inwards; when the arm is fixed, it may assist the Pectoral and Latissimus dorsi muscles in drawing the trunk forwards.

Anterior Humeral Region

Coraco-Brachialis.   Biceps.   Brachialis Anticus.

Dissection. The arm being placed on the table, with the front surface uppermost, make a vertical incision through the integument along the middle line, from the middle of the interval between the folds of the axilla, to about two inches below the elbow joint, where it should be joined by a transverse incision, extending from the inner to the outer side of the fore-arm; the two flaps being reflected on either side, the fascia should be examined.

The Deep Fascia of the arm, continuous with that covering the shoulder and front of the great Pectoral muscle, is attached, above, to the clavicle, acromion, and spine of the scapula; it forms a thin, loose, membranous sheath investing the muscles of this region, sending down septa between them, and composed of fibres disposed in a circular or spiral direction, and these being connected together by vertical fibres. It differs in thickness at different parts, being thin over the Biceps, but thicker where it covers the Triceps and over the condyles of the humerus, and is strengthened by fibrous aponeuroses, which it derives from the Pectoralis major and Latissimus dorsi, on the inner side, and from the Deltoid, externally. On either side it gives off a strong intermuscular septum, which is attached to the condyloid ridge and condyles on either side of the humerus. These septa serve to separate the muscles of the anterior, from those of the posterior brachial region. The external intermuscular septum extends from the lower part of the anterior bicipital ridge, along the external condyloid ridge, to the outer condyle; it is blended with the tendon of the Deltoid; gives attachment to the Triceps behind, to the Brachialis anticus, Supinator longus, and Extensor carpus radialis longior, in front; and is perforated by the musculo-spiral nerve, and superior profunda artery. The internal intermuscular septum, thicker than the preceding, extends from the lower part of the posterior bicipital groove below the Teres major, along the internal condyloid ridge to the inner condyle; it is blended with the tendon of the Coraco-brachialis, and affords attachment to the Triceps, behind, and the brachialis anticus, in front. It is perforated by the ulnar nerve, and the inferior profunda and anastomotic arteries. At the elbow the deep fascia takes attachment to all the prominent points around this joint, and is continuous with the fascia of the fore-arm. On the removal of this fascia the muscles of the anterior humeral region are exposed.

The *Coraco-Brachialis*, the smallest of the three muscles in this region, is situated at the upper and inner part of the arm. It arises from the apex of the coracoid process of the scapula, in common with the short head of the biceps, and from the intermuscular septum between these two muscles; the fibres pass downwards, backwards, and a little outwards, to be inserted by means of a flat tendon into a rough line at the middle of the inner side of the shaft of the humerus. It is perforated by the musculo-cutaneous nerve. The inner border of this muscle forms a guide to the performance of the operation of tying the brachial artery in the upper part of its course.

Relations. By its *anterior surface*, with the Deltoid and Pectoralis major above, at its insertion it is crossed by the brachial artery. By its *posterior surface*, with the tendons of the Subscapularis, Latissimus dorsi, and Teres major, the short head of the Triceps, the humerus, and the anterior circumflex vessels. By its *inner border*, with the brachial artery, and the median and musculo-cutaneous nerves. By its *outer border*, with the short head of the Biceps and Brachialis anticus.

The *Biceps* is a long fusiform muscle, situated along the anterior aspect of the arm its entire length, and divided above into two portions or heads, from which circumstance it has received its name. Its internal or short head arises by a thick flattened tendon from the apex of the coracoid process of the scapula, in common with the Coraco-brachialis. The external or long head, arises from the upper margin of the glenoid cavity of the scapula, by a long rounded tendon, which is continuous with the glenoid ligament. This tendon passes across the head of the humerus, being enclosed in a special sheath of the synovial membrane of the shoulder joint; it then pierces the capsular ligament at its attachment to the humerus, and descends in the bicipital groove which separates the two tuberosities in which it is retained by a sort of fibrous bridge. The fibres from this tendon form a rounded belly, which about the middle of the arm joins with the short portion of the muscle. The belly of the muscle, narrow and somewhat flattened, terminates above the elbow in a flattened tendon, which is inserted into the posterior part of the tuberosity of the radius, a synovial bursa being interposed between the tendon and the anterior part of the tuberosity. The tendon of this muscle is thin and broad; as it approaches the radius it becomes narrowed and twisted upon itself, being applied by a flat surface to the posterior part of the tuberosity, and opposite the bend of the elbow gives off, from its inner side, a broad aponeurosis, which passes obliquely downwards and inwards across the brachial artery, and is continuous with the fascia of the fore-arm. The inner border of this muscle forms a guide to the performance of the operation of tying the brachial artery in the middle of the arm.

Relations. Its *anterior surface* is overlapped above by the Pectoralis major and Deltoid; in the rest of its extent it is covered by the superficial and deep fasciae and the integument. Its *posterior surface* rests on the shoulder-joint and humerus, from which it is separated by the Subscapularis, Teres major, Latissimus dorsi, brachialis anticus, and the musculo-cutaneous nerve. Its *inner border* is in relation with the Coraco-brachialis, the brachial vessels, and median nerve. By its *outer border*, with the Deltoid and Supinator longus.

The *Brachialis Anticus* is a broad muscle, which covers the whole of the anterior surface of the lower part of the humerus. It is somewhat compressed from before backwards, and is broader in the middle than at either extremity. It arises from the lower half of the external and internal surfaces of the shaft of the humerus, commencing above at the insertion of the Deltoid, which it embraces by two well marked angular processes, and extending, below, to within an inch of the margin of the articular surface, and being limited on each side by the external and internal borders. It also arises from the inter-muscular septa on each side, but more extensively from the inner than the outer. Passing down in front of the elbow joint, its fibres converge to a thick tendinous fasciculus, which is inserted into a rough depression on the lower part of the coronoid process of the ulna, being received into a notch at the upper part of the Flexor digitorum profundus.

Relations. By its *anterior surface*, with the Biceps, musculo-cutaneous nerve, the brachial vessels, and median nerve. By its *posterior surface*, with the humerus and anterior ligament of the elbow joint. By its *inner border*, with the Triceps, ulnar nerve, and Pronator radii teres, from which it is separated by the intermuscular septa. By its *outer border*, with the musculo-spiral nerve, radial recurrent artery, the Supinator longus, and Extensor carpi radialis longior.

Nerves. The muscles of this group are supplied by the musculo-cutaneous nerve. The Brachialis anticus receives an additional filament from the musculo-spiral.

Actions. The Coraco-brachialis draws the humerus forwards and inwards, and at the same time assists in elevating it towards the scapula. The Biceps and brachialis anticus are flexors of the fore-arm; the former muscle is also a supinator, and serves to render tense the fascia of the fore-arm by means of the broad aponeurosis given off from its tendon. When the fore-arm is fixed, the Biceps and Brachialis anticus flex the arm upon the fore-arm, as is seen in the efforts of climbing. The Brachialis anticus forms an important defence to the elbow joint.

Posterior Humeral Region
Triceps.   Subanconeus.

The *Triceps* is the only muscle situated on the back of the arm, extending the entire length of the posterior surface of the humerus. It is of large size, and divided above into three portions or heads; hence the name of the muscle. These three portions have been named, the middle or long head, the external, and the internal or short head.

The *middle* or *long head* arises, by a flattened tendon, from a rough triangular depression, immediately below the glenoid cavity of the scapula, being blended at its upper part with the glenoid ligament; the muscular fibres pass downwards between the two other portions of the muscle, and join with them in the common tendon of insertion.

The *external head* arises from the posterior surface of the shaft of the humerus, between the insertion of the Teres minor and the upper part of the musculo-spiral groove, from the external border of the humerus and external intermuscular septum: the fibres from this origin converge towards the common tendon of insertion.

The *internal* or *short head* arises from the whole of the posterior surface of the shaft of the humerus, below the groove for the musculo-spiral nerve, commencing above, narrow and pointed, immediately below the insertion of the Teres major, and extending, below, to within an inch of the trochlear surface; it also arises from the internal border and internal intermuscular septum. The fibres of this portion of the muscle are directed, some downwards to the olecranon, whilst others converge to the common tendon of insertion.

The *common tendon* of the Triceps commences about the middle of the back part of the muscle: it consists of two aponeurotic laminae, one of which is subcutaneous, and covers the posterior surface of the muscle for the lower half of its extent; the other layer is more deeply seated in the substance of the muscle: after receiving the attachment of the muscular fibres, they join together immediately above the elbow, and are inserted into the posterior part of the upper surface of the olecranon process, a small bursa, occasionally multilocular, being interposed between the tendon and the front of this surface.

The long head of the Triceps passes between the Teres minor and Teres major, dividing the triangular space between these two muscles and the humerus into two smaller spaces, one triangular, the other quadrangular (fig. 154). The triangular space transmits the dorsalis scapulae artery and veins, being bounded by the Teres minor above, the Teres major below, and the scapular head of the Triceps externally: the quadrangular space transmits the posterior circumflex vessels and nerve; it is bounded by the Teres minor above, the Teres major below, the scapular head of the Triceps internally, and the humerus externally.

Relations. By its *posterior surface*, with the integument, superficial and deep fascia, and integument. By its *anterior surface*, with the humerus, musculo-spiral nerve, superior profunda artery, and back part of the elbow-joint. Its *middle* or *long head* is in relation, behind, with the Deltoid and Teres minor; in front, with the Subscapularis, Latissimus dorsi, and Teres major.

Subanconeus. This is a small muscle, distinct from the Triceps, and analogous to the Subcrureus in the lower limb. It may be exposed by removing the Triceps from the lower part of the humerus. It consists of one or two slender fasciculi, which arise from the humerus, immediately above the olecranon fossa, and are inserted into the posterior ligament of the elbow-joint.

Nerves. The Triceps and Subanconeus are supplied by the musculo-spiral nerve.

Actions. The Triceps is the great Extensor muscle of the fore-arm; when the fore-arm is flexed, serving to draw it into a right line with the arm. It is the direct antagonist of the Biceps and Brachialis anticus. When the arm is extended, the long head of this muscle may assist the Teres major and Latissimus dorsi in drawing the humerus backwards. The long head of the Triceps protects the under part of the shoulder-joint, and prevents displacement of the head of the humerus downwards and backwards.

Muscles of the Fore-arm

Dissection. To dissect the fore-arm, place the limb in the position indicated in fig. 151; make a vertical incision along the middle line from the elbow to the wrist, and connect each extremity with a transverse incision; the flaps of integument being removed, the fascia of the fore-arm is exposed.

The *Deep Fascia* of the fore-arm, continuous above with that enclosing the arm, is a dense highly glistening aponeurotic investment, which forms a general sheath enclosing all the muscles in this region; it is attached behind to the olecranon and posterior border of the ulna, and gives off from its inner surface numerous intermuscular septa, which enclose each muscle separately. It consists of circular and oblique fibres, connected together at right angles by numerous vertical fibres. It is much thicker on the dorsal than on the palmar surface, and at the lower than at the upper part of the fore-arm, and is strengthened by tendinous fibres, derived from the Brachialis anticus and Biceps in front, and from the Triceps behind. Its inner surface affords extensive origin for muscular fibres, especially at the upper part of the inner and outer sides of the fore-arm, and forms the boundaries of a series of conical-shaped fibrous cavities, in which the muscles in this region are contained. Besides the vertical septa separating each muscle, transverse septa are given off both on the anterior and posterior surfaces of the fore-arm, separating the deep from the superficial layer of muscles. Numerous apertures exist in the fascia for the passage of vessels and nerves; one of these, of large size, situated at the front of the bend of the elbow, serves for the passage of a communicating branch between the superficial and deep veins.

The muscles of the fore-arm may be subdivided into groups corresponding to the region they occupy. The first group occupies the inner and anterior aspect of the fore-arm, and comprises the Flexor and Pronator muscles. The second group occupies the outer side of the fore-arm; and the third, its posterior aspect. The two latter groups include all the Extensor and Supinator muscles.

Anterior Brachial Region
Superficial Layer.

Pronator radii teres.   Flexor carpi radialis.
Palmaris longus.   Flexor carpi ulnaris.
Flexor sublimis digitorum.

All these muscles take origin from the internal condyle by a common tendon.

The *Pronator Radii Teres* arises by two heads. One, the largest and most superficial, from the humerus, immediately above the internal condyle, and from the tendon

155.—Front of the Left Fore-arm. Superficial Muscles.

common to the origin of the other muscles; also from the fascia of the fore-arm, and intermuscular septum between it and the Flexor carpi radialis. The other head is a thin fasciculus, which arises from the inner side of the coronoid process of the ulna, joining the other at an acute angle. Between the two heads passes the median nerve. The muscle passes obliquely across the fore-arm from the inner to the outer side, and terminates in a flat tendon, which turns over the outer margin of the radius, and is inserted into a rough ridge at the middle of the outer surface of the shaft of that bone.

Relations. By its *anterior surface*, with the fascia of the fore-arm, the Supinator longus, and the radial vessels and nerve. By its *posterior surface*, with the Brachialis anticus, Flexor sublimis digitorum, the median nerve, and ulnar artery. Its *upper border* forms the inner boundary of a triangular space, in which is placed the brachial artery, median nerve, and tendon of the Biceps muscle. Its *lower border* is in contact with the Flexor carpi radialis.

The *Flexor Carpi Radialis* lies on the inner side of the preceding muscle. It arises from the internal condyle by the common tendon, from the fascia of the fore-arm, and from the intermuscular septa between it and the Pronator teres, on the inside; the Palmaris longus, externally; and the Flexor sublimis digitorum, beneath. Slender and aponeurotic in structure at its commencement, it increases in size, and terminates in a tendon which forms the lower two-thirds of its structure. This tendon passes through a separate opening on the outer side of the annular ligament, runs through a groove in the os trapezium, converted into a canal by a thin fibrous sheath, lined by a synovial membrane, and is inserted into the base of the metacarpal bone of the index finger. The radial artery lies between the tendon of this muscle and the Supinator longus, and may easily be secured in this situation.

Relations. By its *superficial surface*, with the fascia of the fore-arm and the integument. By its *deep surface*, with the Flexor sublimis digitorum, Flexor longus pollicis, and wrist joint. By its *outer border*, with the Pronator radii teres, and the radial vessels. By its *inner border*, with the Palmaris longus.

The *Palmaris Longus* is a slender fusiform muscle, lying on the inner side of the preceding. It arises from the inner condyle of the humerus by the common tendon, from the fascia of the fore-arm, and inter-muscular septa, between it and the adjacent muscles. It terminates in a slender flattened tendon, which forms the lower two-thirds of its structure, being inserted into the annular ligament, and expanding to be continuous with the palmar fascia.

Variations. This muscle is often found wanting; when it exists, it presents many varieties. Its fleshy belly is sometimes very long, or it may occupy the middle of the muscle, which is tendinous at either extremity; or it may be muscular at its lower extremity, its upper part being tendinous. Occasionally there is a second Palmaris longus placed on the inner side of the preceding, terminating, below, partly in the annular ligament or fascia, and partly in the small muscles of the little finger.

Relations. By its *anterior surface*, with the fascia of the fore-arm. By its *posterior surface*, with the Flexor digitorum sublimis. *Internally*, with the Flexor carpi ulnaris. *Externally*, with the Flexor carpi radialis.

The *Flexor carpi ulnaris* lies along the ulnar side of the fore-arm. It arises by two heads, separated by a tendinous arch, beneath which passes the ulnar nerve, and posterior ulnar recurrent artery. One head arises from the inner condyle of the humerus, by the common tendon; the other, from the inner margin of the olecranon, and by an aponeurosis from the upper two-thirds of the posterior border of the ulna. It also arises from the inter-muscular septum between it and the Flexor sublimis digitorum. The muscular fibres terminate in a tendon, which is inserted on the anterior surface of the pisiform bone, the tendon being prolonged to the annular ligament and base of the metacarpal bone of the little finger. The ulnar artery lies on the outer side of the tendon of this muscle, in the lower two-thirds of the fore-arm; the tendon forming a guide to the operation of including this vessel in a ligature in this situation.

Relations. By its *anterior surface*, with the fascia of the fore-arm, with which it is intimately connected for a considerable extent. By its *posterior surface*, with the Flexor sublimis, the Flexor profundus, the Pronator quadratus, and the ulnar vessels and nerve. By its *outer* or *radial border*, with the Palmaris longus, above; below, with the ulnar vessels and nerve.

The *Flexor Digitorum Sublimis* is placed beneath the preceding muscles; these therefore require to be removed before its entire extent of attachment is brought into view. It is the largest of the muscles of the superficial layer, and arises by three distinct heads. One from the internal condyle of the humerus by the common tendon, from the internal lateral ligament of the elbow joint, and from the inter-muscular septum common to it and the preceding muscles. The second head arises from the coronoid process of the ulna, above the ulnar origin of the Pronator radii teres. The third head arises by tendinous fibres from the oblique line of the radius, extending from the tubercle above, to the insertion of the Pronator radii teres below. The muscular fibres pass vertically downwards, forming a broad and thick muscle, which divides into four tendons about the middle of the fore-arm; as these tendons pass beneath the annular ligament into the palm of the hand, they are arranged in pairs, the anterior pair corresponding to the middle and ring fingers; the posterior pair to the index and little fingers. The tendons diverge from one another as they pass onwards, and are finally inserted into the lateral margins of the second phalanges, about their centre. Opposite the base of the first phalanges, each tendon divides, so as to leave a fissured interval, between which passes one of the tendons of the Flexor profundus, and they both enter an osso-aponeurotic canal, formed by a strong fibrous band which arches across them, and is attached on each side to the margins of the phalanges. The two portions into which the tendon of the Flexor sublimis divides, so as to admit of the passage of the deep flexor, expand somewhat, and form a grooved channel into which the accompanying deep flexor tendon is received; the two divisions then unite, and finally subdivide a second time to be inserted into the fore part and sides of the second phalanges. The tendons whilst contained in the fibro-osseous canals are connected to the phalanges by slender tendinous filaments, called *vincula accessoria tendinum*. A synovial sheath invests the tendons as they pass beneath the annular ligament; a similar membrane surrounds each tendon as it passes along the phalanges.

Relations. In the fore-arm. By its *anterior surface*, with the deep fascia and all the preceding superficial muscles. By its *posterior surface*, with the Flexor profundus digitorum, Flexor longus pollicis, the ulnar vessels and nerves, and the median nerve. In the hand, its tendons are in relation, in front, with the palmar fascia, superficial palmar arch, and the branches of the median nerve. Behind, with the tendons of the deep Flexor and the Lumbricales.

Anterior Brachial Region
Deep Layer.
Flexor Profundus Digitorum. Flexor Longus Pollicis.
Pronator Quadratus.

Dissection. Divide each of the superficial muscles at its centre, and turn either end aside, the deep layer of muscles, together with the median nerve and ulnar artery, will then be exposed.

The *Flexor Profundus Digitorum* (*perforans*) is situated on the ulnar side of the forearm, immediately beneath the superficial Flexors. It arises from the upper two-thirds of the anterior and internal surfaces of the shaft of the ulna, embracing above, the insertion of the Brachialis anticus, and extending, below, to within a short distance of the Pronator quadratus. It also arises from a depression on the inner side of the coronoid process, by an aponeurosis from the upper two-thirds of the posterior border of the ulna, and from the ulnar half of the interosseous membrane. The fibres from these origins pass downwards, forming a fleshy belly of considerable size, which divides into four unequal portions, each of which terminates in a tendon which passes beneath the annular ligament beneath the tendons of the Flexor sublimis. Opposite the first phalanges, the tendons pass between the two slips of the tendons of the Flexor sublimis, and are finally inserted into the bases of the last phalanges. The tendon of the index finger is distinct; the rest are connected together by cellular tissue and tendinous slips, as far as the palm of the hand.

Four small muscles, the Lumbricales, are connected with the tendons of the Flexor profundus in the palm. They will be described with the muscles in that region.

Relations. By its *anterior surface*, in the fore-arm, with the Flexor sublimis digitorum, the Flexor carpi ulnaris, the ulnar vessels and nerve, and the median nerve; and in the hand, with the tendons of the superficial Flexor. By its *posterior surface*, in the forearm, with the ulna, the interosseous ligament, the Pronator quadratus; and in the hand, with the Interossei, Adductor pollicis, and deep palmar arch. By its *ulnar border*, with the Flexor carpi ulnaris. By its *radial border*, with the Flexor longus pollicis, the anterior interosseous artery and nerve being interposed.

The *Flexor Longus Pollicis* is situated on the radial side of the fore-arm, lying on the same plane as the preceding. It arises from the upper two-thirds of the grooved anterior surface of the shaft of the radius; commencing, above, immediately below the tuberosity and oblique line, and extending, below, to within a short distance of the Pronator quadratus. It also arises from the adjacent part of the interosseous membrane, and occasionally by a fleshy slip from the inner side of the base of the coronoid process. The fibres pass downwards and terminate in a flattened tendon, which passes beneath the annular ligament, is then lodged in the interspace between the two heads of the Flexor brevis pollicis, and entering a tendino-osseous canal, similar to those for the other flexor tendons, is inserted into the base of the last phalanx of the thumb.

Relations. By its *anterior surface*, with the Flexor sublimis digitorum, Flexor carpi radialis, Supinator longus, and radial vessels. By its *posterior surface*, with the radius, interosseous membrane, and Pronator quadratus. By its *ulnar border*, with the Flexor profundus digitorum, from which it is separated by the anterior interosseous artery and nerve.

The *Pronator Quadratus* is a small muscle, quadrilateral in form,

156.—Front of the Left Fore-arm. Deep Muscles.

extending transversely across the radius and ulna, immediately above their carpal extremities. It arises from the oblique line on the lower fourth of the anterior surface of the shaft of the ulna, and the surface of bone immediately below it; from the internal border of the ulna; and from a strong aponeurosis which covers the inner third of the muscle. The fibres pass horizontally outwards, to be inserted into the lower fourth of the anterior surface and external border of the shaft of the radius.

Relations. By its *anterior surface*, with the Flexor profundus digitorum, the Flexor longus pollicis, Flexor carpi radialis, and the radial and ulnar vessels, and ulnar nerve. By its *posterior surface*, with the radius, ulna, and interosseous membrane.

Nerves. All the muscles of the superficial layer are supplied by the median nerve, excepting the Flexor carpi ulnaris, which is supplied by the ulnar. Of the deep layer, the Flexor profundus digitorum is supplied conjointly by the ulnar and anterior interosseus nerves, the Flexor longus pollicis and Pronator quadratus by the anterior interosseous nerve.

Actions. These muscles act upon the fore-arm, the wrist, and hand. Those acting on the fore-arm, are the Pronator radii teres and Pronator quadratus, which rotate the radius upon the ulna, rendering the hand prone; when pronation has been fully effected, the Pronator radii teres assists the other muscles in flexing the fore-arm. The flexors of the wrist are the Flexor carpi ulnaris and radialis; and the flexors of the phalanges are the Flexor sublimis and Profundus digitorum; the former flexing the second phalanges, and the latter the last. The Flexor longus pollicis flexes the last phalanx of the thumb. The three latter muscles, after flexing the phalanges by continuing their action, act upon the wrist, assisting the ordinary flexors of this joint; and all assist in flexing the fore-arm upon the arm. The Palmaris longus is a tensor of the palmar fascia; when this action has been fully effected, it flexes the hand upon the fore-arm.

Radial Region

Supinator Longus.   Extensor Carpi Radialis Longior.
Extensor Carpi Radialis Brevior.

Dissection. Divide the integument in the same manner as in the dissection of the anterior brachial region; and after having examined the cutaneous vessels and nerves and deep fascia, they should be removed, when the muscles of this region will be exposed. The removal of the fascia will be considerably facilitated by detaching it from below upwards. Great care should be taken to avoid cutting across the tendons of the muscles of the thumb.

The *Supinator Longus* is the most superficial muscle on the radial side of the fore-arm, fleshy for the upper two-thirds of its extent, tendinous below. It arises from the upper two-thirds of the external condyloid ridge of the humerus, and from the external intermuscular septum being limited above by the musculo-spiral groove. The fibres descend on the anterior and outer side of the fore-arm, and terminate in a flat tendon, which is inserted into the base of the styloid process of the radius.

Relations. By its *superficial surface*, with the integument and fascia for the greater part of its extent; near its insertion it is crossed by the Extensor ossis metacarpi pollicis and the Extensor primi internodii pollicis. By its *deep surface*, with the humerus, the Extensor carpi radialis longior and brevior, the insertion of the Pronator radii teres, and the Supinator brevis. By its *inner border*, above the elbow with the Brachialis anticus, the musculo-spiral nerve, and radial recurrent artery; and in the fore-arm, with the radial vessels and nerve.

157.—Posterior Surface of Fore-arm. Superficial Muscles.

The *Extensor Carpi Radialis Longior* is placed partly beneath the preceding muscle. It arises from the lower third of the external condyloid ridge of the humerus, immediately below the Supinator longus, and from the external inter-muscular septum. The fibres pass downwards, and terminate at the upper third of the fore-arm in a flat tendon, which runs along the outer border of the radius, beneath the extensor tendons of the thumb; it then passes through a groove common to it and the Extensor carpi radialis brevior, immediately behind the styloid process; and is inserted into the base of the metacarpal bone of the index finger, its radial side.

Relations. By its *superficial surface*, with the Supinator longus and fascia of the fore-arm. Its *outer side*, is crossed obliquely by the Extensor ossis metacarpi pollicis and the Extensor primi internodii pollicis; and at the wrist by the Extensor secundi internodii pollicis. By its *deep surface*, with the elbow-joint, the Extensor carpi radialis brevior, and back part of the wrist.

The *Extensor Carpi Radialis Brevior* is shorter, as its name implies, and thicker than the preceding muscle, beneath which it is placed.

It arises from the external condyle of the humerus by a tendon common to it and the other extensor muscles; from the external lateral ligament of the elbow-joint; from a strong aponeurosis which covers its surface; and from the intermuscular septum between it and the adjacent muscles. The fibres pass downwards, and terminate about the middle of the fore-arm in a flat tendon, which is closely connected with that of the preceding muscle, accompanies it to the wrist, lying in the same groove on the posterior surface of the radius; passes beneath the annular ligament, and diverging somewhat from its fellow, is inserted into the base of the metacarpal bone of the middle finger, its radial side.

The tendons of the two preceding muscles, as they pass across the same groove at the back of the radius, are retained in it by a fibrous sheath, lubricated by a single synovial membrane, but separated from each other by a small vertical ridge of bone.

Relations. By its *superficial surface*, with the Extensor carpi radialis longior, and crossed by the Extensor muscles of the thumb. By its *deep surface*, with the Supinator brevis, tendon of the Pronator radii teres, radius and wrist-joint. By its *ulnar border*, with the Extensor communis digitorum.

Posterior Brachial Region
Superficial Layer.

Extensor Communis Digitorum.   Extensor Carpi Ulnaris.
Extensor Minimi Digiti.   Anconeus.

The *Extensor Communis Digitorum* is situated at the back part of the fore-arm. It arises from the external condyle of the humerus by a tendon common to it and the other superficial Extensor muscles, from the deep fascia, and the intermuscular septa between it and the adjacent muscles. Just below the middle of the fore-arm it divides into four tendons, which pass in a separate sheath beneath the posterior annular ligament of the wrist, lubricated by a synovial membrane. The tendons then diverge, the two middle ones passing along the dorsal surface of the corresponding metacarpal bones, the lateral ones crossing obliquely to the metacarpal bones, along which they pass; and are finally inserted into the second and third phalanges of the fingers in the following manner. Each tendon opposite its correspondmg metacarpo-phalangeal articulation becomes narrow and thickened, being reinforced by the tendons of the interossei and lumbricales, gives off a thin fasciculus upon each side of the joint, and spreads out into a broad aponeurosis, which covers the whole of the dorsal surface of the first phalanx. Opposite the first phalangeal joint, this aponeurosis divides into three slips, a middle and two lateral; the former is inserted into the base of the second phalanx, and the two lateral, which are continued onwards along the sides of the second phalanx, unite by their contiguous margins, and are inserted into the upper surface of the last phalanx. The tendons of the middle, ring, and little fingers are connected together as they cross the hand by small oblique tendinous slips. The tendons of the index and little fingers also receive, before their division, the special extensor tendons belonging to them.

Relations. By its *superficial surface*, with the fascia of the fore-arm and hand, the posterior annular ligament and integument. By its *deep surface*, with the Supinator brevis, the Extensor muscles of the thumb and index finger, posterior interosseous artery and nerve, the wrist-joint, carpus, metacarpus, and phalanges. By its *radial border*, with the Extensor carpi radialis brevior. By its *ulnar border*, with the Extensor minimi digiti, and Extensor carpi ulnaris.

The *Extensor Minimi Digiti* is a small slender muscle, placed on the inner side of the Extensor communis, with which it is generally connected. It arises from the common tendon of origin of the Extensor muscles by a thin tendinous slip; and from the inter-muscular septa between it and the adjacent muscles. Passing down to the lower extremity of the ulna, its tendon runs through a separate sheath in the annular ligament, and at the metacarpo-phalangeal articulation unites with the tendon derived from the long Extensor. The common tendon then spreads into a broad aponeurosis, which is inserted into the second and third phalanges of the little finger in a similar manner to the common extensor tendons of the other fingers.

The *Extensor Carpi Ulnaris* is the most superficial muscle on the ulnar side of the fore-arm. It arises by the common tendon from the external condyle of the humerus, from the middle third of the posterior border of the ulna below the Anconeus, and from the fascia of the fore-arm. This muscle terminates in a tendon, which runs through a groove behind the styloid process of the ulna, passes through a separate sheath in the annular ligament, and is inserted into the base of the metacarpal bone of the little finger.

Relations. By its *superficial surface*, with the fascia of the fore-arm. By its *deep surface*, with the ulna, and the muscles of the deep layer.

The *Anconeus* is a small triangular muscle, placed behind and beneath the elbow-joint, and appears to be a continuation of the external portion of the Triceps. It arises by a separate tendon from the back part of the outer condyle of the humerus; the fibres diverge from this origin, the upper ones being directed horizontally, the lower obliquely inwards, to be inserted into the triangular surface at the upper part of the posterior surface of the shaft of the ulna.

Relations. By its *superficial surface*, with a strong fascia derived from the Triceps. By its *deep surface*, with the elbow-joint, the orbicular ligament, the ulna, and a small portion of the Supinator brevis.

Posterior Brachial Region
Deep Layer.

Supinator Brevis.	Extensor Primi Internodii Pollicis.
Extensor Ossis Metacarpi Pollicis.	Extensor Secundi Internodii Pollicis.
	Extensor Indicis.

The *Supinator Brevis* is a broad muscle, of a hollow cylindrical form, curved around the upper third of the radius. It arises from the external condyle of the humerus, from the external lateral ligament of the elbow-joint, from the orbicular ligament of the radius, from the prominent oblique line of the ulna, extending down from the lower extremity of the lesser sigmoid cavity, and the triangular depression in front of it; it also arises from a tendinous expansion which covers its surface. The fibres of the muscle pass obliquely around the upper part of the radius; the most superior fibres forming a sling-like fasciculus, which passes around the neck of the radius above the tuberosity, to be attached to the back part of its inner surface; the middle fibres being attached to the outer edge of the bicipital tuberosity; the lower fibres to the oblique line as low down as the insertion of the Pronator radii teres. This muscle is pierced by the posterior interosseous nerve.

158.—Posterior Surface of the Fore-arm. Deep Muscles.

Relations. By its *superficial surface*, with the Pronator radii teres, all the superficial Extensor and Supinator muscles, the Anconeus, the radial vessels and nerve, and the musculo-spiral nerve. By its *deep surface*, with the elbow joint, the interosseous membrane, and the radius.

The *Extensor Ossis Metacarpi Pollicis* is the most external and the largest of the deep Extensor muscles, lying immediately below the Supinator brevis. It arises from the posterior surface of the shaft of the ulna below the origin of the Supinator brevis, from the interosseous ligament, and from the middle third of the posterior surface of the shaft of the radius. Passing obliquely downwards and outwards, it terminates in a tendon which runs through a groove on the outer side of the styloid process of the radius, accompanied by the tendon of the Extensor primi internodii pollicis, and is inserted into the base of the metacarpal bone of the thumb.

Relations. By its *superficial surface*, with the Extensor communis digitorum, Extensor minimi digiti, and fascia of the fore-arm; being crossed by the branches of the posterior interosseous artery and nerve. By its *deep surface*, with the ulna, interosseous membrane, radius, the tendons of the Extensor carpi radialis longior and brevior, and at the outer side of the wrist with the radial artery. By its *upper border*, with the Supinator brevis. By its *lower border*, with the Extensor primi internodii pollicis.

The *Extensor Primi Internodii Pollicis* is much smaller than the preceding muscle, on the inner side of which it lies. It arises from the posterior surface of the shaft of the radius, immediately below the Extensor ossis metacarpi, and from the interosseous membrane. Its direction is similar to that of the Extensor ossis metacarpi, its tendon passing through the same groove on the outer side of the styloid process, to be inserted into the base of the first phalanx of the thumb.

Relations. The same as those of the Extensor ossis metacarpi pollicis.

The *Extensor Secundi Internodii Pollicis* is much larger than the preceding muscle, the origin of which it partly covers in. It arises from the posterior surface of the shaft of the ulna, below the origin of the Extensor ossis metacarpi pollicis, and from the interosseous membrane. It terminates in a tendon which passes through a distinct canal in the annular ligament, lying in a narrow oblique groove at the back part of the lower end of the radius. It then crosses obliquely the tendons of the Extensor carpi radialis longior and brevior, being separated by a triangular interval from the other Extensor tendons of the thumb, in which space the radial artery is found; and is finally inserted into the base of the last phalanx of the thumb.

Relations. By its *superficial surface*, with the same parts as the Extensor ossis metacarpi pollicis. By its *deep surface*, with the ulna, interosseous membrane, radius, the wrist, the radial artery, and metacarpal bone of the thumb.

The *Extensor Indicis* is a narrow elongated muscle, placed on the inner side of, and parallel with, the preceding. It arises from the posterior surface of the shaft of the ulna below the origin of the Extensor secundi internodii pollicis, and from the interosseous membrane. Its tendon passes with the Extensor communis digitorum through the same canal in the annular ligament, and subsequently joins that tendon of the Extensor communis which belongs to the index finger, opposite the lower end of the corresponding metacarpal bone. It is finally inserted into the second and third phalanges of the index finger, in the manner already described.

Relations. They are similar to those of the preceding muscles.

Nerves. The Supinator longus, Extensor carpi radialis longior, and Anconeus, are supplied by branches from the musculo-spiral nerve. The remaining muscles of the radial and posterior brachial regions, by the posterior interosseous nerve.

Actions. The muscles of the radial and posterior brachial regions, which comprise all the Extensor and Supinator muscles, act upon the fore-arm, wrist and hand; they are the direct antagonists of the Pronator and Flexor muscles. The Anconeus assists the Triceps in extending the fore-arm. The Supinator longus and brevis are the supinators of the fore-arm and hand; the former muscle more especially acting as a supinator when the limb is pronated. When supination has been produced, the Supinator longus, if still continuing to act, flexes the fore-arm. The Extensor carpi radialis longior and brevior, and Extensor carpi ulnaris muscles, are the Extensors of the wrist; continuing their action, they serve to extend the fore-arm upon the arm; they are the direct antagonists of the Flexor carpi radialis and ulnaris. The common Extensor of the fingers, the Extensors of the thumb, and the Extensors of the index and little fingers, serve to extend the phalanges into which they are inserted; and are the direct antagonists of the Flexors. By continuing their action they assist in extending the fore-arm. The Extensors of the thumb may assist in supinating the fore-arm, when this part of the hand has been drawn inwards towards the palm, on account of the oblique direction of the tendons of these muscles.

Muscles and Fasciae of the Hand

Dissection (fig. 130). Make a transverse incision across the front of the wrist, and a second across the heads of the metacarpal bones, connect the two by a vertical incision in the middle line, and continue it through the centre of the middle finger. The anterior and posterior annular ligaments, and the palmar fascia, should first be dissected.

The *Anterior Annular Ligament* is a strong fibrous band, which arches over the front of the carpus, converting the deep groove on the front of these bones into a canal, beneath which the tendons of the muscles of the fore-arm pass, previous to their insertion into the fingers. This ligament is attached, internally, to the pisiform bone, and unciform process of the unciform; and externally, to the tuberosity of the scaphoid, and ridge on the trapezium. It is continuous, above, with the deep fascia of the fore-arm, and below, with the palmar fascia. It is crossed by the tendon of the Palmaris longus, by the ulnar artery and nerve, and the cutaneous branch of the median nerve. It has inserted into its upper and inner part, the tendon of the Flexor carpi ulnaris; and has, arising from it below, the small muscles of the thumb and little finger. It is pierced by the tendon of the Flexor carpi radialis; and, beneath it, pass the tendons of the Flexor sublimis and profundus digitorum, the Flexor longus pollicis, and the median nerve. There are two synovial membranes beneath this ligament; one of large size, enclosing the tendons of the Flexor sublimis and profundus; and a separate one for the tendon of the Flexor longus pollicis; the latter is also large and very extensive, reaching from above the wrist to the extremity of the last phalanx of the thumb.

The *Posterior Annular Ligament* is a strong transverse fibrous band, extending across the back of the wrist, and continuous with the fascia of the fore-arm. It forms a sheath for the extensor tendons in their passage to the fingers, being attached, internally, to the cuneiform and pisiform bones, and palmar fascia; externally, to the margin of the radius; and in its passage across the wrist, to the elevated ridges on the posterior surface of the radius. It presents six compartments for the passage of tendons, each of which is lined by a separate synovial sac. These are, from within outwards, 1. A sheath on the outer side of the radius for the tendons of the Extensor ossis metacarpi, and Extensor primi internodii pollicis. 2. Behind the styloid process, for the tendons of the Extensor carpi radialis longior and brevior. 3. Opposite the middle of the posterior surface of the radius, for the tendon of the Extensor secundi internodii pollicis. 4. For the tendons of the Extensor communis digitorum, and Extensor indicis. 5. For the Extensor minimi digiti. 6. For the tendon of the Extensor carpi ulnaris. The synovial membranes lining these sheaths are usually very extensive, extending from above the annular ligament, down upon the tendons, almost to their insertion.

The Palmar Fascia forms a common sheath which invests the muscles of the hand. It consists of three portions, a central and two lateral. The central portion occupies the middle of the palm, is triangular in shape, of great strength and thickness, and binds down the tendons in this situation. It is narrow above, being attached to the lower margin of the annular ligament, and receives the expanded tendon of the Palmaris longus muscle. Below, it is broad and expanded, and opposite the heads of the metacarpal bones divides into four slips, for the four fingers. Each slip subdivides into two processes which enclose the tendons of the Flexor muscles, and are attached to the sides of the first phalanx, and to the anterior or glenoid ligament; by this arrangement, four arches are formed, under which the Flexor tendons pass. The arched intervals left in the fascia between these four fibrous slips, transmit the digital vessels and nerves, and the tendons of the Lumbricales. At the point of division of the palmar fascia into the slips above mentioned, numerous strong transverse fibres bind the separate processes together. This fascia is intimately adherent to the integument by numerous fibrous bands, and gives origin by its inner margin to the Palmaris brevis; it covers the superficial palmar arch, the tendons of the flexor muscles, and the branches of the median and ulnar nerves; and on each side it gives off a vertical septum, which is continuous with the interosseous aponeurosis, and separates the lateral from the middle palmar region.

The Lateral portions of the palmar fascia are very thin fibrous layers, which cover, on the radial side, the muscles of the ball of the thumb; and on the ulnar side, the muscles of the little finger; they are continuous with the dorsal fascia, and in the palm, with the middle portion of the palmar fascia.

Muscles of the Hand

The muscles of the hand are subdivided into three groups. 1. Those of the thumb, which occupy the radial side. 2. Those of the little finger, which occupy the ulnar side. 3. Those in the middle of the palm and between the interosseous spaces.

Radial Group
Muscles of the Thumb.
Abductor Pollicis.
Opponens Pollicis (Flexor Ossis Metacarpi).
Flexor Brevis Pollicis.
Adductor Pollicis.

The *Abductor Pollicis* is a thin, flat, narrow muscle, placed immediately beneath the integument. It arises from the ridge of the os trapezium and annular ligament; and passing outwards and downwards, is inserted by a thin flat tendon into the radial side of the base of the first phalanx of the thumb.

Relations. By its *superficial surface*, with the palmar fascia. By its *deep surface*, with the Opponens pollicis, from which it is separated by a thin aponeurosis. Its *inner border*, is separated from the Flexor brevis pollicis by a narrow cellular interval.

The *Opponens Pollicis* (*Flexor Ossis Metacarpi*) is a small triangular muscle, placed beneath the preceding. It arises from the palmar surface of the trapezium and annular ligament; the fleshy fibres pass downwards and outwards, to be inserted into the whole length of the metacarpal bone of the thumb on its radial side.

Relations. By its *superficial surface*, with the Abductor pollicis. By its *deep surface*, with the trapezio-metacarpal articulation. By its *inner border*, with the Flexor brevis pollicis.

The *Flexor Brevis Pollicis* is much larger than either of the two preceding muscles, beneath which it is placed. It consists of two distinct portions, in the interval between which lies the tendon of the Flexor longus pollicis. The anterior and more superficial portion arises from the trapezium and outer two-thirds of the annular ligament. The deeper portion from the trapezoides, os magnum, base of the third metacarpal bone, and sheath of the tendon of the Flexor carpi radialis. The fleshy fibres unite to form a single muscle; this divides into two tendons, which are inserted one on either side of the base of the first phalanx of the thumb. A sesamoid bone is developed in each

159.—Muscles of the Left Hand. Palmar Surface.

of these tendons as they pass across the metacarpo-phalangeal joint; the outer one being joined by the tendon of the Abductor, and the inner, by that of the Adductor.

Relations. By its *superficial surface*, with the palmar fascia. By its *deep surface*, with the Adductor pollicis, and tendon of the Flexor carpi radialis. By its *external surface*, with the Opponens pollicis. By its *internal surface*, with the tendon of the Flexor longus pollicis.

The *Adductor Pollicis* (fig. 156), is the most deeply seated, and the largest of this group of muscles. It is of a triangular form, arising, by its broad base, from the whole length of the metacarpal bone of the middle finger on its palmar surface: the fibres, proceeding outwards, converge, to be inserted by a short tendon into the ulnar side of the base of the first phalanx of the thumb, and into the internal sesamoid bone, being blended with the innermost tendon of the Flexor brevis pollicis.

Relations. By its *superficial surface*, with the Flexor brevis pollicis, the tendons of the Flexor profundus digitorum and Lumbricales. Its *deep surface*, covers the two first interosseous spaces, from which it is separated by a strong aponeurosis.

Nerves. The Abductor, Opponens, and outer head of the Flexor brevis pollicis, are supplied by the median nerve; the inner head of the Flexor brevis, and the adductor pollicis, by the ulnar nerve.

Actions. The actions of the muscles of the thumb are almost sufficiently indicated by their names. This segment of the hand is provided with three Extensors, an Extensor of the metacarpal bone, an Extensor of the first, and an Extensor of the second phalanx; these occupy the dorsal surface of the fore-arm and hand. There are, also, three Flexors on the palmar surface, a Flexor of the metacarpal bone, the Flexor ossis metacarpi (Opponens pollicis), the Flexor brevis pollicis, and the Flexor longus pollicis; there is also an Abductor and an Adductor. These muscles give to the thumb that extensive range of motion which it possesses in an eminent degree.

Ulnar Region
Muscles of the Little Finger.

Palmaris Brevis. Flexor Brevis Minimi Digiti.
Abductor Minimi Digiti. Opponens Minimi Digiti.

The *Palmaris Brevis*, is a thin quadrilateral plane of muscular fibres, placed immediately beneath the integument on the ulnar side of the hand. It arises by tendinous fasciculi, from the annular ligament and palmar fascia; the fleshy fibres pass horizontally inwards, to be inserted into the skin on the inner border of the palm of the hand.

Relations. By its *superficial surface*, with the integument to which it is intimately adherent, especially by its inner extremity. By its *deep surface*, with the inner portion of the palmar fascia, which separates it from the ulnar artery and nerve, and from the muscles of the ulnar side of the hand.

The *Abductor Minimi Digiti* is situated on the ulnar border of the palm of the hand. It arises by tendinous fibres from the pisiform bone, and from an expansion of the tendon of the Flexor carpi ulnaris. The muscle terminates in a flat tendon, which is inserted into the base of the first phalanx of the little finger, on its ulnar side.

Relations. By its *superficial surface*, with the inner portion of the palmar fascia, and the Palmaris brevis. By its *deep surface*, with the Flexor ossi metacarpi. By its *inner border*, with the Flexor brevis minimi digiti.

The *Flexor Brevis Minimi Digiti* lies on the same plane as the preceding muscle, on its radial side. It arises from the unciform process of the unciform bone, and anterior surface of the annular ligament, and is inserted into the base of the first phalanx of the little finger, in connection with the preceding. It is separated from the Abductor at its origin, by the communicating branch of the ulnar artery, and deep palmar branch of the ulnar nerve. This muscle is sometimes wanting. The Abductor is then, usually, of large size.

Relations. By its *superficial surface*, with the internal portion of the palmar fascia, and the Palmaris brevis. By its *deep surface*, with the Flexor ossis metacarpi.

The *Opponens Minimi Digiti* (fig. 156), is of a triangular form, and placed immediately beneath the preceding muscles. It arises from the unciform process of the unciform bone, and contiguous portion of the annular ligament; from these points, the fibres pass downwards and inwards, to be inserted into the whole length of the metacarpal bone of the little finger, along its ulnar margin.

Relations. By its *superficial surface*, with the Flexor brevis, and Abductor minimi digiti. By its *deep surface*, with the interossei muscles in the fifth metacarpal space, the metacarpal bone, and the Flexor tendons of the little finger.

Nerves. All the muscles of this group are supplied by the ulnar nerve.

Actions. The actions of the muscles of the little finger are expressed in their names. The Palmaris brevis corrugates the skin on the inner side of the palm of the hand.

MIDDLE PALMAR REGION
Lumbricales.   Interossei Palmares.
Interossei Dorsales.

160.—The Dorsal Interossei of Left Hand.

The *Lumbricales* are four small fleshy fasciculi, accessories to the deep Flexor muscle. They arise by fleshy fibres from the tendons of the deep Flexor, the first and second, from the radial side and palmar surface of the tendons of the index and middle fingers, the third, from the contiguous sides of the tendons of the middle and ring fingers, and the fourth, from the contiguous sides of the tendons of the ring and little fingers. They pass forwards to the radial side of the corresponding fingers, and opposite the Metacarpo-phalangeal articulations, each tendon terminates in a broad aponeurosis, which is inserted into the tendinous expansion from the Extensor communis digitorum, which covers the dorsal aspect of each finger.

The *Interossei Muscles* are so named from their occupying the intervals between the metacarpal bones. They are divided into two sets, a dorsal and palmar, the former are four in number, one in each metacarpal space, the latter, three in number, lie upon the metacarpal bones.

The *Dorsal Interossei* are four in number, larger than the palmar, and occupy the intervals between the metacarpal bones. They are bipenniform muscles, arising by two heads from the adjacent sides of the metacarpal bones, but more extensively from that side of the metacarpal bone, which corresponds to the side of the finger in which the muscle is inserted. They are inserted into the base of the first phalanges, and into the aponeurosis of the common Extensor tendon. Between the double origin of each of these muscles is a narrow triangular interval, through which passes a perforating branch from the deep palmar arch.

The *First Dorsal Interosseous* muscle or Abductor indicis, is larger than the others, and lies in the interval between the thumb and index finger. It is flat, triangular in form, and arises by two heads, separated by a fibrous arch, for the passage of the radial artery into the deep part of the palm of the hand. The outer head arises from the upper half of the ulnar border of the first metacarpal bone, the inner head, from the entire length of the radial border of the second metacarpal bone, the tendon is inserted into the radial side of the index finger. The second and third are inserted into the middle finger, the former into its radial, the latter into its ulnar side. The fourth is inserted in the radial side of the ring finger.

The *Palmar Interossei*, three in number, are smaller than the Dorsal, and placed upon the palmar surface of the metacarpal bones, rather than between them. They arise from the entire length of the metacarpal bone of one finger, and are inserted into the side of the base of the first phalanx and aponeurotic expansion of the common Extensor tendon of the same finger.

161.—The Palmar Interossei of Left Hand.

The first arises from the ulnar side of the second metacarpal bone, and is inserted into the same side of the index finger. The second arises from the radial side of the fourth metacarpal bone, and is inserted into the same side of the ring finger. The third arises from the radial side of the fifth metacarpal bone, and is inserted into the same side of the little finger. From this account it may be seen, that each finger is provided with two Interossei muscles, with the exception of the little finger.

Nerves. The two outer Lumbricales are supplied by the median nerve; the rest of the muscles of this group by the ulnar.

Actions. The Dorsal interossei muscles abduct the fingers from an imaginary line drawn longitudinally through the centre of the middle finger, and the Palmar interossei adduct the fingers towards the same line. They usually assist the Extensor muscles, but when the fingers are slightly bent, assist in flexing the fingers.

SURGICAL ANATOMY

The Student having completed the dissection of the muscles of the upper extremity, should consider the effects likely to be produced by the action of the various muscles in fracture of the bones; the causes of displacement are thus easily recognised, and a suitable treatment in each case may be readily adopted.

In considering the actions of the various muscles upon fractures of the upper extremity, the most common forms of injury have been selected, both for illustration and description.

Fracture of the *clavicle* is an exceedingly common accident, and is usually caused by indirect violence, as a fall upon the shoulder; it occasionally, however, occurs from direct force. Its most usual situation is just external to the centre of the bone, but it may occur at the sternal or acromial ends.

162.—Fracture of the Middle of the Clavicle.

Fracture of the *middle of the clavicle* (fig. 162) is always attended with considerable displacement, the outer fragment being drawn downwards, forwards, and inwards; the inner fragment slightly upwards. The outer fragment is drawn down by the weight of the arm and the action of the Deltoid, and forwards and inwards by the Pectoralis minor and Subclavius muscles; the inner fragment is slightly raised by the Sterno-cleido mastoid, but only to a very limited extent, as the attachment of the costo-clavicular ligament and Pectoralis major below and in front would prevent any very great displacement upwards. The causes of displacement having been ascertained, it is easy to apply the appropriate treatment. The outer fragment is to be drawn outwards, and, together with the scapula, raised upwards to a level with the inner fragment, and retained in that position.

In fracture of the *acromial end of the clavicle* between the conoid and trapezoid ligaments, only slight displacement occurs, as these ligaments, from their oblique insertion, serve to hold both portions of the bone in apposition. Fracture, also, of the *sternal end*, internal to the costo-clavicular ligament, is attended with only slight displacement, this ligament serving to retain the fragments in close apposition.

Fracture of the *acromion process* usually arises from violence applied to the upper and outer part of the shoulder: it is generally known by the rotundity of the shoulder being lost, from the Deltoid drawing downwards and forwards the fractured portion; and the displacement may easily be discovered by tracing the margin of the clavicle outwards, when the fragment will be found resting on the front and upper part of the head of the humerus. In order to relax the anterior and outer fibres of the Deltoid (the opposing muscle), the arm should be drawn forwards across the chest, and the elbow well raised up, so that the head of the bone may press upwards the acromion process, and retain it in its position.

Fracture of the *coracoid process* is an extremely rare accident, and is usually caused by a sharp blow directly on its pointed extremity. Displacement is here produced by the combined actions of the Pectoralis minor, short head of the biceps, and Coraco-brachialis, the former muscle drawing the fragment inwards, the latter directly downwards, the amount of displacement being limited by the connection of this process to the acromion by means of the coraco-acromion ligament. In order to relax these muscles, and replace the fragments in close apposition, the fore-arm should be flexed so as to relax the Biceps, and the arm drawn forwards and inwards across the chest so as to relax the Coraco-brachialis; the action of the Pectoralis minor may be counteracted by placing a pad in the axilla; the humerus should then be pushed upwards against the coraco-acromial ligament, and the arm retained in this position.

Fracture of the *anatomical neck of the humerus* within the capsular ligament is a rare accident, attended with very slight displacement, an impaired condition of the motions of the joint, and crepitus.

Fracture of the *surgical neck* (fig. 163) is very common, is attended with considerable displacement, and its appearances correspond somewhat with those of dislocation of the head of the humerus into the axilla. The upper fragment is slightly elevated under the coraco-acromion ligament by the muscles attached to the greater and lesser tuberosities; the upper end of the lower ligament is drawn inwards by the Pectoralis major, Latissimus dorsi, and Teres major; and the humerus is thrown obliquely outwards from the side by the action of the Deltoid, and occasionally elevated so as to project beneath and in front of the coracoid process. By fixing the shoulder, and drawing the arm outwards and downwards, the existing deformity is at once reduced. To counteract the action of the opposing muscles, and to keep the fragments in position, the arm should be drawn from the side, and pasteboard splints applied on its four sides, a large conical-shaped pad should be placed in the axilla with the base turned upwards, and the elbow approximated to the side, and retained there by a broad roller passed around the chest; by these means, the action of the Pectoralis major, Latissimus dorsi, Teres major, and Deltoid muscles are counteracted: the fore-arm should then be flexed, and the hand supported in a sling, care being taken not to raise the elbow, otherwise the lower fragment may be displaced upwards.

163.—Fracture of the Surgical Neck of the Humerus.

In fracture of the *shaft of the humerus* below the insertion of the Pectoralis major, Latissimus dorsi, and Teres major, and above the insertion of the Deltoid, there is also considerable deformity, the lower end of the upper fragment being drawn inwards by the first mentioned muscles, and the lower fragment drawn upwards and outwards by the Deltoid, producing shortening of the limb, and a considerable prominence at the seat of fracture, from the fractured ends of the bone riding over one another, especially if the fracture takes place in an oblique direction. The fragments may be readily brought into apposition by extension from the elbow, and retained in that position by adopting the same means as in the preceding injury.

In fracture of the *shaft of the humerus* immediately below the insertion of the Deltoid, the amount of deformity depends greatly upon the direction of the fracture. If the fracture occurs in a transverse direction, only slight displacement occurs, the lower extremity of the upper fragment being drawn a little forwards: but in oblique fracture, the combined actions of the Biceps and Brachialis anticus muscles in front, and the Triceps behind, draw upwards the lower fragment, causing it to glide over the lower end of the upper fragment, either backwards or forwards, according to the direction of the fracture. Simple extension reduces the deformity, and the application of splints on the four sides of the arm retain the fragments in apposition. Care should be taken not to raise the elbow, but the fore-arm and hand may be supported in a sling.

164.—Fracture of the Humerus above the Condyles.

Fracture of the *humerus* (fig. 164) immediately above the condyles deserves very attentive consideration, as the general appearances correspond somewhat with those produced by separation of the epiphysis of the humerus, and with those of dislocation of the radius and ulna backwards. If the direction of the fracture is oblique from above, downwards and outwards, the lower fragment is drawn upwards and backwards by the Brachialis anticus and Biceps in front, and the Triceps behind. This injury may be diagnosed from dislocation by the increased mobility in fracture, the existence of crepitus, and the deformity being remedied by extension, by the discontinuance of which it is again reproduced. The age of the patient is of importance in distinguishing this form of injury from separation of the epiphysis. If fracture occurs in the opposite direction to that shewn in the plate, the lower fragment is drawn upwards and forwards, causing a considerable prominence in front, and the lower end of the upper fragment projects backwards beneath the tendon of the Triceps muscle.

Fracture of the *coronoid process of the ulna* is an accident of rare occurrence, and is usually caused by violent action of the brachialis anticus muscle. The amount of displacement varies according to the extent of the fracture. If the tip of the process only is broken off, the fragment is drawn upwards by the Brachialis anticus on a level with the coronoid depression of the humerus, and the power of flexion is partially lost. If the process is broken off near its root, the fragment is still displaced by the same muscle; at the same time, on extending the fore-arm, partial dislocation backwards of the ulna occurs from the action of the Triceps muscle. The appropriate treatment would be to relax the Brachialis anticus by flexing the fore-arm, and to retain the fragments in immediate apposition by keeping the arm in this position. Union is generally ligamentous.

Fracture of the *olecranon process* (fig. 165) is a more frequent accident, and is caused either by violent action of the Triceps muscle, or by a fall or blow upon the point of the elbow. The detached fragment is displaced upwards, by the action of the Triceps muscle, from half an inch to two inches; the prominence of the elbow is consequently lost, and a deep hollow is felt at the back part of the joint, which is much increased on flexing the limb. The patient at the same time loses the power of extending the fore-arm. The treatment consists in relaxing the Triceps by extending the fore-arm, and retaining it in this position by means of a long straight splint applied to the front of the arm; the fragments are thus brought into closer apposition, and may be further approximated by drawing down the upper fragment. Union is generally ligamentous.

165.—Fracture of the Olecranon.

Fracture of the *neck of the radius* is an exceedingly rare accident, and is generally caused by direct violence. Its diagnosis is somewhat obscure, on account of the slight deformity visible from the large number of muscles which surround it; but the movements of pronation and supination are entirely lost. The upper fragment is drawn outwards by the Supinator brevis, its extent of displacement being limited by the attachment of the orbicular ligament. The lower fragment is drawn forwards and slightly upwards by the Biceps, and inwards by the Pronator radii teres, its displacement forwards and upwards being counteracted in some degree by the Supinator brevis. The treatment essentially consists in relaxing the Biceps, Supinator brevis, and Pronator radii teres muscles; by flexing the fore-arm, and placing it in a position midway between pronation and supination, extension having been previously made so as to bring the parts in apposition.

Fracture of the *radius* (fig. 166) is more common than fracture of the ulna, on account of the connection of the former with the wrist. Fracture of the shaft of the radius near its centre may occur from direct violence, but more frequently from a fall forwards, the entire weight of the body being received on the wrist and hand. The upper fragment is drawn upwards by the Biceps, and inwards by the Pronator radii

166.—Fracture of the Shaft of the Radius.

teres, holding a position midway between pronation and supination, and a degree of fulness in the upper half of the fore-arm is thus produced; the lower fragment is drawn downwards and inwards towards the ulna by the Pronator quadratus, and thrown into a state of pronation by the same muscle; at the same time, the Supinator longus, by elevating the styloid process, into which it is inserted, will serve to depress still more the upper end of the lower fragment towards the ulna. In order to relax the opposing muscles the fore-arm should be bent, and the limb placed in a position midway between pronation and supination; the fracture is then easily reduced by extension from the wrist and elbow: well padded splints should then be applied on both sides of the fore-arm from the elbow to the wrist; the hand being allowed to fall, will, by its own weight, counteract the action of the Pronator quadratus and Supinator longus, and elevate this fragment to the level of the upper one.

Fracture of the *shaft of the ulna* is not a common accident; it is usually caused by direct violence. Its more protected position on the inner side of the limb, the greater strength of its shaft, and its indirect connection with the wrist, render it less liable to injury than the radius. It usually occurs a little below the centre, which is the weakest part of the bone. The upper fragment retains its usual position; but the lower fragment is drawn outwards towards the radius by the Pronator quadratus, producing a well marked depression at the seat of fracture, and some fulness on the dorsal and palmar surfaces of the fore-arm. The fracture is easily reduced by extension from the wrist and fore-arm. The fore-arm should be flexed, and placed in a position midway between pronation and supination, and well padded splints applied from the elbow to the ends of the fingers.

Fracture of the *shafts of the radius and ulna together* is not a common accident; it may arise from a direct blow, or from indirect violence. The lower fragments are drawn upwards, sometimes forwards, sometimes backwards, according to the direction of the fracture, by the combined actions of the Flexor and Extensor muscles, producing a degree of fulness on the dorsal or palmar surface of the fore-arm; at the same time the two fragments are drawn into contact by the Pronator quadratus, the radius in a state of pronation: the upper fragment of the radius is drawn upwards and inwards by the Biceps and Pronator radii teres to a higher level than the ulna; the upper portion of the ulna is slightly elevated by the Brachialis anticus. The fracture may be reduced by extension from the wrist and elbow, and the fore-arm should be placed in the same position as in fracture of the ulna.

In the treatment of all cases of fracture of the bones of the fore-arm, the greatest care is requisite to prevent the ends of the bones from being drawn inwards towards the interosseous space: if this is not carefully attended to, the radius and ulna may become anchylosed, and the movements of pronation and supination entirely lost. To obviate this, the splints applied to the limb should be well padded, so as to press the muscles down into their normal situation in the interosseous space, and so prevent the approximation of the fragments.

Fracture of the *lower end of the radius* (fig. 167) is usually called *Colles fracture*, from the name of the eminent Dublin surgeon who first accurately described it. It usually arises from the patient falling from a height, and alighting upon the hand, which receives the entire weight of the body. This fracture usually takes place from

167.—Fracture of the Lower End of the Radius.

half an inch to an inch above the articular surface if it occurs in the adult; but in the child, before the age of sixteen, it is more frequently a separation of the epiphysis from the apophysis. The displacement which is produced is very considerable, and bears some resemblance to dislocation of the carpus backwards, from which it should be carefully distinguished. The lower fragment is drawn upwards and backwards behind the upper fragment by the combined actions of the Supinator longus and the flexors and extensors of the thumb and carpus, producing a well marked prominence on the back of the wrist, with a deep depression behind. The upper fragment projects forwards, often lacerating the substance of the Pronator quadratus, and is drawn by this muscle into close contact with the lower end of the ulna, causing a projection on the anterior surface of the fore-arm, immediately above the carpus, from the flexor tendons being thrust forwards. This fracture may be distinguished from dislocation by the deformity being removed on making sufficient extension, when crepitus may be occasionally detected; at the same time, on extension being discontinued, the parts immediately resume their deformed appearance. The age of the patient will also assist in determining whether the injury is fracture or separation of the epiphysis. The treatment consists in flexing the fore-arm, and making powerful extension from the wrist and elbow, depressing at the same time the radial side of the hand, and retaining the parts in this position by well padded *pistol-shaped* splints.

MUSCLES AND FASCIAE OF THE LOWER EXTREMITY

The Muscles of the Lower Extremity are subdivided into groups, corresponding with the different regions of the limb.

Iliac Region.
Psoas magnus.
Psoas parvus.
Iliacus.

Thigh.
Anterior Femoral Region.
Tensor vaginae femoris.
Sartorius.
Rectus.
Vastus externus.

Vastus internus.
Cruraeus.
Subcruraeus.

Internal Femoral Region.
Gracilis.
Pectineus.
Adductor longus.
Adductor brevis.
Adductor magnus.

HIP.
Gluteal Region.
Gluteus maximus.
Gluteus medius.
Gluteus minimus.
Pyriformis.
Gemellus superior.
Obturator internus.
Gemellus inferior.
Obturator externus.
Quadratus femoris.
Posterior Femoral Region.
Biceps.
Semi-tendinosus.
Semi-membranosus.

LEG.
Anterior Tibio-fibular Region.
Tibialis anticus.
Extensor longus digitorum.
Extensor proprius pollicis.
Peroneus tertius.
Posterior Tibio-fibular Region.
Superficial Layer.
Gastrocnemius.
Plantaris.
Soleus.

Deep Layer.
Popliteus.
Flexor longus pollicis.
Flexor longus digitorum.
Tibialis posticus.
Fibular Region.
Peroneus longus.
Peroneus brevis.

FOOT.
Dorsal Region.
Extensor brevis digitorum.
Interossei dorsales.
Plantar Region.
First Layer.
Abductor pollicis.
Flexor brevis digitorum.
Abductor minimi digiti.
Second Layer.
Musculus accessorius.
Lumbricales.
Third Layer.
Flexor brevis pollicis.
Adductor pollicis.
Flexor brevis minimi digiti.
Transversus pedis.
Fourth Layer.
Interossei plantares.

ILIAC REGION
Psoas Magnus.   Psoas Parvus.   Iliacus.

Dissection. No detailed description is required for the dissection of these muscles. They are exposed after the removal of the viscera from the abdomen, covered by the Peritoneum and a thin layer of fascia, the fascia iliaca.

The *Iliac fascia* is the aponeurotic layer which lines the back part of the abdominal cavity, and encloses the Psoas and Iliacus muscles throughout their whole extent. It is thin above, and becomes gradually thicker below, as it approaches the femoral arch.

The portion investing the Psoas, is attached, above, to the ligamentum arcuatum internum; internally, to the sacrum; and by a series of arched processes to the intervertebral substances, and prominent margins of the bodies of the vertebrae; the intervals left opposite the constricted portions of the bodies, transmitting the lumbar arteries and sympathetic filaments of nerves. Externally, it is continuous with the fascia lumborum.

The portion investing the iliacus is connected, externally, to the whole length of the inner border of the crest of the ilium. Internally, to the brim of the true pelvis, where it is continuous with the periosteum, and receives the tendon of insertion of the Psoas parvus. External to the femoral vessels, this fascia is intimately connected with Poupart's ligament, and is continuous with the fascia transversalis; but corresponding to the point where the femoral vessels pass down into the thigh, it is prolonged down behind them, forming the posterior wall of the femoral sheath. Below this point, the iliac fascia surrounds the Psoas and Iliacus muscles to their termination, and becomes continuous with the iliac portion of the fascia lata. Internal to the femoral vessels the iliac fascia is connected to the ilio-pectineal line, and is continuous with the pubic portion of the fascia lata. The iliac vessels lie in front of the iliac fascia, but all the branches of the lumbar plexus, behind it; it is separated from the peritoneum by a quantity of loose areolar tissue. In abscesses accompanying caries of the lower part of the spine, the matter makes its way to the femoral arch, distending the sheath of the Psoas; and when it accumulates in considerable quantity, this muscle becomes absorbed, and the nervous cords contained in it are dissected out, and lie exposed in the cavity of the abscess; the femoral vessels, however, remain intact, and the peritoneum seldom becomes implicated notwithstanding the extreme thinness of this membrane.

Remove this fascia, and the muscles of the iliac region will be exposed.

The *Psoas Magnus* is a long fusiform muscle, placed on the side of the lumbar region of the spine (fig. 169) and margin of the pelvis. It arises from the sides of the bodies, from the corresponding intervertebral substances, and from the anterior part of the bases of the transverse processes of the last dorsal and all the lumbar vertebrae. The muscle is connected to the bodies of the vertebrae by five slips, each of which is attached to the upper and lower margins of two vertebrae, and to the intervertebral substance between them; the slips themselves being connected by tendinous arches extending across the constricted part of the bodies, beneath which pass the lumbar arteries and sympathetic nervous filaments. These tendinous arches also give origin to muscular fibres and protect the blood-vessels and nerves from pressure during the action of the muscle. The first slip is attached to the contiguous margins of the last dorsal and first lumbar vertebrae; the last, to the contiguous margins of the fourth and fifth lumbar, and intervertebral substance. From these points, the muscle passes down across the brim of the pelvis, and diminishing gradually in size, passes beneath Poupart's ligament, and terminates in a tendon, which after receiving the fibres of the Iliacus, is inserted into the lesser trochanter of the femur.

Relations. In the lumbar region. By its *anterior surface*, which is placed behind the Peritoneum, with the ligamentum arcuatum internum, the kidney, Psoas parvus, renal vessels, ureter, spermatic vessels, genito-crural nerve, the colon, and along its pelvic border, with the common and external iliac artery and vein. By its *posterior surface*, with the transverse processes of the lumbar vertebrae and the quadratus lumborum, from which it is separated by the anterior lamella of the aponeurosis of the Transversalis; the anterior crural nerve is at first situated in the substance of the muscle, and emerges from its outer border at its lower part. The lumbar plexus is situated in the posterior part of the substance of the muscle. By its *inner side*, with the bodies of the lumbar vertebrae, the lumbar arteries, the sympathetic ganglia, and its communicating branches with the spinal nerves. In the thigh it is in relation, in front, with the fascia lata; behind, with the capsular ligament of the hip, from which it is separated by a synovial bursa, which sometimes communicates with the cavity of the joint through an opening of variable size. By its *inner border*, with the Pectineus and the femoral artery, which slightly overlaps it. By its *outer border*, with the crural nerve and Iliacus muscle.

The *Psoas Parvus* is a long slender muscle, placed immediately in front of the preceding. It arises from the sides of the bodies of the last dorsal and first lumbar vertebrae, and from the inter-vertebral substance between them. It forms a small flat muscular bundle, which terminates in a broad flattened tendon, which is inserted into the ilio-pectineal eminence, being continuous, by its outer border, with the iliac fascia. This muscle is most frequently found wanting, being present, according to M. Theile, in one out of every twenty subjects examined.

Relations. It is covered by the peritoneum, and at its origin by the ligamentum arcuatum internum; it rests on the Psoas magnus.

The *Iliacus* is a flat radiated muscle, which fills up the whole of the internal iliac fossa. It arises from the inner concave surface of the ilium, from the inner margin of the crest of that bone; behind, from the ilio-lumbar ligament, and base of the sacrum; in front, from the anterior superior and anterior inferior spinous processes of the ilium, the notch between them, and by a few fibres from the capsular ligament of the hip-joint. The fibres converge to be inserted into the outer side of the tendon common to this muscle and the Psoas magnus, some of them being prolonged down into the oblique line which extends from the lesser trochanter to the linea aspera.

Relations. *Within the pelvis:* by its *anterior surface*, with the iliac fascia, which separates this muscle from the peritoneum, and with the external cutaneous nerve; on the right side, with the caecum; on the left side, with the sigmoid flexure of the colon. By its *posterior surface*, with the iliac fossa. By its *inner border*, with the Psoas magnus, and anterior crural nerve. In the thigh, it is in relation, by its *anterior surface*, with the fascia lata, Rectus and Sartorius; behind, with the capsule of the hip-joint, a synovial bursa common to it, and the Psoas magnus being interposed.

Nerves. The Psoae muscles are supplied by the anterior branches of the lumbar nerves. The Iliacus from the anterior crural.

Actions. The Psoas and Iliacus muscles, acting from above, flex the thigh upon the pelvis, and, at the same time, rotate the femur outwards, from the obliquity of their insertion into the inner and back part of that bone. Acting from below, the femur being fixed, the muscles of both sides bend the lumbar portion of the spine and pelvis forwards. They also serve to maintain the erect position, by supporting the spine and pelvis upon the femur, and assist in raising the trunk when the body is in the recumbent posture.

The Psoas parvus is a tensor of the iliac fascia.

Anterior Femoral Region

Tensor Vaginae Femoris.
Sartorius.
Rectus.
Vastus Externus.
Vastus Internus.
Cruraeus.
Sub-Cruraeus.

Dissection. To expose the muscles and fasciae in this region, an incision should be made along Poupart's ligament, from the spine of the ilium to the pubes, from the centre of this, a vertical incision must be carried along the middle line of the thigh to below the knee-joint, and connected with a transverse incision, carried from the inner to the outer side of the leg. The flaps of integument having been removed, the superficial and deep fasciae should be examined. The more advanced student would commence the study of this region by an examination of the anatomy of femoral hernia, and Scarpa's triangle, the incisions for the dissection of which are marked out in the accompanying figure.

168.—Dissection of Lower Extremity. Front view.

1. *Dissection of* FEMORAL HERNIA, & SCARPA'S TRIANGLE

2. FRONT *of* THIGH

3. FRONT *of* LEG

4. DORSUM *of* FOOT

FASCIAE OF THE THIGH

The *Superficial fascia*, forms a continuous layer over the whole of the lower extremity, consisting of areolar tissue, containing in its meshes much adipose matter, and capable of being separated into two or more layers, between which are found the superficial vessels and nerves. It varies in thickness in different parts of the limb; in the sole of the foot it is so thin, as to be scarcely demonstrable, the integument being closely adherent to the deep fascia beneath, but in the groin it is thicker, and the two layers are separated from one another by the superficial inguinal glands, the internal saphenous vein, and several smaller vessels. Of these two layers, the most superficial is continuous above with the superficial fascia of the abdomen, the deep layer becoming blended with the fascia lata, a little below Poupart's ligament. The deep layer of superficial fascia is intimately adherent to the margins of the saphenous opening in the fascia lata, and pierced in this situation by numerous small blood and lymphatic vessels, hence the name *cribriform fascia*, which has been applied to it. Subcutaneous bursae are found in the superficial fascia over the patella, point of the heel, and phalangeal articulations of the toes.

The *Deep fascia* of the thigh is exposed on the removal of the superficial fascia, and is named, from its great extent, the fascia lata; it forms a uniform investment for the whole of this region of the limb, but varies in thickness in different parts; thus, it is thickest in the upper and outer side of the thigh, where it receives a fibrous expansion from the Gluteus maximus muscle, and the Tensor vaginae femoris is inserted

169.—Muscles of the Iliac and Anterior Femoral Regions.

between its layers, it is very thin behind, and at the upper and inner side, where it covers the Adductor muscles, and again becomes stronger around the knee, receiving fibrous expansions from the tendons of the Biceps externally, and from the Sartorius, Gracilis, Semitendinosus, and Triceps extensor cruris in front. The fascia lata is attached, above; to Poupart's ligament, and crest of the ilium, behind to the margin of the sacrum and coccyx, internally to the pubic arch and pectineal line, and below to all the prominent points around the knee-joint, the condyles of the femur, tuberosities of the tibia, and head of the fibula. That portion which invests the Gluteus medius (the Gluteal aponeurosis) is very thick and strong, and gives origin, by its inner surface, to some of the fibres of that muscle; at the upper border of the Gluteus maximus, it divides into two layers; the most superficial, very thin, covers the surface of the Gluteus maximus, and is continuous below with the fascia lata: the deep layer is thick above, and blends with the great sacro-sciatic ligament, thin below, where it separates the Gluteus maximus from the deeper muscles. From the inner surface of the fascia lata, are given off two strong intermuscular septa, which are attached to the whole length of the linea aspera; the external and stronger one, extending from the insertion of the Gluteus maximus, to the outer condyle, separates the Vastus externus in front from the short head of the Biceps behind, and gives partial origin to these muscles; the inner one, the thinner of the two, separates the Vastus internus from the Adductor muscles. Besides these, there are numerous smaller septa, separating the individual muscles, and enclosing each in a distinct sheath. At the upper and inner part of the thigh, a little below Poupart's ligament, a large oval-shaped aperture is observed in this fascia, it transmits the internal saphenous vein, and other smaller vessels, and is termed the *saphenous opening*. In order more correctly to consider the mode of formation of this aperture, the fascia lata is described as consisting, in this part of the thigh, of two portions, an iliac portion, and a pubic portion.

The *iliac portion* is all that part of the fascia lata placed on the outer side of the saphenous opening. It is attached, externally, to the crest of the ilium, and its anterior superior spine, to the whole length of Poupart's ligament, as far internally as the spine of the pubes, and to the Pectineal line in conjunction with Gimbernat's ligament. From the spine of the pubes, it is reflected downwards and outwards, forming an arched margin, the superior cornu, or outer boundary of the saphenous opening; this margin overlies, and is adherent to the anterior layer of the sheath of the femoral vessels, to its edge is attached the cribriform fascia, and, below, it is continuous with the pubic portion of the fascia lata.

The *pubic portion* is situated at the inner side of the saphenous opening; at the lower margin of this aperture it is continuous with the iliac portion; traced upwards, it is seen to cover the surface of the Pectineus muscle, and passing behind the sheath of the femoral vessels, to which it is closely united, is continuous with the sheath of the Psoas and Iliacus muscles, and is finally lost in the fibrous capsule of the hip-joint. This fascia is attached above, to the pectineal line in front of the insertion of the aponeurosis of the external oblique, and internally, to the margin of the pubic arch. From this description it may be observed, that the iliac portion of the fascia lata passes in front of the femoral vessels, the pubic portion behind them, an apparent aperture consequently exists, between the two, through which the internal saphena joins the femoral vein.

The fascia should now be removed from the surface of the muscles. This may be effected by pinching it up between the forceps, dividing it, and separating it from each muscle in the course of its fibres.

The *Tensor Vaginae Femoris* is a short flat muscle, situated at the upper and outer side of the thigh. It arises by aponeurotic fibres from the anterior part of the outer lip of the crest of the ilium, and from the outer surface of the anterior superior spinous process, between the Gluteus medius, and Sartorius. The muscle passes obliquely downwards, and a little backwards, to be inserted by tendinous fibres between the two layers of the fascia lata, about one fourth down the thigh.

Relations. By its *superficial surface*, with a layer of the fascia lata and the integument. By its *deep surface*, with the deep layer of the fascia lata, the Gluteus medius, Rectus femoris, and, Vastus externus. By its *anterior border*, with the Sartorius, from which it is separated below by a triangular space, in which is seen the Rectus femoris. By its *posterior border*, with the Gluteus medius, being separated from it below by a slight interval.

The *Sartorius*, the longest muscle in the body, is a flat, narrow, riband-like muscle, which arises by tendinous fibres from the anterior superior spinous process of the ilium and upper half of the notch below it; it passes obliquely inwards, across the upper and anterior part of the thigh, then descends vertically, as far as the inner side of the knee, passing behind the inner condyle of the femur, and terminates in a tendon, which curving obliquely forwards, expands into a broad aponeurosis, which is inserted into the upper part of the inner surface of the shaft of the tibia, nearly as far forwards as the crest. This expansion covers in the insertion of the tendons of the Gracilis and Semitendinosus, with which it is partially united, a synovial bursa being interposed between them. An offset is derived from this aponeurosis, which blends with the fibrous capsule of the knee-joint, and another, given off from its lower border, blends with the fascia on the inner side of the leg. The relations of this muscle to the femoral artery should be carefully examined, as its inner border forms the chief guide in the operation of including this vessel in a ligature. In the upper third of the thigh, it forms, with the Adductor longus, the sides of a triangular space, Scarpa's triangle, the base of which, turned upwards, is formed by Poupart's ligament; the femoral artery passes perpendicularly through the centre of this space from its base to its apex. In the middle third of the thigh, the femoral artery lies first along the inner border, and then beneath the Sartorius.

Relations. By its *superficial surface*, with the fascia lata and integument. By its *deep surface* with the Iliacus, Psoas, Rectus, Vastus internus, sheath of the femoral vessels, Adductor longus, Adductor magnus, Gracilis, long saphenous nerve, and internal lateral ligament of the knee-joint.

The *Quadriceps Extensor Cruris*, includes the four remaining muscles on the anterior part of the thigh. They are the great Extensor muscles of the leg, forming a large fleshy mass, which covers the anterior surface and sides of the femur, being united below into a single tendon, attached to the tibia, and above subdividing into separate portions, which have received separate names. Of these, one occupying the middle of the thigh, connected above with the ilium, is called the *Rectus Femoris*, from its straight course. The other divisions lie in immediate connection with the shaft of the femur, which they cover from the condyles to the trochanters. The portion on the outer side of the femur being termed the *Vastus Externus*, that covering the inner side the *Vastus Internus*, and that covering the front of the bone, the *Cruraeus*. The two latter portions are, however, so intimately blended, as to form but one muscle.

The *Rectus Femoris*, is situated in the middle of the anterior region of the thigh; it is fusiform in shape, and its fibres are arranged in a bipenniform manner. It arises by two tendons; one, the straight tendon, from the anterior inferior spinous process of the ilium, the other is flattened and curves outwards, to be attached to a groove above the brim of the acetabulum; this is the reflected tendon of the Rectus, it unites with the straight tendon at an acute angle, and then spreads into an aponeurosis, from which the muscular fibres arise. The muscle terminates in a broad and thick aponeurosis, which occupies the lower two-thirds of its posterior surface, and, gradually becoming narrowed into a flattened tendon, is inserted into the patella in common with the Vasti and Cruraeus.

Relations. By its *superficial surface*, with the anterior fibres of the Gluteus medius, the Tensor vaginae femoris, Sartorius, and the Psoas and Iliacus, by its lower three-fourths with the fascia lata. By its *posterior surface*, with the hip-joint, the anterior circumflex vessels, and the Cruraeus and Vasti muscles.

The three remaining muscles have been described collectively by some anatomists, separate from the Rectus, under the name of the *Triceps Extensor Cruris*; in order to expose them, divide the Sartorius and Rectus muscles across the middle, and turn them aside, when the Triceps extensor will be fully brought into view.

The *Vastus Externus* is the largest part of the Quadriceps extensor. It arises by a broad aponeurosis, which is attached to the anterior border of the great trochanter, to a horizontal ridge on its outer surface, to a rough line, leading from the trochanter major to the linea aspera, and to the whole length of the outer lip of the linea aspera; this aponeurosis covers the upper three-fourths of the muscle, and from its inner surface, many fibres arise. A few additional fibres arise from the tendon of the Gluteus maximus, and from the external intermuscular septum between the Vastus externus, and short head of the Biceps. These fibres form a large fleshy mass, which is attached to a strong aponeurosis, placed on the under surface of the muscle at its lowest part, this becomes contracted and thickened into a flat tendon, which is inserted into the outer part of the upper border of the patella, blending with the great Extensor tendon.

Relations. By its *superficial surface*, with the Rectus, the Tensor vaginae femoris, the fascia lata, and the Gluteus maximus, from which it is separated by a synovial bursa. By its *deep surface*, with the Cruraeus, some large branches of the external circumflex artery being interposed.

The *Vastus internus* and *Cruraeus*, are so inseparably connected together, as to form but one muscle. It is the smallest portion of the Quadriceps extensor. The anterior portion covered by the Rectus, being called the Cruraeus, the internal portion, which lies immediately beneath the fascia lata, the Vastus Internus. It arises by an aponeurosis, which is attached to the lower part of the line that extends from the inner side of the neck of the femur to the linea aspera, from the whole length of the inner lip of the linea aspera, and internal intermuscular septum. It also arises from nearly the whole of the internal, anterior and external surfaces of the shaft of the femur, limited above by the line between the two trochanters, and extending below to within the lower fourth of the bone. From these different origins, the fibres converge to a broad aponeurosis, which covers the anterior surface of the middle portion of the muscle (the Cruraeus), and the deep surface of the inner division of the muscle (the Vastus internus), becoming joined and gradually narrowing, it is inserted into the patella, blending with the other portions of the Quadriceps extensor.

Relations. By their *superficial surface*, with the Psoas and Iliacus, the Rectus, Sartorius, Pectineus, Adductors, and fascia lata, femoral artery, vein, and saphenous nerve. By its *deep surface*, with the femur, Subcruraeus and synovial membrane of the knee-joint.

The student will observe the striking analogy that exists between the Quadriceps extensor, and the Triceps brachialis in the upper extremity. So close is this similarity, that M. Cruvelhier has described it under the name of the *Triceps femoralis*. Like the Triceps brachialis, it consists of three distinct divisions or heads; a middle or long head, analogous to the long head of the Triceps, and of two other portions which have respectively received the names of the external and internal heads of the muscle. These, it will be noticed, are strictly analogous to the outer and inner heads of the Triceps brachialis.

The *Subcruraeus* is a small muscular fasciculus usually distinct from the superficial muscle, which arises from the anterior surface of the lower part of the shaft of the femur, and is inserted into the upper part of the synovial pouch that extends upwards from the knee-joint behind the patella. This fasciculus is occasionally united with the Cruraeus. It sometimes consists of two separate muscular bundles.

The *tendons* of the different portions of the Quadriceps extensor unite at the lower part of the thigh, so as to form a single strong tendon, which is inserted into the upper part of the patella. More properly speaking, the patella may be regarded as a sesamoid bone, developed in the tendon of the Quadriceps extensor, and the ligamentum patellae, which is continued from the lower part of the patella, to the tuberosity of the tibia, as the proper tendon of insertion of this muscle. A small synovial bursa is interposed between the tendon and the upper part of the tuberosity. From the tendons corresponding to the Vasti, a fibrous prolongation is derived, which is attached below to the upper extremities of the tibia and fibula. It serves to protect the knee-joint, which is strengthened on its outer side by the fascia lata.

Nerves. The Tensor vaginae femoris is supplied by the superior gluteal nerve, the other muscles of this region, by branches from the anterior crural.

Actions. The Tensor vaginae femoris is a tensor of the fascia lata; continuing its action, the oblique direction of its fibres enables it to rotate the thigh inwards. In the erect posture, acting from below, it will serve to steady the pelvis upon the head of the femur. The Sartorius flexes the leg upon the thigh, and, continuing to act, the thigh upon the pelvis, at the same time drawing the limb inwards, so as to cross one leg over the other. Taking its fixed point from the leg, it flexes the pelvis upon the thigh, and, if one muscle acts, assists in rotating it. The Quadriceps extensor extends the leg upon the thigh. Taking their fixed point from the leg, as in standing, the Extensor muscles will act upon the femur, supporting it perpendicularly upon the head of the tibia, thus maintaining the entire weight of the body. The Rectus muscle assists the Psoas and Iliacus, in supporting the pelvis and trunk upon the femur, or in bending it forwards.

Internal Femoral Region

Gracilis.   Adductor Longus.
Pectineus.   Adductor Brevis.
Adductor Magnus.

Dissections. These muscles are at once exposed by removing the fascia from the fore part and inner side of the thigh. The limb should be abducted so as to render the muscles tense, and easier of dissection.

The *Gracilis* is the most superficial muscle on the inner side of the thigh. It is a thin, flattened, slender muscle, broad above, narrow and tapering below. It arises by a thin aponeurosis between two and three inches in breadth, from the inner margin of the ramus of the pubes and ischium. The fibres pass vertically downwards, and terminate in a rounded tendon which passes behind the internal condyle of the femur, and curving around the inner tuberosity of the tibia, becomes flattened, and is inserted into the upper part of the inner surface of the shaft of the tibia, beneath the tuberosity. The tendon of this muscle is situated immediately above that of the Semi-tendinosus, and beneath the aponeurosis of the Sartorius, with which it is in part blended. As it passes across the the internal lateral ligament of the knee-joint, it is separated from it by a synovial bursa, common to it and the Semi-tendinosus muscle.

Relations. By its *superficial surface*, with the fascia lata and the Sartorius below; the internal saphena vein crosses it obliquely near its lower part, lying superficial to the fascia lata. By its *deep surface*, with the three Adductors, and the internal lateral ligament of the knee-joint.

170.—Muscles of the Internal Femoral Region.

The *Pectineus* is a flat quadrangular muscle, situated at the anterior part of the upper and inner aspect of the thigh. It arises from the linea ilio-pectinea, from the surface of bone in front of it, between the pectineal eminence and spine of the pubes, and from a tendinous prolongation of Gimbernat's ligament, which is attached to the crest of the pubes, and is continuous with the fascia covering the outer surface of the muscle; the fibres pass downwards, backwards, and outwards, to be inserted into a rough line leading from the trochanter minor to the linea aspera.

Relations. By its *anterior surface*, with the pubic portion of the fascia lata, which separates it from the femoral vessels and internal saphena vein. By its *posterior surface*, with the hip-joint, the Adductor brevis and Obturator externus muscles, the obturator vessels and nerve being interposed. By its *outer border*, with the Psoas, a cellular interval separating them, upon which lies the femoral artery. By its *inner border*, with the margin of the Adductor longus.

The *Adductor Longus*, the most superficial of the three Adductors, is a flat triangular muscle, lying on the same plane as the Pectineus, with which it is often blended above. It arises, by a flat narrow tendon, from the front of the pubes, at the angle of junction of the crest with the symphysis; it soon expands into a broad fleshy belly, which, passing downwards, backwards, and outwards, is inserted, by an aponeurosis, into the middle third of the linea aspera, between the Vastus internus and the Adductor magnus.

Relations. By its *anterior surface*, with the fascia lata, and near its insertion, with the femoral artery and vein. By its *posterior surface*, with the Adductor brevis and magnus, the anterior branches of the obturator vessels and nerve, and with the profunda artery and vein near its insertion. By its *outer border*, with the Pectineus. By its *inner border*, with the Gracilis.

The Pectineus and Adductor longus should now be divided near their origin, and turned downwards, when the Adductor brevis and Obturator externus will be exposed.

The *Adductor Brevis* is situated immediately beneath the two preceding muscles. It is somewhat triangular in form, and arises by a narrow origin from the outer surface of the descending ramus of the pubes, between the Gracilis and Obturator externus. Its fibres passing backwards, outwards, and downwards, are inserted by an aponeurosis into the upper part of the linea aspera, immediately behind the Pectineus and upper part of the Adductor longus.

Relations. By its *anterior surface*, with the Pectineus, Adductor longus, and anterior branches of the obturator vessels and nerve. By its *posterior surface*, with the Adductor magnus, and posterior branches of the obturator vessels and nerves. By its *outer border*, with the Obturator externus, and conjoined tendon of the Psoas and Iliacus. By its *inner border*, with the Gracilis and Adductor magnus. This muscle is pierced, near its insertion, by the middle perforating branch of the profunda artery.

The Adductor brevis should now be cut away near its origin and turned outwards, when the entire extent of the Adductor magnus will be exposed.

The *Adductor Magnus* is a large triangular muscle, forming a septum between the muscles on the inner, and those on the posterior aspect of the thigh. It arises by short tendinous fibres from a small part of the descending ramus of the pubes, from the ascending ramus of the ischium, and from the outer margin and under surface of the tuberosity of the ischium. Those fibres which arise from the ramus of the pubes are very short, horizontal in direction, and are inserted into the rough line leading from the great trochanter to the linea aspera, internal to the Gluteus maximus; those from the ramus of the ischium are directed downwards and outwards with different degrees of obliquity, to be inserted by means of a broad aponeurosis, into the whole length of the interval between the two lips of the linea aspera and upper part of the internal bifurcation below. The internal portion of the muscle, consisting principally of those fibres which arise from the tuberosity of the ischium, forms a thick fleshy mass consisting of coarse bundles which descend almost vertically, and terminate about the lower third of the thigh in a rounded tendon, which is inserted into the tuberosity above the inner condyle of the femur, being connected by a fibrous expansion to the line leading upwards from the condyle to the linea aspera. Between these two portions of the muscle, an angular interval is left, almost entirely tendinous in structure, for the passage of the femoral vessels into the popliteal space. The external portion of the muscle is pierced by four apertures, the three superior for the three perforating arteries; the fourth, for the passage of the profunda. This muscle gives off an aponeurosis, which passes in front of the femoral vessels, and joins with the Vastus internus.

Relations. By its *anterior surface*, with the Pectineus, Adductor brevis, Adductor longus and the femoral vessels. By its *posterior surface*, with the great sciatic nerve, the

Gluteus maximus, Biceps, Semi-tendinosus, and Semi-membranosus. By its *superior* or *shortest border*, it lies parallel with the Quadratus femoris. By its *internal* or *longest border*, with the Gracilis, Sartorius, and fascia lata. By its *external* or *attached border*, it is inserted into the femur behind the Adductor brevis and Adductor longus, which separate it, in front, from the Vastus internus; and in front of the Gluteus maximus and short head of the Biceps, which separate it from the Vastus externus.

Nerves. All the muscles of this group are supplied by the obturator nerve. The Pectineus receives additional branches from the accessory obturator and anterior crural; and the Adductor magnus an additional one from the great sciatic.

Actions. The Pectineus and three Adductors adduct the thigh powerfully; they are especially used in horse-exercise, the flanks of the horse being firmly grasped between the knees by the action of these muscles. From their oblique insertion into the linea aspera, they rotate the thigh outwards, assisting the external Rotators, and when the limb has been abducted, they draw it inwards, carrying the thigh across that of the opposite side. The Pectineus and Adductor brevis and longus assist the Psoas and Iliacus in flexing the thigh upon the pelvis. In progression, also, all these muscles assist in drawing forwards the hinder limb. The Gracilis assists the Sartorius in flexing the leg and drawing it inwards; it is also an Adductor of the thigh. If the lower extremities are fixed, these muscles may take their fixed point from below and act upon the pelvis, serving to maintain the body in the erect posture; or, if their action is continued, to flex the pelvis forwards upon the femur.

GLUTEAL REGION
Gluteus Maximus.
Gluteus Medius.
Gluteus Minimus.
Pyriformis.
Gemellus Superior.
Obturator Internus.
Gemellus Inferior.
Obturator Externus.
Quadratus Femoris.

Dissection (fig. 171). The subject should be turned on its face, a block placed beneath the pelvis to make the buttocks tense, and the limbs allowed to hang over the end of the table, the foot inverted, and the limb abducted. An incision should be made through

171.—Dissection of Lower Extremity. Posterior View.

1. Dissection of GLUTEAL REGION
3. BACK of THIGH
2. POPLITEAL SPACE
4. BACK of LEG
5. SOLE of FOOT

172.—Muscles of the Gluteal and Posterior Femoral Regions.

the integument along the back part of the crest of the ilium and margin of the sacrum to the tip of the coccyx, from which point a second incision should be carried obliquely downwards and outwards to the outer side of the thigh, four inches below the great trochanter. The portion of integument included between these incisions, together with the superficial fascia, should be removed in the direction shewn in the figure, when the Gluteus maximus and the dense fascia covering the Gluteus medius will be exposed.

The *Gluteus Maximus*, the most superficial muscle in the gluteal region, is a very broad and thick fleshy mass, of a quadrilateral shape, which forms the prominence of the nates. Its large size is one of the most characteristic points in the muscular system in man, connected as it is with the power he has of maintaining the trunk in the erect posture. In structure it is remarkably coarse, being made up of muscular fasciculi lying parallel with one another, and collected together into large bundles, separated by deep cellular intervals. It arises from the superior curved line of the ilium, and the portion of bone, including the crest, immediately behind it; from the posterior surface of the last piece of the sacrum, the sides of the coccyx, and posterior surface of the great sacro-sciatic and posterior sacro-iliac ligaments. The fibres are directed obliquely downwards and outwards; those forming the upper and larger portion of the muscle (after converging somewhat) terminate in a thick tendinous lamina, which passes across the great trochanter, and is inserted into the fascia lata covering the outer side of the thigh, the lower portion of the muscle being inserted into the rough line leading from the great trochanter to the linea aspera, between the Vastus externus and Adductor magnus.

Three *synovial bursae* are usually found separating the under surface of this muscle from the eminences which it covers. One of these, of large size, and generally multilocular, separates it from the great trochanter. A second, often wanting, is situated on the tuberosity of the ischium. A third, between the tendon of this muscle and the Vastus externus.

Relations. By its *superficial surface*, with a thin fascia, which separates it from cellular membrane, fat, and the integument. By its *deep surface*, with the ilium, sacrum, coccyx, and great sacro-sciatic ligament, the Gluteus medius, Pyriformis, Gemelli, Obturator internus, Quadratus femoris, the great sacro-sciatic foramen, the tuberosity of the ischium, great trochanter, the Biceps, Semi-tendinosus, Semi-membranosus, and Adductor magnus muscles, the gluteal vessels and nerve issuing from the pelvis above the Pyriformis muscle, the ischiatic and internal pudic vessels and nerves below it. Its *upper border* is thin, and connected with the Gluteus medius by the fascia lata. Its *lower border*, free and prominent, forms the fold of the nates, and is directed towards the perineum.

Dissection. The Gluteus maximus should now be divided near its origin by a vertical incision carried from its upper to its lower border: a cellular interval will be exposed, separating it from the Gluteus medius and External rotator muscles beneath. The upper portion of the muscle should be altogether detached, and the lower portion turned outwards; the loose areolar tissue filling up the interspace between the trochanter major and tuberosity of the ischium being removed, the parts already enumerated as exposed by the removal of this muscle will be seen.

The *Gluteus Medius* is a broad, thick, radiated muscle, situated on the outer surface of the pelvis. Its posterior third is covered by the Gluteus maximus; its anterior two-thirds, is covered by a layer of fascia, which is thick and dense, and separates it from the integument. It arises from the outer surface of the ilium, between the superior and middle curved lines, and from the outer lip of that portion of the crest which is between them; it also arises from the dense fascia covering its outer surface. The fibres gradually converge to a strong flattened tendon, which is inserted into the oblique line which traverses the outer surface of the great trochanter. A synovial bursa separates the tendon of this muscle from the surface of the trochanter in front of its insertion.

Relations. By its *superficial surface*, with the Gluteus maximus, Tensor vaginae femoris, and deep fascia. By its *deep surface*, with the Gluteus minimus and the gluteal vessels and nerve. Its *anterior border* is blended avith the Gluteus minimus and Tensor vaginae femoris. Its *posterior border* lies parallel with the Pyriformis.

This muscle should now be divided near its insertion and turned upwards, when the Gluteus minimus will be exposed.

The *Gluteus Minimus*, the smallest of the three glutei, is placed immediately beneath the preceding. It is a fan-shaped muscle, arising from the external surface of the ilium, between the middle and inferior curved lines, and behind, from the margin of the great sacro-sciatic notch; the fibres converge to the deep surface of a radiated aponeurosis, which, terminating in a tendon, is inserted into an impression on the anterior border of the great trochanter. A synovial bursa is interposed between the anterior part of the tendon and the great trochanter.

Relations. By its *superficial surface*, with the Gluteus medius, and the gluteal vessels and nerves. By its *deep surface*, with the ilium, the reflected tendon of the Rectus femoris, and capsular ligament of the hip-joint. Its *anterior margin* is blended with the Gluteus medius. Its *posterior margin* is often joined with the tendon of the Pyriformis.

The *Pyriformis* is a flat muscle, pyramidal in shape, lying almost parallel with the lower margin of the Gluteus minimus. It is situated partly within the pelvis at its posterior part, and partly at the back part of the hip-joint. It arises from the anterior surface of the sacrum by three fleshy digitations, attached to the portions of bone interposed between the second, third, and fourth anterior sacral foramina, and also from the grooves leading from them: a few fibres also arise from the margin of the great sacro-sciatic foramen, and from the anterior surface of the great sacro-sciatic ligament. The muscle passes out of the pelvis through the great sacro-sciatic foramen, the upper part of which it fills, and is inserted, by a rounded tendon, into the back part of the upper border of the great trochanter, being generally blended with the tendon of the Obturator internus.

Relations. By its *anterior surface, within the pelvis*, with the Rectum (especially on the left side), the sacral plexus of nerves, and the internal iliac vessels; *external to the pelvis*, with the os innominatum and capsular ligament of the hip-joint. By its *posterior surface, within the pelvis*, with the sacrum; and *external to it*, with the Gluteus maximus. By its *upper border*, with the Gluteus medius, from which it is separated by the gluteal vessels and nerves. By its *lower border*, with the Gemellus superior; the ischiatic vessels and nerves, and the internal pudic vessels and nerves, passing from the pelvis in the interval between them.

Dissection. The next muscle, as well as the origin of the Pyriformis, can only be seen when the pelvis is divided, and the viscera contained in this cavity removed.

The *Obturator Internus*, like the preceding muscle, is situated partly within the cavity of the pelvis, partly at the back of the hip-joint. It arises from the inner surface of the anterior and external wall of the pelvis, being attached to the margin of bone around the inner side of the obturator foramen; viz., from the descending ramus of the pubes, and the ascending ramus of the ischium; and, laterally, from the inner surface of the body of the ischium, between the margin of the obturator foramen in front, the great sciatic notch behind, and the brim of the true pelvis above. It also arises from the inner surface of the obturator membrane, and from the tendinous arch which protects the obturator vessels and nerve in passing beneath the sub-pubic arch. The fibres converge downwards and outwards, and terminate in four or five tendinous bands, which are found on its deep surface; these bands are reflected at a right angle over the inner surface of the tuberosity of the ischium, which is covered with cartilage, grooved for their reception, and lined with a synovial bursa. The muscle leaves the pelvis by the lesser sacro-sciatic notch; and the tendinous bands unite into a single flattened tendon, which passes horizontally outwards, and, after receiving the attachment of the Gemelli, is inserted into the upper border of the great trochanter in front of the Pyriformis. A synovial bursa, narrow and elongated in form, is usually found between the tendon of this muscle and the capsular ligament of the hip. It occasionally communicates with that between the tendon and the tuberosity of the ischium, the two forming a single sac.

In order to display the peculiar appearances presented by the tendon of this muscle, it should be divided near its insertion and reflected outwards.

Relations. Within the pelvis, this muscle is in relation, by its *anterior surface*, with the obturator membrane and inner surface of the anterior wall of the pelvis; by its *posterior surface*, with the obturator fascia, which separates it from the Levator ani; and it is crossed by the internal pudic vessels and nerve. This surface forms the outer boundary of the ischio-rectal fossa. *External to the pelvis*, it is covered by the great sciatic nerve and Gluteus maximus, and rests on the back part of the hip-joint.

The *Gemelli* are two small muscular fasciculi, accessories to the tendon of the Obturator internus, which is received into a groove between them. They have received the names *superior* and *inferior* from the position they occupy.

The *Gemellus Superior*, the smaller of the two, is a fleshy fasciculus, which arises from the external surface of the spine of the ischium, and passing horizontally outwards, becomes blended with the upper part of the tendon of the Obturator internus, and is inserted with it into the superior border of the great trochanter. This muscle is sometimes wanting.

Relations. By its *superficial surface*, with the Gluteus maximus and the ischiatic vessels and nerves. By its *deep surface*, with the capsule of the hip-joint. By its *upper border*, with the lower margin of the Pyriformis. By its *lower border*, with the tendon of the Obturator internus.

The *Gemellus Inferior* arises from the upper part of the outer border of the tuberosity of the ischium, and, passing horizontally outwards, is blended with the lower part of the tendon of the Obturator internus, and inserted with it into the upper border of the great trochanter.

Relations. By its *superficial surface*, with the Gluteus maximus, and the ischiatic vessels and nerves. By its *deep surface*, it covers the capsular ligament of the hip-joint. By its *upper border*, with the tendon of the Obturator internus. By its *lower border*, with the tendon of the Obturator externus and Quadratus femoris.

The *Quadratus Femoris* is a short flat muscle, quadrilateral in shape (hence its name), situated immediately below the Gemellus inferior, and above the upper margin of the Adductor magnus. It arises from the external border of the tuberosity of the ischium, and proceeding horizontally outwards, is inserted into the upper part of the linea quadrati, on the posterior surface of the trochanter major. A synovial bursa is often found between the under surface of this muscle and the lesser trochanter, which it covers.

Relations. By its *posterior surface*, with the Gluteus maximus and the ischiatic vessels and nerves. By its *anterior surface*, with the tendon of the Obturator externus and Trochanter minor. By its *upper border*, with the Gemellus inferior. Its *lower border* is separated from the Adductor magnus by the internal circumflex vessels.

Dissection. In order to expose the next muscle (the Obturator externus), it is necessary to remove the Psoas, Iliacus, Pectineus, and Adductor brevis and longus muscles, from the front and inner side of the thigh; and the Gluteus maximus and Quadratus femoris, from the back part. Its dissection should consequently be postponed until the muscles of the anterior and internal femoral regions have been examined.

The *Obturator Externus* is a flat triangular muscle, which covers the outer surface of the anterior wall of the pelvis. It arises from the margin of bone immediately around the inner side of the obturator foramen, viz., from the body and ramus of the pubes, and the ramus of the ischium; it also arises from the inner two-thirds of the outer surface of the obturator membrane, and from the tendinous arch which completes the sub-pubic canal for the passage of the obturator vessel and nerve. The fibres converging pass outwards, and terminate in a tendon which runs across the back part of the hip-joint, and is inserted into the digital fossa of the femur.

Relations. By its *anterior surface*, with the Psoas, Iliacus, Pectineus, Adductor longus. Adductor brevis, and Gracilis; and more externally, with the neck of the femur and capsule of the hip-joint. By its *posterior surface*, with the obturator membrane and Quadratus femoris.

Nerves. The Gluteus maximus is supplied by the inferior gluteal nerve and a branch from the sacral plexus. The Gluteus medius and minimus, by the superior gluteal. The Pyriformis, Gemelli, Obturator internus, and Quadratus femoris, by branches from the sacral plexus. And the Obturator externus, by the obturator nerve.

Actions. The Glutei muscles, when they take their fixed point from the pelvis, are all abductors of the thigh. The Gluteus maximus and the posterior fibres of the Gluteus medius, rotate the thigh outwards; the anterior fibres of the Gluteus medius and the Gluteus minimus rotate it inwards. The Gluteus maximus serves to extend the femur, and the Gluteus medius and minimus draw it forwards. The Gluteus maximus is also a tensor of the fascia lata. Taking their fixed point from the femur, the Glutei muscles act upon the pelvis, supporting it and the whole trunk upon the head of the femur, which is especially obvious in standing on one leg. In order to gain the erect posture after the effort of stooping, these muscles draw the pelvis backwards, assisted by the Biceps, Semi-tendinosus, and Semi-membranosus muscles. The remaining muscles are powerful rotators of the thigh outwards. In the sitting posture, when the thigh is flexed upon the pelvis, their action as rotators ceases, and they become abductors, with the exception of the Obturator externus, which still rotates the femur outwards. When the femur is fixed, the Pyriformis and Obturator muscles serve to draw the pelvis forwards if it has been inclined backwards, and assist in steadying it upon the head of the femur.

Posterior Femoral Region

Biceps. Semi-tendinosus. Semi-membranosus.

Dissection (fig. 171). Make a vertical incision along the middle of the thigh, from the lower fold of the nates to about three inches below the back of the knee-joint, and there connect it with a transverse incision, carried from the inner to the outer side of the leg. A third incision should then be made transversely at the junction of the middle with the lower third of the thigh. The integument having been removed from the back of the knee in the direction indicated in the figure, and the boundaries and contents of the popliteal space examined, the removal of the integument from the remaining part of the thigh should be continued, when the fascia and muscles of this region will be exposed.

The *Biceps* is a large muscle, of considerable length, situated on the posterior and outer aspect of the thigh. It arises by two distinct portions or heads. One, the long head, from an impression at the upper and back part of the tuberosity of the ischium, by a tendon common to it and the Semi-tendinosus. The femoral or short head, from the outer lip of the linea aspera, between the Adductor magnus and Vastus externus, extending from two inches below the Gluteus maximus, to within two inches of the outer condyle; it also arises from the external intermuscular septum. The fibres of the long head form a fusiform belly, which, passing obliquely downwards and a little outwards, terminates in an aponeurosis which covers the posterior surface of the muscle, and receives the fibres of the short head; this aponeurosis becomes gradually contracted into a tendon, which is inserted into the outer side of the head of the fibula. At its insertion, the tendon divides into two portions, which embrace the external lateral ligament of the knee-joint, a strong prolongation being sent forwards to the outer tuberosity of the tibia, which gives off an expansion to the fascia of the leg. The tendon of this muscle forms the outer ham-string.

Relations. By its *superficial surface*, with the Gluteus maximus and fascia lata. By its *deep surface*, with the Semi-membranosus, Adductor magnus, and Vastus externus, the great sciatic nerve, popliteal artery and vein, and near its insertion, with the external head of the Gastrocnemius, Plantaris, and superior external articular artery.

The *Semi-tendinosus*, remarkable for the great length of its tendon, is situated at the posterior and inner aspect of the thigh. It arises from the tuberosity of the ischium by a tendon common to it and the long head of the Biceps; it also arises from an aponeurosis which connects the adjacent surfaces of the two muscles to the extent of about three inches after their origin. It forms a fusiform muscle which, passing downwards and inwards, terminates a little below the middle of the thigh in a long round tendon, which lies along the inner side of the popliteal space, curving around the inner tuberosity of the tibia, to be inserted into the upper part of the inner surface of the shaft of this bone, nearly as far forwards as its anterior border. This tendon lies beneath the expansion of the Sartorius, and below that of the Gracilis, to which it is united. A tendinous intersection is usually observed about the middle of this muscle.

Relations. By its *superficial surface*, with the Gluteus maximus and fascia lata. By its *deep surface*, with the Semi-membranosus, Adductor magnus, inner head of the Gastrocnemius, and internal lateral ligament of the knee-joint.

The *Semi-membranosus*, so called from the tendinous expansion on its anterior and posterior surfaces, is situated at the back part and inner side of the thigh. It arises by a thick tendon from the upper and outer part of the tuberosity of the ischium above and to the outer side of the Biceps and Semi-tendinosus, and is inserted into the posterior part of the inner tuberosity of the tibia, beneath the internal lateral ligament. The tendon of this muscle at its insertion divides into three portions; the middle portion is the fasciculus of insertion into the back part of the inner tuberosity, it sends down an expansion to cover the Popliteus muscle. The internal portion is horizontal, passing forwards beneath the internal lateral ligament, to be inserted into a groove along the inner side of the internal tuberosity. The posterior division passes upwards and backwards, to be inserted into the back part of the outer condyle of the femur, forming the chief part of the posterior ligament of the knee-joint.

The tendon of origin of this muscle expands into an aponeurosis, which covers the upper part of its anterior surface; from this muscular fibres arise, and converge to another aponeurotic expansion, which covers the lower part of its posterior surface, and this contracts into the tendon of insertion. The tendons of the two preceding muscles, with those of the Gracilis and Sartorius, form the inner ham-string.

Relations. By its *superficial surface*, with the Gluteus maximus, Semi-tendinosus, Biceps, and fascia lata. By its *deep surface*, with the Quadratus femoris, Adductor magnus, and inner head of the Gastrocnemius. It covers the popliteal artery and vein, and is separated from the knee-joint by a synovial membrane. By its *inner border*, with the Gracilis. By its *outer border*, with the great sciatic nerve.

Nerves. The muscles of this region are supplied by the great sciatic nerve.

Actions. The three ham-string muscles flex the leg upon the thigh. When the knee is semi-flexed, the Biceps, from its oblique direction downwards and outwards, rotates the leg slightly outwards; and the Semi-membranosus, in consequence of its oblique direction, rotates the leg inwards, assisting the Popliteus. Taking their fixed point from below, these muscles serve to support the pelvis upon the head of the femur, and to draw the trunk directly backwards, as is seen in feats of strength, when the body is thrown backwards in the form of an arch.

Surgical Anatomy. The tendons of these muscles occasionally require subcutaneous division in some forms of spurious anchylosis of the knee-joint, dependent upon permanent contraction and rigidity of the flexor muscles, or from stiffening of the ligamentous and other tissues surrounding the joint, the result of disease. This is easily effected by putting the tendon upon the stretch, and inserting a narrow sharp-pointed knife between it and the skin; the cutting edge being then turned towards the tendon, it should be divided, taking care that the wound in the skin is not at the same time enlarged. This operation has been attended with considerable success in some cases of stiffened knee from rheumatism, gradual extension being kept up for some time after the operation.

Muscles and Fasciae of the Leg

Dissection (fig. 168). The knee should be bent, a block placed beneath it, and the foot kept in an extended position; an incision should then be made through the integument in the middle line of the leg to the ankle, and continued along the dorsum of the foot to the toes. A second incision should be made transversely across the ankle, and a third in the same direction across the bases of the toes: the flaps of integument included between these incisions should be removed, and the fascia of the leg examined.

The *Fascia of the Leg* forms a complete investment to the whole of this region of the limb, excepting to the inner surface of the tibia, to which it is unattached. It is continuous above with the fascia lata, receiving an expansion from the tendon of the Biceps on the outer side, and from the tendons of the Sartorius, Gracilis, and Semi-tendinosus on the inner side; in front it blends with the periosteum covering the tibia and fibula; below, it is continuous with the annular ligaments of the ankle. It is thick and dense in the upper and anterior part of the leg, and gives attachment, by its inner surface, to the Tibialis anticus and Extensor longus digitorum muscles; but thinner behind, where it covers the Gastrocnemius and Soleus muscles. Its inner surface gives off, on the outer side of the leg, two strong intermuscular septa, which enclose the Peronaei muscles, and separate them from those on the anterior and posterior tibial regions, and several smaller and more slender processes enclose the individual muscles in each region; at the same time, a broad transverse intermuscular septum intervenes between the superficial and deep muscles in the posterior tibio-fibular region.

The fascia should now be removed by dividing it in the same direction as the integument, excepting opposite the ankle, where it should be left entire. The removal of the fascia should be commenced from below, opposite the tendons, and detached in the line of direction of the muscular fibres.

Muscles of the Leg

These may be subdivided into three groups: those on the anterior, those on the posterior, and those on the outer side.

Anterior Tibio-Fibular Region
Tibialis Anticus.
Extensor Proprius Pollicis.
Extensor Longus Digitorum
Peroneus Tertius.

The *Tibialis Anticus* is situated on the outer side of the tibia, being thick and fleshy at its upper part, tendinous below. It arises from the outer tuberosity and upper two-thirds of the external surface of the shaft of the tibia; from the adjoining part of the interosseous membrane; from the deep fascia of the leg; and from the intermuscular septum between it and the Extensor communis digitorum: the fibres pass vertically downwards, and terminate in a tendon, which is apparent on the anterior

173.—Muscles of the Front of the Leg.

surface of the muscle at the lower third of the leg. After passing through the innermost compartment of the anterior annular ligament, it is inserted into the inner side of the internal cuneiform bone, and base of the metatarsal bone of the great toe.

Relations. By its *anterior surface*, with the deep fascia, and with the annular ligament. By its *posterior surface*, with the interosseous membrane, tibia, and ankle-joint. By its *inner surface*, with the tibia. By its *outer surface*, with the Extensor longus digitorum, and Extensor proprius pollicis, the anterior tibial vessels and nerve lying between it and the last mentioned muscles.

The *Extensor Proprius Pollicis* is a thin, elongated, and flattened muscle, situated between the Tibialis anticus and Extensor longus digitorum. It arises from the anterior surface of the fibula for about the two middle fourths of its extent, its origin being internal to the Extensor longus digitorum; it also arises from the interosseous membrane to a similar extent. The fibres pass downwards, and terminate in a tendon, which occupies the anterior border of the muscle, passes through a distinct compartment in the annular ligament, and is inserted into the base of the last phalanx of the great toe. Opposite the metatarso-phalangeal articulation, the tendon gives off a thin prolongation on each side, which covers its surface.

Relations. By its *anterior border*, with the deep fascia of the leg, and the anterior annular ligament. By its *posterior border*, with the interosseous membrane, fibula, tibia, ankle-joint, and Extensor brevis digitorum. By its *outer side*, with the Extensor longus digitorum above, the dorsalis pedis artery and anterior tibial nerve below. By its *inner side*, with the Tibialis anticus, and the anterior tibial vessels above.

The *Extensor Longus Digitorum* is an elongated, flattened, semi-penniform muscle, situated the most external of all the muscles on the fore-part of the leg. It arises from the outer tuberosity of the tibia; from the upper three-fourths of the anterior surface of the shaft of the fibula; from the interosseous membrane, deep fascia; and from the intermuscular septa between it and the Tibialis anticus on the inner, and the Peronei on the outer side. The fibres pass downwards, and terminate in four tendons, which pass through a distinct canal in the annular ligament, together with the Peroneus tertius, run across the dorsum of the foot, and are inserted into the second and third phalanges of the four lesser toes. The mode in which these tendons are inserted is the following. Each tendon opposite the metatarso-phalangeal articulation is joined on its outer side by a tendon of the Extensor brevis digitorum (except the fourth), and receives a fibrous expansion from the Interossei and Lumbricales; it then spreads into a broad aponeurosis, which covers the dorsal surface of the first phalanx: this aponeurosis, at the articulation of the first with the second phalanx, divides into three slips, a middle one, which is inserted into the base of the second phalanx, and two lateral slips, which, after uniting on the dorsal surface of the second phalanx, are continued onwards to be inserted into the base of the third.

Relations. By its *anterior surface*, with the deep fascia of the leg, and the annular ligament. By its *posterior surface*, with the fibula, interosseous membrane, ankle-joint, and Extensor brevis digitorum. By its *inner side*, with the Tibialis anticus, Extensor proprius pollicis, and anterior tibial vessels and nerve. By its *outer side*, with the Peroneus longus and brevis.

The *Peroneus Tertius* is but a part of the Extensor longus digitorum, being almost always intimately united with it. It arises from the lower fourth of the anterior surface of the fibula; its outer part, from the lower part of the interosseous membrane; and from an intermuscular septum between it and the Peroneus brevis. Its tendon, after passing through the same canal in the annular ligament as the Extensor longus digitorum, is inserted into the base of the metatarsal bone of the little toe on its dorsal surface.

This muscle is often wanting.

Nerves. These muscles are supplied by the anterior tibial nerve.

Actions. The Tibialis anticus and Peroneus tertius are the direct flexors of the tarsus upon the leg; the former muscle, from the obliquity in the direction of its tendon, raises the inner border of the foot; and the latter, acting with the Peroneus brevis and longus, will draw the outer border of the foot upwards and the sole outwards. The Extensor longus digitorum and Extensor proprius pollicis extend the phalanges of the toes, and continuing their action, flex the tarsus upon the leg. Taking their origin from below, in the erect posture, all these muscles serve to fix the bones of the leg in a perpendicular direction, and give increased strength to the ankle-joint.

174.—Muscles of the Back of the Leg. Superficial Layer.

Posterior Tibio-Fibular Region

Dissection (fig. 171). Make a vertical incision along the middle line of the back of the leg, from the lower part of the popliteal space to the heel, connecting it below by a transverse incision extending between the two malleoli; the flaps of integument being removed, the fascia and muscles should be examined.

The muscles in this region of the leg are subdivided into two layers, superficial and deep. The superficial layer constitutes a powerful muscular mass, forming what is called the calf of the leg. Their large size is one of the most characteristic features of the muscular apparatus in man, and bears a direct connection with his ordinary attitude and mode of progression.

Superficial Layer
Gastrocnemius.   Soleus.
Plantaris.

The *Gastrocnemius* is the most superficial muscle at the back part of the leg, and forms the greater part of the calf. It arises by two heads, which are connected to the condyles of the femur by two strong flat tendons. The inner head, the larger, and a little the most posterior, is attached to a depression at the upper and back part of the inner condyle. The outer head, to the upper and back part of the external condyle, immediately above the origin of the Popliteus. Both heads, also, arise by a few tendinous and fleshy fibres from the ridges which are continued upwards from the condyles to the linea aspera. Each tendon spreads into an aponeurosis, which covers the posterior surface of that portion of the muscle to which it belongs; that covering the inner head being longer and thicker than the outer. From the anterior surface of these tendinous expansions muscular fibres are given off; those in the median line, which correspond to the accessory portion of the muscle derived from the bifurcations of the linea aspera, unite at an angle upon a median tendinous raphe below. The remaining fibres converge to the posterior surface of an aponeurosis which covers the front of the muscle, and this, gradually contracting, unites with the tendon of the Soleus, and forms with it the tendo Achillis.

Relations. By its *superficial surface*, with the fascia of the leg, which separates it from the external saphenous vein and nerve. By its *deep surface*, with the posterior ligament of the knee-joint, the Popliteus, Soleus, Plantaris, popliteal vessels, and internal popliteal nerve. The tendon of the inner head corresponds with the back part of the inner condyle, from which it is separated by a synovial bursa, which in some cases communicates with the cavity of the knee-joint. The tendon of the outer head contains a sesamoid fibro-cartilage (rarely osseous), where it plays over the corresponding outer condyle; and one is occasionally found in the tendon of the inner head.

The Gastrocnemius should be divided across just below its origin, and turned downwards, in order to expose the next muscles.

The *Soleus* is a broad flat muscle, situated immediately beneath the preceding. It has received its name from the fancied resemblance it bears to a sole-fish. It arises by tendinous fibres from the back part of the head, and from the upper half of the posterior surface of the shaft of the fibula, from the oblique line of the tibia, and from the middle third of its internal border; some fibres also arise from a tendinous arch which passes between the tibial and fibular origins of the muscle, and beneath which the posterior tibial vessels and nerve pass into the leg. The fibres pass backwards to an aponeurosis which covers the posterior surface of the muscle, and this, gradually becoming thicker and narrower, joins with the tendon of the Gastrocnemius, and forms with it the tendo Achillis.

Relations. By its *superficial surface*, with the Gastrocnemius and Plantaris. By its *deep surface*, with the Flexor longus digitorum, Flexor longus pollicis, Tibialis posticus, and posterior tibial vessels and nerve; from which it is separated by the transverse intermuscular septum, interposed between the superficial and deep muscles at the back of the leg.

The *Tendo Achillis*, the common tendon of the Gastrocnemius, Soleus, and Plantaris, is the thickest and strongest tendon in the body. It is about six inches in length, and formed by the junction of the aponeuroses of the two preceding muscles. It commences about the middle of the leg, but receives fleshy fibres much lower on its anterior surface. Gradually becoming contracted below, it is inserted into the lower part of the posterior tuberosity of the os calcis, a synovial bursa being interposed between the tendon and the upper part of the tuberosity. Externally it is covered by the fascia and the integument, and it is separated beneath from the deep seated muscles and vessels, by a considerable interval filled up with areolar and adipose tissue.

The *Plantaris* is an extremely diminutive muscle, placed between the Gastrocnemius and Soleus, and remarkable for the long and delicate tendon which it presents. It arises from the lower part of the external bifurcation of the linea aspera, and from the posterior ligament of the knee-joint. It forms a small fusiform belly, about two inches in length, which terminates in a long and slender tendon, which crosses obliquely between the two muscles of the calf, and running along the inner border of the tendo Achillis, is inserted with it into the posterior part of the os calcis. This muscle is occasionally double, it is sometimes wanting. Occasionally its tendon is lost in the subcutaneous adipose tissue, or in the internal annular ligament.

Nerves. These muscles are supplied by the internal popliteal nerve.

Actions. The muscles of the calf possess considerable power, and are constantly called into use in standing, walking, dancing, and leaping, hence the large size they usually present. In walking, these muscles draw powerfully upon the os calcis, raising the heel, and, with it, the entire body, from the ground; the body being thus supported on the raised foot, the opposite limb can be carried forwards. In standing, the Soleus, taking its fixed point from below, steadies the leg upon the foot, and prevents the body from falling forwards, to which there is a constant tendency from the super-incumbent weight. The Gastrocnemius, acting from below, serves to flex the femur upon the tibia, assisted by the Popliteus. The Plantaris is the rudiment of a large muscle which exists in some of the lower animals, and serves as a tensor of the plantar fascia.

Posterior Tibio-Fibular Region
Deep Layer.

Popliteus.	Flexor Longus Digitorum.
Flexor Longus Pollicis.	Tibialis Posticus.

Dissection. Detach the Soleus from its attachment to the fibula and tibia, and turn it downwards, when the deep layer of muscles is exposed, covered by the deep fascia of the leg.

The deep fascia of the leg is a broad, transverse inter-muscular septum, interposed between the superficial and deep muscles, in the posterior tibio-fibular region. On each side it is connected to the margins of the tibia and fibula. Above, where it covers the Popliteus, it is thick and dense, and receives an expansion from the tendon of the Semi-membranosus; it is thinner in the middle of the leg, but, below, where it covers the tendons passing behind the malleoli, it is thickened. It is continued onwards in the interval between the ankle and the heel, where it covers the vessels and is blended with the internal annular ligament.

This fascia should now be removed, commencing from below opposite the tendons, and detaching it from the muscles in the direction of their fibres.

The *Popliteus* is a thin, flat, triangular muscle, which forms the floor of the popliteal space, and is covered in by a tendinous expansion, derived from the Semi-membranosus muscle. It arises by a strong flat tendon, about an inch in length, from a deep depression on the outer side of the external condyle of the femur, and from the posterior ligament of the knee-joint; and is inserted into the inner two-thirds of the triangular surface above the oblique line on the posterior part of the shaft of the tibia, and into the tendinous expansion covering the surface of the muscle. The tendon of this muscle is covered in by that of the Biceps and the external lateral ligament of the knee-joint; it grooves the outer surface of the external semilunar cartilage, and is invested by the synovial membrane of the knee-joint.

175.—Muscles of the Back of the Leg. Deep Layer.

Relations. By its *superficial surface*, with the fascia above mentioned, which separates it from the Gastrocnemius, Plantaris, popliteal vessels and internal popliteal nerve. By its *deep surface*, with the tibio-fibular articulation and back of the tibia.

The *Flexor Longus Pollicis* is situated on the fibular side of the leg, and is the most superficial, and largest of the three next muscles. It arises from the lower two-thirds of the internal surface of the shaft of the fibula, with the exception of an inch below, from the lower part of the interosseous membrane, from an inter-muscular septum between it and the Peroneus longus and brevis, externally; and from the fascia covering the Tibialis posticus. The fibres pass obliquely downwards and backwards, and terminate around a tendon which occupies nearly the whole length of the posterior surface of the muscle. This tendon passes through a groove on the posterior surface of the tibia, external to that for the Tibialis posticus and Flexor longus digitorum; it then passes through a second groove on the posterior extremity of the astragalus, and along a third groove, beneath the tubercle of the os calcis, into the sole of the foot, where it runs forwards between the two heads of the Flexor brevis pollicis, and is inserted into the base of the last phalanx of the great toe. The grooves in the astragalus and os calcis which contain the tendon of this muscle, are converted by tendinous fibres into distinct canals, lined by synovial membrane; and as the tendon crosses the sole of the foot, it is connected to the common Flexor by a tendinous slip.

Relations. By its *superficial surface*, with the Soleus and tendo Achillis, from which it is separated by the deep fascia. By its *deep surface*, with the fibula, Tibialis posticus, the peroneal vessels, the lower part of the interosseous membrane, and the ankle-joint. By its *outer border*, with the Peroneus longus and brevis. By its *inner border*, with the Tibialis posticus, and Flexor longus digitorum.

The *Flexor Longus Digitorum* is situated on the inner or tibial side of the leg. At its origin, it is thin and pointed, but gradually increases in size as it descends. It arises from the posterior surface of the shaft of the tibia, immediately below the oblique line, to within three inches of its extremity, internal to the tibial origin of the Tibialis posticus; some fibres also arise from the intermuscular septum, between it and the Tibialis posticus. The fibres terminate in a tendon, which runs nearly the whole length of the posterior surface of the muscle. This tendon passes, behind the inner Malleolus, in a groove, common to it, and the Tibialis posticus, from which it is separated by a fibrous septum; each tendon is lined by a separate synovial membrane. It then passes, obliquely, forwards and outwards, beneath the arch of the os calcis, into the sole of the foot, where, crossing beneath the tendon of the Flexor longus pollicis, to which it is connected by a strong tendinous slip, it becomes expanded, is joined by the Musculus accessorius, and, finally divides into four tendons, which are inserted into the bases of the last phalanges of the four lesser toes, each tendon passing through a fissure in the tendon of the Flexor brevis digitorum, opposite the middle of the first phalanges.

Relations. In the leg. By its *superficial surface*, with the Soleus, and the posterior tibial vessels and nerve, from which it is separated by the deep fascia. By its *deep surface*, with the Tibia and Tibialis posticus. *In the foot*, it is covered by the Abductor pollicis, and Flexor brevis digitorum, and crosses beneath the Flexor longus pollicis.

The *Tibialis Posticus* lies between the two preceding muscles, and is the most deeply seated of all the muscles in the leg. It commences above, by two pointed processes, separated by an angular interval, through which, the anterior tibial vessels pass forwards to the front of the leg, arising from the posterior surface of the interosseous membrane, its whole length, excepting its lowest part, from the posterior surface of the shaft of the tibia, external to the Flexor longus digitorum, between the commencement of the oblique line above, and the centre of the external border of the bone below, and from the upper two-thirds of the inner surface of the shaft of the fibula; some fibres also arise from the deep fascia, and from the intermuscular septa, separating it from the adjacent muscles on each side. The fibres terminate in a tendon, which passes in front of the Flexor longus digitorum, through a groove behind the inner Malleolus, enclosed in a separate sheath; it then passes through another sheath, over the internal lateral ligament, and beneath the calcaneo-scaphoid articulation, and is inserted into the tuberosity of the scaphoid, and internal cuneiform bones. The tendon of this muscle, contains a sesamoid bone, near its insertion, and gives off fibrous expansions, one of which, passes backwards to the os calcis, others outwards to the middle and external cuneiform, and some forwards to the bases of the third and fourth metatarsal bones.

Relations. By its *superficial surface*, with the Soleus, Flexor longus digitorum, Flexor longus pollicis, the posterior tibial vessels and nerve, and the peroneal vessels, from which it is separated by the deep fascia. By its *deep surface*, with the interosseous ligament, the tibia, fibula, and ankle-joint.

Nerves. The Popliteus is supplied by the internal popliteal nerve, the remaining muscles of this group, by the posterior tibial nerve.

Actions. The Popliteus assists in flexing the leg upon the thigh, and, when flexed, it may rotate the tibia inwards. The Tibialis posticus is a direct Extensor of the tarsus upon the leg; acting in conjunction with the Tibialis anticus, it turns the sole of the foot inwards, antagonizing the Peroneus longus which turns it outwards. The Flexor longus digitorum, and Flexor longus pollicis, are the direct Flexors of the phalanges, and, continuing their action, extend the foot upon the leg; they assist the Gastrocnemius and Soleus in extending the foot, as in the act of walking, or in standing on tiptoe. In consequence of the oblique direction of the tendon of the long Extensor, the toes would be drawn inwards, were it not for the Flexor accessorius muscle, which is inserted into the outer side of that tendon, and draws it to the middle line of foot, during its action. Taking their fixed point from the foot, these muscles serve to maintain the upright posture, by steadying the tibia and fibula, perpendicularly, upon the ankle-joint. They also serve to raise these bones from the oblique position they assume in the stooping posture.

Fibular Region
Peroneus Longus.   Peroneus Brevis.

Dissection. These muscles are readily exposed, by removing the fascia, covering their surface, from below upwards, in the line of direction of their fibres.

The *Peroneus Longus* is situated at the upper part of the outer side of the leg. It arises from the head, and upper two-thirds of the outer surface of the shaft of the fibula, from the deep fascia, and from the intermuscular septa, between it and the muscles on the anterior, and those on the posterior surface of the leg. It terminates in a long tendon, which passes behind the outer malleolus, in a groove, common to it, and the Peroneus brevis, the groove being converted into a canal by a fibrous band, and the tendons, invested by a common synovial membrane; it is then reflected, obliquely forwards, across the outer side of the os calcis, being contained in a separate fibrous sheath, lined by a prolongation of the synovial membrane, from the groove behind the malleolus. Having reached the outer side of the cuboid bone, it runs, in a groove, on its under surface, which is converted into a canal, by the long calcaneo-cuboid ligament, lined by a synovial membrane, and crossing, obliquely, the sole of the foot, is inserted into the outer side of the base of the metatarsal bone of the great toe. The tendon of the muscle has a double reflection, first, behind the external malleolus, secondly, on the outer side of the cuboid bone; in both of these situations, the tendon is thickened, and, in the latter, a sesamoid bone is usually developed in its substance.

Relations. By its *superficial surface*, with the fascia and integument. By its *deep surface*, with the fibula, the Peroneus brevis, os calcis, and cuboid bone. By its *anterior border*, a tendinous septum intervenes between it and the Extensor longus digitorum. By its *posterior border*, an intermuscular septum, separates it from the Soleus above, and the Flexor longus pollicis below.

The *Peroneus Brevis* lies beneath the Peroneus longus, and is shorter and smaller than it. It arises from the lower two-thirds of the external surface of the shaft of the fibula, internal to the Peroneus longus; from the anterior and posterior borders of the bone; and from the intermuscular septa separating it from the adjacent muscles on the front and back part of the leg. The fibres pass vertically downwards, and terminate in a tendon, which runs through the same groove as the preceding muscle, behind the external malleolus, being contained in the same fibrous sheath, and lubricated by the same synovial membrane; it then passes through a separate sheath on the outer side of the os calcis, above that for the tendon of the Peroneus longus, and is finally inserted into the base of the metatarsal bone of the little toe, on its dorsal surface.

Relations. By its *superficial surface*, with the Peroneus longus and the fascia of the leg and foot. By its *deep surface*, with the fibula and outer side of the os calcis.

Nerves. The Peroneus longus and brevis are supplied by the musculo-cutaneous branch of the external popliteal nerve.

Actions. The Peroneus longus and brevis extend the foot upon the leg, in conjunction with the Tibialis posticus, antagonizing the Tibialis anticus and Peroneus tertius, which are flexors of the foot. The Peroneus longus also everts the sole of the foot; hence the extreme eversion observed in fracture of the lower end of the fibula, where that bone offers no resistance to the action of this muscle. Taking their fixed point below, they serve to steady the leg upon the foot. This is especially the case in standing upon one leg, when the tendency of the superincumbent weight is to throw the leg inwards; and the Peroneus longus overcomes this by drawing on the outer side of the leg, and thus maintains the perpendicular direction of the limb.

Surgical Anatomy. The student should now consider the position of the tendons of the various muscles of the leg, their relation with the ankle-joint and surrounding blood-vessels, and especially their action upon the foot, as their rigidity and contraction give rise to one or the other forms of deformity known as *club-foot*. The most simple and common deformity is the *talipes equinus*, the heel being raised from the ground by rigidity and contraction of the Gastrocnemius muscle, and the patient walking upon the ball of the foot. In the *talipes varus*, which is the more common congenital form, the heel is raised by the tendo Achillis, the inner border of the foot drawn upwards by the Tibialis anticus, and the anterior two-thirds of the foot twisted inwards by the Tibialis posticus and Flexor longus digitorum, the patient walking upon the dorsum of the foot and outer ankle. In the *talipes vulgus* the outer edge of the foot is raised by the Peronei muscles, and the patient walks upon the inner ankle. In the *talipes calcaneus* the foot is raised by the Extensor muscles, the heel is depressed, and the patient walks upon it. Each of these deformities may be successfully relieved (after other remedies fail) by division of the opposing tendons; by this means the foot regains its normal position, and the tendons heal by the organization of lymph thrown out between the divided ends. The operation is easily performed by putting the contracted tendon upon the stretch, and dividing it by means of a narrow sharp-pointed knife inserted between it and the skin.

Muscles and Fasciae of the Foot

The fibrous bands which serve to bind down the tendons in front and behind the ankle in their passage to the foot, should now be examined; they are termed the *annular ligaments*, and are three in number, anterior, internal, and external.

The *Anterior Annular Ligament* consists of a superior or vertical portion, which binds down the extensor tendons as they descend on the front of the tibia; and an inferior or horizontal portion, which retains them in connection with the tarsus: the two portions being connected by a thin intervening layer of fascia. The upper and stronger portion is attached externally to the lower end of the fibula, internally to the tibia, and above is continuous with the fascia of the leg: it contains two separate sheaths, one internally, for the tendon of the Tibialis anticus; one externally, for the tendons of the Extensor longus digitorum and Peroneus tertius, the tendon of the Extensor proprius pollicis, and the anterior tibial vessels and nerve pass beneath it. The lower portion is attached externally to the upper surface of the os calcis, in front of the depression for the interosseous ligament, and internally to the inner malleolus and plantar fascia: it contains three sheaths; the most internal for the tendon of the Tibialis anticus, the next in order for the tendon of the Extensor proprius pollicis, and the most external for the Extensor communis digitorum and Peroneus tertius: the anterior tibial vessels and nerve lie altogether beneath it. These sheaths are lined by separate synovial membranes.

The *Internal Annular Ligament* is a strong fibrous band, which extends from the inner malleolus above, to the internal margin of the os calcis below, converting a series of bony grooves in this situation into osteo-fibrous canals, for the passage of the tendons of the Flexor muscles and vessels into the sole of the foot. It is continuous above with the deep fascia of the leg, below with the plantar fascia and the fibres of origin of the Abductor pollicis muscle. The three canals which it forms, transmit from within outwards, first, the tendon of the Tibialis posticus; second, the tendon of the Flexor longus digitorum, then the posterior tibial vessels and nerve, which run through a broad space beneath the ligament; lastly, in a canal formed partly by the astragalus, the tendon of the Flexor longus pollicis. Each of these canals is lined by a separate synovial membrane.

The *External Annular Ligament* extends from the extremity of the outer malleolus to the outer surface of the os calcis, and serves to bind down the tendons of the Peronei muscles in their passage beneath the outer ankle. The two tendons are enclosed in one synovial sac.

Dissection of the Sole of the Foot. The foot should be placed on a high block with the sole uppermost, and firmly secured in that position. Carry an incision around the heel and along the inner and outer borders of the foot to the great and little toes. This incision should divide the integument and thick layer of granular fat beneath, until the fascia is visible; it should then be removed from the fascia in a direction from behind forwards, as seen in fig. 171.

The *Plantar Fascia*, the densest of all the fibrous membranes, consists of three portions, a middle and two lateral.

The *middle portion*, of great strength and thickness, consists of dense glistening fibres, disposed, for the most part, longitudinally; it is narrow and thick behind, and attached to the inner tuberosity on the under surface of the os calcis, behind the origin of the Flexor brevis digitorum, and becoming broader and thinner as it passes forwards, divides opposite the middle of the metatarsal bones into five fasciculi, one for each of the toes. Each of these fasciculi divides opposite the metatarso-phalangeal articulation into two slips, which embrace the sides of the flexor tendons of the toes, and are inserted into the bases of the metatarsal bones, and into the transverse ligaments of the corresponding articulation, thus forming a series of arches through which the tendons of the short and long flexors pass to the toes. The intervals left between the five primary fasciculi allow of the passage of the digital vessels and nerves, and the tendons of the Lumbricales and Interossei muscles. At the point of division of the fascia into fasciculi and slips, numerous transverse fibres are superadded, which serve to increase the strength of the fascia at this part, by binding the processes together and connecting them with the integument. The middle portion of the plantar fascia is continuous with the lateral portions at each side, and sends upwards into the foot, at their point of junction, two strong vertical intermuscular septa, broader in front than behind, which separate the middle from the external and internal plantar group of muscles. From these again thinner transverse septa are derived, which separate the various layers of muscles in this region; the upper surface of this fascia gives attachment behind to the Flexor brevis digitorum muscle.

The *lateral portions* of the plantar fascia cover the sides of the foot.

The *outer portion* covers the under surface of the Abductor minimi digiti; it is very thick behind, thin in front, and extends from the os calcis forwards to the base of the fifth metatarsal bone, into the outer side of which it is inserted; it is continuous internally with the middle portion of the plantar fascia, and externally with the dorsal fascia.

The *inner portion* is very thin, and covers the Abductor pollicis muscle; it is attached behind to the internal annular ligament, is continuous around the side of the foot with the dorsal fascia, and externally with the middle portion of the plantar fascia.

Muscles of the Foot

These are divided into two groups: 1. Those on the dorsum; 2. Those on the plantar surface.

1. Dorsal Region
Extensor Brevis Digitorum.

The *Fascia* on the dorsum of the foot is a thin membranous layer, continuous above with the anterior margin of the annular ligament; it becomes gradually lost in front, opposite the heads of the metatarsal bones, and on each side blends with the lateral portions of the plantar fascia: it forms a sheath for the tendons placed on the dorsum of the foot. On the removal of this fascia, the muscles of the dorsal region of the foot are exposed, covered by their investing fascia.

The *Extensor Brevis Digitorum* is a thin and somewhat broad muscle, which arises by a rounded extremity from the outer side of the os calcis, in front of the groove for the Peroneus brevis, from the astragalo-calcanean ligament, and from the anterior annular ligament of the tarsus: passing obliquely across the dorsum of the foot, it terminates in four tendons. The innermost, which is the largest, is inserted into the first phalanx of the great toe; the other three into the outer sides of the long extensor tendons of the second, third, and fourth toes.

Relations. By its *superficial surface*, with the fascia of the foot, the tendons of the Extensor longus digitorum, and Extensor proprius pollicis. By its *deep surface*, with the tarsal and metatarsal bones, and the Dorsal interossei muscles.

Nerves. It is supplied by the anterior tibial nerve.

Actions. The Extensor brevis digitorum is an accessory to the long Extensor, extending the phalanges of the four inner toes, but acting only on the first phalanx of the great toe. The obliquity of its direction counteracts the oblique movement given to the toes by the long Extensor, so that both muscles acting together, the toes are evenly extended.

2. Plantar Region

The muscles in the plantar region of the foot may be divided into three groups, in a similar manner to those in the hand. Those of the internal plantar region, are connected with the great toe, and correspond with those of the thumb; those of the external plantar region, are connected with the little toe, and correspond with those of the little finger; and those of the middle plantar region, are connected with the tendons intervening between the two former groups. The Interossei are considered separately.

Internal Plantar Group.
Abductor Pollicis.
Flexor Brevis Pollicis.
Adductor Pollicis.
Transversus Pedis.

External Plantar Group.
Abductor Minimi Digiti.
Flexor Brevis Minimi Digiti.

Middle Plantar Group.
Flexor Brevis Digitorum.   Musculus Accessorius.
Lumbricales.

In order to facilitate their dissection, it will be found more convenient to divide them into three layers, as they present themselves, in the order in which they are successively exposed.

First Layer.
Abductor Pollicis.   Flexor Brevis Digitorum.
Abductor Minimi Digiti.

Dissection. Remove the fascia on the inner and outer sides of the foot, commencing in front over the tendons, and proceeding backwards. The central portion should be divided transversely in the middle of the foot, and the two flaps dissected forwards and backwards.

The *Abductor Pollicis* lies along the inner border of the foot. It arises from the inner tuberosity of the os calcis, from the internal annular ligament, from the plantar fascia, and from the intermuscular septum between it and the Flexor brevis digitorum. The fibres terminate in a tendon, which is inserted, together with the innermost tendon of the Flexor brevis pollicis, into the internal sesamoid bone and inner border of the base of the first phalanx of the great toe.

Relations. By its *superficial surface*, with the internal plantar fascia. By its *deep surface*, with the Flexor brevis pollicis, the Musculus accessorius, and the tendons of the Flexor longus digitorum and Flexor longus pollicis, the Tibialis anticus and posticus, the plantar vessels and nerves, and the articulations of the tarsus.

176.—Muscles of the Sole of the Foot. First Layer.

The *Flexor Brevis Digitorum* lies in the middle line of the sole of the foot immediately beneath the plantar fascia, with which it is firmly united. It arises, by a narrow tendinous process, from the inner tuberosity of the os calcis, from the central part of the plantar fascia, and from the intermuscular septa between it and the adjacent muscles. It passes forwards and divides into four tendons. Opposite the middle of the first phalanges, each tendon presents a longitudinal slit, to allow of the passage of the corresponding tendon of the Flexor longus digitorum, the two portions forming a groove for its reception, and after reuniting, divides a second time into two processes, which are inserted into the sides of the second phalanges. The mode of division of the tendons of the Flexor brevis digitorum, and their insertion into the phalanges, is analogous to the Flexor sublimis in the hand.

Relations. By its *superficial surface*, with the plantar fascia. By its *deep surface*, with the Musculus accessorius, the Lumbricales, the tendons of the Flexor longus digitorum, and the plantar vessels and nerves, from which it is separated by a thin layer of fascia. The *outer* and *inner borders* are separated from the adjacent muscles by means of vertical prolongations of the plantar fascia.

The *Abductor Minimi Digiti* lies along the outer border of the foot. It arises, by a very broad origin, from the outer tuberosity of the os calcis, from the under surface of the os calcis in front of both tubercles, from the outer portion of the plantar fascia, and the intermuscular septum between it and the Flexor brevis digitorum. Its tendon, after gliding over a smooth facet on the under surface of the base of the fifth metacarpal bone, is inserted into the outer side of the base of the first phalanx of the little toe.

177.—Muscles of the Sole of the Foot. Second Layer.

Relations. By its *superficial surface*, with the outer portion of the plantar fascia. By its *deep surface*, with the outer head of the Musculus accessorius, the Flexor brevis minimi digiti, the long plantar ligament, and Peroneus longus. Its *inner side* is separated from the Flexor brevis digitorum by a vertical septum of fascia.

Dissection. The muscles of the superficial layer should be divided at their origin, by inserting the knife beneath each, and cutting obliquely backwards, so as to detach them from the bone; they should then be drawn forwards, in order to expose the second layer, but not separated from their insertion. The two layers are separated by a thin membrane, the deep plantar fascia, on the removal of which is seen the tendon of the Flexor longus digitorum, with its accessory muscle, the Flexor longus pollicis and the Lumbricales. The long flexor tendons cross each other at an acute angle, the Flexor longus pollicis running along the inner side of the foot, on a plane superior to that of the Flexor longus digitorum, the direction of which is obliquely outwards.

Second Layer.
Flexor Accessorius.   Lumbricales.

The *Flexor Accessorius* arises posteriorly by two heads, the inner or larger, which is muscular, being attached to the inner concave surface of the os calcis and to the calcaneo-scaphoid ligament; the outer head, flat and tendinous, to the under surface of the os calcis, in front of its outer tuberosity, and to the long plantar ligament: the two portions become united at an acute angle, and are inserted into the outer margin and upper and under surfaces of the tendon of the Flexor longus digitorum, forming a kind of groove, in which the tendon is lodged. A few fibres from the upper surface of the muscle blend with a tendinous expansion from the Flexor longus pollicis.

Relations. By its *superficial surface*, with the muscles of the superficial layer, from which it is separated by the external plantar vessels and nerves. By its *deep surface*, with the os calcis and long calcaneo-cuboid ligament.

The *Lumbricales* are four small muscles, accessory to the tendons of the Flexor longus digitorum: they arise from the tendons of the long Flexor, as far back as their angle of division, each arising from two tendons, except the internal one. Each muscle terminates in a tendon, which passes forwards on the inner side of each of the lesser toes, and is inserted into the expansion of the long Extensor and base of the second phalanx of the corresponding toe.

Dissection. The flexor tendons should be divided at the back part of the foot, and the Musculus accessorius at its origin, and drawn forwards, in order to expose the third layer.

178.—Muscles of the Sole of the Foot. Third Layer.

Third Layer.
Flexor Brevis Pollicis.
Adductor Pollicis.
Flexor Brevis Minimi Digiti
Transversus Pedis.

The *Flexor Brevis Pollicis* arises by a pointed tendinous process, from the inner border of the cuboid bone, from the contiguous portion of the external cuneiform, and from the prolongation of the tendon of the Tibialis posticus, which is attached to that bone. The muscle divides, in front, into two portions, which are inserted into the inner and outer sides of the base of the first phalanx of the great toe, a sesamoid bone being developed in each tendon at its insertion. The inner head of this muscle is blended with the Abductor pollicis previous to its insertion; the outer head, with the Adductor pollicis; and the tendon of the Flexor longus pollicis lies in a groove between them.

Relations. By its *superficial surface*, with the Abductor pollicis, the tendon of the Flexor longus pollicis and plantar fascia. By its *deep surface*, with the tendon of the Peroneus longus, and metatarsal bone of the great toe. By its *inner border*, with the Abductor pollicis. By its *outer border*, with the Adductor pollicis.

The *Adductor Pollicis* is a large, thick, fleshy mass, passing obliquely across the foot, and occupying the hollow space between the four outer metatarsal bones. It arises from the tarsal extremities of the second, third, and fourth metatarsal bones, and from the sheath of the tendon of the Peroneus longus; and is inserted, together with the outer head of the Flexor brevis pollicis, into the outer side of the base of the first phalanx of the great toe.

179.—The Dorsal Interossei. Left Foot.

The *Flexor Brevis Minimi Digiti* is situated along the outer border of the metatarsal bone of the little toe. It arises from the base of the metatarsal bone of the little toe, and from the sheath of the Peroneus longus; its tendon is inserted into the base of the first phalanx of the little toe, on its outer side.

Relations. By its *superficial surface*, with the plantar fascia and tendon of the abductor minimi digiti. By its *deep surface*, with the fifth metatarsal bone.

The *Transversus Pedis* is a narrow, flat, muscular fasciculus, stretched transversely across the heads of the metatarsal bones, between them and the flexor tendons. It arises from the under surface of the head of the fifth metatarsal bone, and from the transverse ligament of the metatarsus; and is inserted into the outer side of the first phalanx of the great toe; its fibres being blended with the tendon of insertion of the Adductor pollicis.

180.—The Plantar Interossei. Left Foot.

Relations. By its *under surface*, with the tendons of the long and short Flexors and Lumbricales. By its *upper surface*, with the Interossei.

The Interossei

The Interossei muscles in the foot are similar to those in the hand. They are seven in number, and consist of two groups, dorsal, and plantar.

The *Dorsal Interossei*, four in number, are situated between the metatarsal bones. They are bipenniform muscles, arising by two heads from the adjacent sides of the metatarsal bones between which they are placed, their tendons being inserted into the bases of the first phalanges, and into the aponeurosis formed by the common extensor tendon. In the angular interval left between each muscle at its posterior extremity, the perforating arteries pass to the dorsum of the foot; except in the first Interosseous muscle, where the interval allows the passage of the communicating branch of the dorsalis pedis artery. The first Dorsal interosseous muscle is inserted into the inner side of the second toe; the other three are inserted into the outer sides of the second, third, and fourth toes. They are all abductors from an imaginary line or axis drawn through the second toe.

The *Plantar Interossei*, three in number, lie beneath, rather than between, the metatarsal bones. They are single muscles, and are each connected with but one metatarsal bone. They arise from the base and inner sides of the shaft of the third, fourth, and fifth metatarsal bones, and are inserted into the inner sides of the bases of the first phalanges of the same toes, and into the aponeurosis of the common extensor tendon. These muscles are all adductors, towards an imaginary line, extending through the second toe.

Nerves. The internal plantar nerve supplies the Abductor pollicis, Flexor brevis digitorum, Flexor brevis pollicis, and the first and second Lumbricales. The external plantar nerve supplies the Abductor minimi digiti, Musculus accessorius, third and fourth Lumbricales, Adductor pollicis, Flexor brevis minimi digiti, Transversus pedis, and all the Interossei.

SURGICAL ANATOMY

The student should now consider the effects produced by the action of the various muscles, in fractures of the bones of the lower extremity. The more common forms of fracture have been especially selected for illustration and description.

Fracture of the *neck of the femur internal to the capsular ligament* (fig. 181) is a very common accident, and is most frequently caused by indirect violence, such as slipping off the edge of the kerbstone, the impetus and weight of the body falling upon the neck of the bone. It usually occurs in females, and seldom under fifty years of age. At this period of life, the neck of the bone, under certain conditions of the system, assumes a horizontal instead of an oblique direction, the head being on a level with the trochanter major; the cancellous tissue of the neck becomes soft and infiltrated with fatty matter, the compact tissue is partially absorbed, and the amount of earthy matter becomes greater in proportion to the animal constituent; hence, the bones are brittle, and more liable to fracture. The characteristic marks of this accident are slight shortening of the limb, and eversion of the foot, neither of which symptoms occur, however, in

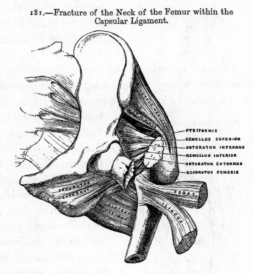

181.—Fracture of the Neck of the Femur within the Capsular Ligament.

some cases until a short time after the injury. The eversion is caused by the combined action of the external rotator muscles, as well as by the Psoas and Iliacus, Pectineus, Adductors, and Glutei muscles. The shortening and retraction of the limb is produced by the action of the Glutei, and by the Rectus femoris in front, and the Biceps, Semi-tendinosus, and Semi-membranosus behind.

182.—Fracture of the Femur below the Trochanter Minor.

Fracture of the *femur below the trochanter minor* (fig. 182), is an accident of not unfrequent occurrence, and is attended with great displacement producing considerable deformity. The upper fragment, the portion chiefly displaced, is tilted forwards almost at right angles with the pelvis by the combined action of the Psoas and Iliacus, and at the same time everted and drawn outwards by the External rotator and Glutei muscles, causing a marked prominence at the upper and outer side of the thigh, and much pain from the bruising and laceration of the muscles. The limb is shortened, from the lower fragment being drawn upwards by the Rectus in front, and the Biceps, Semi-membranosus, and Semi-tendinosus behind; and at the same everted, and the upper end thrown outwards, the lower inwards, by the Pectineus and Adductor muscles. This fracture may be reduced in two different methods: either by direct relaxation of all the opposing muscles, to effect which the limb should be placed on a double inclined plane; or by overcoming the contraction of the muscles by continued extension, which may be effected by means of the long splint.

Oblique fracture of the *femur immediately above the condyles* (fig. 183), is a formidable injury, and attended with considerable displacement. On examination of the limb, the lower fragment may be felt deep in the popliteal space, being drawn backwards by the Gastrocnemius, Soleus, and Plantaris muscles, and upwards by the Posterior femoral, and Rectus muscles. The pointed end of the upper fragment is drawn inwards by the Pectineus and Adductor muscles, and tilted forwards by the Psoas and Iliacus, piercing the Rectus muscle, and occasionally the integument. Relaxation of these muscles, and direct approximation of the broken fragments, is effected by placing the limb on a double inclined plane. The greatest care is requisite in keeping the pointed extremity of the upper fragment in proper apposition; otherwise, after union of the fracture, extension of the limb is partially destroyed from the Rectus muscle being held down by the fractured end of the bone, and from the patella when elevated being drawn upwards against it.

183.—Fracture of the Femur above the Condyles.

Fracture of the *patella* (fig. 184), may be produced by muscular action, or by direct violence. When produced by muscular action, it occurs thus: a person in danger of falling forwards, attempts to recover himself by throwing the body backwards, and the violent action of the Quadriceps extensor upon the patella snaps that bone transversely across. The upper fragment is drawn up the thigh by the Quadriceps extensor, the lower fragment being retained in its position by the ligamentum patellae; the extent of separation of the two fragments depending upon the degree of laceration of the ligamentous structures around the bone. The patient is totally unable to straighten the limb; the prominence of the patella is lost; and a marked but varying interval can be felt between the fragments. The treatment consists in relaxing the opposing muscles, which may be effected by raising the trunk, and slightly elevating the limb, which should be kept in a straight position. Union is usually ligamentous. In fracture from direct violence, the bone is generally comminuted, or fractured obliquely or perpendicularly.

184.—Fracture of the Patella.

Oblique fracture of the *shaft of the tibia* (fig. 185), usually occurs at the lower fourth of the bone, this being the narrowest and weakest part, and is generally accompanied with fracture of the fibula. If the fracture has taken place obliquely from above, downwards, and forwards, the fragments ride over one another, the lower fragment being drawn backwards and upwards by the powerful action of the muscles of the calf; the pointed extremity of the upper fragment projects forwards immediately beneath the integument, often protruding through it, and rendering the fracture a compound one. If the direction of the fracture is the reverse of that shewn in the figure, the pointed extremity of the lower fragment projects forwards, riding upon the lower end of the upper one. By relaxing the opposing muscles (bending the knee), with extension made from the knee and ankle, the fragments may be brought into apposition. It is often necessary, however, in compound fracture, to remove a portion of the projecting bone with the saw before complete adaptation can be effected.

185.—Oblique Fracture of the Shaft of the Tibia.

Fracture of the *fibula, with dislocation of the tibia inwards* (fig. 186), commonly known as "Pott's Fracture," is one of the most frequent injuries of the ankle-joint. The end of the tibia rests upon the inner side of the Astragalus, the internal lateral ligament of the ankle-joint is ruptured, and the inner malleolus projects inwards beneath the integument which is tightly stretched over it, and in danger of bursting. The fibula is broken, usually about three inches above the ankle, and occasionally that portion of the tibia with which it is more directly connected below; the foot is everted by the action of the Peroneus longus, its inner border resting upon the ground, and, at the same time, the heel is drawn up by the muscles of the calf. This injury may be at once reduced by flexing the leg at right angles with the thigh, which relaxes all the opposing muscles, and by making slight extension from the knee and ankle.

186.—Fracture of the Fibula, with Dislocation of the Tibia inwards. 'Pott's Fracture.'

OF THE ARTERIES

The Arteries are cylindrical tubular vessels, which serve to convey blood from the heart to every part of the body. These vessels were named arteries (αηρ τηρειν, to contain air), from the belief entertained by the ancients that they contained air. To Galen is due the honour of refuting this opinion; he showed that these vessels, though for the most part empty after death, contained blood in the living body.

The pulmonary artery, which arises from the right ventricle of the heart, carries venous blood directly into the lungs, from whence it is returned by the pulmonary veins into the left auricle. This constitutes the lesser or pulmonic circulation. The great artery which arises from the left ventricle, the aorta, conveys arterial blood to the body generally; from whence it is brought back to the right side of the heart by means of the veins. This constitutes the greater or systemic circulation.

The distribution of the systemic arteries is like a highly ramified tree, the common trunk of which, formed by the aorta, commences at the left ventricle of the heart, the smallest ramifications corresponding to the circumference of the body and the contained organs. The arteries are found in nearly every part of the animal body, with the exception of the hairs, nails, and epidermis; and the larger trunks usually occupy the most protected situations, running, in the limbs, along the flexor side, where they are less exposed to injury.

There is considerable variation in the mode of division of the arteries; occasionally a short trunk subdivides into several branches at the same point, as we observe in the coeliac and thyroid axes; or the vessel may give off several branches in succession, and still continue as the main trunk, as is seen in the arteries of the limbs; but the usual division is dichotomous, as, for instance, the aorta dividing into the two common iliacs; and the common carotid, into the external and internal.

The branches of arteries arise at very variable angles; some, as the superior intercostal, arise from the aorta at an obtuse angle; others, as the lumbar arteries, at a right angle; or, as the spermatic, at an acute angle. An artery from which a branch is given off is smaller in size than the trunk from which it arises, but retains a uniform diameter until a second branch is derived from it; but if an artery divides into two branches, the combined area of the two vessels is, in nearly every instance, somewhat greater than that of the trunk; and the combined area of all the arterial branches greatly exceeds the diameter of the aorta; so that the arteries collectively may be regarded as a cone, the apex of which corresponds to the aorta; the base, to the capillary system.

The arteries, in their distribution, communicate freely with one another, forming what is called an *anastomosis* (ανα, between; στομα, mouth), or inosculation, and this communication is very free between the large, as well as between the smaller branches. The anastomoses between trunks of equal size is found where great freedom and activity of the circulation is requisite, as in the brain; here the two vertebral arteries unite to form the basilar, and the two internal carotid arteries are connected by a short inter-communicating trunk; it is also found in the abdomen, the intestinal arteries having very free anastomoses between their larger branches. In the limbs, the anastomoses are most frequent and of largest size around the joints; the branches of an artery above, freely inosculating with branches from the vessel below; these anastomoses are of considerable interest to the surgeon, as it is by their enlargement that a collateral circulation is established after the application of a ligature to an artery for the cure of aneurism. The smaller branches of arteries anastomose more frequently than the larger, and between the smallest twigs these inosculations become so numerous as to constitute a close network that pervades nearly every tissue of the body.

Throughout the body generally the larger arterial branches pursue a perfectly straight course, but in certain situations they are tortuous; thus the facial artery in its course over the face, and the labial arteries of the lips, are extremely tortuous in their course, to accommodate themselves to the movements of these parts. The uterine arteries are also tortuous, to accommodate themselves to the increase of size which this organ undergoes during pregnancy. Again, the internal carotid and vertebral arteries, previous to their entering the cavity of the skull, describe a series of curves, which are evidently intended to diminish the velocity of the current of blood, by increasing the extent of surface over which it moves, and adding to the amount of impediment which is produced from friction.

The smaller arterial branches terminate in a system of minute anastomosing vessels which pervade every tissue of the body. These vessels, from their minute size, are termed *capillaries* (capillus, a hair). They are interposed between the smallest branches of the arteries and the commencing veins, constituting a network, the branches of which are of nearly uniform size, their average diameter being about the $1/3000$ of an inch; but the size of the smaller capillaries, and the diameter of the meshes between them, vary in the different organs.

The arteries are dense in structure, of considerable strength, highly elastic, and, when divided, they preserve, although empty, their cylindrical form.

They are composed of three coats, internal, middle, and external.

The *internal* is an epithelial and elastic coat; it consists of two layers, the innermost of which is composed of a single layer of elliptical or spindle-shaped epithelial particles, with round or oval nuclei, resting upon a striated and perforated transparent colourless membrane, highly elastic, but extremely thin and brittle, disposed in one or more layers, and forming the chief substance of the inner coat.

The *middle*, or *contractile coat*, consists of muscular and elastic fibres, it is of a reddish yellow colour, highly elastic, and consists of numerous layers of non-striated muscular fibres, disposed in a circular form around the vessel, having intermixed with them layers of fine elastic or fenestrated membrane; as many as forty layers have been counted in the aorta, twenty-eight in the carotid, and fifteen in the subclavian artery. The muscular tissue exists in greatest abundance in the smallest arteries, whilst in the larger trunks it is blended with much elastic tissue; the great thickness of the walls of the arteries is due chiefly to this coat.

The *external*, or *areolar and elastic coat*, consists of condensed areolar and elastic tissue; in the larger arteries it is composed of two distinct layers; an inner, composed of elastic tissue, most distinct in the larger arteries; and an external layer of condensed areolo-fibrous tissue, the constituent fibres being disposed more or less diagonally or obliquely around the vessel. In the smaller arteries the elastic tissue is wanting, the areolar coat increasing in proportion.

Some arteries have extremely thin coats in proportion to their size; this is especially the case in those situated in the cavity of the cranium and spinal canal the difference depending upon the greater thinness of the external and middle coats.

The arteries in their distribution throughout the body, are included in a thin areolo-fibrous investment, which forms what is called their sheath. In the limbs, this is usually formed by a prolongation of the deep fascia; in the upper part of the thigh it consists of a continuation downwards of the transversalis and iliac fasciae of the abdomen; in the neck, of a prolongation of the deep cervical fascia. The included vessel is loosely connected with its sheath by a delicate areolar tissue; and the sheath usually encloses the accompanying veins and sometimes a nerve. Some arteries, as those in the cranium, are not included in sheaths.

Arteries are supplied with blood-vessels like the other organs of the body, they are called *vasa vasorum*. These nutrient vessels arise from a branch of the artery or from a neighbouring vessel, at some considerable distance from the point at which they are distributed; they ramify in the loose areolar tissue connecting the artery with its sheath, and are distributed to the external and middle coats, and according to Arnold and others, supply the internal coat. Minute veins serve to return the blood from these vessels, they empty themselves into the venae comites in connection with the artery. Arteries are also provided with nerves; they are derived chiefly from the sympathetic, but partly from the cerebro-spinal system. They form intricate plexuses upon the surface of the larger trunks, the smaller branches being usually accompanied by single filaments; their exact mode of distribution is unknown.

In the description of the arteries, we shall first consider the efferent trunk of the systemic circulation, the aorta, and its branches; and then the efferent trunk of the pulmonic circulation, the pulmonary artery.

The Aorta

The *aorta* (αορτη; *arteria magna*) is the main trunk of a series of vessels, which, arising from the heart, conveys the red oxygenated blood to every part of the body for its nutrition. This vessel commences at the upper part of the left ventricle, and after ascending for a short distance, arches backwards to the left side, over the root of the left lung, descends within the thorax on the left side of the vertebral column, passes through the aortic opening in the Diaphragm, and entering the abdominal cavity, terminates, considerably diminished in size, opposite the fourth lumbar vertebra, where it divides into the right and left common iliac arteries. Hence its subdivision into the arch of the aorta, the thoracic aorta and the abdominal aorta, from the direction or position peculiar to each part.

Arch of the Aorta

Dissection. In order to examine the arch of the aorta, the thorax should be opened, by dividing the cartilages of the ribs on each side of the sternum, and raising this bone from below upwards, and then sawing through the sternum on a level with its

187.—The Arch of the Aorta and its Branches.

FIG. 188. Plan of the Branches

articulation with the clavicle. By this means the relations of the large vessels to the upper border of the sternum and root of the neck are kept in view.

The arch of the aorta extends from the origin of the vessel at the upper part of the left ventricle, to the lower border of the body of the third dorsal vertebra. At its commencement, it ascends behind the sternum, obliquely upwards and forwards towards the right side, and opposite the upper border of the second costal cartilage of the right side, passes transversely from right to left, and from before backwards to the left side of the second dorsal vertebra; it then descends upon the left side of the body of the third dorsal vertebra, at the lower border of which it becomes the thoracic aorta. Hence this portion of the vessel is divided into an ascending, a transverse, and a descending portion. The artery in its course describes a curve, the convexity of which is directed upwards and to the right side, the concavity in the opposite direction.

Ascending Part of the Arch

The ascending portion of the arch of the aorta is about two inches in length. It commences at the upper part of the left ventricle, in front of the left auriculo-ventricular orifice, and opposite the middle of the sternum on a line with its junction to the third costal cartilage; it passes obliquely upwards in the direction of the heart's axis, to the right side, as high as the upper border of the second costal cartilage, describing a slight curve in its course, and being situated, when distended, about a quarter of an inch behind the posterior surface of the sternum. A little above its commencement, it is somewhat enlarged, and presents three small dilatations, called the *sinuses of the aorta* (sinuses of Valsalva) opposite to which are attached the three semi-lunar valves, which serve the purpose of preventing any regurgitation of blood into the cavity of the ventricle. A section of the aorta opposite this part has a somewhat triangular figure; but below the attachment of the valves it is circular. This portion of the arch is contained in the cavity of the pericardium, and together with the pulmonary artery, is invested in a tube of serous membrane, continued on to them from the surface of the heart.

Relations. The ascending part of the arch is covered at its commencement by the trunk of the pulmonary artery and the right auricular appendage, and, higher up, is separated from the sternum by the pericardium, some loose areolar tissue, and the remains of the thymus gland; *behind*, it rests upon the right pulmonary vessels and root of the right lung. On the *right side*, it is in relation with the superior vena cava and right auricle; on the *left side*, with the pulmonary artery.

Plan of the Relations of the Ascending Part of the Arch

In front.
Pulmonary artery.
Right auricular appendage.
Pericardium.
Remains of thymus gland.

Right side. Arch of   *Left side.*
Superior cava. Aorta. Pulmonary artery.
Right auricle. Ascending Portion.

Behind.
Right pulmonary vessels.
Root of right lung.

Transverse Part of the Arch

The second or transverse portion of the arch commences at the upper border of the second costo-sternal articulation of the right side in front, and passes from right to left, and from before backwards, to the left side of the second dorsal vertebra behind. Its upper border is usually about an inch below the upper margin of the sternum.

Relations. Its *anterior surface* is covered by the left pleura and lung, and crossed towards the left side by the left pneumogastric and phrenic nerves, and cardiac branches of the sympathetic. Its *posterior surface* lies on the trachea just above its bifurcation, the great cardiac plexus, the oesophagus, thoracic duct, and left recurrent laryngeal nerve. Its *superior border* is in relation with the left innominate vein; and from its upper part are given off the innominate, left carotid, and left subclavian arteries. By its *lower border*, with the bifurcation of the pulmonary artery, and the remains of the ductus arteriosus, which is connected with the left division of that vessel; the left recurrent laryngeal nerve winds round it, whilst the left bronchus passes below it.

Plan of the Relations of the Transverse Part of the Arch

Above.
Left innominate vein.
Arteria innominata.
Left carotid.
Left subclavian.

In front.	Arch of Aorta. Transverse Portion.	*Behind.*
Left pleura and lung.		Trachea.
Left pneumogastric nerve.		Cardiac plexus.
Left phrenic nerve.		Oesophagus.
Cardiac nerves.		Thoracic duct.
		Left recurrent nerve.

Below.
Bifurcation of pulmonary artery.
Remains of ductus arteriosus.
Left recurrent nerve.
Left bronchus.

Descending Part of the Arch

The descending portion of the arch has a straight direction, inclining downwards on the left side of the body of the third dorsal vertebra, at the lower border of which it becomes the thoracic aorta.

Relations. Its *anterior surface* is covered by the pleura and root of the left lung; *behind*, it lies on the left side of the body of the third dorsal vertebra. On its *right side* lies the oesophagus and thoracic duct; on its *left side* it is covered by the pleura.

Plan of the Relations of the Descending Part of the Arch

In front.
Pleura.
Root of left lung.

Right side.	Arch of Aorta. Descending Portion.	*Left side.*
Oesophagus.		Pleura.
Thoracic duct.		

Behind.
Left side of body of third dorsal vertebra.

The ascending, transverse, and descending portions of the arch vary in position according to the movements of respiration, being lowered, together with the trachea, bronchi, and pulmonary vessels, during inspiration by the descent of the diaphragm, and elevated during expiration, when the diaphragm ascends. These movements are greater in the ascending than the transverse, and in the latter than the descending part.

Peculiarities. The height to which the aorta rises in the chest is usually about an inch below the upper border of the sternum; but it may ascend nearly to the top of that bone. Occasionally it is found an inch and a half; more rarely, three inches below this point.

Direction. Sometimes the aorta arches over the root of the right instead of the left lung, as in birds, and passes down on the right side of the spine. In such cases, all the viscera of the thoracic and abdominal cavities are transposed. Less frequently, the aorta, after arching over the right lung, is afterwards directed to its usual position on the left side of the spine, this peculiarity not being accompanied by any transposition of the viscera.

Conformation. The aorta occasionally divides into an ascending and a descending trunk, as in some quadrupeds, the former being directly vertically upwards, and subdividing into three branches, to supply the head and upper extremities. Sometimes the aorta subdivides soon after its origin into two branches, which soon reunite. In one of these cases, the oesophagus and trachea were found to pass through the interval left by their division; this is the normal condition of the vessel in the reptilia.

Surgical Anatomy. Of all the vessels of the arterial system, the aorta, and more especially its arch, is most frequently the seat of disease; hence it is important to consider some of the consequences that may ensue from aneurism of this part.

It will be remembered, that the ascending part of the arch is contained in the pericardium, just behind the sternum, its commencement being crossed by the pulmonary artery and right auricular appendage, having the root of the right lung behind, the vena cava on the right side, and the pulmonary artery and left auricle on the left side.

Aneurism of the ascending aorta, in the situation of the aortic sinuses, in the great majority of cases, affects the right coronary sinus; this is mainly owing to the regurgitation of blood upon the sinuses, taking place chiefly on the right anterior aspect of the vessel. As the aneurismal sac enlarges, it may compress any or all of the structures in immediate proximity with it, but chiefly projects towards the right anterior side; and, consequently, interferes mainly with those structures that have a corresponding relation with the vessel. In the majority of cases, it bursts in the cavity of the pericardium, the patient suddenly drops down dead, and, upon a post-mortem examination, the pericardial bag is found full of blood: or it may compress the right auricle, or the pulmonary artery, and adjoining part of the right ventricle, and open into one or the other of these parts, or it may compress the superior cava.

Aneurism of the ascending aorta, originating above the sinuses, most frequently implicates the right anterior wall of the vessel; this is probably mainly owing to the blood being impelled against this part. Its direction is also chiefly towards the right of the median line. If it attains a large size and projects forwards, it may absorb the sternum and the cartilages of the ribs, usually on the right side, and appear as a pulsating tumour on the front of the chest, just below the manubrium; or it may burst into the pericardium, may compress or even open into the right lung, the trachea, bronchi, or oesophagus.

Regarding the transverse part of the arch, the student is reminded that the vessel lies on the trachea, the oesophagus, and thoracic duct; that the recurrent laryngeal nerve winds around it; and that from its upper part are given off three large trunks, which supply the head, neck, and upper extremities. Now an aneurismal tumour taking origin from the posterior part or right aspect of the vessel, its most usual site, may press upon the trachea, impede the breathing, or produce cough, haemoptysis, or stridulous breathing, or it may ultimately burst into that tube, producing fatal haemorrhage. Again, its pressure on the laryngeal nerves may give rise to symptoms which so accurately resemble those of laryngitis, that the operation of tracheotomy has in some cases been resorted to from the supposition that disease existed in the larynx; or it may press upon the thoracic duct, and destroy life by inanition; or it may involve the oesophagus, producing dysphagia; or may burst into this tube, when fatal haemorrhage will occur. Again, the innominate artery, or the left carotid, or subclavian, may be so obstructed by clots, as to produce a weakness, or even a disappearance, of the pulse in one or the other wrist; or the tumour may present itself at or above the manubrium, generally either in the median line, or to the right of the sternum.

Aneurism affecting the descending part of the arch is usually directed backwards and to the left side, causing absorption of the vertebrae and corresponding ribs; or it may press upon the trachea, left bronchus, oesophagus, and the right and left lungs, generally the latter: when rupture of the sac occurs, this usually takes place in the left pleural cavity; less frequently in the left bronchus, the right pleura, or into the substance of the lungs or trachea. In this form of aneurism, pain is almost a constant and characteristic symptom, existing either in the back or chest, and usually radiating from the spine around the left side. This symptom depends upon the aneurismal sac compressing the intercostal nerves against the bone.

Branches of the Arch of the Aorta (figs. 187, 188)

The branches given off from the arch of the aorta are five in number. Two of small size from the ascending portion, the right and left coronary; and three of large size from the transverse portion, the innominate artery, the left carotid, and the left subclavian.

Peculiarities. Position of the Branches. The branches, instead of arising from the highest part of the arch (their usual position), may be moved more to the right, arising from the commencement of the transverse or upper part of the ascending portion; or the distance from one another at their origin may be increased or diminished, the most frequent change in this respect being the approximation of the left carotid, towards the innominate artery.

The Number of the primary branches may be reduced to two: the left carotid arising from the innominate artery; or (more rarely), the carotid and subclavian arteries of the left side arising from an innominate artery. But the number may be increased to four, from the right carotid and subclavian arteries arising directly from the aorta, the innominate being absent. In most of these latter cases, the right subclavian arose from the left end of the arch; in other cases, it was the second or third branch given off instead of the first. Lastly, the number of trunks from the arch may be increased to five or six; in these instances, the external and internal carotids arose separately from the arch, the common carotid being absent on one or both sides.

Number usual, Arrangement different. When the aorta arches over to the right side, the three branches have an arrangement the reverse of what is usual, the innominate supplying the left side; and the carotid and subclavian (which arise separately), the right side. In other cases, where the aorta takes its usual course, the two carotids may be joined in a common trunk, and the subclavians arise separately from the arch, the right subclavian generally arising from the left end of the arch.

Secondary Branches sometimes arise from the arch; most commonly it is the left vertebral, which usually takes origin between the left carotid, and left subclavian, or beyond them. Sometimes, a thyroid branch is derived from the arch, or the right internal mammary, or left vertebral, or, more rarely, both vertebrals.

The Coronary Arteries

The coronary arteries supply the heart; they are two in number, right and left, arising near the commencement of the aorta immediately above the free margin of the semi-lunar valves.

The *Right Coronary Artery*, about the size of a crow's quill, arises from the aorta immediately above the free margin of the right semi-lunar valve, between the pulmonary artery, and the appendix of the right auricle. It passes forwards to the right side in the groove between the right auricle and ventricle, and curving around the right border of the heart, runs along its posterior surface as far as the posterior interventricular groove, where it divides into two branches, one of which continues onwards in the groove between the left auricle and ventricle, and anastomoses with the left coronary; the other descends along the posterior inter-ventricular furrow, supplying branches to both ventricles, and to the septum, and anastomosing at the apex of the heart with the descending branch of the left coronary.

This vessel sends a large branch along the thin margin of the right ventricle to the apex, and numerous small branches to the right auricle and ventricle, and commencement of the pulmonary artery.

The *Left Coronary*, smaller than the former, arises immediately above the free edge of the left semi-lunar valve, a little higher than the right; it passes forwards between the pulmonary artery and the left appendix auriculae, and descends obliquely towards the anterior inter-ventricular groove, where it divides into two branches. Of these, one passes transversely outwards in the left auriculo-ventricular groove, and winds around the left border of the heart to its posterior surface, where it anastomoses with the superior branch of the right coronary; the other descends along the anterior inter-ventricular groove to the apex of the heart, where it anastomoses with the descending branch of the right coronary. The left coronary supplies the left auricle and its appendix, both ventricles, and numerous small branches to the pulmonary artery, and commencement of the aorta.

Peculiarities. These vessels occasionally arise by a common trunk, or their number may be increased to three; the additional branch being of small size. More rarely, there are two additional branches.

Arteria Innominata

The innominate artery is the largest branch given off from the arch of the aorta. It arises from the commencement of the transverse portion in front of the left carotid, and ascending obliquely to the upper border of the right sterno-clavicular articulation, divides into the right carotid and subclavian arteries. This vessel varies from an inch and-a-half to two inches in length.

Relations. In front, it is separated from the first bone of the sternum by the Sterno-hyoid and Sterno-thyroid muscles, the remains of the thymus gland, and by the left innominate and right inferior thyroid veins which cross its root. *Behind,* it lies upon the trachea which it crosses obliquely. On the *right side* is the right vena innominata, right pneumogastric nerve, and the pleura; and on the *left side,* the remains of the thymus gland, and origin of the left carotid artery.

Plan of the Relations of the Innominate Artery

In front.
Sternum.
Sterno-hyoid and Sterno-thyroid.
Remains of thymus gland.
Left innominate and right inferior thyroid veins.

Right side.	Innominate Artery.	*Left side.*
Right vena innominata.		Remains of thymus.
Right pneumogastric nerve.		Left carotid.
Pleura.		

Behind.
Trachea.

Peculiarities in point of division. When the bifurcation of the innominate artery varies from the point above mentioned, it sometimes ascends a considerable distance above the sternal end of the clavicle; less frequently it divides below it. In the former class of cases, its length may exceed two inches; and, in the latter, be reduced to an inch or less. These are points of considerable interest for the surgeon to remember in connection with the operation of including this vessel in a ligature.

Branches. The arteria innominata occasionally supplies a thyroid branch (middle thyroid artery), which ascends along the front of the trachea to the thyroid gland; and sometimes, a thymic or bronchial branch. The left carotid is frequently joined with the innominate artery at its origin. Sometimes, there is no innominate artery, the right subclavian and right carotid arising directly from the arch of the aorta.

Position. When the aorta arches over to the right side, the innominate is directed to the left side of the neck, instead of to the right.

Surgical Anatomy. Although the operation of tying the innominate artery, has been performed by several surgeons, for aneurism of the right subclavian extending inwards as far as the scalenus, in no instance has it been attended with success. An important fact has, however, been established; viz., that the circulation in the parts supplied by the artery, can be supported after the operation; a fact which cannot but encourage surgeons to have recourse to it whenever the urgency of the case may require it, notwithstanding that it must be regarded as peculiarly hazardous.

The failure of the operation in those cases where it has been performed, has depended on subsequent repeated secondary haemorrhage, or on inflammation of the adjoining pleural sac and lung. The main obstacles to its performance are, as the student will perceive from his dissection of this vessel, its deep situation behind and beneath the sternum, and the number of important structures which surround it in every part.

In order to apply a ligature to this vessel, the patient is placed upon his back, with the shoulders raised, and the head bent a little backwards, so as to draw out the artery from behind the sternum into the neck. An incision two inches long is then made along the anterior border of the Sterno-mastoid muscle, terminating at the sternal end of the clavicle. From this joint, a second incision is to be carried about the same length along the upper border of the clavicle. The skin is to be dissected back, and the platysma being exposed, must be divided on a director: the sternal end of the Sterno-mastoid is now brought into view, and a director being passed beneath it, and close to its under surface, so as to avoid any small vessels, it must be divided transversely throughout the greater part of its attachment.

Pressing aside any loose cellular tissue or vessels that may now appear, the Sterno-hyoid and Sterno-thyroid muscles will be exposed, and must be divided, a director being previously passed beneath them. The inferior thyroid veins now come into view, and must be carefully drawn either upwards or downwards, by means of a blunt hook. On no account should these vessels be divided, as it would add much to the difficulty of the operation, and endanger its ultimate success. After tearing through a strong fibro-cellular lamina, the right carotid is brought into view, and being traced downwards, the arteria innominata is arrived at. The left vena innominata should now be depressed, the right vena innominata, the internal jugular vein, and pneumogastric nerve drawn to the right side; and a curved aneurism needle may then be passed around the vessel, close to its surface, and in a direction from below upwards and inwards; care being taken to avoid the right pleural sac, the trachea, and cardiac nerves. The ligature should be applied to the artery as high as possible, in order to allow room between it and the aorta for the formation of a coagulum.

It has been seen that the failure of this operation depends either upon repeated secondary haemorrhage, or inflammation of the pleural sac and lung. The importance of avoiding the thyroid plexus of veins during the primary steps of the operation, and the pleural sac whilst including the vessel in the ligature, should be most carefully attended to.

Common Carotid Arteries

The common carotid arteries, although occupying a nearly similar position in the neck, differ in position, and, consequently, in their relations at their origin. The right carotid arises from the arteria innominata, behind the right sterno-clavicular articulation; the left from the highest part of the arch of the aorta. The left carotid is, consequently, longer and placed more deeply in the thorax. It will, therefore, be more convenient to describe first the course and relations of that portion of the left carotid which intervenes between the arch of the aorta and the left sterno-clavicular articulation (fig. 187, color illustration p. A1).

The left carotid within the thorax passes obliquely outwards from the arch of the aorta to the root of the neck. In *front*, it is separated from the first piece of the sternum by the Sterno-hyoid and Sterno-thyroid muscles, the left innominate vein, and the remains of the thymus gland; *behind*, it lies on the trachea, oesophagus, and thoracic duct. *Internally*, it is in relation with the arteria innominata; *externally*, with the left pneumogastric nerve, cardiac branches of the sympathetic, and left subclavian artery.

PLAN OF THE RELATIONS OF THE LEFT COMMON CAROTID
THORACIC PORTION

In front.
Sternum.
Sterno-hyoid and Sterno-thyroid muscles.
Left innominate vein.
Remains of thymus gland.

Internally.	Left Common Carotid. Thoracic Portion.	*Externally.*
Arteria innominata.		Left pneumogastric nerve. Cardiac nerves. Left subclavian artery.

Behind.
Trachea.
Oesophagus.
Thoracic duct.

In the neck, the two common carotids resemble each other so closely, that one description will apply to both. Starting from each side of the neck, these vessels pass obliquely upwards, from behind the sterno-clavicular articulations, to a level with the upper border of the thyroid cartilage, where they divide into the external and internal carotids; these names being derived, the former from their distribution to the external parts of the head and face, the latter from their distribution to the internal parts of the cranium. The course of each vessel is indicated by a line drawn from the sternal end of the clavicle below, to a point midway between the angle of the jaw and the mastoid process above.

At the lower part of the neck the two common carotid arteries are separated from each other by a very small interval, which corresponds to the trachea; but at the upper part, the thyroid body, the larynx and pharynx project forwards between these

189.—Surgical Anatomy of the Arteries of the Neck. Right Side.

FIG. 190. Plan of the Branches of the EXTERNAL CAROTID.

vessels, and give the appearance of their being placed further back in this situation. The common carotid artery is contained in a sheath, derived from the deep cervical fascia, which also encloses the internal jugular vein and pneumogastric nerve, the vein lying on the outer side of the artery, and the nerve between the artery and vein, on a plane posterior to both. On opening the sheath, these three structures are seen to be separated from one another, each being enclosed in a separate fibrous investment.

Relations. At the lower part of the neck the common carotid artery is very deeply seated, being covered by the Platysma, superficial and deep fasciae, the Sterno-mastoid, Sterno-hyoid, and Sterno-thyroid muscles, and by the Omo-hyoid opposite the cricoid cartilage; but in the upper part of its course, near its termination, it is more superficial, being covered merely by the integument, Platysma, the superficial and deep fascia, and inner margin of the Sterno-mastoid, and is contained in a triangular space, bounded behind by the Sterno-mastoid, above by the posterior belly of the Digastric, and below by the anterior belly of the Omo-hyoid. This part of the artery is crossed obliquely from within outwards by the sterno-mastoid artery; it is also crossed by the superior thyroid veins, which terminate in the internal jugular, and, descending on its sheath in front, is seen the descendens noni nerve, this filament being joined with branches from the cervical nerves, which cross the vessel from without inwards. Sometimes the descendens noni is contained within the sheath. The middle thyroid vein crosses it about its centre, and the anterior jugular vein below, the latter vessel being usually placed beneath the Sterno-mastoid. *Behind*, the artery lies in front of the cervical portion of the spine, resting first on the Longus colli muscle, then on the Rectus anticus major, from which it is separated by the sympathetic nerve. The recurrent laryngeal nerve and inferior thyroid artery cross behind the vessel at its lower part. *Internally*, it is in relation with the trachea and thyroid gland, the inferior thyroid artery and recurrent laryngeal nerve being interposed; higher up, with the larynx and pharynx. On its *outer side* are placed the internal jugular vein and pneumogastric nerve.

At the lower part of the neck, the internal jugular vein on the right side recedes from the artery, but on the left side it approaches it, and often crosses its lower part. This arises from the circumstance of the veins on both sides having to pass towards the right side of the thorax. This is an important fact to bear in mind during the performance of any operation on the lower part of the left common carotid artery.

Plan of the Relations of the Common Carotid Artery

In front.

Integument and fasciae.
Platysma.
Sterno-mastoid.
Sterno-hyoid.
Sterno-thyroid.
Omo-hyoid.
Descendens noni nerve.
Sterno-mastoid artery.
Superior and middle thyroid veins.
Anterior jugular vein.

Externally.
Internal jugular vein.
Pneumogastric nerve.

Common Carotid.

Internally.
Trachea.
Thyroid gland.
Recurrent laryngeal nerve.
Inferior thyroid artery.
Larynx.
Pharynx.

Behind.

Longus colli.
Rectus anticus major.
Sympathetic nerve.
Inferior thyroid artery.
Recurrent laryngeal nerve.

Peculiarities as to Origin. The *right common carotid* may arise above or below its usual point (the upper border of the sterno-clavicular articulation). This variation occurs in one out of about eight cases and a half, and is more frequently above than below the point stated; or its origin may be transferred to the arch of the aorta, or it may arise in conjunction with the left carotid. The *left common carotid* varies more frequently in its origin than the right. In the majority of cases it arises with the innominate artery, or where the innominate artery was absent, the two carotids arose usually by a single trunk. This vessel has a tendency towards the right side of the arch, occasionally being the first branch given off from the transverse portion. It rarely joins with the left subclavian, except in cases of transposition of the viscera.

Point of Division. The most important peculiarities of this vessel, in a surgical point of view, relate to its place of division in the neck. In the majority of cases, this occurs higher than usual, the artery dividing into two branches opposite the hyoid bone, or even higher; more rarely, it occurs below its usual place, opposite the middle of the larynx, or the lower border of the cricoid cartilage; and one case is related by Morgagni, where this vessel, only an inch and a half in length, divided at the root of the neck. Very rarely, the common carotid ascends in the neck without any subdivision, the internal carotid being wanting; and in two cases the common carotid has been found to be absent, the external and internal carotids arising directly from the arch of the aorta. This peculiarity existed on both sides in one subject, on one side in another.

Occasional Branches. The common carotid usually gives off no branches, but it occasionally gives origin to the superior thyroid, or a laryngeal branch, the inferior thyroid, or, more rarely, the vertebral artery.

Surgical Anatomy. The operation of tying the common carotid artery may be necessary in a wound of that vessel or its branches, in an aneurism, or in a case of pulsating tumour of the orbit or skull. If the wound involves the trunk of the common carotid itself, it will be necessary to tie the artery above and below the wounded part. If, however, one of the branches of that vessel is wounded, or has an aneurismal tumour connected with it, a ligature may be applied to any part of it, excepting its origin and termination. When the case is such as to allow of a choice being made, the lower part of the carotid should never be selected as the spot upon which a ligature should be placed, for not only is the artery in this situation placed very deeply in the neck, but it is covered by three layers of muscles, and on the left side the jugular vein, in the great majority of cases, passes obliquely over its front surface. Neither should the upper end be selected, for here the superior thyroid veins would give rise to very considerable difficulty in the application of a ligature. The point most favourable for the operation is opposite the lower part of the larynx, and here a ligature may be applied on the vessel, either above or below the point where it is crossed by the Omo-hyoid muscle. In the former situation the artery is most accessible, and it may be tied there in cases of wounds, or aneurism of any of the large branches of the carotid; whilst in cases of aneurism of the upper part of the carotid, that part of the vessel may be selected which is below the Omo-hyoid. It occasionally happens that the carotid artery bifurcates below its usual position: if the artery be exposed at its point of bifurcation, both divisions of the vessel should be tied near their origin, in preference to tying the trunk of the artery near its termination; and if, in consequence of the entire absence of the common carotid, or from its early division, two arteries, the external and internal carotids, are met with, the ligature should be placed on that vessel which is found on compression to be connected with the disease.

In this operation, the direction of the vessel and the inner margin of the Sterno-mastoid are the chief guides to its performance.

To tie the Common Carotid above the Omo-hyoid. The patient should be placed on his back with the head thrown back; an incision is to be made, three inches long, in the direction of the anterior border of the Sterno-mastoid, from a little below the angle of the jaw to a level with the cricoid cartilage: after dividing the integument, Platysma, and superficial fascia, the deep fascia must be cut through on a director, so as to avoid wounding numerous small veins that are usually found beneath. The head may now be brought forwards so as to relax the parts somewhat, and the margins of the wound must be held asunder by copper spatulae. The descendens noni nerve is now exposed, and must be avoided, and the sheath of the vessel having been raised by forceps, is to be opened over the artery to a small extent. The internal jugular vein will now present itself alternately distended and relaxed; this should be compressed both above and below, and drawn outwards, in order to facilitate the operation. The aneurism needle is now passed from the outside, care being taken to keep the needle in close contact with the artery, and thus avoid the risk of injuring the jugular vein, or including the vagus nerve. Before the ligature is secured, it should be ascertained that nothing but the artery is included in it.

To tie the Common Carotid, below the Omo-hyoid. The patient should be placed in the same situation as before. An incision about three inches in length is to be made parallel to the inner edge of the Sterno-mastoid, commencing on a level with the cricoid cartilage. The inner border of the Sterno-mastoid having been exposed, the sterno-mastoid artery and a large vein, the Middle thyroid will be seen, and must be carefully avoided; the Sterno-mastoid is to be turned outwards, and the Sterno-hyoid and thyroid muscles inwards. The deep fascia must now be divided below the Omo-hyoid muscle, and the sheath having been exposed, must be opened, care being taken to avoid the descendens noni, which here runs on the inner or tracheal side. The jugular vein and vagus nerve being then pressed to the outer side, the needle must be passed around the artery from without inwards, great care being taken to avoid the inferior thyroid artery, the recurrent laryngeal, and sympathetic nerves which lie behind it.

EXTERNAL CAROTID ARTERY

The external carotid artery (fig. 189, color illustration p. A2), arises opposite the upper border of the thyroid cartilage, and taking a slightly curved course, ascends upwards and forwards, and then inclines backwards, to the space between the neck of the condyle of the lower jaw, and the meatus auditorius, where it divides into the temporal and internal maxillary arteries. It rapidly diminishes in size as it ascends the neck, owing to the number and large size of the branches given off from it. In the child, it is somewhat smaller than the internal carotid; but in the adult, the two vessels are of nearly equal size. At its commencement, this artery is more superficial, and placed nearer the middle line than the internal carotid, and is contained in the triangular space bounded by the Sterno-mastoid behind, the Omo-hyoid below, and the posterior belly of the Digastric and Stylo-hyoid above; it is covered by the skin, platysma, deep fascia, and anterior margin of the Sterno-mastoid, crossed by the hypoglossal nerve, and by the lingual and facial veins; it is afterwards crossed by the Digastric and Stylo-hyoid muscles, and higher up passes deeply into the substance of the parotid gland, where it lies beneath the facial nerve, and by the junction of the temporal and internal maxillary veins.

Internally is the hyoid bone, the wall of the pharynx, and the ramus of the jaw, from which it is separated by a portion of the parotid gland.

Behind it, near its origin, is the superior laryngeal nerve; and, higher up, it is separated from the internal carotid by the Stylo-glossus and Stylo-pharyngeus muscles, the glosso-pharyngeal nerve, and part of the parotid gland.

PLAN OF THE RELATIONS OF THE EXTERNAL CAROTID

In front.		*Behind.*
Integument, platysma.		Superior Laryngeal nerve.
Superficial and deep fasciae.	External	Stylo-glossus.
Hypoglossal nerve.	Carotid.	Stylo-pharyngeus.
Lingual and facial veins.		Glosso-pharyngeal nerve.
Digastric and Stylo-hyoid muscles.		Parotid gland.
Facial nerve and parotid gland.		
Temporal and maxillary veins.		

Internally.
Hyoid bone.
Pharynx.
Parotid gland.
Ramus of jaw.

Surgical Anatomy. The application of a ligature to the external carotid may be required in cases of wounds of this vessel, or of its branches when these cannot be tied; this, however, is an operation very rarely performed, ligature of the common carotid being preferable, on account of the number of branches given off from the external. To tie this vessel near its origin, below the point where it is crossed by the Digastric, an incision about three inches in length should be made along the margin of the Sterno-mastoid, from the angle of the jaw to the cricoid cartilage, as in the operation for tying the common carotid. To tie the vessel above the Digastric, between it and the parotid gland, an incision should be made from the lobe of the ear to the great cornu of the Os-hyoides, dividing successively the skin, platysma, and fascia. By separating the posterior belly of the Digastric and Stylo-hyoid muscles which are seen at the lower part of the wound, from the parotid gland, the vessel will be exposed, and a ligature may be applied to it.

Branches. The external carotid artery gives off eight branches, which, for convenience of description, may be divided into four sets. (See fig. 190, Plan of the Branches.)

Anterior.	*Posterior.*	*Ascending.*	*Terminal.*
Superior thyroid.	Occipital.	Ascending	Temporal.
Lingual.	Posterior auricular.	pharyngeal.	Internal maxillary.
Facial.			

The student is here reminded that many variations are met with in the number, origin, and course of these branches in different subjects; but the above arrangement is that which is found in the great majority of cases.

The SUPERIOR THYROID ARTERY (figs. 189 and 194), is the first branch given off from the external carotid, being derived from that vessel just below the greater cornu of the hyoid bone. At its commencement, it is quite superficial, being covered by the integument, fasciae, and platysma, and is contained in the triangular space bounded by the Sterno-mastoid, Digastric, and Omo-hyoid muscles. After ascending upwards and inwards for a short distance, it curves downwards and forwards in an arched and tortuous manner to the upper part of the thyroid gland, passing beneath the Omo-hyoid, Sterno-hyoid, and Sterno-thyroid muscles; and distributes numerous branches to its anterior surface, anastomosing with its fellow of the opposite side, and with the inferior thyroid arteries. Besides the arteries distributed to the muscles and substance of the gland, its branches are the following.

> Hyoid.
> Superficial descending branch.
> Superior laryngeal.
> Crico-thyroid.

The *Hyoid* is a small branch which runs along the lower border of the os-hyoides, beneath the Thyro-hyoid muscle; and, after supplying the muscles connected to that bone, forms an arch, by anastomosing with the vessel of the opposite side.

The *Superficial Descending Branch* runs downwards and outwards across the sheath of the common carotid artery, and supplies the Sterno-mastoid and neighbouring muscles and integument. It is of importance that the situation of this vessel be remembered, in the operation for tying the common carotid artery.

The *Superior Laryngeal*, larger than either of the preceding, accompanies the superior laryngeal nerve, beneath the Thyro-hyoid muscle; it pierces the thyro-hyoidean membrane, and supplies the muscles, mucous membrane, and glands of the larynx and epiglottis, anastomosing with the branch from the opposite side.

The *Crico-thyroid* (Inferior laryngeal) is a small branch which runs transversely across the crico-thyroid membrane, communicating with the artery of the opposite side. The position of this vessel should be remembered, as it may prove the source of troublesome haemorrhage during the operation of laryngotomy.

Surgical Anatomy. The superior thyroid, or some of its branches, are occasionally divided in cases of cut throat, giving rise to considerable haemorrhage. In such cases, the artery should be secured, the wound being enlarged for that purpose, if necessary. The operation may be easily performed, the position of the artery being very superficial, and the only structures of importance covering it, being a few small veins. The operation of tying the superior thyroid artery, in bronchocele, has been performed in numerous instances with partial or temporary success. When, however, the collateral circulation between this vessel with the artery of the opposite side, and with the inferior thyroid is completely re-established, the tumour usually regains its former size.

The LINGUAL ARTERY (fig. 194) arises from the external carotid between the superior thyroid and facial; it runs obliquely upwards and inwards to the greater cornu of the hyoid bone, then passes horizontally forwards parallel with the great cornu, and, ascending perpendicularly to the under surface of the tongue, turns forwards on its under surface as far as the tip of that organ, under the name of the ranine artery.

Relations. Its first, or oblique portion, is superficial, being contained in the triangular inter-muscular space already described, resting upon the Middle constrictor muscle of the pharynx, and covered in by the Platysma and fascia of the neck. Its second, or horizontal portion, also lies upon the Middle constrictor, being covered at first by the tendon of the Digastric, and the Stylo-hyoid muscle, and afterwards by the Hyo-glossus, the latter muscle separating it from the hypoglossal nerve. Its third, or ascending portion, lies between the Hyo-glossus and Genio-hyo-glossus muscles. The fourth, or terminal part, under the name of the ranine, runs along the under surface of the tongue to its tip, it is very superficial, being covered only by the mucous membrane, and rests on the Lingualis on the outer side of the Genio-hyo-glossus. The hypoglossal nerve lies nearly parallel with the lingual artery, separated from it, in the second part of its course, by the Hyo-glossus muscle.

The branches of the lingual artery are the

        Hyoid.   Sublingual.
        Dorsalis Linguae.   Ranine.

The *Hyoid* branch runs along the upper border of the hyoid bone, supplying the muscles attached to it, and anastomosing with its fellow of the opposite side.

The *Dorsalis Linguae* (fig. 194) arises from the lingual artery beneath the Hyo-glossus muscle; ascending to the dorsum of the tongue, it supplies its mucous membrane, the tonsil, soft palate, and epiglottis; anastomosing with its fellow from the opposite side.

The *Sublingual*, a branch of bifurcation of the lingual artery, arises at the anterior margin of the Hyo-glossus muscle, and running forwards and outwards beneath the Mylo-hyoid to the sublingual gland, supplies its substance, giving branches to the Mylo-hyoid and neighbouring muscles, the mucous membrane of the mouth and gums.

The *Ranine* may be regarded as the continuation of the lingual artery; it runs along the under surface of the tongue, resting on the Lingualis, and covered by the mucous membrane of the mouth; it lies on the outer side of the Genio-hyo-glossus, and is covered in by the Hyo-glossus and Stylo-glossus, accompanied by the gustatory nerve. On arriving at the tip of the tongue, it anastomoses with the artery of the opposite side. These vessels in the mouth are placed one on each side of the fraenum.

Surgical Anatomy. The lingual artery may be divided near its origin in cases of cut throat, a complication that not unfrequently happens in this class of wounds, or severe haemorrhage which cannot be restrained by ordinary means may ensue from a wound, or deep ulcer of the tongue. In the former case, the primary wound may be enlarged if necessary, and the bleeding vessel at once secured. In the latter case, it has been suggested that the lingual artery should be tied near its origin. If the student, however, will observe the depth at which this vessel is placed from the surface, the number of important parts which surround it on every side, and its occasional irregularity of origin, the great difficulty of such an operation will be apparent; under such circumstances, it is more advisable that the external or common carotid should be tied.

Troublesome haemorrhage may occur in the division of the fraenum in children, if the ranine artery, which lies on each side of it, is cut through. The student should remember that the operation is always to be performed with a pair of blunt pointed scissors, which should be so held as to divide the part in the direction downwards and backwards; the ranine artery and veins are then avoided.

The FACIAL ARTERY (fig. 191) arises a little above the lingual, and ascends obliquely forwards and upwards, beneath the body of the lower jaw, to the submaxillary gland, in which it is imbedded; this may be called the cervical part of the artery. It then curves upwards over the body of the jaw at the anterior inferior angle of the Masseter muscle, ascends forwards and upwards across the cheek to the angle of the mouth, passes up along the side of the nose, and terminates at the inner canthus of the eye, under the name of the angular artery. This vessel, both in the neck, and on the face, is remarkably tortuous; in the former situation, to accommodate itself to the movements of the pharynx in deglutition; and in the latter, to the movements of the jaw, and of the lips and cheeks.

Relations. *In the neck* its origin is superficial, being covered by the integument, Platysma, and fascia; it then passes beneath the Digastric and Stylo-hyoid muscles, and the submaxillary gland. *On the face*, where passing over the body of the lower jaw, it is comparatively superficial, being covered by the Platysma. In this situation its pulsation may be distinctly felt, and compression of the vessel effectually made against the bone. In its course over the face, it is covered by the integument, the fat of the cheek, and, near the angle of the mouth, by the Platysma and Zygomatic muscles. It rests on the Buccinator, the Levator anguli oris, and the Levator labii superioris alaeque nasi. It is accompanied by the facial vein throughout its entire course; the vein is not tortuous like the artery, and, on the face, is separated from that vessel by a considerable interval. The branches of the facial nerve cross this vessel, and the infraorbital nerve lies beneath it.

The branches of this vessel may be divided into two sets, those given off below the jaw (cervical), and those on the face (facial).

Cervical Branches.	*Facial Branches.*
Inferior or Ascending Palatine.	Muscular.
Tonsillitic.	Inferior Labial.
Submaxillary.	Inferior Coronary.
Submental.	Superior Coronary.
	Lateralis Nasi.
	Angular.

The *Inferior* or *Ascending Palatine* (fig. 194) ascends between the Stylo-glossus and Stylo-pharyngeus to the outer side of the pharynx. After supplying these muscles,

191.—The Arteries of the Face and Scalp.

the tonsil, and Eustachian tube, it divides near the Levator palati into two branches, one follows the course of the Tensor palati, supplies the soft palate and the palatine glands; the other passes to the tonsil, which it supplies; anastomosing with the tonsillitic artery. These vessels inosculate with the posterior palatine branch of the internal maxillary artery.

The *Tonsillitic* branch (fig. 194) passes up along the side of the pharynx, and perforating the Superior constrictor, ramifies in the substance of the tonsil and root of the tongue.

The *Submaxillary* consists of three or four large branches, which supply the submaxillary gland, some being prolonged to the neighbouring muscles, lymphatic glands, and integument.

The *Submental*, the largest of the cervical branches, is given off from the facial artery just as that vessel quits the submaxillary gland; it runs forwards upon the Mylo-hyoid muscle, just below the body of the jaw, and beneath the Digastric; after supplying the muscles attached to the jaw, and anastomosing with the sublingual artery, it arrives at the symphysis of the chin, where it divides into a superficial and a deep branch; the former turns round the chin, and passing between the integument and Depressor labii inferioris, supplies both, and anastomoses with the inferior Labial. The deep branch passes between that muscle and the bone, supplies the lip; anastomosing with the inferior labial and mental arteries.

The muscular branches are distributed to the internal Pterygoid, Masseter, and Buccinator.

The *Inferior Labial* passes beneath the Depressor anguli oris, to supply the muscles and integument of the lower lip, anastomosing with the inferior coronary and submental branches of the facial, and with the mental branch of the inferior dental artery.

The *Inferior Coronary*, is derived from the facial artery near the angle of the mouth; it passes upwards and inwards beneath the Depressor anguli oris, and, penetrating the Orbicularis muscle, runs in a tortuous course along the edge of the lower lip between this muscle and the mucous membrane, inosculating with the artery of the opposite side. This artery supplies the labial glands, the mucous membrane, and muscles of the lower lip; and anastomoses with the inferior labial, and mental branch of the inferior dental artery.

The *Superior Coronary* is larger, and more tortuous in its course than the preceding. It follows the same course along the edge of the upper lip, lying between the mucous membrane and the Orbicularis, and anastomoses with the artery of the opposite side. It supplies the textures of the upper lip, and gives off in its course two or three vessels which ascend to the nose. One, named the artery of the septum, ramifies on the septum of the nares as far as the point of the nose; another supplies the ala of the nose.

The *Lateralis Nasi* is derived from the facial, as this vessel is ascending along the side of the nose; it supplies the ala and dorsum of the nose, anastomosing with its fellow, the nasal branch of the ophthalmic, the artery of the septum, and the infra-orbital.

The *Angular Artery* is the termination of the trunk of the facial; it ascends to the inner angle of the orbit, accompanied by a large vein, the angular; it distributes some branches on the cheek which anastomose with the infra-orbital, and after supplying the lachrymal sac, and Orbicularis muscle, terminates by anastomosing with the nasal branch of the ophthalmic artery.

The *anastomoses* of the facial artery are very numerous, not only with the vessel of the opposite side, but with other vessels from different sources; viz., with the sub-lingual branch of the lingual, with the mental branch of the inferior dental as it emerges from the dental foramen, with the ascending pharyngeal and posterior palatine, and with the ophthalmic, a branch of the internal carotid; it also inosculates with the transverse facial, and with the infra-orbital.

Peculiarities. The facial artery not unfrequently arises by a common trunk with the lingual. This vessel also is subject to some variations in its size, and in the extent to which it supplies the face. It occasionally terminates as the submental, and not unfrequently supplies the face only as high as the angle of the mouth or nose. The deficiency is then supplied by enlargement of one of the neighbouring arteries.

Surgical Anatomy. The passage of the facial artery over the body of the jaw would appear to afford a favourable position for the application of pressure in cases of haemorrhage from the lips, the result either of an accidental wound, or from an operation; but its application is useless, on account of the free communication of this vessel with its fellow, and with numerous branches from different sources. In a wound involving the lip, it is better to seize the part between the fingers and evert it, when the bleeding vessel may be at once secured with a tenaculum. In order to prevent haemorrhage in cases of excision, or in the removal of diseased growths from the part, the lip should be compressed on each side between the finger and thumb, whilst the surgeon excises the diseased part. In order to stop haemorrhage where the lip has been divided in an operation, it is necessary in uniting the edges of the wound, to pass the sutures through the cut edges, almost as deep as its mucous surface; by these means, not only are the cut surfaces more neatly adapted to each other, but the possibility of haemorrhage is prevented by including in the suture the divided artery. If the suture is, on the contrary, passed through merely the cutaneous portion of the wound, haemorrhage occurs into the cavity of the mouth. The student should, lastly, observe the relation of the angular artery to the lachrymal sac, and it will be seen that, as the vessel passes up along the inner margin of the orbit, it ascends on its nasal side. In operating for fistula lachrymalis, the sac should always be opened on its outer side, in order that this vessel should be avoided.

The OCCIPITAL ARTERY arises from the posterior part of the external carotid, opposite the facial, near the lower margin of the Digastric muscle. At its origin, it is covered by the posterior belly of the Digastric and Stylo-hyoid muscles, and part of the parotid gland, the hypo-glossal nerve winding around it from behind forwards; higher up, it passes across the internal carotid artery, the internal jugular vein, and the pneumogastric and spinal accessory nerves; it then ascends to the interval between the transverse process of the atlas, and the mastoid process of the temporal bone, passes horizontally backwards, grooving the surface of the latter bone, being covered by the Sterno-mastoid, Splenius, Digastric, and Trachelo-mastoid muscles, resting upon the Complexus, Superior oblique, and Rectus posticus major muscles; it then passes vertically upwards, piercing the cranial attachment of the Trapezius, and ascends in a tortuous course on the occiput as high as the vertex, where it divides into numerous branches.

The branches given off from this vessel are,

Muscular.   Inferior meningeal.
Auricular.   Arteria princeps cervicis.

The *Muscular Branches* supply the Digastric, Stylo-hyoid, Sterno-mastoid, Splenius, and Trachelo-mastoid muscles. The branch distributed to the Sterno-mastoid is of a large size.

The *Auricular Branch* supplies the back part of the concha.

The *Meningeal Branch*, ascends with the internal jugular vein, and enters the skull through the foramen lacerum posterius, to supply the dura mater in the posterior fossa.

The *Arteria Princeps Cervicis* (fig. 194), is a large vessel which descends along the back part of the neck, and divides into a superficial and deep branch. The former runs beneath the Splenius, giving off branches which perforate that muscle to supply the Trapezius, anastomosing with the superficial cervical artery; the latter passes beneath the Complexus, between it and the Semi-spinalis colli, and anastomoses with the vertebral, and deep cervical branch of the superior intercostal. The anastomosis between this vessel and the profunda cervicis serves mainly to establish the collateral circulation after ligature of the carotid or subclavian artery.

The cranial branches of the occipital artery are distributed upon the occiput;

they are very tortuous, and lie between the integument and Occipito-frontalis, anastomosing with their fellows of the opposite side, the posterior auricular, and temporal arteries. They supply the posterior part of the Occipito-frontalis muscle, the integument, pericranium, and one or two branches occasionally pass through the parietal or mastoid foramina, to supply the dura mater.

The POSTERIOR AURICULAR ARTERY (fig. 191) is a small vessel, which arises from the external carotid, above the Digastric and Stylo-hyoid muscles, opposite the apex of the styloid process. It ascends, under cover of the parotid gland, to the groove between the cartilage of the ear and the mastoid process, immediately above which it divides into two branches, an anterior, which passes forwards to anastomose with the posterior division of the temporal, and a posterior, which communicates with the occipital. Just before arriving at the mastoid process, this artery is crossed by the portio dura, and has beneath it the spinal accessory nerve.

Besides several small branches to the Digastric, Stylo-hyoid, and Sterno-mastoid muscles, and to the parotid gland, this vessel gives off two branches.

    Stylo-mastoid.   Auricular.

The *Stylo-mastoid Branch* enters the stylo-mastoid foramen, and supplies the tympanum, mastoid cells, and semi-circular canals. In the young subject, a branch from this vessel forms, with the tympanic branch from the internal maxillary, a delicate vascular circle, which surrounds the auditory meatus, and from which delicate vessels ramify on the membrana tympani.

The *Auricular Branch* is distributed to the posterior part of the cartilage of the ear, upon which it minutely ramifies, some branches curving round its margin, others perforating the fibre-cartilage, to supply its anterior surface.

The ASCENDING PHARYNGEAL ARTERY (fig. 194), the smallest branch of the external carotid, is a long slender vessel, deeply seated in the neck, beneath the other branches of the external carotid and Stylo-pharyngeus muscle. It arises from the posterior part of the external carotid, and ascends the neck to the under surface of the base of the skull. It lies upon the Rectus capitis anticus major, between the internal carotid, and the outer wall of the pharynx. Its branches may be subdivided into three sets: 1. Those directed outwards to supply muscles and nerves. 2. Those directed inwards to the pharynx. 3. Meningeal branches.

    External.   Pharyngeal.
        Meningeal.

The *External Branches* are numerous small vessels, which supply the Recti antici muscles, the sympathetic, lingual, and pneumogastric nerves, and the lymphatic glands of the neck, anastomosing with the ascending cervical branch of the subclavian artery.

The *Pharyngeal Branches* are three or four in number. Two of these descend to supply the Middle and Inferior constrictors and the Stylo-pharyngeus, ramifying in their substance and in the mucous membrane lining them. The largest of the pharyngeal branches passes inwards, running upon the Superior constrictor, and sending ramifications to the soft palate, Eustachian tube, and tonsil, taking the place of the ascending palatine branch of the facial artery, when that vessel is of small size.

The *Meningeal Branches* consist of several small vessels, which pass through foramina in the base of the skull, to supply the dura mater. One, the posterior meningeal, enters the cranium through the foramen lacerum posterius with the internal jugular vein. A second passes through the foramen lacerum basis cranii; and occasionally a third through the anterior condyloid foramen. They are all distributed to the dura mater.

The Temporal Artery (fig. 191), the smaller of the two terminal branches of the external carotid, appears, from its direction, to be the continuation of that vessel. It commences in the substance of the parotid gland, in the interspace between the neck of the condyle of the lower jaw and the external meatus; crossing over the root of the Zygoma, immediately beneath the integument, it divides about two inches above the zygomatic arch into two branches, an anterior and a posterior.

The *Anterior Temporal* inclines forwards over the forehead, supplying the muscles, integument, and pericranium in this region, and anastomoses with the supra-orbital and frontal arteries, its branches being directed from before backwards.

The *Posterior Temporal*, larger than the anterior, curves upwards and backwards along the side of the head, lying above the temporal fascia, and inosculates with its fellow of the opposite side, and with the posterior auricular and occipital arteries.

The temporal artery, as it crosses the zygoma, is covered by the Attrahens aurem muscle, and by a dense fascia given off from the parotid gland; it is also usually crossed by one or two veins, and accompanied by branches of the facial nerve. Besides some twigs to the parotid gland, the articulation of the jaw, and to the Masseter muscle, its branches are the

Transverse facial.	Middle temporal.
Anterior auricular.	

The *Transverse Facial* is given off from the temporal before that vessel quits the parotid gland; running forwards through its substance, it passes transversely across the face, between Stenon's duct and the lower border of the zygoma, and divides on the side of the face into numerous branches, which supply the parotid gland, the Masseter muscle, and the integument, anastomosing with the facial and infra-orbital arteries. This vessel rests on the Masseter, and is accompanied by one or two branches of the facial nerve.

The *Middle Temporal Artery* arises immediately above the zygomatic arch, and perforating the temporal fascia, supplies the Temporal muscle, anastomosing with the deep temporal branches of the internal maxillary. It occasionally gives off an orbital branch, which runs along the upper border of the zygoma, between the two layers of the temporal fascia, to the outer angle of the orbit; it supplies the Orbicularis, and anastomoses with the lachrymal and palpebral branches of the ophthalmic artery.

The *Anterior Auricular Branches* are distributed to the anterior portion of the pinna, the lobule, and part of the external meatus, anastomosing with branches of the posterior auricular.

Surgical Anatomy. It occasionally happens that the surgeon is called upon to perform the operation of arteriotomy upon this vessel in cases of inflammation of the eye or brain. Under these circumstances, the anterior branch is the one usually selected. If the student will consider the relations of the trunk of this vessel with the surrounding structures, as it crosses the zygomatic arch, he will observe that it is covered by a thick and dense fascia, crossed by one or two veins, and accompanied by branches of the facial and temporo-auricular nerves. Bleeding should not be performed in this situation, as considerable difficulty may arise from the dense fascia covering this vessel preventing a free flow of blood, and considerable pressure is requisite afterwards to repress it. Again, a varicose aneurism may be formed by the accidental opening of one of the veins covering it; or severe neuralgic pain may arise from the operation implicating one of the nervous filaments which accompany the artery.

The anterior branch, is, on the contrary, subcutaneous, is a large vessel, and as readily compressed as any other portion of the artery; it should consequently always be selected for the operation.

The INTERNAL MAXILLARY, the larger of the two terminal branches of the external carotid, passes inwards, at right angles from that vessel, behind the neck of the lower jaw, to supply the deep structures of the face. At its origin, it is imbedded in the substance of the parotid gland, being on a level with the lower extremity of the lobe of the ear.

192.—The Internal Maxillary Artery, and its Branches.

FIG. 193. *Plan of the Branches*

In the first part of its course (maxillary portion), the artery passes horizontally forwards and inwards, between the ramus of the jaw, and the internal lateral ligament. The artery here lies parallel with the auricular nerve; it crosses the inferior dental nerve, and lies beneath the narrow portion of the External pterygoid muscle.

In the second part of its course (pterygoid portion), it ascends obliquely forwards and upwards upon the outer surface of the External pterygoid muscle, being

In the third part of its course (spheno-maxillary portion), it approaches the superior maxillary bone, crosses the interval between the two heads of the External pterygoid, and enters the spheno-maxillary fossa, where it lies in relation with Meckel's ganglion, and gives off its terminal branches.

Peculiarities. Occasionally, this artery passes between the two Pterygoid muscles. The vessel in this case passes forwards to the interval between the two heads of the External pterygoid, in order to reach the maxillary bone. Sometimes, the vessel escapes from beneath the External pterygoid by perforating the middle of this muscle.

The branches of this vessel may be divided into three groups, corresponding with its three divisions.

Branches from the Maxillary Portion

Tympanic.   Small meningeal.
Middle meningeal.   Inferior dental.

The *Tympanic Branch* passes upwards behind the articulation of the lower jaw, enters the tympanum through the fissura Glaseri, supplies the Laxator tympani, and ramifies upon the membrana tympani, anastomosing with the stylo-mastoid and Vidian arteries.

The *Middle Meningeal* is the largest of the branches which supply the dura mater. It arises from the internal maxillary between the internal lateral ligament, and the neck of the jaw, and ascends vertically upwards to the foramen spinosum in the spinous process of the sphenoid bone. On entering the cranium, it divides into two branches, an anterior, and a posterior. The anterior branch, the larger, crosses the great ala of the sphenoid, and reaches the groove, or canal, in the anterior inferior angle of the parietal bone; it then divides into branches which spread out between the dura mater and internal surface of the cranium, some passing upwards over the parietal bone as far as the vertex, and others backwards to the occipital bone. The posterior branch crosses the squamous portion of the temporal, and on the inner surface of the parietal bone divides into branches which supply the posterior part of the dura mater and cranium. The branches of this vessel are distributed to the dura mater, but chiefly to the bones; they anastomose with the arteries of the opposite side, and with the anterior and posterior meningeal.

The middle meningeal, on entering the cranium, gives off the following collateral branches: 1. Numerous small vessels to the ganglion of the fifth nerve, and to the dura mater in this situation. 2. A branch to the facial nerve, which enters the hiatus Fallopii, supplies the facial nerve, and anastomoses with the stylo-mastoid branch of the occipital artery. 3. Orbital branches, which pass through the sphenoidal fissure, or through separate canals in the great wing of the sphenoid, to anastomose with the lachrymal or other branches of the ophthalmic artery. 4. Temporal branches, which pass through foramina in the great wing of the sphenoid, and anastomose in the temporal fossa with the deep temporal arteries.

The *Small Meningeal* is sometimes derived from the preceding. It enters the skull through the foramen ovale, and supplies the Casserian ganglion and dura mater. Before entering the cranium, it gives off a branch to the nasal fossa and soft palate.

The *Inferior Dental* descends with the dental nerve, to the foramen on the inner side of the ramus of the jaw. It runs along the dental canal in the substance of the bone, accompanied by the nerve, and opposite the bicuspid tooth divides into two branches, an incisor, which is continued forwards beneath the incisor teeth as far as the symphysis, where it anastomoses with the artery of the opposite side; the other, the mental, escapes with the nerve at the mental foramen, supplies the structures composing the chin, and anastomoses with the submental, inferior labial, and inferior coronary arteries. As the dental artery enters the foramen, it gives off a mylo-hyoid branch, which runs in the mylo-hyoid groove, and ramifies on the under surface of the Mylo-hyoid muscle. The dental and incisor arteries during their course through the substance of the bone, give off a few twigs which are lost in the diploë, and a series of branches which correspond in number to the roots of the teeth; these enter the minute apertures at the extremities of the fangs, and ascend to supply the pulp of the teeth.

Branches of the Second, or Pterygoid Portion

Deep temporal.	Masseteric.
Pterygoid.	Buccal.

These branches are distributed, as their names imply, to the muscles in the maxillary region.

The *Deep Temporal Branches*, two in number, anterior, and posterior, each occupy that part of the temporal fossa indicated by its name. Ascending between the Temporal muscle and pericranium, they supply that muscle, and anastomose with the other temporal arteries. The anterior branch communicating with the lachrymal through small branches which perforate the malar bone.

The *Pterygoid Branches*, irregular in their number and origin, supply the Pterygoid muscles.

The *Masseteric* is a small branch which passes outwards above the sigmoid notch of the lower jaw, to the deep surface of the Masseter. It supplies that muscle, and anastomoses with the masseteric branches of the facial and transverse facial arteries.

The *Buccal* is a small branch which runs obliquely forwards between the Internal pterygoid, and the ramus of the jaw, to the outer surface of the Buccinator, to which it is distributed, anastomosing with branches of the facial artery.

Branches of the Third, or Spheno-maxillary Portion

Alveolar.	Vidian.
Infra-orbital.	Pterygo-palatine.
Posterior or Descending palatine.	Nasal or Spheno-palatine.

The *Alveolar* is given off from the internal maxillary by a common branch with the infra-orbital, and just as the trunk of the vessel is passing into the spheno-maxillary fossa. Descending upon the tuberosity of the superior maxillary bone, it divides into numerous branches; one, the superior dental, larger than the rest, supplies the molar and bicuspid teeth; its branches entering the foramina in the alveolar process, some branches pierce the bone to supply the lining of the antrum, and others are continued forwards on the alveolar process to supply the gums.

The *Infra-orbital* appears, from its direction, to be the continuation of the trunk of the internal maxillary. It arises from that vessel by a common trunk with the preceding branch, and runs along the infra-orbital canal with the superior maxillary nerve, emerging upon the face at the infra-orbital foramen, beneath the Levator labii superioris. Whilst contained in the canal, it gives off branches which ascend into the orbit, and supply the Inferior rectus, and Inferior oblique muscles, and the lachrymal gland. Other branches descend through canals in the bone, to supply the mucous membrane of the antrum, and the front teeth of the upper jaw. On the face, it supplies the lachrymal sac, and inner angle of the orbit, anastomosing with the facial and nasal branch of the ophthalmic arteries; and other branches descend beneath the elevator of the upper lip, and anastomose with the transverse facial and buccal branches.

The four remaining branches arise from that portion of the internal maxillary which is contained in the spheno-maxillary fossa.

The *Descending Palatine* passes down along the posterior palatine canal with the posterior palatine branches of Meckel's ganglion, and emerging from the posterior palatine foramen, runs forwards in a groove on the inner side of the alveolar border of the hard palate, to be distributed to the gums, the mucous membrane of the hard palate, and palatine glands. Whilst it is contained in the palatine canal, it gives off branches, which descend in the accessory palatine canals to supply the soft palate, anastomosing with the ascending palatine artery; and anteriorly it terminates in a small vessel, which ascends in the anterior palatine canal, and anastomoses with the artery of the septum, a branch of the spheno-palatine.

The *Vidian Branch* passes backwards along the Vidian canal with the Vidian nerve. It is distributed to the upper part of the pharynx and Eustachian tube, sending a small branch into the tympanum.

The *Pterygo-Palatine* is also a very small branch, which passes backwards through the pterygo-palatine canal with the pharyngeal nerve, and is distributed to the upper part of the pharynx and Eustachian tube.

The *Nasal* or *Spheno-Palatine* passes through the spheno-palatine foramen into the cavity of the nose, at the back part of the superior meatus, and divides into two branches; one internal, the artery of the septum, passes obliquely downwards and forwards along the septum nasi, supplies the mucous membrane, and anastomoses in front with the ascending branch of the descending palatine. The external branches, two or three in number, supply the mucous membrane covering the lateral wall of the nares, the antrum, and the ethmoid and sphenoid cells.

Surgical Anatomy of the Triangles of the Neck

The student having studied the relative anatomy of the large arteries of the neck and their branches, and the relations they bear to the veins and nerves, should now examine these structures collectively, as they present themselves in certain regions of the neck, in each of which important operations are being constantly performed.

For this purpose, the Sterno-mastoid, or any other muscles that have been divided in the dissection of these vessels, should be replaced in their normal position, the head should be supported by placing a block at the back of the neck, and the face turned to the side opposite to that which is being examined.

The side of the neck presents a somewhat quadrilateral outline, limited, above, by the lower border of the body of the jaw, and an imaginary line extending from the angle of the jaw to the mastoid process; below, by the prominent upper border of the clavicle; in front, by the median line of the neck; behind, by the anterior margin of the Trapezius muscle. This space is subdivided into two large triangles by the Sterno-mastoid muscle, which passes obliquely across the neck, from the sternum and clavicle, below, to the mastoid process, above. The anterior margin of this muscle forms a prominent ridge beneath the skin, which serves as a guide to the surgeon in the operation for applying a ligature to the common carotid artery, or in oesophagotomy. The triangular space in front of this muscle is called the *anterior triangle*, and that behind it, the *posterior triangle*.

Anterior Triangular Space

The anterior triangle is limited, in front, by a line extending from the chin to the sternum; behind, by the anterior margin of the Sterno-mastoid; its base, directed upwards, is formed by the lower border of the body of the jaw, and a line extending from the angle of the jaw to the mastoid process; its apex is formed below by the sternum. This space is covered in by the integument, superficial fascia, Platysma, deep fascia, crossed by branches of the facial and superficial cervical nerves; and subdivided into three smaller triangles by the Digastric muscle, above, and the anterior belly of the Omo-hyoid, below. These are named, from below upwards, the inferior carotid triangle, the superior carotid triangle, and the submaxillary triangle.

Each of these spaces must now be separately examined.

The *Inferior Carotid Triangle* is limited, in front, by the median line of the neck; behind, by the anterior margin of the Sterno-mastoid; above, by the anterior belly of the Omo-hyoid; and it is covered in by the integument, Platysma, superficial and deep fasciae; ramifying between which, is seen the cutaneous descending branch of the superficial cervical nerve. Beneath these superficial structures, are the Sterno-hyoid and Sterno-thyroid muscles, which, together with the anterior margin of the Sterno-mastoid, conceal the lower part of the common carotid artery. This vessel is enclosed within its sheath, together with the internal jugular vein and pneumogastric nerve; the vein lying on the outer side of the artery on the right side of the neck, but overlapping it, or passing directly across it on the left side; the nerve lying between the artery and vein, on a plane posterior to both. In front of the sheath are a few filaments descending from the loop of communication between the descendens and communicans noni; behind the sheath is seen the inferior thyroid artery, the recurrent laryngeal and sympathetic nerves; and on its inner side, the trachea, the thyroid gland, much more prominent in the female than in the male, and the lower part of the larynx. In the upper part of this space, the common carotid artery may be tied below the Omo-hyoid muscle.

The *Superior Carotid Triangle* is bounded, behind, by the Sterno-mastoid; below, by the anterior belly of the Omo-hyoid; and above, by the posterior belly of the Digastric muscle. Its floor is formed by parts of the Thyro-hyoid, Hyo-glossus, and the inferior and middle Constrictor muscles of the pharynx; and it is covered in by the integument, Platysma, superficial and deep fasciae; ramifying between which, are branches of the facial and superficial cervical nerves. This space contains the upper part of the common carotid artery, which bifurcates opposite the upper border of the thyroid cartilage into the external and internal carotid. These vessels are concealed from view by the anterior margin of the Sterno-mastoid muscle, which overlaps them. The external and internal carotids lie side by side, the external being the most anterior of the two. The following branches of the external carotid are also met with in this space; the superior thyroid, which runs forwards and downwards; the lingual, which passes directly forwards; the facial, forwards and upwards; the occipital is directed backwards; and the ascending pharyngeal runs directly upwards on the inner side of the internal carotid. The veins met with are, the internal jugular which lies on the outer side of the common and internal carotid vessels, and veins corresponding to the above mentioned branches of the external carotid, viz., the superior and middle thyroid, the lingual, facial, ascending pharyngeal, and sometimes the occipital; all of which accompany their corresponding arteries, and terminate in the internal jugular. In front of the sheath of the common carotid is the descendens noni, the hypo-glossal, from which it is derived, crossing both carotids above, curving around the occipital artery at its origin. Within the sheath, between the artery and vein, and behind both, is the pneumogastric nerve; behind the sheath, the sympathetic. On the outer side of the vessels the spinal accessory nerve runs for a short distance before it pierces the Sterno-mastoid muscle; and on the inner side of the internal carotid, just below the hyoid bone, may be seen the superior laryngeal nerve; and still more inferiorly, the external laryngeal nerve. The upper part of the larynx and the pharynx, are also found in the front part of this space.

The *Submaxillary Triangle* corresponds to that part of the neck immediately beneath the body of the jaw. It is bounded above, by the lower border of the body of the jaw, the parotid gland, and mastoid process; behind, by the posterior belly of the Digastric and Stylo-hyoid muscles; in front, by the middle line of the neck. The floor of this space is formed by the anterior belly of the Digastric, the Mylo-hyoid, and Hyo-glossus muscles; and it is covered in by the integument, Platysma, superficial and deep fasciae; ramifying between which, are branches of the facial and ascending filaments of the superficial cervical nerve. This space contains, in front, the submaxillary gland, imbedded in which, are the facial artery and vein, and their glandular branches; beneath this gland, on the surface of the Mylo-hyoid muscle, are the submental artery, and the mylo-hyoid artery and nerve. The posterior part of this space is separated from the anterior part, by the stylo-maxillary ligament, it contains the external carotid artery, ascending deeply in the substance of the parotid gland; this vessel here lies in front of and superficial to the internal carotid, being crossed by the facial nerve, and giving off in its course the posterior auricular, temporal and internal maxillary branches; more deeply seated is the internal carotid, the internal jugular vein, and the pneumogastric nerve, separated from the external carotid, by the Stylo-glossus and Stylo-pharyngeus muscles, and the glosso-pharyngeal nerve.

Posterior Triangular Space

The posterior triangular space is bounded in front by the Sterno-mastoid muscle, behind, by the anterior margin of the Trapezius; its base corresponds to the upper border of the clavicle, its apex to the occiput. This space is crossed about an inch above the clavicle by the posterior belly of the Omo-hyoid, which divides it unequally into two, an upper or occipital, and a lower or subclavian.

The *Occipital*, the larger of the two posterior triangles, is bounded in front by the Sterno-mastoid; behind by the Trapezius; below by the Omo-hyoid. Its floor is formed from above downwards by the Splenius, Levator anguli scapulae, and the Middle and Posterior scaleni muscles. It is covered in by the integument, the Platysma below, the superficial and deep fasciae, and crossed, above, by the ascending branches of the cervical plexus, the spinal accessory nerve is directed obliquely across the space from the Sterno-mastoid, which it pierces, to the under surface of the Trapezius; below, it is crossed by the ascending branches of the same plexus and transversalis colli artery and vein. A chain of lymphatic glands is also found running along the posterior border of the Sterno-mastoid, from the mastoid process to the root of the neck.

The *Subclavian*, the smaller of the two posterior triangles, is bounded, above, by the posterior belly of the Omo-hyoid; below, by the clavicle; its base, directed forwards, being formed by the Sterno-mastoid. The size of this space varies according to the extent of attachment of the clavicular portion of the Sterno-mastoid and Trapezius muscles, and also according to the height at which the Omo-hyoid crosses the neck above the clavicle. The height also of this space varies much, according to the position of the arm, being much diminished on raising the limb, on account of the ascent of the clavicle, and increased on drawing the arm downwards, when this bone is consequently depressed. This space is covered in by the integument, the superficial and deep fasciae, and crossed by the descending branches of the cervical plexus. Passing across it, just above the level of the clavicle, is the third portion of the subclavian artery, which curves outwards and downwards from the outer margin of the Scalenus anticus, across the first rib, to the axilla. Sometimes this vessel rises as high as an inch and a half above the clavicle, or to any point intermediate between this and its usual level. Occasionally it passes in front of the Scalenus anticus, or pierces the fibres of this muscle. The subclavian vein lies beneath the clavicle, and is usually not seen in this space; but it occasionally rises as high up as the artery, and has even been seen to pass with that vessel behind the Scalenus anticus. The brachial plexus of nerves lies above the artery, and in close contact with it. Passing transversely across the clavicular margin of the space are the supra-scapular vessels, and traversing its upper angle in the same direction, the transverse cervical vessels. The external jugular vein descends vertically downwards behind the posterior border of the Sterno-mastoid, to terminate in the Subclavian; it receives the transverse cervical and supra-scapular veins, which occasionally form a plexus in front of the artery, and a small vein which crosses the clavicle from the cephalic. The small nerve to the Subclavius also crosses this space about its centre.

Internal Carotid Artery

The internal carotid artery commences at the bifurcation of the common carotid, opposite the upper border of the thyroid cartilage, and ascends perpendicularly upwards, in front of the transverse processes of the three upper cervical vertebræ, to the carotid foramen in the petrous portion of the temporal bone. After ascending in it for a short distance, it passes forwards and inwards through the carotid canal, and ascending a little by the side of the sella Turcica, curves upwards by the anterior clinoid process, where it pierces the dura mater, and divides into its terminal branches.

This vessel supplies the anterior part of the brain, the eye, and its appendages. Its size in the adult is equal to that of the external carotid; in the child it is larger

194— The Internal Carotid and Vertebral Arteries. Right Side.

than that vessel. It is remarkable for the number of curvatures that it presents in different parts of its course. In its cervical portion it occasionally presents one or two flexures near the base of the skull, whilst through the rest of its extent it describes a double curvature, which resembles the italic letter ∫ placed horizontally ∽. These curvatures most probably diminish the velocity of the current of blood, by increasing the extent of surface over which it moves, and adding to the amount of impediment produced by friction. In considering the course and relations of this vessel, it may be conveniently divided into four portions, a cervical, petrous, cavernous, and cerebral.

Cervical Portion. This portion of the internal carotid at its commencement is very superficial, being contained in the superior carotid triangle, on the same level but behind the external carotid, overlapped by the Sterno-mastoid, and covered by the Platysma, deep fascia, and integument; it then passes beneath the parotid gland, being crossed by the hypo-glossal nerve, the Digastric and Stylo-hyoid muscles, and the external carotid and occipital arteries. Higher up it is separated from the external carotid by the Stylo-glossus and Stylo-pharyngeus muscles and the glosso-pharyngeal nerve. It is in relation, posteriorly, with the Rectus anticus major, the superior cervical ganglion of the sympathetic and superior laryngeal nerve; externally, with the internal jugular vein, and pneumogastric nerve; internally, with the pharynx, the tonsil, and the ascending pharyngeal artery.

Petrous Portion. When the internal carotid artery enters the canal in the petrous portion of the temporal bone it first ascends a short distance, then curves forwards and inwards, and again ascends as it leaves the canal to enter the cavity of the skull. In this canal, the artery lies at first anterior to the tympanum, from which it is separated by a thin bony lamella, which is cribriform in the young subject, and often absorbed in old age. It is separated from the bony wall of the carotid canal by a prolongation of dura mater, and is surrounded by filaments of the carotid plexus.

Cavernous Portion. The internal carotid artery, in this part of its course, ascends by the side of the body of the sphenoid bone, being situated on the inner wall of the cavernous sinus, in relation, externally, with the sixth nerve, and covered by the lining membrane of the sinus. The third, fourth, and ophthalmic nerves are placed on the outer wall of the cavernous sinus, being separated from its cavity by the lining membrane.

Cerebral Portion. On the inner side of the anterior clinoid process the internal carotid perforates the dura mater, and is received into a sheath of the arachnoid. This portion of the artery is on the outer side of the optic nerve; it lies at the inner extremity of the fissure of Sylvius, having the third nerve externally.

Peculiarities. The length of the internal carotid varies according to the length of the neck, and also according to the point of bifurcation of the common carotid. Its origin sometimes takes place from the arch of the aorta; this vessel, in such rare instances, was placed nearer the middle line of the neck than the external carotid, as far upwards as the larynx, when the latter vessel crossed the internal carotid. The course of the vessel, instead of being straight, may be very tortuous. A few instances are recorded in which this vessel may be altogether absent: in one of these the common carotid ascended the neck, and gave off the usual branches of the external carotid; the cranial portion of the vessel being replaced by two branches of the internal maxillary, which entered the skull through the foramen rotundum and ovale, and joined to form a single vessel.

Surgical Anatomy. The cervical part of the internal carotid is sometimes wounded by a stab or gun-shot wound in the neck, or even occasionally by a stab from within the mouth, as when a person receives a thrust from the end of a parasol, or falls down with a tobacco-pipe in his mouth. In such cases a ligature should be applied to the common carotid.

The branches given off from the internal carotid are:

From Petrous Portion	Tympanic.
From Cavernous Portion	{ Arteria receptaculi. Anterior meningeal. Ophthalmic.
From Cerebral Portion	{ Anterior cerebral. Middle cerebral. Posterior communicating. Anterior choroid.

The *Cervical Portion* of the internal carotid gives off no branches.

The *Tympanic* is a small branch which enters the cavity of the tympanum, through a minute foramen in the carotid canal, and anastomoses with the tympanic branch of the internal maxillary, and stylo-mastoid arteries.

The *Arteriae receptaculi* are numerous small vessels, derived from the carotid artery in the cavernous sinus; they supply the pituitary body, the Casserian ganglion, and the walls of the cavernous and inferior petrosal sinuses. One of these branches, distributed to the dura mater, is called the *anterior meningeal*; it anastomoses with the middle meningeal.

The *Ophthalmic Artery* arises from the internal carotid, just as that vessel is emerging from the cavernous sinus, on the inner side of the anterior clinoid process, and enters the orbit through the optic foramen, below and on the outer side, of the optic nerve. It then crosses above, and to the inner side of this nerve, to the inner wall of the orbit, and, passing horizontally forwards, beneath the lower border of the Superior oblique muscle, to the inner angle of the eye, divides into two terminal branches, the frontal, and nasal.

Branches. The branches of this vessel may be divided into an orbital group, which are distributed to the orbit and surrounding parts; and an ocular group, which supply the muscles and globe of the eye.

Orbital Group.	*Ocular Group.*
Lachrymal.	Muscular.
Supra-orbital.	Anterior ciliary.
Posterior ethmoidal.	Short ciliary.
Anterior ethmoidal.	Long ciliary.
Palpebral.	Arteria centralis retinae.
Frontal.	
Nasal.	

195.—The Ophthalmic Artery and its Branches, the Roof of the Orbit having been removed.

The *Lachrymal* is the first, and one of the largest branches, derived from the ophthalmic, arising close to the optic foramen, and not unfrequently from that vessel before entering the orbit. It accompanies the lachrymal nerve along the upper border of the External rectus muscle, and is distributed to the lachrymal gland. Its terminal branches, escaping from the gland, are distributed to the upper eyelid and conjunctiva, anastomosing with the palpebral arteries. The lachrymal artery gives off one or two malar branches; one of which passes through a foramen in the malar bone to reach the temporal fossa and anastomoses with the deep temporal arteries. The other appears on the cheek, and anastomoses with the transverse facial. A branch is also sent backwards, through the sphenoidal fissure, to the dura mater, which anastomoses with a branch of the middle meningeal artery.

Peculiarities. The lachrymal artery is sometimes derived from one of the anterior branches of the middle meningeal artery.

The *Supra-orbital artery*, the largest branch of the ophthalmic, arises from that vessel above the optic nerve. Ascending so as to rise above all the muscles of the orbit, it passes forwards, with the frontal nerve, between the periosteum and Levator palpebrae; and, passing through the supra-orbital foramen, divides into a superficial and deep branch, which supply the muscles and integument of the forehead and pericranium, anastomosing with the temporal, angular branch of the facial, and the artery of the opposite side. This artery in the orbit supplies the Superior rectus and the Levator palpebrae, sends a branch inwards, across the pulley of the Superior oblique muscle, to supply the parts of the inner canthus; and at the supra-orbital foramen, frequently transmits a branch to the diploë.

The *Ethmoidal branches* are two in number; posterior, and anterior. The former, which is the smaller, passes through the posterior ethmoidal foramen, supplies the posterior ethmoidal cells, and, entering the cranium, gives off a meningeal branch, which supplies the adjacent dura mater, and nasal branches, which descends into the nose through apertures in the cribriform plate, anastomosing with branches of the spheno-palatine. The anterior ethmoidal artery accompanies the nasal nerve through the anterior ethmoidal foramen, supplies the anterior ethmoidal cells, and frontal sinuses, and, entering the cranium, divides into a meningeal branch, which supplies the adjacent dura mater, and a nasal branch which descends into the nose, through an aperture in the cribriform plate.

The *Palpebral arteries*, two in number, superior and inferior, arise from the ophthalmic, opposite the pulley of the Superior oblique muscle; they encircle the eyelids near their free margin, forming a superior and an inferior arch, which lie between the Orbicularis muscle and tarsal cartilage. The superior palpebral, inosculating at the outer angle of the orbit with the orbital branch of the temporal artery. The inferior branch anastomoses with the orbital branch of the infra-orbital artery, at the inner side of the lid; from this anastomosis a branch passes to the nasal duct, ramifying, in its mucous membrane, as far as the inferior meatus.

The *Frontal artery*, one of the terminal branches of the ophthalmic, passes from the orbit at its inner angle, and, ascending on the forehead, supplies the muscles, integument, and pericranium, anastomosing with the supra-orbital artery.

The *Nasal artery*, the other terminal branch of the ophthalmic, emerges from the orbit above the tendo oculi, and, after giving a branch to the lachrymal sac, divides into two, one of which anastomoses with the angular artery, the other branch, the dorsalis nasi, runs along the dorsum of the nose, supplies its entire surface, and anastomoses with the artery of the opposite side.

The *Ciliary arteries* are divisible into three groups, the short, long, and anterior.

The *Short Ciliary arteries*, from twelve to fifteen in number, arise from the ophthalmic, or some of its branches; they surround the optic nerve as they pass forwards to the posterior part of the eyeball, pierce the sclerotic coat around the entrance of this nerve, and supply the choroid coat and ciliary processes.

The *Long Ciliary arteries*, two in number, also pierce the posterior part of the sclerotic, and run forwards, along each side of the eyeball, between the sclerotic and choroid, to the ciliary ligament, where they divide into two branches; these form an arterial circle around the circumference of the iris, from which numerous radiating branches pass forwards, in its substance, to its free margin, where they form a second arterial circle around its pupillary margin.

The *Anterior Ciliary arteries* are derived from the muscular branches; they pierce the sclerotic a short distance from the cornea, and terminate in the great arterial circle of the iris.

The *Arteria Centralis Retinae*, is one of the smallest branches of the ophthalmic artery. It arises near the optic foramen, pierces the optic nerve obliquely, and runs forwards, in the centre of its substance, to the retina, in which its branches are distributed as far forwards as the ciliary processes. In the human foetus, a small vessel passes forwards, through the vitreous humour, to the posterior surface of the capsule of the lens.

The *Muscular branches*, two in number, superior and inferior, supply the muscles of the eyeball. The superior the smaller, often wanting, supplies the Levator palpebrae, Superior rectus, and Superior oblique. The inferior, more constant in its existence, passes forwards, between the optic nerve and Inferior rectus, and is distributed to the External and Inferior recti, and Inferior oblique. This vessel gives off most of the anterior ciliary arteries.

The *Cerebral Branches* of the internal carotid are, the anterior cerebral, the middle cerebral, the posterior communicating, and the anterior choroid.

The *Anterior Cerebral* arises from the internal carotid, at the inner extremity of the fissure of Sylvius. It passes forwards in the great longitudinal fissure between the

196.—The Arteries at the Base of the Brain. The Right Half of the Cerebellum and Pons have been removed.

two anterior lobes of the brain, being connected soon after its origin with the vessel of the opposite side by a short anastomosing trunk, about two lines in length, the anterior communicating. The two anterior cerebral arteries, lying side by side, curve around the anterior border of the corpus callosum, and run along its upper surface to its posterior part, where they terminate by anastomosing with the posterior cerebral arteries. They supply the olfactory and optic nerves, the under surface of the anterior lobes, the third ventricle, the anterior perforated space, the corpus callosum, and the inner surface of the hemispheres.

The *Anterior communicating Artery* is a short branch, about two lines in length, but of moderate size, connecting together the two anterior cerebral arteries across the longitudinal fissure. Sometimes this vessel is wanting, the two arteries joining together to form a single trunk, which afterwards subdivides. Or the vessel may be wholly or partially subdivided into two; frequently, it is longer and smaller than usual.

The *Middle Cerebral Artery*, the largest branch of the internal carotid, passes obliquely outwards along the fissure of Sylvius, within which it divides into three branches: an anterior, which supplies the pia mater, investing the surface of the anterior lobe; a posterior, which supplies the middle lobe; and a median branch, which supplies the small lobe at the outer extremity of the Sylvian fissure. Near its origin, this vessel gives off numerous small branches, which enter the substantia perforata, to be distributed to the corpus striatum.

The *Posterior communicating Artery* arises from the back part of the internal carotid, runs directly backwards, and anastomoses with the posterior cerebral, a branch of the basilar. This artery varies considerably in size, being sometimes small, and occasionally so large that the posterior cerebral may be considered as arising from the internal carotid rather than from the basilar. It is frequently larger on one than on the other side.

The *Anterior Choroid* is a small but constant branch which arises from the back part of the internal carotid, near the posterior communicating artery. Passing backwards and outwards, it enters the descending horn of the lateral ventricle, beneath the edge of the middle lobe of the brain. It is distributed to the hippocampus major, corpus fimbriatum, and choroid plexus.

ARTERIES OF THE UPPER EXTREMITY

The artery which supplies the upper extremity, continues as a single trunk from its commencement, as far as the elbow; but different portions of it have received different names, according to the region through which it passes. Thus, that part of the vessel which extends from its origin, as far as the outer border of the first rib, is termed the subclavian; beyond this point to the lower border of the axilla, it is termed the axillary; and from the lower margin of the axillary space to the bend of the elbow, it is termed brachial; here, the single trunk terminates by dividing into two branches, the radial, and ulnar, an arrangement precisely similar to what occurs in the lower limb.

Subclavian Arteries

The subclavian artery on the right side arises from the arteria innominata, opposite the right sterno-clavicular articulation; on the left side, it arises from the arch of the aorta. It follows, therefore, that these two vessels must, in the first part of their course, differ in their length, their direction, and in their relation with neighbouring parts.

In order to facilitate the description of these vessels, more especially in a surgical point of view, each subclavian artery has been divided into three parts. The first portion, on the right side, ascends obliquely outwards, from the origin of the vessel to the inner border of the Scalenus anticus. On the left side, it ascends perpendicularly to the inner border of this muscle. The second part passes outwards, behind the Scalenus anticus; and the third part passes from the outer margin of this muscle, beneath the clavicle, to the lower border of the first rib, where it becomes the axillary artery. The first portions of these two vessels differ so much in their course, and in their relation with neighbouring parts, that they will be described separately. The second and third parts are precisely alike on both sides.

FIRST PART OF THE RIGHT SUBCLAVIAN ARTERY (figs. 188, 189)

It arises from the arteria innominata, opposite the right sterno-clavicular articulation, passes upwards and outwards across the root of the neck, and terminates at the inner margin of the Scalenus anticus muscle. In this part of its course, it ascends a little above the clavicle, the extent to which it does so varying in different cases. It is covered, *in front*, by the integument, superficial and deep fasciae, Platysma, the clavicular origin of the Sterno-mastoid, the Sterno-hyoid and Sterno-thyroid muscles, and another layer of the deep fascia. It is crossed by the internal jugular and vertebral veins, and by the pneumogastric, the cardiac branches of the sympathetic, and phrenic nerves. *Beneath*, the artery is invested by the pleura, and *behind*, it is separated by a cellular interval from the Longus colli, the transverse process of the

PLAN OF RELATIONS OF THE FIRST PORTION OF RIGHT SUBCLAVIAN ARTERY

In front.

Integument, superficial and deep fasciae.
Platysma.
Clavicular origin of Sterno-mastoid.
Sterno-hyoid and Sterno-thyroid.
Internal jugular and vertebral veins.
Pneumogastric, cardiac, and phrenic nerves.

Behind.

Recurrent laryngeal nerve.
Sympathetic.
Longus colli.
Transverse process of seventh cervical vertebra.

seventh cervical vertebra, and the sympathetic; the recurrent laryngeal nerve winding around the lower and back part of this vessel. The subclavian vein lies below the subclavian artery, immediately behind the clavicle.

First Part of the Left Subclavian Artery (fig. 187).

It arises from the end of the transverse portion of the arch of the aorta, opposite the second dorsal vertebra, and ascends to the inner margin of the first rib, behind the insertion of the Scalenus anticus muscle. This vessel is, therefore, longer than the right, situated more deeply in the cavity of the chest, and directed almost vertically upwards, instead of arching outwards like the vessel of the opposite side.

It is in relation, *in front*, with the pleura, the left lung, the pneumogastric, phrenic, and cardiac nerves, which lie parallel with it, the left carotid artery, left internal jugular and innominate veins, and is covered by the Sterno-thyroid, Sterno-hyoid, and Sterno-mastoid muscles; *behind*, with the oesophagus, thoracic duct, inferior cervical ganglion of the sympathetic, Longus colli, and vertebral column. To its *inner side* is the oesophagus, trachea, and thoracic duct; to its *outer side*, the pleura.

Plan of Relations of First Portion of Left Subclavian Artery

In front.
Pleura and left lung.
Pneumogastric, cardiac, and phrenic nerves.
Left carotid artery.
Left internal jugular and innominate veins.
Sterno-thyroid, Sterno-hyoid, and Sterno-mastoid muscles.

Inner side.
Oesophagus.
Trachea.
Thoracic duct.

Left Subclavian Artery. First portion.

Outer side.
Pleura.

Behind.
Oesophagus and thoracic duct.
Inferior cervical ganglion of sympathetic.
Longus colli and vertebral column.

The relations of the second and third portions of the subclavian arteries are precisely similar on both sides.

The *Second Portion of the Subclavian Artery* lies between the two Scaleni muscles; it is very short, and forms the highest part of the arch described by that vessel.

Relations. It is covered, *in front*, by the integument, Platysma, Sterno-mastoid, cervical fascia, and by the phrenic nerve, which is separated from the artery by the Scalenus anticus muscle. *Behind*, it is in relation with the Middle scalenus. *Above*, with the brachial plexus of nerves. *Below*, with the pleura. The subclavian vein lies below the artery, separated from it by the Scalenus anticus.

PLAN OF THE RELATIONS OF THE SECOND PORTION OF THE SUBCLAVIAN ARTERY

In front.

Platysma and Sterno-mastoid. Phrenic nerve.
Cervical fascia. Scalenus anticus.

Above. — Subclavian Artery. Second portion. — *Below.*
Brachial plexus. Pleura.

Behind.
Middle scalenus.

The *Third Portion of the Subclavian Artery* passes downwards and outwards from the outer margin of the Scalenus anticus to the lower border of the first rib, where it becomes the axillary artery. This portion of the vessel is the most superficial, and is contained in a triangular space, the base of which is formed in front by the Anterior scalenus, and the two sides by the Omo-hyoid above and the clavicle below.

Relations. It is covered, *in front*, by the integument, the superficial and deep fasciae, the Platysma; and by the clavicle, the Subclavius muscle, and the supra-scapular artery and vein *below*; the clavicular descending branches of the cervical plexus and the nerve to the Subclavius pass vertically downwards in front of the artery. The external jugular vein crosses it at its inner side, and receives the supra-scapular and transverse cervical veins, which occasionally form a plexus in front of it. The subclavian vein is below the artery, lying close behind the clavicle. *Behind*, it lies on the Middle scalenus muscle. *Above* it, and to its outer side, is the brachial plexus, and Omo-hyoid muscle. *Below*, it rests on the outer surface of the first rib.

PLAN OF THE RELATIONS OF THE THIRD PORTION OF THE SUBCLAVIAN ARTERY

In front.

Integument, fasciae, and Platysma.
The external jugular, supra-scapular, and transverse cervical veins.
Descending branches of cervical plexus.
Subclavius muscle, supra-scapular vessels, and clavicle.

Above. — Subclavian Artery. Third portion. — *Below.*
Brachial plexus. First rib.
Omo-hyoid.

Behind.
Scalenus medius.

Peculiarities. The subclavian arteries vary in their origin, their course, and in the height to which they rise in the neck.

The origin of the right subclavian from the innominate takes place, in some cases, above the sterno-clavicular articulation; more frequently in the cavity of the thorax, below that joint. Or the artery may arise as a separate trunk from the arch of the aorta; in such cases it may be either the first, second, third, or even the last branch derived from that vessel: in the majority of cases, it is the first or last, rarely the second or third.

When it is the first branch, it occupies the ordinary position of the innominate artery; when the second or third, it gains its usual position by passing behind the right carotid; and when the last branch, it arises from the left extremity of the arch, at its upper or back part, and passes obliquely towards the right side, behind the oesophagus and right carotid, sometimes between the oesophagus and trachea, to the upper border of the first rib, where it follows its ordinary course. In very rare instances, this vessel arises from the thoracic aorta, as low down as the fourth dorsal vertebra. Occasionally it perforates the anterior scalenus; more rarely it passes in front of this muscle: sometimes the subclavian vein passes with the artery behind the Scalenus. The artery sometimes ascends as high as an inch and a half above the clavicle, or to any intermediate point between this and the upper border of this bone, the right subclavian usually ascending higher than the left.

The left subclavian is occasionally joined at its origin with the left carotid.

Surgical Anatomy. The relations of the subclavian arteries of the two sides having been examined, the student should direct his attention to consider the best position in which compression of the vessel may be effected, or in what situation a ligature may be best applied in cases of aneurism or wounds.

Compression of the subclavian artery is required in cases of operations about the shoulder, in the axilla, or at the upper part of the arm; and the student will observe that there is only one situation in which it can be effectually applied, viz., where the artery passes across the outer surface of the first rib. In order to compress the vessel in this situation, the shoulder should be depressed, and the surgeon, grasping the side of the neck, may press with his thumb in the hollow behind the clavicle downwards against the rib; if from any cause the shoulder cannot be sufficiently depressed, pressure may be made from before backwards, so as to compress the artery against the Middle scalenus and transverse process of the seventh cervical vertebra.

Ligature of the subclavian artery may be required in cases of wounds of the axillary artery, or in aneurism of that vessel; and the third part of the artery is consequently that which is most favourable for such an operation, on account of its being comparatively superficial, and most remote from the origin of the large branches. In those cases where the clavicle is not displaced, this operation may be performed with comparative facility; but where the clavicle is elevated from the presence of a large aneurismal tumour in the axilla, the artery is placed at a great depth from the surface, which materially increases the difficulty of the operation. Under these circumstances, it becomes a matter of importance to consider the height to which this vessel reaches above the bone. In ordinary cases its arch is about half an inch above the clavicle, occasionally as high as an inch and a half, and sometimes so low as to be on a level with its upper border. If displacement of the clavicle occurs, these variations will necessarily make the operation more or less difficult, according as the vessel is more or less accessible.

The chief points in the operation of tying the third portion of the subclavian artery are as follows. The patient being placed on a table in the horizontal position, and the shoulder depressed as much as possible, the integument should be drawn downwards upon the clavicle and an incision made through it upon that bone from the anterior border of the Trapezius to the posterior border of the Sterno-mastoid, to which may be added a short vertical incision meeting the centre of the preceding; the Platysma myoides and cervical fascia should be divided upon a director, and if the interval between the Trapezius and Sterno-mastoid muscles be insufficient for the performance of the operation, a portion of one or both may be divided. The external jugular vein will now be seen towards the inner side of the wound; this and the scapular and transverse cervical veins which terminate in it should be held aside, and if divided both ends should be included in a ligature: the supra-scapular artery should be avoided, and the Omo-hyoid muscle must now be looked for, and held aside if necessary. In the space beneath this muscle, careful search must be made for the vessel; the deep fascia having been divided with the finger-nail or silver scalpel, the outer margin of the Scalenus muscle must be felt for, and the finger being guided by it to the first rib, the pulsation of the subclavian artery will be felt as it passes over its surface. The aneurism needle may then be passed around the vessel from before backwards, by which means the vein will be avoided, care being taken not to include a branch of the brachial plexus instead of the artery in the ligature. If the clavicle is so raised by the tumour that the application of the ligature cannot be effected in this situation, the artery may be tied above the first rib, or even behind the Scalenus muscle: the difficulties of the operation in such a case will be materially increased, on account of the greater depth of the artery, and alteration of the position of the surrounding parts.

The second division of the subclavian artery, from being that portion which rises highest in the neck, has been considered favourable for the application of the ligature, where it is difficult to apply it in the third part of its course. There are, however, many objections to the operation in this situation. It is necessary to divide the Scalenus anticus muscle, upon which lies the phrenic nerve, and at the inner side of which is situated the internal jugular vein; a wound of either of these structures might lead to the most dangerous consequences. Again, the artery is in contact, below, with the pleura, which must also be avoided; and lastly, the proximity of so many of its larger branches arising internal to this point, must be a still further objection to the operation. If, however, it has been determined upon to perform the operation in this situation, it should be remembered that it occasionally happens that the artery passes in front of the Scalenus anticus, or through the fibres of that muscle; or that the vein sometimes passes with the artery behind the Scalenus anticus.

In those cases of aneurism of the axillary or subclavian artery which encroach upon the outer portion of the Scalenus muscle to such an extent that a ligature cannot be applied in that situation, it may be deemed advisable, as a last resource, to tie the first portion of the subclavian artery. On the left side this operation is quite impracticable, the great depth of the artery from the surface, its intimate relation with the pleura, and its close proximity with so many important veins and nerves, present a series of difficulties which it is impossible to overcome. On the right side the operation is practicable, and has been performed, though not with success. The main objection to the operation in this situation is the smallness of the interval which usually exists between the commencement of the vessel and the origin of the nearest branch. This operation may be performed in the following manner. The patient being placed on a table in the horizontal position, with the neck extended, an incision should be made parallel with the inner part of the clavicle, and a second along the inner border of the Sterno-mastoid, meeting it at right angles. The sternal attachment of the Sterno-mastoid may now be divided on a director, and turned outwards; a few small arteries and veins, and occasionally the anterior jugular must be avoided, and the Sterno-hyoid and thyroid muscles divided in the same manner as the preceding muscle. After tearing through the deep fascia with the finger-nail, the internal jugular vein will be seen crossing the artery; this should be pressed aside, and the artery secured by passing the needle from below upwards, by which the pleura is more effectually avoided. The exact position of the vagus nerve, the recurrent laryngeal, the phrenic and sympathetic nerves, should be remembered, and the ligature should be applied near the origin of the vertebral, in order to afford as much room as possible for the formation of a coagulum between the ligature and the origin of the vessel. It should be remembered that the right subclavian artery is occasionally deeply placed in the first part of its course, when it arises from the left side of the aortic arch, and passes in such cases behind the oesophagus, or between it and the trachea.

BRANCHES OF THE SUBCLAVIAN ARTERY (fig. 197, color illustration p. A3)

These are four in number. Three arising from the first portion of the vessel, the vertebral, the internal mammary, and the thyroid axis; and one from the second portion, the superior intercostal. The vertebral arises from the upper and back part of the first portion of the artery; the thyroid axis from the front, and the internal mammary from the under part of this vessel. The superior intercostal is given off from the upper and back part of the second portion of the artery. On the left side the second portion usually gives off no branch, the superior intercostal arising to the inner side of the Scalenus anticus. On both sides of the body, the first three branches arise close together at the inner margin of the Scalenus anticus in the majority of cases a free interval of half an inch to an inch existing between the commencement of the artery and the origin of the nearest branch; in a smaller number of cases, an interval of more than an inch existed, never exceeding an inch and three-quarters. In a very few instances the interval was less than half an inch.

197.—Plan of the Branches of the Right Subclavian Artery.

VERTEBRAL ARTERY (fig. 194)

The vertebral artery is generally the first and largest branch of the subclavian; it arises from the upper and back part of the first portion of the vessel, and passing upwards, enters the foramen in the transverse process of the sixth cervical vertebra, and ascends through the foramina in the transverse processes of all the vertebrae. Above the upper border of the axis, it inclines outwards and upwards to the foramen in the transverse process of the atlas, through which it passes; it then winds backwards behind its articular process, runs in a deep groove on the surface of the posterior arch of this bone, and piercing the posterior occipito-atloid ligament and dura mater, enters the skull through the foramen magnum. It then passes in front of the medulla oblongata, and unites with the vessel of the opposite side at the lower border of the pons Varolii, to form the basilar artery.

At its origin, it is situated behind the internal jugular vein and inferior thyroid artery, and, near the spine, lies between the Longus colli and Scalenus anticus muscles, having the thoracic duct in front of it on the left side. Within the foramina formed by the transverse processes of the vertebrae, it is accompanied by a plexus of nerves from the sympathetic, and lies between the vertebral vein, which is in front, and the cervical nerves, which issue from the intervertebral foramina behind it. Whilst winding around the articular process of the atlas, it is contained in a triangular space formed by the Rectus posticus minor, the Superior and Inferior oblique muscles; and is covered by the Rectus posticus major and Complexus. And within the skull, as it winds around the medulla oblongata, it is placed between the hypoglossal and anterior root of the sub-occipital nerves.

Branches. These may be divided into two sets, those given off in the neck, and those within the cranium.

Cervical Branches. { Lateral spinal.
Muscular.

Cranial Branches. { Posterior meningeal.
Anterior spinal.
Posterior spinal.
Inferior cerebellar.

The *Lateral Spinal Branches* enter the spinal canal through the intervertebral foramina, each dividing into two branches. Of these, one passes along the roots of the nerves to supply the spinal cord and its membranes anastomosing with the other spinal arteries; the other is distributed to the posterior surface of the bodies of the vertebrae.

Several large *Muscular Branches* are given off to the deep muscles of the neck where the vertebral artery curves around the articular process of the atlas. They anastomose with the occipital and deep cervical arteries.

The *Posterior Meningeal* are one or two small branches given off from the vertebral opposite the foramen magnum. They ramify between the bone and dura mater in the cerebellar fossae, and supply the falx cerebelli.

The *Anterior Spinal* is a small branch, larger than the posterior spinal, which arises near the termination of the vertebral, and unites with its fellow of the opposite side in front of the medulla oblongata. The single trunk thus formed, descends a short distance on the anterior aspect of the spinal cord, and joins with a succession of small branches which enter the spinal canal through some of the intervertebral foramina;

these branches are derived from the vertebral and ascending cervical, in the neck; from the intercostal, in the dorsal region; and from the lumbar, ilio-lumbar, and lateral sacral arteries in the lower part of the spine. They unite, by means of ascending and descending branches, to form a single anterior median artery, which extends as low down as the lower part of the spinal cord. This vessel is placed beneath the pia mater along the anterior median fissure, it supplies that membrane and the substance of the cord, and sends off branches at its lower part to be distributed to the Cauda equina.

The *Posterior Spinal* arises from the vertebral, at the side of the medulla oblongata, passing backwards to the posterior aspect of the spinal cord, it descends on either side, lying behind the posterior roots of the spinal nerves, and is reinforced by a succession of small branches, which enter the spinal canal through the intervertebral foramina, and by which it is continued to the lower part of the cord, and to the Cauda equina. Branches from these vessels form a free anastomosis around the posterior roots of the spinal nerves, and communicate, by means of very tortuous transverse branches, with the vessel of the opposite side. At its commencement, it gives off an ascending branch, which terminates on the sides of the fourth ventricle.

The *Inferior Cerebellar artery*, the largest branch of the vertebral, winds backwards around the upper part of the medulla oblongata, passing between the origin of the spinal accessory and pneumogastric nerves, over the restiform body, to the under surface of the cerebellum, where it divides into two branches; an internal one, which is continued backwards to the notch between the two hemispheres of the cerebellum; and an external one, which supplies the under surface of the cerebellum, as far as its outer border, where it anastomoses with the superior cerebellar. Branches from this artery supply the choroid plexus of the fourth ventricle.

The *Basilar artery*, so named from its position at the base of the skull, is a single trunk, formed by the junction of the two vertebral arteries; it extends from the posterior to the anterior border of the pons Varolii, where it divides into two terminal branches, the posterior cerebral arteries. Its branches are, on each side, the following:

Transverse.	Superior cerebellar.
Anterior cerebellar.	Posterior cerebral.

The *Transverse* branches supply the pons Varolii and adjacent parts of the brain; one accompanies the auditory nerve into the internal auditory meatus; and another, of larger size, passes along the crus cerebelli, to be distributed to the anterior border of the under surface of the cerebellum. It is called the *anterior (inferior) cerebellar artery*.

The *Superior Cerebellar arteries* arise near the termination of the basilar. They wind round the crus cerebri, close to the fourth nerve, and, arriving at the upper surface of the cerebellum, divide into branches which supply the pia mater, covering its surface, anastomosing with the inferior cerebellar. It gives several branches to the pineal gland, and also to the velum interpositum.

The *Posterior Cerebral arteries*, the two terminal branches of the basilar, are larger than the preceding, from which they are separated near their origin by the third nerves. Winding around the crus cerebri, they pass to the under surface of the posterior lobes of the cerebrum, which they supply, anastomosing with the anterior and middle cerebral arteries. Near their origin they give off a number of parallel branches, which enter the posterior perforated spot, and receive the posterior communicating arteries from the internal carotid. They also give off a branch, the posterior choroid, which supplies the velum interpositum and choroid plexus, entering the interior of the brain, beneath the posterior border of the corpus callosum.

Circle of Willis. The remarkable anastomosis which exists between the branches of the internal carotid, and vertebral arteries at the base of the brain, constitutes the circle of Willis. It is formed, in front, by the anterior cerebral and anterior communicating arteries; on each side, by the trunk of the internal carotid, and the posterior communicating; behind, by the posterior cerebral, and point of the basilar. It is by this anastomosis that the cerebral circulation is equalized, and provision made for effectually carrying it on if one or more of the branches are obliterated. The parts of the brain included within this arterial circle are, the lamina cinerea, the commissure of the optic nerves, the infundibulum, the tuber cinereum, the corpora albicantia, and the pars perforata postica.

Thyroid Axis

The thyroid axis is a short, thick trunk, which arises from the fore part of the first portion of the subclavian artery, close to the inner side of the Scalenus anticus muscle, and divides, almost immediately after its origin, into three branches, the inferior thyroid, supra-scapular, and transversalis colli.

The INFERIOR THYROID ARTERY passes upwards, in a serpentine course, behind the sheath of the carotid vessel and sympathetic nerve, the middle cervical ganglion resting upon this vessel, and is distributed to the under surface of the thyroid gland, anastomosing with the superior thyroid, and with the corresponding artery of the opposite side. Its branches are the

  Laryngeal.   Oesophageal.
  Tracheal.   Ascending cervical.

The *Laryngeal* branch ascends upon the trachea to the back part of the larynx, and supplies the muscles and the mucous membrane of this part.

The *Tracheal* branches are distributed over the trachea, anastomosing below with the bronchial arteries.

The *Oesophageal* branches are distributed to the oesophagus.

The *Ascending cervical* artery is a small branch which arises from the inferior thyroid, just where that vessel is passing behind the carotid artery, and runs up the neck in the interval between the Scalenus anticus, and Rectus anticus major. It gives branches to the muscles of the neck, which communicate with those sent out from the vertebral, and sends one or two through the intervertebral foramina, along the cervical nerves, to supply the bodies of the vertebrae, the spinal cord, and its membranes.

The SUPRA-SCAPULAR ARTERY, smaller than the transverse cervical, passes obliquely from within outwards, across the root of the neck. It at first lies on the lower part of the Scalenus anticus, being covered by the Sterno-mastoid; it then crosses the subclavian artery, and runs outwards behind, and parallel with, the clavicle and

Subclavius muscle, and beneath the posterior belly of the Omo-hyoid, to the superior border of the scapula, where it passes over the ligament of the supra-scapular notch to the supra-spinous fossa. In this situation it lies close to the bone, and ramifies between it and the Supra-spinatus muscle to which it is mainly distributed, giving off a communicating branch, which crosses the neck of the scapula, to reach the infra-spinous fossa, where it anastomoses with the dorsal branch of the subscapular artery. Besides distributing branches to the Sterno-mastoid, and neighbouring muscles, it gives off a supra-acromial branch, which, piercing the Trapezius muscle, supplies the cutaneous surface of the acromion, anastomosing with the acromial thoracic artery. As the artery passes across the supra-scapular notch, a branch descends into the subscapular fossa, ramifies beneath that muscle, and anastomoses with the posterior and subscapular arteries. It also supplies the shoulder joint.

The TRANSVERSALIS COLLI passes transversely outwards, across the upper part of the subclavian triangle, to the anterior margin of the Trapezius muscle, beneath which it divides into two branches, the superficial cervical, and the posterior scapular. In its course across the neck, it passes in front of the Scaleni muscles and the brachial plexus, between the divisions of which it sometimes passes, and is covered in by the Platysma, Sterno-mastoid, Omo-hyoid, and Trapezius muscles.

198.—The Scapular and Circumflex Arteries.

The *Superficial Cervical* ascends beneath the anterior margin of the Trapezius, distributing branches to it, and to the neighbouring muscles and glands in the neck.

The *Posterior Scapular*, the continuation of the transverse cervical, passes beneath the levator anguli scapulae to the superior angle of the scapula, and descends along the posterior border of that bone as far as the inferior angle, where it anastomoses with the subscapular branch of the axillary. In its course it is covered in by the Rhomboid muscles, supplying these, the Latissimus dorsi and Trapezius, and anastomoses with the supra-scapular and subscapular arteries, and with the posterior branches of some of the intercostal arteries.

Peculiarities. The *Superficial Cervical* frequently arises as a separate branch from the thyroid axis; and the posterior scapular, from the third, more rarely from the second, part of the subclavian.

The INTERNAL MAMMARY arises from the under surface of the first portion of the subclavian artery, opposite the thyroid axis. It descends behind the clavicle, to the inner surface of the anterior wall of the chest, resting upon the costal cartilages, a short distance from the margin of the sternum; and, at the interval between the sixth and seventh cartilages, divides into two branches, the musculophrenic, and superior epigastric.

At its origin, it is covered by the internal jugular and subclavian veins, and crossed by the phrenic nerve. In the upper part of the thorax, it lies upon the costal cartilages, and internal Intercostal muscles in front, covered by the pleura behind. At the lower part of the thorax, the Triangularis sterni separates this vessel from the pleura. It is accompanied by two veins, which join at the upper part of the thorax into a single trunk.

The branches of the internal mammary are,

Comes nervi phrenici (superior phrenic).	Anterior intercostal.
Mediastinal.	Perforating.
Pericardiac.	Musculo-phrenic.
Sternal.	Superior epigastric.

The *Comes Nervi Phrenici (Superior Phrenic)*, is a long slender branch, which accompanies the phrenic nerve, between the pleura and pericardium, to the Diaphragm, to which it is distributed; anastomosing with the other phrenic arteries from the internal mammary, and abdominal aorta.

The *Mediastinal Branches* are small vessels, which are distributed to the areolar tissue in the anterior mediastinum, and the remains of the thymus gland.

The *Pericardiac Branches* supply the upper part of the pericardium, the lower part receiving branches from the musculo-phrenic artery. Some *sternal* branches are distributed to the Triangularis sterni, and both surfaces of the sternum.

The *Anterior Intercostal Arteries* supply the five or six upper intercostal spaces. The branch corresponding to each space passes outwards, and soon divides into two, which run along the opposite edges of the ribs, and inosculate with the intercostal arteries from the aorta. They are at first situated between the pleura and the internal Intercostal muscles, and then between the two layers of these muscles. They supply the Intercostal and Pectoral muscles, and the mammary gland.

The *Anterior* or *Perforating Arteries* correspond to the five or six upper intercostal spaces. They arise from the internal mammary, pass forwards through the intercostal spaces, and, curving outwards, supply the Pectoralis major, and the integument. Those which correspond to the first three spaces, are distributed to the mammary gland. In females, during lactation, these branches are of large size.

The *Musculo-phrenic Artery* is directed obliquely downwards and outwards, behind the cartilages of the false ribs, perforating the Diaphragm at the eighth or ninth rib, and terminating, considerably reduced in size, opposite the last intercostal space. It gives off anterior intercostal arteries to each of the intercostal spaces across which it passes;

they diminish in size as the spaces decrease in length, and are distributed in a manner precisely similar to the anterior intercostals from the internal mammary. It also gives branches backwards to the Diaphragm, and downwards to the abdominal muscles.

The *Superior Epigastric* continues in the original direction of the internal mammary, descends behind the Rectus muscle, and, perforating its sheath, divides into branches which supply the Rectus, anastomosing with the epigastric artery from the external iliac. Some vessels perforate the sheath of the Rectus, and supply the muscles of the abdomen and the integument, and a small branch which passes inwards upon the side of the ensiform appendix, anastomoses in front of that cartilage with the artery of the opposite side.

The SUPERIOR INTERCOSTAL arises from the upper and back part of the second portion of the subclavian artery on the right side, and to the inner side of the Scalenus anticus on the left side. Passing backwards, it gives off the deep cervical branch, and descends behind the pleura in front of the necks of the first two ribs, and inosculates with the first aortic intercostal. In the first intercostal space, it gives off a branch which is distributed in a similar manner with the aortic intercostals. The branch for the second intercostal space usually joins with one from the first aortic intercostal. Each intercostal gives off a branch to the posterior Spinal muscles, and a small one, which passes through the corresponding intervertebral foramen to the spinal cord and its membranes.

The *Deep Cervical Branch (Profunda Cervicis)* arises, in most cases, from the superior intercostal, and is analogous to the posterior branch of an aortic intercostal artery. Passing backwards, between the transverse process of the seventh cervical vertebra and the first rib, it ascends the back part of the neck, between the Complexus and Semi-spinalis colli muscles, as high as the axis, supplying these and adjacent muscles, and anastomosing with the arteria princeps cervicis of the occipital, and with branches which pass outwards from the vertebral.

SURGICAL ANATOMY OF THE AXILLA

The *Axilla* is a conical space, situated between the upper and lateral parts of the chest, and inner side of the arm.

Boundaries. Its *apex*, which is directed upwards towards the root of the neck, corresponds to the interval between the first rib internally, the superior border of the scapula externally, and the clavicle and Subclavius muscle in front. The *base*, directed downwards, is formed by the integument, and a thick layer of fascia, extending between the lower border of the Pectoralis major in front, and the lower border of the Latissimus dorsi behind; it is broad internally, at the chest, but narrow and pointed externally, at the arm. Its *anterior boundary* is formed by the Pectoralis major and Minor muscles, the former covering in the whole of the anterior wall of the axilla, the latter covering only its central part. Its *posterior boundary*, which extends somewhat lower than the anterior, is formed by the Subscapularis above, the Teres major and Latissimus dorsi below. On the *inner side* are the first four ribs and their corresponding Intercostal muscles, and part of the Serratus magnus. On the *outer side*, where the anterior and posterior boundaries converge, the space is narrow, and bounded by the humerus, the Coraco-brachialis and Biceps muscles.

199.—The Axillary Artery, and its Branches.

Contents. This space contains the axillary vessels, and brachial plexus of nerves with their branches, some branches of the intercostal nerves, a large number of lymphatic glands, all connected together by a quantity of fat and loose areolar tissue.

Their Position. The axillary artery (fig. 199, color illustration p. A3) and vein, with the brachial plexus of nerves, extend obliquely along the outer boundary of the axillary space, from its apex to its base, and are placed much nearer the anterior than the posterior wall, the vein lying to the inner or thoracic side of the artery, and altogether concealing it. At the fore part of the axillary space, in contact with the Pectoral muscles, are the thoracic branches of the axillary artery, and along the anterior margin of the axilla, the long thoracic artery extends to the side of the chest. At the back part, in contact with the lower margin of the Subscapularis muscle, are the subscapular vessels and nerves; winding around the lower border of this muscle, is the dorsalis scapulae artery and veins; and towards the outer extremity of the muscle, the posterior circumflex vessels and nerve are seen curving backwards to the shoulder.

Along the inner or thoracic side, no vessel of any importance exists, its upper part being crossed by a few small branches from the superior thoracic artery. There are some important nerves, however, in this situation; the posterior thoracic or external respiratory nerve, descending on the surface of the Serratus magnus, to which it is distributed; and perforating the upper and anterior part of this wall, are the intercosto-humeral nerves, which pass across the axilla to the inner side of the arm.

The cavity of the axilla is filled by a quantity of loose areolar tissue, a large number of small arteries and veins, all of which are, however, of inconsiderable size, and numerous lymphatic glands; these are from ten to twelve in number, and situated chiefly on the thoracic side, and lower and back part of this space.

The student should attentively consider the relation of the vessels and nerves in the several parts of the axilla; for it not unfrequently happens that the surgeon is called upon to extirpate diseased glands, or to remove a tumour from this situation. In performing such an operation, it will be necessary to proceed with much caution in the direction of the outer wall and apex of the space, as here the axillary vessels will be in danger of being wounded. Towards the posterior wall, it will be necessary to avoid the subscapular, dorsalis scapulae, and posterior circumflex vessels, and, along the anterior wall, the thoracic branches. It is only along the inner or thoracic wall, and in the centre of the axillary cavity, that there are no vessels of any importance; a most fortunate circumstance, for it is in this situation more especially that tumours requiring removal, are most frequently situated.

THE AXILLARY ARTERY

The axillary artery, the continuation of the subclavian, commences at the lower border of the first rib, and terminates at the lower border of the tendons of the Latissimus dorsi and Teres major muscles, when it becomes the brachial. Its direction varies with the position of the limb: where the arm lies by the side of the chest, the vessel forms a gentle curve, the convexity being upwards and outwards; when it is directed at right angles with the trunk, the vessel is nearly straight; and if elevated still higher, it describes a curve, the concavity of which is directed upwards. At its commencement the artery is very deeply situated, but near its termination is superficial, being covered only by the skin and fascia. The description of the relations of this vessel may be facilitated by its division into three portions. The first portion being that above the Pectoralis minor; the second portion, beneath; and the third, below that muscle.

The *first portion* of the axillary artery is in relation, *in front*, with the clavicular portion of the Pectoralis major, the costo-coracoid membrane, and the cephalic vein; *behind*, with the first intercostal space, the corresponding Intercostal muscle, the first serration of the Serratus magnus, and the posterior thoracic nerve; on its *outer side* with the brachial plexus, from which it is separated by a little cellular interval; on its *inner*, or thoracic side, with the axillary vein.

RELATIONS OF FIRST PORTION OF THE AXILLARY ARTERY

In front.
Pectoralis major.
Costo-coracoid membrane.
Cephalic vein.

Outer side.	Axillary Artery. First portion.	*Inner side.*
Brachial plexus.		Axillary vein.

Behind.
First intercostal space, and Intercostal muscle.
First serration of Serratus magnus.
Posterior thoracic nerve.

The *second portion* of the axillary artery lies beneath the Pectoralis minor. It is covered, *in front*, by the Pectoralis major and minor muscles; *behind*, it is separated from the Subscapularis by a cellular interval; on the *inner side*, it is in contact with the axillary vein. The brachial plexus of nerves surrounds the artery, and separates it from direct contact with the vein and adjacent muscles.

Relations of Second Portion of the Axillary Artery

In front.
Pectoralis major and minor.

Outer side.	Axillary Artery. Second portion.	Inner side.
Brachial plexus.		Axillary vein.

Behind.
Subscapularis.

The *third portion* of the axillary artery lies below the Pectoralis minor. It is in relation, *in front*, with the lower border of the Pectoralis major above, being covered only by the integument and fascia below; *behind*, with the lower part of the Subscapularis, and the tendons of the Latissimus dorsi and Teres major; on its *outer side*, with the Coraco-brachialis; on its *inner*, or thoracic side, with the axillary vein. The brachial plexus of nerves bears the following relation to the artery in this part of its course: on the *outer side* is the median nerve, and the musculo-cutaneous for a short distance; on the *inner side*, the ulnar, the internal, and lesser internal cutaneous nerves; and *behind*, the musculo-spiral, and circumflex, the latter extending only to the lower border of the Subscapularis muscle.

Relations of Third Portion of the Axillary Artery

In front.
Integument and fascia.
Pectoralis major.

Outer side.	Axillary Artery. Third portion.	Inner side.
Coraco-brachialis.		Ulnar nerve.
Median nerve.		Internal cutaneous nerves.
Musculo-cutaneous nerve.		Axillary vein.

Behind.
Subscapularis.
Tendons of Latissimus dorsi, and Teres major.
Musculo-spiral, and circumflex nerves.

Peculiarities. The axillary artery, in about one case out of every ten, gives off a large branch, which forms either one of the arteries of the forearm, or a large muscular trunk. In the first set of cases this artery is most frequently the radial (1 in 33), sometimes the ulnar (1 in 72), and very rarely, the interosseous (1 in 506). In the second set of cases, the trunk gave origin to the subscapular, circumflex, and profunda arteries of the arm. Sometimes only one of the circumflex, or one of the profunda arteries, arose from the trunk. In these cases the brachial plexus surrounded the trunk of the branches, and not the main vessel.

Surgical Anatomy. The student having carefully examined the relations of the axillary artery in its various parts, should now consider in what situation compression of this vessel may be most easily effected, and the best position for the application of a ligature to it when necessary.

Compression of this vessel is required in the removal of tumours, or in amputation of the upper part of the arm; and the only situation in which this can be effectually made, is in the lower part of its course, just beneath the integument and fascia; and on compressing it from within outwards upon the humerus, the circulation may be efficiently suspended.

The *application of a ligature to the axillary artery* may be required, in cases of aneurism of the upper part of the brachial; and there are only two situations in which it may be secured, viz., in the upper, or in the lower part of its course.

The *axillary artery at its central part* is so deeply seated, and, at the same time, so closely surrounded with large nervous trunks, that the application of a ligature to it would be almost impracticable.

In the *lower part* of its course the operation is more simple, and may be performed in the following manner:—The patient being placed on a bed, and the arm separated from the side, with the hand supinated, the head of the humerus is felt for, and an incision made through the integument over it, about two inches in length, a little nearer to the posterior than the anterior fold of the axilla. After carefully dissecting through the areolar tissue and fascia, the median nerve and axillary vein are exposed; the former having been displaced to the outer, and the latter to the inner side of the arm, the elbow being at the same time bent so as to relax these structures, and facilitate their separation, the ligature may be passed around the artery from the ulnar to the radial side.

The *upper portion* of the axillary artery may be tied, in cases of aneurism encroaching so far upwards that a ligature cannot be applied in the lower part of its course. Notwithstanding that this operation has been performed in some few cases, and with success, its performance is attended with much difficulty and danger. The student will remark, that in this situation, it would be necessary to divide a thick muscle, and, after separating the costo-coracoid membrane, the artery would be exposed at the bottom of a more or less deep space, with the cephalic and axillary veins in such relation with it as must render the application of a ligature to this part of the vessel particularly hazardous. Under such circumstances it is an easier, and, at the same time, more advisable operation, to tie the subclavian artery in the third part of its course.

In a case of wound of this vessel, the general practice of cutting down upon, and tying the vessel above and below the wounded point, should be adopted in all cases.

The branches of the axillary artery are, the

From 1st Part.	{ Superior thoracic. Acromial thoracic.
From 2nd Part.	{ Thoracica longa. Thoracica alaris.
From 3rd Part.	{ Subscapular. Anterior circumflex. Posterior circumflex.

The *Superior Thoracic* is a small artery, which arises from the axillary, or by a common trunk with the acromial thoracic. Running forwards and inwards along the upper border of the Pectoralis minor, it passes between it and the Pectoralis major to the side of the chest. It supplies these muscles, and the parietes of the thorax, anastomosing with the internal mammary and intercostal arteries.

The *Acromial Thoracic* is a short trunk, which arises from the fore part of the axillary artery. Projecting forwards to the upper border of the Pectoralis minor, it divides into three sets of branches, thoracic, acromial, and descending. The thoracic branches, two or three in number, are distributed to the Serratus magnus and Pectoral muscles, anastomosing with the intercostal branches of the internal mammary. The acromial branches are directed outwards towards the acromion, supplying the Deltoid muscle, and anastomosing, on the surface of the acromion, with the suprascapular and posterior circumflex arteries. The descending branch passes in the interspace between the Pectoralis major and Deltoid, accompanying the cephalic vein, and supplying both muscles.

The *Long Thoracic* passes downwards and inwards along the lower border of the Pectoralis minor to the side of the chest, supplying the Serratus magnus, the Pectoral muscles, and mammary gland, and sending branches across the axilla to the axillary glands and Subscapularis, which anastomose with the internal mammary and intercostal arteries.

The *Thoracica Alaris* is a small branch, which supplies the glands and areolar tissue of the axilla. Its place is frequently supplied by branches from some of the other thoracic arteries.

The *Subscapular*, the largest branch of the axillary artery, arises opposite the lower border of the Subscapularis muscle, and passes downwards and backwards along its lower margin to the inferior angle of the scapula, where it anastomoses with the posterior scapular, a branch of the subclavian. It distributes branches to the Subscapularis, the Serratus magnus, Teres major, and Latissimus dorsi muscles, and gives off, about an inch and a-half from its origin, a large branch, the dorsalis scapulae. This vessel curves round the inferior border of the scapula, leaving the axilla in the interspace between the Teres minor above, the Teres major below, and the long head of the Triceps in front; and divides into three branches, a subscapular, which enters the subscapular fossa beneath the Subscapularis which it supplies, anastomosing with the subscapular and suprascapular arteries; an infra-spinous branch (dorsalis scapulae), which turns round the axillary border of the scapula, between the Teres minor and the bone, enters the infra-spinous fossa, supplies the Infra-spinatus muscle, and anastomoses with the supra-scapular and posterior scapular arteries; and a median branch, which is continued along the axillary border of the scapula, between the Teres major and minor, and, at the dorsal surface of the inferior angle of the bone, anastomoses with the supra-scapular.

The *Circumflex Arteries* wind around the neck of the humerus.

The *Posterior*, the larger of the two, arises from the back part of the axillary, opposite the lower border of the Subscapularis muscle, and, passing backwards with the circumflex veins and nerve, through the quadrangular space bounded by the Teres major and minor, the scapular head of the Triceps and the humerus, winds around the neck of that bone, is distributed to the Deltoid muscle and shoulder-joint, anastomosing with the anterior circumflex, supra-scapular, and acromial thoracic arteries.

The *Anterior Circumflex*, considerably smaller than the preceding, arises just below that vessel, from the outer side of the axillary artery. It passes horizontally outwards, beneath the coraco-brachialis and short head of the Biceps, lying upon the fore part of the neck of the humerus, and, on reaching the bicipital groove, gives off an ascending branch, which passes upwards along it, to supply the head of the bone and the shoulder-joint. The trunk of the vessel is then continued outwards beneath the Deltoid which it supplies, and anastomoses with the posterior circumflex, and acromial thoracic arteries.

200.—The Surgical Anatomy of the Brachial Artery.

BRACHIAL ARTERY (fig. 200, color illustration p. A3)

The brachial artery commences at the lower margin of the tendon of the Teres major, and, passing down the inner and anterior aspect of the arm, terminates about half an inch below the bend of the elbow, where it divides into the radial and ulnar arteries.

The direction of this vessel is marked by a line drawn from the outer side of the axillary space between the folds of the axilla, to a point midway between the condyles of the humerus, which corresponds to the depression along the inner border of the Coraco-brachialis and Biceps muscles. In the upper part of its course, this vessel lies internal to the humerus; but below, it is in front of that bone.

Relations. This artery is superficial throughout its entire extent, being covered, *in front*, by the integument, the superficial and deep fasciae; the bicipital fascia separates it opposite the elbow from the median basilic vein, the median nerve crosses it at its centre, and the basilic vein lies in the line of the artery for the lower half of its course. *Behind*, it is separated from the inner side of the humerus above, by the long and inner heads of the Triceps, the musculo-spiral nerve and superior profunda artery intervening; and from the front of the bone below, by the insertion of the Coraco-brachialis and the Brachialis anticus muscles. By its *outer side*, it is in relation with the commencement of the median nerve, and the Coraco-brachialis and Biceps muscles, which slightly overlap the artery. By its *inner side*, with the internal cutaneous and ulnar nerves, its upper half; the median nerve, its lower half. It is accompanied by two veins, the venae comites; they lie in close contact with the artery, being connected together at intervals by short transverse communicating branches.

PLAN OF THE RELATIONS OF THE BRACHIAL ARTERY
In front.
Integument and fasciae.
Bicipital fascia, median basilic vein.
Median nerve.

Outer side.		*Inner side.*
Median nerve.	Brachial Artery.	Inner cutaneous.
Coraco-brachialis.		Ulnar and median nerves.
Biceps.		

Behind.
Triceps.
Musculo-spiral nerve.
Superior profunda artery.
Coraco brachialis.
Brachialis anticus.

BEND OF THE ELBOW

At the bend of the elbow, the brachial artery sinks deeply into a triangular interval, the base of which is directed upwards towards the humerus, and the sides of which are bounded, externally, by the Supinator longus, internally, by the Pronator radii teres; its floor is formed by the Brachialis anticus, and Supinator brevis. This space contains the brachial artery, with its accompanying veins; the radial and ulnar arteries, the median and musculo-spiral nerves, and the tendon of the biceps. The brachial artery occupies the middle line of this space, and divides opposite the coronoid process of the ulna into the radial and ulnar arteries; it is covered, *in front*, by the integument, the superficial fascia, and the median basilic vein, the vein being separated from direct contact with the artery by the bicipital fascia. *Behind*, it lies on the Brachialis anticus, which separates it from the elbow-joint. The median nerve lies on the inner side of the artery, but separated from it below by an interval of half an inch. The tendon of the Biceps lies to the outer side of the space, and the musculo-spiral nerve still more externally, lying upon the Supinator brevis, and partly concealed by the Supinator longus.

Peculiarities of the Artery as regards its Course. The brachial artery, accompanied by the median nerve, may leave the inner border of the Biceps, and descend towards the inner condyle of the humerus, where it usually curves around a prominence of bone, to which it is connected by a fibrous band; it then inclines outwards, beneath or through the substance of the Pronator teres muscle, to the bend of the elbow. This variation bears considerable analogy with the normal condition of the artery in some of the carnivora.

As regards its Division. Occasionally, the artery is divided for a short distance at its upper part into two trunks, which are united above and below. A similar peculiarity occurs in the main vessel of the lower limb.

The point of bifurcation may be above or below the usual point, the former condition being by far the most frequent. Out of 481 examinations recorded by Mr. Quain, some made on the right and some on the left side of the body, in 386 the artery bifurcated in its normal position. In one case only was the place of division lower than usual, being two or three inches below the elbow-joint. "In ninety cases out of 481, or about 1 in $5^{1}/_{9}$, there were two arteries instead of one in some part, or in the whole of the arm."

There appears, however, to be no correspondence between the arteries of the two arms, with respect to their irregular division; for in sixty-one bodies it occurred in one side only in forty-three; on both sides, in different positions, in thirteen; on both sides, in the same position, in five.

The point of bifurcation takes place at different parts of the arm, being most frequent in the upper part, less so in the lower part, and least so in the middle, the most usual point for the application of a ligature; under any of these circumstances two large arteries would be found in the arm instead of one. The most frequent (in three out of four) of these peculiarities is the high division of the radial. It often arises from the inner side of the brachial, and runs parallel with the main trunk to the elbow, where it crosses it, lying beneath the fascia; or it may perforate the fascia, and pass over the artery, immediately beneath the integument.

The ulnar sometimes arises from the brachial high up, and then occasionally leaves that vessel at the lower part of the arm, and descends towards the inner condyle. In the fore-arm it generally lies beneath the deep fascia, superficial to the Flexor muscles; occasionally between the integument and deep fascia, and very rarely beneath the Flexor muscles.

The interosseous artery sometimes arises from the upper part of the brachial or axillary: as it descends the arm it lies behind the main trunk, and at the bend of the elbow regains its usual position.

In some cases of high division of the radial, the remaining trunk (ulnar interosseous) occasionally passes, together with the median nerve, along the inner margin of the arm to the inner condyle, and then passing from within outwards, beneath or through the Pronator teres, regains its usual position at the bend of the elbow.

Occasionally, the two arteries representing the brachial are connected at the bend of the elbow by a short transverse branch, and are even sometimes reunited.

Sometimes long slender vessels, *vasa aberrantia*, connect the brachial or axillary arteries with one of the arteries of the fore-arm, or a branch from them. These vessels usually join the radial.

In some subjects the brachial artery is covered by an additional slip from the Biceps or brachialis anticus muscles; and occasionally a slip from the latter muscle covers the whole extent of the ulnar interosseous trunk in cases of high division of the radial.

Surgical Anatomy. Compression of the brachial artery is required in cases of amputation of the arm or fore-arm, in resection of the elbow-joint, and the removal of tumours; and it will be observed, that it may be effected in almost any part of its course; if pressure is made in the upper part of the limb it should be directed from within outwards, and if in the lower part, from before backwards, as the artery lies on the inner side of the humerus above, and in front of it below. The most favourable situation is either above or below the insertion of the Coraco-brachialis.

The application of a ligature to the brachial artery may be required in cases of wounds of the vessel, or of aneurism of the brachial, the radial, ulnar, or interosseous arteries; and this vessel may be secured in any part of its course. The chief guides in determining its position are the surface-markings produced by the inner margin of the Coraco-brachialis and Biceps, the known course of the vessel, and its pulsation, which should be carefully felt for before any operation is performed, as the vessel occasionally deviates from its usual position in the arm. In whatever situation the operation is performed, great care is necessary on account of the extreme thinness of the parts covering the artery, and the intimate connection which the vessel has throughout its whole course with important nerves and veins. Sometimes a thin layer of muscular fibre is met with concealing the artery; if such is the case, it must be divided across, in order to expose it.

In the upper third of the arm the artery may be exposed in the following manner. The patient being placed horizontally upon a table, the affected limb should be raised from the side, and the hand supinated. An incision about two inches in length should be made on the ulnar side of the Coraco-brachialis muscle, and the subjacent fascia cautiously divided so as to avoid wounding the internal cutaneous nerve or basilic vein, which sometimes runs on the surface of the artery as high as the axilla. The fascia having been divided, it should be remembered, that the ulnar and internal cutaneous nerves lie on the inner side of the artery, the median on the outer side, the latter nerve being occasionally superficial to the artery in this situation, and that the venae comites are also in relation with the vessel, one on either side. These being carefully separated, the aneurism needle should be passed around the artery from the ulnar to the radial side.

If two arteries are present in the arm in consequence of a high division, they are usually placed side by side; and if they are exposed in an operation, the surgeon should endeavour to ascertain, by alternately pressing on one or the other vessel, which of the two communicates with the wound or aneurism, when a ligature may be applied accordingly; or if pulsation or haemorrhage ceases only when both vessels are compressed, both vessels may be tied, as it may be concluded that the two communicate above the seat of disease or are reunited.

It should also be remembered, that two arteries may be present in the arm in a case of high division, and that one of these may be found along the inner intermuscular septum, in a line towards the inner condyle of the humerus, or in its usual position, but deeply placed, beneath the common trunk: a knowledge of these facts will at once suggest the precautions necessary in every case, and indicate the necessary measure to be adopted when met with.

In the middle of the arm the brachial artery may be exposed by making an incision along the inner margin of the Biceps muscle. The fore-arm being bent so as to relax the muscle, it should be drawn slightly aside, and the fascia being carefully divided, the median nerve will be exposed lying upon the artery (sometimes beneath); this being drawn inwards and the muscle outwards, the artery should be separated from its accompanying veins and secured. In this situation the inferior profunda may be mistaken for the main trunk, especially if enlarged, from the collateral circulation having become established; this may be avoided by directing the incision externally towards the Biceps, rather than inwards or backwards towards the Triceps.

The lower part of the brachial artery is of extreme interest in a surgical point of view, on account of the relation which it bears to those veins most commonly opened in venesection. Of these vessels, the median basilic is the largest and most prominent, and, consequently, the one usually selected for the operation. It should be remembered, that this vein runs parallel with the brachial artery, from which it is separated by the bicipital fascia, and that in no case should this vessel be selected for the operation, except in a part which is not in contact with the artery.

The branches of the brachial artery are the

Superior profunda.	Inferior profunda.
Nutrient artery.	Anastomotica magna.
Muscular.	

The *Superior Profunda* arises from the inner and back part of the brachial, opposite the lower border of the Teres major, and passes backwards to the interval between the outer and inner heads of the Triceps muscle, accompanied by the musculo-spiral nerve; it winds around the back part of the shaft of the humerus in the spiral groove, between the Triceps and the bone, and descends on the outer side of the arm to the space between the Brachialis anticus, and Supinator longus, as far as the elbow, where it anastomoses with the recurrent branch of the radial artery. It supplies the Deltoid, Coraco-brachialis, and Triceps muscles, and whilst in the groove, between the Triceps and the bone, it gives off the posterior articular artery, which descends perpendicularly between the Triceps and the bone, to the back part of the elbow-joint, where it anastomoses with the interosseous recurrent branch, and, on the inner side of the arm, with the ulnar recurrent, and with the anastomotica magna or inferior profunda (fig. 203).

The *Nutrient artery* of the shaft of the humerus arises from the brachial, about the middle of the arm. Passing downwards, it enters the nutritious canal of that bone, near the insertion of the Coraco-brachialis muscle.

The *Inferior Profunda*, of small size, arises from the brachial, a little below the middle of the arm; piercing the internal intermuscular septum, it descends on the surface of the inner head of the Triceps muscle, to the space between the inner condyle and olecranon, accompanied by the ulnar nerve, and terminates by anastomosing with the posterior ulnar recurrent, and anastomotica magna.

The *Anastomotica Magna* arises from the brachial, about two inches above the elbow-joint. It passes transversely inwards upon the Brachialis anticus, and, piercing the internal intermuscular septum, winds around the back part of the humerus, between the Triceps and the bone, forming an arch above the olecranon fossa, by its junction with the posterior articular branch of the superior profunda. As this vessel lies on the Brachialis anticus, an offset passes between the internal condyle and olecranon, where it anastomoses with the inferior profunda and posterior ulnar recurrent arteries. Other branches ascend to join the inferior profunda; and some descend in front of the inner condyle, to anastomose with the anterior ulnar recurrent.

201.—The Surgical Anatomy of the Radial and Ulnar Arteries.

The *Muscular* are three or four large branches, which are distributed to the muscles in the course of the artery. They supply the Coraco-brachialis, Biceps, and Brachialis anticus muscles.

Radial Artery

The Radial artery appears, from its direction, to be the continuation of the brachial, but, in size, it is smaller than the ulnar. It commences at the bifurcation of the brachial, just below the bend of the elbow, and passes along the radial side of the fore-arm to the wrist; it then winds backwards, round the outer side of the carpus, beneath the extensor tendons of the thumb, and, running forwards, passes between the two heads of the first Dorsal interosseous muscle, into the palm of the hand. It then crosses the metacarpal bones to the ulnar border of the hand, forming the deep palmar arch, and, at its termination, inosculates with the deep branch of the ulnar artery. The relations of this vessel may thus be conveniently divided into three parts, viz., in front of the fore-arm, at the back of the wrist, and in the hand.

Relations. In the fore-arm, this vessel extends from opposite the neck of the radius, to the fore part of the styloid process, being placed to the inner side of the shaft of that bone above, and in front of it below. It is superficial throughout its entire extent, being covered by the integument, the superficial and deep fasciae, and slightly over-lapped superiorly by the Supinator longus. In its course downwards it lies upon the tendon of the Biceps, the Supinator brevis, the Pronator radii teres, radial origin of the Flexor sublimis digitorum, the Flexor longus pollicis, Pronator quadratus, and the lower extremity of the radius. In the upper third of its course, it lies between the Supinator longus and the Pronator radii teres; in its lower two thirds, between the tendons of the Supinator longus and the Flexor carpi radialis. The radial nerve lies along the outer side of the artery, in the middle third of its course; and some filaments of the musculo-cutaneous nerve, after piercing the deep fascia, run along the lower part of the artery as it winds around the wrist. The vessel is accompanied by venae comites throughout its whole course.

PLAN OF THE RELATIONS OF THE RADIAL ARTERY IN THE FORE-ARM
In front.
Integument—superficial and deep fasciae.
Supinator longus.

Inner side.	Radial artery in Fore-arm.	*Outer side.*
Pronator radii teres.		Supinator longus.
Flexor carpi radialis.		Radial nerve (middle third).

Behind.
Tendon of Biceps.
Supinator brevis.
Pronator radii teres.
Flexor sublimis digitorum.
Flexor longus pollicis.
Pronator quadratus.
Radius.

At the wrist, as it winds around the outer side of the Carpus, from the styloid process to the first interosseous space, it lies upon the external lateral ligament being covered by the extensor tendons of the thumb, subcutaneous veins, some filaments of the radial nerve, and the integument. It is accompanied by two veins, and a filament of the musculo-cutaneous nerve.

In the hand, it passes from the upper end of the first interosseous space, between the heads of the Abductor indicis, transversely across the palm, to the base of the metacarpal bone of the little finger, where it inosculates with the communicating branch from the ulnar artery, forming the deep palmar arch. It lies upon the carpal extremities of the metacarpal bones and the Interossei muscles, being covered by the flexor tendons of the fingers, the Lumbricales, the muscles of the little finger, and the Flexor brevis pollicis, and is accompanied by the deep branch of the ulnar nerve.

Peculiarities. The origin of the radial artery varies in the proportion nearly of one in eight cases. In one case the origin was lower than usual. In the other cases, the upper part of the brachial was a more frequent source of origin than the axillary. The variations in the position of this vessel in the arm, and at the bend of the elbow, have been already mentioned. In the fore-arm it deviates less frequently from its position than the ulnar. It has been found lying over the fascia, instead of beneath it. It has also been observed on the surface of the Supinator longus, instead of along its inner border; and in turning round the wrist, it has been seen lying over, instead of beneath, the Extensor tendons.

Surgical Anatomy. The operation of tying the radial artery is required in cases of wounds either of its trunk, or of some of its branches, or for aneurism: and it will be observed, that the vessel may be easily exposed in any part of its course through the fore-arm. This operation in the middle or inferior third of this region is easily performed; but in the upper third, near the elbow, the operation is attended with some difficulty, from the greater depth of the vessel, and from its being overlapped by the Supinator longus and Pronator teres muscles.

To tie the artery in this situation, an incision three inches in length should be made through the integument, from the bend of the elbow obliquely downwards and outwards, on the radial side of the fore-arm, avoiding the branches of the median vein; the fascia of the arm being divided, and the Supinator longus drawn a little outwards, the artery will be exposed. The venae comites should be carefully separated from the vessel, and the ligature passed from the radial to the ulnar side.

In the middle third of the fore-arm the artery may be exposed by making an incision of similar length on the inner margin of the Supinator longus. In this situation the radial nerve lies in close relation with the outer side of the artery, and should, as well as the veins, be carefully avoided.

In the inferior third, the artery is easily secured by dividing the integument and fasciae in the interval between the tendons of the Supinator longus and Flexor carpi radialis muscles.

The branches of the radial artery may be divided into three groups, corresponding with the three regions in which this vessel is situated.

In the Fore-arm.
{ Radial recurrent.
  Muscular.
  Superficialis volae.
  Anterior carpal. }

Wrist.
{ Posterior carpal.
  Metacarpal.
  Dorsales pollicis.
  Dorsalis indicis. }

Hand.
{ Princeps pollicis.
  Radialis indicis.
  Perforantes.
  Interossei. }

The *Radial Recurrent* is given off immediately below the elbow. It ascends between the branches of the musculo-spiral nerve, lying on the Supinator brevis, and then between the Supinator longus and Brachialis anticus, supplying these muscles, the elbow-joint, and anastomosing with the terminal branches of the superior profunda.

The *Muscular Branches* are distributed to the muscles on the radial side of the fore-arm.

The *Superficialis Volae* arises from the radial artery, just where this vessel is about to wind around the wrist. Running forwards, it passes between the muscles of the thumb, which it supplies, and anastomoses with the termination of the ulnar artery, completing the superficial palmar arch. This vessel varies considerably in size, usually it is very small, and terminates in the muscles of the thumb; sometimes it is as large as the continuation of the radial.

202.—Ulnar and Radial Arteries. Deep View.

The Carpal branches supply the joints of the wrist. The *Anterior Carpal* is a small vessel which arises from the radial artery near the lower border of the Pronator quadratus, and running inwards in front of the radius, anastomoses with the anterior carpal branch of the ulnar artery. From the arch thus formed, branches descend to supply the articulations of the wrist.

The *Posterior Carpal* is a small vessel which arises from the radial artery beneath the extensor tendons of the thumb; crossing the carpus transversely to the inner border of the hand, it anastomoses with the posterior carpal branch of the ulnar.

Superiorly it sends branches upwards, which anastomose with the termination of the anterior interosseous artery. Other branches descend to the metacarpal spaces; they are the dorsal interosseous arteries for the third and fourth interosseous spaces; they anastomose with the posterior perforating branches from the deep palmar arch.

The *Metacarpal (First Dorsal Interosseous Branch)* arises beneath the extensor tendons of the thumb, sometimes with the posterior carpal artery; running forwards on the second dorsal interosseous muscle; it communicates, behind, with the corresponding perforating branch of the deep palmar arch; and, in front, inosculates with the digital branch of the superficial palmar arch, and supplies the adjoining sides of the index and middle fingers.

The *Dorsales Pollicis* are two small vessels which run along the sides of the dorsal aspect of the thumb. They sometimes arise separately, or occasionally by a common trunk, near the base of the first metacarpal bone.

The *Dorsalis Indicis*, also a small branch, runs along the radial side of the back of the index finger, sending a few branches to the Abductor indicis.

The *Princeps Pollicis* arises from the radial just as it turns inwards to the deep part of the hand; it descends between the Abductor indicis and Adductor pollicis, along the ulnar side of the metacarpal bone of the thumb, to the base of the first phalanx, where it divides into two branches, which run along the sides of the palmar aspect of the thumb, and form an arch on the under surface of the last phalanx, from which branches are distributed to the integument and cellular membrane of the thumb.

The *Radialis Indicis* arises close to the preceding, descends between the Abductor indicis and Adductor pollicis, and runs along the radial side of the index finger to its extremity, where it anastomoses with the collateral digital artery from the superficial palmar arch. At the lower border of the Adductor pollicis this vessel anastomoses with the princeps pollicis, and gives a communicating branch to the superficial palmar arch.

The *Perforantes*, three in number, pass backwards between the heads of the last three Dorsal interossei muscles, to inosculate with the dorsal interosseous arteries.

The *Palmar Interossei*, three or four in number, are branches of the deep palmar arch; they run forwards upon the Interossei muscles, and anastomose at the clefts of the fingers with the digital branches of the superficial arch.

Ulnar Artery

The *Ulnar Artery*, the larger of the two sub-divisions of the brachial, commences a little below the bend of the elbow, and crosses the inner side of the fore-arm obliquely to the commencement of its lower half; it then runs along its ulnar side to the wrist, crosses the annular ligament on the radial side of the pisiform bone, and passes across the palm of the hand, forming the superficial palmar arch, which terminates by inosculating with the superficialis volae.

Relations in the Fore-arm. In its *upper half*, it is deeply seated, being covered by all the superficial Flexor muscles, excepting the Flexor carpi ulnaris; crossed by the median nerve, which, at its origin, for about an inch lies to its inner side; and it lies upon the Brachialis anticus and Flexor profundus digitorum muscles. In the *lower half* of the fore-arm, it lies upon the Flexor profundus, being covered by the integument, the superficial and deep fasciae, and is placed between the Flexor carpi ulnaris and Flexor sublimis digitorum muscles. It is accompanied by two veins, which lie one on each side of the vessel; the ulnar nerve lies on its inner side for the lower two thirds of its extent, and a small branch from it descends on the lower part of the vessel to the palm of the hand.

Plan of Relations of the Ulnar Artery in the Fore-arm

In front.

Superficial flexor muscles. } Upper half.
Median nerve.
Superficial and deep fasciae. Lower half.

Inner side.	Ulnar Artery in Fore-arm.	*Outer side.*
Flexor carpi ulnaris.		Flexor sublimis digitorum.
Ulnar nerve (lower two-thirds).		

Behind.
Brachialis anticus.
Flexor profundus digitorum.

At the wrist, the ulnar artery is covered by the integument and fascia, and lies upon the anterior annular ligament. On its inner side is the pisiform bone. The ulnar nerve lies at the inner side, and somewhat behind the artery.

In the palm of the hand, the continuation of the ulnar artery is called the superficial palmar arch; it passes obliquely outwards to the interspace between the ball of the thumb and the index finger, where it anastomoses with the superficialis volae, and a branch from the radialis indicis, thus completing the superficial palmar arch. The convexity of this arch is directed towards the fingers, its concavity towards the muscles of the thumb.

The superficial palmar arch is covered by the Palmaris brevis, the palmar fascia, and integument; and lies upon the annular ligament, the muscles of the little finger, the tendons of the superficial Flexor, and the divisions of the median and ulnar nerves, the latter accompanying the artery a short part of its course.

Relations of the Superficial Palmar Arch

In front.	Ulnar Artery in Hand.	*Behind.*
Integument.		Annular ligament.
Palmaris brevis.		Origin of muscles of little finger.
Palmar fascia.		Superficial flexor tendons.
		Divisions of median and ulnar nerves.

Peculiarities. The ulnar artery was found to vary in its origin nearly in the proportion of one in thirteen cases, in one case arising lower than usual, about two or three inches below the elbow, and in all the other cases much higher, the brachial being a more frequent source of origin than the axillary.

Variations in the position of this vessel are more frequent than in the radial. When its origin is normal, the course of the vessel is rarely changed. When it arises high up, its position in the fore-arm is almost invariably superficial to the Flexor muscles, lying commonly beneath the fascia, more rarely between the fascia and integument.

In a few cases, its position was subcutaneous in the upper part of the fore-arm, sub-aponeurotic in the lower part.

Surgical Anatomy. The application of a ligature to this vessel is required in cases of wound of the artery, or of its branches, or in consequence of aneurism. In the upper half of the fore-arm, the artery is deeply seated beneath the superficial Flexor muscles, and their division would be requisite in a case of recent wound of the artery in this situation, in order to secure it, but under no other circumstances. In the middle and inferior thirds of the fore-arm, this vessel may be easily secured by making an incision on the radial side of the tendon of the Flexor carpi ulnaris; the deep fascia being divided, and the Flexor carpi ulnaris and its companion muscle, the Flexor sublimis, being separated from each other, the vessel will be exposed, accompanied by its venae comites, the ulnar nerve lying on its inner side. The veins being separated from the artery, the ligature should be passed from its ulnar to its radial side, taking care to avoid the ulnar nerve.

The branches of the ulnar artery may be arranged into three groups,

Fore-arm.	Anterior ulnar recurrent.	
	Posterior ulnar recurrent.	
	Interosseous	Anterior interosseous.
		Posterior interosseous.
	Muscular.	
Wrist.	Anterior carpal.	
	Posterior carpal.	
Hand.	Deep or communicating branch.	
	Digital.	

The *Anterior Ulnar Recurrent* arises immediately below the elbow-joint, passes upwards and inwards between the Brachialis anticus and Pronator radii teres, supplies these muscles, and, in front of the inner condyle, anastomoses with the anastomotica magna and inferior profunda.

The *Posterior Ulnar Recurrent* is much larger, and arises somewhat lower than the preceding. It passes backwards and inwards, beneath the Flexor sublimis, and ascends behind the inner condyle of the humerus. In the interval between this eminence and the olecranon, it lies beneath the Flexor carpi ulnaris, ascending between the heads of that muscle, beneath the ulnar nerve; it supplies the neighbouring muscles and joint, and anastomoses with the inferior profunda, anastomotica magna, and interosseous recurrent arteries.

The *Interosseous Artery* is a short trunk, about an inch in length, and of considerable size, which arises immediately below the bicipital tuberosity of the radius, and, passing backwards to the upper border of the interosseous membrane, divides into two branches, the anterior, and posterior interosseous.

The *Anterior Interosseous* passes down the fore-arm on the anterior surface of the interosseous membrane, to which it is connected by a thin aponeurotic arch. It is accompanied by the interosseous branch of the median nerve, and overlapped by the contiguous margins of the Flexor profundus digitorum and Flexor longus pollicis muscles, giving off in this situation muscular branches, and the nutrient arteries of the radius and ulna. At the upper border of the Pronator quadratus, a branch descends in front of that muscle, to anastomose in front of the carpus with the branches from the anterior carpal and deep palmar arch. The continuation of the artery passes behind the Pronator quadratus, and, piercing the interosseous membrane, descends to the back of the wrist, where it anastomoses with the posterior carpal branches of the radial and ulnar arteries. The anterior interosseous gives off a long, slender branch, which accompanies the median nerve, and gives off-sets to its substance. This, the median artery, is sometimes much enlarged.

The *Posterior Interosseous Artery* passes backwards through the interval between the oblique ligament and the upper border of the interosseous membrane, and passes down the back part of the fore-arm, between the superficial and deep layer of muscles, to both of which it distributes branches. Descending to the back of the wrist, it anastomoses with the termination of the anterior interosseous, and with the posterior carpal branches of the radial and ulnar arteries. This artery gives off, near its origin, the posterior interosseous recurrent branch, a large vessel, which ascends to the interval between the external condyle and olecranon, beneath the Anconeus and Supinator brevis, anastomosing with a branch from the supinator profunda, and with the posterior ulnar recurrent arteries.

The *Muscular Branches* are distributed to the muscles along the ulnar side of the fore-arm.

203.—Arteries of the Back of the Fore-arm and Hand.

The *Carpal Branches* are intended for the supply of the wrist-joint.

The *Anterior Carpal* is a small vessel, which crosses the anterior surface of the carpus beneath the tendons of the Flexor profundus, and inosculates with a corresponding branch of the radial artery.

The *Posterior Carpal* arises immediately above the pisiform bone, winding backwards beneath the tendon of the Flexor carpi ulnaris; it gives off a branch which passes across the dorsal surface of the carpus beneath the extensor tendons, anastomosing with a corresponding branch of the radial artery, and forming the posterior carpal arch; it is then continued along the metacarpal bone of the little finger, forming its dorsal branch.

The *Deep* or *Communicating Branch* arises at the commencement of the palmar arch, passing deeply inwards between the Abductor minimi digiti and Flexor brevis minimi digiti, near their origins; it anastomoses with the termination of the radial artery, completing the deep palmar arch.

The *Digital Branches*, four in number, are given off from the convexity of the superficial palmar arch. They supply the ulnar side of the little finger, and the adjoining sides of the ring, middle, and index fingers; the radial side of the index finger and thumb being supplied from the radial artery. The digital arteries at first lie superficial to the flexor tendons, but as they pass forwards with the digital nerves to the clefts between the fingers, they lie between them, and are there joined by the interosseous branches from the deep palmar arch. The digital arteries on the sides of the fingers lie beneath the digital nerves; and, about the middle of the last phalanx, the two branches for each finger, form an arch, from the convexity of which branches pass to supply the matrix of the nail.

The Descending Aorta

The descending aorta is divided into two portions, the thoracic, and abdominal, in correspondence with the two great cavities of the trunk in which it is situated.

The *Thoracic Aorta* commences at the lower border of the third dorsal vertebra, on the left side, and terminates at the aortic opening in the Diaphragm in front of the last dorsal vertebra. At its commencement, it is situated on the left side of the spine; it approaches the median line as it descends, and, at its termination, lies directly in front of the column. The direction of this vessel being influenced by the spine, upon which it rests, it is concave forwards in the dorsal region, and, as the branches given off from it are small, the diminution in the size of the vessel is inconsiderable. It is contained in the back part of the posterior mediastinum, being in relation in front, from above downwards, with the left pulmonary artery, the left bronchus, the pericardium, and the oesophagus; *behind*, with the vertebral column, and the vena azygos minor; on the *right side*, with the vena azygos major, and thoracic duct; on the *left side*, with the left pleura and lung. The oesophagus, with its accompanying nerves, lies on the right side of the aorta *above*; in front of this vessel, in the middle of its course; whilst, at its *lower part*, it is on the left side, on a plane anterior to it.

Plan of the Relations of the Thoracic Aorta

In front.
Left pulmonary artery.
Left bronchus.
Pericardium.
Oesophagus.

Right side.
Oesophagus (above).
Vena azygos major.
Thoracic duct.

Thoracic Aorta.

Left side.
Pleura.
Left lung.
Oesophagus (below).

Behind.
Vertebral column.
Vena azygos minor.

Surgical Anatomy. The student should now consider the effects likely to be produced by aneurism of the thoracic aorta, a disease of common occurrence. When we consider the great depth of the vessel from the surface, and the number of important structures which surround it on every side, it may be easily conceived what a variety of obscure symptoms may arise, from disease of this part of the arterial system, and how they may be liable to be mistaken for those of other affections. Aneurism of the thoracic aorta most usually extends backwards, along the left side of the spine, producing absorption of the bodies of the vertebrae, causing extensive curvature of the spine; whilst the irritation or pressure on the cord, will give rise to pain, either in the chest, back, or loins, with radiating pain in the left upper intercostal spaces, from pressure on the intercostal nerves; at the same time the tumour may project back on each side of the spine, beneath the integument, as a pulsating swelling, simulating abscess connected with diseased bone; or it may displace the oesophagus, and compress the lung on one or the other side. If the tumour extend forward, it may press upon and displace the heart, giving rise to palpitation, and other symptoms of disease of that organ; or it may displace, or even compress, the oesophagus, causing pain and difficulty of swallowing, as in stricture of that tube, and ultimately even open into it by ulceration, producing fatal haemorrhage. If the disease make way to either side, it may press upon the thoracic duct; or it may burst into the pleural cavity, or into the trachea or lung; and lastly, it may open into the posterior mediastinum.

Branches of the Thoracic Aorta

Pericardiac.
Bronchial.
Intercostal.
Oesophageal.
Posterior mediastinal.

The *Pericardiac* are a few small vessels, irregular in their origin, distributed to the pericardium.

The *Bronchial* arteries are the nutrient vessels of the lungs, and vary in number, size, and origin. That of the right side arises from the first aortic intercostal, or by a common trunk with the left bronchial, from the anterior part of the thoracic aorta. Those of the left side, usually two in number, arise from the thoracic aorta, one a little lower than the other. Each vessel is directed forwards to the back part of the corresponding bronchus, along which they run, dividing and subdividing, upon the bronchial tubes, supplying them, the cellular tissue of the lungs, the bronchial glands and the oesophagus.

The *Oesophageal Arteries*, usually four or five in number, arise from the anterior part of the aorta, and pass obliquely downwards to the oesophagus, forming a chain of anastomoses along that tube, anastomosing with the oesophageal branches of the inferior thyroid arteries above, and with ascending branches from the phrenic and gastric arteries below.

The *Posterior Mediastinal Arteries* are numerous small vessels which supply the glands and loose areolar tissue in the mediastinum.

The *Intercostal Arteries* arise from the posterior part of the aorta. They are usually ten in number on each side, the superior intercostal space (and occasionally the second one) being supplied by the superior intercostal, a branch of the subclavian. The right intercostals are longer than the left, on account of the position of the aorta to the left side of the spine. They pass outwards, across the bodies of the vertebrae, to the intercostal spaces, being covered by the pleura, and crossed by the oesophagus, thoracic duct, sympathetic nerve, and the vena-azygos major, the left passing beneath the superior intercostal vein, the vena-azygos minor, and sympathetic. In the intercostal spaces each artery divides into two branches, an anterior, or proper intercostal branch, which passes outwards; and a posterior, or dorsal branch which passes backwards. The anterior branch passes outwards, at first lying upon the external Intercostal muscle, covered in front by the pleura and a thin fascia. It then passes between the two layers of Intercostal muscles, and, having ascended obliquely to the lower border of the rib above, divides, near the angle of that bone, into two branches; of these the larger runs in the groove, on the lower border of the rib above; the smaller branch along the upper border of the rib below; passing forward, they supply the Intercostal muscles, and anastomose with the anterior intercostal arteries, branches of the internal mammary, and with the thoracic branches of the axillary artery. The first aortic intercostal anastomoses with the superior intercostal, and the last three, pass between the Abdominal muscles, inosculating with the epigastric in front, and with the phrenic, and lumbar arteries. Each intercostal artery is accompanied by a vein and nerve, the former being above, and the latter below, except in the upper intercostal spaces, where the nerve is at first above the artery. The arteries are protected from pressure during the action of the Intercostal muscles, by fibrous arches thrown across, and attached by each extremity to the bone.

The *Posterior*, or *Dorsal Branch*, of each intercostal artery, passes backwards to the inner side of the anterior costo-transverse ligament, and divides into a spinal branch, which supplies the vertebrae, the spinal cord and its membranes, and a muscular branch, which is distributed to the muscles and integument of the back.

THE ABDOMINAL AORTA (fig. 204, color illustration p. A4)

The *Abdominal Aorta* commences at the aortic opening of the Diaphragm, in front of the body of the last dorsal vertebra, and descending a little to the left side of the vertebral column, terminates on the left side of the body of the fourth lumbar vertebra, where it divides into the two common iliac arteries. At it lies upon the bodies of the vertebrae it is convex forwards, the greatest convexity corresponding to the third lumbar vertebra, which is a little above and to the left side of the umbilicus.

Relations. It is covered, *in front*, by the lesser omentum and stomach, behind which are the branches of the coeliac axis and the solar plexus; below these, by the splenic vein, the pancreas, the left renal vein, the transverse portion of the duodenum, the mesentery, and aortic plexus. *Behind*, it is separated from the lumbar vertebrae by the left lumbar veins, the receptaculum chyli, and thoracic duct. On the *right side*, with the inferior vena cava (the right crus of the Diaphragm being interposed above), the vena azygos, thoracic duct, and right semilunar ganglion. On the *left side*, with the sympathetic nerve and left semilunar ganglion.

PLAN OF THE RELATIONS OF THE ABDOMINAL AORTA

In front.

Lesser omentum and stomach.	Left renal vein.
Branches of coeliac axis and solar plexus.	Transverse duodenum.
Splenic vein.	Mesentery.
Pancreas.	Aortic plexus.

Right side.	Abdominal Aorta.	*Left side.*
Right crus of diaphragm.		Sympathetic nerve.
Inferior vena cava.		Left semilunar ganglion.
Vena agyzos.		
Thoracic duct.		
Right semilunar ganglion.		

Behind.

Left lumbar veins.	Thoracic duct.
Receptaculum chyli.	Vertebral column.

Surgical Anatomy. Aneurisms of the abdominal aorta near the coeliac axis communicate in nearly equal proportion with the anterior and posterior parts of this vessel.

When an aneurismal sac is connected with the back part of the aorta, it usually produces absorption of the bodies of the vertebrae, and forms a pulsating tumour, that presents itself in the left hypochondriac or epigastric regions, accompanied by symptoms of disturbance of the alimentary canal. Pain is invariably present, and is usually of two kinds, a fixed and constant pain in the back, caused by the tumour pressing on or displacing the branches of the solar plexus and splanchnic nerves, and a sharp lancinating pain, radiating along those branches of the lumbar nerves pressed on by the tumour; hence the pain in the loins, the testes, the hypogastrium, and in the lower limb (usually of the left side). This form of aneurism usually bursts into

204.—The Abdominal Aorta and its Branches.

the peritoneal cavity, or behind the peritoneum, in the left hypochondriac region; or it may form a large aneurismal sac, extending down as low as Poupart's ligament; haemorrhage in these cases being generally very extensive, but slowly produced, and never rapidly fatal.

When an aneurismal sac is connected with the front of the aorta near the coeliac axis, it forms a pulsating tumour in the left hypochondriac or epigastric regions, usually attended with symptoms of disturbance of the alimentary canal, as sickness, dyspepsia, or constipation, and accompanied by pain, which is constant but nearly always fixed in the loins, epigastrium, or some part of the abdomen; the radiating pain being rare, as the lumbar nerves are seldom implicated. This form of aneurism may burst into the peritoneal cavity, or behind the peritoneum, between the layers of the mesentery, or, more rarely, into the duodenum; it rarely extends backwards so as to affect the spine.

Branches of the Abdominal Aorta

Phrenic.

Coeliac axis. { Gastric. Hepatic. Splenic. }

Superior mesenteric.

Supra-renal.

Renal.

Spermatic.

Inferior mesenteric.

Lumbar.

Sacra media.

The branches may be divided into two sets: 1. Those supplying the viscera. 2. Those distributed to the walls of the abdomen.

Visceral Branches.

Supplying viscera of digestion. { Coeliac axis { Gastric. Hepatic. Splenic. } Superior mesenteric. Inferior mesenteric. }

Supra-renal glands.
Kidneys.
Testes.

Supra-renal.
Renal.
Spermatic.

Parietal Branches.

Phrenic.
Lumbar.
Sacra media.

COELIAC AXIS (figs. 205, 206, color illustrations p. A5)

To expose this artery, raise the liver, draw down the stomach, and then tear through the layers of the lesser omentum.

It is a short thick trunk, about half an inch in length, arising from the aorta, opposite the margin of the Diaphragm, and passing nearly horizontally forwards (in the erect posture), divides into three large branches, the gastric, hepatic, and splenic, occasionally giving off one of the phrenic arteries.

Relations. It is covered, *in front*, by the lesser omentum. On the *right side*, it is in relation with the right semilunar ganglion, and the lobus Spigelii of the liver. On the *left side*, with the left semilunar ganglion and cardiac end of the stomach. *Below*, it rests upon the upper border of the pancreas.

The GASTRIC ARTERY (*Coronaria ventriculi*), the smallest of the three branches of the coeliac axis, passes upwards and to the left side, to the cardiac orifice of the stomach, distributing branches to the oesophagus, which anastomose with the aortic oesophageal arteries; others supply the cardiac end of the stomach, inosculating with branches of the splenic artery: it then passes from left to right, along the lesser curvature of the stomach to the pylorus, lying in its course between the layers of the lesser omentum, and giving branches to both surfaces of the organ; at its termination it anastomoses with the pyloric branch of the hepatic.

The HEPATIC ARTERY in the adult is intermediate in size between the gastric and splenic; in the foetus, it is the largest of the three branches of the coeliac axis. It passes upwards to the right side, between the layers of the lesser omentum, and in front of the foramen of Winslow, to the transverse fissure of the liver, where it divides into two branches (right and left), which supply the corresponding lobes of that organ, accompanying the ramifications of the vena portae and hepatic duct. The hepatic artery, in its course along the right border of the lesser omentum, is in relation with the ductus communis choledocus and portal vein, the former lying to the right of the artery, and the vena portae behind.

Its branches are the

      Pyloric.

   Gastro-duodenalis.   { Gastro-epiploica dextra.
              Pancreatico-duodenalis.

      Cystic.

205.—The Cœliac Axis and its Branches, the Liver having been raised, and the Lesser Omentum removed.

The *Pyloric Branch* arises from the hepatic, above the pylorus, descends to the pyloric end of the stomach; and passes from right to left along its lesser curvature, supplying it with branches, and inosculating with the gastric artery.

The *Gastro-Duodenalis* is a short but large branch, which descends behind the duodenum, near the pylorus, and divides at the lower border of the stomach into two branches, the gastro-epiploica dextra and the pancreatico-duodenalis. Previous to its division, it gives off two or three small inferior pyloric branches to the pyloric end of the stomach and pancreas.

The *Gastro-Epiploica Dextra* runs from right to left along the greater curvature of the stomach, between the layers of the great omentum, anastomosing about the middle of the lower border of this organ with the gastro-epiploica sinistra from the splenic artery. This vessel gives off numerous branches, some of which ascend to supply both surfaces of the stomach, whilst others descend to supply the great omentum.

The *Pancreatico Duodenalis* descends along the contiguous margins of the duodenum and pancreas. It supplies both these organs, and anastomoses with the inferior pancreatico-duodenal branch of the superior mesenteric artery.

In ulceration of the duodenum, which frequently occurs in connexion with severe burns, this artery is often involved, and death may occur from sudden haemorrhage into the intestinal canal.

206.—The Cæliac Axis and its Branches, the Stomach having been raised, and the Transverse Meso-Colon removed.

The *Cystic Artery*, usually a branch of the right hepatic, passes upwards and forwards along the neck of the gall bladder, and divides into two branches, one of which ramifies on its free surface, the other, between it and the substance of the liver.

The SPLENIC ARTERY, in the adult, is the largest of the three branches of the caeliac axis, and is remarkable for the extreme tortuosity of its course. It passes horizontally to the left side behind the upper border of the pancreas, accompanied by the splenic vein, which lies below it; and on arriving near the spleen, divides into branches, some of which enter the hilus of that organ to be distributed to its structure, whilst others are distributed to the great end of the stomach.

The branches of this vessel are:

    Pancreaticae parvae.   Gastric (Vasa brevia).
    Pancreatica magna.   Gastro-epiploica sinistra.

The *Pancreatic* are numerous small branches derived from the splenic as it runs behind the upper border of the pancreas, supplying its middle and left parts. One of these, larger than the rest, is given off from the splenic near the left extremity of the pancreas; it runs from left to right near the posterior surface of the gland following the course of the pancreatic duct, and is called the *pancreatica magna*. These vessels anastomose with the pancreatic branches of the pancreatico-duodenal arteries.

The *Gastric* (*Vasa brevia*) consist of from five to seven small branches, which arise either from the termination of the splenic artery, or from its terminal branches; and passing from left to right, between the layers of the gastro-splenic omentum, are distributed to the great curvature of the stomach; anastomosing with branches of the gastric and gastro-epiploica sinistra arteries.

The *Gastro-Epiploica Sinistra*, the largest branch of the splenic, runs from left to right along the great curvature of the stomach, between the layers of the great omentum; and anastomoses with the gastro-epiploica dextra. In its course, it distributes several branches to the stomach, which ascend upon both surfaces; others descend to supply the omentum.

Superior Mesenteric Artery

In order to expose this vessel, raise the great omentum and transverse colon, draw down the small intestines, and if the peritoneum is divided where the transverse meso-colon and mesentery join, this artery will be exposed just as it issues beneath the lower border of the pancreas.

The Superior Mesenteric Artery (fig. 207, color illustration p. A6) supplies the whole length of the small intestine, except the first part of the duodenum; it also supplies the caecum, ascending and transverse colon; it is a vessel of large size arising from the fore part of the aorta, about a quarter of an inch below the caeliac axis; being covered, at its origin, by the splenic vein and pancreas. It passes forwards, between the pancreas and transverse portion of the duodenum, crosses in front of this portion of the intestine, and descends between the layers of the mesentery to the right iliac fossa, where it terminates considerably diminished in size. In its course it forms an arch, the convexity being directed forwards and downwards to the left side, the concavity backwards and upwards to the right. It is accompanied by the superior mesenteric vein, and surrounded by the superior mesenteric plexus of nerves. Its branches are the

    Inferior pancreatico-duodenal.   Ileo-colic.
    Vasa intestini tenuis.   Colica dextra.
          Colica media.

The *Inferior Pancreatico-Duodenal* is given off from the superior mesenteric below the pancreas, and is distributed to its right extremity and the transverse and descending portions of the duodenum; anastomosing with the pancreatico-duodenal artery.

The *Vasa Intestini Tenuis* arise from the convex side of the superior mesenteric artery. They are usually from twelve to fifteen in number, and are distributed to the jejunum and ileum. They run parallel with one another between the layers of the mesentery; each vessel dividing into two branches, which unite with a similar branch on each side, forming a series of arches, the convexities of which are directed towards the intestine. From this first set of arches branches arise, which again unite with similar branches from either side, and thus a second series of arches is formed; and from these latter, a third, and even a fourth or fifth series of arches are constituted, diminishing in size the nearer they approach the intestine. From the terminal arches numerous small straight vessels arise which encircle the intestine, upon which they are minutely distributed, ramifying between its coats.

The *Ileo-Colic Artery* is the lowest branch given off from the concavity of the superior mesenteric artery. It descends between the layers of the mesentery to the right iliac fossa, where it divides into two branches. Of these, the inferior one inosculates with the lowest branches of the vasa intestini tenuis, from the convexity of which branches proceed to supply the termination of the ileum, the caecum and appendix coeci, and the ileo-coecal and ileo-colic valves. The superior division inosculates with the colica dextra, and supplies the commencement of the colon.

207.—The Superior Mesenteric Artery and its Branches.

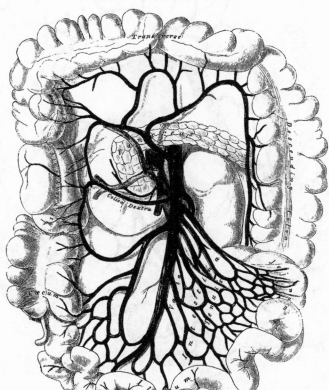

The *Colica Dextra* arises from about the middle of the concavity of the superior mesenteric artery, and passing beneath the peritoneum to the middle of the ascending colon, divides into two branches; a descending branch, which inosculates with the ileo-colic; and an ascending branch, which anastomoses with the colica media. These branches form arches, from the convexity of which vessels are distributed to the ascending colon. The branches of this vessel are covered with peritoneum only on their anterior aspect.

The *Colica Media* arises from the upper part of the concavity of the superior mesenteric, and, passing forwards between the layers of the transverse meso-colon, divides into two branches; the one on the right side inosculating with the colica dextra; that on the left side, with the colica sinistra, a branch of the inferior mesenteric. From the arches formed by their inosculation, branches are distributed to the transverse colon. The branches of this vessel lie between two layers of peritoneum.

208.—The Inferior Mesenteric Artery and its Branches.

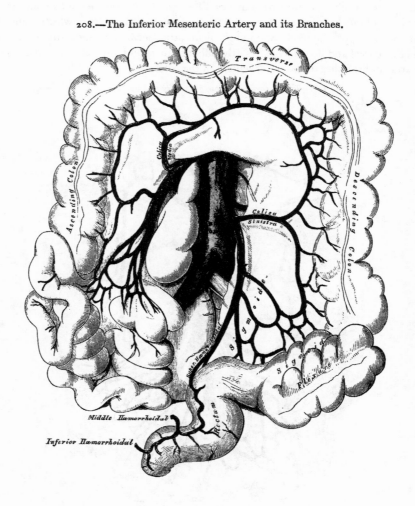

Inferior Mesenteric Artery

In order to expose this vessel, draw the small intestines and mesentery over to the right side of the abdomen, raise the transverse colon towards the thorax, and divide the peritoneum covering the left side of the aorta.

The INFERIOR MESENTERIC ARTERY (fig. 208) supplies the descending and sigmoid flexure of the colon, and greater part of the rectum. It is smaller than the superior mesenteric; and arises from the left side of the aorta, between one and two inches above its division into the common iliacs. It passes downwards to the left iliac fossa, and then descends, between the layers of the meso-rectum, into the pelvis, under the name of the *superior haemorrhoidal artery*. It lies at first in close relation with the left side of the aorta, and then passes in front of the left common iliac artery. Its branches are the

Colica sinistra.   Sigmoidea.
Superior haemorrhoidal.

The *Colica Sinistra* passes behind the peritoneum, in front of the left kidney, to reach the descending colon, and divides into two branches; an ascending branch, which inosculates with the colica media; and a descending branch, which anastomoses with the sigmoid artery. From the arches formed by these inosculations, branches are distributed to the descending colon.

The *Sigmoid Artery* runs obliquely downwards across the Psoas muscle to the sigmoid flexure of the colon, and divides into branches which supply this part of the intestine; anastomosing above, with the colica sinistra; and below, with the superior haemorrhoidal artery. This vessel is sometimes replaced by three or four small branches.

The *Superior Haemorrhoidal Artery*, the continuation of the inferior mesenteric, descends into the pelvis between the layers of the meso-rectum, crossing, in its course, the ureter, and left common iliac artery and vein. Opposite the middle of the sacrum it divides into two branches, which descend one on each side of the rectum, where they divide into several small branches, which are distributed between the mucous and muscular coats of this tube, to near its lower end; anastomosing with each other, with the middle haemorrhoidal arteries, branches of the internal iliac, and with the inferior haemorrhoidal branches of the internal pudic.

The student should especially remark, that the trunk of this vessel descends along the back part of the intestine as far as the middle of the sacrum before it divides; this is about a finger's length or four inches from the anus. In disease of this tube, the rectum should never be divided beyond this point in that direction, for fear of involving this artery.

The *Supra-Renal Arteries* are two small vessels which arise, one on each side of the aorta, opposite the superior mesenteric artery. They pass obliquely upwards and outwards, to the under surface of the supra-renal capsules, to which they are distributed, anastomosing with capsular branches from the phrenic and renal arteries. In the adult these arteries are of small size; in the foetus they are as large as the renal arteries.

The *Renal Arteries* are two large trunks, which arise from the sides of the aorta, immediately below the superior mesenteric artery. Each is directed outwards, so as to form nearly a right angle with the aorta. The right one longer than the left, on account of the position of the aorta, passes behind the inferior vena cava. The left is somewhat higher than the right. Previously to entering the kidney, each artery divides into four or five branches, which are distributed to its substance. At the hilus, these branches lie between the renal vein and ureter, the vein being usually in front, the ureter behind. Each vessel gives off some small branches to the supra-renal capsules, the ureter, and to the surrounding cellular membrane and muscles.

The *Spermatic Arteries* are distributed to the testes in the male, and to the ovaria in the female. They are two small slender vessels, of considerable length, which arise from the front of the aorta, a little below the renal arteries. Each artery passes obliquely outwards and downwards, behind the peritoneum, crossing the ureter, and resting on the Psoas muscle, the right spermatic lying in front of the inferior vena cava, the left behind the sigmoid flexure of the colon. On reaching the margin of the pelvis, each vessel passes in front of the corresponding external iliac artery, and takes a different course in the two sexes.

In the Male, it is directed outwards, to the internal abdominal ring, and accompanies the other constituents of the spermatic cord along the spermatic canal to the testis, where it becomes tortuous, and divides into several branches, two or three of which accompany the vas deferens, and supply the epididymis, anastomosing with the deferential artery; others pierce the back part of the tunica albuginea, and supply the substance of the testis.

In the Female, the spermatic arteries (ovarian) are shorter than in the male, and do not pass out of the abdominal cavity. On arriving at the margins of the pelvis, they pass inwards, between the two laminae of the broad ligament of the uterus, to be distributed to the ovaries. One or two small branches supply the Fallopian tubes; another passes on to the side of the uterus, and anastomoses with the uterine arteries. Other offsets are continued along the round ligament, through the inguinal canal, to the integument of the labium and groin. At an early period of foetal life, when the testes lie by the side of the spine, below the kidneys, the spermatic arteries are short; but as these organs descend from the abdomen into the scrotum, they become gradually lengthened.

The *Phrenic Arteries* are two small vessels, which present much variety in their origin. They may arise separately from the front of the aorta, immediately below the caeliac axis, or by a common trunk, which may spring either from the aorta, or from the caeliac axis. Sometimes one is derived from the aorta, and the other from one of the renal arteries. In only one out of thirty-six cases, did these arteries arise as two separate vessels from the aorta. They diverge from one another across the crura of the Diaphragm, and then pass obliquely upwards and outwards upon its under surface. The left phrenic passes behind the oesophagus, and runs forwards on the left side of the oesophageal opening. The right phrenic, passing behind the liver and inferior vena cava, ascends along the right side of the aperture for transmitting that vein. Near the back part of the central tendon, each vessel divides into two branches. The internal branch runs forwards to the anterior margin of the thorax, supplying the Diaphragm, and anastomosing with its fellow of the opposite side, and with the musculo-phrenic, a branch of the internal mammary. The external branch passes towards the side of the thorax, and inosculates with the intercostal arteries. The internal branch of the right phrenic gives off a few vessels to the inferior vena cava; and the left one some branches to the oesophagus. Each vessel also sends capsular branches to the supra-renal capsule of its own side. The spleen on the left side, and the liver on the right, also receive a few branches from these vessels.

The *Lumbar Arteries* are analogous to the intercostal. They are usually four in number on each side, and arise from the back part of the aorta, nearly at right angles with that vessel. They pass outwards and backwards, around the sides of the body of the corresponding lumbar vertebra, behind the sympathetic nerve and the Psoas muscle; those on the right side being covered by the vena cava, and the two upper ones on each side by the crura of the Diaphragm. In the interval between the transverse processes of the vertebrae, each artery divides into a dorsal and an abdominal branch.

The *dorsal branch* gives off, immediately after its origin, a spinal branch, which enters the spinal canal; it then continues its course backwards, between the transverse processes, and is distributed to the muscles and integument of the back, anastomosing with each other, and with the posterior branches of the intercostal arteries.

The *spinal branch*, besides supplying offsets which run along the nerves to the dura mater and cauda equina, anastomosing with the other spinal arteries, divides into two branches, one of which ascends on the posterior surface of the body of the vertebra above, and the other descends on the posterior surface of the body of the vertebra below, both vessels anastomosing with similar branches from neighbouring spinal arteries. The inosculations of these vessels on each side, throughout the whole length of the spine, form a series of arterial arches behind the bodies of the vertebrae, which are connected with each other, and with a median longitudinal vessel, extending along the centre of the bodies of the vertebrae, by transverse branches. From these vessels offsets are distributed to the periosteum and bones.

The *abdominal branches* pass outwards, behind the quadratus lumborum, the lowest branch occasionally in front of that muscle, and, being continued between the abdominal muscles, anastomose with branches of the epigastric and internal mammary *in front*, the intercostals *above*, and those of the ilio-lumbar, and circumflex iliac, *below*.

The *Middle Sacral Artery* is a small vessel, about the size of a crow-quill, which arises from the posterior part of the aorta, just at its bifurcation. It descends upon the last lumbar vertebra, and along the middle line of the anterior surface of the sacrum, to the upper part of the coccyx, where it terminates by anastomosing with the lateral sacral arteries. From it branches arise which run through the meso-rectum, to supply the posterior surface of the rectum. Other branches are given off on each side, which anastomose with the lateral sacral arteries, and send off small offsets which enter the anterior sacral foramina.

Common Iliac Arteries

The abdominal aorta terminates by dividing into the two common iliac arteries. The bifurcation of this vessel usually takes place on the left side of the body of the fourth lumbar vertebra. This point corresponds to the left side of the umbilicus, and is

209.—Arteries of the Pelvis.

on a level with a line drawn across from one crista-ilii to the other. The common iliac arteries are about two inches in length; diverging from the termination of the aorta, they pass downwards and outwards to the margin of the pelvis, and divide opposite the intervertebral substance, between the last lumbar vertebra and the sacrum, into two branches, the external and internal iliac arteries; the former supplying the lower extremity, the latter the viscera and parietes of the pelvis (fig. 209, color illustration p. A6).

The *Right Common Iliac* is somewhat longer than the left, and passes more obliquely across the body of the last lumbar vertebra. It is covered *in front* by the peritoneum, the intestines, the branches of the sympathetic nerve, and crossed, at its point of division, by the ureter. *Behind*, it is separated from the last lumbar vertebra by the two common iliac veins. On its *outer side* it is in relation with the vena cava, and right common iliac vein above, and the Psoas magnus muscle below.

The *Left Common Iliac* is in relation in front with the peritoneum, branches of the sympathetic nerve, the rectum and superior haemorrhoidal artery, and crossed, at its point of bifurcation, by the ureter. The left common iliac vein lies partly on the inner side, and part beneath the artery; on its outer side, it is in relation with the Psoas magnus.

Branches. The common iliac arteries give off small branches to the peritoneum, Psoae muscles, ureters, and to the surrounding cellular membrane, and occasionally give origin to the ilio-lumbar, or renal arteries.

Peculiarities. Its *point of origin* varies according to the bifurcation of the aorta. In three-fourths of a large number of cases, the aorta bifurcated either upon the fourth lumbar vertebra, or upon the inter-vertebral disc, between it and the fifth; one case in nine being below, and one in eleven above this point. In ten out of every thirteen cases, the vessel bifurcated within half an inch above or below the level of the crest of the ilium; more frequently below than above.

The *point of division* is subject to great variety. In two-thirds of a large number of cases, it was between the last lumbar vertebra and the upper border of the sacrum; in one case in eight being above, and in one in six below that point. The left common iliac artery divides lower down more frequently than the right.

The *relative length*, also, of the two common iliac arteries varies. The right common iliac was longest in sixty-three cases; the left, in fifty-two; whilst they were both equal in fifty-three. The length of the arteries varied in five-sevenths of the cases examined, from an inch and a-half to three inches; in about half of the remaining cases, the artery was longer; and in the other half, shorter; the minimum length being less than half an inch, the maximum, four and a-half inches. In one instance, the right common iliac was found wanting, the external and internal iliacs arising directly from the aorta.

Surgical Anatomy. The application of a ligature to the common iliac artery may be required on account of aneurism or haemorrhage, implicating the external or internal iliacs, or on account of secondary haemorrhage after amputation of the thigh high up. It has been seen that the commencement of this vessel corresponds to the left side of the umbilicus on a level with a line drawn from the highest point of one iliac crest to the opposite one, and its course to a line extending from this point downwards towards the middle of Poupart's ligament. The line of incision required in the first steps of an operation for securing this vessel, would materially depend upon the nature of the disease. If the surgeon select the iliac region, a curved incision, about five inches in length, may be made, commencing on the left side of the umbilicus, carried outwards towards the anterior superior iliac spine, and then along the upper border of Poupart's ligament, as far as its middle. But if the aneurismal tumour should extend high up in the abdomen, along the external iliac, it is better to select the side of the abdomen, approaching the artery from above, by making an incision from four to five inches in length, from about two inches above and to the left of the umbilicus, carried outwards in a curved direction towards the lumbar region, and terminating a little below the anterior superior iliac spine. The abdominal muscles (in either case) having been cautiously divided in succession, the transversalis fascia must be carefully cut through, and the peritoneum, together with the ureter, separated from it and from the iliac fascia, and pushed aside; the sacro-iliac articulation must be felt for, and upon it the vessel will be felt pulsating, and may be fully exposed in close connection with its accompanying vein. On the right side, both common iliac veins, as well as the inferior vena cava, are in close connection with the artery, and must be carefully avoided. On the left side, the vein usually lies on the inner side, and behind the artery; but it occasionally happens that the two common iliac veins are joined on the left instead of the right side, which would add much to the difficulty of an operation in such a case. If the common iliac artery is so short that danger is to be apprehended from secondary haemorrhage if a ligature is applied to it, it would be preferable, in such a case, to tie both the external and internal iliac near their origin. This operation has been performed in several instances; in a few with success.

INTERNAL ILIAC ARTERY

The internal iliac artery supplies the walls and viscera of the pelvic cavity, the generative organs, and inner side of the thigh. It is a short, thick vessel, smaller than the external iliac, and about an inch and a-half in length, which arises at the point of bifurcation of the common iliac; and, passing downwards to the upper margin of the great sacro-sciatic foramen, divides into two large trunks, an anterior, and posterior; a partially obliterated cord, the hypogastric artery, extending from the extremity of the vessel forwards to the bladder.

Relations. In front, with the ureter, which separates it from the peritoneum. *Behind,* it is in relation with the internal iliac vein, the lumbo-sacral nerve, and Pyriformis muscle. By its *outer side,* near its origin, with the Psoas muscle.

PLAN OF THE RELATIONS OF THE INTERNAL ILIAC ARTERY

In front.
Peritoneum.
Ureter.

Outer side.
Psoas magnus.

(Internal Iliac.)

Behind.
Internal iliac vein.
Lumbo-sacral nerve.
Pyriformis muscle.

In the foetus, the internal iliac artery (hypogastric), is twice as large as the external iliac, and appears the continuation of the common iliac. Passing forwards to the bladder, it ascends along the side of this viscus to its apex, to which it gives branches (superior vesical); it then passes upwards along the posterior part of the abdomen to the umbilicus, converging towards its fellow of the opposite side. Having passed through the umbilical opening, the two arteries twine around the umbilical vein, forming with it the umbilical cord; and, ultimately, ramify in the substance of the placenta. That portion of the vessel placed within the abdomen, is called the hypogastric artery; and that external to that cavity, the umbilical artery.

At birth, when the placental circulation ceases, that portion of the hypogastric artery which extends from the umbilicus to the apex of the bladder, contracts, and ultimately dwindles to a solid fibrous cord; the portion of the same vessel extending from the apex of the bladder to within an inch and a-half of its origin, is not totally impervious, though it becomes considerably reduced in size; and serves to convey blood to the bladder, under the name of the superior vesical artery.

Peculiarities, as regards its length. In two-thirds of a large number of cases, the length of the internal iliac varied between an inch and an inch and a-half; in the remaining third, it was more frequently longer than shorter, the maximum length being three inches, the minimum, half an inch.

The lengths of the common and internal iliac arteries bear an inverse proportion to each other, the internal iliac artery being long when the common iliac is short, and *vice versa*.

As regards its place of division. The place of division of the internal iliac varies between the upper margin of the sacrum, and the upper border of the sacro-sciatic foramen.

The arteries of the two sides in a series of cases often differed in length, but neither seemed constantly to exceed the other.

Surgical Anatomy. The application of a ligature to the internal iliac artery may be required in cases of aneurism or haemorrhage affecting one of its branches. This vessel may be secured by making an incision through the abdominal parieties in the iliac region, in a direction and to an extent similar to that for securing the common iliac; the transversalis fascia having been cautiously divided, and the peritoneum pushed inwards from the iliac fossa towards the pelvis, the finger may feel the pulsation of the external iliac at the bottom of the wound; and, by tracing this vessel upwards, the internal iliac is arrived at, opposite the sacro-iliac articulation. It should be remembered that the vein lies behind, and on the right side a little external to the artery, and in close contact with it; the ureter and peritoneum, which lie in front, must also be avoided. The degree of facility in applying a ligature to this vessel, will mainly depend upon its length. It has been seen that, in the great majority of the cases examined, the artery was short, varying from an inch to an inch and a-half; in these cases, the artery is deeply seated in the pelvis; when, on the contrary, the vessel is longer, it is found partly above that cavity. If the artery is very short, which occasionally happens, it would be preferable to apply a ligature to the common iliac, or upon the external and internal iliacs at their origin.

Branches of the Internal Iliac

From the Anterior Trunk.	*From the Posterior Trunk.*
Superior vesical.	Gluteal.
Inferior vesical.	Ileo-lumbar.
Middle haemorrhoidal.	Lateral sacral.
Obturator.	
Internal pudic.	
Sciatic.	
In female. { Uterine. Vaginal.	

The *Superior Vesical* is that part of the foetal hypogastric artery which remains pervious after birth. It extends to the side of the bladder, distributing numerous branches to the body and fundus of this organ. From one of these, a slender vessel is derived which accompanies the vas deferens in its course to the testis, where it anastomoses with the spermatic artery. This is the artery of the vas deferens. Other branches supply the ureter.

The *Middle Vesical*, usually a branch of the superior, is distributed to the base of the bladder, and under surface of the vesiculae seminales.

The *Inferior Vesical* arises from the anterior division of the internal iliac, in common with the middle haemorrhoidal, and is distributed to the base of the bladder, the prostate gland, and vesiculae seminales. Those branches distributed to the prostate, communicate with the corresponding vessel of the opposite side.

The *Middle Haemorrhoidal Artery* usually arises together with the preceding vessel. It supplies the rectum, anastomosing with the other haemorrhoidal arteries.

The *Uterine Artery* passes downwards from the anterior trunk of the internal iliac to the neck of the uterus. Ascending, in a tortuous course, on the side of this viscus, between the layers of the broad ligament, it distributes branches to its substance, anastomosing, near its termination, with a branch from the ovarian artery. Branches from this vessel are also distributed to the bladder and ureter.

The *Vaginal Artery* is analogous to the inferior vesical in the male; it descends upon the vagina, supplying its mucous membrane, and sending branches to the neck of the bladder, and contiguous part of the rectum.

The *Obturator Artery* usually arises from the anterior trunk of the internal iliac, frequently from the posterior. It passes forwards below the brim of the pelvis, to the groove in the upper border of the obturator foramen, and escaping from the pelvic cavity through this aperture, divides into an internal and an external branch. In the pelvic cavity, this vessel lies upon the pelvic fascia, beneath the peritoneum, and a little below the obturator nerve. And whilst passing through the obturator foramen, is contained in an oblique canal, formed by the horizontal branch of the pubes, above, and the arched border of the obturator membrane, below.

Branches. Within the pelvis, the obturator artery gives off an *iliac branch* to the iliac fossa, which supplies the bone and the Iliacus muscle, and anastomoses with the ilio-lumbar artery; a *vesical branch* which runs backwards to supply the bladder; and a *pubic branch*, which is given off from the vessel just before it leaves the pelvic cavity. It ascends upon the back of the pubes, communicating with offsets from the epigastric artery, and with the corresponding vessel of the opposite side. This branch is placed on the inner side of the femoral ring. *External to the pelvis*, the obturator artery divides into an external and an internal branch, which are deeply situated beneath the External obturator muscle, skirting the circumference of the obturator foramen, and anastomosing at the lower part of this aperture with each other, and with branches of the internal circumflex artery.

The *internal branch* curves inwards along the inner margin of the obturator foramen, distributing branches to the Obturator muscles, Pectineus, Adductors, and Gracilis, and anastomoses with the external branch, and with the internal circumflex artery.

The *external branch* curves around the outer margin of the foramen, to the space between the Gemellus inferior and Quadratus femoris, where it anastomoses with the sciatic artery. It supplies the Obturator muscles, anastomoses, as it passes backwards, with the internal circumflex, and sends a branch to the hip-joint through the cotyloid notch, which ramifies on the round ligament as far as the head of the femur.

Peculiarities. In two out of every three cases this vessel arises from the internal iliac. In one case in 3½ from the epigastric; and in about one in seventy-two cases by two roots from both vessels. It arises in about the same proportion from the external iliac artery. The origin of the obturator from the epigastric is not commonly found on both sides of the same body.

When the obturator artery arises at the front of the pelvis from the epigastric, it descends almost vertically downwards to the upper part of the obturator foramen. The artery in this course usually descends in contact with the external iliac vein, and lies on the outer side of the femoral ring; in such cases it would not be endangered in the operation for femoral hernia. Occasionally, however, it curves inwards along the free margin of Gimbernat's ligament, and under such circumstances it would almost completely encircle the neck of a hernial sac (supposing a hernia to exist in such a case), and would be in great danger of being wounded if an operation was necessary.

The *Internal Pudic* is the smaller of the two terminal branches of the anterior trunk of the internal iliac, and supplies the external organs of generation. It passes downwards and outwards to the lower border of the great sacro-sciatic foramen, and emerges from the pelvis between the Pyriformis and Coccygeus muscles; it then crosses the spine of the ischium, and enters that cavity through the lesser sacro-sciatic foramen. The artery now crosses the Internal obturator muscle, to the ramus of the ischium, being covered by the obturator fascia, and situated about an inch and a half from the margin of the tuberosity; it then ascends forwards and upwards along the ramus of the ischium, pierces the posterior layer of the deep perinaeal fascia, and runs forwards along the inner margin of the ramus of the pubes; finally it perforates the anterior layer of the deep perinaeal fascia, and divides into its two terminal branches, the dorsal artery of the penis, and the artery of the corpus cavernosum.

Relations. In the first part of its course, within the pelvis, it lies in front of the Pyriformis muscle and sacral plexus of nerves, and on the outer side of the rectum (on the left side). As it crosses the spine of the ischium, it is covered by the Gluteus maximus, and great sacro-sciatic ligament. And when it enters the pelvis, it lies on the outer side of the ischio-rectal fossa, upon the surface of the Obturator internus muscle, contained in a fibrous canal formed by the obturator fascia and the falciform process of the great sacro-sciatic ligament. It is accompanied by the pudic veins, and the internal pudic nerve.

Peculiarities. The internal pudic is sometimes smaller than usual, or fails to give off one or two of its usual branches; in such cases, the deficiency is supplied by branches derived from an additional vessel, the *accessory pudic*, which generally arises from the pudic artery before its exit from the great sacro-sciatic foramen, and passes forwards near the base of the bladder, on the upper part of the prostate gland, to the perinaeum, where it gives off those branches usually derived from the pudic artery itself. The deficiency most frequently met with, is that in which the internal pudic ends as the artery of the bulb; the artery of the corpus cavernosum and arteria dorsalis penis being derived from the accessory pudic. Or the pudic may terminate as the superficial perinaeal, the artery of the bulb being derived, with the other two branches, from the accessory vessel.

The relation of the accessory pudic to the prostate gland and urethra, is of the greatest interest in a surgical point of view, as this vessel is in danger of being wounded in the lateral operation of lithotomy.

Branches. Within the pelvis, the internal pudic gives off several small branches, which supply the muscles, sacral nerves, and viscera in this cavity. *In the perinaeum* the following branches are given off.

Inferior or external haemorrhoidal.	Artery of the bulb.
Superficial perinaeal.	Artery of the corpus cavernosum.
Transverse perinaeal.	Dorsal artery of the penis.

The *External Haemorrhoidal* are two or three small arteries, which arise from the internal pudic as it passes above the tuberosity of the ischium. Crossing the ischio-rectal fossa, they are distributed to the muscles and integument of the anal region.

The *Superficial Perinaeal Artery* supplies the scrotum, and muscles and integument of the perinaeum. It arises from the internal pudic, in front of the preceding branches, and piercing the lower border of the deep perinaeal fascia, runs across the Transversus perinaei, and through the triangular space between the Accelerator urinae and Erector penis, both of which it supplies, and is finally distributed to the skin of the scrotum and dartos. In its passage through the perinaeum it lies beneath the superficial perinaeal fascia.

The *Transverse Perinaeal* is a small branch which arises either from the internal pudic, or from the superficial perinaeal artery as it crosses the Transversus perinaei muscle. Piercing the lower border of the deep perinaeal fascia, it runs transversely inwards along the cutaneous surface of the Transversus perinaei muscle, which it supplies, as well as the structures between the anus and bulb of the urethra.

The *Artery of the Bulb* is a large but very short vessel, arising from the internal pudic between the two layers of the deep perinaeal fascia, and passing nearly transversely inwards, pierces the bulb of the urethra, in which it ramifies. It gives off a small branch which descends to supply Cowper's gland. This artery is of considerable importance in a surgical point of view, as it is in danger of being wounded in the lateral operation of lithotomy, an accident usually attended with severe and alarming haemorrhage. This vessel is sometimes very small, occasionally wanting, or even double. It sometimes arises from the internal pudic earlier than usual, and crosses the perinaeum to reach the back part of the bulb. In such a case the vessel could hardly fail to be wounded in the performance of the lateral operation for lithotomy. If, on the contrary, it should arise from an accessory pudic, it lies more forward than usual, and is out of danger in the operation.

The *Artery of the Corpus Cavernosum*, one of the terminal branches of the internal pudic, arises from that vessel while it is situated between the crus penis and the ramus of the pubes; piercing the crus penis obliquely, it runs forwards in the corpus cavernosum by the side of the septum pectiniforme, to which its branches are distributed.

The *Dorsal Artery of the Penis* ascends between the crus and pubic symphysis, and piercing the suspensory ligament, runs forwards on the dorsum of the penis to the glans, where it divides into two branches, which supply the glans and prepuce. On the dorsum of the penis, it lies immediately beneath the integument, parallel with the dorsal vein and corresponding artery of the opposite side. It supplies the integument and fibrous sheath of the corpus cavernosum.

The *Internal Pudic Artery in the Female*, is smaller than in the male. Its origin and course are similar, and there is considerable analogy in the distribution of its branches. The superficial artery supplies the labia pudenda; the artery of the bulb supplies the erectile tissue of the bulb of the vagina, whilst the two terminal branches supply the clitoris; the artery of the corpus cavernosum, the cavernous body of the clitoris; and the arteria dorsalis clitoridis, the dorsum of that organ.

The *Sciatic Artery* (fig. 210), the larger of the two terminal branches of the anterior trunk of the internal iliac, is distributed to the muscles on the back of the pelvis. It passes downwards to the lower part of the great sacro-sciatic foramen, behind the internal pudic, resting on the sacral plexus of nerves and Pyriformis muscle, and escapes from the pelvis between the Pyriformis and Coccygeus. It then descends in the interval between the Trochanter major and tuberosity of the ischium, accompanied by the sciatic nerves, and covered in by the Gluteus maximus, and divides into branches, which supply the deep muscles at the back of the hip.

Within the pelvis it distributes branches to the Pyriformis, Coccygeus, and Levator ani muscles; some haemorrhoidal branches, which supply the rectum, and occasionally take the place of the middle haemorrhoidal artery; and vesical branches to the base and neck of the bladder, vesiculae seminales, and prostate gland.

210.—The Arteries of the Gluteal and Posterior Femoral Regions.

External to the pelvis it gives off the coccygeal, inferior gluteal, comes nervi ischiadici, muscular, and articular branches.

The *Coccygeal Branch* runs inwards, pierces the great sacro-sciatic ligament, and supplies the Gluteus maximus, the integument, and other structures on the posterior surface of the coccyx.

The *Inferior Gluteal Branches*, three or four in number, supply the Gluteus maximus muscle.

The *Comes Nervi Ischiadici* is a long slender vessel, which accompanies the great sciatic nerve for a short distance; it then penetrates it, and runs in its substance to the lower part of the thigh.

The *Muscular Branches* supply the muscles on the back part of the hip, anastomosing with the gluteal, internal and external circumflex, and superior perforating arteries.

Some articular branches are also distributed to the capsule of the hip-joint.

The *Gluteal Artery* is the largest branch of the internal iliac, and appears to be the continuation of the posterior division of that vessel. It is a short thick trunk, which passes downwards to the upper part of the great sacro-sciatic foramen, escapes from the pelvis above the upper border of the Pyriformis muscle, and immediately divides into a superficial and deep branch. Within the pelvis, it gives off a few muscular branches to the Iliacus, Pyriformis, and Obturator internus, and just previous to quitting that cavity a nutritious artery, which enters the ilium.

The *superficial branch* passes beneath the Gluteus maximus, and divides into numerous branches, some of which supply this muscle, whilst others perforate its tendinous origin, and supply the integument of the posterior surface and side of the sacrum, anastomosing with the posterior branches of the sacral arteries.

The *deep branch* runs between the Gluteus medius and minimus, and subdivides into two. Of these, the *superior division*, continuing the original course of the vessel, passes along the upper border of the Gluteus minimus to the anterior superior spine of the ilium, anastomosing with the circumflex iliac and ascending branches of the external circumflex artery. The *inferior division* crosses the Gluteus minimus obliquely to the Trochanter major, distributing branches to the Glutei muscles, and inosculates with the external circumflex artery. Some branches pierce the Gluteus minimus to supply the hip-joint.

The *Ilio-Lumbar Artery* ascends beneath the Psoas muscle and external iliac vessels, to the upper part of the iliac fossa, where it divides into a lumbar and an iliac branch.

The *lumbar branch* supplies the Psoas and Quadratus lumborum muscles, anastomosing with the last lumbar artery, and sends a small spinal branch through the intervertebral foramen, between the last lumbar vertebra and the sacrum, into the spinal canal, to supply the spinal cord and its membranes.

The *iliac branch* descends to supply the Iliacus internus, some offsets running between the muscle and the bone, one of which enters an oblique canal to supply the diploë, whilst others run along the crest of the ilium, distributing branches to the Gluteal and Abdominal muscles, and anastomosing in their course with the gluteal, circumflexa ilii, external circumflex, and epigastric arteries.

The *Lateral Sacral Arteries* are usually two in number on each side, superior and inferior.

The *superior*, which is of large size, passes inwards, and after anastomosing with branches from the middle sacral, enters the first or second sacral foramen, is distributed to the contents of the sacral canal, and escaping by the corresponding posterior sacral foramen, supplies the skin and muscles on the dorsum of the sacrum.

The *inferior branch* passes obliquely across the front of the Pyriformis muscle and sacral nerves to the inner side of the anterior sacral foramina, descends on the front of the sacrum, and anastomoses over the coccyx with the sacra-media and opposite lateral sacral arteries. In its course, it gives off branches, which enter the anterior sacral foramina, these after supplying the bones and membranes of the interior of the spinal canal, escape by the posterior sacral foramina, and are distributed to the muscles and skin on the dorsal surface of the sacrum.

External Iliac Artery

The external iliac artery is the chief vessel which supplies the lower limb. It is larger in the adult than the internal iliac, and passes obliquely downwards and outwards along the inner border of the Psoas muscle, from the bifurcation of the common iliac to the femoral arch, where it enters the thigh, and becomes the femoral artery. The course of this vessel would be indicated by a line drawn from the left side of the umbilicus to a point midway between the anterior superior spinous process of the ilium and the symphysis pubis.

Relations. In front, with the peritoneum, sub-peritoneal areolar tissue, the intestines, and a thin layer of fascia, derived from the iliac fascia, which surrounds the artery and vein. At its origin it is occasionally crossed by the ureter. The spermatic vessels descend for some distance upon it near its termination, and it is crossed in this situation by a branch of the genito-crural nerve and the circumflexa ilii vein; the vas deferens curves down along its inner side. *Behind*, it is in relation with the external iliac vein, which, at the femoral arch, lies at its inner side; on the left side the vein is altogether internal to the artery. *Externally*, it rests against the Psoas muscle, from which it is separated by the iliac fascia. The artery rests upon this muscle near Poupart's ligament. Numerous lymphatic vessels and glands are found lying on the front and inner side of the vessel.

Plan of the Relations of the External Iliac Artery

In front.
Peritoneum, intestines, and iliac fascia.

Near Poupart's Ligament.
{ Spermatic vessels.
Genito-crural nerve.
Circumflexa ilii vein.
Lymphatic vessels and glands. }

Outer side.
Psoas magnus.
Iliac fascia.

(External Iliac.)

Inner side.
External iliac vein and vas deferens at femoral arch.

Behind.
External iliac vein.

Surgical Anatomy. The application of a ligature to the external iliac may be required in cases of aneurism of the femoral artery, or in cases of secondary haemorrhage, after the latter vessel has been tied for popliteal aneurism. This vessel may be secured in any part of its course, excepting near its upper end, on account of the circulation through the internal iliac, and near its lower end, on account of the origin of the epigastric and circumflex vessels. One of the chief points in the performance of the operation is to secure the vessel without injury to the peritoneum. The patient having been placed in the recumbent position, an incision should be made, commencing about an inch above and to the inner side of the anterior superior spinous process of the ilium, and running downwards and outwards to the outer end of Poupart's ligament, and parallel with its outer half, to a little above its middle. The abdominal muscles and transversalis fascia having been cautiously divided, the peritoneum should be separated from the iliac fossa and pushed towards the pelvis; and on introducing the finger to the bottom of the wound the artery may be felt pulsating along the inner border of the Psoas muscle. The external iliac vein is situated along the inner side of the artery, and must be cautiously separated from it by the finger-nail, or point of the knife, and the aneurism needle should be introduced on the inner side, between the artery and vein.

Branches. Besides several small branches to the Psoas muscle and the neighbouring lymphatic glands, the external iliac gives off two branches of considerable size, the

      Epigastric.   Circumflexa ilii.

The *Epigastric Artery* arises from the external iliac, a few lines above Poupart's ligament. It at first descends to reach this ligament, and then ascends obliquely upwards and inwards between the peritoneum and transversalis fascia, to the margin of the sheath of the Rectus muscle. Having perforated the sheath near its lower third, it ascends vertically upwards behind the Rectus, to which it is distributed, dividing into numerous branches, which anastomose above the umbilicus with the terminal branches of the internal mammary and inferior intercostal arteries. It is accompanied by two veins, which usually unite into a single trunk before their termination in the external iliac vein. As this artery ascends from Poupart's ligament to the Rectus, it lies behind the inguinal canal, to the inner side of the internal abdominal ring, and immediately above the femoral ring, the vas deferens in the male, and the round ligament in the female, crossing behind the artery in descending into the pelvis.

Branches. The branches of this vessel are the *cremasteric*, which accompanies the spermatic cord, and supplies the Cremaster muscle, anastomosing with the spermatic artery. A *pubic branch*, which runs across Poupart's ligament, and then descends behind the pubes to the inner side of the crural ring, and anastomoses with offsets from the obturator artery. *Muscular branches*, some of which are distributed to the abdominal muscles and peritoneum, anastomosing with the lumbar and circumflexa ilii arteries; others perforate the tendon of the external Oblique and supply the integument, anastomosing with branches of the external epigastric.

Peculiarities. The origin of the epigastric may take place from any part of the external iliac between Poupart's ligament and two inches and a half above it; or it may arise below this ligament, from the femoral, or from the deep femoral.

Union with Branches. It frequently arises from the external iliac by a common trunk with the obturator. Sometimes the epigastric arises from the obturator, the latter vessel being furnished by the internal iliac, or the epigastric may be formed of two branches, one derived from the external iliac, the other from the internal iliac.

The *Circumflex Iliac Artery* arises from the outer side of the external iliac, nearly opposite the epigastric artery. It ascends obliquely outwards behind Poupart's ligament, and runs along the inner surface of the crest of the ilium to about its middle, where it pierces the Transversalis, and runs backwards between this muscle and the Internal oblique, to anastomose with the ilio-lumbar and gluteal arteries. Opposite the anterior superior spine of the ilium, it gives off a large branch, which ascends between the Internal oblique and Transversalis muscles, supplying them and anastomosing with the lumbar and epigastric arteries. The circumflex iliac artery is accompanied by two veins, which, uniting into a single trunk, crosses the external iliac artery just above Poupart's ligament, and enters the external iliac vein.

211.—Surgical Anatomy of the Femoral Artery.

Femoral Artery

The femoral artery (fig. 211, color illustration p. A7) is the continuation of the external iliac. It commences immediately beneath Poupart's ligament, midway between the anterior superior spine of the ilium and the symphysis pubis, and passing down the fore part and inner side of the thigh, terminates at the opening in the Adductor magnus, at the junction of the middle with the lower third of the thigh, where it becomes the popliteal artery. A line drawn from a point midway between the anterior superior spine of the ilium and the symphysis of the pubes to the inner side of the internal condyle of the femur, will be nearly parallel with the course of this artery. This vessel, at the upper part of the thigh, lies a little internal to the head of the femur; in the lower part of its course, on the inner side of the shaft of this bone, and between these two points the vessel is separated from the bone by a considerable interval.

In the upper third of the thigh the femoral artery is very superficial, being covered by the integument, inguinal glands, and by the superficial and deep fasciae, and is contained in a triangular space, called "Scarpa's triangle."

Scarpa's Triangle. Scarpa's triangle corresponds to the depression seen immediately below the fold of the groin. It is a triangular space, the apex of which is directed downwards, and the sides of which are formed externally by the Sartorius, internally by the Adductor longus, and the base, by Poupart's ligament. The floor of this space is formed from without inwards by the Iliacus, Psoas, Pectineus, Adductor longus, and a small part of the Adductor brevis muscles; and it is divided into two nearly equal parts by the femoral artery and vein, which extend from the middle of its base to its apex: the artery giving off in this situation its cutaneous and profunda branches, the vein receiving the deep femoral and internal saphena veins. In this space, the femoral artery rests on the inner margin of the Psoas muscle, which separates it from the capsular ligament of the hip-joint. The artery in this situation is crossed in front by the crural branch of the genitocrural nerve, and behind by the branch to the Pectineus from the anterior crural. The femoral vein lies at its inner side, between the margins of the Pectineus and Psoas muscles. The anterior crural nerve lies about half an inch to the outer side of the femoral artery, deeply imbedded between the Iliacus and Psoas muscles; and on the Iliacus muscle, internal to the anterior superior spinous process of the ilium, is the external cutaneous nerve. The femoral artery and vein are enclosed in a strong fibrous sheath, formed by fibrous and cellular tissue, and by a process of fascia sent inwards from the fascia lata; the vessels are separated, however, from one another by thin fibrous partitions.

In the middle third of the thigh, the femoral artery is more deeply seated, being covered by the integument, the superficial and deep fasciae, and the Sartorius, and is contained in an aponeurotic canal; formed by a dense fibrous band, which extends transversely from the Vastus internus to the tendons of the Adductor longus and magnus muscles. In this part of its course it lies in a depression, bounded externally by the Vastus internus, internally by the Adductor longus and Adductor magnus. The femoral vein lies on the outer side of the artery, in close apposition with it, and, still more externally, is the internal (long) saphenous nerve.

Relations. From above downwards, the femoral artery rests upon the Psoas muscle, which separates it from the margin of the pelvis and capsular ligament of the hip; it is next separated from the Pectineus, by the profunda vessels and femoral vein; it then lies upon the Adductor longus; and lastly, upon the tendon of the Adductor magnus, the femoral vein being interposed. To its *inner side*, it is in relation, above, with the femoral vein, and, lower down, with the Adductor longus, and Sartorius. To its *outer side*, the Vastus internus separates it from the femur, in the lower part of its course.

The *femoral vein,* at Poupart's ligament, lies close to the inner side of the artery, separated from it by a thin fibrous partition, but, as it descends, gets behind it, and then to its outer side.

The *internal saphenous nerve,* is situated on the outer side of the artery, in the middle third of the thigh, beneath the aponeurotic covering, but not within the sheath of the vessels. Small cutaneous nerves cross the front of the sheath.

Peculiarities. Double femoral re-united. Four cases are at present recorded, in which the femoral artery divided into two trunks below the origin of the Profunda, and became re-united near the opening in the Adductor magnus, so as to form a single popliteal artery. One of them occurred in a patient operated upon for popliteal aneurism.

Change of Position. A similar number of cases have been recorded, in which the femoral artery was situated at the back of the thigh, the vessel being continuous above with the internal iliac, escaping from the pelvis through the great sacro-sciatic foramen, and accompanying the great sciatic nerve to the popliteal space, where its division occurred in the usual manner.

Position of the Vein. The femoral vein is occasionally placed along the inner side of the artery, throughout the entire extent of Scarpa's triangle; or it may be slit, so that a large vein is placed on each side of the artery for a greater or less extent.

Origin of the Profunda. This vessel occasionally arises from the inner side, and more rarely, from the back of the common trunk; but the more important peculiarity, in a surgical point of view, is that which relates to the height at which the vessel arises from the femoral. In three-fourths of a large number of cases it arose between one and two inches below Poupart's ligament; in a few cases, the distance was less than an inch; more rarely, opposite the ligament; and in one case, above Poupart's Ligament, from the external iliac. Occasionally, the distance between the origin of the vessel and Poupart's ligament, exceeds two inches, and in one case, it was found to be as much as four inches.

Surgical Anatomy. Compression of the femoral artery, which is constantly requisite in amputations, or other operations on the lower limb, is most effectually made immediately below Poupart's ligament. In this situation, the artery is very superficial, and is merely separated from the margin of the acetabulum and front of the head of the femur, by the Psoas muscle; so that the surgeon, by means of his thumb, or any other resisting body, may effectually control the circulation through it. This vessel may also be compressed in the middle third of the thigh, by placing a compress over the artery, beneath the tourniquet, and directing the pressure from within outwards, so as to compress the vessel on the inner side of the shaft of the femur.

The *application of a ligature* to the femoral artery may be required in cases of wound or aneurism of the arteries of the leg, of the popliteal or femoral; and the vessel may be exposed and tied in any part of its course. The great depth of this vessel in the middle of the thigh, its close connection with important structures, and the density of its sheath, render the operation in this situation one of much greater difficulty than the application of a ligature in the upper part of its course, where it is more superficial.

Ligature of the femoral artery, within two inches of its origin, is usually considered very unsafe, on account of the connection of large branches with it, the epigastric and circumflex iliac arising just above its origin; the profunda, from one to two inches below; occasionally, also, one of the circumflex arteries arises from the vessel in the interspace between these. The profunda sometimes arises higher than the point above-mentioned, and rarely between two or three inches (in one case four) below Poupart's ligament. It would appear, then, that the most favourable situation for the application of a ligature to this vessel, is between four and five inches from its point of origin. In order to expose the artery in this situation, an incision, between two and three inches long, should be made in the course of the vessel, the patient lying in the recumbent position, with the limb slightly flexed and abducted. A large vein is frequently met with, passing in the course of the artery to join the saphena; this must be avoided, and the fascia lata having been cautiously divided, and the Sartorius exposed, this muscle must be drawn outwards, in order to fully expose the sheath of the vessels. The finger being introduced into the wound, and the pulsation of the artery felt, the sheath should be divided over it to a sufficient extent to allow of the introduction of the ligature, but no further; otherwise the nutrition of the coats of the vessel may be interfered with, or muscular branches which arise from the vessel at irregular intervals may be divided. In this part of the operation, a small nerve which crosses the sheath should be avoided. The aneurism needle must be carefully introduced and kept close to the artery, to avoid the femoral vein, which lies behind the vessel in this part of its course.

To expose the artery in the middle of the thigh, an incision should be made through the integument, between three and four inches in length, over the inner margin of the Sartorius, taking care to avoid the internal saphena vein, the situation of which may be previously known by compressing it higher up in the thigh. The fascia lata having been divided, and the Sartorius muscle exposed, it should be drawn outwards, when the strong fascia which is stretched across from the Adductors to the Vastus internus, will be exposed, and must be freely divided; the sheath of the vessels is now seen, and must be opened, and the artery secured by passing the aneurism needle between the vein and artery, in the direction from within outwards. The femoral vein in this situation lies on the outer side of the artery, the long saphenous nerve on its anterior and outer side.

It has been seen that the femoral artery occasionally divides into two trunks, below the origin of the profunda. If, in the operation for tying the femoral, two vessels are met with, the surgeon should alternately compress each, in order to ascertain which vessel is connected with the aneurismal tumour, or with the bleeding from the wound, and that one only tied which controls it. If, however, it is necessary to compress both vessels before the circulation in the tumour is controlled, both should be tied, as it would be probable that they became re-united, as is mentioned above.

Branches. The branches of the femoral artery are the:

- Superficial epigastric.
- Superficial circumflex iliac.
- Superficial external pudic.
- Deep external pudic.
- Profunda { External circumflex. Internal circumflex. Three perforating. }
- Muscular.
- Anastomotica magna.

The *Superficial Epigastric* arises from the femoral, about half an inch below Poupart's ligament, and, passing through the saphenous opening in the fascia lata, ascends on to the abdomen, in the superficial fascia covering the External oblique muscle, nearly as high as the umbilicus. It distributes branches to the inguinal glands, the superficial fascia and integument, anastomosing with branches of the deep epigastric, and internal mammary arteries.

The *Superficial Circumflex Iliac*, the smallest of the cutaneous branches, arises close to the preceding, and, piercing the fascia lata, runs outwards, parallel with Poupart's ligament, as far as the crest of the Ilium, dividing into branches which supply the integument of the groin, the superficial fascia, and inguinal glands, anastomosing with the deep circumflex iliac, and with the gluteal and external circumflex arteries.

The *Superficial External Pudic* (superior), arises from the inner side of the femoral artery, close to the preceding vessels, and, after piercing the fascia lata at the saphenous opening, passes inwards, across the spermatic cord, to be distributed to the integument on the lower part of the abdomen, and of the penis and scrotum in the male, and to the labia in the female, anastomosing with branches of the internal pudic.

The *Deep External Pudic* (inferior), more deeply seated than the preceding, passes inwards on the Pectineus muscle, covered by the fascia lata, which it pierces opposite the ramus of the pubes, its branches being distributed, in the male, to the integument of the scrotum and perinaeum, and in the female, to the labium, anastomosing with branches of the superficial perinaeal artery.

The *Profunda Femoris* (deep femoral artery), nearly equals the size of the superficial femoral. It arises from the outer and back part of the femoral artery, from one to two inches below Poupart's ligament. It at first lies on the outer side of the superficial femoral, and then passes beneath it and the femoral vein to the inner side of the femur, and terminates at the lower third of the thigh in a small branch, which pierces the Adductor magnus, to be distributed to the Flexor muscles, on the posterior part of the thigh, anastomosing with branches of the popliteal and inferior perforating arteries.

Relations. Behind, it lies first upon the Iliacus, and then on the Adductor brevis and Adductor magnus muscles. *In front*, it is separated from the femoral artery; *above*, by the femoral and profunda veins; and *below*, by the Adductor longus. On its *outer side*, the insertion of the Vastus internus separates it from the femur.

PLAN OF THE RELATIONS OF THE PROFUNDA ARTERY
In front.
Femoral and profunda veins.
Adductor longus.

Outer side.
Vastus internus.

Behind.
Iliacus.
Adductor brevis.
Adductor magnus.

The *External Circumflex Artery* supplies the muscles on the front of the thigh. It arises from the outer side of the profunda, passes horizontally outwards, between the divisions of the anterior crural nerve, and beneath the Sartorius and Rectus muscles, and divides into three sets of branches, ascending, transverse, and descending.

The *ascending branches* pass upwards, beneath the Tensor vaginae femoris muscle, to the outer side of the hip, anastomosing with the terminal branches of the gluteal, and circumflex iliac arteries.

The *descending branches*, three or four in number, pass downwards, beneath the Rectus, upon the Vasti muscles, to which they are distributed, one or two passing beneath the Vastus externus as far as the knee, anastomosing with the superior articular branches of the popliteal artery.

The *transverse branches*, the smallest and least numerous, pass outwards over the Cruraeus, pierce the Vastus externus, and wind around the femur to its back part, just below the great trochanter, anastomosing at the back of the thigh with the internal circumflex, sciatic, and superior perforating arteries.

The *Internal Circumflex Artery*, smaller than the external, arises from the inner and back part of the profunda, and winds around the inner side of the femur, between the Pectineus and Psoas muscles. On reaching the tendon of the Obturator externus, it divides into two branches; one, ascending, is distributed to the Adductor muscles, the Gracilis, and Obturator externus, anastomosing with the obturator artery, a descending branch which passes beneath the Adductor brevis, to supply it and the great Adductor; the continuation of the vessel passing backwards, between the Quadratus femoris and upper border of the Adductor magnus, anastomosing with the sciatic, external circumflex, and superior perforating arteries. Opposite the hip-joint, this branch gives off an articular vessel, which enters the joint beneath the transverse ligament; and, after supplying the adipose tissue, passes along the round ligament to the head of the bone.

The *Perforating Arteries* (fig. 210), usually three in number, are so called from their perforating the tendons of the Adductor brevis and magnus muscles to reach the back of the thigh.

The *First* or *Superior Perforating Artery* passes backwards between the Pectineus and Adductor brevis (sometimes perforates the latter); it then pierces the Adductor magnus close to the linea aspera, and divides into branches which supply both Adductors, the Biceps, and Gluteus maximus muscle; anastomosing with the sciatic, internal circumflex, and middle perforating arteries.

The *Second* or *Middle Perforating Artery*, larger than the first, passes through the tendons of the Adductor brevis and Adductor magnus muscles, divides into ascending and descending branches, which supply the flexor muscles of the thigh; anastomosing with the superior and inferior perforantes. The nutrient artery of the femur is usually given off from this branch.

The *Third* or *Inferior Perforating Artery* is given off below the Adductor brevis; it pierces the Adductor magnus, and divides into branches which supply the flexor muscles of the thigh; anastomosing with the perforating arteries, above, and with the terminal branches of the profunda, below.

Muscular Branches are given off from this vessel throughout its entire course.

They vary from two to seven in number, and supply chiefly the Sartorius and Vastus internus.

The *Anastomotica Magna* arises from the femoral artery just before it passes through the tendinous opening in the Adductor magnus muscle, and divides into a superficial and deep branch.

The *superficial branch* accompanies the long saphenous nerve, beneath the Sartorius, and piercing the fascia lata, is distributed to the integument.

The *deep branch* descends in the substance of the Vastus internus, lying in front of the tendon of the Adductor magnus, to the inner side of the knee, where it anastomoses with the superior internal articular artery and recurrent branch of the anterior tibial. A branch from this vessel crosses outwards above the articular surface of the femur, forming an anastomotic arch with the superior external articular artery, and supplies branches to the knee-joint.

Popliteal Artery

The popliteal artery commences at the termination of the femoral, at the opening in the Adductor magnus, and passing obliquely downwards and outwards behind the knee-joint, to the lower border of the Popliteus muscle, divides into the anterior and posterior tibial arteries. Through this extent the artery lies in the popliteal space.

The Popliteal Space

Dissection. A vertical incision about eight inches in length should be made along the back part of the knee-joint, connected above and below by a transverse incision passing from the inner to the outer side of the limb. The flaps of integument included between these incisions should be reflected in the direction shown in fig. 171.

On removing the integument, the superficial fascia is exposed, and ramifying in it along the middle line are found some filaments of the small sciatic nerve, and towards the inner part, some offsets from the internal cutaneous nerve.

The superficial fascia having been removed, the fascia lata is brought into view. In this region it is strong and dense, being strengthened by transverse fibres, and firmly attached to the tendons on the inner and outer sides of the space. It is perforated below by the external saphena vein. This fascia having been reflected back in the same direction as the integument, the small sciatic nerve and external saphena vein are seen immediately beneath it, in the middle line. If the loose adipose tissue is now removed, the boundaries and contents of the space may be examined.

Boundaries. The popliteal space or the ham, occupies the lower third of the thigh and the upper fifth of the leg; extending from the aperture in the Adductor magnus, to the lower border of the Popliteus muscle. It is a lozenge-shaped space, being widest at the back part of the knee-joint, and deepest above the articular end of the femur. It is bounded, externally, above the joint, by the Biceps, and below the articulation, by the Plantaris and external head of the Gastrocnemius. Internally, above the joint, by the Semi-membranosus, Semi-tendinosus, Gracilis, and Sartorius; below the joint, by the inner head of the Gastrocnemius.

Above, it is limited by the apposition of the inner and outer hamstring muscles; below, by the junction of the two heads of the Gastrocnemius. The floor is formed by the lower part of the posterior surface of the shaft of the femur, the posterior ligament of the knee-joint, the upper end of the tibia, and the fascia covering the Popliteus muscle, and the space is covered in by the fascia lata.

Contents. It contains the popliteal vessels and their branches, together with the termination of the external saphena vein, the internal and external popliteal nerves and their branches, the small sciatic nerve, the articular branch from the obturator nerve, a few small lymphatic glands, and a considerable quantity of loose adipose tissue.

212.—The Popliteal, Posterior Tibial, and Peroneal Arteries.

Position of contained parts. The internal popliteal nerve descends in the middle line of the space, lying superficial, and a little external to the vein and artery. The external popliteal nerve descends on the outer side of the space, lying close to the tendon of the Biceps muscle. More deeply at the bottom of the space are the popliteal vessels, the vein lying superficial and a little external to the artery, to which it is closely united by dense areolar tissue; sometimes the vein is placed on the inner instead of the outer side of the artery; or the vein may be double, the artery then lies between them, the two veins being usually connected by short transverse branches. More deeply, and close to the surface of the bone, is the popliteal artery, and passing off from it at right angles are its articular branches. The articular branch from the obturator nerve descends upon the popliteal artery to supply the knee; and deeply in the space is an articular filament occasionally derived from the great sciatic nerve. The popliteal lymphatic glands, four or five in number, are found surrounding the artery; one usually lies superficial to the vessel, another is situated between it and the bone, and the rest are placed on either side of it. In health, these glands are small; but when enlarged and indurated from inflammation, the pulsation communicated to them from the popliteal artery makes them resemble so closely an aneurismal tumour, that it requires a very careful examination to discriminate between them.

The POPLITEAL ARTERY (fig. 212. color

illustration p. A7), in its course downwards from the aperture in the Adductor magnus, to the lower border of the Popliteus muscle, rests first on the inner, and then on the posterior surface of the femur; in the middle of its course, on the posterior ligament of the knee-joint; and below, on the fascia covering the Popliteus muscle. *Superficially*, it is covered, above, by the Semi-membranosus; in the middle of its course, by a quantity of fat, which separates it from the deep fascia and integument; and below, it is overlapped by the Gastrocnemius, Plantaris, and Soleus muscles, the popliteal vein, and the internal popliteal nerve. The popliteal vein, which is intimately attached to the artery, lies superficial and external to it, until near its termination, when it crosses it and lies to its inner side. The popliteal nerve is still more superficial and external, crossing, however, the artery below the joint, and lying on its inner side. *Laterally*, it is bounded by the muscles which form the boundaries of the popliteal space.

Peculiarities in point of division. Occasionally the popliteal artery divides prematurely into its terminal branches; this division occurs most frequently opposite the knee-joint.

Unusual branches. This artery sometimes divides into the anterior tibial and peroneal, the posterior tibial being wanting, or very small. In a single case, this artery divided into three branches, the anterior and posterior tibial, and peroneal.

Surgical Anatomy. Ligature of the popliteal artery is required in cases of wound of that vessel, but for aneurism of the posterior tibial, it is preferable to tie the superficial femoral. The popliteal may be tied in the upper or lower part of its course; but in the middle of the space the operation is attended with considerable difficulty, from the great depth of the artery, and from the extreme degree of tension of its lateral boundaries.

In order to expose the vessel in the upper part of its course, the patient should be placed in the prone position, with the limb extended. An incision about three inches in length should then be made through the integument, along the posterior margin of the Semi-membranosus, and the fascia lata having been divided, this muscle must be drawn inwards, when the pulsation of the vessel will be detected with the finger; the nerve lies on the outer or fibular side of the artery, the vein, superficial and also to its outer side; having cautiously separated it from the artery, the aneurism needle should be passed around the latter vessel from without inwards.

To expose the vessel in the lower part of its course, where the artery lies between the two heads of the Gastrocnemius, the patient should be placed in the same position as in the preceding operation. An incision should then be made through the integument in the middle line, commencing opposite the bend of the knee-joint, care being taken to avoid the external saphena vein and nerve. After dividing the deep fascia and separating some dense cellular membrane, the artery, vein, and nerve will be exposed, descending between the two heads of the Gastrocnemius. Some muscular branches of the popliteal should be avoided if possible, or if divided, tied immediately. The leg being now flexed, in order the more effectually to separate the two heads of the Gastrocnemius, the nerve should be drawn inwards and the vein outwards, and the aneurism needle passed between the artery and vein from without inwards.

The branches of the popliteal artery are

Muscular { Superior.
{ Inferior or Sural.
Superior external articular.
Superior internal articular.
Azygos articular.
Inferior external articular.
Inferior internal articular.

The *Superior Muscular Branches*, two or three in number, arise from the upper part of the popliteal artery, and are distributed to the Flexor muscles of the leg and the Vastus externus; anastomosing with the inferior perforating, and terminal branches of the profunda.

The *Inferior Muscular* (*Sural*) are two large branches which are distributed to the two heads of the Gastrocnemius and Plantaris muscles. They arise from the popliteal artery opposite the knee-joint.

Cutaneous branches descend on each side and in the middle of the limb, between the Gastrocnemius and integument; they arise separately from the popliteal artery, or from some of its branches, and supply the integument of the calf.

The *Superior Articular Arteries*, two in number, arise one on either side of the popliteal, and wind around the femur immediately above its condyles to the front of the knee-joint.

The *internal branch* passes beneath the tendon of the Adductor magnus, and divides into two, one of which supplies the Vastus internus, inosculating with the anastomotica magna and inferior internal articular; the other ramifies close to the surface of the femur, supplying it and the knee-joint, and anastomosing with the superior external articular artery.

The *external branch* passes above the outer condyle, beneath the tendon of the Biceps, and divides into a superficial and deep branch: the superficial branch supplies the Vastus externus, and anastomoses with the descending branch of the external circumflex artery; the deep branch supplies the lower part of the femur and knee-joint, and forms an anastomotic arch across the bone with the anastomotic artery.

The *Azygos Articular* is a small branch, arising from the popliteal artery opposite the bend of the joint. It pierces the posterior ligament, and supplies the ligaments and synovial membrane in the interior of the articulation.

The *Inferior Articular Arteries*, two in number, arise from the popliteal, beneath the Gastrocnemius, and wind round the head of the tibia, below the joint.

The *internal* one passes below the inner tuberosity, beneath the internal lateral ligament, at the anterior border of which it ascends to the front and inner side of the joint, to supply the head of the tibia and the articulation of the knee.

The *external* one passes outwards above the head of the fibula, to the front of the knee-joint, lying in its course beneath the outer head of the Gastrocnemius, the external lateral ligament, and the tendon of the Biceps muscle, and divides into branches, which anastomose with the artery of the opposite side, the superior articular, and the recurrent branch of the anterior tibial.

Anterior Tibial Artery

The anterior tibial artery commences at the bifurcation of the popliteal, at the lower border of the Popliteus muscle, passes forwards between the two heads of the Tibialis posticus, and through the aperture left between the bones at the upper part of the interosseous membrane, to the deep part of the front of the leg; it then descends on the anterior surface of the interosseous ligament and of the tibia to the front of the ankle-joint, where it lies more superficial, and becomes the dorsalis pedis. A line drawn from the inner side of the head of the fibula to midway between the two malleoli, will be parallel with the course of this artery.

Relations. In the upper two-thirds of its extent, it rests upon the interosseous ligament, to which it is connected by delicate fibrous arches thrown across it. In the lower third, upon the front of the tibia, and the anterior ligament of the ankle-joint. In the upper third of its course, it lies between the Tibialis anticus and Extensor longus digitorum: in the middle third, between the Tibialis anticus and Extensor proprius pollicis. In the lower third it is crossed by the tendon of the Extensor proprius pollicis, and lies between it and the innermost tendon of the Extensor longus digitorum. It is covered, in the upper two-thirds of its course, by the muscles which lie on either side of it, and by the deep fascia: in the lower third, by the integument, annular ligament, and fascia.

The anterior tibial artery is accompanied by two veins (venae comites), which lie one on either side of the artery; the anterior tibial nerve lies at first to its outer side, and about the middle of the leg is placed superficial to it; at the lower part of the artery the nerve is on the outer side.

Plan of the Relations of the Anterior Tibial Artery

In front.
Integument, superficial and deep fasciae.
Tibialis anticus.
Extensor longus digitorum.
Extensor proprius pollicis.
Anterior tibial nerve.

Inner side.
Tibialis anticus.
Extensor proprius pollicis.

Outer side.
Anterior tibial nerve.
Extensor longus digitorum.
Extensor proprius pollicis.

Behind.
Interosseous membrane.
Tibia.
Anterior ligament of ankle-joint.

Peculiarities in Size. This vessel may be diminished in size, or it may be deficient to a greater or less extent, or it may be entirely wanting, its place being supplied by perforating branches from the posterior tibial, or by the anterior division of the peroneal artery.

Course. This artery occasionally deviates in its course towards the fibular side of the leg, regaining its usual position beneath the annular ligament at the front of the ankle. In two instances, this vessel has approached the surface in the middle of the leg, from this point onwards being covered merely by the integument and fascia.

Surgical Anatomy. The anterior tibial artery may be tied in the upper or lower part of the leg. In the upper part, the operation is attended with great difficulty, on account of the depth of the vessel from the surface. An incision, about four inches in length, should be made through the integument, midway between the spine of the tibia and the outer margin of the fibula, the fascia and intermuscular septum between the Tibialis anticus and Extensor communis digitorum being divided to the same extent. The foot must be flexed to relax these muscles, and they must be separated from each other by the finger. The artery is then exposed, deeply seated, lying upon the interosseous membrane, the nerve lying externally, and one of the venae comites on either side; these must be separated from the artery before the aneurism needle is passed around it.

213.—Surgical Anatomy of the Anterior Tibial and Dorsalis Pedis Arteries.

To tie this vessel in the lower third of the leg above the ankle-joint, an incision about three inches in length should be made through the integument between the tendons of the Tibialis anticus and Extensor proprius pollicis muscles, the deep fascia being divided to the same extent; the tendon on either side should be held aside, when the vessel will be seen lying upon the tibia, with the nerve superficial to it, and one of the venae comites on either side.

In order to secure this vessel over the instep, an incision should be made on the fibular side of the tendon of the Extensor proprius pollicis, between it and the innermost tendon of the long Extensor: the deep fascia having been divided, the artery will be exposed, the nerve lying either superficial to it or to its outer side.

The branches of the anterior tibial artery are

    Recurrent tibial.   Muscular.
    Internal malleolar.   External malleolar.

The *Recurrent Branch* arises from the anterior tibial as soon as that vessel has passed through the interosseous space; it ascends in the Tibialis anticus muscle, and ramifies on the front and sides of the knee-joint, anastomosing with the articular branches of the popliteal.

The *Muscular Branches* are numerous; they are distributed to the muscles which lie on either side of this vessel, some piercing the deep fascia to supply the integument, others passing through the interosseous membrane, and anastomosing with branches of the posterior tibial and peroneal arteries.

The *Malleolar Arteries* supply the ankle-joint.

The *internal*, arises about two inches above the articulation, passes beneath the tendon of the Tibialis anticus to the inner ankle, upon which it ramifies, anastomosing with branches of the posterior tibial and internal plantar arteries.

The *external* passes beneath the tendons of the Extensor longus digitorum and Extensor proprius pollicis, and supplies the outer ankle, anastomosing with the anterior peroneal artery and with ascending branches from the tarsea branch of the dorsalis pedis.

DORSALIS PEDIS ARTERY

The dorsalis pedis, the continuation of the anterior tibial, passes forwards from the bend of the ankle along the tibial side of the foot to the back part of the first interosseous space, where it divides into two branches, the dorsalis hallucis and communicating.

Relations. This vessel in its course forwards rests upon the astragalus, scaphoid, and internal cuneiform bones, and the ligaments connecting them, being covered by the integument and fascia, and crossed near its termination by the innermost tendon of the Extensor brevis digitorum. On its *tibial side* is the tendon of the Extensor proprius pollicis; on its *fibular side*, the innermost tendon of the Extensor longus digitorum. It is accompanied by two veins, and by the anterior tibial nerve, which lies on its outer side.

PLAN OF THE RELATIONS OF THE DORSALIS PEDIS ARTERY

In front.
Integument and fascia.
Innermost tendon of Extensor brevis digitorum.

Tibial side.		*Fibular side.*
Extensor proprius pollicis.	Dorsalis Pedis.	Extensor longus digitorum.
		Anterior tibial nerve.

Behind.
Astragalus.
Scaphoid.
Internal cuneiform,
 and their ligaments.

Peculiarities in Size. The dorsal artery of the foot may be larger than usual, to compensate for a deficient plantar artery; or it may be deficient in its terminal branches to the toes, which are then derived from the internal plantar; or its place may be supplied altogether by a large anterior peroneal artery.

Position. This artery frequently curves outwards, lying external to the line between the middle of the ankle and the back of the first interosseous space.

Surgical Anatomy. This artery may be tied, by making an incision through the integument, between two and three inches in length, on the fibular side of the tendon of the Extensor proprius pollicis, in the interval between it and the inner border of the short Extensor muscle. The incision should not extend further forwards than the back part of the first interosseous space, as the artery divides in this situation. The deep fascia being divided to the same extent, the artery will be exposed, the nerve lying upon its outer side.

Branches. The branches of the dorsalis pedis are, the

Tarsea.	Interosseae.
Metatarsea.	Dorsalis pollicis.
Communicating.	

The *Tarsea Artery* arises from the dorsalis pedis, as that vessel crosses the scaphoid bone; it passes in an arched direction outwards, lying upon the tarsal bones, and covered by the Extensor brevis digitorum: it supplies that muscle and the articulations of the tarsus, and anastomoses with branches from the metatarsea, external malleolar, peroneal, and external plantar arteries.

The *Metatarsea* arises a little anterior to the preceding; it passes outwards to the outer part of the foot, over the bases of the metatarsal bones, beneath the tendons of the short Extensor, its direction being influenced by its point of origin; and it anastomoses with the tarsea and external plantar arteries. This vessel gives off three branches, the interosseae, which pass forwards upon the three outer Dorsal interossei muscles, and, in the clefts between the toes, divide into two dorsal collateral branches for the adjoining toes. At the back part of each interosseous space these vessels receive the posterior perforating branches from the plantar arch; and at the fore part of each interosseous space, they are joined by the anterior perforating branches, from the digital arteries. The outermost interosseous artery gives off a branch which supplies the outer side of the little toe.

The *Dorsalis Hallucis* runs forwards along the outer surface of the first metatarsal bone, and, at the cleft between the first and second toes, divides into two branches, one of which passes inwards, beneath the tendon of the Flexor longus pollicis, and is distributed to the inner border of the great toe; the other branch bifurcating to supply the adjoining sides of the great and second toes.

The *Communicating Artery* dips down into the sole of the foot, between the two heads of the first Dorsal interosseous muscle, and inosculates with the termination of the external plantar artery, to complete the plantar arch. It here gives off two digital branches; one runs along the inner side of the great toe, on its plantar surface; the other passes forwards along the first metatarsal space, and bifurcates for the supply of the adjacent sides of the great and second toes.

Posterior Tibial Artery

The posterior tibial is an artery of large size, which extends obliquely downwards from the lower border of the Popliteus muscle, along the tibial side of the leg, to the fossa between the inner ankle and the heel, where it divides beneath the origin of the Abductor pollicis, into the internal and external plantar arteries. At its origin it lies opposite the interval, between the tibia and fibula; as it descends, it approaches the inner side of the leg, lying behind the tibia, and, in the lower part of its course, is situated midway between the inner malleolus and the tuberosity of the os calcis.

Relations. It lies successively upon the Tibialis posticus, the Flexor longus digitorum, and below, upon the tibia and back part of the ankle-joint. It is *covered* by the intermuscular fascia, which separates it above from the Gastrocnemius and Soleus muscles. In the lower third, where it is more superficial, it is covered only by the integument and fascia, and runs parallel with the inner border of the tendo Achillis. It is accompanied by two veins, and by the posterior tibial nerve, which lies at first to the inner side of the artery, but soon crosses it, and is, in the greater part of its course, on its outer side.

Plan of the Relations of the Posterior Tibial Artery

In front.
Tibialis posticus.
Flexor longus digitorum.
Tibia.
Ankle-joint.

Inner side.	*Outer side.*
Posterior tibial nerve, upper third.	Posterior tibial nerve, lower two-thirds.

(Posterior Tibial.)

Behind.
Gastrocnemius.
Soleus.
Deep fascia and integument.

Behind the Inner Ankle, the tendons and blood-vessels are arranged in the following order, from within outwards: First, the tendons of the Tibialis posticus and Flexor longus digitorum, lying in the same groove, behind the inner malleolus, the former being the most internal. External to these is the posterior tibial artery, having a vein on either side; and, still more externally, the posterior tibial nerve. About half an inch nearer the heel is the tendon of the Flexor longus pollicis.

Peculiarities in Size. The posterior tibial is not unfrequently smaller than usual, or absent, its place being compensated for by a large peroneal artery, which passes inwards at the lower end of the tibia, and either joins the small tibial artery, or continues alone to the sole of the foot.

Surgical Anatomy. The *application of a ligature* to the posterior tibial may be required in cases of wound of the sole of the foot, attended with great haemorrhage, when the vessel should be tied at the inner ankle. In cases of wound of the posterior tibial itself, it will be necessary to enlarge the wound so as to expose the vessel at the wounded point (excepting where the vessel is injured by a punctured wound from the front of the leg). In cases of aneurism from wound of the artery low down, the vessel should be tied in the middle of the leg. But in aneurism of the posterior tibial high up, it would be better to tie the femoral artery.

To tie the posterior tibial artery at the ankle, a semi-lunar incision should be made through the integument, about two inches and a half in length, midway between the heel and inner ankle, but a little nearer the latter. The subcutaneous cellular membrane having been divided, a strong and dense fascia, the internal annular ligament, is exposed. This ligament is continuous above with the deep fascia of the leg, covers the vessels and nerves, and is intimately adherent to the sheaths of the tendons. This having been cautiously divided upon a director, the sheath of the vessels is exposed, and being opened, the artery is seen with one of the vena comites on each side. The aneurism needle should be passed around the vessel from the heel towards the ankle, in order to avoid the posterior tibial nerve, care being at the same time taken not to include the venae comites.

The vessel may also be tied in the lower third of the leg, by making an incision about three inches in length, parallel with the inner margin of the tendo Achillis. The internal saphena vein being carefully avoided, the two layers of fascia must be divided upon a director, when the artery is exposed along the inner margin of the Flexor longus digitorum, with one of its venae comites on either side, and the nerve lying external to it.

To tie the posterior tibial in the middle of the leg, is a very difficult operation, on account of the great depth of the vessel from the surface, and from its being covered in by the Gastrocnemius and Soleus muscles. The patient being placed in the recumbent position, the injured limb should rest on its outer side, the knee being partially bent, and the foot extended, so as to relax the muscles of the calf. An incision about four inches in length should then be made through the integument, along the inner margin of the tibia, taking care to avoid the internal saphena vein. The deep fascia having been divided, the margin of the Gastrocnemius is exposed, and must be drawn aside, and the tibial attachment of the Soleus divided, a director being previously passed beneath it. The artery may now be felt pulsating beneath the deep fascia, about an inch from the margin of the tibia. The fascia having been divided, and the limb placed in such a position as to relax the muscles of the calf as much as possible, the veins should be separated from the artery, and the aneurism needle passed around the vessel from without inwards, so as to avoid wounding the posterior tibial nerve.

The branches of the posterior tibial artery are, the

Peroneal.	Nutritious.
Muscular.	Communicating.
Internal calcanean.	

The *Peroneal Artery* lies, deeply seated, along the back part of the fibular side of the leg. It arises from the posterior tibial, about an inch below the lower border of the Popliteus muscle, passes obliquely outwards to the fibula, and then descends along the inner border of this bone to the lower third of the leg, where it gives off the anterior peroneal. It then descends across the articulation, between the tibia and fibula, to the outer side of the os calcis, supplying the neighbouring muscles and back of the ankle, and anastomosing with the external malleolar, tarsal, and external plantar arteries.

Relations. This vessel rests at first upon the Tibialis posticus, and, in the greater part of its course, in the fibres of the Flexor longus pollicis, in a groove between the interosseous ligament and the bone. It is *covered*, in the upper part of its course, by the Soleus and deep fascia; *below*, by the Flexor longus pollicis.

PLAN OF THE RELATIONS OF THE PERONEAL ARTERY

In front.
Tibialis posticus.
Flexor longus pollicis.

Outer side.   Peroneal Artery.
Fibula.

Behind.
Soleus.
Deep fascia.
Flexor longus pollicis.

Peculiarities in Origin. The peroneal artery may arise three inches below the popliteus, or from the posterior tibial high up, or even from the popliteal.

Its Size is more frequently increased than diminished, either reinforcing the posterior tibial by its junction with it, or by altogether taking the place of the posterior tibial, in the lower part of the leg and foot, the latter vessel only existing as a short muscular branch. In those rare cases, where the peroneal artery is smaller than usual, a branch from the posterior tibial supplies its place, and a branch from the anterior tibial compensates for the diminished anterior peroneal artery. In one case, the peroneal artery has been found entirely wanting.

The anterior peroneal is sometimes enlarged, and takes the place of the dorsal artery of the foot.

The peroneal artery, in its course, gives off branches to the Soleus, Tibialis posticus, Flexor longus pollicis, and Peronei muscles, and a nutrient branch to the fibula.

The *Anterior Peroneal* pierces the interosseous membrane, about two inches above the outer malleolus, to reach the fore part of the leg, and, passing down beneath the peroneus tertius to the outer ankle, ramifies on the front and outer side of the tarsus, anastomosing with the external malleolar and tarsal arteries.

The *Nutritious Artery* of the tibia arises from the posterior tibial near its origin, and after supplying a few muscular branches, enters the nutritious canal of that bone, which it traverses obliquely from above downwards. This is the largest nutrient artery of bone in the body.

The *Muscular Branches* are distributed to the Soleus and deep muscles along the back of the leg.

The *Communicating Branch* to the peroneal passes transversely across the back of the tibia, about two inches above its lower end, passing beneath the Flexor longus pollicis.

The *Internal Calcanean* consists of several large branches, which arise from the posterior tibial just before its division; they are distributed to the fat and integument behind the tendo Achillis and about the heel, and to the muscles on the inner side of the sole, anastomosing with the posterior peroneal and internal malleolar arteries.

The *Internal Plantar Artery* (figs. 214, 215), much smaller than the external, passes forwards along the inner side of the foot. It is at first situated above the Abductor pollicis, and then between it and the Flexor brevis digitorum, both of which it supplies. At the base of the first metatarsal bone, where it has become much diminished in size, it passes along the inner border of the great toe, inosculating with its digital branches.

The *External Plantar Artery*, much larger than the internal, passes obliquely outwards and forwards to the base of the fifth metatarsal bone. It then turns obliquely inwards to the interval between the bases of the first and second metatarsal bones, where it inosculates with the communicating branch from the dorsalis pedis artery, thus completing the plantar arch. As this artery passes outwards it is at first placed between the os calcis and Abductor pollicis, and then between the Flexor brevis digitorum and Flexor accessorius; and as it passes forwards to the base of the little toe,

214.—The Plantar Arteries. Superficial View.

215.—The Plantar Arteries. Deep View.

it lies more superficial between the Flexor brevis digitorum and Abductor minimi digiti, covered by the deep fascia and integument. The remaining portion of the vessel is deeply situated: it extends from the base of the metatarsal bone of the little toe to the back part of the first interosseous space, and forms the plantar arch; it is convex forwards, lies upon the Interossei muscles, opposite the tarsal ends of the metatarsal bones, and is covered by the Adductor pollicis, the flexor tendons of the toes, and the Lumbricales.

Branches. The plantar arch, besides distributing numerous branches to the muscles, integument, and fasciae in the sole, gives off the following branches:

   Posterior perforating.   Digital—Anterior perforating.

The *Posterior Perforating* are three small branches, which ascend through the back part of the three outer interosseous spaces, between the heads of the dorsal Interossei muscles, and anastomose with the interosseous branches from the metatarsal artery.

The *Digital Branches* are four in number, and supply the three outer toes and half the next. The *first* passes outwards from the outer side of the plantar arch, and is distributed to the outer side of the little toe, passing in its course beneath the Abductor and short Flexor muscles. The *second*, *third*, and *fourth* run forwards along the metatarsal spaces, and on arriving at the clefts between the toes, divide into collateral branches, which supply the adjacent sides of the three outer toes and the outer side of the second. At the bifurcation of the toes, each digital artery sends upwards, through the fore part of the corresponding metatarsal space, a small branch, which inosculates with the interosseous branches of the metatarsal artery. These are the anterior perforating arteries.

From the arrangement already described of the distribution of the vessels to the toes, it will be seen that both sides of the three outer toes, and the outer side of the second toe, are supplied by branches from the plantar arch; both sides of the great toe, and the inner side of the second, being supplied by the dorsal artery of the foot.

Pulmonary Artery

The pulmonary artery conveys the dark impure venous blood from the right side of the heart to the lungs. It is a short wide vessel, about two inches in length, arising from the left side of the base of the right ventricle, in front of the ascending aorta. It ascends obliquely upwards, backwards, and to the left side, as far as the under surface of the arch of the aorta, where it divides into two branches of nearly equal size, the right and left pulmonary arteries.

Relations. The greater part of this vessel is contained, together with the aorta, in the pericardium, being enclosed with it in a tube of serous membrane, continued upwards from the base of the heart, and has attached to it, above, the fibrous layer of this membrane. Behind, it rests at first upon the ascending aorta, and higher up in front of the left auricle. On either side of its origin is the appendix of the corresponding auricle and a coronary artery; and higher up it passes to the left side of the ascending aorta. A little to the left of its point of bifurcation it is connected to the under surface of the arch by a short fibrous cord, the remains of a vessel peculiar to foetal life, the ductus arteriosus.

The *Right Pulmonary Artery*, longer and larger than the left, runs horizontally outwards, behind the ascending aorta and superior vena cava, to the root of the right lung, where it divides into two branches, of which the lower, the larger, supplies the lower lobe, the upper giving a branch to the middle lobe.

The *Left Pulmonary Artery*, shorter but somewhat smaller than the right, passes horizontally in front of the descending aorta and left bronchus to the root of the left lung, where it divides into two branches for the two lobes.

OF THE VEINS

The Veins are the vessels which serve to return the blood from the capillaries of the different parts of the body to the heart. They consist of two distinct sets of vessels, the pulmonary and systemic.

The *Pulmonary Veins*, unlike other vessels of this kind, contain arterial blood, which they return from the lungs to the left auricle of the heart.

The *Systemic Veins* return the venous blood from the body generally to the right auricle of the heart.

The *Portal Vein*, an appendage to the systemic venous system, is confined to the abdominal cavity, returning the venous blood from the viscera of digestion, and carrying it to the liver by a single trunk of large size, the vena portae. From this organ the same blood is conveyed to the inferior vena cava by means of the hepatic veins.

The veins, like the arteries, are found in nearly every tissue of the body; they commence by minute plexuses, which communicate with the capillaries, the branches from which, uniting together, constitute trunks, which increase in size as they pass towards the heart, from the termination of larger branches in them. The veins are larger and altogether more numerous than the arteries; hence the entire capacity of the venous system is much greater than the arterial; the pulmonary veins excepted, which do not exceed in capacity the pulmonary arteries. From the combined area of the smaller venous branches being greater than the main trunks, it results that the venous system represents a cone, the summit of which corresponds to the heart, its base to the circumference of the body. In form, the veins are not perfectly cylindrical, like the arteries, their walls being collapsed when empty, and the uniformity of their surface being interrupted at intervals by slight contractions, which indicate the existence of valves in their interior. They usually retain, however, the same calibre as long as they receive no neighbouring branches.

The veins communicate very freely with one another, especially in certain regions of the body; and this communication exists between the larger trunks as well as between the smaller branches. Thus, in the cavity of the cranium, and between the veins of the neck, where obstruction of the cerebral venous system would be attended with imminent danger, we find that the sinuses and larger veins have large and very frequent anastomoses. The same free communication exists between the veins throughout the whole extent of the spinal canal, and between the veins composing the various venous plexuses in the abdomen and pelvis, as the spermatic, uterine, vesical, prostatic, etc.

The veins are subdivided into three sets; superficial, deep, and sinuses.

The *Superficial* or *Cutaneous Veins* are found between the layers of superficial fascia, immediately beneath the integument: they return the blood from these structures, and communicate with the deep veins by perforating the deep fascia.

The *Deep Veins* accompany the arteries, and are usually enclosed in the same sheath with those vessels. In the smaller arteries, as the radial, ulnar, brachial, tibial, peroneal, they exist generally in pairs, one lying on each side of the vessel, and are called *venae comites*. The larger arteries, as the axillary, subclavian, popliteal, and femoral, have usually only one accompanying vein. In certain organs of the body, however, the deep veins do not accompany the arteries; for instance, the veins in the skull and spinal canal, the hepatic veins in the liver, and the larger veins returning blood from the osseous tissue.

Sinuses are venous channels, which, in their structure and mode of distribution, differ altogether from the veins. They are found only in the interior of the skull, and are formed by a subdivision of the layers of the dura mater; their outer coat consisting of fibrous tissue, their inner of a serous membrane continuous with the serous membrane of the veins.

Veins are thinner in structure than the arteries, and possessed of considerable strength. The superficial veins usually have thicker coats than the deep veins, and the veins of the lower limb are thicker than those of the upper.

Veins are composed of three coats, internal, middle, and external.

The *Internal* is an epithelial and elastic coat, consisting of an epithelial lining supported on several laminae of longitudinal elastic fibres. It is less brittle in structure than the same coat in the arteries, and its laminae seldom fenestrated.

The *Middle* or *Contractile Coat*, thinner than that of the arteries, consists of numerous alternating layers of muscular and elastic fibres. The muscular fibres are disposed in a circular form around the vessel, intermixed with areolar tissue and elastic fibres. The elastic coat consists of well developed elastic fibres, reticulating in a longitudinal direction. This coat is best marked in the splenic and portal veins, and appears to be wanting in the hepatic part of the vena cava, in the hepatic and subclavian veins.

Muscular tissue is also wanting in the veins: 1. Of the maternal part of the placenta. 2. In most of the cerebral veins and sinuses of the dura mater. 3. In the veins of the retina. 4. In the veins of the cancellous tissue of bones. 5. In the venous spaces of the corpora cavernosa. The veins of the above mentioned parts consist of an internal epithelial lining, supported on one or more layers of areolar tissue. On the other hand, muscular tissue is abundantly developed in the veins of the gravid uterus, being found in all three coats; and in the vena cava and pulmonary veins, it is prolonged on to them from the auricles of the heart.

The *External* or *Areolar Fibrous Coat* consists of areolar tissue and longitudinal elastic fibres, it also contains in some of the larger veins a longitudinal network of non-striated muscular fibres, as in the whole length of the inferior vena cava, the renal, azygos, and external iliac veins, and in all the large trunks of the portal venous system, and in the trunks of the hepatic veins.

Most veins are provided with valves which serve to prevent the reflux of the blood. They are formed by a reduplication of the lining membrane, strengthened by a little fibrous tissue; their form is semilunar, they are attached by their convex edge to the wall of the vein, the concave margin is free, directed in the course of the venous current, and lies in close apposition with the wall of the vein as long as the current of blood takes its natural course; if, however, any regurgitation takes place, the valves become distended, their opposed edges are brought into contact, and the current is intercepted. Most commonly two such valves are found, placed opposite one another, more especially in the smaller veins, or in the larger trunks at the point where they

216.—Veins of the Head and Neck.

are joined by small branches; occasionally there are three, and sometimes only one. The wall of the vein immediately above the point of attachment of each segment of the valve, is expanded into a pouch or sinus, which gives to the vessel, when injected or distended with blood, a knotted appearance. The valves are very numerous in the veins of the extremities, especially the lower ones, these vessels having to conduct the blood against the force of gravity. They are absent in the very small veins, also in the vena cava, the hepatic vein, portal vein and its branches, the renal, uterine, and ovarian veins. A few valves are found in the spermatic veins, and one also at their point of junction with the renal vein and inferior cava in both sexes. The cerebral and spinal veins, the veins of the cancelled tissue of bone, the pulmonary veins, and the umbilical vein and its branches, are also destitute of valves. They are occasionally found, few in number, in the venae azygos and intercostal veins.

The veins are supplied with nutrient vessels, *vasa vasorum*, like the arteries; but nerves are not generally found distributed upon them, the only vessels upon which they have at present been traced are the inferior vena cava and cerebral veins.

The veins may be arranged into three groups. 1. Those of the head and neck, upper extremity, and thorax, which terminate in the superior vena cava. 2. Those of the lower limb, pelvis, and abdomen, which terminate in the inferior vena cava. 3. The cardiac veins, which open directly into the right auricle of the heart.

Veins of the Head and Neck

The veins of the head and neck may be subdivided into three groups. 1. The veins of the exterior of the head. 2. The veins of the neck. 3. The veins of the diploë and interior of the cranium.

The veins of the exterior of the head are the

    Facial.
    Temporal.
    Internal Maxillary.
    Temporo-maxillary.
    Posterior auricular.
    Occipital.

The FACIAL VEIN passes obliquely across the side of the face, extending from the inner angle of the orbit, downwards and outwards, to the anterior margin of the Masseter muscle. It lies to the outer side of the facial artery, and is not so tortuous as that vessel. It commences in the frontal region, where it is called the *frontal vein*; at the inner angle of the eye it has received the name of the *angular vein*; and from this point to its termination, the *facial vein*.

The *frontal vein* commences on the anterior part of the skull, by a venous plexus, which communicates with the anterior branches of the temporal vein; the veins converge to form a single trunk, which descends along the middle line of the forehead parallel with the vein of the opposite side, and unites with it at the root of the nose by a transverse trunk, called the *nasal arch*. Occasionally the frontal veins join to form a single trunk which bifurcates at the root of the nose into the two angular veins. At the nasal arch the branches diverge, and run along the side of the root of the nose. The frontal vein as it descends upon the forehead, receives the supra-orbital vein; the dorsal veins of the nose terminate in the nasal arch; and the angular vein receives, on its inner side, the veins of the ala nasi; on its outer side, the superior palpebral vein; it moreover communicates with the ophthalmic vein, which establishes an important anastomosis between this vessel and the cavernous sinus.

The *facial vein* commences at the inner angle of the orbit, being a continuation of the angular vein. It passes obliquely downwards and outwards, beneath the great Zygomatic muscle, descends along the anterior border of the Masseter, crosses over the body of the lower jaw, with the facial artery, and, passing obliquely outwards and backwards, beneath the Platysma and cervical fascia, unites with a branch of communication from the temporo-maxillary vein, to form a trunk of large size which enters the internal jugular.

Branches. The facial vein receives, near the angle of the mouth, communicating branches from the pterygoid-plexus. It is also joined by the inferior palpebral, the superior and inferior labial veins, the buccal veins from the cheek, and the masseteric veins. Below the jaw, it receives the submental, the inferior palatine, which returns the blood from the plexus around the tonsil and soft palate; the submaxillary vein, which commences in the submaxillary gland; and lastly, the ranine vein.

The *Temporal Vein* commences by a minute plexus on the side and vertex of the skull, which communicates with the frontal vein in front, the corresponding vein of the opposite side, and the posterior auricular and occipital veins behind. From this network, anterior and posterior branches are formed which unite above the zygoma, forming the trunk of the vein. This trunk is joined in this situation by a large vein, the middle temporal, which receives the blood from the substance of the Temporal muscle and pierces the fascia at the upper border of the zygoma. The temporal vein then descends between the external auditory meatus and the condyle of the jaw, enters the substance of the parotid gland, and unites with the internal maxillary vein, to form the temporo-maxillary.

Branches. The temporal vein receives in its course some parotid veins, an articular branch from the articulation of the jaw, anterior auricular veins from the external ear, and a vein of large size, the transverse facial, from the side of the face.

The *Internal Maxillary Vein* is a vessel of considerable size, receiving branches which correspond with those derived from the internal maxillary artery. Thus it receives the middle meningeal veins, the deep temporal, the pterygoid, masseteric, and buccal, some palatine veins, and the inferior dental. These branches form a large plexus, the pterygoid, which is placed between the Temporal and External pterygoid, and partly between the Pterygoid muscles. This plexus communicates very freely with the facial vein, and with the cavernous sinus, by branches through the base of the skull. The trunk of the vein then passes backwards, behind the neck of the lower jaw, and unites with the temporal vein, forming the temporo-maxillary.

The *Temporo-Maxillary Vein*, formed by the union of the temporal and internal maxillary veins, descends in the substance of the parotid gland, between the ramus of the jaw and the Sterno-mastoid muscle, and divides into two branches, one of which passes inwards to join the facial vein, the other is continuous with the external jugular. It receives near its termination the posterior auricular vein.

The *Posterior Auricular Vein* commences upon the side of the head, by a plexus which communicates with the branches of the temporal and occipital veins, descending behind the external ear. It joins the temporo-maxillary, just before that vessel terminates in the external jugular. This vessel receives the stylo-mastoid vein, and some branches from the back part of the external ear.

The *Occipital Vein* commences at the back part of the vertex of the skull, by a plexus in a similar manner with the other veins. It follows the course of the occipital artery, passing deeply beneath the muscles of the back part of the neck, and terminates in the internal jugular, occasionally in the external jugular. As this vein passes opposite the mastoid process, it receives the mastoid vein, which establishes a communication with the lateral sinus.

Veins of the Neck

The veins of the neck, which return the blood from the head and face, are the

External jugular. Anterior jugular.
Posterior external jugular. Internal jugular.
Vertebral.

The *External Jugular Vein* receives the greater part of the blood from the exterior of the cranium and deep parts of the face, being a continuation of the temporo-maxillary and posterior auricular veins. It commences in the substance of the parotid gland, on a level with the angle of the lower jaw, and runs perpendicularly down the neck, in the direction of a line drawn from the angle of the jaw to the middle of the clavicle. In its course, it crosses the Sterno-mastoid muscle, and runs parallel with its posterior border as far as its attachment to the clavicle, where it perforates the deep fascia, and terminates in the subclavian vein, on the outer side of the internal jugular. As it descends the neck, it is separated from the Sterno-mastoid by the anterior layer of the deep cervical fascia, and is covered by the Platysma, the superficial fascia, and the integument. This vein is crossed about its centre by the superficial cervical nerve, and its upper half is accompanied by the auricularis magnus nerve. The external jugular vein varies in size, bearing an inverse proportion to that of the other veins of the neck: it is occasionally double. It is provided with two valves, one being placed at its entrance into the subclavian vein, and a second in most cases about the middle of its course. These valves do not prevent the regurgitation of the blood, or the passage of injection from below upwards.

Branches. This vein receives the occipital, the posterior external jugular, and, near its termination, the supra-scapular and transverse cervical veins. It communicates with the anterior jugular, and, in the substance of the parotid, receives a large branch of communication from the internal jugular.

The *Posterior External Jugular Vein* returns the blood from the integument and superficial muscles in the upper and back part of the neck, lying between the Splenius and Trapezius muscles. It descends the back part of the neck, and opens into the external jugular just below the middle of its course.

The *Anterior Jugular Vein* collects the blood from the integument and muscles in the middle of the anterior region of the neck. It passes down between the median line and the anterior border of the Sterno-mastoid, and, at the lower part of the neck, passes beneath that muscle to open into the subclavian vein, near the termination of the external jugular. This vein varies considerably in size, bearing almost always an inverse proportion to the external jugular. Most frequently there are two anterior jugulars, a right, and left; but occasionally only one. This vein receives some laryngeal branches, and occasionally an inferior thyroid vein. Just above the sternum, the two anterior jugular veins communicate by a transverse trunk, which receives branches from the inferior thyroid veins. It also communicates with the external and with the internal jugular.

The *Internal Jugular Vein* collects the blood from the interior of the cranium, from the superficial parts of the face, and from the neck. It commences at the jugular foramen, in the base of the skull, being formed by the coalescence of the lateral and inferior petrosal sinuses. At its origin it is somewhat dilated, and this dilatation is called the sinus, or gulf of the internal jugular vein. It runs down the side of the neck in a vertical direction, lying at first on the outer side of the internal carotid, and then on the outer side of the common carotid, and at the root of the neck unites with the subclavian vein, to form the vena innominata. The internal jugular vein, at its commencement, lies upon the Rectus lateralis, behind, and at the outer side of the internal carotid, and the eighth and ninth pairs of nerves; lower down, the vein and artery lie upon the same plane, the glosso-pharyngeal and hypoglossal nerves passing forwards between them; the pneumogastric descends between and behind them, in the same sheath; and the spinal accessory passes obliquely outwards, behind the vein. At the root of the neck the vein of the right side is placed at a little distance from the artery; on the left side, it usually crosses it at its lower part. This vein is of considerable size, but varying in different individuals, the left one being usually the smallest. It is provided with two valves, which are placed at its point of termination, or from one to two inches above it.

Branches. This vein receives in its course, the facial, lingual, pharyngeal, superior and middle thyroid veins, and the occipital. At its point of junction with the branch common to the temporal and facial veins, it becomes greatly increased in size.

The *Lingual Veins* commence on the dorsum, sides, and under surface of the tongue, and, passing backwards, following the course of the lingual artery and its branches, terminate in the internal jugular.

The *Pharyngeal Vein* commences in a minute plexus, the pharyngeal, at the back part and sides of the pharynx, and after receiving meningeal branches, and some from the vidian and spheno-palatine veins, terminates in the internal jugular. It occasionally opens into the facial, lingual, or superior thyroid vein.

The *Superior Thyroid Vein* commences in the substance and on the surface of the thyroid gland, by branches corresponding with those of the superior thyroid artery, and terminates in the upper part of the internal jugular vein.

The *Middle Thyroid Vein* collects the blood from the lower part of the lateral lobe of the thyroid gland, and, being joined by some branches from the larynx and trachea, terminates in the lower part of the internal jugular vein.

Veins of the Diploë

The diploë of the cranial bones is channelled, in the adult, with a number of tortuous canals, which are lined by a more or less complete layer of compact tissue. The veins they contain are large and capacious, their walls being thin, and formed only of epithelium, resting upon a layer of elastic tissue, and they present, at irregular

217.—Veins of the Diploë, as displayed by the Removal of the Outer Table of the Skull.

intervals, pouch-like dilatations, or *culs de sac*, which serve as reservoirs for the blood. These are the veins of the diploë, and can only be displayed by removing the outer table of the skull.

In adult life, as long as the cranial bones are distinct and separable, these veins are confined to the particular bones; but in old age, when the sutures are united, they communicate with each other, and increase in size. These vessels communicate, in the interior of the cranium, with the meningeal veins, and with the sinuses of the dura mater; and on the exterior of the skull, with the veins of the pericranium. In the cranium, they are divided into the *frontal*, which opens into the supra-orbital vein, by an aperture at the supra-orbital notch, the *anterior temporal*, which is confined chiefly to the frontal bone, and opens into one of the deep temporal veins, after escaping by an aperture in the great wing of the sphenoid, the *posterior temporal*, which is confined to the parietal bone, terminates in the lateral sinus, by an aperture at the posterior inferior angle of the parietal bone, and the *occipital* which is confined to the occipital bone, and opens either into the occipital vein, or the occipital sinus.

Cerebral Veins

The *Cerebral Veins* are remarkable for the extreme thinness of their coats, from the muscular tissue in them being wanting, and for the absence of valves. They may be divided into two sets, the superficial, which are placed on the surface, and the deep veins, which occupy the interior of the organ.

The *Superficial Cerebral Veins* ramify upon the surface of the brain, being lodged in the sulci, between the convolutions, a few running across the convolutions. They receive branches from the substance of the brain, and terminate in the sinuses. They are named from the position they occupy, superior, inferior, internal, or external.

The *Superior Cerebral Veins*, seven or eight in number on each side, pass forwards

218.—Vertical Section of the Skull, showing the Sinuses of the Dura Mater.

and inwards towards the great longitudinal fissure, where they receive the internal cerebral veins, which return the blood from the convolutions of the flat surface of the corresponding hemisphere; passing obliquely forwards, they become invested with a tubular sheath of the arachnoid membrane, and open into the superior longitudinal sinus, in the opposite direction to the course of the blood.

The *Inferior Anterior Cerebral Veins* commence on the under surface of the anterior lobes of the brain, and terminate in the cavernous sinuses.

The *Inferior Lateral Cerebral Veins* commence on the lateral parts of the hemispheres and at the base of the brain: they unite to form from three to five veins, which open into the lateral sinus from before backwards.

The *Inferior Median Cerebral Veins*, which are very large, commence at the fore part of the under surface of the cerebrum, and from the convolutions of the posterior lobe, and terminate in the straight sinus behind the venae Galeni.

The *Deep Cerebral*, or *Ventricular Veins* (venae Galeni), are two in number, one from the right, the other from the left, ventricle. They are each formed by two veins, the vena corporis striati, and the choroid vein. They pass backwards, parallel with one another, enclosed within the velum interpositum, and pass out of the brain at the great transverse fissure, between the under surface of the corpus callosum and the tubercula quadrigemina, and enter the straight sinus.

The *Vena Corporis Striati* commences in the groove between the corpus striatum and thalamus opticus, receives numerous veins from both of these parts, and unites behind the anterior pillar of the fornix with the choroid vein, to form one of the venae Galeni.

The *Choroid Vein* runs along the whole length of the outer border of the choroid plexus, receiving veins from the hippocampus major, the fornix and corpus callosum, and unites, at the anterior extremity of the choroid plexus, with the vein of the corpus striatum.

The *Cerebellar Veins* occupy the surface of the cerebellum, and are disposed in three sets, superior, inferior, and lateral. The superior pass forwards and inwards, across the superior vermiform process, and terminate in the straight sinus: some open into the venae Galeni. The inferior cerebellar veins, of large size, run transversely outwards, and terminate by two or three trunks in the lateral sinuses. The lateral anterior cerebellar veins, terminate in the superior petrosal sinuses.

Sinuses of the Dura Mater

The sinuses of the dura mater (fig. 218) are venous channels, analogous to the veins, their outer coat being formed by the dura mater; their inner, by a continuation of the serous membrane of the veins. They are twelve in number, and are divided into two sets. 1. Those situated at the upper and back part of the skull. 2. The sinuses at the base of the skull. The former are the

> Superior longitudinal.   Straight sinus.
> Inferior longitudinal.   Lateral sinuses.
> Occipital sinuses.

The *Superior Longitudinal Sinus* occupies the attached margin of the falx cerebri. Commencing at the crista Galli, it runs from before backwards, grooving the inner surface of the frontal, the adjacent margins of the two parietal and the superior division of the crucial ridge of the occipital bone, and terminates by dividing into the two lateral sinuses. This sinus is triangular in form, narrow in front, and gradually increasing in size as it passes backwards. On examining its inner surface, it presents the internal openings of the cerebral veins, the apertures of which are, for the most part, directed from behind forwards, and chiefly open at its back part, their orifices being concealed by fibrous areolae; numerous fibrous bands are also seen (*chordae Willisi*), which extend transversely across its inferior angle; and lastly, some small, white, projecting bodies, the glandulae Pacchioni. This sinus receives the superior cerebral veins, numerous veins from the diploë and dura mater, and, at the posterior extremity of the sagittal suture, the parietal veins from the pericranium.

The point where the superior longitudinal and lateral sinuses are continuous is called the *confluence of the sinuses*, or the *torcular Herophili*. It presents a considerable dilatation, of very irregular form, and is the point of meeting of six sinuses, the superior longitudinal, the two lateral, the two occipital, and the straight.

The *Inferior Longitudinal Sinus*, more correctly described as the *inferior longitudinal vein*, is contained in the posterior part of the free margin of the falx cerebri. It is of a circular form, increases in size as it passes backwards, and terminates in the straight sinus. It receives several veins from the falx cerebri, and occasionally a few from the flat surface of the hemispheres.

The *Straight Sinus* is situated at the line of junction of the falx cerebri with the tentorium. It is triangular in form, increases in size as it proceeds backwards, and runs

219.—The Sinuses at the Base of the Skull.

obliquely downwards and backwards from the termination of the inferior longitudinal sinus to the torcular Herophili. Besides the inferior longitudinal sinus, it receives the vena Galeni, the inferior median cerebral veins, and the superior cerebellar. A few transverse bands cross its interior.

The *Lateral Sinuses* are of large size, and situated in the attached margin of the tentorium cerebelli. They commence at the torcular Herophili, and passing horizontally outwards to the base of the petrous portion of the temporal bone, curve downwards and inwards on each side to reach the jugular foramen, where they terminate in the internal jugular vein. Each sinus rests, in its course, upon the inner surface of the occipital, the posterior inferior angle of the parietal, the mastoid portion of the temporal, and on the occipital again just before its termination. These sinuses are of unequal size, the right being the larger, and they increase in size as they proceed from behind forwards. The horizontal portion is of a triangular form, the curved portion semi-cylindrical; their inner surface is smooth, and not crossed by the fibrous bands found in the other sinuses. These sinuses receive blood from the superior longitudinal, the straight, and the occipital sinuses; and in front they communicate with the superior and inferior petrosal. They communicate with the veins of the pericranium by means of the mastoid and posterior condyloid veins, and they receive the inferior cerebral and inferior cerebellar veins, and some from the diploë.

The *Occipital* are the smallest of the cranial sinuses. They are usually two in number, and situated in the attached margin of the falx cerebelli. They commence by several small veins around the posterior margin of the foramen magnum, which communicate with the posterior spinal veins, and terminate by separate openings (sometimes by a single aperture) in the torcular Herophili.

The sinuses at the base of the skull (fig. 219) are the

        Cavernous.   Inferior petrosal.
        Circular.   Superior petrosal.
                Transverse.

The *Cavernous Sinuses* are named from their presenting a reticulated structure. They are two in number, of large size, and placed one on each side of the sella Turcica, extending from the sphenoidal fissure to the apex of the petrous portion of the temporal bone: they receive anteriorly the ophthalmic vein through the sphenoidal fissure, communicate behind with the petrosal sinuses, and with each other by the circular and transverse sinuses. On the inner wall of each sinus is found the internal carotid artery, accompanied by filaments of the carotid plexus, and by the sixth nerve; and on its outer wall, the third, fourth, and ophthalmic nerves. These parts are separated from the blood flowing along the sinus by the lining membrane, which is continuous with the inner coat of the veins. The cavity of the sinus, which is larger behind than in front, is intersected by filaments of fibrous tissue and small vessels. The cavernous sinuses receive the inferior anterior cerebral veins; they communicate with the lateral sinuses by means of the superior and inferior petrosal, and with the facial vein through the ophthalmic.

The *Ophthalmic* is a large vein, which connects the frontal at the inner angle of the orbit with the cavernous sinus; it pursues the same course as the ophthalmic artery, and receives branches corresponding to those derived from that vessel. Forming a short single trunk, it passes through the inner extremity of the sphenoidal fissure, and terminates in the cavernous sinus.

220.—The Superficial Veins of the Upper Extremity.

The *Circular Sinus* completely surrounds the pituitary body, and communicates on each side with the cavernous sinuses. Its posterior half is larger than the anterior; and in old age it is more capacious than at an early period of life. It receives veins from the pituitary body, and from the adjacent bone and dura mater.

The *Inferior Petrosal Sinus* is situated in the groove formed by the junction of the inferior border of the petrous portion of the temporal with the basilar process of the occipital. It commences in front at the termination of the cavernous sinus, and opens behind, at the jugular foramen, forming with the lateral sinus the commencement of the internal jugular vein. These sinuses are semi-cylindrical in form.

The *Transverse Sinus* is placed transversely across the fore part of the basilar process of the occipital bone serving to connect the two inferior petrosal and cavernous sinuses. A second is occasionally found opposite the foramen magnum.

The *Superior Petrosal Sinus* is situated along the upper border of the petrous portion of the temporal bone, in the front part of the attached margin of the tentorium. It is small and narrow, and connects together the cavernous and lateral sinuses at each side. It receives a cerebral vein (inferior lateral cerebral) from the under part of the middle lobe, and a cerebellar vein (anterior lateral cerebellar) from the anterior border of the cerebellum.

VEINS OF THE UPPER EXTREMITY

The veins of the upper extremity are divided into two sets: 1. The superficial veins. 2. The deep veins.

The *Superficial Veins* are placed immediately

beneath the integument between the two layers of superficial fascia; they commence in the hand chiefly on its dorsal aspect, where they form a more or less complete arch.

The *Deep Veins* accompany the arteries, and constitute the venae comites of those vessels.

Both sets of vessels are provided with valves, which are more numerous in the deep than in the superficial.

The superficial veins of the upper extremity are the

    Anterior ulnar.   Posterior ulnar.
    Basilic.   Radial.
    Cephalic.   Median.
    Median basilic.   Median cephalic.

The *Anterior Ulnar Vein* commences on the anterior surface of the wrist and ulnar side of the hand, and ascends along the inner side of the fore-arm to the bend of the elbow, where it joins with the posterior ulnar vein, to form the basilic. It communicates with branches of the median vein in front, and with the posterior ulnar behind.

The *Posterior Ulnar Vein* commences on the posterior surface of the ulnar side of the hand, and from the vein of the little finger (vena salvatella), situated over the fourth metacarpal space. It ascends on the posterior surface of the ulnar side of the fore-arm, and just below the elbow unites with the anterior ulnar vein to form the basilic.

The *Basilic* is a vein of considerable size, formed by the coalescence of the anterior and posterior ulnar veins; ascending along the inner side of the elbow, it receives the median basilic vein, and passing upwards along the inner side of the arm, pierces the deep fascia, and ascends in front of the brachial artery, terminating either in one of the venae comites of that vessel, or in the axillary vein.

The *Radial Vein* commences from the dorsal surface of the thumb, index finger, and radial side of the hand, by branches which communicate with the vena salvatella. They form by their union a large vessel, which ascends along the radial side of the fore-arm, receiving numerous branches from both its surfaces. At the bend of the elbow it receives the median cephalic, when it becomes the cephalic vein.

The *Cephalic Vein* ascends along the outer border of the Biceps muscle, to the upper third of the arm; it then passes in the interval between the Pectoralis major and Deltoid muscles, accompanied by the descending branch of the thoracica acromialis artery, and terminates in the axillary vein just below the clavicle. This vein is occasionally connected with the external jugular or subclavian, by a branch which passes from it upwards in front of the clavicle.

The *Median Vein* collects the blood from the superficial structures in the palmar surface of the hand and middle line of the fore-arm, communicating with the anterior ulnar and radial veins. At the bend of the elbow, it receives a branch of communication from the deep veins, accompanying the brachial artery, and divides into two branches, the median cephalic and median basilic, which diverge from each other as they ascend.

The *Median Cephalic*, the smaller of the two, passes outwards in the groove between the Supinator longus and Biceps muscles, and joins with the cephalic vein. The branches of the external cutaneous nerve pass behind this vessel.

The *Median Basilic* vein passes obliquely inwards, in the groove between the biceps and Pronator radii teres, and joins with the basilic. This vein passes in front of the brachial artery, from which it is separated by a fibrous expansion, given off from the tendon of the Biceps to the fascia covering the Flexor muscles. Filaments of the internal cutaneous nerve pass in front as well as behind this vessel.

The *Deep Veins of the Upper Extremity* follow the course of the arteries, forming their venae comites. They are generally two in number, one lying on each side of the corresponding artery, and they are connected at intervals by short transverse branches.

There are two digital veins accompanying each artery along the sides of the fingers; these, uniting at their base, pass along the interosseous spaces in the palm, and terminate in the two superficial palmar veins. Branches from these vessels on the radial side of the hand accompany the superficialis volae, and on the ulnar side, terminate in the deep ulnar veins. The deep ulnar veins, as they pass in front of the wrist, communicate with the interosseous and superficial veins, and unite at the elbow, with the deep radial veins, to form the venae comites of the brachial artery.

The *Interosseous Veins* accompany the anterior and posterior interosseous arteries. The anterior interosseous veins commence in front of the wrist, where they communicate with the deep radial and ulnar veins; at the upper part of the fore-arm they receive the posterior interosseous veins, and terminate in the venae comites of the ulnar artery.

The *Deep Palmar Veins* accompany the deep palmar arch, being formed by branches which accompany the ramifications of this vessel. They communicate with the superficial palmar veins at the inner side of the hand; and on the outer side, terminate in the venae comites of the radial artery. At the wrist, they receive a dorsal and a palmar branch from the thumb, and unite with the deep radial veins. Accompanying the radial artery, these vessels terminate in the venae comites of the brachial artery.

The *Brachial Veins* are placed one on each side of the brachial artery, receiving branches corresponding with those given off from this vessel; at the lower margin of the axilla they unite with the basilic to form the axillary vein.

The deep veins have numerous anastomoses, not only with each other, but also with the superficial veins.

Axillary Vein

The axillary vein is of large size and formed by the continuation upwards of the basilic vein. It commences at the lower part of the axillary space, and increasing in size as it ascends, by receiving branches corresponding with those of the axillary artery, terminates immediately beneath the clavicle at the outer margin of the first rib, and becomes the subclavian vein. This vessel is covered in front by the Pectoral muscles and costo-coracoid membrane, and lies on the thoracic side of the axillary artery. Opposite the Subscapularis, it is joined by a large vein, formed by the junction of the venae comites of the brachial; and near its termination it receives the cephalic vein. This vein is provided with valves at the point where it is joined by the cephalic and basilic veins.

Subclavian Vein

The subclavian vein, the continuation of the axillary, extends from the outer margin of the first rib to the inner end of the sterno-clavicular articulation, where it unites with the internal jugular, to form the vena innominata. It is in relation, in front, with the clavicle and Subclavius muscle; behind, with the subclavian artery, from which it is separated internally by the Scalenus anticus and phrenic nerve. Below, it rests in a depression on the first rib and upon the pleura. Above, it is covered by the cervical fascia and integument.

221.—The Venæ Cavæ and Azygos Veins, with their Formative Branches.

The subclavian vein occasionally rises in the neck to a level with the third part of the subclavian artery, and in two instances, has been seen passing with this vessel behind the Scalenus anticus. This vessel is provided with valves near its termination in the innominate.

Branches. It receives the external and anterior jugular veins and a small branch from the cephalic, outside the Scalenus; and on the inner side of this muscle, the vertebral and internal jugular veins.

The *Vertebral Vein* commences by numerous small branches in the occipital region, from the deep muscles at the upper and back part of the neck, passes outwards, and enters the foramen in the transverse process of the atlas, and descends by the side of the vertebral artery, in the canal formed by the transverse processes of the cervical vertebrae. Emerging from the foramen in the transverse process of the sixth cervical, it terminates at the root of the neck in the subclavian vein near its junction with the vena innominata. This vein, in the lower part of its course, occasionally divides into two branches, one emerges with the artery at the sixth cervical vertebra; the other escapes through the foramen in the seventh cervical.

Branches. This vein receives in its course the posterior condyloid vein, muscular branches from the muscles in the prevertebral region; dorsi-spinal veins, from the back part of the cervical portion of the spine; meningo-rachidian veins, from the interior of the spinal canal; and lastly, the ascending and deep cervical veins.

Innominate Veins

The venae innominatae (fig. 221) are two large trunks, placed one on each side of the root of the neck, and formed by the union of the internal jugular and subclavian veins of the corresponding side.

The *Right Vena Innominata* is a short vessel, about an inch and a half in length, which commences at the inner end of the clavicle, and passing almost vertically downwards, joins with the left vena innominata just below the cartilage of the first rib, to form the superior vena cava. It lies superficial and external to the arteria innominata; on its right side the pleura is interposed between it and the apex of the lung. This vein at its angle of junction with the subclavian, receives the right lymphatic duct; and lower down, the right internal mammary, right inferior thyroid, and right superior intercostal veins.

The *Left Vena Innominata*, about three inches in length, and larger than the right, passes obliquely from left to right across the upper and front part of the chest, to unite with its fellow of the opposite side, forming the superior vena cava. It is in relation in front, with the sternal end of the left clavicle, the left sterno-clavicular articulation, and with the first piece of the sternum, from which it is separated by the Sterno-hyoid and Sterno-thyroid muscles, the thymus glands or its remains, and some loose areolar tissue. Behind, it lies across the roots of the three large arteries arising from the arch of the aorta. This vessel, at its commencement, receives the thoracic duct; it is joined also by the left inferior thyroid, left internal mammary, and the left superior intercostal veins, and occasionally some thymic and pericardiac veins. There are no valves in the venae innominatae.

Peculiarities. Sometimes the innominate veins open separately into the right auricle; in such cases the right vein takes the ordinary course of the superior vena cava, but the left vein, after communicating by a small branch with the right one, passes in front of the root of the left lung, and turning to the back of the heart, receives the cardiac veins, and terminates in the back of the right auricle. This occasional condition of the veins in the adult, is a regular one in the foetus at an early period, and the two vessels are persistent in birds and some mammalia. The subsequent changes which take place in these vessels are the following. The communicating branch between the two trunks enlarges and forms the future left innominate vein; the remaining part of the left trunk is obliterated as far as the heart, where it remains pervious, and forms the coronary sinus; a remnant of the obliterated vessel is seen in adult life as a fibrous band passing along the back of the left auricle and in front of the root of the left lung, called by Mr. Marshall, the vestigial fold of the pericardium.

The *Internal Mammary Veins*, two in number to each artery, follow the course of that vessel, and receive branches corresponding with those derived from it. The two veins unite into a single trunk, which terminates in the innominate vein.

The *Inferior Thyroid Veins*, two, frequently three or four in number, arise in the venous plexus, on the thyroid body, communicating with the middle and superior thyroid veins. The left one, descends in front of the trachea, behind the Sterno-thyroid muscles, communicating with its fellow by transverse branches, and terminates in the left vena innominata. The right one, which is placed a little to the right of the median line, opens into the right vena innominata, just at its junction with the superior cava. These veins receive tracheal and inferior laryngeal branches, and are provided with valves at their termination in the innominate veins.

The Superior Intercostal Veins return the blood from the upper intercostal spaces.

The *right superior intercostal*, much smaller than the left, closely corresponds with the superior intercostal artery, receiving the blood from the first, or first and second intercostal spaces, and terminates in the right vena innominata. Sometimes it passes down, and opens into the vena azygos major.

The *left superior intercostal* is always larger than the right, but varies in size in different subjects, being small when the left upper azygos vein is large, and vice versa. It is usually formed by branches from the two or three upper intercostal spaces, and, passing across the arch of the aorta, terminates in the left vena innominata. The left bronchial vein opens into it.

Superior Vena Cava

The *Superior Vena Cava* receives the blood which is conveyed to the heart from the whole of the upper half of the body. It is a short trunk, varying from two inches and a half to three inches in length, formed by the junction of the two venae innominata. It commences immediately below the cartilage of the first rib on the right side, and, descending vertically downwards, enters the pericardium, and terminates in the upper part of the right auricle. In its course, it describes a slight curve, the convexity of which is turned to the right side.

Relations. The part external to the pericardium is in relation, *in front*, with the thoracic fascia, which separates it from the thymus gland, and from the sternum; *behind*, with the trachea, from which it is separated by numerous lymphatic glands. On its *right side*, with the right phrenic nerve and the right lung; on its *left side*, with the arch of the aorta. The portion contained within the pericardium, is covered by the serous layer of that membrane, in its anterior three-fourths. It is in relation, *behind*, with the right pulmonary artery, and right superior pulmonary veins; on the *left side*, with the aorta. It receives the vena azygos major, just before it enters the pericardium, and several small veins from the mediastinum and pericardium. The superior vena cava has no valves.

Azygos Veins

The *Azygos Veins* connect together the superior and inferior venae cavae, supplying the place of these vessels in that part of the trunk in which they are deficient, on account of their connection with the heart.

The larger, or *Right Azygos Vein*, commences opposite the first or second lumbar vertebra, by receiving a branch from the right lumbar veins; sometimes by a branch from the renal vein, or from the inferior vena cava. It enters the thorax through the aortic opening in the Diaphragm, and passes along the right side of the vertebral column to the third dorsal vertebra, where it arches forward, over the root of the right lung, and terminates in the superior vena cava, just before that vessel enters the pericardium. Whilst passing through the aortic opening of the Diaphragm, it lies with the thoracic duct on the right side of the aorta; and in the thorax it lies upon the intercostal arteries, on the right side of the aorta and thoracic duct, and covered by the pleura.

Branches. It receives nine or ten lower intercostal veins of the right side, the vena azygos minor, several oesophageal, mediastinal, and vertebral veins; near its termination, the right bronchial vein; and it is occasionally connected with the right superior intercostal vein. A few imperfect valves are found in this vein, but its branches are provided with complete valves.

The intercostal veins on the left side, below the two or three upper intercostal spaces, usually form two trunks, named the left lower, and the left upper, azygos veins.

The *Left lower*, or *Smaller Azygos Vein*, commences in the lumbar region, by a branch from one of the lumbar veins, or from the left renal. It passes into the thorax, through the left crus of the Diaphragm, and, ascending on the left side of the spine, as high as the sixth or seventh vertebra, passes across the column, behind the aorta and thoracic duct, to terminate in the right azygos vein. It receives the four or five lower intercostal veins of the left side, and some oesophageal and mediastinal veins.

The *Left Upper Azygos*, varies according to the size of the left superior intercostal. It receives veins from the intercostal spaces between the superior intercostal vein, and highest branch of the left lower azygos. They are usually two or three in number, and join to form a trunk which ends in the right azygos vein, or in the left lower azygos. When this vein is small, or altogether wanting, the left superior intercostal vein will extend as low as the fifth or sixth intercostal space.

The *Bronchial Veins* return the blood from the substance of the lungs; that of the right side opens into the vena azygos major, near its termination; that of the left side, in the left superior intercostal vein.

The Spinal Veins

The numerous venous plexuses placed upon and within the spine, may be arranged into four sets.

1. Those placed on the exterior of the spinal column, the dorsi-spinal veins.
2. Those situated in the interior of the spinal canal, between the vertebra and the theca vertebralis (meningo-rachidian veins).
3. The veins of the bodies of the vertebrae.
4. The veins of the spinal cord (medulli spinal).

1. The *Dorsi-Spinal Veins* (figs. 222, 223, color illustrations p. A8) commence by small branches, which receive their blood from the integument of the back of the spine, and from the muscles in the vertebral grooves. They form a complicated network, which surrounds the spinous processes, laminae, and the transverse and articular processes of all the vertebrae. At the bases of the transverse processes, they communicate, by means of ascending and descending branches, with the veins surrounding the contiguous vertebrae, and with the veins in the interior of the spine, in the intervals between

the arches of the vertebrae, perforating the ligamenta subflava, and terminate in the vertebral vein in the neck, in the intercostal veins in the thorax, in the lumbar and sacral veins in the loins and pelvis.

2. The veins contained in the interior of the spinal canal, are situated between the theca vertebralis and the vertebrae. They consist of two longitudinal plexuses, one of which runs along the posterior surface of the bodies of the vertebrae, throughout the entire length of the spinal canal (anterior longitudinal spinal veins), receiving the veins belonging to the bodies of the vertebrae (venae basis vertebrarum). The other plexus is placed on the inner, or anterior surface of the lamina of the vertebrae. They also extend along the entire length of the spinal canal, and are called the posterior longitudinal spinal veins.

222.—Transverse Section of a Dorsal Vertebra, showing the Spinal Veins.

The *Anterior Longitudinal Spinal Veins* consist of two large, tortuous venous canals, which extend along the whole length of the vertebral column, from the foramen magnum to the base of the coccyx, being placed one on each side of the posterior surface of the bodies of the vertebra, external to the posterior common ligament. These veins communicate together opposite each vertebra, by transverse trunks, which pass beneath the ligament, and receive the large venae basis vertebrarum, from the interior of the body of each vertebra. The anterior longitudinal spinal veins are least developed in the cervical and sacral regions. They are not of uniform size throughout, being alternately enlarged and constricted. At the intervertebral foramina they communicate with the dorsi-spinal veins, and with the vertebral vein in the neck,

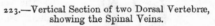

223.—Vertical Section of two Dorsal Vertebræ, showing the Spinal Veins.

with the intercostal veins in the dorsal region, and with the lumbar and sacral veins in the corresponding regions.

The *Posterior Longitudinal Spinal Veins*, smaller than the anterior, are situated one on either side, between the inner surface of the laminae and the theca vertebralis. They communicate (like the anterior), opposite each vertebra, by transverse trunks; and with the anterior longitudinal veins, by lateral transverse branches, which pass from behind forwards. These veins, at the intervertebral foramina, join with the dorsi-spinal veins.

3. The *Veins of the Bodies of the Vertebrae* (venae basis vertebrarum), emerge from the foramina on their posterior surface, and join the transverse trunk connecting the anterior longitudinal spinal veins. They are contained in large, tortuous channels, in the substance of the bones, similar in every respect to those found in the diploë of the cranial bones. These canals lie parallel to the upper and lower surface of the bones, arise from the entire circumference of the vertebra, communicate with veins which enter through the foramina, on the anterior surface of the bodies, and converge to the principal canal, which is sometimes double towards its posterior part. They become greatly developed in advanced age.

224.—The Internal or Long Saphenous Vein and its Branches.

4. The *Veins of the Spinal Cord* (medulli spinal), consist of a minute tortuous venous plexus, which covers the entire surface of the cord, being situated between the pia-mater and arachnoid. These vessels emerge chiefly from the posterior median furrow, and are largest in the lumbar region. Near the base of the skull they unite, and form two or three small trunks, which communicate with the vertebral veins, and then terminate in the inferior cerebellar veins, or in the petrosal sinuses. Each of the spinal nerves is accompanied by a branch as far as the intervertebral foramina, where they join the other veins from the spinal canal. There are no valves in the spinal veins.

VEINS OF THE LOWER EXTREMITY

The veins of the lower extremity are divided, like those of the upper, into two sets, superficial and deep: the superficial veins being placed beneath the integument, between the two layers of superficial fascia; the deep veins accompanying the arteries, and forming the venae comites of those vessels. Both sets of veins are provided with valves, which are more numerous in the deep than in the superficial set. These valves are also more numerous in the lower than in the upper limbs.

The *Superficial Veins* of the lower extremity are the internal or long saphenous, and the external or short saphenous.

The *Internal Saphenous Vein* (fig. 224) commences from a minute plexus, which covers the dorsum and inner side of the foot; it ascends in front of the inner ankle, and along the inner side of the leg, behind the inner margin of the tibia, accompanied by the internal saphenous nerve. At the knee it passes backwards behind the inner condyle of the femur, ascends along the inside of the thigh, and, passing through the saphenous opening in the fascia lata, terminates in the femoral vein, an inch and a half below Poupart's ligament. This vein receives in its course cutaneous branches from the leg and thigh, and at the saphenous opening, the superficial epigastric, superficial circumflex iliac, and external pudic veins. The veins from the inner and back part of the thigh frequently unite to form a large vessel, which enters the main trunk near the saphenous opening, and sometimes those on the outer side of the thigh join to form a large branch; so that occasionally three large veins are seen converging from different parts of the thigh towards the saphenous opening. The internal saphena communicates in the foot with the internal plantar vein; in the leg, with the posterior tibial veins, by branches which perforate the tibial origin of the Soleus muscle, and also with the anterior tibial veins; at the knee, with the articular veins; in the thigh, with the femoral vein by one or more branches. The valves in this vein vary from two to six in number; they are more numerous in the thigh than in the leg.

The *External* or *Short Saphenous Vein* is formed by branches which collect the blood from the dorsum and outer side of the foot; it passes behind the outer ankle, and along the outer border of the tendo Achillis, across which it passes at an acute angle to reach the middle line of the posterior aspect of the leg. Ascending directly upwards, it perforates the deep fascia in the lower part of the popliteal space, and terminates in the popliteal vein, between the heads of the Gastrocnemius muscle. It is accompanied by the external saphenous nerve. It receives numerous large branches from the back part of the leg, and communicates with the deep veins on the dorsum of the foot, and behind the outer malleolus. This vein has only two valves, one of which is always found near its termination in the popliteal vein.

225.—External, or Short Saphenous Vein.

The *Deep Veins* of the lower extremity accompany the arteries and their branches, and are called the *venae comites* of those vessels.

The external and internal plantar veins unite to form the posterior tibial. They accompany the posterior tibial artery, and are joined by the peroneal veins.

The *Anterior Tibial Veins* are formed by a continuation upwards of the venae dorsales pedis. They perforate the interosseous membrane at the upper part of the leg, and form, by their junction with the posterior tibial, the popliteal vein.

The valves in the deep veins are very numerous.

Popliteal Vein

The popliteal vein is formed by the junction of the venae comites of the anterior and posterior tibial vessels; it ascends through the popliteal space to the tendinous aperture in the Adductor magnus, where it becomes the femoral vein. In the lower part of its course it is placed internal to the artery; between the heads of the Gastrocnemius it is superficial to that vessel, but above the knee-joint it is close to its outer side. It receives the sural veins from the Gastrocnemius muscle, the articular veins, and the external saphenous. The valves in this vein are usually four in number.

Femoral Vein

The femoral vein accompanies the femoral artery through the upper two-thirds of the thigh. In the lower part of its course it lies external to the artery; higher up it is behind it; and beneath Poupart's ligament it lies to its inner side, and on the same plane as that vessel. It receives numerous muscular branches; the profunda femoris joins it about an inch and a half below Poupart's ligament, and near its termination the internal saphenous vein. The valves in this vein are four or five in number.

External Iliac Vein

The external iliac vein commences at the termination of the femoral, beneath the crural arch, and passing upwards along the brim of the pelvis, terminates opposite the sacro-iliac symphysis, by uniting with the internal iliac to form the common iliac vein. On the right side, it lies at first along the inner side of the external iliac artery; but as it passes upwards, gradually inclines behind it. On the left side, it lies altogether on the inner side of the artery. It receives, immediately above Poupart's ligament, the epigastric and circumflex iliac veins. It has no valves.

Internal Iliac Vein

The internal iliac vein is formed by the venae comites of the branches of the internal iliac artery, the umbilical arteries excepted. It receives the blood from the exterior of the pelvis by the gluteal, sciatic, internal pudic, and obturator veins; and from the organs in the cavity of the pelvis by the haemorrhoidal and vesico-prostatic plexuses in the male, and the uterine and vaginal plexuses in the female. The vessels forming these plexuses are remarkable for their large size, their frequent anastomoses, and the number of valves which they contain. The internal iliac vein lies at first on the inner side and then behind the internal iliac artery, and terminates opposite the sacro-iliac articulation, by uniting with the external iliac, to form the common iliac vein. This vessel has no valves.

The *Haemorrhoidal Plexus* surrounds the lower end of the rectum, being formed by the superior haemorrhoidal veins, branches of the inferior mesenteric, and the middle and inferior haemorrhoidal, which terminate in the internal iliac. The portal and general venous systems have a free communication by means of the branches composing this plexus.

The *Vesico-prostatic Plexus* surrounds the neck and base of the bladder and prostate gland. It communicates with the haemorrhoidal plexus behind, and receives the great dorsal vein of the penis, which enters the pelvis beneath the sub-pubic ligament.

This plexus is supported upon the sides of the bladder by a reflection of the pelvic fascia. These veins are very liable to become varicose, and often contain hard earthy concretions, called *phlebolites*.

The *Dorsal Vein of the Penis* is a vessel of large size, which returns the blood from the body of this organ. At first it consists of two branches, which are contained in the groove on the dorsum of the penis, and receives veins from the glans, the corpus spongiosum, and numerous superficial veins; these unite near the root of the penis into a single trunk, which pierces the triangular ligament beneath the pubic arch, and divides into two branches, which enter the prostatic plexus.

The *Vaginal Plexus* surrounds the mucous membrane of the vagina, being especially developed at the orifice of this canal; it communicates with the vesical plexus in front, and with the haemorrhoidal plexus behind.

The *Uterine Plexus* is situated along the sides and superior angles of the uterus, receiving large venous canals (the uterine sinuses) from its substance. The veins composing this plexus anastomose frequently with each other, and some of them communicate with the ovarian veins. They are not tortuous like the arteries.

Common Iliac Vein

Each common iliac vein is formed by the union of the external and internal iliac veins in front of the sacro-vertebral articulation; passing obliquely upwards towards the right side, they terminate upon the intervertebral substance between the fourth and fifth lumbar vertebrae, where they unite at an acute angle to form the inferior vena cava. The *right common iliac* is shorter than the left, nearly vertical in its direction, and ascends behind and then to the outer side of its corresponding artery. The *left common iliac*, longer and more oblique in its course, is at first situated at the inner side of the corresponding artery, and then behind the right common iliac. Each common iliac receives the ilio-lumbar, and sometimes the lateral sacral veins. The left one receives, in addition, the middle sacral vein. No valves are found in these veins.

The *Middle Sacral Vein* accompanies its corresponding artery along the front of the sacrum, and terminates in the left common iliac vein; occasionally in the commencement of the inferior vena cava.

Peculiarities. The left common iliac vein, instead of joining with the right one in its usual position, occasionally ascends on the left side of the aorta as high as the kidney, where, after receiving the left renal vein, it crosses over the aorta, and then joins with the right vein to form the vena cava. In these cases, the two common iliacs are connected by a small communicating branch at the spot where they are usually united.

Inferior Vena Cava

The inferior vena cava returns to the heart the blood from all the parts below the Diaphragm. It is formed by the junction of the two common iliac veins on the right side of the intervertebral substance, between the fourth and fifth lumbar vertebrae. It passes upwards along the front of the spine, on the right side of the aorta, and having reached the under surface of the liver, is contained in a groove in its posterior border. It then perforates the tendinous centre of the Diaphragm, enters the pericardium, where it is covered by its serous layer, and terminates in the lower and back part of the right auricle. At its termination in the auricle, it is provided with a valve, the Eustachian, which is of large size during foetal life.

Relations. In front, from below upwards, with the mesentery, transverse portion of the duodenum, the pancreas, portal vein, and the posterior border of the liver, which partly and occasionally completely surrounds it; *behind,* it rests upon the vertebral column, the right crus of the Diaphragm, the right renal and lumbar arteries; on the *left side,* it is in relation with the aorta. It receives in its course the following branches:

Lumbar.   Supra-renal.
Right spermatic.   Phrenic.
Renal.   Hepatic.

Peculiarities. In Position. This vessel is sometimes placed on the left side of the aorta, as high as the left renal vein, after receiving which, it crosses over to its usual position on the right side; or it may be placed altogether on the left side of the aorta, as far upwards as its termination in the heart: in such cases, the abdominal and thoracic viscera, together with the great vessels, are all transposed.

Point of Termination. Occasionally the inferior vena cava joins the right azygos vein, which is then of large size. In such cases, the superior cava receives the whole of the blood from the body before transmitting it to the right auricle, the blood from the hepatic veins excepted, these vessels terminating directly in the right auricle.

The *Lumbar Veins,* three or four in number on each side, collect the blood by dorsal branches from the muscles and integument of the loins, and by abdominal branches from the walls of the abdomen, where they communicate with the epigastric veins. At the spine, they receive branches from the spinal plexuses, and then pass forwards round the sides of the bodies of the vertebrae beneath the Psoas magnus, and terminate at the back part of the inferior cava. The left lumbar veins are longer than the right, and pass behind the aorta. The lumbar veins communicate with each other by branches which pass in front of the transverse processes. Occasionally two or more of these veins unite to form a single trunk, the ascending lumbar, which serves to connect the common iliac, ilio-lumbar, lumbar, and azygos veins of the corresponding side of the body.

The *Spermatic Veins* emerge from the back of the testis, and receive branches from the epididymis; they form a branched and convoluted plexus, called the *spermatic plexus* (plexus pampiniformis), below the abdominal ring: the vessels composing this plexus are very numerous, and ascend along the cord in front of the vas deferens; having entered the abdomen, they coalesce to form two branches, which ascend on the Psoas muscle, behind the peritoneum, lying one on each side of the spermatic artery, and unite to form a single vessel, which opens on the right side in the inferior vena cava, piercing this vessel obliquely; on the left side in the left renal vein, terminating at right angles with this vein. The spermatic veins are provided with valves. The left spermatic vein passes behind the sigmoid flexure of the colon; this circumstance, as well as the indirect communication of the vessel with the vena cava, may serve to explain the more frequent occurrence of varicocele on the left side.

The *Ovarian Veins* are analogous to the spermatic in the male; they form a plexus near the ovary, and in the broad ligament and Fallopian tube, communicating with the uterine plexus. They terminate as in the male. Valves are occasionally found in these veins. These vessels, like the uterine veins, become much enlarged during pregnancy.

The *Renal Veins* are of large size, and placed in front of the divisions of the renal arteries. The left is longer than the right, and passes in front of the aorta, just below the origin of the superior mesenteric artery. It receives the left spermatic vein. It usually opens into the vena cava, a little higher than the right.

The *Supra-Renal Vein* terminates, on the right side, in the vena cava; on the left side, in the left renal or phrenic vein.

The *Phrenic Veins* follow the course of the phrenic arteries. The *two superior*, of small size, accompany the corresponding nerve and artery; the right terminating opposite the junction of the two venae innominatae, the left in the left superior intercostal or left internal mammary. The *two inferior phrenic veins* follow the course of the inferior phrenic arteries, and terminate, the right in the inferior vena cava, the left in the left renal vein.

The *Hepatic Veins* commence in the substance of the liver, in the capillary terminations of the vena portae: these branches, gradually uniting, form three large veins, which converge towards the posterior border of the liver, and open into the inferior vena cava, whilst that vessel is situated in the groove at the back part of this organ. Of these three veins, one from the right, and another from the left lobes, open obliquely into the vena cava; that from the middle of the organ and lobus Spigelii having a straight course. The hepatic veins run singly, and are in direct contact with the hepatic tissue. They are destitute of valves.

Portal System of Veins

The portal venous system is composed of four large veins, which collect the venous blood from the viscera of digestion. The trunk formed by their union (vena portae) enters the liver, ramifies throughout its substance, and its branches again emerging from that organ as the hepatic veins, terminate in the inferior vena cava. The branches of this vein are in all cases single, and destitute of valves.

The veins forming the portal system are the

Inferior mesenteric.   Splenic.
Superior mesenteric.   Gastric.

The *Inferior Mesenteric Vein* returns the blood from the rectum, sigmoid flexure, and descending colon, corresponding with the ramifications of the branches of the inferior mesenteric artery. Ascending beneath the peritoneum in the lumbar region, it passes behind the transverse portion of the duodenum and pancreas, and terminates in the splenic vein. Its haemorrhoidal branches inosculate with those of the internal iliac, and thus establish a communication between the portal and the general venous system.

The *Superior Mesenteric Vein* returns the blood from the small intestines, and from the caecum and ascending and transverse portions of the colon, corresponding with the distribution of the branches of the superior mesenteric artery. The large trunk formed by the union of these branches ascends along the right side and in front of the corresponding artery, passes in front of the transverse portion of the duodenum, and unites behind the upper border of the pancreas with the splenic vein, to form the vena portae.

The *Splenic Vein* commences by five or six large branches, which return the blood from the substance of the spleen. These uniting form a single vessel, which passes from left to right behind the upper border of the pancreas, and terminates at its greater end by uniting at a right angle with the superior mesenteric to form the vena portae. The splenic vein is of large size, and not tortuous like the artery. It receives the vasa brevia from the left extremity of the stomach, the left gastro-epiploic vein, pancreatic branches from the pancreas, the pancreatico-duodenal vein, and the inferior mesenteric vein.

The *Gastric* is a vein of small size, which accompanies the gastric artery from left to right along the lesser curvature of the stomach, and terminates in the vena portae.

225.—Portal Vein and its Branches.

The *Portal Vein* is formed by the junction of the superior mesenteric and splenic veins, their union taking place in front of the vena cava, and behind the upper border of the great end of the pancreas. Passing upwards through the right border of the lesser omentum to the under surface of the liver, it enters the transverse fissure, where it is somewhat enlarged, forming the sinus of the portal vein, and divides into two branches, which accompany the ramifications of the hepatic artery and hepatic duct throughout its substance. Of these two branches the right is the larger but the shorter of the two. The portal vein is about four inches in length, and, whilst contained in the lesser omentum, lies behind and between the hepatic duct and artery, the former being to the right, the latter to the left. These structures are accompanied by filaments of the hepatic plexus and numerous lymphatics, surrounded by a quantity of loose areolar tissue, the capsule of Glisson, and placed between the layers of the lesser omentum. The vena portae receives the gastric and cystic veins; the latter vein sometimes terminates in the right branch of the vena portae. Within the liver, the portal vein receives the blood from the branches of the hepatic artery.

Cardiac Veins

The veins which return the blood from the substance of the heart are, the
- Great cardiac vein.
- Anterior cardiac veins.
- Posterior cardiac vein.
- Venae Thebesii.

The *Great Cardiac Vein* is a vessel of considerable size, which commences at the apex of the heart, and ascends along the anterior ventricular groove to the base of the ventricles. It then curves to the left side, around the auriculo-ventricular groove, between the left auricle and ventricle, to the back part of the heart, and opens into the coronary sinus, its aperture being guarded by two valves. It receives the posterior cardiac vein, and the left cardiac veins from the left auricle and ventricle, one of which, ascending along the left margin of the ventricle, is of large size. The branches joining it are provided with valves.

The *Posterior Cardiac Vein* commences, by small branches, at the apex of the heart, communicating with those of the preceding. It ascends along the groove between the ventricles, on the posterior surface of the heart, to its base, and terminates in the coronary sinus, its orifice being guarded by a valve. It receives the veins from the posterior surface of both ventricles.

The *Anterior Cardiac Veins* are three or four small branches, which collect the blood from the anterior surface of the right ventricle. One of these, larger than the rest, runs along the right border of the heart, the vein of Galen. They open separately into the lower part of the right auricle.

The *Venae Thebesii* are numerous minute veins, which return the blood directly from the muscular substance, without entering the venous current. They open, by minute orifices, (*foramina Thebesii*), on the inner surface of the right auricle.

The Coronary Sinus is that portion of the coronary vein which is situated in the posterior part of the left auriculo-ventricular groove. It is about an inch in length, presents a considerable dilatation, and is covered by the muscular fibres of the left auricle. It receives the great cardiac vein, the posterior cardiac vein, and an oblique vein from the back part of the left auricle, the remnant of the obliterated left innominate trunk of the foetus, described by Mr. Marshall. The coronary sinus terminates in the right auricle, between the inferior vena cava and the auriculo-ventricular aperture, its orifice being guarded by a semilunar fold of the lining membrane of the heart, the coronary valve. All the branches joining this vessel, excepting the oblique vein, above-mentioned, are provided with valves.

The Pulmonary Veins

The *Pulmonary Veins* return the arterial blood from the lungs to the left auricle of the heart. They are four in number, two for each lung. The pulmonary differ from other veins in several respects. 1. They carry arterial, instead of venous, blood. 2. They are destitute of valves. 3. They are only slightly larger than the arteries they accompany. 4. And they accompany those vessels singly. They commence in a capillary network, upon the parietes of the bronchial cells, where they are continuous with the ramifications of the pulmonary artery, and, uniting together, form a single trunk for each lobule. These branches, successively uniting, form a single trunk for each lobe, three for the right, and two for the left, lung. The vein of the middle lobe of the right lung unites with that from the upper lobe, in most cases, forming two trunks on each side, which open separately into the left auricle. Occasionally they remain separate; there are then three veins on the right side. Not unfrequently, the two left pulmonary veins terminate by a common opening.

Within the lung, the branches of the pulmonary artery are *in front*, the veins *behind*, and the bronchi between the two.

At the root of the lung, the veins are *in front*, the artery *in the middle*, and the bronchus *behind*.

Within the pericardium, their anterior surface is invested by the serous layer of this membrane, the right pulmonary veins pass behind the right auricle and ascending aorta; the left pass in front of the thoracic aorta, with the left pulmonary artery.

OF THE LYMPHATICS

The Lymphatics have derived their name from the appearance of the fluid contained in their interior (*lympha*, water). They are also called *absorbents*, from the property these vessels possess of absorbing foreign matters into the system, and carrying them into the circulation.

The lymphatic system includes not only the lymphatic vessels and the glands through which they pass, but also the *lacteal*, or *chyliferous* vessels. The lacteals are the lymphatic vessels of the small intestine, and differ in no respect from the lymphatics generally, excepting that they carry a milk-white fluid, the chyle, during the process of digestion, and convey it into the blood through the thoracic duct.

The lymphatics are exceedingly delicate vessels, the coats of which are so transparent, that the fluid they contain is readily seen through them. They retain a nearly uniform size, being interrupted at intervals by constrictions, which give to them a knotted or beaded appearance, owing to the presence of valves in their interior. They are found in nearly every texture and organ of the body, with the exception of the substance of the brain and spinal cord, the eyeball, cartilage, tendon, membranes of the ovum, the placenta, and umbilical cord. Their existence in the substance of bone is doubtful.

The lymphatics are arranged into a superficial and deep set. The superficial vessels, on the surface of the body, are placed immediately beneath the integument, accompanying the superficial veins; they join the deep lymphatics in certain situations by perforating the deep fascia. In the interior of the body, they lie in the sub-mucous areolar tissue, throughout the whole length of the gastro-pulmonary and genito-urinary tracts; or in the sub-serous areolar tissue, beneath the serous membrane covering the various organs in the cranial, thoracic, and abdominal cavities. In each of these situations these vessels arise in the form of a dense plexiform network, consisting of several strata; the vessels composing which, as well as the meshes between them, are much larger than the capillary plexus. From these networks small vessels emerge, which pass, either to a neighbouring gland, or to join some larger lymphatic trunk. The deep lymphatics, fewer in number, and larger than the superficial, accompany the deep blood-vessels. Their mode of origin is not known; it is, however, probable, that it is similar to that of the superficial vessels. The lymphatics of any part or organ exceed, in number, the veins; but in size, they are much smaller. Their anastomoses also, especially of the large trunks, is more frequent, and is effected by vessels equal in diameter to those which they connect, the continuous trunks retaining the same diameter.

The lymphatic vessels, like arteries and veins, are composed of three coats, internal, middle, and external.

The *internal* is an epithelial and elastic coat. It is thin, transparent, slightly elastic, and ruptures sooner than the other coats. It is composed of a layer of scaly epithelium, supported on one or more laminae of longitudinal elastic fibres.

The *middle*, or muscular coat, is thin, extensile, and elastic, consisting internally of a layer of longitudinal muscular fibres of the involuntary kind, intermixed with some areolar tissue, external to which, in the larger lymphatics, is a layer of circular fibres.

The *external*, or areolar-fibrous coat, is similar to that of the blood-vessels. It is thin, but very extensile and elastic, composed of filaments of areolar tissue, intermixed with some muscular fibres, longitudinally or obliquely disposed. It forms a protective covering to the other coats, and serves to connect the vessel with the neighbouring structures.

The lymphatics are supplied by nutrient vessels, which are distributed to their outer and middle coats; but no nerves have at present been traced into them.

The lymphatics are very generally provided with valves, which assist very materially in effecting the circulation of the fluid they contain. They are formed of a thin layer of fibrous tissue, coated on both surfaces with scaly epithelium. Their form is semilunar; they are attached by their convex edge to the sides of the vessel, the concave edge being free, and directed in the course of the contained current. Most usually, two such valves, of equal size, are found placed opposite one another; but occasionally exceptions occur, especially at or near the anastomoses of lymphatic vessels. Thus one valve may be of very rudimentary size, the other increased in proportion. In other cases, the semilunar flaps have been found directed transversely across the vessel, instead of obliquely, so as to impede the circulation in both directions, but not to completely arrest it in either; or the semilunar flaps, taking the same direction, have been united on one side, so that they formed, by their union, a transverse septum, having a partial transverse slit; and sometimes the flap was constituted of a circular fold, attached to the entire circumference of the vessel, and having in its centre a circular or elliptical aperture, the arrangements of the flaps being similar to those composing the ilio-caecal valve.

The valves in the lymphatic vessels are placed at much shorter intervals than in the veins. They are most numerous near the lymphatic glands, and they are found more frequently in the lymphatics of the neck and upper extremity, than in the lower. The wall of the lymphatics, immediately above the point of attachment of each segment of a valve, is expanded into a pouch or sinus, which gives to these vessels, when distended, the knotted or beaded appearance which they present. Valves are wanting in the vessels composing the plexiform network in which the lymphatics originate.

There is no satisfactory evidence to prove that any natural communication exists between the lymphatics of glandular organs and their ducts, or between the lymphatics and the capillary vessels.

The lymphatic or absorbent glands, named also *conglobate glands*, are small solid glandular bodies, situated in the course of the lymphatic and lacteal vessels. They are found in the neck and on the external parts of the head; in the upper extremity, in the axilla and front of the elbow; in the lower extremity, in the groin and popliteal space. In the abdomen, they are found in large numbers in the mesentery, and along the side of the aorta, vena cava, and iliac vessels; and in the thorax, in the anterior and posterior mediastina. They are somewhat flattened, and of a round or oval form. In size, they vary from a hemp-seed to an almond, and their colour, on section, is of a pinkish grey tint, excepting the bronchial glands, which in the adult are mottled with black. The lymphatic and lacteal vessels pass through these bodies in their passage to the thoracic and lymphatic ducts. A lymphatic or lacteal, previous to entering a gland, divides into several small branches, which are named *inferent* or *afferent vessels* (vasa inferentia or afferentia); and those which emerge from it are called *efferent vessels* (vasa efferentia). In structure they are composed of a superficial or cortical spongy substance about two or three lines in thickness, containing numerous small cavities or loculi, filled with a whitish pulpy matter; the afferent vessels pour their contents into these loculi, which communicate by minute vessels with the lymphatic plexus composing the centre or medullary portion of the gland, from which the efferent vessels emerge. These plexuses of lymphatic vessels are intermixed with a capillary plexus, and the whole enclosed in a thin fibro-areolar capsule.

Thoracic Duct

The thoracic duct (fig. 227, color illustration p. A8) conveys the great mass of the lymph and chyle into the blood. It is the common trunk of all the lymphatic vessels of the body, excepting those of the right side of the head, neck, and thorax, and right upper extremity, the right lung, right side of the heart, and the convex surface of the liver. It varies from eighteen to twenty inches in length in the adult, and extends from the second lumbar vertebra to the root of the neck. It commences in the abdomen by a triangular dilatation, the receptaculum chyli (reservoir or cistern of Pecquet), which is situated upon the front of the body of the second lumbar vertebra, to the right side and behind the aorta, by the side of the right crus of the Diaphragm. It ascends into the thorax through the aortic opening in the Diaphragm, and is placed in the postetrior mediastinum in front of the vertebral column, lying between the aorta and vena azygos. Opposite the fourth dorsal vertebra it inclines towards the left side and ascends behind the arch of the aorta, on the left side of the oesophagus, and

227.—The Thoracic and Right Lymphatic Ducts.

behind the first portion of the left subclavian artery, to the upper orifice of the thorax. Opposite the upper border of the seventh cervical vertebra it curves downwards above the subclavian artery, and in front of the Scalenus muscle, so as to form an arch; and terminates near the angle of junction of the left internal jugular and subclavian veins. The thoracic duct, at its commencement, is about equal in size to the diameter of a goose-quill, diminishes considerably in its calibre in the middle of the thorax, and is again dilated just before its termination. It is generally flexuous in its course, and constricted at intervals so as to present a varicose appearance. The thoracic duct not unfrequently divides in the middle of its course into two branches of unequal size which soon re-unite, or into several branches which form a plexiform interlacement. It occasionally bifurcates, at its upper part, into two branches, the left terminating in the usual manner, the right opening into the left subclavian vein, in connection with the right lymphatic duct. The thoracic duct has numerous valves throughout its whole course, but they are more numerous in the upper than in the lower part; at its termination it is provided with a pair of valves, the free borders of which are turned towards the vein, so as to prevent the regurgitation of venous blood into the duct.

Branches. The thoracic duct at its commencement receives four or five large trunks from the abdominal lymphatic glands, and also the trunk of the lacteal vessels. Within the thorax, it is joined by the lymphatic vessels from the left half of the wall of the thoracic cavity; and the lymphatics from the sternal and intercostal glands, those of the left lung, left side of the heart, trachea, and oesophagus; and just before its termination, receives the lymphatics of the left side of the head and neck, and left upper extremity.

The *Right Lymphatic Duct* is a short trunk, about an inch in length, and a line or a line and a half in diameter, which receives the lymph from the right side of the head and neck, the right upper extremity, and right side of the thorax; and terminates at the angle of union of the right subclavian and right internal jugular veins. Its orifice is guarded by two semilunar valves, which prevent the entrance of blood from the veins.

Branches. In addition to those already mentioned, it receives the lymphatics of the right lung and right side of the heart, and some from the convex surface of the liver.

Lymphatics of the Head, Face, and Neck

The *Superficial Lymphatic Glands of the Head* (fig. 228) are of small size, few in number, and confined to its posterior region. They are the *occipital*, placed at the back of the head along the attachment of the Occipito-frontalis; and the *posterior auricular*, near the upper end of the Sterno-mastoid. These glands become considerably enlarged in cutaneous affections and other diseases of the scalp. In the face, the superficial lymphatic glands are more numerous: they are the *parotid*, some of which are superficial and others deeply placed in its substance; the *zygomatic*, situated under the zygoma; the *buccal*, on the surface of the buccinator muscle; and the *submaxillary*, the largest, beneath the body of the lower jaw.

The *Superficial Lymphatics of the Head* are divided into an anterior and a posterior set, which follow the course of the temporal and occipital vessels. The temporal set accompany the temporal artery in front of the ear, to the parotid lymphatic glands, from which they proceed to the lymphatic glands of the neck. The occipital set follow the course of the occipital artery, descend to the occipital and posterior auricular lymphatic glands, and from thence join the cervical glands.

The *Superficial Lymphatics of the Face* are more numerous than those of the head. They commence over its entire surface, those from the frontal region accompanying the frontal vessels; they then pass obliquely across the face, accompanying the facial vein, pass through the buccal glands on the surface of the Buccinator muscle, and join the submaxillary lymphatic glands. These glands receive the lymphatic vessels from the lips, and are often found enlarged in cases of malignant disease of this part.

The *Deep Lymphatics of the Face* are derived from the pituitary membrane of the nose, the mucous membrane of the mouth and pharynx, and the contents of the temporal and orbital fossae; they accompany the branches of the internal maxillary artery, and terminate in the deep parotid and cervical lymphatic glands.

The *Deep Lymphatics of the Cranium* consist of two sets, the meningeal and cerebral. The meningeal lymphatics accompany the meningeal vessels, escape through foramina at the base of the skull, and join the deep cervical lymphatic glands. The cerebral lymphatics are described by Fohmann as being situated between the arachnoid and pia mater, as well as in the choroid plexuses of the lateral ventricles; they accompany the trunks of the carotid and vertebral arteries, and probably pass through foramina at the base of the skull, to terminate in the deep cervical glands. They have not at present been demonstrated in the dura mater, or in the substance of the brain.

228.—The Superficial Lymphatics and Glands of the Head, Face, and Neck.

The *Lymphatic Glands of the Neck* are divided into two sets, superficial and deep.

The *superficial cervical glands* are placed in the course of the external jugular vein, between the Platysma and Sterno-mastoid. They are most numerous at the root of the neck, in the triangular interval between the clavicle, the Sterno-mastoid, and the Trapezius, where they are continuous with the axillary glands. A few small glands are also found on the front and sides of the larynx.

The *deep cervical glands* (fig. 229) are numerous and of large size; they form an uninterrupted chain along the sheath of the carotid artery and internal jugular vein, lying by the side of the pharynx, oesophagus, and trachea, and extending from the base of the skull to the thorax, where they communicate with the lymphatic glands in this cavity.

The *Superficial and Deep Cervical Lymphatics* are a continuation of those already described on the cranium and face. After traversing the glands in those regions, they pass through the chain of glands which lie along the sheath of the carotid vessels, being joined by the lymphatics from the pharynx, oesophagus, larynx, trachea, and thyroid gland. At the lower part of the neck, after receiving some lymphatics from the thorax, they unite into a single trunk, which terminates on the left side, in the thoracic duct; on the right side, in the right lymphatic duct.

229.—The Deep Lymphatics and Glands of the Neck and Thorax.

Lymphatics of the Upper Extremity

The *Lymphatic Glands* of the upper extremity (fig. 230) may be subdivided into two sets, superficial and deep.

The *superficial lymphatic glands* are few, and of small size. There are occasionally two or three in front of the elbow, and one or two above the internal condyle of the humerus, near the basilic vein.

230.—The Superficial Lymphatics and Glands of the Upper Extremity.

The *deep lymphatic glands* are also few in number. In the fore-arm a few small ones are occasionally found in the course of the radial and ulnar vessels; and in the arm, there is a chain of small glands along the inner side of the brachial artery.

The *Axillary Glands* are of large size, and usually ten or twelve in number. A chain of these glands surrounds the axillary vessels imbedded in a quantity of loose areolar tissue; they receive the lymphatic vessels from the arm: others are dispersed in the

231.—The Superficial Lymphatics and Glands of the Lower Extremity.

Superficial Inguinal Glands

areolar tissue of the axilla: the remainder are arranged in two series, a small chain running along the lower border of the Pectoralis major, as far as the mammary gland, receiving the lymphatics from the front of the chest and mamma; and others are placed along the lower margin of the posterior wall of the axilla, which receive the lymphatics from the integument of the back. Two or three subclavian lymphatic glands are placed immediately beneath the clavicle; it is through these that the axillary and deep cervical glands communicate with each other. One is figured by Mascagni near the umbilicus. In malignant diseases, tumours or other affections implicating the upper part of the back and shoulder, the front of the chest and mamma, the upper part of the front and side of the abdomen, or the hand, fore-arm, and arm, these glands are usually found enlarged.

The *Superficial Lymphatics* of the upper extremity arise from the skin of the hand, and run along the sides of the fingers chiefly on the dorsal surface of the hand; they then pass up the fore-arm, and subdivide into two sets, which take the course of the subcutaneous veins. Those from the inner border of the hand accompany the ulnar veins along the inner side of the fore-arm to the bend of the elbow, where they join with some lymphatics from the outer side of the fore-arm, follow the course of the basilic vein, communicate with the glands immediately above the elbow, and terminate in the axillary glands, joining with the deep lymphatics. The superficial lymphatics from the outer and back part of the hand accompany the radial veins to the bend of the elbow, being less numerous than the preceding. Here the greater number join the basilic group; the rest ascend with the cephalic vein on the outer side of the arm, some crossing obliquely the upper part of the Biceps to terminate in the axillary glands, whilst one or two accompany the cephalic vein in the cellular interval between the Pectoralis major and Deltoid, and enter the subclavian lymphatic glands.

The *Deep Lymphatics* of the upper extremity accompany the deep blood-vessels. In the fore-arm they consist of three sets, corresponding with the radial, ulnar, and interosseous arteries; they pass through the glands occasionally found in the course of these vessels, and communicate at intervals with the superficial lymphatics. In their ascent upwards, some of them pass through the glands which lie upon the brachial artery; they then enter the axillary and subclavian glands, and at the root of the neck terminate, on the left side in the thoracic duct, and on the right side in the right lymphatic duct.

Lymphatics of the Lower Extremity

The *Lymphatic Glands* of the lower extremity may be subdivided into two sets, superficial and deep.

The *superficial lymphatic glands* of the lower extremity are confined to the inguinal region.

The *superficial inguinal glands*, placed immediately beneath the integument, are of large size, and vary from eight to ten in number. They are divisible into two groups; an upper, disposed irregularly along Poupart's ligament, receiving the lymphatic vessels from the integument of the scrotum, penis, parietes of the abdomen, perinaeum, and gluteal regions; and an inferior group, which surround the saphenous opening in the fascia lata, a few being sometimes continued along the saphena vein to a variable extent. These receive the superficial lymphatic vessels from the lower extremity. These glands frequently become enlarged in diseases implicating the parts from which their efferent lymphatics originate. Thus, in malignant or syphilitic affections of the prepuce and penis, the labia majora in the female, in cancer scroti, in abscess in the perinaeum, or in any other disease affecting the integument and superficial structures in these parts, or the sub-umbilical part of the abdomen or gluteal region, the upper chain of glands is almost invariably enlarged, the lower chain being implicated in diseases affecting the lower limb.

The *Deep Lymphatic Glands* are the anterior tibial, popliteal, deep inguinal, gluteal, and ischiatic.

The *Anterior Tibial Gland* is not constant in its existence. It is generally found by the side of the anterior tibial artery, upon the interosseous membrane at the upper part of the leg. Occasionally two glands are found in this situation.

The *Deep Popliteal Glands*, four or five in number, are of small size; they surround the popliteal vessels, imbedded in the cellular tissue and fat of the popliteal space.

The *Deep Inguinal Glands* are placed beneath the deep fascia around the femoral artery and vein. They are of small size, and communicate with the superficial inguinal glands through the saphenous opening.

The *Gluteal and Ischiatic Glands* are placed, the former above, the latter below the Pyriformis muscle, resting on their corresponding vessels as they pass out of the great sacro-sciatic foramen.

The *Lymphatics* of the lower extremity, like the veins, may be divided into two sets, superficial and deep.

The *Superficial Lymphatics* are placed between the integument and superficial fascia, and are divisible into two groups, an internal group, which follow the course of the internal saphena vein, and an external group, which accompany the external saphena.

The *internal group*, the largest, commence on the inner side and dorsum of the foot; they pass, some in front and some behind the inner ankle, ascend the leg with the internal saphenous vein, pass with it behind the inner condyle of the femur, and accompany it to the groin, where they terminate in the group of inguinal glands which surround the saphenous opening. Some of the efferent vessels from these glands pierce the cribriform fascia and sheath of the femoral vessels, and terminate in a lymphatic gland contained in the femoral canal, thus establishing a communication between the lymphatics of the lower extremity and those of the trunk; others pierce the fascia lata, and join the deep inguinal glands.

The *external group* arise from the outer side of the foot, ascend in front of the anterior region of the leg, and just below the knee cross the tibia from without inwards, to join the lymphatics on the inner side of the thigh. Others commence on the outer side of the foot, pass behind the outer malleolus, and accompany the external saphenous vein along the back of the leg, where they enter the popliteal glands.

The *Deep Lymphatics* of the lower extremity are few in number, and accompany the deep blood-vessels. In the leg they consist of three sets, the anterior tibial, peroneal, and posterior tibial, which accompany the corresponding vessels, being two or three in number to each: they ascend with the blood-vessels, and enter the lymphatic glands in the popliteal space: the efferent vessels from these glands accompany the femoral vein, and join the deep inguinal glands; from these the vessels pass beneath Poupart's ligament, and communicate with the chain of glands surrounding the external iliac vessels.

The deep lymphatics of the gluteal and ischiatic regions follow the course of the blood-vessels, and join the gluteal and ischiatic glands at the great sacro-sciatic foramen.

Lymphatics of the Pelvis and Abdomen

The *Deep Lymphatic Glands in the Pelvis* are the external iliac, the internal iliac, and the sacral. Those of the abdomen are the lumbar glands.

The *External Iliac Glands* form an uninterrupted chain around the external iliac vessels, three being placed around the commencement of the vessel just behind the crural arch. They communicate by one extremity with the femoral lymphatics, and by the other with the lumbar glands.

The *Internal Iliac Glands* surround the internal iliac vessels; they receive the lymphatics corresponding to the branches of the internal iliac artery, and communicate with the lumbar glands.

The *Sacral Glands* occupy the sides of the anterior surface of the sacrum, some being situated in the folds of the meso-rectum. These and the internal iliac glands become greatly enlarged in malignant disease of the bladder, rectum, or uterus.

232.—The Deep Lymphatic Vessels and Glands of the Abdomen and Pelvis.

The *Lumbar Glands* are very numerous; they are situated on the front of the lumbar vertebrae, surrounding the common iliac vessels, the aorta, and vena cava; they receive the lymphatic vessels from the lower extremities and pelvis, as well as from the testes and some of the abdominal viscera: the efferent vessels from these glands unite into a few large trunks, which, with the lacteals, form the commencement of the thoracic duct. In some cases of malignant disease, these glands become enormously enlarged, completely surrounding the aorta and vena cava, and occasionally greatly contracting the calibre of these vessels. In all cases of malignant disease of the testis, and in malignant disease of the lower limb, before any operation is attempted, careful examination of the abdomen should be made in order to ascertain if any enlargement exists, and if any should be detected, all operative measures are fruitless.

The *Lymphatics of the Pelvis and Abdomen* may be divided into two sets, superficial and deep.

The *Superficial Lymphatics* of the walls of the abdomen and pelvis follow the course of the superficial blood-vessels. Those derived from the integument of the lower part of the abdomen below the umbilicus, follow the course of the superficial epigastric vessels, and converge to the superior group of the superficial inguinal glands, the deep set accompany the deep epigastric vessels, and communicate with the external iliac glands. The superficial lymphatics from the sides and lumbar part of the abdominal wall wind round the crest of the ilium, accompanying the superficial circumflex iliac vessels, to join the superior group of the superficial inguinal glands; the greater number, however, accompany the ilio-lumbar and lumbar vessels backwards to join the lumbar glands.

The *Superficial Lymphatics of the Gluteal Region* turn horizontally round the outer side of the nates, and join the superficial inguinal glands.

The *Superficial Lymphatics of the Scrotum and Perinaeum* follow the course of the external pudic vessels, and terminate in the superficial inguinal glands.

The *Superficial Lymphatics of the Penis* occupy the sides and dorsum of the organ, the latter receiving the lymphatics from the skin covering the glans penis; they all converge to the superior group of the superficial inguinal glands. The deep lymphatic vessels of the penis follow the course of the internal pudic vessels, and join the internal iliac glands.

In the female, the lymphatic vessels of the mucous membrane of the labia, nymphae, and clitoris, terminate in the superior group of the inguinal lymphatic glands.

The *Deep Lymphatics of the Pelvis and Abdomen* take the course of the principal blood-vessels. Those of the parietes of the pelvis, which accompany the gluteal, ischiatic, and obturator vessels, follow the course of the internal iliac artery, and ultimately join the lumbar lymphatics.

The efferent vessels from the inguinal glands enter the pelvis beneath Poupart's ligament, where they lie in close relation with the femoral vein; they then pass through the chain of glands surrounding the external iliac vessels, and finally terminate in the lumbar glands. They receive the deep epigastric, circumflex ilii, and ilio-lumbar lymphatics.

The *Lymphatics of the Bladder* arise from the entire surface of the organ; the greater number run beneath the peritoneum on its posterior surface, and, after passing through the lymphatic glands in this situation, join with those from the prostate and vesiculae seminales, and enter the internal iliac glands.

The *Lymphatics of the Rectum* are of large size; after passing through some small glands that lie upon its outer wall and in the meso-rectum, they pass to the sacral or lumbar glands.

The *Lymphatics of the Uterus* consist of two sets, superficial and deep; the former being placed beneath the peritoneum, the latter in the substance of the organ. The lymphatics of the cervix uteri, together with those from the vagina, enter the internal iliac and sacral glands: those from the body and fundus of the uterus pass outwards in the broad ligaments, and being joined by the lymphatics from the ovaries, broad ligaments, and Fallopian tubes, ascend with the ovarian vessels to open into the lumbar glands. In the unimpregnated uterus they are small, but during gestation they become very greatly enlarged.

The *Lymphatics of the Testicle* consist of two sets, superficial and deep; the former commence on the surface of the tunica vaginalis, the latter in the epididymis and body of the gland. They form several large trunks, which ascend with the spermatic cord, and accompanying the spermatic vessels into the abdomen, open into the lumbar glands; hence the enlargement of these glands in malignant disease of this organ.

The *Lymphatics of the Kidney* arise on the surface, and also in the interior of the organ; they unite together at the hilus, and after receiving the lymphatic vessels from the ureters and supra-renal capsules, open into the lumbar glands.

The *Lymphatics of the Liver* are divisible into two sets, superficial and deep. The former arise in the sub-peritoneal areolar tissue over the entire surface of the organ. Those on the convex surface may be divided into four groups: 1. Those which pass from behind forwards, consisting of three or four trunks, which ascend in the longitudinal ligament, and unite to form a single trunk, which passes up between the fibres of the Diaphragm, behind the ensiform cartilage, to enter the anterior mediastinal glands, and finally ascend to the root of the neck, to terminate in the right lymphatic duct. 2. Another group, which also incline from behind forwards, are reflected over the anterior margin of the liver to its concave surface, and from thence pass along the longitudinal fissure to the glands in the gastro-hepatic omentum. 3. A third group incline outwards to the right lateral ligament, and uniting into one or two large trunks, pierce the Diaphragm, and run along its upper surface to enter the anterior mediastinal glands; or, instead of entering the thorax, turn inwards across the crus of the Diaphragm, and open into the commencement of the thoracic duct. 4. The fourth group incline outwards from the surface of the left lobe of the liver to the left lateral ligament, pierce the Diaphragm, and passing forwards, terminate in the glands in the anterior mediastinum.

The *Superficial Lymphatics on the under surface of the Liver* may be divided into three sets: 1. Those on the right side of the gall-bladder enter the lumbar glands. 2. Those surrounding the gall-bladder form a remarkable plexus, which accompanies the hepatic vessels, and open into the glands in the gastro-hepatic omentum. 3. Those on the left of the gall-bladder pass to the oesophageal glands, and to those placed along the lesser curvature of the stomach.

The *Deep Lymphatics* accompany the branches of the portal vein and the hepatic artery and duct through the substance of the gland; passing out at the transverse fissure, they enter the lymphatic glands along the lesser curvature of the stomach and behind the pancreas, or join with one of the lacteal vessels previous to its termination in the thoracic duct.

The *Lymphatic Glands of the Stomach* are of small size; they are placed along the lesser and greater curvatures, some within the gastro-splenic omentum, whilst others surround its cardiac and pyloric orifices.

The *Lymphatics of the Stomach* consist of two sets, superficial and deep; the former originating in the subserous, and the latter in the submucous coats. They follow the course of the blood-vessels, and may consequently be arranged into three groups.

The *first group* accompany the coronary vessels along the lesser curvature, receiving branches from both surfaces of the organ, and pass to the glands around the pylorus. The *second group* pass from the great end of the stomach, accompany the vasa brevia, and enter the splenic lymphatic glands. The *third group* run along the greater curvature with the right gastro-epiploic vessels, and terminate at the root of the mesentery in one of the principal lacteal vessels.

The *Lymphatic Glands of the Spleen* occupy the hilus. Its *lymphatic vessels* consist of two sets, superficial and deep; the former being placed beneath its peritoneal covering, the latter in the substance of the organ: they accompany the blood-vessels, passing through a series of small glands, and after receiving those derived from the pancreas, ultimately pass into the thoracic duct.

The Lymphatic System of the Intestines

The *Lymphatic Glands of the Small Intestines* are placed between the layers of the mesentery, occupying the meshes formed by the superior mesenteric vessels, and hence called *mesenteric glands*. They vary in number from a hundred and thirty to about a hundred and fifty, and are about the size of an almond. These glands are most numerous, and largest, superiorly near the duodenum, and inferiorly opposite the termination of the ileum in the colon. This latter group becomes greatly enlarged and infiltrated with deposit in cases of fever accompanied with ulceration of the intestines.

The *Lymphatic Glands of the Large Intestine* are much less numerous than the mesenteric glands; they are situated along the vascular arches formed by the arteries previous to their distribution, and even sometimes upon the intestine itself. They are fewest in number along the transverse colon, where they form an uninterrupted chain with the mesenteric glands.

The *Lymphatics of the Small Intestine* are called *lacteals*, from the milk-white fluid they usually contain: they consist of two sets, superficial and deep; the former lie beneath the peritoneal coat, taking a longitudinal course along the outer side of the intestine; the latter occupy the submucous tissue, and course transversely round the intestine, accompanied by the branches of the mesenteric vessels: they pass between the layers of the mesentery, enter the mesenteric glands, and finally unite to form two or three large trunks, which terminate in the thoracic duct.

The *Lymphatics of the Great Intestine* consist of two sets: those of the caecum, ascending and transverse colon, which, after passing through their proper glands, enter the mesenteric glands; and those of the descending colon and rectum, which pass to the lumbar glands.

The Lymphatics of the Thorax

The *Deep Lymphatic Glands of the Thorax* are the intercostal, internal mammary, anterior mediastinal, and posterior mediastinal.

The *Intercostal Glands* are small, irregular in number, and situated on each side of the spine, near the costo-vertebral articulations, some being placed between the two planes of intercostal muscles.

The *Internal Mammary Glands* are placed at the anterior extremity of each intercostal space, by the side of the internal mammary vessels.

The *Anterior Mediastinal Glands* are placed in the loose areolar tissue of the anterior mediastinum, some lying upon the Diaphragm in front of the pericardium, and others around the great vessels at the base of the heart.

The *Posterior Mediastinal Glands* are situated in the areolar tissue in the posterior mediastinum, forming a continuous chain by the side of the aorta and oesophagus; they communicate on each side with the intercostal, below with the lumbar glands, and above with the deep cervical.

The *Superficial Lymphatics of the front of the Thorax* run across the great Pectoral muscle, and those on the back part of this cavity lie upon the Trapezius and Latissimus dorsi; they all converge to the axillary glands. The lymphatics from the mamma run along the lower border of the Pectoralis major, through a chain of small lymphatic glands, and communicate with the axillary glands.

The *Deep Lymphatics of the Thorax* are the intercostal, internal mammary, and diaphragmatic.

The *Intercostal Lymphatics* follow the course of the intercostal vessels, receiving lymphatics from the Intercostal muscles and pleura; they pass backwards to the spine, and unite with lymphatics from the back part of the thorax and spinal canal. After traversing the intercostal glands, they incline down the spine, and terminate in the thoracic duct.

The *Internal Mammary Lymphatics* follow the course of the internal mammary vessels: they commence in the muscles of the abdomen above the umbilicus, communicating with the epigastric lymphatics, ascend between the fibres of the Diaphragm at its attachment to the ensiform appendix, and in their course behind the costal cartilages are joined by the intercostal lymphatics, terminating on the right side in the right lymphatic duct, on the left side in the thoracic duct.

The *Lymphatics of the Diaphragm* follow the course of their corresponding vessels, and terminate, some in front in the inferior mediastinal and internal mammary glands, some behind in the intercostal and hepatic lymphatics.

The *Bronchial Glands* are situated around the bifurcation of the trachea and roots of the lungs. They are ten or twelve in number, the largest being placed opposite the bifurcation of the trachea, the smallest around the bronchi and their primary divisions for some little distance within the substance of the lungs. In infancy, they present the same appearance as lymphatic glands in other situations, in the adult they assume a brownish tinge, and in old age a deep black colour. Occasionally they become sufficiently enlarged to compress and narrow the canal of the bronchi; and they are often the seat of tubercle or deposits of phosphate of lime.

The *Lymphatics of the Lung* consist of two sets, superficial and deep: the former are placed beneath the pleura, forming a minute plexus, which covers the outer surface of the lung; the latter accompany the blood-vessels, and run along the bronchi: they both terminate at the root of the lungs in the bronchial glands. The efferent vessels from these glands, two or three in number, ascend upon the trachea to the root of the neck, traverse the tracheal and oesophageal glands, and terminate on the left side in the thoracic duct, on the right side in the right lymphatic duct.

The *Cardiac Lymphatics* consist of two sets, superficial and deep; the former arise in the subserous areolar tissue of the surface, and the latter beneath the internal lining membrane of the heart. They follow the course of the coronary vessels; those of the right side unite into a trunk at the root of the aorta, which, ascending across the arch of that vessel, passes backwards to the trachea, upon which it ascends, to terminate at the root of the neck in the right lymphatic duct. Those of the left side unite into a single vessel at the base of the organ, which passing along the pulmonary artery, and traversing some glands at the root of the aorta, ascends on the trachea to terminate in the thoracic duct.

The *Thymic Lymphatics* arise from the spinal surface of the thymus gland, and terminate on each side in the internal jugular veins.

The *Thyroid Lymphatics* arise from either lateral lobe of this organ; they converge to form a short trunk, which terminates, on the right side in the right lymphatic duct, on the left side in the thoracic duct.

The *Lymphatics of the Oesophagus* form a plexus around that tube, traverse the glands in the posterior mediastinum, and, after communicating with the pulmonary lymphatic vessels near the root of the lungs, terminate in the thoracic duct.

NERVOUS SYSTEM

The Nervous System consists of a series of connected central organs, called, collectively, the *cerebro-spinal centre* or *axis*, of the *ganglia*, and of the *nerves*.

The *Cerebro-Spinal Axis* consists of two portions, the brain or encephalon, which is contained within the cranium, and the spinal cord, continuous with the brain, which is enclosed in the spinal canal. The cerebro-spinal centre consists of two lateral symmetrical halves, which correspond in their structure in every respect; they are partially separated by longitudinal fissures, and connected together by broad transverse bands of nervous substance, called *commissures*.

The cerebro-spinal axis consists of two substances, which differ from each other in density and colour; they are called the *grey cineritious* or *cortical substance*, and the *white* or *medullary*.

The *grey* or *cortical substance* is disposed in the form of a thin layer upon the outer surface of the convolutions of the cerebrum and laminas of the cerebellum; it is not confined, however, to the external surface, for it exists in the interior of the spinal cord throughout its entire length, and from this part may be traced up through the medulla oblongata, pons Varolii, and crura cerebri, to the central parts of the hemispheres, the optic thalami, and corpora striata. It also forms at the base of the brain, the lamina cinerea, the tuber cinereum, and the grey matter in the anterior and posterior perforated spaces. The grey matter may be traced from the anterior perforated space into the olfactory nerve as far as the bulb, and from the posterior space as forming part of the infundibulum and pituitary body. The grey matter in this situation is continued upon the sides of the thalami, forms the soft commissure, surrounds the anterior pillars of the fornix, enters below into the substance of the corpus albicans; and, above, forms part of the lateral walls of the septum lucidum. It is also found in the centre of each of the corpora quadrigemina, in the pineal gland, and corpora geniculata. It forms also the corpus dentatum in the centre of each lateral lobe of the cerebellum.

The *white* or *medullary portion* of the cerebro-spinal axis consists of fibres, which are arranged chiefly in a longitudinal direction, or interlace at various angles with transverse fibres; they may be arranged into three classes, ascending, transverse, and longitudinal. The *ascending fibres* pass up from the medulla oblongata, increase in number as they ascend through the pons, the optic thalami, and striated bodies, and then diverge to every part of the surface of the hemispheres. They were called by Gall the *diverging fibres*. The *transverse* or *commissural fibres* commence at the surface of the hemispheres, and proceed inwards towards the centre, connecting the two hemispheres together; these were named by Gall the *converging fibres*. The *longitudinal fibres*, also commissural, connect together different parts of the same hemisphere, being confined to the same side of the middle line.

Chemical Composition. The following analysis by Lassaigne represents the relative proportion of the different constituents composing the grey and white matter of the brain.

	Grey	White
Water	85.2	73.0
Albuminous matter	7.5	9.9
Colourless fat	1.0	13.9
Red fat	3.7	0.9
Osmazome and lactates	1.4	1.0
Phosphates	1.2	1.3
	100.0	100.0

It appears from this analysis, that the cerebral substance consists of albumen, dissolved in water, combined with fatty matters and salts. The fatty matters, according to Fremy, consist of cerebric acid, which is most abundant, cholesterin, oleophosphoric acid, and olein, margarin, and traces of their acids. The same analyst states, that the fat contained in the brain is confined almost exclusively to the white substance, and that its colour becomes lost when the fatty matters are removed. According to Vauquelin, the cord contains a larger proportion of fat than the brain; and according to L'Heritier, the nerves contain more albumen and more soft fat than the brain.

The *Ganglia* may be regarded as separate and independent nervous centres, of smaller size and less complex structure than the brain, connected with each other, with the cerebro-spinal axis, and with the nerves in various situations. They are found on the posterior root of each of the spinal nerves; on the posterior or sensory root of the fifth cranial nerve; on the seventh nerve; on the two sensory divisions of the eighth pair (the glosso-pharyngeal and pneumogastric); in a connected series along each side of the vertebral column, forming the trunk of the sympathetic; on the branches of this nerve in the head, neck, thorax, and abdomen; or at the point of junction of branches of this nerve with the cerebro-spinal nerves. On section, they are seen to consist of a reddish grey substance, traversed by numerous white nerve-fibres: they vary considerably in form and size; the largest are those found in the cavity of the abdomen; the smallest, the microscopic ganglia, which exist in considerable numbers upon the nerves distributed to the different viscera. The ganglia are invested by a smooth and firm closely-adhering membranous envelope, consisting of dense areolar tissue; this sheath is continuous with the neurilemna of the nerves, and sends numerous processes into the interior of the ganglia, which support the blood-vessels supplying its substance.

The *Nerves* are round or flattened white cords, communicating on the one hand with the cerebro-spinal centre or the ganglia, and by the other distributed to the various textures of the body, forming the medium of communication between the two. One class of nerve-fibres, the *afferent or centripetal*, serve to convey impressions to the brain, the great centre of sensation and volition, where they are rendered cognizable to the mind; whilst another class of nerve-fibres, the *efferent or centrifugal*, convey the stimulus of volition to the organs of motion. The brain and spinal cord are also capable of receiving impressions by means of the afferent nerve-fibres, which results in a motorial stimulus being propagated along the efferent nerves, quite independent of the efforts of volition, and without even consciousness. The movements of this kind are called *reflex* or *excito-motory*.

The nerves are subdivided into two great classes, the cerebro-spinal, which proceed from the cerebro-spinal axis, and the sympathetic or ganglionic nerves, which proceed from the sympathetic ganglia; the cerebro-spinal are the *nerves of animal life*, being distributed to the organs of the senses, the skin, and to the active organs of locomotion, the muscles. The sympathetic or ganglionic nerves are distributed chiefly to the viscera and blood-vessels, and are termed the *nerves of organic life*.

The *cerebro-spinal nerves* consist of numerous nerve-fibres, collected together and enclosed in a membranous sheath. A small bundle of primitive fibres enclosed in a tubular sheath is called a *funiculus*: if the nerve is of small size, it may consist only of a single funiculus, but if large, the funiculi are collected together into larger bundles or fasciculi; and one or more fasciculi bound together in a common membranous investment, termed the *sheath*, constitutes a nerve. In structure, the common sheath investing the whole nerve, as well as the septa given off from it, which separates the fasciculi, consists of areolar tissue, composed of the white and yellow elastic fibres, the latter existing in greatest abundance. The tubular sheath of the funiculi, or *neurilemma*, consists of a fine smooth transparent membrane, which may be easily separated, in the form of a tube, from the fibres it encloses; in structure, it is, for the most part, a simple and homogeneous transparent film, occasionally composed of numerous minute reticular fibres.

The nerve-fibres, as far as is at present known, do not coalesce, but pursue an uninterrupted course from the centre to the periphery. In dissecting a nerve, however, into its component funiculi, it may be seen that they do not pursue a perfectly insulated course, but occasionally join at a very acute angle with other funiculi proceeding in the same direction; from these again branches are given off, which join again in like manner with other funiculi. It must be remembered, however, that in these communications the nerve-fibres do not coalesce, but merely pass into the sheath of the adjacent nerve, become intermixed with the nerve-fibres, and again pass on to become blended with the nerve-fibres in some adjoining fasciculus.

The cerebro-spinal nerves consist almost exclusively of the tubular nerve-fibres, the gelatinous fibres existing in very small proportion.

The blood-vessels supplying a nerve terminate in a minute capillary plexus, the vessels composing which run, for the most part, parallel with the funiculi; they are connected together by short transverse vessels, forming narrow oblong meshes, similar to the capillary system of muscle.

Nerves in their course subdivide into branches, and these frequently communicate with branches of a neighbouring nerve. In the subdivision of a nerve, the filaments of which it is composed are continued from the trunk into the branches, and at their junction with the branches of neighbouring nerves, the filaments pass to become intermixed with those of the other nerve in their further progress; in no instance, however, do the separate nerve-fibres either subdivide or inosculate.

The communications which take place between two or more nerves form what is called a *plexus*. Sometimes a plexus is formed by the primary branches of the trunks of the nerves, as the cervical, brachial, lumbar, and sacral plexuses, and occasionally by the terminal fasciculi, as in the plexuses formed at the periphery of the body. In the formation of a plexus, the component nerves divide, then join, and again subdivide in such a complex manner that the individual fasciculi become interlaced most intricately; so that each branch leaving a plexus may contain filaments from each of the primary nervous trunks which form it. In the formation also of the smaller plexuses at the periphery of the body, there is a free interchange of the fasciculi and primitive fibrils. In each case, however, the individual filaments remain separate and distinct, neither subdividing nor inosculating.

Some nerve-fibres have no peripheral termination. Gerber has shewn, that nerve-fibres occasionally form loops by their junction with a neighbouring fibre in the same fasciculus, and return to the cerebro-spinal centre without having any peripheral termination. These he considers to be sentient nerves, appropriated exclusively to the nerve itself, the *nervi nervorum*, upon which the sensibility of the nerve depends, and quite exclusive of the sensation produced by an impression made at the peripheral end of the nerve. These fibres bear some analogy to those met with in the posterior part of the optic commissure, where a set of fibres pass from one optic tract across the commissure to the opposite tract, having no communication with the optic nerve; also in the communications formed between the cervical nerves and spinal accessory and descendens noni, the nerve-fibres form an arch connected by each extremity with the cerebro-spinal centre, and have no peripheral termination.

Again, some nerve-fibres would appear to have no central connection with the cerebro-spinal centre, as those forming the most anterior part of the optic commissure. These *inter-retinal fibres*, as they are called, commence in the retina, pass along the optic nerve, and across the commissure to the optic nerve and retina of the opposite side.

The point of connection of a nerve with the brain or spinal cord is called, for convenience of description, its *origin* or *root*. If the fasciculi of which the nerve is composed should all arise at or near one point, or along one tract, the root is called *single*. If, on the contrary, the fasciculi divide into two separate bundles, which are connected at two different points with any part of the cerebro-spinal centre, such nerve is said to have a *double origin*, or to arise by *two roots*, each of which may have a separate function, as in the spinal nerves. The point where the separate fasciculi of a nerve are connected to the surface of the cerebro-spinal centre is called the *apparent origin* of a nerve; the term *real* or *deep origin* being given to that part of the centre from which a nerve actually springs.

The nerve-fibres at their periphery terminate in a varied manner. Occasionally the elementary fibres are disposed in terminal loops or plexuses, which, for a considerable period, was supposed to be their usual mode of termination, but later investigations have shewn that such is not the case. Nerve-fibres most commonly terminate by blunted and slightly-swollen ends, such as is observed in those which enter into the Paccinian bodies, or they may become gradually lost to view in the tissue in which they are distributed, becoming diminished in size, and their tubular sheath and white substance being wanting. Occasionally the elementary nerve-fibres, as in the nerves of special sense, may be brought into connection at their periphery with cells similar to those met with in the grey matter of the brain and ganglia.

The *Sympathetic System* consists of numerous parts, which may be arranged as follows. 1. A connected series of ganglia placed along both sides of the spinal column, from the cranium above to the coccyx below. 2. Branches of communication passing between the ganglia. 3. Branches of connection between the ganglia and the cranial and spinal nerves. 4. Primary branches of distribution, remarkable for their plexiform communications on the vessels, glands, and neighbouring viscera to which they are distributed, or for passing to other larger ganglia, situated in each of the great cavities of the body, and usually placed on the roots of origin of the larger blood-vessels. 5. Plexuses of nerves proceeding from these secondary ganglia, accompanying the blood-vessels, and receiving branches from the spinal or cerebral nerves. The sympathetic nerves consist of tubular and gelatinous fibres, intermixed with a varying proportion of filamentous areolar tissue, and enclosed in a sheath formed of fibro-areolar tissue. The tubular fibres are, for the most part, smaller than those composing the cerebro-spinal nerves; their double contour is less distinct, and, according to Remak, they present nuclei similar to those found in the gelatinous nerve-fibres. Those branches of the sympathetic which present a well-marked grey colour, are composed more especially of gelatinous nerve-fibres, intermixed with few tubular fibres; whilst those of a white colour contain more of the tubular fibres, and few gelatinous. Occasionally the grey and white cords run together in a single nerve, without any intermixture, as in the branches of communication between the sympathetic ganglia and the spinal nerves, or in the communicating cords between the ganglia.

The Cerebro-Spinal Centre consists of two parts, the spinal cord and the encephalon: the latter may be subdivided into the cerebrum or brain proper, the cerebellum or little brain, the tuber annulare or pons Varolii, and the medulla oblongata.

The Spinal Cord and its Membranes

Dissection. To dissect the cord and its membranes, it will be necessary to lay open the whole length of the spinal canal. For this purpose, the muscles must be separated from the vertebral grooves, so as to expose the spinous processes and laminae of the vertebrae; and the latter must be sawn through on each side, close to the roots of the transverse processes, from the third or fourth cervical vertebra, above, to the sacrum below. The vertebral arches having been displaced, by means of a chisel, and the separate fragments removed, the dura mater will be exposed, covered by a plexus of veins and a quantity of loose areolar tissue, often infiltrated with serous fluid. The arches of the upper vertebrae are best divided by means of a strong pair of forceps.

Membranes of the Cord

The membranes which envelope the spinal cord are three in number. The most external is the dura mater, a strong fibrous membrane, which forms a loose sheath around the cord. The most internal is the pia mater, a cellulo-vascular membrane, which closely invests the entire surface of the cord. Between the two, is the arachnoid membrane, an intermediate serous sac, which envelopes the cord, and is then reflected on the inner surface of the dura mater.

The Dura Mater of the cord, continuous with that which invests the brain, is a loose sheath which surrounds it, being separated from the bony walls of the spinal canal by a quantity of loose areolar adipose tissue, and a plexus of veins. It is attached, above, to the circumference of the foramen magnum, and extends, below, as far as the top of the sacrum; but, beyond this point, it is impervious, being continued, in the form of a slender cord, to the back of the coccyx, where it blends with the periosteum. This sheath is much larger than is necessary for its contents, and its size is greater in the cervical and lumbar regions, than in the dorsal. Its inner surface is smooth, being lined by the arachnoid membrane; and on each side may be seen the double openings which transmit the two roots of the corresponding spinal nerve, the fibrous layer of the dura mater being continued in the form of a tubular prolongation on them as they issue from these apertures, and becoming lost upon them. These prolongations of the dura mater are short in the upper part of the spine, but become gradually longer below, forming a number of tubes of fibrous membrane, which enclose the sacral nerves, and are contained in the spinal canal.

The chief peculiarities of the dura mater of the cord, as compared with that investing the brain, are the following:

The dura mater of the cord is not adherent to the bones of the spinal canal, which have an independent periosteum.

It does not send partitions into the fissures of the cord, as in the brain.

Its fibrous laminae do not separate, to form venous sinuses, as in the brain.

Structure. The dura mater consists of white fibrous tissue, arranged in bands, which intersect one another. It is sparingly supplied with vessels, as compared with the dura mater of the brain; and no nerves have as yet been traced into it.

233.—The Spinal Cord and its Membranes.

The *Arachnoid* is exposed by slitting up the dura mater, and reflecting this membrane on either side (fig. 233). It is a thin, delicate, serous membrane, which invests the outer surface of the cord, and is then reflected upon the inner surface of the dura mater, to which it is intimately adherent. That portion which surrounds the cord, is called the *visceral layer of the arachnoid*; and that which lines the inner surface of the dura mater, *the parietal layer*; the interval between the two, is called the *cavity of the arachnoid*. The visceral layer forms a loose sheath around the cord, so as to leave a considerable interval between the two, which is called the *sub-arachnoidean space*. This space is largest at the lower part of the spinal canal, and encloses the mass of nerves which form the cauda equina. It contains an abundant serous secretion, the cerebro-spinal fluid, and usually communicates with the general ventricular cavity of the brain, by means of an opening in the fibrous layer of the inferior boundary of the fourth ventricle. This secretion is sufficient in amount to expand the arachnoid membrane, so as to completely fill up the whole of the space included in the dura mater. The sub-arachnoidean space is crossed, at the back part of the cord, by numerous fibrous bands, which stretch from the arachnoid to the pia mater, especially in the cervical region, and is partially sub-divided by a longitudinal membranous partition, which serves to connect the arachnoid with the pia mater, opposite the posterior median fissure. This partition is incomplete, and cribriform in structure, consisting of bundles of white fibrous tissue, interlacing with each other. The visceral layer of the arachnoid surrounds the spinal nerves where they arise from the cord, and encloses them in a tubular sheath as far as their point of exit from the dura mater, where it becomes continuous with the parietal layer.

The arachnoid is not very vascular. No nerves have as yet been traced into this membrane.

The *Pia Mater* of the cord is exposed on the removal of the arachnoid (fig. 233). It is less vascular in structure than the pia mater of the brain, with which it is continuous, being thicker, more dense in structure, and composed of fibrous tissue, arranged in longitudinal bundles. It covers the entire surface of the cord, to which it is very intimately adherent, forming its neurilemma, and sends a process downwards into its anterior fissure, and another, extremely delicate, into the posterior fissure. It also forms a sheath for each of the filaments of the spinal nerves, and invests the nerves themselves. A longitudinal fibrous band extends along the middle line on its anterior surface, called by Haller, the linea splendens; and a somewhat similar band, the ligamentum denticulatum, is situated on each side. At the point where the cord terminates, the pia mater becomes contracted, and is continued down as a long, slender filament, which descends through the centre of the mass of nerves forming the cauda equina, and is blended with the impervious sheath of dura mater (before mentioned), on a level with the top of the sacral canal. It assists in maintaining the cord in its position during the movements of the trunk, and is, from this circumstance, called the central ligament of the spinal cord. It contains a little nervous substance, which may be traced for some distance into its upper part, and is accompanied by a small artery and vein.

234.—Transverse Section of the Spinal Cord and its Membranes.

Structure. The pia mater of the cord, though less vascular than that which invests the brain, contains a network of delicate vessels in its substance. It is also supplied with nerves, which, according to Purkinje, are derived from the sympathetic; but Remak states that they are chiefly supplied from the posterior roots of the spinal nerves. At the upper part of the cord, it presents a greyish, mottled tint, which is owing to yellowish or brown pigment cells being scattered within its tissue.

The *Ligamentum Denticulatum* (fig. 233) is a narrow, fibrous band, situated on each side of the spinal cord, throughout its entire length, and separating the anterior from the posterior roots of the spinal nerves, having received its name from the serrated appearance which it presents. Its inner border is continuous with the pia mater, at the side of the cord. Its outer border presents a series of triangular, dentated serrations, the points of which are fixed, at intervals, to the dura mater, serving to unite together the two layers of the arachnoid membrane. These serrations are about twenty in number, on each side, the first being attached to the dura mater, opposite the margin of the foramen magnum, between the vertebral artery and the hypoglossal nerve; and the last corresponds to nearly the lower end of the cord. Its use is to support the cord in the fluid by which it is surrounded.

The Spinal Cord

The spinal cord (*medulla spinalis*) is that elongated part of the cerebro-spinal axis, which is contained in the spinal canal. It weighs, when divested of its membranes and nerves, about one ounce and a half, its proportion to the encephalon being about 1 to 33. It does not nearly fill the canal in which it is contained, its investing membranes being separated from the bony walls of the canal by areolar tissue and a plexus of veins. It occupies, in the adult, the upper two-thirds of the spinal canal, extending from the foramen magnum to the lower border of the body of the first lumbar vertebra, where it terminates in a jointed extremity which is concealed among the leash of nerves forming the cauda equina. In the foetus, before the third month, it reaches to the bottom of the vertebral canal; but, after this period, it gradually recedes from below, as the growth of the bones composing the canal is more rapid in proportion than the cord; so that, in the child at birth, it extends as far as the third lumbar vertebra. Its position varies according to the degree of curvature of the spinal column, being raised somewhat in flexion of the spine forwards. Its length varies from fifteen to eighteen inches, and it presents a difference in its diameter in different parts, being marked by two enlargements, an upper or cervical, and a lower, or lumbar. The cervical enlargement, which is the larger, extends from the third cervical to the first dorsal vertebra: its greatest diameter is in the transverse direction, and it corresponds with the origin of the nerves which supply the upper extremities. The lower, or lumbar enlargement, is situated opposite the last dorsal vertebra, its greatest diameter being from before backwards. It corresponds with the origin of the nerves which supply the lower extremities. In form, the spinal cord is a flattened cylinder. Its anterior surface presents, along the middle line, a longitudinal fissure, the anterior median fissure; and, on its posterior surface, another fissure exists, which also extends along the entire length of the cord, the posterior median fissure. These fissures serve to divide the cord into two equal and symmetrical portions, which are connected together throughout their entire length, by a transverse band of nervous substance, the commissure.

235.—Spinal Cord. Side View. Plan of the Fissures and Columns.

The *Anterior median fissure* is wider, but of less depth than the posterior, extending into the cord for about one-third of its thickness, and is deepest at the lower part of the cord. It contains a prolongation from the pia mater; and its floor is formed by the anterior white commissure, which is perforated by numerous blood-vessels, which pass to the centre of the cord.

The *Posterior median fissure* is much more delicate than the anterior, and more distinct at the upper part of the cord and below. It extends into the cord to about one half of its depth. It contains a very slender process of the pia mater and numerous blood-vessels, and its floor is formed by a thin layer of white substance, the posterior white commissure. Some anatomists state, that the bottom of this fissure corresponds to the grey matter, except in the cervical region, and at a point corresponding to the enlargement in the lumbar region.

On either side of the anterior median fissure, a linear series of foramina may be observed, indicating the points where the anterior roots of the spinal nerves emerge from the cord. This is called, by some anatomists, the anterior lateral fissure of the cord, although no actual fissure exists in this situation. And on either side of the posterior median fissure, along the line of attachment of the posterior roots of the nerves, a delicate fissure may be seen, leading down to the grey matter which approaches the surface in this situation: this is called the posterior lateral fissure of the spinal cord. On the posterior surface of the spinal cord, on either side of the posterior median fissure, is a slight longitudinal furrow, marking off two slender tracts, the posterior median columns. These are most distinct in the cervical region, but are stated by Foville to exist throughout the whole length of the cord. The fissures divide each half of the spinal cord into four columns, an anterior column, a lateral column, a posterior column, and a posterior median column.

The *Anterior column* includes all the portion of the cord between the anterior median fissure and the anterior lateral fissure, from which the anterior roots of the nerves arise, and is continuous with the anterior pyramid of the medulla oblongata.

The *Lateral column*, the largest segment of the cord, includes all the portion between the anterior and posterior lateral fissures. It is continuous with the lateral column of the medulla. By some anatomists, the anterior and lateral columns are included together, under the name of the antero-lateral column, which forms rather more than two-thirds of the entire circumference of the cord.

The *Posterior column* is situated between the posterior median and posterior lateral fissures. It is continued, above, into the restiform body of the medulla.

The *Posterior median column* is that narrow segment of the cord which is seen on each side of the posterior median fissure, usually included with the preceding, as the posterior column.

If a transverse section of the spinal cord be made, it will be seen to consist of white and grey nervous matter. The white matter is situated at the circumference, the grey matter in the interior.

The Grey matter presents two crescentic masses, placed one in each lateral half of the cord, with their convexities towards one another, and joined by a transverse band of grey matter, the grey commissure. Each crescentic mass has an anterior and posterior horn.

The posterior horn is long and narrow, and approaches the surface at the posterior lateral fissure, near which it presents a slight enlargement. The grey matter, in this situation, is pale and soft, and was called by Rolando, the *substantia cinerea gelatinosa*, being surrounded by a layer of reddish-brown substance.

The anterior horn is short and thick, and does not quite reach the surface, but extends towards the point of attachment of the anterior roots of the nerves. Its margin presents a dentate, or stellate appearance. Owing to this peculiar arrangement of the grey matter, the anterior and posterior horns projecting towards the surface, each half of the cord is divided, more or less completely, into three columns, anterior, middle, and posterior; the anterior and middle being joined, as the anterior horn does not quite reach the surface, to form the antero-lateral column.

The *Grey commissure*, which connects the two crescentic masses of grey matter, is separated from the bottom of the anterior median fissure by a thick layer of white substance, the anterior white commissure; and, from the bottom of the posterior fissure by the posterior white commissure. The existence of the latter commissure is doubted by some anatomists. The grey commissure consists of a transverse band of grey matter, and of white fibres, derived from the opposite half of the cord and the posterior roots of the nerves. The white commissure is formed, partly of fibres from the anterior column, and partly from the fibrils of the anterior roots of the spinal nerves, which decussate as they pass across from one to the other side.

The mode of arrangement of the grey matter, and its amount in proportion to the white, vary in different parts of the cord. Thus, the posterior horns are long and narrow, in the cervical region; short and narrower, in the dorsal; short, but wider, in the lumbar region. In the cervical region, the crescentic portions are small, the white matter more abundant than in any other region of the cord. In the dorsal region, the grey matter is least developed, the white matter being also small in quantity. In the lumbar region, the grey matter is more abundant than in any other region of the cord. Towards the lower end of the cord, the white matter gradually ceases. The crescentic portions of the grey matter gradually blend into a single mass, which forms the only constituent of its extreme point.

236.—Transverse Sections of the Cord.

Opposite Middle of Cervical regⁿ

Opposite Middle of Dorsal regⁿ

Opposite Lumbar region

The *white matter of the cord* forms about seven-eighths of its entire substance. It is composed of parallel fibres collected into compressed, longitudinal bundles between which blood-vessels, supported by a delicate process of pia mater, pass transversely into the substance of the cord.

In the foetus, until after the sixth month, a canal, continuous with the general ventricular cavity of the brain, extends throughout the entire length of the spinal cord, formed by the closing-in of a previously open groove.

In the adult, this canal can only be seen at the upper part of the cord, extending from the point of the calamus scriptorius, in the floor of the fourth ventricle, for about half an inch down the centre of the cord, where it terminates in a *cul de sac*, the remnant of the canal being just visible in a section of the cord, as a small, pale spot, corresponding to the centre of the grey commissure, its cavity having become obliterated. In some cases this canal remains pervious throughout the whole length of the cord.

The Brain and its Membranes

Dissection. To examine the brain with its membranes, the skull cap should first be removed. This may be effected by sawing through the external table, commencing, in front, about an inch above the margin of the orbit, and extending, behind, to a level with the occipital protuberance. The internal table must then be broken through with the chisel and hammer to prevent injury to the investing membranes or brain, and after having been loosened, it should be forcibly detached, when the dura mater will be exposed. The adhesion between the bone and the dura mater is very intimate, and much more so in the young subject than in the adult.

The membranes of the brain are the dura mater, arachnoid membrane, and pia mater.

Dura Mater

The dura mater is a thick and dense inelastic fibrous membrane, which lines the interior of the skull. Its outer surface is rough and fibrillated, and adheres closely to the inner surface of the bones, forming their internal periosteum; this adhesion being more intimate opposite the sutures and at the base of the skull, where it is attached to the margin of the foramen magnum, and is here continuous with the dura mater lining the spinal canal. Its inner surface is smooth and epitheliated, being lined by the parietal layer of the arachnoid. The dura mater is therefore a fibro-serous membrane, composed of an external fibrous lamella, and an internal serous layer. It sends numerous processes inwards, into the cavity of the skull, for the support and protection of the different parts of the brain; it is also prolonged to the outer surface of the skull, through the various foramina which exist at its base, where it is continuous with the pericranium, and its fibrous layer forms sheaths for the nerves which pass through these apertures. At the base of the skull, it sends a fibrous prolongation into the foramen caecum; it lines the olfactory groove, and sends a series of tubular prolongations around the filaments of the olfactory nerves as they pass through the cribriform foramina; a prolongation is also continued through the sphenoidal fissure into the orbit, and another is continued into the same cavity through the optic foramen, forming a sheath for the optic nerve, which is continued as far as the eye-ball. In certain situations in the skull already mentioned, the fibrous layer of this membrane subdivides into two, to form the sinuses for the passage of venous blood. Upon the upper surface of the dura mater, in the situation of the longitudinal sinus, may be seen numerous small whitish bodies, the glandulae Pacchioni.

Structure. The dura mater consists of white fibrous and elastic tissues, arranged in flattened laminae, which intersect one another in every direction.

Its *arteries* are very numerous, but are chiefly distributed to the bones. Those found in the anterior fossa, are the anterior meningeal, from the anterior and posterior ethmoidal, and internal carotid. In the middle fossa are the middle and small meningeal, from the internal maxillary, and a third branch from the ascending pharyngeal, which enters the skull through the foramen lacerum basis cranii. In the posterior fossa, are the posterior meningeal branch of the occipital, which enters the skull through the jugular foramen, the posterior meningeal, from the vertebral, and occasionally meningeal branches from the ascending pharyngeal, which enter the skull, one at the jugular foramen, the other at the anterior condyloid foramen.

The *veins* which return the blood from the dura mater and partly from the bones, anastomose with the diploic veins. These vessels terminate in the various sinuses, with the exception of two which accompany the middle meningeal artery: these pass from the skull at the foramen spinosum.

The *nerves* of the dura mater, are the recurrent branch of the fourth, and filaments from the Gasserian ganglion, the ophthalmic nerve, and sympathetic.

The so-called Glandulae Pacchioni are numerous small whitish granulations, usually collected into clusters of variable size, which are found in the following situations: 1. Upon the outer surface of the dura mater, in the vicinity of the superior longitudinal sinus, being received into little depressions on the inner surface of the calvarium. 2. On the inner surface of the dura mater. 3. In the superior longitudinal sinus. 4. On the pia mater near the margin of the hemispheres.

These bodies are not glandular in structure, but consist of a fibro-cellular matrix originally developed from the pia mater: by their growth they produce absorption or separation of the fibres of the dura mater; in a similar manner they make their way into the superior longitudinal sinus, where they are covered by the lining membrane. The cerebral layer of the arachnoid in the situation of these growths is usually thickened and opaque, and adherent to the parietal portion.

These bodies are not found in infancy, and very rarely until the third year. They are usually found after the seventh year; and from this period they increase in number as age advances. Occasionally they are wanting.

Processes of the Dura Mater

The processes of the dura mater, sent inwards into the cavity of the skull, are three in number, the falx cerebri, the tentorium cerebelli, and the falx cerebelli.

The *falx cerebri*, so named from its sickle-like form, is a strong arched process of the dura mater, which descends vertically in the longitudinal fissure between the two hemispheres of the brain. It is narrow in front, where it is attached to the crista galli process of the ethmoid bone, and broad behind, where it is connected with the upper surface of the tentorium. Its upper margin is convex, and attached to the inner surface of the skull as far back as the internal occipital protuberance. In this situation it is broad, and contains the superior longitudinal sinus. Its lower margin is free, concave, and presents a sharp curved edge which contains the inferior longitudinal sinus.

The *tentorium cerebelli*, so named from its tent-like form, is a roof of dura mater, elevated in the middle, and inclining downwards towards its circumference. It covers the upper surface of the cerebellum, supporting the posterior lobes of the brain, and preventing their pressure upon it. It is attached behind, by its convex border, to the transverse ridges upon the inner surface of the occipital bone, and there encloses the lateral sinuses; in front, to the superior margin of the petrous portion of the temporal bone, enclosing the superior petrosal sinuses, and from the apex of this bone, on each side, is continued into the anterior and posterior clinoid processes. Along the middle line of its upper surface, the posterior border of the falx cerebri is attached, the straight sinus being placed at their point of junction. Its anterior border is free and concave, and presents a large oval opening for the transmission of the crura cerebri.

The *falx cerebelli* is a small triangular process of dura mater, received into the indentation between the two lateral lobes of the cerebellum behind. Its base is attached, above, to the under and back part of the tentorium; its posterior margin, to the lower division of the vertical crest on the inner surface of the occipital bone. As it descends, it sometimes divides into two smaller folds, which are lost on the sides of the foramen magnum.

Arachnoid Membrane

The arachnoid (ἀράχνη, εἶδος, like a spider's web), so named from its extreme thinness, is the serous membrane which envelopes the brain, and is then reflected on the inner surface of the dura mater. Like other serous membranes, it is a shut sac, and consists of a parietal and a visceral layer.

The *parietal layer* covers the inner surface of the dura mater, to which it is very intimately adherent, and gives this membrane the smooth and polished surface which it presents; it is also reflected over those processes which separate the hemispheres of the brain and cerebellum.

The *visceral layer* invests the brain more loosely, being separated from direct contact with the cerebral matter by the pia mater, and a quantity of loose areolar tissue, the sub-arachnoidean. On the upper surface of the cerebrum the arachnoid is thin and transparent, and may be easily demonstrated by injecting a stream of air beneath it by means of a blowpipe; it passes over the convolutions without dipping down into the sulci between them. At the base of the brain, the arachnoid is thicker, and slightly opaque towards the central part; it covers the anterior lobes, is extended across between the two middle lobes, so as to leave a considerable interval between it and the brain, the *anterior sub-arachnoidean space*; it is closely adherent to the pons and under surface of the cerebellum, but between the hemispheres of the cerebellum and the medulla oblongata another considerable interval is left between it and the brain, called the *posterior sub-arachnoidean space*. These two spaces communicate together across the crura cerebri. The arachnoid membrane surrounds the nerves which arise from the brain, and encloses them in loose sheaths as far as their point of exit from the skull, where it becomes continuous with the parietal layer.

The *Sub-arachnoid Space* is the interval left between the arachnoid and pia mater: this space is narrow on the surface of the hemispheres, but at the base of the brain a wide interval is left between the two middle lobes, and behind, between the hemispheres of the cerebellum and the medulla oblongata. This space is the seat of an abundant serous secretion, the cerebro-spinal fluid, which fills up the interval between the arachnoid and pia mater. The sub-arachnoid space usually communicates with the general ventricular cavity of the brain, by means of an opening in the inferior boundary of the fourth ventricle.

The sac of the arachnoid also contains serous fluid; this is, however, small in quantity compared with the cerebro-spinal fluid.

Structure. The arachnoid consists of bundles of white fibrous and elastic tissues intimately blended together. The visceral portion is covered with a layer of scaly epithelium. It is almost destitute of vessels, and the existence of nerves in it has not been satisfactorily demonstrated.

The *Cerebro-spinal Fluid* fills up the sub-arachnoid space, keeping the opposed surfaces of the arachnoid membrane in contact. It is a clear limpid fluid, having a saltish taste, and a slightly alkaline reaction. According to Lassaigne, it consists of 98.5 parts of water, the remaining 1.5 percent being solid matters, animal and saline.

It varies in quantity from two to ten ounces, being most abundant in old persons, and is quickly reproduced. Its chief use is probably to afford mechanical protection to the nervous centres, and to prevent the effects of concussion communicated from without.

Pia Mater

The pia mater is a vascular membrane, and derives its blood from the internal carotid and vertebral arteries. It consists of a minute plexus of blood-vessels, held together by an extremely fine areolar tissue. It invests the entire surface of the brain, dipping down between the convolutions and laminae, and is prolonged into the interior, forming the velum interpositum and choroid plexuses of the fourth ventricle. Upon the surface of the hemispheres, where it covers the grey matter of the convolutions, it is very vascular, and gives off from its inner surface a multitude of minute vessels, which extend perpendicularly for some distance into the cerebral substance. At the base of the brain, in the situation of the substantia perforata and locus perforatus, a number of long straight vessels are given off, which pass through the white matter to reach the grey substance in the interior. On the cerebellum, the membrane is more delicate, and the vessels from its inner surface are shorter. Upon the crura cerebri and pons Varolii its characters are altogether changed; it here presents a dense fibrous structure, marked only by slight traces of vascularity.

According to Fohmann and Arnold, this membrane contains numerous lymphatic vessels. Its nerves are derived from the sympathetic, and also from the third, sixth, seventh, eighth, and accessorius. They accompany the branches of the arteries.

The Brain

The brain (*encephalon*) is that portion of the cerebro-spinal axis that is contained in the cranial cavity. It is divided into four principal parts: viz., the cerebrum, the cerebellum, the pons Varolii, and medulla oblongata.

The *Cerebrum* forms the largest portion of the encephalic mass, and occupies a considerable part of the cavity of the cranium, resting in the anterior and middle fossa of the base of the skull, and separated posteriorly from the cerebellum by the tentorium cerebelli. About the middle of its under surface is a narrow constricted portion, part of which, the crura cerebri, is continued onwards into the pons Varolii below, and through it to the medulla oblongata and spinal cord; whilst another portion, the crura cerebelli, pass down into the cerebellum.

The *Cerebellum* (little brain or after brain) is situated in the inferior occipital fossae, being separated from the under surface of the posterior lobes of the cerebrum by the tentorium cerebelli. It is connected to the rest of the encephalic mass by means of connecting bands, called *crura*; of these, two ascend to the cerebrum, two descend to the medulla oblongata, and two blend together in front, forming the pons Varolii.

The *Pons Varolii* is that portion of the encephalic mass which rests upon the upper part of the basilar process. It constitutes a sort of centre to the various segments above named, receiving, above, the crura from the cerebrum; at the sides, the crura from the cerebellum; and, being connected, below, with the medulla oblongata.

The *Medulla Oblongata* extends from the lower border of the pons Varolii to the upper part of the spinal cord. It lies beneath the cerebellum, resting on the lower part of the basilar groove of the occipital bone.

Weight of the Encephalon. The average weight of the brain in the adult male is 49½ oz., or a little more than 3 lb. avoirdupois, that of the female 44 oz., the average difference between the two being from 5 to 6 oz. The prevailing weight of the brain in the male ranges between 46 oz. and 53 oz., and in the female, between 41 oz. and 47 oz. In the male, the maximum weight out of 278 cases was 65 oz., and the minimum weight 34 oz. The maximum weight of the adult female brain, out of 191 cases, was 56 oz., and the minimum weight 31 oz. It appears that the weight of the brain increases rapidly up to the seventh year, more slowly to between sixteen and twenty, and still more slowly to between thirty and forty, when it reaches its maximum. Beyond this period, as age advances and the mental faculties decline, the brain diminishes slowly in weight about an ounce for each subsequent decennial period. These results apply alike to both sexes.

The size of the brain appears to bear a general relation to the intellectual capacity of the individual. Cuvier's brain weighed rather more than 64 oz., that of the late Dr. Abercrombie 63 oz., and that of Dupuytren 62½ oz. On the other hand, the brain of an idiot seldom weighs more than 23 oz.

The human brain is heavier than that of all the lower animals excepting the elephant and whale. The brain of the former weighs from 8 lb. to 10 lb., and that of the whale, in a specimen seventy-five feet long, weighed rather more than 5 lb.

Medulla Oblongata

The medulla oblongata is the upper enlarged part of the spinal cord, and extends from the upper border of the atlas to the lower border of the pons Varolii. It is directed obliquely downwards and backwards, its anterior surface resting on the basilar groove of the occipital bone, its posterior surface being received into the fossa between the hemispheres of the cerebellum, forming the floor of the fourth ventricle. It is pyramidal in form, its broad extremity directed upwards, its lower end being narrow at its point of connection with the cord. It measures an inch and a quarter in length, three quarters of an inch in breadth at its widest part, and half an inch in thickness. Its surface is marked in the median line, in front and behind, by an anterior and posterior median fissure, which are continuous with those of the spinal cord. The anterior fissure contains a fold of pia mater, and terminates below the pons in a *cul-de-sac*, the foramen caecum. The posterior is a deep but narrow fissure, continued upwards along the floor of the fourth ventricle, where it is finally lost. These two fissures divide the medulla into

237.—Medulla Oblongata and Pons Varolii. Anterior Surface.

two symmetrical halves, each lateral half being subdivided by minor grooves into four columns, which, from before backwards, are named, the *anterior pyramid, lateral tract and olivary body*, the *restiform body*, the *posterior pyramid*.

238.—Posterior Surface of Medulla Oblongata.

The *Anterior Pyramids* are two pyramidal-shaped bundles of white matter, placed one on either side of the anterior median fissure, and separated from the olivary body, which is external to them, by a slight depression. At the lower border of the pons they are somewhat constricted; they then become enlarged, and taper slightly as they descend, being continuous below with the anterior columns of the cord. On separating the pyramids below, it will be observed that the innermost fibres of the two form from four to five bundles on each side, which decussate with one another; this decussation, however, is not formed entirely of fibres from the pyramids, but mainly from the deep portion of the lateral columns of the cord which pass forwards to the surface between the diverging anterior columns. The outermost fibres do not decussate; they are derived from the anterior columns of the cord, and are continued directly upwards through the pons Varolii.

Lateral Tract and *Olivary Body*. The lateral tract is continuous with the lateral column of the cord. Below, it is broad, and includes that part of the medulla between the anterior pyramid and restiform body; but, above, it is pushed a little backwards, and narrowed by the projection forwards of the olivary body.

The *Olivary Bodies* are two prominent, oval masses, situated behind the anterior pyramids, from which they are separated by slight grooves. They equal, in breadth, the anterior pyramids, are a little broader above than below, and are about half an inch in length, being separated, above, from the pons Varolii, by a slight depression. Numerous white fibres (*fibrae arciformes*) are seen winding around the lower end of each body; sometimes crossing their surface.

The *Restiform Bodies* are the largest columns of the medulla, and continuous, below, with the posterior columns of the cord. They are two rounded, cord-like eminences, placed between the lateral tracts, in front, and the posterior pyramids, behind; from both of which they are separated by slight grooves. As they ascend, they diverge from each other, assist in forming the lateral boundaries of the fourth ventricle, and then enter the corresponding hemisphere of the cerebellum, forming its inferior peduncle.

The *Posterior Pyramids* (*fasciculi graciles*) are two narrow, white cords, placed one on each side of the posterior median fissure, and separated from the restiform bodies by a narrow groove. They consist entirely of white fibres, and are continuous with the posterior median columns of the spinal cord. These bodies lie at first, in close contact. Opposite the apex of the fourth ventricle, they form an enlargement (*processus clavatus*), and then, diverging, are lost in the corresponding restiform body.

The upper part of the posterior pyramids form the lateral boundaries of the calamus scriptorius.

The *Posterior surface of the Medulla Oblongata* forms part of the floor of the fourth ventricle. It is of a triangular form, bounded on each side by the diverging posterior pyramids, and is that part of the ventricle which, from its resemblance to the point of a pen, is called the *calamus scriptorius*. The divergence of these columns, and the restiform bodies, opens to view the grey matter of the medulla, which is continuous, below, with the grey commissure of the cord. In the middle line is seen a longitudinal furrow, continuous with the posterior median fissure of the cord, terminating, below, at the point of the ventricle, in a *cul-de-sac*, the ventricle of Arantius, which descends into the medulla for a slight extent. It is the remains of a canal, which, in the foetus, extends throughout the entire length of the cord.

Structure. The columns of the cord are directly continuous with those of the medulla oblongata, below; but, higher up, both the white and grey constituents are re-arranged before they are continued upwards to the cerebrum and cerebellum.

The *Anterior Pyramid* is composed of fibres derived from the anterior column of the cord of its own side, and from the lateral column of the opposite half of the cord, and is continued upwards into the cerebrum and cerebellum. The cerebellar fibres form a superficial and deep layer, which pass beneath the olive to the restiform body, and spread out into the structure of the cerebellum. A deeper fasciculus encloses the olivary body, and, receiving fibres from it, enters the pons as the olivary fasciculus or fillet; but the chief mass of fibres from the pyramid, the cerebral fibres, enter the pons in their passage upwards to the cerebrum. The anterior pyramids contain no grey matter.

The *Lateral Tract* is continuous, below, with the lateral column of the cord. Its fibres pass in three different directions. The most external join the restiform body, and pass to the cerebellum. The internal, more numerous, pass forwards, pushing aside the fibres of the anterior column, and form part of the opposite anterior pyramid. The middle fibres ascend, beneath the olivary body, to the cerebrum, passing along the back of the pons, and form, together with fibres from the restiform body, the *fasciculi teretes*, in the floor of the fourth ventricle.

Olivary Body. If a transverse section is made through either olivary body, it will be found to be a small ganglionic mass, deeply imbedded in the medulla, partly appearing on the surface as a smooth, olive-shaped eminence (fig. 239). It consists, externally, of white substance; and internally, of a grey nucleus, the corpus dentatum. The grey matter is arranged in the form of a hollow capsule, open at its upper and inner part and presenting a zig-zag, or dentated outline. White fibres originate from the interior of this body, by the aperture in the posterior part of the capsule. They join with those fibres of the anterior column which ascend on the outer side, and beneath the olive, to form the olivary fasciculus, which ascends to the cerebrum.

239.—Transverse Section of Medulla Oblongata.

The *Restiform Body* is formed chiefly of fibres from the posterior column of the cord; but it receives some from the lateral column, and a fasciculus from the anterior, and is continued, upwards, to the cerebrum and cerebellum. On entering the pons, it divides into two fasciculi, above the point of the fourth ventricle. The most external one enters the cerebellum: the inner one joins the posterior pyramid, is continued up along the fourth ventricle, and, joining the fasciculi teretes, passes up to the cerebrum.

Septum of the Medulla Oblongata. Above the decussation of the anterior pyramids, numerous white fibres extend, from behind forwards, in the median line, forming a septum, which subdivides the medulla into two lateral halves. Some of these fibres emerge at the anterior median fissure, and form a band which curves around the lower border of the olivary body, or passes transversely across it, and round the sides of the medulla, forming the *arciform fibres* of Rolando. Others appear in the floor of the fourth ventricle, issuing from the posterior median fissure, and form the white striae in that situation.

Grey Matter of the Medulla Oblongata. The grey matter of the medulla, is a continuation of that contained in the interior of the spinal cord, besides a series of special deposits, or nuclei.

In the lower part of the medulla the grey matter is arranged as in the cord, but, at the upper part, it becomes more abundant, and is disposed with less apparent regularity, becoming blended with all the white fibres, except the anterior pyramids. The part corresponding to the transverse grey commissure of the cord, is exposed to view in the floor of the medulla oblongata, by the divergence of the restiform bodies, and posterior pyramids, becoming blended with the ascending fibres of the lateral column, and thus forming the fasciculi teretes. The lateral crescentic portions but especially the posterior horns, become enlarged, blend with the fibres of the restiform bodies, and form the *tuberculo cinereo* of Rolando.

Special deposits of grey matter are found both in the anterior and posterior parts of the medulla; in the former situation, forming the corpus dentatum within the olivary body, and in the latter, a series of special masses, or nuclei, connected with the roots of origin of the spinal accessory, vagus, glosso-pharyngeal, and hypo-glossal nerves.

Pons Varolii

The pons Varolii (*mesocephale*, Chaussier) is the bond of union of the various segments of the encephalon, connecting the cerebrum above, the medulla oblongata below, and the cerebellum behind. It is situated above the medulla oblongata, below the crura cerebri, and between the hemispheres of the cerebellum.

Its *under surface* presents a broad transverse band of white fibres, which arches like a bridge across the upper part of the medulla, extending between the two hemispheres of the cerebellum. This surface projects considerably beyond the level of these parts, is of a quadrangular form, rests upon the basilar groove of the occipital bone, and is limited before and behind by very prominent margins. It presents along the middle line a longitudinal groove, wider in front than behind, which lodges the basilar artery; numerous transverse striae are also observed on each side, which indicate the course of its superficial fibres.

Its *upper surface* forms part of the floor of the fourth ventricle, and at each side it becomes contracted into a thick rounded cord, the crus cerebelli, which enters the substance of the cerebellum, constituting its middle peduncle.

Structure. The pons Varolii consists of alternate layers of transverse and longitudinal fibres intermixed with grey matter (fig. 240).

240.—The Columns of the Medulla Oblongata, and their Connection with the Cerebrum and Cerebellum.

The *transverse fibres* connect together the two lateral hemispheres of the cerebellum, and constitute its great transverse commissure. They consist of a superficial and a deep layer. The superficial layer passes uninterruptedly across the surface of the pons, forming a uniform layer, consisting of fibres derived from the crus cerebelli on each side, which meet in the median line. The deep layer of transverse fibres decussate with the longitudinal fibres continued up from the medulla; they also connect the hemispheres of the cerebellum.

The *longitudinal fibres* are continued up through the pons. 1. From the anterior pyramidal body. 2. From the olivary body. 3. From the lateral and posterior columns of the cord, receiving special fibres from the grey matter of the pons itself.

1. The fibres from the anterior pyramid ascend through the pons, embedded between two layers of transverse fibres, being subdivided in their course into smaller bundles; at the upper border of the pons they enter the crus cerebri, forming its fasciculated portion.

2. The olivary fasciculus divides in the pons into two bundles, one of which ascends to the corpora quadrigemina; the other is continued to the cerebrum with the fibres of the lateral column.

3. The fibres from the lateral and posterior columns of the cord, with a bundle from the olivary fasciculus, are intermixed with much grey matter, and appear in the floor of the fourth ventricle as the fasciculi teretes, they ascend to the deep or cerebral part of the crus cerebri.

Septum. The pons is subdivided into two lateral halves by a median septum, which extends through its posterior half. The septum consists of antero-posterior and transverse fibres. The former are derived from the floor of the fourth ventricle and from the transverse fibres of the pons, which bend backwards before passing across to the opposite side. The latter are derived from the floor of the fourth ventricle, they pierce the longitudinal fibres, and are then continued across from one to the other side of the medulla, piercing the antero-posterior fibres.

The two halves of the pons, in front, are connected together by transverse commissural fibres.

Cerebrum: Upper Surface

The cerebrum, in man, constitutes the largest portion of the encephalon. Its upper surface is of an ovoidal form, broader behind than in front, convex in its general outline, and divided into two lateral halves or hemispheres, right and left, by the great longitudinal fissure. This fissure extends throughout the entire length of the cerebrum in the middle line, reaching down to the base of the brain in front and behind, but interrupted in the middle by a broad transverse commissure of white matter, the corpus callosum, which connects the two hemispheres together. This fissure lodges the falx cerebri, and indicates the original development of the brain by two lateral halves.

Each hemisphere presents an outer surface, which is convex to correspond with the vault of the cranium; an inner surface, flattened, and in contact with the opposite

241.—Upper Surface of the Brain, the Pia Mater having been removed.

hemisphere, the two forming the sides of the longitudinal fissure; and an under surface or base, of more irregular form, which rests, in front, in the anterior and middle fossae at the base of the skull, and behind, upon the tentorium.

Convolutions. If the pia mater is removed with the forceps, the entire surface of each hemisphere will present a number of convoluted eminences, the convolutions, separated from each other by depressions (*sulci*) of various depths. The outer surface of each convolution, as well as the sides and bottom of the sulci between them, are composed of grey matter, which is here called the *cortical substance*. The interior of each convolution is composed of white matter, white fibres also blend with the grey matter at the sides and bottom of the sulci. By this arrangement the convolutions are admirably adapted to increase the amount of grey matter without occupying much additional space, and also afford a greater extent of surface for the fibres to terminate in it. On closer examination, however, the grey matter of the cortical substance is found subdivided into four layers, two of which are composed of grey and two of white substance. The most external is an outer white stratum, not equally thick over all parts of the brain, being most marked on the convolutions in the longitudinal fissure and on the under part of the brain, especially on the middle lobe, near the descending horn of the lateral ventricle. Beneath the latter is a thick reddish grey lamina, and then another thin white stratum; lastly, a thin stratum of grey matter, which lies in close contact with the white fibres of the hemispheres: consequently white and grey laminae alternate with one another in the grey matter of the convolutions. In certain convolutions, however, the cortical substance consists of no less than six layers, three grey and three white, an additional white stratum dividing the most superficial grey one into two; this is especially marked in those convolutions which are situated near the corpus callosum.

A perfect resemblance between the convolutions does not exist in all brains, nor are they symmetrical on the two sides of the same brain. Occasionally the free borders or the sides of a deep convolution present a fissured or notched appearance.

The *sulci* are generally an inch in depth; they also vary in different brains, and in different parts of the same brain; they are usually deepest on the outer convex surface of the hemispheres; the deepest is situated on the inner surface of the hemisphere, on a level with the corpus callosum, and corresponds to the projection in the posterior horn of the lateral ventricle, the hippocampus minor.

The number and extent of the convolutions, as well as their depth, appear to bear a close relation to the intellectual power of the individual, as is shown in their increasing complexity of arrangement as we ascend from the lowest mammalia up to man. Thus they are absent in some of the lower orders of this class, and they increase in number and extent through the higher orders. In man they present the most complex arrangement. Again, in the child at birth before the intellectual faculties are exercised, the convolutions have a very simple arrangement, presenting few undulations; and the sulci between them are less deep than in the adult. In old age, when the mental faculties have diminished in activity, the convolutions become much less prominently marked.

Those convolutions which are the largest and most constantly present, are the convolution of the corpus callosum, the convolution of the longitudinal fissure, the supra-orbital convolution, and the convolutions of the outer surface of the hemisphere.

The *Convolution of the Corpus Callosum* (*gyrus fornicatus*) is always well marked. It lies parallel with the upper surface of the corpus callosum, commencing, in front, on the under surface of the brain in front of the anterior perforated space; it winds round the curved border of the corpus callosum, and passes along its upper surface as far as its posterior extremity, where it is connected with the convolutions of the posterior lobe; it then curves downwards and forwards, embracing the cerebral peduncle, passes into the middle lobe, forming the hippocampus major and terminates just behind the point from whence it arose.

The *Supra-orbitar Convolution* on the under surface of the anterior lobe is well marked.

The *Convolution of the Longitudinal Fissure* bounds the margin of the fissure on the upper surface of the hemisphere. It commences on the under surface of the brain, at the anterior perforated spot, passes forwards along the inner margin of the anterior lobe, being here divided by a deep sulcus, in which the olfactory nerve is received; it then curves over the anterior and upper surface of the hemisphere, along the margin of the longitudinal fissure, to its posterior extremity, where it curves forwards along the under surface of the hemisphere as far as the middle lobe.

The convolutions on the outer convex surface of the hemisphere, the general direction of which is more or less oblique, are the largest and the most complicated convolutions of the brain, frequently becoming branched like the letter Y in their course upwards and backwards towards the longitudinal fissure: these convolutions attain their greatest development in man, and are especially characteristic of the human brain. They are seldom symmetrical on the two sides.

Cerebrum: Under Surface or Base

The under surface of each hemisphere presents a subdivision, as already mentioned, into three lobes, named, from their position, anterior, middle, and posterior.

The *anterior lobe*, of a triangular form, with its apex backwards, is somewhat concave, and rests upon the convex surface of the roof of the orbit, being separated from the middle lobe by the fissure of Sylvius. The *middle lobe*, which is more prominent, is received into the middle fossa of the base of the skull. The *posterior lobe* rests upon the tentorium, its extent forwards being limited by the anterior margin of the cerebellum.

The various objects exposed to view on the under surface of the cerebrum in the middle line are here arranged in the order in which they are met with from before backwards.

Longitudinal fissure.	Tuber cinereum.
Corpus callosum and its peduncles.	Infundibulum.
Lamina cinerea.	Pituitary body.
Olfactory nerve.	Corpora albicantia.
Fissure of Sylvius.	Posterior perforated space.
Anterior perforated space.	Crura cerebri.
Optic commissure.	

The *Longitudinal Fissure* separates the two hemispheres from one another; it divides the two anterior lobes in front; and on raising the cerebellum and pons, it will be seen completely separating the two posterior lobes, the intermediate portion of the fissure being arrested by the great transverse band of white matter, the corpus callosum. Of these two portions of the longitudinal fissure, that which separates the posterior lobes is the longest. In the fissure between the two anterior lobes the anterior cerebral arteries may be seen ascending to the corpus callosum; and at the back part of this portion of the fissure, the anterior curved portion of the corpus callosum descends to the base of the brain.

The *Corpus Callosum* terminates at the base of the brain by a concave margin, which is connected with the tuber cinereum through the intervention of a thin layer of grey substance, the lamina cinerea. This may be exposed by gently raising and drawing back the optic commissure. A broad white band may be observed on each

242.—Base of the Brain.

side, passing from the under surface of the corpus callosum in front, backwards and outwards, to the commencement of the fissure of Sylvius; these bands are called the *peduncles of the corpus callosum*. Laterally, the corpus callosum extends into the anterior lobe.

The *Lamina Cinerea* is a thin layer of grey substance, extending backwards from the termination of the corpus callosum above the optic commissure to the tuber cinereum; it is continuous on either side with the grey matter of the anterior perforated space, and forms the anterior part of the inferior boundary of the third ventricle.

The *Olfactory Nerve*, with its bulb, is seen on either side of the longitudinal fissure, upon the under surface of each anterior lobe.

The *Fissure of Sylvius* separates the anterior and middle lobes, and lodges the middle cerebral artery. At its entrance is seen a point of medullary substance, corresponding to a subjacent band of white fibres, connecting the anterior and middle lobes, and called the *fasciculus unciformis*; on following this fissure outwards, it divides into two branches, which enclose a triangular-shaped prominent cluster of isolated convolutions, the *island of Reil*. These convolutions, from being covered in by the sides of the fissure, are called the *gyri operti*.

The *Anterior Perforated Space* is situated at the inner side of the fissure of Sylvius. It is of a triangular shape, bounded in front by the convolution of the anterior lobe and roots of the olfactory nerve; behind, by the optic tract; externally, by the middle lobe and commencement of the fissure of Sylvius; internally, it is continuous with the lamina cinerea, and crossed by the peduncle of the corpus callosum. It is of a greyish colour, and corresponds to the under surface of the corpus striatum, a large mass of grey matter, situated in the interior of the brain; it has received its name from being perforated by numerous minute apertures for the transmission of small straight vessels into the substance of the corpus striatum.

The *Optic Commissure* is situated in the middle line, immediately behind the lamina cinerea. It is the point of junction between the two optic nerves.

Immediately behind the diverging optic tracts, and between them and the peduncles of the cerebrum (crura cerebri) is a lozenge-shaped interval, the interpeduncular space, in which are found the following parts, arranged in the following order from before backwards: the tuber cinereum, infundibulum, pituitary body, corpora albicantia, and the posterior perforated space.

The *Tuber Cinereum* is an eminence of grey substance, situated between the optic tracts and the corpora albicantia; it is connected with the surrounding parts of the cerebrum, forms part of the floor of the third ventricle, and is continuous with the grey substance in that cavity. From the middle of its under surface, a conical tubular process of grey matter, about two lines in length, is continued downwards and forwards to be attached to the posterior lobe of the pituitary body; this is the infundibulum. Its canal, funnel-shaped in form, communicates with that of the third ventricle.

The *Pituitary Body* is a small reddish-grey vascular mass, weighing from five to ten grains, and of an oval form, situated in the sella Turcica, in connection with which it is retained by the dura mater which forms the inner wall of the cavernous sinus. It is very vascular, and consists of two lobes, separated from one another by a fibrous lamina. Of these, the anterior is the larger, of an oblong form, and somewhat concave behind, where it receives the posterior lobe, which is round. The anterior lobe consists externally of firm yellowish-grey substance, and internally of a soft pulpy substance of a yellowish-white colour. The posterior lobe is darker than the anterior. In the foetus it is larger proportionally than in the adult, and contains a cavity which communicates through the infundibulum with the third ventricle. In the adult it is firmer and more solid, and seldom contains any cavity. Its structure, especially the anterior lobe, is similar to that of the ductless glands.

The *Corpora Albicantia* are two small round white masses, each about the size of a pea, placed side by side immediately behind the tuber cinereum. They are formed by the anterior crura of the fornix, hence called the *bulbs of the fornix*, which, after descending to the base of the brain, are folded upon themselves, before passing upwards to the thalami optici. They are composed externally of white substance, and internally of grey matter; the grey matter of the two being connected by a transverse commissure of the same material. At an early period of foetal life they are blended together into one large mass, but become separated about the seventh month.

The *Posterior Perforated Space* (*Pons Tarini*) corresponds to a whitish-grey substance, placed between the corpora albicantia in front, the pons Varolii behind, and the crura cerebri on either side. It forms the back part of the floor of the third ventricle, and is perforated by numerous small orifices for the passage of blood-vessels, to the thalami optici.

The *Crura Cerebri* (*Peduncles of the Cerebrum*) are two thick cylindrical bundles of white matter, which emerge from the anterior border of the pons, and diverge as they pass forwards and outwards to enter the under part of either hemisphere. Each crus is about three-quarters of an inch in length, and somewhat broader in front than behind. They are marked upon their surface with longitudinal striae, and each is crossed, just before entering the hemisphere, by a flattened white band, the optic tract, which is adherent by its upper border to the peduncle. In its interior is contained a mass of dark grey matter, called *locus niger*. The third nerves may be seen emerging from the inner side of either crus; and the fourth nerve winding around its outer side from above.

Each crus consists of a superficial and deep layer of longitudinal white fibres, continued upwards from the pons, separated by a mass of grey matter, the locus niger.

The *Superficial Longitudinal Fibres* are continued upwards, from the anterior pyramids to the cerebrum. They consist of coarse fasciculi, which form the free part of the crus, and have received the name of the fasciculated portion of the peduncle, or crust.

The *Deep Layer of Longitudinal Fibres* are continued upwards, to the cerebrum, from the lateral and posterior columns of the medulla, and from the olivary fasciculus, these fibres consisting of some derived from the same, and others from the opposite lateral tract of the medulla. More deeply, are a layer of finer fibres, mixed with grey matter, derived from the cerebellum, blended with the former. The cerebral surface of the crus cerebri is formed of these fibres, and is named the tegmentum.

The *Locus Niger* is a mass of grey matter, situated between the superficial and deep layer of fibres above described. It is placed nearer the inner than the outer side of this body.

The posterior lobes of the cerebrum are concealed from view by the upper surface of the cerebellum, and pons Varolii. When these parts are removed, the two hemispheres are seen to be separated by the great longitudinal fissure, this fissure being arrested, in front, by the posterior rounded border of the corpus callosum.

General Arrangement of the Parts composing the Cerebrum

As the peduncles of the cerebrum enter the hemispheres, they diverge from one another, so as to leave an interval between them, the interpeduncular space. As they ascend, the component fibres of each pass through two large masses of grey matter called the ganglia of the brain, the thalami optici, and corpora striata, which project as rounded eminences from the upper and inner side of each peduncle. The hemispheres are connected together, above these masses, by the great transverse commissure, the corpus callosum, and the interval left between its under surface, the upper surface of the ganglia, and the parts closing the interpeduncular space, forms the general ventricular cavity. The upper part of this cavity is subdivided into two, by a vertical septum, the septum lucidum; and thus the two lateral ventricles are formed. The lower part of this cavity forms the third ventricle, which communicates with the lateral ventricles, above, and with the fourth ventricle, behind. The fifth ventricle is the interval left between the two layers composing the septum lucidum.

Interior of the Cerebrum

If the upper part of either hemisphere is removed with a scalpel, about half an inch above the level of the corpus callosum, its internal white matter will be exposed. It is an oval-shaped centre, of white substance, surrounded on all sides by a narrow, convoluted margin of grey matter, which presents an equal thickness in nearly every part. This white, central mass, has been called the *centrum ovale minus*. Its surface is studded with numerous minute red dots (*puncta vasculosa*), produced by the escape of blood from divided blood-vessels. In inflammation, or great congestion of the brain, these are very numerous, and of a dark colour. If the remaining portion of the hemispheres are slightly separated from one another, a broad band of white substance will be observed connecting them, at the bottom of the longitudinal fissure: this is the corpus callosum. The margins of the hemispheres, which overlap this portion of the brain, are called the *labia cerebri*. It is a part of the convolution of the corpus callosum (*gyrus fornicatus*), already described; and the space between it and the upper surface of the corpus callosum, has been termed the *ventricle of the corpus callosum*.

The hemispheres should now be sliced off, to a level with the corpus callosum, when the white substance of that structure will be seen connecting together both hemispheres. The large expanse of medullary matter now exposed, surrounded by the convoluted margin of grey substance, is called the *centrum ovale majus* of Vieussens.

The *Corpus Callosum* is a thick stratum of transverse fibres, exposed at the bottom of the longitudinal fissure. It connects the two hemispheres of the brain, forming their great transverse commissure; and forms the roof of a space in the interior of each hemisphere, the lateral ventricle. It is about four inches in length, extending to within an inch and a half of the anterior, and to within two inches and a half of the posterior, part of the brain. It is somewhat broader behind than in front, and it is thicker at either end than in its central part, being thickest behind. It presents a somewhat arched form, from before backwards, terminating anteriorly in a rounded border, which curves downwards and backwards, between the anterior lobes to the base of the brain. In its course, it forms a distinct bend, named the *knee*, or *genu*, and the reflected portion, named the *beak* (*rostrum*), becoming gradually narrower, is attached to the anterior cerebral lobe, and is connected, through the lamina cinerea, with the optic commissure. The reflected portion of the corpus callosum gives off, near its termination, two bundles of white substance, which, diverging from one another, pass backwards, across the anterior perforated space, to the entrance of the fissure of Sylvius. They are called the *peduncles* of the corpus callosum. Posteriorly, the corpus callosum forms a thick, rounded fold, which is free for a little distance,

243.—Section of the Brain. Made on a Level with the Corpus Callosum.

as it curves forwards, and is then continuous with the fornix. On its upper surface, its fibrous structure is very apparent to the naked eye, being collected into coarse, transverse bundles. Along the middle line, is a linear depression, the raphe, bounded laterally by two or more slightly elevated longitudinal bands, called the *striae longitudinales*, or nerves of Lancisi; and, still more externally, other longitudinal striae are seen, beneath the convolution, which rests on the corpus callosum. These are the striae longitudinales laterales. The under surface of the corpus callosum is continuous behind with the fornix, being separated from it in front by the septum lucidum, which forms a vertical partition between the two ventricles. On either side, the fibres of the corpus callosum penetrate into the substance of the hemispheres, and connect together the anterior, middle, and part of the posterior lobes. It is the increased aggregation of fibres derived from the anterior and posterior lobes, which explains the great thickness of the two extremities of this commissure.

An incision should now be made through the corpus callosum, on either side of the raphe, when two large irregular cavities will be exposed, which extend throughout the entire length of each hemisphere. These are the lateral ventricles.

244.—The Lateral Ventricles of the Brain.

The *Lateral Ventricles* are serous cavities, formed by the upper part of the general ventricular space in the interior of the brain. They are lined by a thin diaphanous lining membrane, covered with ciliated epithelium, and moistened by a serous fluid, which is sometimes, even in health, secreted in considerable quantity. These cavities are two in number, one in each hemisphere, and they are separated from each other by a vertical septum, the septum lucidum.

Each lateral ventricle consists of a central cavity, or body, and three smaller cavities, or cornua, which extend from it in different directions. The anterior cornu, curves forwards and outwards, into the substance of the anterior lobe. The posterior cornu, called the *digital cavity*, curves backwards into the posterior lobe. The middle cornu, descends into the middle lobe.

The *Central Cavity*, or body of the lateral ventricle, is triangular in form. It is bounded, above, by the under surface of the corpus callosum, which forms the roof of the cavity. Internally, is a vertical partition, the septum lucidum, which separates it from the opposite ventricle, and connects the under surface of the corpus callosum with the fornix. Its floor is formed by the following parts, enumerated in their order of position, from before backwards, the corpus striatum, taenia semicircularis, thalamus opticus, choroid plexus, corpus fimbriatum, and fornix.

The *Anterior Cornu* is triangular in form, passing outwards into the anterior lobe, and curving round the anterior extremity of the corpus striatum. It is bounded, above and in front, by the corpus callosum; behind, by the corpus striatum.

The *Posterior Cornu*, or digital cavity, curves backwards into the substance of the posterior lobe, its direction being backwards and outwards, and then inwards. On its floor is seen a longitudinal eminence, which corresponds with a deep sulcus between two convolutions: this is called the *hippocampus minor*. Between the middle and posterior horns, a smooth eminence is observed, which varies considerably in size in different subjects. It is called the *eminentia collateralis*.

The *Corpus Striatum* (superior ganglion of the cerebrum), has received its name from the striated appearance which its section presents, from white fibres diverging through its substance. The intra-ventricular portion is a large pear-shaped mass, of a grey colour externally; its broad extremity is directed forwards, into the fore-part of the body, and anterior cornu of the lateral ventricle; its narrow end is directed outwards and backwards, being separated from its fellow by the thalami optici; it is covered by the serous lining of the cavity, and crossed by some veins of considerable size. The extra-ventricular portion is imbedded in the white substance of the hemisphere.

The *Taenia Semicircularis* is a narrow, whitish, semi-transparent band, of medullary substance, situated in the depression between the corpus striatum and thalamus opticus. Anteriorly, it descends in connexion with the anterior pillar of the fornix; behind, it is continued into the descending horn of the ventricle, where it becomes lost. Its surface, especially at its fore-part, is transparent, and dense in structure, and was called by Tarinus the *horny band*. It consists of longitudinal white fibres, the deepest of which run between the corpus striatum and thalamus opticus. Beneath it is a large vein (*vena corporis striati*), which receives numerous smaller veins from the surface of the corpus striatum, and thalamus opticus, and terminates in the venae Galeni.

The *Choroid Plexus* is a highly vascular, fringe-like membrane, occupying the margin of the fold of pia mater (*velum interpositum*), in the interior of the brain. It extends, in a curved direction, across the floor of the lateral ventricle. In front, where it is small and tapering, it communicates with the choroid plexus of the opposite side, through a large oval aperture, the foramen of Monro. Posteriorly, it descends into the middle horn of the lateral ventricle, where it joins with the pia mater through the transverse fissure. In structure, it consists of minute, and highly vascular villous processes, the villi being covered by a single layer of epithelium, composed of large, round corpuscles, containing, besides a central nucleus, a bright yellow spot. The arteries of the choroid plexus enter the ventricle at the descending cornu, and, after ramifying through its substance, send branches into the substance of the brain. The veins of the choroid plexuses terminate in the venae Galeni.

The *Corpus Fimbriatum*, or *Taenia Hippocampi*, is a narrow, white, tape-like band, situated immediately behind the choroid plexus. It is the lateral edge of the posterior pillar of the fornix, and is attached along the inner border of the hippocampus major as it descends into the middle horn of the lateral ventricle. It may be traced as far as the pes hippocampi.

The *Thalami Optici* and *Fornix* will be described when more completely exposed, in a later stage of the dissection of the brain.

The middle cornu should now be exposed, throughout its entire extent, by introducing the little finger gently into it, and cutting through the hemisphere, between it and the surface, in the direction of the cavity.

The *Middle*, or *Descending Cornu*, the largest of the three, traverses the middle lobe of the brain, forming in its course a remarkable curve round the back of the optic thalamus. It passes, at first, backwards, outwards, and downwards, and then curves around the crus cerebri, forwards and inwards, nearly to the point of the middle lobe, close to the fissure of Sylvius. Its superior boundary is formed by the medullary substance of the middle lobe, and the under surface of the thalamus opticus. Its inferior boundary presents for examination the following parts: The hippocampus major, pes hippocampi, pes accessorius, corpus fimbriatum, choroid plexus, fascia dentata, transverse fissure.

The *Hippocampus Major*, or *Cornu Ammonis*, so called from its resemblance to a ram's horn, is a white eminence, of a curved elongate form, extending along the entire length of the floor of the middle horn of the lateral ventricle. At its lower extremity it becomes enlarged, and presents a number of rounded elevations with intervening depressions, which, from presenting some resemblance to the claw of an animal, is called the *pes hippocampi*. If a transverse section is made through the hippocampus major, it will be seen that this eminence is the inner surface of the convolution of the corpus callosum, doubled upon itself like a horn, the white convex portion projecting into the cavity of the ventricle; the grey portion being on the surface of the cerebrum, the edge of which, slightly indented, forms the fascia dentata. The white matter of the hippocampus major is continuous through the corpus fimbriatum, with the fornix and corpus callosum.

The *Pes Accessorius*, or *Eminentia Collateralis*, has been already mentioned, as a white eminence, varying in size, placed between the hippocampus major and minor, at the junction of the posterior with the descending cornu. Like the hippocampi, it is formed by white matter corresponding to one of the sulci, between two convolutions protruding into the cavity of the ventricle.

The *Corpus Fimbriatum* (*Taenia Hippocampi*), is a narrow, tape-like band, attached along the inner concave border of the hippocampus major, and reaching down as far as the pes hippocampi. It is a continuation of the posterior pillar of the fornix, prolonged from the central cavity of the lateral ventricle.

Fascia Dentata. On separating the inner border of the corpus fimbriatum from the choroid plexus, and raising the edge of the former, a serrated band of grey substance, the edge of the grey substance of the middle lobe, will be seen beneath it: this is the fascia dentata. Correctly speaking, it is placed external to the cavity of the descending cornu.

The *Transverse Fissure* is seen on separating the corpus fimbriatum from the thalamus opticus. It is situated beneath the fornix, extending from the middle line behind, downwards on either side, to the end of the descending cornu, being bounded

245.—The Fornix, Velum Interpositum, and Middle or Descending Cornu of the Lateral Ventricle.

on one side by the fornix and the hemisphere, and on the other by the thalamus opticus. Through this fissure the pia mater passes from the exterior of the brain into the ventricles, to form the choroid plexuses. Where the pia mater projects into the lateral ventricle, beneath the edge of the fornix, it is covered by a prolongation of the lining membrane, which excludes it from the cavity.

The *Septum Lucidum* forms the internal boundary of the lateral ventricle. It is a thin, semi-transparent septum, attached, above, to the under surface of the corpus callosum; below, to the anterior part of the fornix; and, in front of this, to the prolonged portion of the corpus callosum. It is triangular in form, broad in front, and narrow behind, its surfaces looking towards the cavities of the ventricles. The septum consists of two laminae, separated by a narrow interval, the fifth ventricle.

Each lamina consists of an internal layer of white substance, covered by the lining membrane of the fifth ventricle; and an outer layer of grey matter, covered by the lining membrane of the lateral ventricle. The cavity of the ventricle is lined by a serous membrane, covered with epithelium, and contains fluid. In the foetus, and in some animals, this cavity communicates, below, with the third ventricle; but in the adult, it forms a separate cavity. In cases of serous effusion into the ventricles, the septum is often found softened and partially broken down.

The fifth ventricle may be exposed by cutting through the septum, and attached portion of the corpus callosum, with the scissors; after examining which, the corpus callosum should be cut across, towards its anterior part, and the two portions carefully dissected, the one forwards, the other backwards, when the fornix will be exposed.

The *Fornix* is a longitudinal lamella, of fibrous matter, situated beneath the corpus callosum, with which it is continuous behind, but separated from it in front by the septum lucidum. It may be divided along the middle line into two symmetrical halves, one for either hemisphere. These two portions are joined together in the middle line, where they form the body, but are separated from one another in front and behind; in front, forming the anterior crura, and behind, the posterior crura.

The *body* of the fornix is triangular in form; narrow in front, broad behind. Its upper surface is connected, in the median line, to the septum lucidum in front, and the corpus callosum behind. Its under surface rests upon the velum interpositum, which separates it from the third ventricle, and the inner portion of the optic thalami. Its lateral edges form, on each side, part of the floor of the lateral ventricles, and are in contact with the choroid plexuses.

The *anterior crura* arch downwards towards the base of the brain, separated from each other by a narrow interval. They are composed of white fibres, which descend through a quantity of grey matter in the lateral walls of the third ventricle, and are placed immediately behind the anterior commissure. At the base of the brain, the white fibres of each crus form a sudden curve upon themselves, spread out and form the outer part of the corresponding corpus albicans, from which point they may be traced upwards into the substance of the corresponding thalamus opticus. The anterior crura of the fornix are connected in their course with the optic commissure, the white fibres covering the optic thalamus, the peduncle of the pineal gland, and the superficial fibres of the taenia semicircularis.

The *posterior crura*, at their commencement, are intimately connected by their upper surfaces with the corpus callosum; diverging from one another, they pass downwards into the descending horn of the lateral ventricle, being continuous with the concave border of the hippocampus major. The lateral thin edges of the posterior crura have received the name *corpus fimbriatum*, already described. On the under surface of the fornix, towards its posterior part, between the diverging posterior crura, may be seen some transverse lines, and others longitudinal or oblique. This appearance has been termed the *lyra*, from the fancied resemblance it bears to the strings of a harp.

Between the anterior pillars of the fornix and the anterior extremities of the thalami optici, an oval aperture is seen on each side, the foramen of Monro. The two openings descend towards the middle line, and joining together, lead into the upper part of the third ventricle. These openings form a transverse communication between the lateral ventricles, and below with the third ventricle.

Divide the fornix across anteriorly, and reflect the two portions, the one forwards, the other backwards, when the velum interpositum will be exposed.

The *Velum Interpositum* is a vascular membrane, reflected from the pia mater into the interior of the brain through the transverse fissure, passing beneath the posterior rounded border of the corpus callosum and fornix, and above the corpora quadrigemina, pineal gland, and optic thalami. It is of a triangular form, and separates the under surface of the body of the fornix from the cavity of the third ventricle. Its posterior border forms an almost complete investment for the pineal gland. Its anterior extremity, or apex, is bifid; each bifurcation being continued into the corresponding lateral ventricle, behind the anterior crura of the fornix, forming the anterior extremity of the choroid plexus. On its under surface are two vascular fringes, which diverge from each other behind, and project into the cavity of the third ventricle. These are the choroid plexuses of the third ventricle. To its lateral margins are connected the choroid plexuses of the lateral ventricles. The arteries of the velum interpositum enter from behind, beneath the corpus callosum. Its veins, the venae Galeni, two in number, run along its under surface; they are formed by the venae corporis striati and the venae plexus choroides: the venae Galeni unite posteriorly into a single trunk, which terminates in the straight sinus.

The velum interpositum should now be removed. This must be effected carefully, especially at its posterior part, where it invests the pineal gland; the thalami optici will then be exposed with the cavity of the third ventricle between them (fig. 246).

The *Thalami Optici* (*Superior Ganglia of the Cerebrum*) are two large oblong masses, placed between the diverging portions of the corpora striata; they are of a white colour superficially, internally they are composed of white fibres intermixed with grey matter. Each thalamus rests upon its corresponding crus cerebri, which it embraces. *Externally*, it is bounded by the corpus striatum and taenia semicircularis, and is continuous with the hemisphere. *Internally*, it forms the lateral boundary of the third ventricle; and running along its upper border is seen the peduncle of the pineal gland. Its *upper surface* is free, being partly seen in the lateral ventricle; it is partly covered by the fornix, and marked in front by an eminence, the anterior tubercle. Its *under surface* forms the roof of the descending cornu of the lateral ventricle; into it the crus cerebri passes. Its *posterior and inferior part*, which projects into the descending horn of the lateral ventricle, presents two small round eminences, the internal and external geniculate bodies. Its *anterior extremity*, which is narrow, forms the posterior boundary of the foramen of Monro.

The *Third Ventricle* is the narrow oblong fissure placed between the thalami optici, and extending to the base of the brain. It is bounded above by the under surface of the velum interpositum, from which are suspended the choroid plexuses of the third ventricle, and laterally by two white tracts, one on either side, the peduncles of the pineal gland. Its floor, somewhat oblique in its direction, is formed, from before backwards, by the parts which close the interpeduncular space, viz., the lamina cinerea, the tuber cinereum and infundibulum, the corpora albicantia, and the locus perforatus; its sides, by the optic thalami; in front, by the anterior crura of the fornix and part of the anterior commissure; behind, by the posterior commissure and the *iter a tertio ad quartum ventriculum*.

The cavity of the third ventricle is crossed by three commissures, named, from their position, *anterior*, *middle*, and *posterior*.

246.—The Third and Fourth Ventricles.

The *Anterior Commissure* is a rounded cord of white fibres, placed in front of the anterior crura of the fornix. It perforates the corpus striatum on either side, and spreads out into the substance of the hemispheres, over the roof of the descending horn of the lateral ventricle.

The *Middle* or *Soft Commissure* consists almost entirely of grey matter. It connects together the thalami optici, and is continuous with the grey matter lining the anterior part of the third ventricle.

The *Posterior Commissure*, smaller than the anterior, is a flattened white band of fibres, connecting together the two thalami optici posteriorly. It bounds the third ventricle posteriorly, and is placed in front of and beneath the pineal gland, above the opening leading to the fourth ventricle.

The third ventricle has four openings connected with it. In front are two oval apertures, one on either side, the foramina of Monro, through which the third communicates with the lateral ventricles. Behind, is a third opening leading into the fourth ventricle by a canal, the aqueduct of Sylvius, or *iter a tertio ad quartum ventriculum*. The fourth, situated in the anterior part of the floor of the ventricle, is a deep pit, which leads downwards to the funnel-shaped cavity of the infundibulum (*iter ad infundibulum*).

The lining membrane of the lateral ventricles is continued through the foramina of Monro into the third ventricle, and extends along the *iter a tertio* into the fourth ventricle; at the bottom of the *iter ad infundibulum* it ends in a *cul-de-sac*.

Grey Matter of the Third Ventricle. A layer of grey matter covers the greater part of the surface of the third ventricle. In the floor of this cavity it exists in great abundance, and is prolonged upwards on the sides of the thalami, extending across the cavity as the soft commissure; below, it enters into the corpora albicantia, surrounds in part the anterior pillars of the fornix, and ascends on the sides of the septum lucidum.

Behind the third ventricle, and in front of the cerebellum, are the corpora quadrigemina, and resting upon these the pineal gland.

The *Pineal Gland* (*Conarium*), so named from its peculiar shape (*pinus*, the fruit of the fir), is a small reddish-grey body, conical in form, placed immediately behind the posterior commissure, and between the nates, upon which it rests. It is retained in its position by a duplicature of pia mater, derived from the under surface of the velum interpositum, which almost completely invests it. The pineal gland is about four lines in length, and from two to three in width at its base, and is said to be larger in the child than in the adult, and in the female than in the male. Its base is connected with the cerebrum by some transverse commissural fibres derived from the posterior commissure, and by four slender peduncles, formed of medullary fibres. Of these, the two superior pass forwards upon the upper and inner margin of the optic thalami, to the anterior crura of the fornix, with which they become blended. The inferior peduncles pass vertically downwards from the base of the pineal body, along the back part of the inner surface of the thalami, and are only seen on a longitudinal vertical section through the gland. The pineal gland is very vascular, and consists chiefly of grey matter, with a few medullary fibres. In its base is a small cavity, said by some to communicate with that of the third ventricle. It contains a transparent viscid fluid, and occasionally a quantity of sabulous matter, named *acervulus cerebri*, composed of phosphate and carbonate of lime, phosphate of magnesia and ammonia, with a little animal matter. These concretions are almost constant in their existence, and are found at all periods of life. When this body is solid, the sabulous matter is found upon its surface, and occasionally upon its peduncles.

On the removal of the pineal body and adjacent portion of pia mater, the corpora quadrigemina are exposed.

The *Corpora* or *Tubercula Quadrigemina* (*optic lobes*) are four rounded eminences placed in pairs, two in front, two behind, and separated from another by a crucial depression. They are situated immediately behind the third ventricle and posterior commissure, beneath the posterior border of the corpus callosum, and above the *iter a tertio ad quartum ventriculum*. The anterior pair, the nates, are the larger, oblong from before backwards, and of a grey colour. The posterior pair, the testes, are hemispherical in form, and lighter in colour than the preceding. They are connected on each side with the thalamus opticus and commencement of the optic tracts, by means of two white prominent bands, termed *brachia*. Those connecting the nates with the thalamus (*brachia anteriora*) are the larger, and pass obliquely outwards. Those connecting the testes with the thalamus, are called the *brachia posteriora*. Both pairs, in the adult, are quite solid, being composed of white matter externally, and grey matter within. These bodies are larger in the lower animals than in man. In fishes, reptiles, and birds, they are only two in number, and called the *optic lobes*, from their connection with the optic nerves; and are hollow in their interior; but in mammalia they are four in number, as in man, and quite solid. In the human foetus they are developed at a very early period, and form a large proportion of the cerebral mass; at first they are only two in number, as in the lower mammalia, and hollow in their interior.

These bodies, from below, receive white fibres from the olivary fasciculus or fillet; they are also connected with the cerebellum, by means of a large white cord on each side, the *processus ad testes*, or superior peduncles of the cerebellum, from the corpora quadrigemina, these tracts pass upwards to the thalami.

The *Valve of Vieussens* is a thin translucent lamina of medullary substance, stretched between the two *processus e cerebello ad testes*; it covers in the canal leading from the third to the fourth ventricle, forming part of the roof of the latter cavity. It is narrow in front, where it is connected with the testes; and broader behind, at its connection with the vermiform process of the cerebellum. A slight elevated ridge (the frenulum) descends upon the upper part of the valve from the corpora quadrigemina, and on either side of it may be seen the fibres of origin of the fourth nerve. Its lower half is covered by a thin transversely grooved lobule of grey matter prolonged from the anterior border of the cerebellum; this is called the *linguetta laminosa*.

The *Corpora Geniculata* are two small flattened oblong masses, placed on the outer side of the corpora quadrigemina, and on the under and back part of each optic thalamus, and are named from their position, *corpus geniculatum externum* and *internum*. They are placed one on the outer and one on the inner side of each optic tract. In this situation, the optic tract may be seen dividing into two bands, one of which is connected with the external geniculate body and nates, the other being connected with the internal geniculate body and testis.

Structure of the Cerebrum. The white matter of each hemisphere consists of three kinds of fibres. 1. Diverging or peduncular fibres, which connect the hemisphere with the cord and medulla oblongata. 2. Transverse commissural fibres which connect together the two hemispheres. 3. Longitudinal commissural fibres, which connect distant parts of the same hemisphere.

The *diverging* or *peduncular fibres* consist of a main body and of certain accessory fibres. The main body originate in the columns of the cord and medulla oblongata, and enter the cerebrum through the crus cerebri, where they are arranged in two bundles, separated by the locus niger. Those fibres which form the inferior or fasciculated portion of the crus are derived from the pyramid, and ascending, pass mainly through the centre of the striated body; those on the opposite surface of the crus, which form the tegmentum, are derived from the posterior pyramid and fasciculi teretes; ascending, they pass, some through the under part of the thalamus, and others through both thalamus and corpus striatum, decussating in these bodies with each other and with the fibres of the corpus callosum. The optic thalami also receive accessory fibres from the *processus ad testes*, the olivary fasciculus, the corpora quadrigemina, and corpora geniculata. Some of the diverging fibres end in the cerebral ganglia, whilst others pass through and receive additional fibres from them, and as they emerge, radiate into the anterior, middle, and posterior lobes of the hemisphere, decussating again with the fibres of the corpus callosum, before passing to the convolutions.

The *transverse commissural fibres* connect together the two hemispheres across the middle line. They are formed by the corpus callosum and the anterior and posterior commissures.

The *longitudinal commissural fibres* connect together distant parts of the same hemisphere, the fibres being disposed in a longitudinal direction. They form the fornix, the taenia semicircularis, and peduncles of the pineal gland, the striae longitudinales, the fibres of the gyrus fornicatus, and the fasciculus uncinatus.

The Cerebellum

The cerebellum or little brain, is that portion of the encephalon which is contained in the inferior occipital fossae. It is situated beneath the posterior lobes of the cerebrum, from which it is separated by the tentorium. Its average weight in the male is 5 oz. 4 drs. It attains its maximum weight between the twenty-fifth and fortieth years; its increase in weight after the fourteenth year being relatively greater in the female than in the male. The proportion between the cerebellum and cerebrum is, in the male, as 1 to $8^{4}/_{7}$, and in the female, as 1 to $8\frac{1}{4}$. In the infant, it is proportionally much smaller than in the adult, the relation between them being, according to Chaussier, between 1 to 13, and 1 to 26; by Cruvelhier it was found to be 1 to 20. In form the cerebellum is oblong, flattened from above downwards, its greatest diameter being from side to side. It measures from three and a half to four inches transversely, from two to two and a half inches from before backwards, being about two inches thick in the centre, and about six lines at its circumference, the thinnest part. It consists of grey and white matter, the former, darker than that of the cerebrum, occupies the surface; the latter, the interior. The surface of the cerebellum is not convoluted

247.—Upper Surface of the Cerebellum.

like the cerebrum, but traversed by numerous curved furrows or sulci, which vary in depth at different parts, and correspond to the intervals between the laminae of which its exterior is composed.

Its *upper surface* (fig. 247) is somewhat elevated in the median line, and depressed towards its circumference; it consists of two lateral hemispheres, connected together by an elevated median portion or lobe, the superior vermiform process. The median lobe is the fundamental part, and in some animals, as fishes and reptiles, the only part which exists, the hemispheres being additions, and attaining their maximum in man. The hemispheres are separated in front by a deep notch, the *incisura cerebelli anterior*, which encircles the corpora quadrigemina behind; they are also separated by a similar notch behind, the *incisura cerebelli posterior*, in which is received the upper part of the falx cerebelli. The superior vermiform process (upper part of the median lobe of the cerebellum), extends from the notch on the anterior to that on the posterior border. It is divided into three lobes; the *lobulus centralis*, a small lobe, situated in the incisura anterior; the *monticulus cerebelli*, the central projecting part of the process; and the *commissura simplex*, a small lobe near the incisura posterior.

The *under surface* of the cerebellum (fig. 248) is subdivided into two well marked convex lateral hemispheres by a depression, the valley, which extends from before backwards in the middle line. The lateral hemispheres are lodged in the inferior occipital fossae; the median depression, or valley, receives the back part of the medulla oblongata, is broader in the centre than at either extremity, and has, projecting from its floor, part of the median lobe of the cerebellum, called the *inferior vermiform process*. The parts entering into the composition of this body are, from behind forwards, the *commissura brevis*, situated in the incisura posterior; in front of this, a laminated conical projection, the *pyramid*; more anterior, a larger eminence, the *uvula*; placed between the two rounded lobes which occupy the sides of the valley, the *amygdalae*; and connected with them by a commissure of grey matter, indented on the surface, and called the *furrowed band*. In front of the uvula is the *nodulus*; it is the

248.—Under Surface of the Cerebellum.

anterior pointed termination of the inferior vermiform process, and projects into the cavity of the fourth ventricle; it has been named by Malacarne the *laminated tubercle*. On each side of the nodule is a thin layer of white substance, attached externally to the flocculus, and internally to the nodule, and to a corresponding part on the opposite side; they form together the *posterior medullary velum*, or commissure of the flocculus. They are usually covered in and concealed by the amygdalae, and cannot be seen until these are drawn aside. This band is of a semilunar form on each side, its anterior margin being free and concave, its posterior being attached just in front of the furrowed band. Between it and the nodulus and uvula behind, is a deep fossa, called the *swallow's nest* (*nidus hirundinis*).

Lobes of the Cerebellum. Each hemisphere is divided into an upper and a lower portion by the great horizontal fissure, which commences in front at the pons, and passes horizontally round the free margin of either hemisphere, backwards to the middle line. From this primary fissure numerous secondary fissures proceed, which separate the cerebellum into lobes.

Upon the upper surface of either hemisphere there are two lobes, separated from each other by a fissure. These are the anterior or square lobe, which extends as far back as the posterior edge of the vermiform process, and the posterior or semi-lunar lobe, which passes from the termination of the preceding to the great horizontal fissure.

Upon the under surface of either hemisphere there are five lobes, separated by sulci; these are from before backwards; the *flocculus* or *sub-peduncular lobe*, a prominent tuft, situated behind and below the middle peduncle of the cerebellum; its surface is composed of grey matter, subdivided into a few small laminae: it is sometimes called the *pneumogastric lobule*, from being situated behind the pneumogastric nerve. The *amygdala* or *tonsil* is situated on either side of the great median fissure or valley, and projects into the fourth ventricle. The *digastric lobe* is situated on the outside of the tonsil, being connected in part with the pyramid. Behind the digastric is the *slender lobe*, which is connected with the back part of the pyramid and the commissura brevis: and most posteriorly is the *inferior posterior lobe*, which also joins the commissura brevis in the valley.

Fourth Ventricle

The fourth ventricle, or ventricle of the cerebellum, is the space between the posterior surface of the medulla oblongata and pons in front, and the cerebellum behind. It is lozenge-shaped, being contracted above and below, and broadest across its central part. It is bounded laterally by the *processus e cerebello ad testes* above, and by the diverging posterior pyramids and restiform bodies below.

The *roof* is arched; it is formed by the valve of Vieussens and the under surface of the cerebellum, which presents in this situation four small eminences or lobules, two occupying the median line, the nodulus and uvula, the remaining two, the amygdala, being placed on either side of the uvula.

The anterior boundary, or *floor*, is formed by the posterior surface of the medulla oblongata and pons. In the median line is seen the posterior median fissure; it becomes gradually obliterated above, and terminates below in the point of the calamus scriptorius, formed by the convergence of the posterior pyramids. At this point is the orifice of a short canal terminating in a *cul-de-sac*, the remains of the canal which extends in foetal life through the centre of the cord. On each side of the median fissure are two slightly convex longitudinal eminences, the fasciculi teretes; they extend the entire length of the floor, being indistinct below and of a greyish colour, but well marked and whitish above. Each eminence consists of fibres derived from the lateral tract and restiform body, which ascend to the cerebrum. Opposite the crus cerebelli, on the outer side of the fasciculi teretes, is a small eminence of dark grey substance, which presents a blueish tint through the thin stratum covering it; this is called the *locus coeruleus*; and a thin streak of the same colour continued up from this on either side of the fasciculi teretes, as far as the top of the ventricle, is called the *taenia violacea*. The lower part of the floor of the ventricle is crossed by several white transverse lines, *lineae transversae*; they emerge from the posterior median fissure; some enter the crus cerebelli, others enter the roots of origin of the auditory nerve, whilst some pass upwards and outwards on the floor of the ventricle.

249.—Vertical Section of the Cerebellum.

The *Lining Membrane* of the fourth ventricle is continuous with that of the third, through the aqueduct of Sylvius, and its cavity communicates below with the sub-arachnoid space of the brain and cord through an aperture in the layer of pia mater extending between the cerebellum and medulla oblongata. Laterally, this membrane is reflected outwards a short distance between the cerebellum and medulla.

The *Choroid Plexuses* of the fourth ventricle are two in number; they are delicate vascular fringes, which project into the ventricle on each side, passing from the point of the inferior vermiform process to the outer margin of the restiform bodies.

The *Grey Matter* in the floor of the ventricle consists of a tolerably thick stratum, continuous below with the grey commissure of the cord, and extending up as high as the aqueduct of Sylvius, besides some special deposits connected with the roots of origin of certain nerves. In the upper half of the ventricle is a projection situated over the nucleus, from which the sixth and facial nerves take a common origin. In the lower half are three eminences on each side for the roots of origin of the eighth and ninth nerves.

Structure. If a vertical section is made through either hemisphere of the cerebellum, midway between its centre and the superior vermiform process; it will be found to consist of a central stem of white matter, which contains in its interior a dentate body. From the surface of each hemisphere, a series of plates of medullary matter are detached, which, covered with grey matter, form the laminae; and from its anterior part arise three large processes or peduncles, superior, middle, and inferior, by which it is connected with the rest of the encephalon.

The *Laminae* are about ten or twelve in number, including those on both surfaces of the organ, those in front being detached at a right angle, and those behind at an acute angle; as each lamina proceeds outwards, other secondary laminae are detached from it, and from these tertiary laminae. The arrangement thus described gives to the cut surface of the organ a foliated appearance, to which the name *arbor vitae* has been given. Each lamina consists of white matter, covered externally by a layer of grey substance.

The white matter of each lamina is derived partly from the central stem; in addition to which white fibres pass from one lamina to another.

The grey matter resembles somewhat the cortical substance of the convolutions, consisting of two layers, the external one, soft and of a greyish colour, the internal one, firmer and of a rust colour.

The *Corpus Dentatum*, or *Ganglion of the Cerebellum*, is situated a little to the inner side of the centre of the stem of white matter. It consists of an open bag or capsule of grey matter, the section of which presents a grey dentated outline, being open at its anterior part. It is surrounded by white fibres; white fibres are also contained in its interior, which issue from it to join the superior peduncles.

The *Peduncles* of the cerebellum, superior, middle, and inferior, serve to connect it with the rest of the encephalon.

The *Superior Peduncles* (*Processus e Cerebello ad Testes*) connect the cerebellum with the cerebrum; they pass forwards and upwards to the testes, beneath which they ascend to the crura cerebri and optic thalami, forming part of the diverging cerebral fibres: each peduncle forms part of the lateral boundary of the fourth ventricle, and is connected with its fellow of the opposite side by the valve of Vieussens. Behind, it is continuous with the folia of the inferior vermiform process, and with the white fibres in the interior of the corpus dentatum. Beneath the corpora quadrigemina, the innermost fibres of each peduncle decussate with each other, so that some fibres from the right half of the cerebellum are continued to the left half of the cerebrum.

The *Inferior Peduncles* (*Processus ad Medullam*), connect the cerebellum with the medulla oblongata. They pass downwards, to the back part of the medulla, and form part of the restiform bodies. Above, the fibres of each process are connected chiefly with the laminae, on the upper surface of the cerebellum; and below, they are connected with all three tracts of the half of the medulla, and, through these, with the corresponding half of the cord, excepting the posterior median columns.

The *Middle Peduncles* (*Processus ad Pontem*), the largest of the three, connect together the two hemispheres of the cerebellum, forming their great transverse commissure. They consist of a mass of curved fibres, which arise in the lateral parts of the cerebellum, and pass across to the same points on the opposite side. They form the transverse fibres of the pons Varolii.

CRANIAL NERVES

The Cranial Nerves, nine in number on each side, include all those which arise from some part of the cerebro-spinal centre, and are transmitted through foramina in the base of the cranium. They have been named numerically, according to the order in which they pass out of this cavity. Their names are also derived from the part to which each is distributed, or from the special function appropriated to each. Taken in their order, from before backwards, they are as follows:

1st. Olfactory.
2nd. Optic.
3rd. Motores oculorum.
4th. Pathetic.
5th. Trifacial, Trigemini.
6th. Abducentes.
7th. {Facial (Portio dura). / Auditory (Portio mollis).}
8th. {Glosso-pharyngeal. / Pneumogastric, or Par vagum. / Spinal accessory.}
9th. Hypo-glossal.

The cranial nerves may be subdivided into three groups, according to the peculiar function possessed by each, viz., nerves of special sense; nerves of motion; and compound nerves, that is, the function of which is both motor and sensitive. These groups may be thus arranged:

Nerves of Special Sense.
1st. Olfactory.
2nd. Optic.
7th. Auditory (Portio mollis).

Nerves of Motion.
3rd. Motores oculorum.
4th. Pathetic.
6th. Abducentes.
7th. Facial (Portio dura).
9th. Hypo-glossal.

Compound Nerves.
5th. Trifacial.
8th. {Glosso-pharyngeal. / Pneumogastric. / Spinal accessory.}

All the cranial nerves are connected to some part of the surface of the brain. This is termed their *superficial*, or *apparent origin*. But the fibres may, in all cases, be traced deeply into the substance of the organ. This would form their *deep*, or *real origin*.

Nerves of Special Sense

The First, or Olfactory Nerve, the special nerve of the sense of smell, may be regarded as a portion of the cerebral substance, pushed forward in direct relation with the organ to which it is distributed. It arises by three roots.

The *external, or long root*, is a narrow, white, medullary band, which passes outwards across the fissure of Sylvius, into the substance of the middle lobe of the cerebrum. Its deep origin may be traced to the corpus striatum*, the superficial fibres of the optic thalamus†, the anterior commissure‡, and the convolutions of the island of Reil.

The *middle*, or *grey root*, arises from a papilla of grey matter (caruncula mammillaris), imbedded in the anterior lobe. This root is prolonged into the nerve from the adjacent part of the brain, and contains white fibres in its interior, which are connected with the corpus striatum.

The *internal*, or *short root*, is composed of white fibres, which arise from the inner and back part of the anterior lobe, being connected, according to Foville, with the longitudinal fibres of the gyrus fornicatus.

These three roots unite and form a flat band, narrower in the middle than at either extremity, and its section of a somewhat prismoid form. It is of soft texture, and contains a considerable amount of grey matter in its substance. As it passes forwards, it is contained in a deep sulcus, between two convolutions, lying on the under surface of the anterior lobe, on either side of the longitudinal fissure, and is retained in position by the arachnoid membrane which covers it. On reaching the cribriform plate of the ethmoid bone, it expands into an oblong mass of greyish-white substance, the olfactory bulb. From the under part of this bulb are given off numerous filaments, about twenty in number, which pass through the cribriform foramina, and are distributed to the mucous membrane of the nose. Each filament is surrounded by a tubular prolongation from the dura mater, and pia mater, the former being lost on the periosteum lining the nose; the latter, in the neurilemma of the nerve. The filaments, as they enter the nares, are divisible into three groups, an inner group, larger than those on the outer wall, spread out over the upper third of the septum; a middle set, confined to the roof of the nose; and an outer set, which are distributed over the superior and middle turbinated bones, and the surface of the ethmoid in front of them. As the filaments descend, they unite in a plexiform network, and become gradually lost in the lining membrane. Their mode of termination is unknown.

The olfactory differs in structure from other nerves, in containing grey matter in its interior, being soft and pulpy in structure, and destitute of neurilemma. Its filaments are deficient in the white substance of Schwann, are not divisible into fibrillae, and resemble the gelatinous fibres in being nucleated, and of a finely-granular texture.

250.—The Optic Nerves and Optic Tracts.

Optic Nerve

The SECOND, or OPTIC NERVE, the special nerve of the sense of sight, is distributed exclusively to the eyeball. The nerves of opposite sides are connected together at the commissure; and from the back of the commissure they may be traced to the brain, under the name of the optic tracts.

---

* Vieussens, Winslow, Monro, Mayo.   † Valentin.   ‡ Cruvelhier.

The *optic tract*, at its connection with the brain, divides into two bands which are continued into the optic thalami, the corpora geniculata, and the corpora quadrigemina. The fibres of origin from the thalamus may be traced partly from its surface, and partly from its interior. From this origin, the tract winds obliquely across the under surface of the crus cerebri, in the form of a flattened band, destitute of neurilemma, and is attached to it by its anterior margin. It now assumes cylindrical form, and, as it passes forwards, is connected with the tuber cinereum, and lamina cinerea, from both of which it receives fibres. According to Foville, it is also connected with the taenia semicircularis, and the anterior termination of the gyrus fornicatus. It finally joins with the nerve of the opposite side to form the optic commissure.

251.—Course of the Fibres in the Optic Commissure.

The *commissure*, somewhat quadrilateral in form, rests upon the olivary process of the sphenoid bone, being bounded, in front, by the lamina cinerea; behind, by the tuber cinereum; on either side, by the substantia perforata antica; within the commissure, the optic nerves of the two sides undergo a partial decussation. The fibres which form the inner margin of each tract, are continued across from one to the other side of the brain, and have no connection with the optic nerves. These may be regarded as commissural fibres between the thalami of opposite sides. Some fibres are continued across the anterior border of the chiasma, and connect the optic nerves of the two sides, having no relation with the optic tracts. They may be regarded as commissural fibres between the two retinas. The outer fibres of each tract are continued into the optic nerve of the same side. The central fibres of each tract are continued into the optic nerve of the opposite side, decussating in the commissure with similar fibres of the opposite tract.

The *optic nerves* arise from the fore part of the commissure, and, diverging from one another, become rounded in form, firm in texture, and are enclosed in a sheath derived from the arachnoid. As each nerve passes through the corresponding optic foramen, it receives a sheath from the dura mater; and as it enters the orbit, this sheath subdivides into two layers, one of which becomes continuous with the periosteum of the orbit; the other forms a sheath for the nerve, and becomes lost in the sclerotic. The nerve passes through the cavity of the orbit, pierces the sclerotic and choroid coats at the back part of the eyeball, a little to the nasal side of its centre, and expands into the retina. A small artery, the arteria centralis retinae, perforates the optic nerve a little behind the globe, and runs along its interior in a tubular canal of fibrous tissue. It supplies the internal surface of the retina, and is accompanied by corresponding veins.

Auditory Nerve

The AUDITORY NERVE (portio mollis of the seventh pair), is the special nerve of the sense of hearing, being distributed exclusively to the internal ear. The hard portion of the seventh pair (portio dura), or facial nerve, is the motor nerve of the face. It will be described with the motor cranial nerves.

The auditory nerve arises from numerous white striae, the lineae transversae, which emerge from the posterior median fissure in the anterior wall, or floor, of the fourth ventricle. It is also connected with the grey matter of the medulla, which corresponds to the locus caeruleus. According to Foville, the roots of this nerve are connected, on the under surface of the middle peduncle, with the grey substance of the cerebellum, with the flocculus, and with the grey matter at the borders of the calamus scriptorius. The nerve winds round the restiform body, from which it receives fibres, and passes forwards across the posterior border of the crus cerebelli, in company with the facial nerve, from which it is partially separated by a small artery. It then enters the meatus auditorius, in company with the facial nerve, and, at the bottom of the meatus, divides into two branches, cochlear and vestibular. The auditory nerve is very soft in texture, (hence the name, *portio mollis*), destitute of neurilemma, and within the meatus, receives one or two filaments from the facial.

THE MOTOR CRANIAL NERVES

The THIRD NERVE (*Motor Oculi*), is the chief motor nerve of the muscles of the eyeball. It is a rather large nerve, of rounded form and firm texture, having its apparent origin from the inner surface of the crus cerebri, immediately in front of the pons Varolii.

The *deep origin* may be traced into the substance of the crus, where some of its fibres are connected with the locus niger; others run downwards, among the longitudinal fibres of the pons; whilst others ascend, to be connected with the tubercula quadrigemina, and valve of Vieussens. According to Stilling, the fibres of the nerve pierce the peduncle and locus niger, and arise from a grey nucleus in the floor of the aqueduct of Sylvius. On emerging from the brain, it is invested in a sheath of pia mater, and enclosed in a prolongation from the arachnoid. It then pierces the dura mater on the outer side of the anterior clinoid process, where its serous covering is reflected from it, and passes along the outer wall of the cavernous sinus, above the other orbital nerves, receiving in its course one or two filaments from the cavernous plexus of the sympathetic. It then divides into two branches, which enter the orbit through the sphenoidal fissure, between the two heads of

252.—Nerves of the Orbit. Seen from above.

the External rectus muscle. On passing through this fissure, this nerve is placed below the fourth, and the frontal and lachrymal branches of the ophthalmic nerve.

The *superior division*, the smaller, passes inwards across the optic nerve, and supplies the Superior rectus and Levator palpebrae.

The *inferior division*, the larger, divides into three branches. One passes beneath the optic nerve to the Internal rectus; another to the Inferior rectus; and the third, the largest of the three, passes forwards between the Inferior and External recti, to the Inferior oblique. From the latter, a short, thick branch is given off to the lower part of the lenticular ganglion, forming its inferior root, as well as two filaments to the Inferior rectus. All these branches enter the muscles on their ocular surface.

Fourth Nerve

The FOURTH, or trochlear nerve, is the smallest of the cranial nerves. It arises from the upper part of the valve of Vieussens, immediately behind the testis, and divides beneath the corpora quadrigemina, into two fasciculi; the anterior one arising from a nucleus of grey matter, close to the middle line of the floor of the Sylvian aqueduct; the posterior one from a grey nucleus, at the upper part of the floor of the fourth ventricle, close to the origin of the fifth nerve. The two nerves are connected together at their origin, by a transverse band of white fibres, which crosses the surface of the velum. The nerve winds round the outer side of the crus cerebri, immediately above the pons Varolii, pierces the dura mater in the free border of the tentorium cerebelli, near the posterior clinoid process, above the oval opening for the fifth nerve, and passes forwards through the outer wall of the cavernous sinus, below the third; but, as it enters the orbit, through the sphenoidal fissure, it becomes the highest of all the nerves. In the orbit, it passes inwards, above the origin of the Levator palpebrae, and finally enters the orbital surface of the Superior oblique muscle.

In the outer wall of the cavernous sinus, this nerve receives some filaments from the carotid plexus of the sympathetic. It is not unfrequently blended with the ophthalmic division of the fifth; and occasionally gives off a branch to assist in the formation of the lachrymal nerve. It also gives off a recurrent branch, which passes backwards between the layers of the tentorium, dividing into two or three filaments, which may be traced as far back as the wall of the lateral sinus.

Sixth Nerve

The SIXTH NERVE (*Abducens*), takes its apparent origin by several filaments from the constricted part of the corpus pyramidale, close to the pons, or from the lower border of the pons itself.

The *deep origin* of this nerve has been traced, by Mayo, between the fasciculi of the corpus pyramidale, to the posterior part of the medulla, where Stilling has shown its connection with a grey nucleus in the floor of the fourth ventricle. The nerve pierces the dura mater, immediately below the posterior clinoid process, lying in a groove by the side of the body of the sphenoid bone. It passes forwards through the cavernous sinus, lying on the outer side of the internal carotid artery, where it is joined by several filaments from the carotid plexus, by one from Meckel's ganglion (Böck), and another from the ophthalmic nerve. It enters the orbit through the sphenoidal fissure,

253.—Nerves of the Orbit and Ophthalmic Ganglion. Side view.

and lies above the ophthalmic vein, from which it is separated by a lamina of dura mater. It then passes between the two heads of the External rectus, and is distributed to that muscle on its ocular surface.

The above-mentioned nerves, as well as the ophthalmic division of the fifth, as they pass to the orbit, bear a certain relation to each other in the cavernous sinus, at the sphenoidal fissure, and in the cavity of the orbit, which will be now described.

In the *Cavernous Sinus*, the third, fourth, and ophthalmic division of the fifth, are placed in the dura mater, forming the outer wall of the sinus in numerical order, both from above downwards, and from within outwards. The sixth nerve lies at the outer side of the internal carotid artery. As these nerves pass forwards to the sphenoidal fissure, the third and fifth nerves become divided: the sixth approaches the rest; so that their relative position becomes considerably changed.

In the *Sphenoidal Fissure*, the fourth, and the frontal and lachrymal divisions of the ophthalmic, lie upon the same plane, the former being most internal, the latter external; and they enter the cavity of the orbit above the muscles. The remaining nerves enter that cavity between the two heads of the External rectus. The superior division of the third is the highest; beneath this, the nasal branch of the fifth; then the inferior division of the third; and the sixth lowest of all.

In the Orbit, the fourth, and the frontal and lachrymal divisions of the ophthalmic, lie on the same plane immediately beneath the periosteum, the fourth nerve being internal and resting on the Superior oblique, the frontal resting on the Levator palpebrae, and the lachrymal on the External rectus. Next in order comes the superior division of the third nerve lying immediately beneath the Superior rectus, and then the nasal division of the fifth crossing the optic nerve from the outer to the inner side of this cavity. Beneath these is found the optic nerve, surrounded in front by the ciliary nerves, and having the lenticular ganglion on its outer side, between it and the External rectus. Below the optic is the inferior division of the third, and the sixth, which lies on the outer side of the cavity.

Facial Nerve

The FACIAL NERVE, the hard portion of the seventh pair, is the motor nerve of the face. It arises from the lateral tract of the medulla oblongata, in the groove between the olivary and restiform bodies. Its deep origin may be traced to the floor of the fourth ventricle, where it is connected with the same nucleus as the sixth nerve. This nerve is situated a little nearer to the middle line than the portio mollis, close to the lower border of the pons Varolii, from which some of its fibres are derived.

Connected with this nerve, and lying between it and the portio mollis, is a small fasciculus (*portio inter duram et mollem* of Wrisberg). This accessory portion arises from the lateral column of the cord.

The nerve passes forwards and outwards upon the crus cerebelli, and enters the internal auditory meatus with the auditory nerve. Within the meatus, the facial nerve lies first to the inner side of, and then in, a groove upon the auditory, and is connected to it by one or two filaments.

254.—The Course and Connections of the Facial Nerve in the Temporal Bone.

At the bottom of the meatus, it enters the aqueductus Fallopii, and follows the serpentine course of that canal through the petrous portion of the temporal bone, from its commencement at the internal meatus to its termination at the stylo-mastoid foramen. It is at first directed outwards towards the hiatus Fallopii, where it forms a reddish gangliform swelling (intumescentia ganglioformis), and is joined by several nerves; bending suddenly backwards, it runs in the internal wall of the cavity of the tympanum, above the fenestra ovalis, and at the back of this cavity passes vertically downwards to the stylo-mastoid foramen.

On emerging from this aperture, it runs forwards in the substance of the parotid gland, crosses the external jugular vein and external carotid artery, and divides behind the ramus of the lower jaw into two primary branches, temporo-facial and cervico-facial, from which numerous offsets are distributed over the side of the head, face, and upper part of the neck, supplying the superficial muscles in this region.

The communications of the facial nerve may be thus arranged:

In the internal auditory meatus.   With the auditory nerve.

In the aqueductus Fallopii. {
    With Meckel's ganglion by the large petrosal nerve.
    With the otic ganglion by the small petrosal nerve.
    With the sympathetic on the middle meningeal by the external petrosal nerve.

At its exit from the stylo-mastoid foramen.	With the pneumogastric. „ glosso-pharyngeal. „ carotid plexus. „ auricularis magnus. „ auriculo-temporal.
On the face.	With the three divisions of the fifth.

In the internal auditory meatus, some minute filaments pass between the facial and auditory nerves.

Opposite the hiatus Fallopii, the gangliform enlargement on the facial nerve communicates, by means of the large superficial petrosal nerve, with Meckel's ganglion; by a filament from the smaller superficial petrosal, with the otic ganglion; and by the external superficial petrosal, with the sympathetic filaments accompanying the middle meningeal artery (Bidder). From the gangliform enlargement, according to Arnold, a twig is sent back to the auditory nerve.

At its exit from the stylo-mastoid foramen, it sends a twig to the pneumogastric, another to the glosso-pharyngeal nerve, and communicates with the carotid plexus of the sympathetic, with the great auricular branch of the cervical plexus, with the auriculo-temporal branch of the inferior maxillary nerve in the parotid gland, and on the face with the terminal branches of the three divisions of the fifth.

Branches of Distribution

Within aqueductus Fallopii.	Tympanic. Chorda tympani.	
At exit from stylo-mastoid foramen.	Posterior auricular. Digastric. Stylo-hyoid.	
On the face.	Temporo-facial.	Temporal. Malar. Infra-orbital.
	Cervico-facial.	Supra-maxillary. Infra-maxillary. Cervical.

The *Tympanic Branch* is a small filament, which supplies the Stapedius muscle. It arises from the nerve opposite the pyramid.

The *Chorda Tympani* is given off from the facial as it passes vertically downwards at the back of the tympanum, about a quarter of an inch before its exit from the stylo-mastoid foramen. It ascends from below upwards in a distinct canal, parallel with the aqueductus Fallopii, and enters the cavity of the tympanum through an opening between the base of the pyramid and the attachment of the membrana tympani, and becomes invested with mucous membrane. It passes forwards through the cavity of the tympanum, between the handle of the malleus and vertical ramus of the incus, to its anterior inferior angle, and emerges from that cavity through a distinct foramen at the inner side of the Glasserian fissure. It then descends between the two Pterygoid muscles, and meets the gustatory nerve at an acute angle; after communicating with this nerve, it accompanies it to the submaxillary gland; it then joins the submaxillary ganglion, and terminates in the Lingualis muscle.

The *Posterior Auricular Nerve* arises close to the stylo-mastoid foramen, and passes upwards in front of the mastoid process, where it is joined by a filament from the auricular branch of the pneumogastric, and communicates with the deep branch of the auricularis magnus; as it ascends between the meatus and mastoid process it divides into two branches. The *auricular branch* supplies the Retrahens aurem, and the integument at the back part of the auricle. The *occipital branch*, the larger, passes backwards along the superior curved line of the occipital bone, and supplies the occipital portion of the Occipito-frontalis and the integument.

The *Stylo-hyoid* is a long slender branch, which passes inwards, entering the Stylo-hyoid muscle about its middle; it communicates with the sympathetic filaments on the external carotid artery.

The *Digastric Branch* usually arises by a common trunk with the preceding; it divides into several filaments, which supply the posterior belly of the Digastric; one of these perforates that muscle to join the glosso-pharyngeal nerve.

The *Temporo-facial*, the larger of the two terminal branches, passes upwards and forwards through the parotid gland, crosses the neck of the condyle of the jaw, being connected in this situation with the auriculo-temporal branch of the inferior maxillary nerve, and divides into branches, which are distributed over the temple and upper part of the face; these may be divided into three sets, temporal, malar, and infra-orbital.

The *temporal branches* cross the zygoma to the temporal region, supplying the Attrahens aurem and the integument, and join with the temporal branch of the superior maxillary, and with the auriculo-temporal branch of the inferior maxillary. The more anterior branches supply the frontal portion of the Occipito-frontalis, and the Orbicularis palpebrarum muscle, joining with the supra-orbital branch of the ophthalmic.

The *malar branches* pass across the malar bone to the outer angle of the orbit, where they supply the Orbicularis and Corrugator supercilii muscles, joining with filaments from the lachrymal and supra-orbital nerves: others supply the lower eyelid, joining with filaments of the malar branches of the superior maxillary nerve.

The *infra-orbital*, of larger size than the rest, pass horizontally forwards to be distributed between the lower margin of the orbit and the mouth. The *superficial branches* run beneath the skin and above the superficial muscles of the face, which they supply, being distributed to the integument and hair follicles; some supply the lower eyelid and Pyramidalis nasi, joining, at the inner angle of the orbit, with the infra-trochlear and nasal branches of the ophthalmic. The *deep branches* pass beneath the Levator labii superioris, supply it and the Levator anguli oris, and form a plexus (infra-orbital) by joining with the infra-orbital branch of the superior maxillary nerve.

The *Cervico-facial*, the other division of the facial nerve, passes obliquely downwards and forwards through the parotid gland, where it is joined by branches from the great auricular nerve; opposite the angle of the lower jaw it divides into branches, which are distributed on the lower half of the face and upper part of the neck. These may be divided into three sets, buccal, supra-maxillary, and infra-maxillary.

255.—The Nerves of the Scalp, Face, and Side of the Neck.

The *buccal branches* cross the Masseter muscle, join the infra-orbital branches of the cervico-facial division of the nerve, and with filaments of the buccal branch of the inferior maxillary nerve. They supply the Buccinator and Orbicularis oris.

The *supra-maxillary branches* pass forwards beneath the Platysma and Depressor anguli oris, supplying the muscles and the integument of the lip and chin, anastomosing with the mental branch of the inferior dental nerve.

The *infra-maxillary branches* run forward beneath the Platysma, and form a series of arches across the side of the neck over the supra-hyoid region. One of these branches descends vertically to join with the superficial cervical nerve from the cervical plexus; others supply the Platysma and Levator labii superioris.

Ninth, or Hypo-glossal Nerve

The Ninth Nerve (*Hypo-glossal*) is the motor nerve of the tongue. It arises by several filaments, from ten to fifteen in number, from the groove between the pyramidal and olivary bodies, in a continuous line with the anterior roots of the spinal nerves. According to Stilling, these roots may be traced to a grey nucleus in the floor of the medulla oblongata, between the posterior median furrow and the nuclei of the glosso-pharyngeal and vagus nerves. The filaments of this nerve are collected into two bundles, which perforate the dura mater separately, opposite the anterior condyloid foramen, and unite together after their passage through it. The nerve descends almost vertically downwards to a point corresponding with the angle of the jaw. It is at first deeply seated beneath the internal carotid and jugular vein, and intimately connected with the pneumogastric nerve; it then passes forwards between the vein and artery, and descending the neck, becomes superficial below the Digastric muscle. The nerve then loops round the occipital artery, and crosses the external carotid below the tendon of the Digastric muscle. It passes beneath the Mylo-hyoid muscle, lying between it and the Hyo-glossus, and is connected at the anterior border of the latter muscle with the gustatory nerve; it is then continued forwards into the Genio-hyo-glossus muscle as far as the tip of the tongue, distributing branches to its substance.

The communicating branches of this nerve are with the

Pneumogastric. First and second cervical nerves.
Sympathetic. Gustatory.

The communication with the pneumogastric takes place close to the exit of the nerve from the skull, numerous filaments passing between the Hypo-glossal and second ganglion of the pneumogastric, or both being united so as to form one mass.

It communicates with the sympathetic opposite the atlas, by branches derived from the superior cervical ganglion, and in the same situation it is joined by a filament with the loop connecting the two first cervical nerves.

The communication with the gustatory takes place near the anterior border of the Hyo-glossus muscle by numerous filaments, which ascend upon it.

The branches of distribution are the

Descendens noni. Thyro-hyoid.
Muscular.

The *Descendens Noni* is a long slender branch, which quits the hypo-glossal where it turns round the occipital artery. It descends obliquely across the sheath of the carotid vessels, and joins just below the middle of the neck, to form a loop with the communicating branches from the second and third cervical nerves. From the convexity of this loop, branches pass forwards to supply the Sterno-hyoid, Sterno-thyroid, and both bellies of the Omo-hyoid. According to Arnold, another filament descends in front of the vessels into the chest, which joins the cardiac and phrenic nerves. The descendens noni is occasionally contained in the sheath of the carotid vessels, being sometimes placed over and sometimes beneath the internal jugular vein.

The *Thyro-hyoid* is a small branch, arising from the hypo-glossal near the posterior

256.—Hypo-glossal Nerve, Cervical Plexus, and their Branches.

border of the Hyo-glossus; it passes obliquely across the great cornu of the hyoid bone, and supplies the Thyro-hyoid muscle.

The *Muscular Branches* are distributed to the Stylo-glossus, Hyo-glossus, Genio-hyoid, and Genio-hyo-glossus muscles. At the under surface of the tongue, numerous slender branches pass upwards into the substance of the organ.

Compound Cranial Nerves

The FIFTH NERVE (*Trifacial*, *Trigeminus*) is the largest cranial nerve, and is somewhat analogous to a spinal nerve, in its origin by two roots, and in the existence of a ganglion on its posterior root. The functions of this nerve are various. It is a nerve of special sense, of common sensation, and of motion. It is the nerve of the special sense of taste, the great sensitive nerve of the head and face, and the motor nerve of the muscles of mastication. It arises by two roots, a posterior larger or sensory, and an anterior smaller or motor root. Its *superficial origin* is from the side of the pons

Varolii, a little nearer to its upper than its lower border. The smaller root consists of three or four bundles; in the larger, the bundles are more numerous, varying in number from seventy to a hundred: the two roots are separated from one another by a few of the transverse fibres of the pons. The *deep origin* of the larger, or sensory root, may be traced between the transverse fibres of the pons Varolii to the lateral tract of the medulla oblongata, immediately behind the olivary body. According to some anatomists, it is connected with the grey nucleus at the back part of the medulla, between the fasciculi teretes and restiform columns. By others, it is said to be continuous with the fasciculi teretes and lateral column of the cord; and, according to Foville, some of its fibres are connected with the transverse fibres of the pons; whilst others enter the cerebellum, spreading out on the surface of its middle peduncle. The motor root has been traced by Bell and Retzius to be connected with the pyramidal body. The two roots of the nerve pass forwards through an oval opening in the dura mater, at the apex of the petrous portion of the temporal bone: here the fibres of the larger root enter a large semilunar ganglion (Gasserian), while the smaller root passes beneath it without having any connection with it, and joins outside the cranium with one of the trunks derived from it.

The Gasserian, or Semilunar Ganglion, is lodged in a depression near the apex of the petrous portion of the temporal bone. It is of a somewhat crescentic form, with its convexity turned forwards. Its upper surface is intimately adherent to the dura mater.

Branches. This ganglion receives, on its *inner side*, filaments from the carotid plexus of the sympathetic; and from it some minute branches are given off to the tentorium cerebelli, and the dura mater, in the middle fossa of the cranium. From its *anterior border*, which is directed forwards and outwards, three large branches proceed, the ophthalmic, superior maxillary, and inferior maxillary. The two first divisions of this nerve consist exclusively of fibres derived from the larger root and ganglion, and are solely nerves of common sensation. The third, or inferior maxillary, is composed of fibres from both roots. This, therefore, strictly speaking, is the only portion of the fifth nerve which is compound, and which can be said to bear analogy with a spinal nerve.

The Ophthalmic, or first division of the fifth, is a sensory nerve. It supplies the eyeball, the lachrymal gland, the mucous lining of the eye and nose, and the integument and muscles of the eyebrow and forehead (fig. 252). It is the smallest of the three divisions of the fifth, arising from the upper part of the Gasserian ganglion. It is a short, flattened band, about an inch in length, which passes forwards along the outer wall of the cavernous sinus, below the other nerves, and just before entering the orbit, through the sphenoidal fissure, divides into three branches, frontal, lachrymal, and nasal. The ophthalmic nerve is joined by filaments from the cavernous plexus of the sympathetic, and gives off recurrent filaments which pass between the layers of the tentorium, with a branch from the fourth nerve.

Its branches are, the

                Lachrymal.   Frontal.   Nasal.

The *Lachrymal* is the smallest of the three branches of the ophthalmic. Not unfrequently, it arises by two filaments, one from the ophthalmic, the other from the fourth, and this, Swan considers as the usual condition. It passes forwards in a separate tube of dura mater, and enters the orbit through the narrowest part of the sphenoidal fissure. In this cavity, it runs along the upper border of the External rectus muscle, with the lachrymal artery, and is connected with the orbital branch of the superior maxillary nerve. Within the lachrymal gland it gives off several filaments, which supply it and the conjunctiva. Finally, it pierces the palpebral ligament, and terminates in the integument of the upper eyelid, joining with filaments of the facial nerve.

The *Frontal* (fig. 252), is the largest division of the ophthalmic, and may be regarded, both from its size and direction, as the continuation of this nerve. It enters the orbit above the muscles, through the highest and broadest part of the sphenoidal fissure, and runs forwards along the middle line, between the Levator palpebrae and the periosteum. Midway between the apex and base of this cavity, it divides into two branches, supra-trochlear and supra-orbital.

The *supra-trochlear branch*, the smaller of the two, passes inwards, above the pulley of the Superior oblique muscle, and gives off a descending filament, which joins with the infra-trochlear branch of the nasal nerve. It then escapes from the orbit between the pulley of the Superior oblique and the supra-orbital foramen, curves up on to the forehead close to the bone, and ascends behind the Corrugator supercilii, and Occipito-frontalis muscles, to both of which it is distributed. Finally, it is lost in the integument of the forehead.

The *supra-orbital branch* passes forwards through the supra-orbital foramen, and gives off, in this situation, palpebral filaments to the upper eyelid. It then ascends upon the forehead, and terminates in muscular, cutaneous, and pericranial branches. The *muscular branches* supply the Corrugator supercilii, Occipito-frontalis, and Orbicularis palpebrarum, joining in the substance of the latter muscle with the facial nerve. The *cutaneous branches*, two in number, an inner and an outer, supply the integument of the cranium as far back as the occiput. They are at first situated beneath the Occipito-frontalis, the former perforating the frontal portion of the muscle, the latter its tendinous aponeurosis. The *pericranial branches* are distributed to the pericranium, over the frontal and parietal bones. They are derived from the cutaneous branches whilst beneath the muscle.

The *Nasal Nerve*, is intermediate in size between the frontal and lachrymal, and more deeply placed than the other branches of the ophthalmic. It enters the orbit between the two heads of the External rectus, passes obliquely inwards across the optic nerve, beneath the Levator palpebrae and Superior rectus muscles, to the inner wall of this cavity, where it enters the anterior ethmoidal foramen, immediately below the Superior oblique. It now enters the cavity of the cranium, traverses a shallow groove on the front of the cribriform plate of the ethmoid bone, and passes down, through the slit by the side of the crista galli, into the nose, where it divides into two branches, an internal and an external. The *internal branch* supplies the mucous membrane near the fore part of the septum of the nose. The *external branch* descends in a groove on the inner surface of the nasal bone, and supplies a few filaments to the mucous membrane covering the front part of the outer wall of the nares as far as the inferior spongy bone; it then leaves the cavity of the nose, between the lower border of the nasal bone and the upper lateral cartilage of the nose, and, passing down beneath the Compressor nasi, supplies the integument of the ala and tip of the nose, joining with the facial nerve.

The branches of the nasal nerve are, the ganglionic, ciliary, and infra-trochlear.

The *ganglionic* is a long, slender branch, about half an inch in length, which usually arises from the nasal, between the two heads of the External rectus. It passes forwards on the outer side of the optic nerve, and enters the superior and posterior angle of the ciliary ganglion, forming its superior, or long root. It is sometimes joined by a filament from the cavernous plexus of the sympathetic, or from the superior division of the third nerve.

The *long ciliary nerves*, two or three in number, are given off from the nasal as it crosses the optic nerve. They join the short ciliary nerves from the ciliary ganglion, pierce the posterior part of the sclerotic, and, running forwards between it and the choroid, are distributed to the Ciliary muscle and iris.

The *infra-trochlear branch* is given off just as the nasal nerve passes through the anterior ethmoidal foramen. It runs forwards along the upper border of the Internal rectus, and is joined, beneath the pulley of the Superior oblique, by a filament from the supra-trochlear nerve. It then passes to the inner angle of the eye, and supplies the Orbicularis palpebrarum, the integument of the eyelids, and side of the nose, the conjunctiva, lachrymal sac, and caruncula lachrymalis.

Ganglia of the Fifth Nerve

Connected with the three divisions of the fifth nerve are four small ganglia, which form the whole of the cephalic portion of the sympathetic. With the first division is connected the ophthalmic ganglion; with the second division, the spheno-palatine or Meckel's ganglion; and with the third, the otic and sub-maxillary ganglia. These ganglia receive sensitive filaments from the fifth, and motor filaments from other sources; these filaments are called the *roots of the ganglia*. They are also connected with each other, and with the cervical portion of the sympathetic.

The OPHTHALMIC, LENTICULAR, or CILIARY GANGLION (fig. 253), is a small quadrangular flattened ganglion, of a reddish-grey colour, and about the size of a pin's head, situated at the back part of the orbit between the optic nerve and the External rectus muscle, generally lying on the outer side of the ophthalmic artery. It is enclosed in a quantity of loose fat, which makes its dissection somewhat difficult.

Its *branches of communication*, or its roots, are three, all of which enter its posterior border. One, the long root, is derived from the nasal branch of the ophthalmic, and joins its superior angle. Another branch, the short root, is a short thick nerve, occasionally divided into two parts; it is derived from that branch of the third nerve which supplies the Inferior oblique muscle, and is connected with the inferior angle of the ganglion. A third branch, the sympathetic root, is a slender filament from the cavernous plexus of the sympathetic. This is occasionally blended with the long root, and sometimes passes to the ganglion by itself. According to Tiedemann, this ganglion receives a filament of communication from the spheno-palatine ganglion.

Its *branches of distribution* are the short ciliary nerves. These consist of from ten to twelve delicate filaments, which arise from the fore part of the ganglion in two bundles, connected with its superior and inferior angles; the upper bundle consisting of four filaments, and the lower of six or seven. They run forwards with the ciliary arteries in a wavy course, one set above and the other below the optic nerve, pierce the sclerotic at the back part of the globe, pass forwards in delicate grooves on its inner surface, and are distributed to the ciliary muscle and iris. A small filament is described by Tiedemann, penetrating the optic nerve with the arteria centralis retinae.

Superior Maxillary Nerve (fig. 257)

The superior maxillary, or second division of the fifth, is a sensory nerve. It is intermediate, both in position and size, between the ophthalmic and inferior maxillary. It commences at the middle of the Gasserian ganglion as a flattened plexiform band, passes forwards through the foramen rotundum, where it becomes more cylindrical in form and firmer in texture. It then crosses the spheno-maxillary fossa, traverses the infra-orbital canal in the floor of the orbit, emerging upon the face at the infra-orbital foramen. At its termination, the nerve lies beneath the Levator labii superioris muscle, and divides into a leash of branches, which spread out upon the side of the nose, the lower eyelid, and upper lip, joining with filaments of the facial nerve.

257.—Distribution of the Second and Third Divisions of the Fifth Nerve and Sub-maxillary Ganglion.

The branches of this nerve may be divided into three groups: 1. Those given off in the spheno-maxillary fossa. 2. Those in the infra-orbital canal. 3. Those on the face.

Spheno-maxillary fossa.	Orbital. Spheno-palatine. Posterior dental.
Infra-orbital canal.	Anterior dental.
On the face.	Palpebral. Nasal. Labial.

The *Orbital Branch* arises in the spheno-maxillary fossa, enters the orbit by the spheno-maxillary fissure, and divides into two branches, temporal and malar.

The *temporal branch* runs in a groove along the outer wall of the orbit (in the malar bone), receives a branch of communication from the lachrymal, and passing through a foramen in the malar bone, enters the temporal fossa. It ascends between the bone and substance of the Temporal muscle, pierces this muscle and the temporal fascia about an inch above the zygoma, and is distributed to the integument covering the temple and side of the forehead, communicating with the facial and auriculo-temporal branch of the inferior maxillary nerve.

The *malar branch* passes along the external inferior angle of the orbit, emerges upon the face through a foramen in the malar bone, and perforating the Orbicularis palpebrarum muscle on the prominence of the cheek, joins with the facial.

The *Spheno-palatine Branches*, two in number, descend to the spheno-palatine ganglion.

The *Posterior Dental Branches* arise from the trunk of the nerve just as it is about to enter the infra-orbital canal; they are two in number, posterior and anterior.

The *posterior branch* passes from behind forwards in the substance of the superior maxillary bone, and joins opposite the canine fossa with the anterior dental. Numerous filaments are given off from the lower border of this nerve, which form a minute plexus in the outer wall of the superior maxillary bone immediately above the alveolus. From this plexus, filaments are distributed to the pulps of the molar and bicuspid teeth, the lining membrane of the antrum, and corresponding portion of the gums.

The *anterior branch* is distributed to the gums and Buccinator muscle.

The *Anterior Dental*, of large size, is given off from the superior maxillary nerve just before its exit from the infra-orbital foramen; it enters a special canal in the anterior wall of the antrum, and anastomoses with the posterior dental. From this branch filaments are distributed to the incisor, canine, and first bicuspid teeth; others are lost upon the lining membrane covering the front part of the inferior meatus.

The *Palpebral Branches* pass upwards beneath the Orbicularis palpebrarum. They supply this muscle, the integument, and conjunctiva of the lower eyelid, joining at the outer angle of the orbit with the facial nerve and malar branch of the orbital.

The *Nasal Branches* pass inwards; they supply the muscles and integument of the side of the nose, and join with the nasal branch of the ophthalmic.

The *Labial Branches*, the largest and most numerous, descend beneath the Levator labii superioris, and are distributed to the integument and muscles of the upper lip, the mucous membrane of the mouth, and labial glands.

All these branches are joined, immediately beneath the orbit, by filaments from the facial nerve, forming an intricate plexus, the infra-orbital.

Spheno-Palatine Ganglion

The *Spheno-Palatine Ganglion* (Meckel's) (fig. 258), the largest of the cranial ganglia, is deeply placed in the spheno-maxillary fossa, close to the spheno-palatine foramen. It is triangular, or heart-shaped in form, of a reddish-grey colour, and placed mainly behind the palatine branches of the superior maxillary nerve, at the point where the sympathetic root joins the ganglion. It consequently does not involve those nerves which pass to the palate and nose. Like other ganglia, it possesses a motor, a sensory, and a sympathetic root. Its motor root is derived from the facial, through the Vidian; its sensory root from the fifth; and its sympathetic root from the carotid plexus, through the Vidian. Its branches are divisible into four groups: ascending, which pass to the orbit; descending, to the palate; internal, to the nose; and posterior branches to the pharynx.

The *Ascending Branches* are two or three delicate filaments, which enter the orbit by the spheno-maxillary fissure, and supply the periosteum. Arnold describes and delineates these branches as ascending to the optic nerve; one, to the sixth nerve (Böck); and one, to the ophthalmic ganglion (Tiedemann).

The *Descending* or *Palatine Branches* are distributed to the roof of the mouth, the soft palate, tonsil, and lining membrane of the nose. They are almost a direct continuation of the spheno-palatine branches of the superior maxillary nerve, and are three in number, anterior, middle, and posterior.

The *anterior*, or large palatine nerve, descends through the posterior palatine canal,

258.—The Spheno-Palatine Ganglion and its Branches.

emerges upon the hard palate, at the posterior palatine foramen, and passes forwards through a groove in the hard palate, extending nearly to the incisor teeth. It supplies the gums, the mucous membrane and glands of the hard palate, and communicates in front with the termination of the naso-palatine nerve. While in the posterior palatine canal, it gives off inferior nasal branches, which enter the nose through openings in the palate-bone, and ramify over the middle meatus, and the middle and inferior spongy bones; and, at its exit from the canal, a palatine branch is distributed to both surfaces of the soft palate.

The *middle*, or external palatine nerve, descends in the same canal as the preceding, to the posterior palatine foramen, distributing branches to the uvula, tonsil, and soft palate. It is occasionally wanting.

The *posterior*, or small palatine nerve, descends with a small artery through the small posterior palatine canal, emerging by a separate opening behind the posterior palatine foramen. It supplies the Levator palati muscle, the soft palate, tonsil, and uvula.

The *Internal Branches* are distributed to the septum, and outer wall of the nasal fossae. They are the superior nasal (anterior), and the naso-palatine.

The *superior nasal branches* (anterior), four or five in number, enter the back part of the nasal fossa by the spheno-palatine foramen. They supply the mucous membrane, covering the superior and middle spongy bones, and that lining the posterior ethmoidal cells, a few being prolonged to the upper and back part of the septum.

The *naso-palatine nerve* (Cotunnius), enters the nasal fossa with the other nasal nerves, and passes inwards across the roof of the nose, below the orifice of the sphenoidal sinus, to reach the septum; and then obliquely downwards and forwards along its lower part, lying between the periosteum and pituitary membrane, to the anterior palatine foramen. It descends to the roof of the mouth by a distinct canal, which opens below in the anterior palatine fossa; the right nerve, also in a separate canal, being posterior to the left one. In the mouth, they become united, supply the mucous membrane behind the incisor teeth, joining with the anterior palatine nerve. It occasionally furnishes a few small filaments to the mucous membrane of the septum.

The *Posterior Branches* are the Vidian and pharyngeal (pterygo-palatine).

The *Vidian* arises from the back part of the spheno-palatine ganglion, passes through the Vidian canal, enters the cartilage filling in the foramen lacerum basis cranii, and divides into two branches, the superficial petrosal, and the carotid. In its course along the Vidian canal, it distributes a few filaments to the lining membrane at the back part of the roof of the nose and septum, and that covering the end of the Eustachian tube. These are upper posterior nasal branches.

The *petrosal branch* (*nervus petrosus superficialis major*), enters the cranium through the foramen lacerum basis cranii, having pierced the cartilaginous substance, filling in this aperture. It runs beneath the Gasserian ganglion and dura mater, contained in a groove in the anterior surface of the petrous portion of the temporal bone, enters the hiatus Fallopii, and, being continued through it, into the aqueductus Fallopii, joins the gangliform enlargement on the facial nerve. Properly speaking, this nerve passes from the facial to the spheno-palatine ganglion, forming its motor root.

The *carotid branch* is shorter, but larger than the petrosal, of a reddish-grey colour, and soft in texture. It crosses the foramen lacerum, surrounded by the cartilaginous substance which fills in that aperture, and enters the carotid canal, on the outer side of the carotid artery, to join the carotid plexus.

The *Pharyngeal Nerve (pterygo-palatine)*, is a small branch arising from the back part of the ganglion, occasionally springing from the Vidian nerve. It passes through the pterygo-palatine canal with the pterygo-palatine artery, and is distributed to the lining membrane of the pharynx, behind the Eustachian tube.

Inferior Maxillary Nerve

The Inferior Maxillary Nerve distributes branches to the teeth and gums of the lower jaw, the integument of the temple and external ear, lower part of the face and lower lip, and the muscles of mastication: it also supplies the tongue with its special nerve of the sense of taste. It is the largest of the three divisions of the fifth, and consists of two portions, the larger, or sensory root, proceeding from the inferior angle of the Gasserian ganglion; and the smaller, or motor root, which passes beneath the ganglion, and unites with the inferior maxillary nerve, just after its exit through the foramen ovale. Immediately beneath the base of the skull, this nerve divides into two trunks, anterior and posterior.

The anterior, and smaller division, which receives nearly the whole of the motor root, divides into five branches, which supply the muscles of mastication. They are the masseteric, deep temporal, buccal, and pterygoid.

The *Masseteric Branch* passes outwards, above the External pterygoid muscle, in front of the temporo-maxillary articulation, and crosses the sigmoid notch, with the masseteric artery, to the Masseter muscle, in which it ramifies nearly as far as its anterior border. It occasionally gives a branch to the Temporal muscle, and a filament to the articulation of the jaw.

The *Deep Temporal Branches*, two in number, anterior and posterior, supply the deep surface of the Temporal muscle. The *posterior branch*, of small size, is placed at the back of the temporal fossa. It is sometimes joined with the masseteric branch. The *anterior branch* is reflected upwards, at the pterygoid ridge of the sphenoid, to the front of the temporal fossa. It is occasionally joined with the buccal nerve.

The *Buccal Branch* pierces the External pterygoid, and passes downwards beneath the inner surface of the coronoid process of the lower jaw, or through the fibres of the Temporal muscle to reach the surface of the Buccinator, upon which it divides into a superior and an inferior branch. It gives a branch to the External pterygoid during its passage through this muscle, and a few ascending filaments to the Temporal muscle, one of which occasionally joins with the anterior branch of the deep temporal nerve. The *upper branch* supplies the integument and upper part of the Buccinator muscle, joining with the facial nerve around the facial vein. The *lower branch* passes forwards to the angle of the mouth; supplies the integument and Buccinator muscle, as well as the mucous membrane lining its inner surface, joining with the facial nerve.

The *Pterygoid Branches* are two in number, one for each Pterygoid muscle. The *branch to the Internal pterygoid* is long and slender, and passes inwards to enter the deep surface of the muscle. This nerve is intimately connected at its origin with the otic ganglion.

The *branch to the External pterygoid* is most frequently derived from the buccal, but it may be given off separately from the anterior trunk of the nerve.

The posterior and larger division of the inferior maxillary nerve also receives a few filaments from the motor root. It divides into three branches, auriculo-temporal, gustatory, and inferior dental.

The AURICULO-TEMPORAL NERVE generally arises by two roots, between which passes the middle meningeal artery. It passes backwards beneath the External pterygoid muscle to the inner side of the articulation of the lower jaw. It then turns upwards with the temporal artery, between the external ear and condyle of the jaw, under cover of the parotid gland, and escaping from beneath this structure, divides into two temporal branches. The *posterior temporal*, the smaller of the two, supplies the Attrahens aurem muscle, and is distributed to the upper part of the pinna and the neighbouring integument. The *anterior temporal* accompanies the temporal artery to the vertex of the skull, and supplies the integument of the temporal region, communicating with the facial nerve.

The auriculo-temporal nerve has *branches of communication* with the facial and otic ganglion. Those joining the facial nerve, usually two in number, pass forwards behind the neck of the condyle of the jaw, and join this nerve at the posterior border of the Masseter muscle. They form one of the principal branches of communication between the facial and the fifth nerve. The filaments of communication with the otic ganglion are derived from the commencement of the auriculo-temporal nerve.

The *Auricular Branches* are two in number, inferior and superior. The *inferior auricular* arises behind the articulation of the jaw, and is distributed to the ear below the external meatus; other filaments twine around the internal maxillary artery, and communicate with the sympathetic. The *superior auricular* arises in front of the internal ear, and supplies the integument covering the tragus and pinna.

Branches to the Meatus Auditorius, two in number, arise from the point of communication between the temporo-auricular and facial nerves, and are distributed to the meatus.

The *Branch to the Temporo-maxillary Articulation* is usually derived from the auriculo-temporal nerve.

The *Parotid Branches* supply the parotid gland.

The GUSTATORY or LINGUAL NERVE (fig. 257), the special nerve of the sense of taste, supplies the papillae and mucous membrane of the tongue. It is deeply placed throughout the whole of its course. It lies at first beneath the External pterygoid muscle, together with the inferior dental nerve, being placed to the inner side of the latter nerve, and is occasionally joined to it by a branch which crosses the internal maxillary artery. The chorda tympani also joins it at an acute angle in this situation. The nerve then passes between the Internal pterygoid muscle and the inner side of the ramus of the jaw, and crosses obliquely to the side of the tongue over the Superior constrictor muscle of the pharynx, and between the Stylo-glossus muscle and deep part of the sub-maxillary gland; the nerve lastly runs across Wharton's duct, and along the side of the tongue to its apex, being covered by the mucous membrane of the mouth.

Its *branches of communication* are with the sub-maxillary ganglion and hypo-glossal nerve. The branches to the sub-maxillary ganglion are two or three in number; those connected with the hypo-glossal nerve form a plexus at the anterior margin of the hyo-glossus muscle.

Its *branches of distribution* are few in number. They supply the mucous membrane of the mouth, the gums, the sub-lingual gland, and the conical and fungiform papillae and mucous membrane of the tongue, the terminal filaments anastomosing at the tip of this organ with the hypo-glossal nerve.

The INFERIOR DENTAL is the largest of the three branches of the inferior maxillary nerve. It passes downwards with the inferior dental artery, at first beneath the External pterygoid muscle, and then between the internal lateral ligament and the ramus of the jaw to the dental foramen. It then passes forwards in the dental canal in the inferior maxillary bone, lying beneath the teeth, as far as the mental foramen, where it divides into two terminal branches, incisor and mental. The *incisor branch* is continued onwards within the bone to the middle line, and supplies the canine and incisor teeth. The *mental branch* emerges from the bone at the mental foramen, and divides beneath the Depressor anguli oris into an external branch, which supplies this muscle, the Orbicularis oris, and the integument, communicating with the facial nerve; and an inner branch, which ascends to the lower lip beneath the Quadratus menti; it supplies this muscle and the mucous membrane and integument of the lip, communicating with the facial nerve.

The branches of the inferior dental are the mylo-hyoid and dental.

The *Mylo-hyoid* is derived from the inferior dental just as that nerve is about to enter the dental foramen. It descends in a groove on the inner surface of the ramus of the jaw, in which it is retained by a process of fibrous membrane. It supplies the cutaneous surface of the Mylo-hyoid muscle, and the anterior belly of the Digastric, occasionally sending one or two filaments to the sub-maxillary gland.

The *Dental Branches* supply the molar and bicuspid teeth. They correspond in number to the fangs of those teeth; each nerve entering the orifice at the point of the fang, and supplying the pulp of the tooth.

Two small ganglia are connected with the inferior maxillary nerve: the otic, with the trunk of the nerve; and the submaxillary, with its lingual branch, the gustatory.

OTIC GANGLION

The *Otic Ganglion* (Arnold's) (fig. 259), is a small oval-shaped, flattened ganglion, of a reddish-grey colour, situated immediately below the foramen ovale, on the inner surface of the inferior maxillary nerve, and around the origin of the internal pterygoid nerve. It is in relation, *externally*, with the trunk of the inferior maxillary nerve, at the point where the motor root joins the sensory portion; *internally*, with the cartilaginous part of the Eustachian tube, and the origin of the Tensor palati muscle; *behind* it, is the middle meningeal artery.

Branches of Communication. This ganglion is connected with the inferior maxillary nerve, and its internal pterygoid branch, by two or three short, delicate filaments, and also with the auriculo-temporal nerve: from the former, it obtains its motor, from the latter its sensory root; its communication with the sympathetic being effected by a filament from the plexus surrounding the middle meningeal artery. This ganglion

259.—The Otic Ganglion and its Branches

also communicates with the glosso-pharyngeal and facial nerves, through the small petrosal nerve continued from the tympanic plexus.

Its *Branches of Distribution* are a filament to the tensor tympani, and one to the tensor palati. The former passes backwards, on the outer side of the Eustachian tube; the latter arises from the ganglion, near the origin of the internal pterygoid nerve, and passes forwards.

Submaxillary Ganglion

The *Submaxillary Ganglion* (fig. 257), is of small size, circular in form, and situated above the deep portion of the submaxillary gland, near the posterior border of the Mylo-hyoid muscle, being connected by filaments with the lower border of the gustatory nerve.

Branches of Communication. This ganglion is connected with the gustatory nerve by a few filaments which join it separately, at its fore and back part. It also receives a branch from the chorda tympani, by which it communicates with the facial; and communicates with the sympathetic by filaments from the nervi molles, surrounding the facial artery.

Branches of Distribution. These are five or six in number; they arise from the lower part of the ganglion, and supply the mucous membrane of the mouth and Wharton's duct, some being lost in the submaxillary gland. According to Meckel, a branch from this ganglion occasionally descends in front of the Hyo-glossus muscle, and, after joining with one from the hypo-glossal, passes to the Genio-hyo-glossus muscle.

Eighth Pair

The *Eighth Pair* consists of three nerves, the glosso-pharyngeal, pneumo-gastric, and spinal accessory.

260.—Origin of the Eighth Pair, their Ganglia and Communications.

The GLOSSO-PHARYNGEAL NERVE is distributed, as its name implies, to the tongue and pharynx, being the nerve of sensation to the mucous membrane of the fauces and root of the tongue; and of motion, to the Pharyngeal muscles. It arises by three or four filaments, closely connected together, from the upper part of the medulla oblongata, immediately behind the olivary body.

Its *deep origin* may be traced through the fasciculi of the lateral tract, to a nucleus of grey matter at the lower part of the floor of the fourth ventricle, external to the fasciculi teretes. From its superficial origin, it passes outwards across the flocculus, and leaves the skull at the central part of the jugular foramen, in a separate sheath of the dura mater and arachnoid, in front of the pneumogastric and spinal accessory nerves. In its passage through the jugular foramen, it grooves the lower border of the petrous portion of the temporal bone, and, at its exit from the skull, passes forwards between the jugular vein and internal carotid artery, and descends in front of the latter vessel and beneath the styloid process and the muscles connected with it, to the lower border of the Stylo-pharyngeus. The nerve now curves inwards, forming an arch on the side of the neck, lying upon the Stylo-pharyngeus, and the Middle constrictor of the pharynx, above the superior laryngeal nerve. It then passes beneath the Hyo-glossus, and is finally distributed to the mucous membrane of the fauces, and base of the tongue, the mucous glands of the mouth and tonsil.

In passing through the jugular foramen, the nerve presents, in succession, two gangliform enlargements. The superior one, the smaller, is called the *jugular ganglion*, the inferior, and larger one, the *petrous ganglion*, or the ganglion of Andersch.

The *Superior*, or *Jugular Ganglion*, is situated in the upper part of the groove in which the nerve is lodged during its passage through the jugular foramen. It is of very small size, and involves only the outer side of the trunk of the nerve, a small fasciculus passing beyond it, which is not connected directly with it.

The *Inferior*, or *Petrous Ganglion*, is situated in a depression in the lower border of the petrous portion of the temporal bone; it is larger than the former, and involves the whole of the fibres of the nerve. From this ganglion arise those filaments which connect the glosso-pharyngeal with other nerves at the base of the skull.

Its *Branches of Communication* are with the pneumogastric, sympathetic, and facial, and the tympanic branch.

The branches to the pneumogastric, are two filaments, one to its auricular branch, and one to the upper ganglion of the pneumogastric.

The branch to the sympathetic is connected with the superior cervical ganglion.

The branch of communication with the facial, perforates the posterior belly of the digastric. It arises from the trunk of the nerve below the petrous ganglion, and joins the facial just after its exit from the stylo-mastoid foramen.

The *Tympanic Branch* (Jacobson's nerve), arises from the petrous ganglion, and enters a small bony canal on the base of the petrous portion of the temporal bone.

(This opening is placed on the bony ridge which separates the carotid canal from the jugular fossa.) It ascends to the tympanum, enters this cavity by an aperture in its floor close to the inner wall, and divides into three branches, which are contained in grooves upon the surface of the promontory.

Its *branches of distribution* are, one to the fenestra rotunda, one to the fenestra ovalis, and one to the lining membrane of the Eustachian tube and tympanum.

Its *branches of communication* are three, and occupy separate grooves on the surface of the promontory. One of these passes forwards and downwards to the carotid canal to join the carotid plexus. A second runs vertically upwards to join the greater superficial petrosal nerve, as it lies in the hiatus Fallopii. The third branch runs upwards and forwards towards the anterior surface of the petrous bone, and passes through a small aperture in the sphenoid and temporal bones, to the exterior of the skull, where it joins the otic ganglion. This nerve, in its course through the temporal bone, passes by the gangliform enlargement of the facial, and has a connecting filament with it.

261.—Course and Distribution of the Eighth Pair of Nerves.

The branches of the glossopharyngeal nerve are the carotid, pharyngeal, muscular, tonsillitic, and lingual.

The *Carotid Branches* descend along the trunk of the internal carotid artery as far as its point of bifurcation, communicating with the pharyngeal branch of the pneumogastric, and with branches of the sympathetic.

The *Pharyngeal Branches* are three or four filaments which unite opposite the

Middle constrictor of the pharynx with the pharyngeal branches of the pneumogastric, superior laryngeal, and sympathetic nerves, to form the pharyngeal plexus, branches from which perforate the muscular coat of the pharynx to supply the mucous membrane.

The *Muscular Branches* are distributed to the Stylo-pharyngeus and Constrictors of the pharynx.

The *Tonsillitic Branches* supply the tonsil, forming a plexus (circulus tonsillaris) around this body, from which branches are distributed to the soft palate and fauces, where they anastomose with the palatine nerves.

The *Lingual Branches* are two in number; one supplies the mucous membrane covering the surface of the base of the tongue, the other perforates its substance, and supplies the mucous membrane and papillae of the side of the organ.

The SPINAL ACCESSORY NERVE consists of two parts; one, the accessory part to the vagus, and the other the spinal portion.

The *accessory part*, the smaller of the two, arises by four or five delicate filaments from the lateral tract of the cord below the roots of the vagus; these filaments may be traced to a nucleus of grey matter at the back of the medulla, below the origin of the vagus. It joins, in the jugular foramen, with the upper ganglion of the vagus by one or two filaments, and is continued into the vagus below the second ganglion. It gives branches to the pharyngeal and superior laryngeal branches of the vagus.

The *spinal portion*, firm in texture, arises by several filaments from the lateral tract of the cord, as low down as the sixth cervical nerve; the fibres pierce the tract, and are connected with the anterior horn of the grey crescent of the cord. This portion of the nerve ascends between the ligamentum denticulatum and the posterior roots of the spinal nerves, enters the skull through the foramen magnum, and is then directed outwards to the jugular foramen, through which it passes, lying in the same sheath as the pneumogastric, separated from it by a fold of the arachnoid, and is here connected with the accessory portion. At its exit from the jugular foramen, it passes backwards behind the internal jugular vein, and descends obliquely behind the Digastric and Stylo-hyoid muscles to the upper part of the Sterno-mastoid. It pierces this muscle, and passes obliquely across the sub-occipital triangle, to terminate in the deep surface of the Trapezius. This nerve gives several branches to the Sterno-mastoid during its passage through it, and joins in its substance with branches from the third cervical. In the sub-occipital triangle it joins with the second and third cervical nerves, assists in the formation of the cervical plexus, and occasionally of the great auricular nerve. On the front of the Trapezius, it is reinforced by branches from the third, fourth, and fifth cervical nerves, joins with the posterior branches of the spinal nerves, and is distributed to the Trapezius, some filaments ascending and others descending in its substance as far as its inferior angle.

The PNEUMOGASTRIC, or VAGUS, one of the three divisions of the eighth pair, has a more extensive distribution than any of the other cranial nerves, passing through the neck and cavity of the chest to the upper part of the abdomen. It is composed of both motor and sensitive filaments. It supplies the organs of voice and respiration with motor and sensitive fibres; and the pharynx, oesophagus, stomach, and heart with motor influence. Its *superficial origin* is by eight or ten filaments from the lateral tract immediately behind the olivary body and below the glosso-pharyngeal; its fibres may, however, be traced deeply through the fasciculi of the medulla, to terminate in a grey nucleus near the lower part of the floor of the fourth ventricle. The filaments become united, and form a flat cord, which passes outwards across the flocculus to the jugular foramen, through which it emerges from the cranium. In passing through this opening, the pneumogastric accompanies the spinal accessory, being contained in the same sheath of dura mater with it, a membranous septum separating it from the glosso-pharyngeal, which lies in front. The nerve in this situation presents a well-marked ganglionic enlargement, which is called the *ganglion jugulare*, or the *ganglion of the root of the pneumogastric:* to it the accessory part of the spinal accessory nerve is connected. After the exit of the nerve from the jugular foramen, a second gangliform swelling is formed upon it, called the *ganglion inferius*, or the *ganglion of the trunk of the nerve*; below which it is again joined by filaments from the *accessory* nerve. The nerve descends the neck in a straight direction within the sheath of the carotid vessels, lying between the internal carotid artery and internal jugular vein as far as the thyroid cartilage, and then between the same vein and the common carotid to the root of the neck. Here the course of the nerve becomes different on the two sides of the body.

On the *right side*, the nerve passes across the subclavian artery between it and the subclavian vein, and descends by the side of the trachea to the back part of the root of the lung, where it spreads out in a plexiform network (posterior pulmonary), from the lower part of which two cords descend upon the oesophagus, on which they divide, forming, with branches from the opposite nerve, the oesophageal plexus; below, these branches are collected into a single cord, which runs along the back part of the oesophagus, enters the abdomen, and is distributed to the posterior surface of the stomach, joining the left side of the caeliac plexus, and the splenic plexus.

On the *left side*, the pneumogastric nerve enters the chest, between the left carotid and subclavian arteries, behind the left innominate vein. It crosses the arch of the aorta, and descends behind the root of the left lung and along the anterior surface of the oesophagus to the stomach, distributing branches over its anterior surface, some extending over the great *cul-de-sac*, and others along the lesser curvature. Filaments from these latter branches enter the gastro-hepatic omentum, and join the left hepatic plexus.

The *Ganglion of the Root* is of a greyish colour, circular in form, about two lines in diameter, and resembles the ganglion on the large root of the fifth nerve.

Connecting Branches. To this ganglion the accessory portion of the spinal accessory nerve is connected by several delicate filaments; it also has an anastomotic twig with the petrous ganglion of the glosso-pharyngeal, with the facial nerve by means of the auricular branch, and with the sympathetic by means of an ascending filament from the superior cervical ganglion.

The *Ganglion of the Trunk* (inferior) is a plexiform cord, cylindrical in form, of a reddish colour, and about an inch in length; it involves the whole of the fibres of the nerve, except the portion of the accessory nerve derived from the spinal accessory, which blends with the nerve beyond the ganglion.

Connecting Branches. This ganglion is connected with the hypo-glossal, the superior cervical ganglion of the sympathetic, and with the loop between the first and second cervical nerves.

The *branches of the Pneumogastric* are

In the jugular fossa.	Auricular.
In the neck.	{ Pharyngeal. Superior laryngeal. Recurrent laryngeal. Cervical Cardiac.
In the thorax.	{ Thoracic Cardiac. Anterior pulmonary. Posterior pulmonary. Oesophageal.
In the abdomen.	Gastric.

The *Auricular Branch* arises from the ganglion of the root, and is joined soon after its origin by a filament from the glosso-pharyngeal; it crosses the jugular fossa to an opening near the root of the styloid process. Traversing the substance of the temporal bone, it crosses the aqueductus Fallopii about two lines above its termination at the stylo-mastoid foramen; it here gives off an ascending branch, which joins the facial, and a descending branch, which anastomoses with the posterior auricular branch of the same nerve: the continuation of the nerve reaches the surface between the mastoid process and the external auditory meatus, and supplies the integument at the back part of the pinna.

The *Pharyngeal Branch* arises from the upper part of the inferior ganglion of the pneumogastric, receiving a filament from the accessory portion of the spinal accessory; it passes across the internal carotid artery (in front or behind), to the upper border of the Middle constrictor, where it divides into numerous filaments, which anastomose with those from the glosso-pharyngeal, superior laryngeal, and sympathetic, to form the pharyngeal plexus, from which branches are distributed to the muscles and mucous membrane of the pharynx. As this nerve crosses the internal carotid, some filaments are distributed, together with those from the glosso-pharyngeal, upon the wall of this vessel.

The *Superior Laryngeal Nerve*, larger than the preceding, arises from the middle of the inferior ganglion of the pneumogastric. It descends, by the side of the pharynx, behind the internal carotid, where it divides into two branches, the external and internal laryngeal.

The *external larynygeal branch*, the smaller, descends by the side of the larynx, beneath the Sterno-thyroid, to supply the crico-thyroid and the thyroid gland. It gives branches to the pharyngeal plexus, and the Inferior constrictor, and communicates with the superior cardiac nerve, behind the common carotid.

The *Internal Laryngeal Branch* descends to the opening in the thyro-hyoid membrane, through which it passes with the superior laryngeal artery, and is distributed to the mucous membrane of the larynx, and the Arytenoid muscle, anastomosing with the recurrent laryngeal.

The *branches to the mucous membrane* are distributed, some in front, to the epiglottis, the base of the tongue, and epiglottidean gland; and others pass backwards, in the aryteno-epiglottidean fold, to supply the mucous membrane surrounding the superior orifice of the larynx, as well as the membrane which lines the cavity of the larynx as low down as the vocal chord.

The *filament to the Arytenoid muscle* is distributed partly to it, and part to the mucous lining of the larynx.

The *filament* which joins with the recurrent laryngeal, descends beneath the mucous membrane on the posterior surface of the larynx, behind the lateral part of the thyroid cartilage, where the two nerves become united.

The *Inferior*, or *Recurrent Laryngeal*, so called from its reflected course, arises, on the right side, in front of the subclavian artery: it winds round this vessel, and ascends obliquely to the side of the trachea, behind the common carotid and inferior thyroid arteries. On the left side, it arises in front of the arch of the aorta, and winds round it at the point where the obliterated remains of the ductus arteriosus are connected with this vessel, and then ascends to the side of the trachea. The nerves on both sides ascend in the groove between the trachea and oesophagus, and, piercing the lower fibres of the Inferior constrictor muscle, enter the larynx behind the articulation of the inferior cornu of the thyroid cartilage with the cricoid, being distributed to all the muscles of the larynx, excepting the Crico-thyroid, and joining with the superior laryngeal.

The recurrent laryngeal, as it winds round the subclavian artery and aorta, gives off several cardiac filaments, which unite with the cardiac branch from the pneumogastric and sympathetic. As it ascends the neck, it gives off oesophageal branches, more numerous on the left than on the right side; tracheal branches to the posterior membranous portion of the trachea; and some pharyngeal filaments to the Inferior constrictor of the pharynx.

The *Cervical Cardiac Branches*, two or three in number, arise from the pneumogastric, at the upper and lower part of the neck.

The *superior branches* are small, and communicate with the cardiac branches of the sympathetic, and with the great cardiac plexus.

The *inferior cardiac branches*, one on each side, arise at the lower part of the neck, just above the first rib. On the right side, this branch passes in front of the arteria innominata, and anastomoses with the superior cardiac nerve. On the left side, it passes in front of the arch of the aorta, and anastomoses either with the superior cardiac nerve, or with the cardiac plexus.

The *Thoracic Cardiac Branches*, on the right side, arise from the trunk of the pneumogastric, as it lies by the side of the trachea: passing inwards, they terminate in the deep cardiac plexus. On the left side, they arise from the left recurrent laryngeal nerve.

The *Anterior Pulmonary Branches*, two or three in number, and of small size, are distributed on the anterior aspect of the root of the lungs. They join with filaments from the sympathetic, and form the anterior pulmonary plexus.

The *Posterior Pulmonary Branches*, more numerous and larger than the anterior, are distributed on the posterior aspect of the root of the lung: they are joined by filaments from the third and fourth thoracic ganglia of the sympathetic, and form the posterior pulmonary plexus. Branches from both plexuses accompany the ramifications of the air tubes through the substance of the lungs.

The *Oesophageal Branches* are given off from the pneumo-gastric, both above and below the pulmonary branches. The latter are the most numerous and largest. They form, together with branches from the opposite nerve, the oesophageal plexus.

The *Gastric Branches* are the terminal filaments of the pneumo-gastric nerve. The nerve on the right side is distributed to the posterior surface of the stomach, and joins the left side of the coeliac plexus, and the splenic plexus. The nerve on the left side is distributed over the anterior surface of the stomach, some filaments passing across the great *cul-de-sac*, and others along the lesser curvature. They unite with branches of the right nerve and sympathetic, some filaments passing through the lesser omentum to the left hepatic plexus.

THE SPINAL NERVES

The *Spinal Nerves* are so called, from taking their origin from the spinal cord, and from being transmitted through the intervertebral foramina on either side of the spinal column. There are thirty-one pairs of spinal nerves, which are arranged into the following groups, corresponding to the region of the spine through which they pass:

Cervical.	8 pairs.
Dorsal.	12 "
Lumbar.	5 "
Sacral.	5 "
Coccygeal.	1 "

It will be observed, that each group of nerves corresponds in number with the vertebrae in each region, excepting in the cervical and coccygeal.

Each spinal nerve arises by two roots, an anterior, or motor root, and a posterior, or sensitive root.

Roots of the Spinal Nerves

The *anterior roots* arise somewhat irregularly from a linear series of foramina, on the antero-lateral column of the spinal cord, gradually approaching towards the anterior median fissure as they descend.

The fibres of the anterior roots pass between the anterior and lateral columns, and enter the grey matter of the anterior horn, where they divide into two bundles, the larger of which is connected with the lateral column of the same side, while the smaller bundle passes to the anterior column of the opposite side. The component fibres of the latter bundle form part of the anterior white commissure, and decussate with the roots of the nerve of the opposite side of the cord.

The *posterior roots* arise, in a perfectly straight line, from the posterior lateral fissure, opposite the corresponding horn of grey matter.

The fibres of the posterior roots pass directly into the grey matter of the posterior horn, at the bottom of the posterior lateral fissure, where they subdivide, some passing into the lateral and posterior columns of the same side; while others enter the grey commissure, and form the transverse commissural fibres connecting these roots with the opposite side of the cord.

The *posterior roots* of the nerves are larger, and the individual filaments more numerous, and thicker, than those of the anterior. As their component fibrils pass outwards, towards the aperture in the dura mater, they coalesce into two bundles, receive a tubular sheath from this membrane, and enter the intervertebral ganglion which is developed upon each root.

The posterior root of the first cervical nerve forms an exception to these characters. It is smaller than the anterior, has frequently no ganglion developed upon it, and, when the ganglion exists, it is often situated within the dura mater.

The *anterior roots* are the smaller of the two, devoid of any ganglionic enlargement, and their component fibrils are collected into two bundles, near the intervertebral foramina.

Ganglia of the Spinal Nerves

A ganglion is developed upon each posterior root of the spinal nerves. These ganglia are of an oval form, of a reddish colour, bear a proportion in size to the nerves upon which they are formed, and are placed in the intervertebral foramina, external to the point where the nerves perforate the dura mater. Each ganglion is bifid internally, where it is joined by the two bundles of the posterior root, the two portions being united into a single mass externally. The ganglia upon the first and second cervical nerves form an exception to these characters, being placed on the arches of the vertebrae over which they pass. The ganglia, also, of the sacral nerves are placed within the spinal canal; and that on the coccygeal nerve, also in the canal about the middle of its posterior root. Immediately beyond the ganglion, the two roots unite, and the trunk thus formed passes out of the intervertebral foramen, and divides into an anterior branch, for the supply of the anterior part of the body; and a posterior branch, for the posterior part.

Anterior Branches of the Spinal Nerves

The *anterior branches of the Spinal Nerves* supply the parts of the body in front of the spine, including the limbs. They are for the most part larger than the posterior branches; this increase of size being proportioned to the larger extent of structures they are required to supply. Each branch is connected by slender filaments with the sympathetic. In the dorsal region, the anterior branches of the spinal nerves are completely separate from each other, and are uniform in their distribution; but in the cervical, lumbar, and sacral regions, they form intricate plexuses previous to their distribution.

Posterior Branches of the Spinal Nerves

The *posterior branches of the Spinal Nerves* are generally smaller than the anterior: they arise from the trunk, resulting from the union of the nerves in the intervertebral foramina, and, passing backwards, divide into external and internal branches, which are distributed to the muscles and integument behind the spine. The first cervical and lower sacral nerves are exceptions to these characters.

Cervical Nerves

The *roots of the Cervical Nerves* increase in size from the first to the fifth, and then maintain the same size to the eighth. The posterior roots bear a proportion to the anterior as 3 to 1, which is much greater than in any other region; the individual filaments being also much larger than those of the anterior roots. In direction, they are less oblique than those of the other spinal nerves. The first is directed a little upwards and outwards; the second is horizontal; the others are directed obliquely downwards and outwards, the lowest being the most oblique, and consequently longer than the upper, the distance between their place of origin and their point of exit from the spinal canal, never exceeding the depth of one vertebra.

The *trunk of the first Cervical Nerve* (*Sub-occipital*), leaves the spinal canal, between the occipital bone and the posterior arch of the atlas; the second between the posterior arch of the atlas and the lamina of the axis; and the eighth (the last), between the last cervical and first dorsal vertebrae.

Each nerve, at its exit from the intervertebral foramen, divides into an anterior and a posterior branch. The anterior branches of the four upper cervical nerves, form the cervical plexus. The anterior branches of the four lower cervical nerves, together with the first dorsal, form the brachial plexus.

Anterior Branches of the Cervical Nerves

The *anterior branch of the first*, or *Sub-occipital Nerve*, is of small size. It escapes from the vertebral canal, through a groove upon the posterior arch of the atlas. In this groove it lies beneath the vertebral artery, to the inner side of the Rectus lateralis. As it crosses the foramen in the transverse process of the atlas, it receives a filament from the sympathetic. It then descends, in front of this process, to communicate with an ascending branch from the second cervical nerve.

Communicating filaments from this nerve join the pneumogastric, the hypoglossal and sympathetic, and some branches are distributed to the Rectus lateralis, and the two Anterior recti. According to Valentin, it also distributes filaments to the occipito-atloid articulation, and mastoid process of the temporal bone.

The *anterior branch of the second Cervical Nerve* escapes from the spinal canal, between the posterior arch of the atlas and the lamina of the axis, and, passing forwards on the outer side of the vertebral artery, divides in front of the Intertransverse muscle, into an ascending branch, which joins the first cervical; and two descending branches which join the third.

The *anterior branch of the third Cervical Nerve* is double the size of the preceding. At its exit from the intervertebral foramen, it passes downwards and outwards beneath the Sterno-mastoid, and divides into two branches. The ascending branch joins the anterior division of the second cervical, communicates with the sympathetic and spinal accessory nerves, and subdivides into the superficial cervical, and great auricular nerves. The descending branch passes down in front of the Scalenus anticus, anastomoses with the fourth cervical nerve, and becomes continuous with the clavicular nerves.

The *anterior branch of the fourth Cervical* is of the same size as the preceding. It receives a branch from the third, sends a communicating branch to the fifth cervical, and, passing downwards and outwards, divides into numerous filaments, which cross the posterior triangle of the neck, towards the clavicle and acromion. It usually gives a branch to the phrenic nerve whilst it is contained in the intertransverse space.

The *anterior branches of the fifth, sixth, seventh, and eighth Cervical Nerves*, are remarkable for their large size. They are much larger than the preceding nerves, and are all of equal size. They assist in the formation of the brachial plexus.

Cervical Plexus

The cervical plexus (fig. 256) is formed by the anterior branches of the four upper cervical nerves. It is situated in front of the four upper vertebrae, resting upon the

Levator anguli scapulae, and Scalenus medius muscles, and covered in by the Sternomastoid.

Its branches may be divided into two groups, superficial and deep, which may be thus arranged:

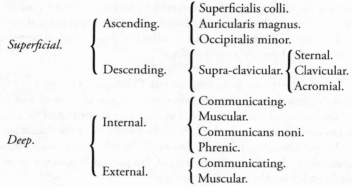

Superficial Branches of the Cervical Plexus

The *Superficialis Colli* arises from the second and third cervical nerves, turns round the posterior border of the Sterno-mastoid about its middle, and passing obliquely forwards behind the external jugular vein to the anterior border of that muscle, perforates the deep cervical fascia, and divides beneath the Platysma into two branches, which are distributed to the anterior and lateral parts of the neck.

The *ascending branch* gives a filament, which accompanies the external jugular vein; it then passes upwards to the sub-maxillary region, and divides into branches, some of which form a plexus with the cervical branches of the facial nerve beneath the Platysma; others pierce this muscle, supply it, and are distributed to the integument of the upper half of the neck, at its fore part, as high up as the chin.

The *descending branch* pierces the Platysma, and is distributed to the integument of the side and front of the neck, as low as the sternum.

This nerve is occasionally represented by two or more filaments.

The *Auricularis Magnus* is the largest of the ascending branches. It arises from the second and third cervical nerves, winds round the posterior border of the Sternomastoid, and, after perforating the deep fascia, ascends upon that muscle beneath the Platysma to the parotid gland, where it divides into numerous branches.

The *facial branches* pass across the parotid, and are distributed to the integument of the face; others penetrate the substance of the gland, and communicate with the facial nerve.

The *posterior* or *auricular branches* ascend vertically to supply the integument of the posterior part of the pinna, communicating with the auricular branches of the facial and pneumogastric nerves.

The *mastoid branch* joins the posterior auricular branch of the facial, and crossing the mastoid process, is distributed to the integument behind the ear.

The *Occipitalis Minor* arises from the second cervical nerve; it curves round the posterior border of the Sterno-mastoid above the preceding, and ascends vertically along the posterior border of this muscle to the back part of the side of the head. Near the cranium it perforates the deep fascia, and is continued upwards along the side of the head behind the ear, supplying the integument and Occipito-frontalis muscle, and communicating with the occipitalis major, auricularis magnus, and posterior auricular branch of the facial.

This nerve gives off an *auricular branch*, which supplies the Attollens aurem and the integument of the upper and back part of the auricle. This branch is occasionally derived from the great occipital nerve. The occipitalis minor varies in size; it is occasionally double.

The *Descending* or *Supra-clavicular Branches* arise from the third and fourth cervical nerves; emerging beneath the posterior border of the Sterno-mastoid, they descend in the interval between this muscle and the Trapezius, and divide into branches, which are arranged, according to their position, into three groups.

The *inner* or *sternal branch* crosses obliquely over the clavicular and sternal attachments of the Sterno-mastoid, and supplies the integument as far as the median line.

The *middle* or *clavicular branch* crosses the clavicle, and supplies the integument over the Pectoral and Deltoid muscles, communicating with the cutaneous branches of the upper intercostal nerves. Not unfrequently, the supra-clavicular nerve passes through a foramen in the clavicle, at the junction of the outer with the inner two-thirds of the bone.

The *external* or *acromial branch* passes obliquely across the outer surface of the Trapezius and the acromion, and supplies the integument of the upper and back part of the shoulder.

Deep Branches of the Cervical Plexus: Internal Series

The *Communicating Branches* consist of several filaments, which pass from the loop between the first and second cervical nerves in front of the atlas to the pneumogastric, hypo-glossal, and sympathetic.

Muscular Branches supply the Anterior recti and Rectus lateralis muscles; they proceed from the first cervical nerve, and from the loop formed between it and the second.

The *Communicans Noni* (fig. 256) consists usually of two filaments, one being derived from the second, and the other from the third cervical. These filaments descend vertically downwards on the outer side of the internal jugular vein, cross in front of the vein a little below the middle of the neck, and form a loop with the descendens noni in front of the sheath of the carotid vessels. Occasionally, the junction of these nerves takes place within the sheath.

The *Phrenic Nerve* (*Internal Respiratory of Bell*) arises from the third and fourth cervical nerves, and receives a communicating branch from the fifth. It descends to the root of the neck, lying obliquely across the front of the Scalenus anticus, passes over the first part of the subclavian artery, between it and the subclavian vein, and, as it enters the chest, crosses the internal mammary artery near its root. Within the chest, it descends nearly vertically in front of the root of the lung, and by the side of the pericardium, between it and the mediastinal portion of the pleura, to the

Diaphragm, where it divides into branches, which separately pierce that muscle, and are distributed to its under surface.

The two phrenic nerves differ in their length, and also in their relations at the upper part of the thorax.

The *right nerve* is situated more deeply, and is shorter and more vertical in direction than the left; it lies on the outer side of the right vena innominata and superior vena cava.

The *left nerve* is rather larger than the right, from the inclination of the heart to the left side, and from the Diaphragm being lower in this than on the opposite side. At the upper part of the thorax, it crosses in front of the arch of the aorta to the root of the lung.

Each nerve supplies filaments to the pericardium and pleura, and near the chest is joined by a filament from the sympathetic; by another derived from the fifth and sixth cervical nerves; and occasionally by one from the union of the descendens noni with the spinal nerves, which, Swan states, occurs only on the left side.

From the *right nerve*, one or two filaments pass to join in a small ganglion with phrenic branches of the solar plexus; and branches from this ganglion are distributed to the hepatic plexus, the supra-renal capsule, and inferior vena cava. From the *left nerve* filaments pass to join the phrenic plexus, but without any ganglionic enlargement.

Deep Branches of the Cervical Plexus: External Series

Communicating Branches. The cervical plexus communicates with the spinal accessory nerve, in the substance of the Sterno-mastoid muscle, in the sub-occipital triangle, and beneath the Trapezius.

Muscular Branches are distributed to the Sterno-mastoid, Levator anguli scapulae, Scalenus medius, and Trapezius.

The branch for the Sterno-mastoid is derived from the second cervical; the Levator anguli scapulae receiving branches from the third; and the Trapezius branches from the third and fourth.

Posterior Branches of the Cervical Nerves

The *Posterior Branches* of the cervical nerves, with the exception of the first two, pass backwards, and divide, behind the posterior Intertransverse muscles, into external and internal branches.

The *external branches* supply the muscles at the side of the neck, viz., the Cervicalis ascendens, Transversalis colli, and Trachelo-mastoid.

The external branch of the second cervical nerve is the largest; it is often joined with the third, and supplies the Complexus, Splenius, and Trachelo-mastoid muscles.

The *internal branches*, the larger, are distributed differently in the upper and lower part of the neck. Those derived from the third, fourth, and fifth nerves pass between the Semi-spinalis and Complexus muscles, and having reached the spinous processes, perforate the aponeurosis of the Splenius and Trapezius, and are continued outwards to the integument over the Trapezius; whilst those derived from the three lowest cervical nerves are the smallest, and are placed beneath the Semi-spinalis, which they supply, and do not furnish any cutaneous filaments. These internal branches supply the Complexus, Semi-spinalis colli, Inter-spinales, and Multifidus spinae.

The posterior branches of the three first cervical nerves require a separate description.

The *posterior branch of the First Cervical Nerve (Sub-occipital)* is larger than the anterior, and escapes from the vertebral canal between the occipital bone and the posterior arch of the atlas, lying behind the vertebral artery, and enters the triangular space formed by the Rectus posticus major, the Obliquus superior, and Obliquus inferior. It supplies the Recti and Obliqui muscles, and the complexus. From the branch which supplies the Inferior oblique a filament is given off, which joins the second cervical nerve. It also occasionally gives off a cutaneous filament, which accompanies the occipital artery, and communicates with the occipitalis major and minor nerves.

The posterior division of the first cervical has no branch analogous to the external branch of the other cervical nerves.

The *posterior branch of the Second Cervical Nerve* is three or four times greater than the anterior branch, and the largest of all the other posterior cervical nerves. It emerges from the spinal canal between the posterior arch of the atlas and lamina of the axis, below the Inferior oblique. It supplies this muscle, and receives a communicating filament from the first cervical. It then divides into external and internal branches.

The *internal branch*, called, from its size and distribution, the *occipitalis major*, ascends obliquely inwards between the Obliquus inferior and Complexus, and pierces the latter muscle and the Trapezius near their attachments to the cranium. It is now joined by a filament from the third cervical nerve, and ascending on the back part of the head with the occipital artery, divides into two branches, which supply the integument of the scalp as far forwards as the vertex, communicating with the occipitalis minor. It gives off an auricular branch to the back part of the ear, and muscular branches to the Complexus.

The *posterior branch of the Third Cervical* is smaller than the preceding, but larger than the fourth; it differs from the posterior branches of the other cervical nerves in its supplying an additional filament to the integument of the occiput. This occipital branch arises from the internal or cutaneous branch beneath the Trapezius; it pierces that muscle, and supplies the skin on the lower and back part of the head. It lies to the inner side of the occipitalis major, with which it is connected.

The internal branches of the posterior divisions of the three first cervical nerves are occasionally joined beneath the Complexus by communicating branches. This communication has been described by Cruvelhier as the *posterior cervical plexus*.

THE BRACHIAL PLEXUS (fig. 262)

The brachial plexus is formed by the union of the anterior branches of the four lower cervical and first dorsal nerves. It extends from the lower part of the side of the neck to the axilla, being very broad, and presenting but little of a plexiform arrangement at its commencement, narrow opposite the clavicle, broad and presenting a more dense interlacement in the axilla, and dividing opposite the coracoid process into numerous branches for the supply of the upper limb. These nerves are all similar in

262.—Plan of the Brachial Plexus.

size, and their mode of union in the formation of the plexus is the following. The fifth and sixth nerves unite near their exit from the spine into a common trunk; the seventh nerve joins this trunk near the outer border of the Middle scalenus; and the three nerves thus form one large single cord. The eighth cervical and first dorsal nerves unite beneath the Anterior scalenus into a common trunk. Thus two large trunks are formed, the upper one by the union of the fifth, sixth, and seventh cervical; and the lower one by the eighth cervical and first dorsal. These two trunks accompany the subclavian artery to the axilla, lying upon its outer side, that formed by the union of the last cervical and first dorsal being nearest to the vessel. Opposite the clavicle, and sometimes in the axilla, each of these cords gives off a fasciculus, which uniting, a third trunk is formed, so that in the centre of the axilla three cords are found, one lying on the outer side of the axillary artery, one on its inner side, and one behind. The brachial plexus communicates with the cervical plexus by a branch from the fourth to the fifth nerve, and with the phrenic by a branch from the fifth cervical, which joins that nerve on the anterior scalenus muscle: the cervical and first dorsal nerves are also joined by filaments from the middle and inferior cervical ganglia of the sympathetic, close to their exit from the intervertebral foramina.

Relations. In the neck, the brachial plexus lies at first between the Anterior and Middle scaleni muscles, and then above and to the outer side of the subclavian artery; it then passes beneath the clavicle and Subclavius muscle, lying upon the first serration of the Serratus magnus and Subscapularis muscles. *In the axilla*, it is placed on the outer side of the first portion of the axillary artery: it surrounds the artery in the second part of its course, one cord lying upon the outer side of that vessel, one on the inner side, and one behind it; and at the lower part of the axillary space gives off its terminal branches to the upper extremity.

Branches. The branches of the brachial plexus may be arranged into two groups, viz., those given off above the clavicle, and those below that bone.

Branches above the Clavicle

Communicating.	Posterior thoracic.
Muscular.	Supra-scapular.

The *communicating branch* with the phrenic is derived from the fifth cervical nerve; it joins the phrenic on the Anterior scalenus muscle.

The *muscular branches* supply the Longus colli, Scaleni, Rhomboidei, and Subclavius muscles. Those for the Scaleni and Longus colli arise from the lower cervical nerves at their exit from the intervertebral foramina. The rhomboid branch arises from the fifth cervical, pierces the Scalenus medius, and passes beneath the Levator anguli scapulae, which it occasionally supplies, to the Rhomboid muscles. The subclavian branch is a small filament, which arises from the trunk formed by the junction of the fifth and sixth cervical nerves; it descends in front of the subclavian artery to the Subclavius muscle, and is usually connected by a filament with the phrenic nerve.

The *Posterior Thoracic Nerve* (long thoracic, external respiratory of Bell), supplies the Serratus magnus, and is remarkable for the length of its course. It arises by two roots, from the fifth and sixth cervical nerves, immediately after their exit from the intervertebral foramina. These unite in the substance of the Middle scalenus muscle, and, after emerging from it, the nerve passes down behind the brachial plexus and the axillary vessels, resting on the outer surface of the Serratus magnus. It extends along the side of the chest to the lower border of this muscle, and supplies it with numerous filaments.

The *Supra-Scapular Nerve* arises from the cord formed by the fifth, sixth, and seventh cervical nerves: passing obliquely outwards beneath the Trapezius, it enters the supra-spinous fossa, through the notch in the upper border of the scapula; and, passing beneath the Supra-spinatus muscle, curves in front of the spine of the scapula to the infra-spinous fossa. In the supra-spinous fossa, it gives off two branches to the Supra-spinatus muscle, and an articular filament to the shoulder-joint; and in the infra-spinous fossa, it gives off two branches to the Infra-spinatus muscle, besides some filaments to the shoulder-joint and scapula.

Branches Below the Clavicle

To chest.	Anterior thoracic.
To shoulder.	{ Subscapular. { Circumflex.
To arm, Fore-arm, and Hand.	{ Musculo-cutaneous. { Internal cutaneous. { Lesser internal cutaneous. { Median. { Ulnar. { Musculo-spiral.

The branches given off below the clavicle, are derived from the three cords of the brachial plexus, in the following manner:

From the outer cord, arises the external of the two anterior thoracic nerves, the musculo-cutaneous nerve, and the outer head of the median.

From the inner cord, arises the internal of the two anterior thoracic nerves, the internal cutaneous, the lesser internal cutaneous (nerve of Wrisberg), the ulnar, and inner head of the median.

263.—Cutaneous Nerves of Right Upper Extremity. Anterior View.

From the posterior cord, arises the subscapular; and it then subdivides into the musculo-spiral and circumflex nerves.

The *Anterior Thoracic Nerves*, two in number, supply the Pectoral muscles.

The *external*, or superficial branch, the larger of the two, arises from the outer cord of the brachial plexus, passes inwards, across the axillary artery and vein, and is distributed to the under surface of the Pectoralis major. It sends down a communicating filament to join the internal branch.

The *internal*, or deep branch, arises from the inner cord, and passes upwards between the axillary artery and vein (sometimes perforates the vein), and joins with the filament from the superficial branch. From the loop thus formed, branches are distributed to the under surface of the Pectoralis minor and major muscles.

The *Subscapular Nerves* are three in number, and supply the Subscapularis, Teres major, and Latissimus dorsi muscles.

The *upper subscapular nerve*, the smallest, enters the upper part of the Subscapularis muscle.

The *lower subscapular nerve* enters the axillary border of the Subscapularis, and terminates in the Teres major. The latter muscle is sometimes supplied by a separate branch.

The *long subscapular*, the largest of the three, descends along the lower border of the Subscapularis to the Latissimus dorsi, through which it may be traced as far as its lower border.

The *Circumflex Nerve* supplies some of the muscles, and the integument of the shoulder, and the shoulder-joint. It arises from the posterior cord of the brachial plexus, in common with the musculo-spiral nerve. It passes down behind the axillary artery, and in front of the Subscapularis; and, at the lower border of this muscle, passes backwards, and divides into two branches.

The *superior branch* winds round the neck of the humerus, beneath the Deltoid, with the posterior circumflex vessels, as far as the anterior border of this muscle, supplying it and giving off cutaneous branches, which pierce it to ramify in the integument covering its lower part.

264.—Cutaneous Nerves of Right Upper Extremity. Posterior View.

The *inferior branch*, at its origin, distributes filaments to the Teres minor and back part of the Deltoid muscles; upon the filament to the former muscle a gangliform enlargement usually exists. The nerve then pierces the deep fascia, and supplies the integument over the lower two-thirds of the posterior surface of the Deltoid, as well as that covering the long head of the Triceps.

The circumflex nerve, before its division, gives off an articular filament, which enters the shoulder-joint below the Sub-scapularis.

The *Musculo-Cutaneous Nerve* (external cutaneous, *perforans Gasserii*), supplies some of the muscles of the arm, and the integument of the fore-arm. It arises from the outer cord of the brachial plexus, opposite the lower border of the Pectoralis minor. It then perforates the Coraco-brachialis muscle, and passes obliquely between the Biceps and Brachialis anticus, to the outer side of the arm, a little above the elbow, where it perforates the deep fascia and becomes cutaneous. This nerve, in its course through the arm, supplies the Coraco-brachialis, Biceps, and brachialis anticus muscles, besides some filaments to the elbow-joint and humerus.

The cutaneous portion of the nerve passes behind the median cephalic vein, and divides, opposite the elbow-joint, into an anterior and a posterior branch.

The *anterior branch* descends along the radial border of the fore-arm to the wrist. It is here placed in front of the radial artery, and, piercing the deep fascia, accompanies that vessel to the back of the wrist. It communicates with a branch from the radial nerve, and distributes filaments to the integument of the ball of the thumb, and to the wrist-joint.

The *posterior branch* is given off about the middle of the fore-arm, and passes downwards, along the back part of its radial side, to the wrist. It supplies the integument of the lower third of the fore-arm, communicating with the radial nerve, and the external cutaneous branch of the musculo-spiral.

The *Internal Cutaneous Nerve* is one of the smallest branches of the brachial plexus. It arises from the inner cord, in common with the ulnar and internal head of the median, and, at its commencement, is placed on the inner side of the brachial artery. It passes down the inner side of the arm, pierces the deep fascia with the basilic vein, about the middle of the limb, and, becoming cutaneous, divides into two branches.

This nerve gives off, near the axilla, a cutaneous filament, which pierces the fascia, and supplies the integument covering the Biceps muscles, nearly as far as the elbow. This filament lies a little external to the common trunk from which it arises.

The *anterior branch*, the larger of the two, passes in front of, occasionally behind, the median basilic vein. It then descends on the anterior surface of the ulnar side of the fore-arm, distributing filaments to the integument as far as the wrist, and communicating with a cutaneous branch of the ulnar nerve.

The *posterior branch*, passes obliquely downwards on the inner side of the basilic vein, winds over the internal condyle of the humerus to the back of the fore-arm, and descends, on the posterior surface of its ulnar side, to a little below the middle, distributing filaments to the integument. It anastomoses above the elbow, with the lesser internal cutaneous, and above the wrist, with the dorsal branch of the ulnar nerve (Swan).

The *Lesser Internal Cutaneous Nerve* (nerve of Wrisberg), is distributed to the integument on the inner side of the arm. It is the smallest of the branches of the brachial plexus, and usually arises from the inner cord, with the internal cutaneous and ulnar nerves. It passes through the axillary space, at first lying beneath, and then on the inner side, of the axillary vein, and communicates with the intercosto-humeral nerve. It then descends along the inner side of the brachial artery, to the middle of the arm, where it pierces the deep fascia, and is distributed to the integument of the back part of the lower third of the arm, extending as far as the elbow, where some filaments are lost in the integument in front of the inner condyle, and others over the olecranon. It communicates with the inner branch of the internal cutaneous nerve.

In some cases, the nerve of Wrisberg and intercosto-humeral, are connected by two or three filaments, which form a kind of plexus at the back part of the axilla. In other cases, the intercosto-humeral is of large size, and takes the place of the nerve of Wrisberg, receiving merely a filament of communication from the brachial plexus, which represents this nerve. In other cases, this filament is wanting, the place of the nerve of Wrisberg being supplied entirely from the intercosto-humeral.

The *Median Nerve* (fig. 265) has received its name from the course it takes along the middle line of the arm and fore-arm to the hand, lying between the ulnar and musculo-spiral and radial nerves. It arises by two roots, one from the outer, and one from the inner cord of the brachial plexus; these embrace the lower part of the axillary artery, uniting either in front or on the outer side of that vessel. As it descends through the arm, it lies at first on the outer side of the brachial artery, crosses that vessel in the middle of its course, usually in front, but occasionally behind it, and lies on its

265.—Nerves of the Left Upper Extremity. Front View.

inner side to the bend of the elbow, where it is placed beneath the bicipital fascia, and is separated from the elbow-joint by the Brachialis anticus. *In the fore-arm*, it passes between the two heads of the Pronator radii teres, and descends beneath the Flexor sublimis, to within two inches above the annular ligament, where it becomes more superficial, lying between the Flexor sublimis and Flexor carpi radialis, covered by the integument and fascia. It then passes beneath the annular ligament into the hand.

Branches. No branches are given off from the median nerve in the arm. *In the fore-arm*, its branches are, muscular, anterior interosseous, and palmar cutaneous.

The *muscular branches* supply all the superficial layer of muscles on the anterior surface of the fore-arm, except the Flexor carpi ulnaris. These branches are derived from the nerve near the elbow. The branch furnished to the Pronator radii teres often arises above the joint.

The *anterior interosseous* supplies the deep muscles on the anterior surface of the fore-arm. It accompanies the anterior interosseous artery along the interosseous membrane, in the interval between the Flexor longus pollicis and Flexor profundus digitorum muscles, both of which it supplies, and terminates below in the Pronator quadratus.

The *palmar cutaneous branch* arises from the median nerve at the lower part of the fore-arm. It pierces the fascia above the annular ligament, and divides into two branches; the *outer one* supplies the skin over the ball of the thumb, and communicates with the external cutaneous nerve; the *inner one* supplies the integument of the palm of the hand, anastomosing with the cutaneous branch of the ulnar. Both nerves cross the annular ligament previous to their distribution.

In the palm of the hand, the median nerve is covered by the integument and palmar fascia, and rests upon the tendons of the Flexor muscles. In this situation it becomes enlarged, somewhat flattened, of a reddish colour, and divides into two branches. Of these, the *external one* supplies a muscular branch to some of the muscles of the thumb, and digital branches to the thumb and index finger; the *internal branch* supplying digital branches to the middle finger and part of the index and ring fingers.

The *branch to the muscles of the thumb* is a short nerve, which subdivides to supply the Abductor, Opponens, and outer head of the Flexor brevis pollicis muscles; the remaining muscles of this group being supplied by the ulnar nerve.

The *digital branches* are five in number. The *first* and *second* pass along the borders of the thumb, the most external one communicating with branches of the radial nerve. The *third* passes along the radial side of the index finger, and supplies the first Lumbrical muscle. The *fourth* subdivides to supply the adjacent sides of the index and middle fingers, and sends a branch to the second Lumbrical muscle. The *fifth* supplies the adjacent sides of the middle and ring fingers, and communicates with a branch from the ulnar nerve.

Each digital nerve, opposite the base of the first phalanx, gives off a dorsal branch, which joins the dorsal digital nerve, and runs along the side of the dorsum of the finger, ending in the integument over the last phalanx. At the end of the finger, the digital nerve divides into a palmar and a dorsal branch; the former supplies the extremity of the finger, and the latter ramifies around and beneath the nail. The digital nerves, as they run along the fingers, are placed superficial to the digital arteries.

The ULNAR NERVE is placed along the inner or ulnar side of the upper limb, and is distributed to the muscles and integument of the fore-arm and hand. It is smaller than the median, behind which it is placed, diverging from it in its course down the arm. It arises from the inner cord of the brachial plexus, in common with the internal head of the median and the internal cutaneous nerves. At its commencement, it lies at the inner side of the axillary artery, and holds the same relation with the brachial artery to the middle of the arm. From this point, it runs obliquely across the internal head of the Triceps, pierces the internal intermuscular septum, and descends to the groove between the internal condyle and olecranon, accompanied by the inferior profunda artery. *At the elbow*, it rests upon the inner condyle, and passes into the fore-arm between the two heads of the Flexor carpi ulnaris. *In the fore-arm*, it descends in a perfectly straight course along its ulnar side, lying upon the Flexor profundus digitorum, its upper half being covered by the Flexor carpi ulnaris, its lower half lying on the outer side of this muscle, covered by the integument and fascia. The ulnar artery, in the upper part of its course, is separated from the ulnar nerve by a considerable interval; in the lower half of its course, the nerve lies to its inner side. *At the wrist*, the ulnar nerve crosses the annular ligament on the outer side of the pisiform bone, a little behind the ulnar artery, and immediately beyond this bone divides into two branches, superficial and deep palmar.

The branches of the ulnar nerve are

In fore-arm.
- Articular (elbow).
- Muscular.
- Cutaneous.
- Dorsal branch.
- Articular (wrist).

In hand.
- Superficial palmar.
- Deep palmar.

The *Articular branches* distributed to the elbow-joint consist of several small filaments. They arise from the nerve as it lies in the groove between the inner condyle and olecranon.

The *Muscular branches* are two in number; one supplying the Flexor carpi ulnaris; the other, the inner half of the Flexor profundus digitorum. They arise from the trunk of the nerve near the elbow.

The *Cutaneous branch* arises from the ulnar nerve about the middle of the fore-arm, and divides into a superficial and deep branch.

The *superficial branch* (frequently absent) pierces the deep fascia near the wrist, and is distributed to the integument, communicating with a branch of the internal cutaneous nerve.

The *deep branch* lies on the ulnar artery, which it accompanies to the hand, some filaments entwining around the vessel, which end in the integument of the palm, communicating with branches of the median nerve.

The *Dorsal cutaneous branch* arises about two inches above the wrist; it passes backwards beneath the Flexor carpi ulnaris, perforates the deep fascia, and running along the ulnar side of the wrist and hand, supplies the inner side of the little finger, and the adjoining sides of the little and ring fingers; it also sends a communicating filament to that branch of the radial nerve which supplies the adjoining sides of the middle and ring fingers.

Articular filaments to the wrist are also supplied by the ulnar nerve.

The *Superficial palmar branch* supplies the Palmaris brevis, and the integument on the inner side of the hand, and terminates in two digital branches, which are distributed, one to the ulnar side of the little finger, the other to the adjoining sides of the little and ring fingers, the latter communicating with a branch from the median.

266.—The Supra-Scapular, Circumflex, and Musculo-Spiral Nerves.

The *Deep palmar branch* passes between the Abductor and Flexor brevis minimi digiti muscles, and follows the course of the deep palmar arch beneath the flexor tendons. At its origin, it supplies the muscles of the little finger. As it crosses the deep part of the hand it sends two branches to each interosseous space, one for the Dorsal and one for the Palmar interosseous muscle, the branches to the second and third Palmar interossei supplying filaments to the two inner Lumbrical muscles. At its termination between the thumb and index finger, it supplies the Adductor pollicis and the inner head of the Flexor brevis pollicis.

The MUSCULO-SPIRAL NERVE (fig. 266), the largest branch of the brachial plexus, supplies the muscles of the back part of the arm and forearm, and the integument of the same parts, as well as that of the hand. It arises from the posterior cord of the brachial plexus by a common trunk with the circumflex nerve. At its commencement, it is placed behind the axillary and upper part of the brachial arteries, passing down in front of the tendons of the Latissimus dorsi and Teres major. It winds round the humerus in the spiral groove with the superior profunda artery and vein, passing from the inner to the outer side of the bone, beneath the Triceps muscle. At the outer side of the arm, it descends between the Brachialis anticus and Supinator longus to the front of the external condyle, where it divides into the radial and posterior interosseous nerves.

The branches of the musculo-spiral nerve are:

    Muscular.   Radial.
    Cutaneous.   Posterior interosseous.

The *Muscular branches* supply the Triceps, Anconeus, Supinator longus, Extensor carpi radialis longior, and Brachialis anticus. These branches are derived from the nerve, at the inner side, back part, and outer side of the arm.

The *internal muscular branches* supply the inner and middle heads of the Triceps muscle. That to the inner head of the Triceps, is a long, slender filament, which lies close to the ulnar nerve, as far as the lower third of the arm.

The *posterior muscular branch*, of large size, arises from the nerve in the groove between the Triceps and the humerus. It divides into branches which supply the outer head of the Triceps and Anconeus muscles. The branch for the latter muscle is a long, slender filament, which descends in the substance of the Triceps to the Anconeus.

The *external muscular branches* supply the Supinator longus, Extensor carpi radialis longior, and Brachialis anticus.

The *Cutaneous branches* are three in number, one internal, and two external.

The *internal cutaneous branch* arises in the axillary space, with the inner muscular branch. It is of small size, and passes across the axilla to the inner side of the arm, supplying the integument on its posterior aspect nearly as far as the olecranon.

The *two external cutaneous branches* perforate the outer head of the Triceps, at its attachment to the humerus. The upper and smaller one follows the course of the cephalic vein to the front of the elbow, supplying the integument of the lower half of the upper arm on its anterior aspect. The *lower branch* pierces the deep fascia below the insertion of the Deltoid, and passes down along the outer side of the arm and elbow, and along the radial side of the fore-arm to the wrist, supplying the integument in its course, and joining, near its termination, with a branch of the external cutaneous nerve.

The *Radial Nerve* passes along the front of the radial side of the fore-arm, to the commencement of its lower third. It lies at first a little to the outer side of the radial artery, concealed beneath the Supinator longus. In the middle third of the fore-arm, it lies beneath the same muscle, in close relation with the outer side of that vessel. It quits the artery about three inches above the wrist, passes beneath the tendon of the Supinator longus, and, piercing the deep fascia at the outer border of the fore-arm, divides into two branches.

The *external branch*, the smaller of the two, supplies the integument of the radial side, and ball of the thumb, joining with the posterior branch of the external cutaneous nerve.

The *internal branch* communicates, above the wrist, with a branch from the external cutaneous, and, on the back of the hand, forms an arch with the dorsal branch of the ulnar nerve. It then divides into digital nerves, which supply, the first, the ulnar side of the thumb; the second, the radial side of the index finger; the third, the adjoining sides of the index and middle fingers; and the fourth, the adjacent borders of the middle and ring fingers. The latter nerve communicates with a filament from the dorsal branch of the ulnar nerve.

The *Posterior Interosseous Nerve* pierces the Supinator brevis, winds to the back of the fore-arm, in the substance of this muscle, and, emerging from its lower border, passes down between the superficial and deep layer of muscles, to the middle of the fore-arm. Considerably diminished in size, it descends on the interosseous membrane, beneath the Extensor secundi internodii pollicis, to the back of the carpus, where it presents a gangliform enlargement, from which filaments are distributed to the ligaments and articulations of the carpus. It supplies all the muscles of the radial and posterior brachial regions, excepting the Anconeus, Supinator longus, and Extensor carpi radialis longior.

Dorsal Nerves

The *Dorsal Nerves* are twelve in number on each side. The first appears between the first and second dorsal vertebrae, and the last between the last dorsal and first lumbar.

The *roots of origin* of the dorsal nerves are few in number, of small size, and vary but slightly from the second to the last. Both roots are very slender; the posterior ones exceeding in thickness those of the anterior only in a slight degree. These roots gradually increase in length from above downwards, and remain in contact with the spinal cord for a distance equal to the height of, at least, two vertebrae, in the lower part of the dorsal region. They then join in the intervertebral foramen, and, at their exit, divide into two branches, a posterior, or dorsal, and an anterior, or intercostal branch.

The first and last dorsal nerves are exceptions to these characters.

The *Posterior primary branches of the Dorsal Nerves*, which are smaller than the intercostal, pass backwards between the transverse processes, and divide into external and internal branches.

The *external branches* increase in size from above downwards. They pass through the Longissimus dorsi, corresponding to the cellular interval between it and the Sacro-lumbalis, supplying these muscles, as well as those by which they are continued upwards to the head, and the Levatores costarum; the five or six lower ones giving off cutaneous filaments.

The *internal branches* of the six upper nerves pass inwards to the interval between the Multifidus spinae, and Semi-spinalis dorsi muscles, which they supply; then, piercing the origin of the Rhomboideus and Trapezius, become cutaneous by the side of the spinous processes. The internal branches of the six lower nerves are distributed to the Multifidus spinae, without giving off any cutaneous filaments.

The *cutaneous branches* of the dorsal nerves are twelve in number, the six upper being derived from the internal branches, and the six lower from the external branches. The former pierce the Rhomboid and Trapezius muscles, close to the spinous processes, and ramify in the integument. They are frequently furnished with gangliform enlargements. The six lower cutaneous branches pierce the Serratus posticus inferior, and Latissimus dorsi, in a line with the angles of the ribs.

Intercostal Nerves

The *Intercostal Nerves* (anterior primary branches of the dorsal nerves), are twelve in number on each side. They are distributed to the parietes of the thorax and abdomen, separately from each other, without being joined in a plexus, in which respect they differ from all the other spinal nerves. Each nerve is connected with the adjoining ganglia of the sympathetic by one or two filaments. The intercostal nerves may be divided into two sets, from the difference they present in their distribution. The six upper, with the exception of the first, are limited in their distribution to the parietes of the chest. The six lower supply the parietes of the chest and abdomen.

Upper Intercostal Nerves

The *Upper Intercostal Nerves* pass forwards in the intercostal spaces with the intercostal vessels, lying below the veins and artery. At the back of the chest, they lie between the pleura and the External intercostal muscle, but are soon placed between the two planes of Intercostal muscles as far as the costal cartilages, where they lie between the pleura and the Internal intercostal muscles. Near the sternum, they cross the internal mammary artery, and Triangularis sterni, pierce the Internal intercostal and Pectoralis major muscles, and supply the integument of the mamma and front of the chest, forming the anterior cutaneous nerves of the thorax; that from the second nerve becoming joined with the supra-clavicular nerves.

Branches. Numerous slender muscular filaments supply the Intercostal and Triangularis sterni muscles. Some of these branches, at the front of the chest, cross the costal cartilages from one to another intercostal space.

Lateral Cutaneous Nerves. These are derived from the intercostal nerves, midway between the vertebrae and sternum, pierce the External intercostal and Serratus magnus muscles, and divide into two branches, anterior and posterior.

The *anterior branches* are reflected forwards to the side and fore part of the chest, supplying the integument of the chest and mamma, and the upper digitations of the External oblique.

The *posterior branches* are reflected backwards, to supply the integument over the scapula and Latissimus dorsi.

The first intercostal nerve has no lateral cutaneous branch. The lateral cutaneous branch of the second intercostal nerve is of large size, and named from its origin and distribution, the intercosto-humeral nerve.

The *Intercosto-humeral Nerve* is of large size. It pierces the External intercostal muscle, crosses the axillary space to the inner side of the arm, and joins with a filament from the nerve of Wrisberg. It then pierces the fascia, and supplies the integument of the upper half of the inner and posterior side of the arm, communicating with the internal cutaneous branch of the musculo-spiral nerve. The size of this nerve is in inverse proportion to the size of the other cutaneous nerves, especially the nerve of Wrisberg.

LOWER INTERCOSTAL NERVES

The *Lower Intercostal Nerves* (excepting the last) have the same arrangement as the upper ones as far as the anterior extremities of the intercostal spaces, where they pass behind the costal cartilages, and between the Internal oblique and Transversalis muscles, to the sheath of the Rectus, which they perforate. They supply the Rectus muscle, and terminate in branches which become subcutaneous near the linea alba (anterior cutaneous nerves of the abdomen), and supply the integument in front of the abdomen, being directed outwards to the lateral cutaneous nerves. The lower intercostal nerves supply the Intercostal and Abdominal muscles, and about the middle of their course give off lateral cutaneous branches, which pierce the External intercostal and External oblique muscles, and are distributed to the integument of the abdomen, the anterior branches passing nearly as far forwards as the margin of the Rectus, the posterior branches passing to supply the skin over the Latissimus dorsi, where they join the dorsal cutaneous nerves.

PECULIAR DORSAL NERVES

First Dorsal Nerve. Its *roots of origin* are similar to those of a cervical nerve. Its *posterior* or *dorsal branch* resembles, in its mode of distribution, the dorsal branches of the cervical nerves. Its *anterior branch* enters almost wholly into the formation of the brachial plexus, giving off, before it leaves the thorax, a small intercostal branch, which runs along the first intercostal space, and terminates on the front of the chest, by forming the first anterior cutaneous nerve of the thorax. The first intercostal nerve gives off no lateral cutaneous branch.

The *Last Dorsal* is larger than the other dorsal nerves. Its anterior branch runs along the lower border of the last rib in front of the Quadratus lumborum, perforates the aponeurosis of the Transversalis, and passes forwards between it and the Internal oblique, to be distributed in the same manner as the preceding nerves. It communicates with the ilio-hypogastric branch of the lumbar plexus, and is occasionally connected with the first lumbar nerve by a slender branch, the dorsi-lumbar nerve, which descends in the substance of the Quadratus lumborum.

The *lateral cutaneous branch* of the last dorsal is remarkable for its large size; it perforates the Internal and External oblique muscles, passes downwards over the crest of the ilium, and is distributed to the integument of the front of the hip, some of its filaments extending as low down as the trochanter major.

LUMBAR NERVES

The *Lumbar Nerves* are five in number on each side; the first appears between the first and second lumbar vertebrae, and the last between the last lumbar and the base of the sacrum.

The *roots* of the lumbar nerves are the largest, and their filaments the most numerous, of all the spinal nerves, and they are closely aggregated together upon the lower end of the cord. The anterior roots are smaller, but there is not the same disproportion between them and the posterior roots as in the cervical nerves. The roots of these nerves have a vertical direction, and are of considerable length, more especially the lower ones, as the spinal cord does not extend beyond the first lumbar vertebra. The roots become joined in the intervertebral foramina, and at their exit divide into two branches, anterior and posterior.

The *Posterior branches* of the lumbar nerves diminish in size from above downwards; they pass backwards between the transverse processes, and divide into external and internal branches.

The *external branches* supply the Erector spinae and Intertransverse muscles. From the three upper branches cutaneous nerves are derived, which pierce the Sacro-lumbalis and Latissimus dorsi muscles, and descend over the back part of the crest of the ilium to be distributed to the integument of the gluteal region, some of the filaments passing as far as the trochanter major.

The *internal branches*, the smaller, pass inwards close to the articular processes of the vertebrae, and supply the Multifidus spinae and Interspinales muscles.

The *Anterior branches* of the lumbar nerves increase in size from above downwards. At their origin, they communicate with the lumbar ganglia of the sympathetic by long slender filaments, which accompany the lumbar arteries around the sides of the bodies of the vertebrae, beneath the Psoas muscle. The nerves pass obliquely outwards behind the Psoas magnus, or between its fasciculi, distributing filaments to it and the Quadratus lumborum. The anterior branches of the four upper nerves are connected together in this situation by anastomotic loops, and form the lumbar plexus. The anterior branch of the fifth lumbar, joined with a branch from the fourth, descends across the base of the sacrum to join the anterior branch of the first sacral nerve, and assist in the formation of the sacral plexus. The cord resulting from the union of these two nerves is called the *lumbo-sacral nerve*.

Lumbar Plexus

The *Lumbar Plexus* is formed by the loops of communication between the anterior branches of the four upper lumbar nerves. The plexus is narrow above, and occasionally connected with the last dorsal by a slender branch, the dorsi lumbar nerve; it is broad below, where it is joined to the sacral plexus by the lumbo-sacral. It is situated in the substance of the Psoas muscle near its posterior part, in front of the transverse processes of the lumbar vertebrae.

The mode in which the plexus is formed is the following. The first lumbar nerve gives off the ilio-hypogastric and ilio-inguinal nerves, and a communicating branch to the second. The second gives off the external cutaneous and genito-crural, and a communicating branch to the third nerve. The third nerve gives a descending filament to the fourth, and divides into two branches, which assist in forming the anterior crural and obturator nerves. The fourth nerve completes the formation of the anterior crural, and the obturator; furnishes part of the accessory obturator, and gives off a communicating branch to the fifth lumbar.

The branches of the lumbar plexus are the

Ilio-hypogastric.	Obturator.
Ilio-inguinal.	Accessory obturator.
Genito-crural.	Anterior crural.
External cutaneous.	

These branches may be divided into two groups, according to their mode of distribution. One group, including the ilio-hypogastric, ilio-inguinal, and part of the genito-crural nerves, supplies the lower part of the parietes of the abdomen; the other group, which includes the remaining nerves, supplies the fore part of the thigh and inner side of the leg.

267.—The Lumbar Plexus and its Branches.

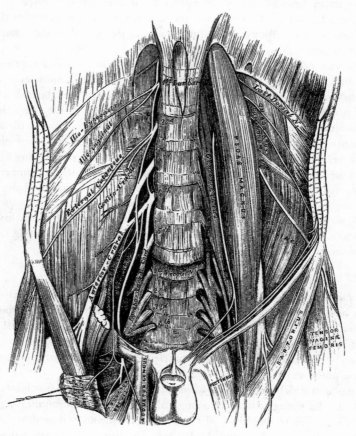

The ILIO-HYPOGASTRIC BRANCH (*superior musculo-cutaneous*) arises from the first lumbar nerve. It pierces the outer border of the Psoas muscle at its upper part, and crosses obliquely over the Quadratus lumborum to the crest of the ilium. It then perforates the Transversalis muscle, and divides between it and the Internal oblique into two branches, iliac and hypogastric.

The *iliac branch* pierces the Internal and External oblique muscles immediately above the crest of the ilium, and is distributed to the integument of the gluteal region, behind the lateral cutaneous branch of the last dorsal nerve (fig. 270). The size of this nerve bears an inverse proportion to that of the cutaneous branch of the last dorsal nerve.

The *hypogastric branch* continues onwards between the Internal oblique and Transversalis muscles. It first pierces the Internal oblique, and near the middle line perforates the External oblique, and is distributed to the integument covering the hypogastric region.

The ILIO-INGUINAL BRANCH (*inferior musculo-cutaneous*), smaller than the preceding, also arises from the first lumbar nerve. It pierces the outer border of the Psoas just below the ilio-hypogastric, and passes obliquely downwards and outwards across the Quadratus lumborum and Iliacus muscles, perforates the Transversalis, and communicates with the ilio-hypogastric nerve between that muscle and the Internal oblique, near the fore part of the crest of the ilium. The nerve then pierces the Internal oblique, distributing filaments to it, and accompanying the spermatic cord, escapes at the external abdominal ring, and is distributed to the integument of the scrotum and upper and inner part of the thigh in the male, and to the labium in the female. The size of this nerve is in inverse proportion to that of the ilio-hypogastric. Occasionally it is very small, and ends by joining it; in such cases, a branch from the ilio-hypogastric takes the place of that nerve, or the nerve may be altogether absent.

The GENITO-CRURAL NERVE arises from the second lumbar, and by a few fibres from the cord of communication between it and the first. It passes obliquely through the substance of the Psoas, descends on its surface to near Poupart's ligament, and divides into a genital and a crural branch.

The *genital branch* descends on the external iliac artery, sending a few filaments around that vessel; it then pierces the fascia transversalis, and passing through the internal abdominal ring, descends along the back part of the spermatic cord to the scrotum, and supplies, in the male, the Cremaster muscle. In the female, it accompanies the round ligament, and is lost upon it.

The *crural branch* passes along the inner margin of the Psoas muscle, beneath Poupart's ligament, into the thigh, where it pierces the fascia lata, and is distributed to the integument of the upper and anterior aspect of the thigh, communicating with the middle cutaneous nerve.

A few filaments from this nerve may be traced on to the femoral artery; they are derived from the nerve as it passes beneath Poupart's ligament.

The EXTERNAL CUTANEOUS NERVE arises from the second lumbar, or from the loop between it and the third. It perforates the outer border of the Psoas muscle about its middle, and crosses the Iliacus muscle obliquely, to the notch immediately beneath the anterior superior spine of the ilium, where it passes beneath Poupart's ligament into the thigh, and divides into two branches of nearly equal size, anterior and posterior.

The *anterior branch* descends in an aponeurotic canal formed in the fascia lata, becomes superficial about four inches below Poupart's ligament, and divides into branches, which are distributed to the integument along the anterior and outer part of the thigh, as far down as the knee. This nerve occasionally communicates with the long saphenous nerve.

The *posterior branch* pierces the fascia lata, and subdivides into branches which pass across the outer and posterior surface of the thigh, supplying the integument in this region as far as the middle of the thigh.

GRAY'S ANATOMY

268.—Cutaneous Nerves of Lower Extremity. Front View.

269.—Nerves of the Lower Extremity. Front View.

The OBTURATOR NERVE supplies the Obturator externus and Adductor muscles of the thigh, the articulations of the hip and knee, and occasionally the integument of the thigh and leg. It arises by two branches; one from the third, the other from the fourth lumbar nerve. It descends through the inner fibres of the Psoas muscle, and emerges from its inner border near the brim of the pelvis; it then runs along the lateral wall of the pelvis, above the obturator vessels, to the upper part of the obturator foramen, where it enters the thigh, and divides into an anterior and a posterior branch, separated by the Adductor brevis muscle.

The *anterior branch* passes down in front of the Adductor brevis, being covered by the Pectineus and Adductor longus; and at the lower border of the latter muscle, communicates with the internal cutaneous and internal saphenous nerves, forming a kind of plexus. It then descends upon the femoral artery, upon which it is finally distributed.

This nerve, near the obturator foramen, gives off an articular branch to the hip-joint. Behind the Pectineus, it distributes muscular branches to the Adductor longus and Gracilis, and occasionally to the Adductor brevis and Pectineus, and receives a communicating branch from the accessory obturator nerve.

Occasionally this communicating branch is continued down, as a cutaneous branch, to the thigh and leg; emerging from the lower border of the Adductor longus, it descends along the posterior margin of the Sartorius to the inner side of the knee, where it pierces the deep fascia, communicates with the long saphenous nerve, and is distributed to the integument of the inner side of the leg, as low down as its middle. When this branch is small, its place is supplied by the internal cutaneous nerve.

The *posterior branch* of the obturator nerve pierces the Obturator externus, and passes behind the Adductor brevis to the front of the Adductor magnus, where it divides into numerous muscular branches, which supply the Obturator externus, the Adductor magnus, and occasionally the Adductor brevis.

The *articular branch* for the knee-joint perforates the lower part of the Adductor magnus, and enters the upper part of the popliteal space; descending upon the popliteal artery, as far as the back part of the knee-joint, it perforates the posterior ligament, and is distributed to the synovial membrane. It gives filaments to the artery in its course.

The ACCESSORY OBTURATOR NERVE is of small size, and arises either from the obturator nerve near its origin, or by separate filaments from the third and fourth lumbar nerves. It descends along the inner border of the Psoas muscle, crosses the body of the pubes, and passes beneath the Pectineus muscle, where it divides into numerous branches. One of these supplies the Pectineus, penetrating its under surface; another is distributed to the hip-joint; while a third communicates with the anterior branch of the obturator nerve. This branch, when of large size, is prolonged (as already mentioned), as a cutaneous branch, to the leg. The accessory obturator nerve is not constantly found; when absent, the hip-joint receives branches from the obturator nerve. Occasionally it is very small, and becomes lost in the capsule of the hip-joint.

The ANTERIOR CRURAL NERVE is the largest branch of the lumbar plexus. It supplies muscular branches to the Iliacus, Pectineus, and all the muscles on the front of the thigh, excepting the Tensor vaginae femoris; cutaneous filaments to the front and inner side of the thigh, and to the leg and foot; and articular branches to the knee. It arises from the third and fourth lumbar nerves, receiving also a fasciculus from the second. It descends through the fibres of the Psoas muscle, emerging from it at the lower part of its outer border; and passes down between it and the Iliacus, and beneath Poupart's ligament, into the thigh, where it becomes somewhat flattened, and divides into an anterior or cutaneous, and a posterior or muscular part. Beneath Poupart's ligament, it is separated from the femoral artery by the Psoas muscle, and lies beneath the iliac fascia.

Within the pelvis, the anterior crural nerve gives off some small branches to the Iliacus, and a branch to the femoral artery, which is distributed upon the upper part of that vessel. The origin of this branch varies; it occasionally arises higher than usual, or it may arise lower down in the thigh.

External to the pelvis, the following branches are given off:

From the Anterior Division.	*From the Posterior Division.*
Middle cutaneous.	Muscular.
Internal cutaneous.	Articular.
Long Saphenous.	

The *Middle Cutaneous Nerve* pierces the fascia lata (occasionally the Sartorius also), about three inches below Poupart's ligament, and divides into two branches, which descend in immediate proximity along the fore part of the thigh, distributing numerous branches to the integument as low as the front of the knee, where it joins a branch of the internal saphenous nerve. Its outer branch communicates, above, with the crural branch of the genito-crural nerve; and the inner branch with the internal cutaneous nerve below. The Sartorius muscle is supplied by this or the following nerve.

The *Internal Cutaneous Nerve* passes obliquely across the upper part of the sheath of the femoral artery, and divides in front, or at the inner side, of that vessel, into two branches, anterior and internal.

The *anterior branch* perforates the fascia lata at the lower third of the thigh, and divides into two branches, one of which supplies the integument as low down as the inner side of the knee; the other crosses the patella to the outer side of the joint, communicating in its course with the long saphenous nerve. A cutaneous filament is occasionally given off from this nerve, which accompanies the long saphenous vein; and it sometimes communicates with the internal branch of the nerve.

The *inner branch* descends along the posterior border of the Sartorius muscle to the knee, where it pierces the fascia lata, communicates with the long saphenous nerve, and gives off several cutaneous branches. The nerve then passes down the inner side of the leg, to the integument of which it is distributed. This nerve, beneath the fascia lata, joins in a plexiform network, by uniting with branches of the long saphenous and obturator nerves. When the communicating branch from the latter nerve is large, and continued to the integument of the leg, the inner branch of the internal cutaneous is small, and terminates at the plexus, occasionally giving off a few cutaneous filaments.

This nerve, before subdividing, gives off a few filaments, which pierce the fascia lata, to supply the integument of the inner side of the thigh, accompanying the long saphena vein. One of these filaments passes through the saphenous opening; a second becomes subcutaneous about the middle of the thigh; and a third pierces the fascia at its lower third.

The *Long*, or *Internal Saphenous Nerve*, is the largest of the cutaneous branches of the anterior crural nerve. It approaches the femoral artery where this vessel passes beneath the Sartorius, and lies on its outer side, beneath the aponeurotic covering, as far as the opening in the lower part of the Adductor magnus. It then quits the artery, and descends vertically along the inner side of the knee, beneath the Sartorius, pierces the deep fascia between the tendons of the Sartorius and Gracilis, and becomes subcutaneous. The nerve then passes along the inner side of the leg, accompanied by the internal saphenous vein, descends behind the internal border of the tibia, and, at the lower third of the leg, divides into two branches: one continues its course along the margin of the tibia, terminating at the inner ankle; the other passes in front of the ankle, and is distributed to the integument along the inner side of the foot, as far as the great toe.

Branches. The long saphenous nerve, *about the middle of the thigh*, gives off a communicating branch, which joins the plexus formed by the obturator and internal cutaneous nerves.

At *the inner side of the knee*, it gives off a large branch (*n. cutaneus patellae*), which pierces the Sartorius and fascia lata, and is distributed to the integument in front of the patella. This nerve communicates *above* the knee with the anterior branch of the internal cutaneous; *below* the knee, with other branches of the long saphenous; and, on the *outer side* of the joint, with branches of the middle and external cutaneous nerves, forming a plexiform network, the plexus patellae. This nerve is occasionally small, and terminates by joining the internal cutaneous, which supplies its place in front of the knee.

Below the knee, the branches of the long saphenous nerve are distributed to the integument of the front and inner side of the leg, communicating with the cutaneous branches from the internal cutaneous, or obturator nerve.

The DEEP GROUP of branches of the anterior crural nerve are muscular and articular.

The *Muscular branches* supply the Pectineus, and all the muscles on the front of the thigh, except the Tensor vaginae femoris, which is supplied from the gluteal nerve, and the Sartorius, which is supplied by filaments from the middle or internal cutaneous nerves.

The branches to the *Pectineus*, usually two in number, pass inwards behind the femoral vessels, and enter the muscle on its anterior surface.

The branch to the *Rectus* muscle enters its under surface high up.

The branch to the *Vastus externus*, of large size, follows the course of the descending branch of the external circumflex artery, to the lower part of the muscle. It gives off an articular filament.

The branches to the *Vastus internus* and crureus, enter the middle of those muscles.

The *Articular branches*, two in number, supply the knee-joint. One, a long, slender filament, is derived from the nerve to the Vastus externus. It penetrates the capsular ligament of the joint on its anterior aspect. The other is derived from the nerve to the Vastus internus. It descends along the internal intermuscular septum, accompanying the deep branch of the anastomotica magna, pierces the capsular ligament of the joint on its inner side, and supplies the synovial membrane.

The Sacral and Coccygeal Nerves

The *Sacral Nerves* are five in number on each side. The four upper ones pass from the sacral canal, through the sacral foramina; the fifth escaping with the coccygeal nerve, from the sacral canal at its termination.

The *roots of origin* of the upper sacral (and lumbar) nerves, are the largest of all the spinal nerves; whilst those of the lowest sacral and coccygeal nerve are the smallest.

The *length of the roots* of these nerves is very considerable, being longer than those of any of the other spinal nerves, on account of the spinal cord not extending beyond the first lumbar vertebra. From their great length, and the appearance they present in connection with the spinal cord, the roots of origin of these nerves are called collectively the *cauda equina*. Each sacral and coccygeal nerve divides into two branches, anterior and posterior.

The *Posterior Sacral Nerves* are small, diminish in size from above downwards, and emerge, except the last, from the sacrum by the posterior sacral foramina.

The *three upper* ones are covered, at their exit from the sacrum, by the Multifidus spinae, and divide into external and internal branches.

The *internal branches* are small, and supply the Multifidus spinae.

The *external branches* communicate with one another, and with the last lumbar and fourth sacral nerves, by means of anastomosing loops. These branches pass outwards, to the outer surface of the great sacro-sciatic ligament, where they form a second series of loops beneath the Gluteus maximus. Cutaneous branches from these second series of loops, usually three in number, pierce this muscle, one near the posterior inferior spine of the ilium; another opposite the end of the sacrum; and the third, midway between these two. They supply the integument over the posterior part of the gluteal region.

The *two lower posterior Sacral Nerves* are situated below the Multifidus spinae. They are of small size, and join with each other, and with the coccygeal nerve, so as to form loops on the back of the sacrum, filaments from which supply the integument over the coccyx.

The *posterior branch of the Coccygeal Nerve* is small. It separates from the anterior in the sacral canal, and receives, as already mentioned, a communicating branch from the last sacral. It is lost in the fibrous structure on the back of the coccyx.

The *Anterior Sacral Nerves* diminish in size from above downwards. The four upper ones emerge from the anterior sacral foramina; the anterior branch of the fifth, together with the coccygeal nerve, between the sacrum and the coccyx. All the anterior sacral nerves communicate with the sacral ganglia of the sympathetic, at their exit from the sacral foramina. The *first* nerve, of large size, unites with the lumbo-sacral nerve. The *second* equals in size the preceding, with which it joins. The *third*, about one-fourth the size of the second, unites with the preceding nerves, to form the sacral plexus.

The *fourth anterior Sacral Nerve* sends a branch to join the sacral plexus. The remaining portion of the nerve divides into visceral and muscular branches: and a communicating filament descends to join the fifth sacral nerve. The *visceral branches* are distributed to the viscera of the pelvis, communicating with the sympathetic nerve. These branches ascend upon the rectum and bladder: in the female, upon the vagina and bladder, communicating with branches of the sympathetic to form the hypogastric plexus. The *muscular branches* are distributed to the Levator ani, Coccygeus, and Sphincter ani. Cutaneous filaments arise from the latter branch, which supply the integument between the anus and coccyx.

The *fifth anterior Sacral Nerve*, after passing from the lower end of the sacral canal, pierces the Coccygeus muscle, and descends upon its anterior surface to the tip of the Coccyx, where it perforates that muscle, to be distributed to the integument over the back part and side of the Coccyx. This nerve communicates above with the fourth, and below with the coccygeal nerve, and supplies the Coccygeus muscle.

The *anterior branch* of the coccygeal nerve is a delicate filament which escapes at the termination of the sacral canal. It pierces the sacro-sciatic ligament and Coccygeus muscle, is joined by a branch from the fifth anterior sacral, and becomes lost in the integument at the back part and side of the Coccyx.

Sacral Plexus

The sacral plexus is formed by the lumbo-sacral, the anterior branches of the three upper, and part of the fourth sacral nerves. These nerves proceed in different directions; the upper ones obliquely outwards, the lower ones nearly horizontally, and unite into a single, broad, flat cord. The sacral plexus is triangular in form, its base corresponding with the exit of the nerves from the sacrum, its apex with the lower part of the great sacro-sciatic foramen. It rests upon the anterior surface of the Pyriformis, and is covered in front by the pelvic fascia, which separates it from the sciatic and pudic branches of the internal iliac artery, and from the viscera of the pelvis.

The branches of the sacral plexus are:

    Muscular.   Pudic.
    Superior gluteal.   Small sciatic.
        Great sciatic.

The *Muscular branches* supply the Pyriformis, Obturator Internus, the two Gemelli, and the Quadratus femoris. The branch to the Pyriformis arises either from the plexus, or from the upper sacral nerves: the branch to the Obturator internus arises at the junction of the lumbo-sacral and first sacral nerves; it crosses behind the spine of the ischium, and passes through the lesser sacro-sciatic foramen to the inner surface of the Obturator internus: the branch to the Gemellus superior arises from the lower part of the plexus, near the pudic nerve: the small branch to the Gemellus inferior and Quadratus femoris also arises from the lower part of the plexus; it passes beneath the Gemelli and tendon of the Obturator internus, and supplies an articular branch to the hip-joint. This branch is occasionally derived from the upper part of the great sciatic nerve.

The SUPERIOR GLUTEAL NERVE arises from the back part of the lumbo-sacral; it passes from the pelvis through the great sacro-sciatic foramen above the Pyriformis muscle, accompanied by the gluteal artery, and divides into a superior and an inferior branch.

The *superior branch* follows the line of origin of the Gluteus minimus, and supplies it and the Gluteus medius.

The *inferior branch* crosses obliquely between the Gluteus minimus and medius, distributing filaments to both these muscles, and terminates in the Tensor vaginae femoris, extending nearly to its lower end.

The PUDIC NERVE arises from the lower part of the sacral plexus, and leaves the pelvis, through the great sacro-sciatic foramen, below the Pyriformis. It then crosses the spine of the ischium, and re-enters the pelvis through the lesser sacro-sciatic foramen. It accompanies the pudic artery upwards and forwards along the outer wall of the ischio-rectal fossa, being covered by the obturator fascia, and divides into two terminal branches, the perineal nerve, and the dorsal nerve of the penis. Near its origin, it gives off the inferior haemorrhoidal nerve.

The *Inferior Haemorrhoidal Nerve* is occasionally derived from the sacral plexus. It passes across the ischio-rectal fossa, with its accompanying vessels, towards the lower end of the rectum, and is distributed to the External sphincter and the integument around the anus. Branches of this nerve communicate with the inferior pudendal and superficial perineal nerves on the inner margin of the thigh.

The *Perineal Nerve*, the most inferior and largest of the two terminal branches of the pudic, is situated below the pudic artery. It accompanies the superficial perineal artery in the perineum, dividing into cutaneous and muscular branches.

The *cutaneous branches* (superficial perineal) are two in number, posterior and anterior. The *posterior branch* passes to the back part of the ischio-rectal fossa, distributing filaments to the Sphincter ani and integument in front of the anus, which communicate with the inferior haemorrhoidal nerve; it then passes forwards, with the anterior branch, to the back of the scrotum, communicating with this nerve and the inferior pudendal. The *anterior branch* passes to the fore part of the ischio-rectal fossa, in front of the preceding, and accompanies it to the scrotum and under part of the penis. This branch gives one or two filaments to the Levator ani.

The *muscular branches* are distributed to the Transversus perinei, Accelerator urinae, Erector penis, and Compressor urethrae. The nerve of the bulb supplies the corpus spongiosum; some of its filaments run for some distance on the surface, before penetrating its interior.

The *Dorsal Nerve of the Penis* is the superior division of the pudic nerve; it accompanies the pudic artery along the ramus of the ischium, and between the two layers of the deep perineal fascia; it then pierces the suspensory ligament of the penis, and accompanies the arteria dorsalis penis to the glans, to which it is distributed. On the penis, this nerve gives off a cutaneous branch, which runs along the side of the organ; it is joined with branches of the sympathetic, and supplies the integument of the upper surface and sides of the penis and prepuce, giving a large branch to the corpus cavernosum.

In the female, the pudic nerve is distributed to the parts analogous to those of the

270.—Cutaneous Nerves of Lower Extremity. Posterior View.

271.—Nerves of the Lower Extremity. Posterior View.

male; its superior division terminating in the clitoris, the inferior in the external labia and perineum.

The SMALL SCIATIC NERVE supplies the integument of the perineum and back part of the thigh and leg, and one muscle, the Gluteus maximus. It is usually formed by the union of two branches, which arise from the lower part of the sacral plexus. It arises below the Pyriformis muscle, descends beneath the Gluteus maximus with the sciatic artery, and at the lower border of that muscle passes along the back part of the thigh, beneath the fascia lata, to the lower part of the popliteal region, where it pierces the fascia and becomes cutaneous. It then accompanies the external saphenous vein below the middle of the leg, its terminal filaments communicating with the external saphenous nerve.

The branches of the small sciatic nerve are muscular (inferior gluteal) and cutaneous.

The *inferior gluteal* consist of several large branches given off to the under surface of the Gluteus maximus, near its lower part.

The *cutaneous branches* consist of two groups; internal and ascending.

The *internal cutaneous branches* are distributed to the skin at the upper and inner side of the thigh, on its posterior aspect. One branch longer than the rest, the inferior pudendal, curves forward below the tuber ischii, pierces the fascia lata on the outer side of the ramus of that bone, and is distributed to the integument of the scrotum, communicating with the superficial perineal nerve.

The *ascending cutaneous branches* consist of two or three filaments, which turn upwards round the lower border of the Gluteus maximus, to supply the integument covering its surface. One or two filaments occasionally descend along the outer side of the thigh, supplying the integument as far as the middle of this region.

Two or three branches are given off from the lesser sciatic nerve as it descends beneath the fascia of the thigh; they supply the integument of the back part of the thigh, popliteal region, and upper part of the leg.

The GREAT SCIATIC NERVE supplies nearly the whole of the integument of the leg, the muscles of the back of the thigh, and of the leg and foot. It is the largest nervous cord in the body, measuring three-quarters of an inch in breadth, and is the continuation of the lower part of the sacral plexus. It passes out of the pelvis through the great sacro-sciatic foramen, below the Pyriformis muscle. It descends between the trochanter major and tuberosity of the ischium, along the back part of the thigh, to about its lower third, where it divides into two large branches, the internal and external popliteal nerves.

This division may take place at any point between the sacral plexus and the lower third of the thigh. When the division occurs at the plexus, the two nerves descend together, side by side; or they may be separated, at their commencement, by the interposition of part or the whole of the Pyriformis muscle. As the nerve descends along the back of the thigh, it rests at first upon the External rotator muscles, together with the small sciatic nerve and artery, being covered by the Gluteus maximus; lower down, it lies upon the Adductor magnus, being covered by the long head of the Biceps.

The *branches* of the nerve, before its division, are articular and muscular.

The *articular branches* arise from the upper part of the nerve; they supply the hip-joint, perforating its fibrous capsule posteriorly. These branches are sometimes derived from the sacral plexus.

The *muscular branches* are distributed to the Flexors of the leg, viz. the biceps, Semi-tendinosus and Semi-membranosus, and a branch to the Adductor magnus. These branches are given off beneath the Biceps muscle.

The INTERNAL POPLITEAL NERVE, the larger of the two terminal branches of the great sciatic nerve, descends along the back part of the thigh through the middle of the popliteal space, to the lower part of the Popliteus muscle, where it passes with the artery beneath the arch of the Soleus, and becomes the posterior tibial. It lies at first very superficial, and at the outer side of the popliteal artery; opposite the knee-joint it is in close relation with these vessels, and crosses the artery to its inner side.

The *branches* of this nerve are articular, muscular, and a cutaneous branch, the external or short saphenous nerve.

The *articular branches*, usually three in number, supply the knee-joint; two of these branches accompany the superior and inferior internal articular arteries, and a third the azygos.

The *muscular branches*, four or five in number, arise from the nerve as it lies between the two heads of the Gastrocnemius muscle; they supply this muscle, the Plantaris, Soleus, and Popliteus.

The *External* or *Short Saphenous Nerve* descends between the two heads of the Gastrocnemius muscle, and about the middle of the back of the leg pierces the deep fascia, and receives a communicating branch (communicans peronei) from the external popliteal nerve. The nerve then continues its course down the leg near the outer margin of the tendo Achillis, in company with the external saphenous vein, winds round the outer malleolus, and is distributed to the integument along the outer side of the foot and little toe, communicating on the dorsum of the foot with the musculo-cutaneous nerve.

The *Posterior Tibial Nerve* commences at the lower border of the Popliteus muscle, and passes along the back part of the leg with the posterior tibial vessels to the interval between the inner malleolus and the heel, where it divides into the external and internal plantar nerves. It lies upon the deep muscles of the leg, and is covered by the deep fascia, the superficial muscles, and integument. In the upper part of its course, it lies to the inner side of the posterior tibial artery; but it soon crosses that vessel, and lies to its outer side as far as the ankle. In the lower third of the leg, it is placed parallel with the inner margin of the tendo Achillis.

The *branches* of the posterior tibial nerve are muscular and plantar-cutaneous.

The *muscular branches* arise either separately, or by a common trunk from the upper part of the nerve. They supply the Tibialis posticus, Flexor longus digitorum, and Flexor longus pollicis muscles; the branch to the latter muscle accompanies the peroneal artery.

The *plantar cutaneous branch* perforates the internal annular ligament, and supplies the integument of the heel and inner side of the sole of the foot.

The *Internal Plantar Nerve* (fig. 272), the larger of the two terminal branches of the posterior tibial, accompanies the internal plantar artery along the inner side of the foot. From its origin at the inner ankle it passes forwards between the Abductor

272.—The Plantar Nerves.

pollicis and Flexor brevis digitorum, divides opposite the bases of the metatarsal bones, into four digital branches, and communicates with the external plantar nerve.

Branches. In its course, the internal plantar nerve gives off *cutaneous branches*, which pierce the plantar fascia, and supply the integument of the sole of the foot; *muscular branches*, which supply the Abductor pollicis and Flexor brevis digitorum; *articular branches* to the articulations of the tarsus and metatarsus; and *four digital branches*. These pierce the plantar fascia in the clefts between the toes, and are distributed in the following manner. The *first* supplies the inner border of the great toe, and sends a filament to the Flexor brevis pollicis muscle; the *second* bifurcates to supply the adjacent sides of the great and second toes, sending a filament to the first Lumbrical muscle; the *third* digital branch supplies the adjacent sides of the second and third toes and the second Lumbrical muscle; and the *fourth* the corresponding sides of the third and fourth toes. This nerve receives a communicating branch from the external plantar nerve. It will be observed that the distribution of these branches is precisely similar to that of the median. Each digital nerve gives off cutaneous and articular filaments; and opposite the last phalanx sends a dorsal branch, which supplies the structures around the nail, the continuation of the nerve being distributed to the ball of the toe.

The *External Plantar Nerve*, the smaller of the two, completes the nervous supply to the structures of the foot, being distributed to the little toe and one half of the fourth, as well as to some of the deep muscles. It passes obliquely forwards with the external plantar artery to the outer side of the foot, lying between the Flexor brevis digitorum and Flexor accessorius; and in the interval between the former muscle and Abductor minimi digiti, divides into a superficial and deep branch. Before its division, it supplies the Flexor accessorius and Abductor minimi digiti.

The *superficial branch* separates into two digital nerves: one, the smaller of the two, supplies the outer side of the little toe, the Flexor brevis minimi digiti, and the two interosseous muscles of the fourth metatarsal space; the other, and larger digital branch, supplies the adjoining sides of the fourth and fifth toes, and communicates with the internal plantar nerve.

The *deep* or *muscular branch* accompanies the external plantar artery into the deep part of the sole of the foot, beneath the tendons of the Flexor muscles and Adductor pollicis, and supplies all the Interossei (except those in the fourth metatarsal space), the two outer Lumbricales, the Adductor pollicis, and the Transversus pedis.

The *External Popliteal* or *Peroneal Nerve*, about one-half the size of the internal popliteal, descends obliquely along the outer side of the popliteal space, close to the margin of the Biceps muscle, to the fibula; and, about an inch below the head of this bone, pierces the origin of the Peroneus longus, and divides beneath this muscle into the anterior tibial and musculo-cutaneous nerves.

The *branches* of the peroneal nerve, previous to its division, are articular and cutaneous.

The *articular branches*, two in number, accompany the superior and inferior external articular arteries to the outer side of the knee. The upper one occasionally arises from the great sciatic nerve before its bifurcation. A third (recurrent) articular nerve is given off at the point of division of the peroneal nerve; it ascends with the tibial recurrent artery through the Tibialis anticus muscle to the front of the knee, which it supplies.

The *cutaneous branches*, two or three in number, supply the integument along the back part and outer side of the leg, as far as its middle or lower part; one of these, larger than the rest, the communicans peronei, arises near the head of the fibula, crosses the external head of the Gastrocnemius to the middle of the leg, where it joins with the external saphenous. This nerve occasionally exists as a separate branch, which is continued down as far as the heel.

The *Anterior Tibial Nerve* commences at the bifurcation of the peroneal nerve, between the fibula and upper part of the Peroneus longus, passes obliquely forwards beneath the Extensor longus digitorum to the fore part of the interosseous membrane, and reaches the outer side of the anterior tibial artery above the middle of the leg; it then descends with the artery to the front of the ankle-joint, where it divides into an external and an internal branch. This nerve lies at first on the outer side of the anterior tibial, then in front of it, and again at its outer side at the ankle-joint.

The *branches* of the anterior tibial, in its course through the leg, are muscular: these supply the Tibialis anticus, the Extensor longus digitorum, and Extensor proprius pollicis muscles.

The *external*, or *tarsal branch* of the anterior tibial, passes outwards across the tarsus, beneath the Extensor brevis digitorum, and, having become ganglionic, like the posterior interosseous nerve at the wrist, supplies the Extensor brevis digitorum and the articulations of the tarsus and metatarsus.

The *internal branch*, the continuation of the nerve, accompanies the dorsalis pedis artery along the inner side of the dorsum of the foot, and, at the first interosseous space, divides into two branches, which supply the adjacent sides of the great and second toes, communicating with the internal division of the musculo-cutaneous nerve.

The *Musculo-Cutaneous branch* supplies the muscles on the fibular side of the leg, and the integument of the dorsum of the foot. It passes forwards between the Peronei muscles and the Extensor longus digitorum, pierces the deep fascia at the lower third of the leg, on its front and outer side, and divides into two branches. This nerve, in its course between the muscles, gives off muscular branches to the Peroneus longus and brevis, and cutaneous filaments to the integument of the lower part of the leg.

The *internal branch* of the musculo-cutaneous nerve, passes in front of the ankle-joint, and along the dorsum of the foot, it supplies the inner side of the great toe, and the adjoining sides of the second and third toes. It also supplies the integument of the inner ankle and inner side of the foot, communicating with the internal saphenous nerve, and joins with the anterior tibial nerve, between the great and second toes.

The *external branch*, the larger, passes along the outer side of the dorsum of the foot, to be distributed to the adjoining sides of the third, fourth, and fifth toes. It also supplies the integument of the outer ankle and outer side of the foot, communicating with the short saphenous nerve. The distribution of these nerves will be found to vary; together, they supply all the toes excepting the outer side of the little toe, and the adjoining sides of the great and second toes.

THE SYMPATHETIC NERVE

The Sympathetic Nerve consists of a series of ganglia connected together by intervening cords, extending on each side of the vertebral column from the base of the skull to the coccyx. It may, moreover, be traced up into the head, where the ganglia occupy spaces between the cranial and facial bones. These two gangliated cords lie parallel with one another as far as the sacrum, on which bone they converge, communicating together in front of the coccyx, through a single ganglion (*ganglion impar*), placed in front of this bone. Some anatomists also state that the two cords are joined at their cephalic extremity, through a small ganglion (the ganglion of Ribes), situated upon the anterior communicating artery. Moreover, the chains of opposite sides communicate together between these two extremities in several parts, by means of the nervous cords that arise from them.

The ganglia are somewhat less numerous than the vertebrae: thus there are only three in the cervical region, twelve in the dorsal, four in the lumbar, five in the sacral, and one in the coccygeal.

The sympathetic nerve, for convenience of description, may be divided into several parts, according to the position occupied by each; and the number of ganglia of which each part is composed, may be thus arranged:

Cephalic portion	4	ganglia.	
Cervical	"	3	"
Dorsal	"	12	"
Lumbar	"	4	"
Sacral	"	5	"
Coccygeal	"	1	"

Each ganglion may be regarded as a distinct centre, from, or to, which, branches pass in various directions. These branches may be thus arranged: 1. Branches of communication between the ganglia. 2. Branches of communication with the cerebral or spinal nerves. 3. Primary branches passing to be distributed to the arteries in the vicinity of the ganglia, and to the viscera, or proceeding to other ganglia placed in the thorax, abdomen, or pelvis.

1. The branches of communication between the ganglia are composed of grey and white nerve-fibres, the latter being continuous with those fibres of the spinal nerves which pass to the ganglia.

2. The branches of communication between the ganglia and the cerebral or spinal nerves, also consist of a white and a grey portion; the former proceeding from the spinal nerve *to* the ganglion, the latter passing *from* the ganglion to the spinal nerve.

3. The primary branches of distribution also consist of two kinds of nerve-fibres, the sympathetic and spinal. They have a remarkable tendency to form intricate plexuses, which encircle the blood-vessels, and are conducted by them to the

273.—The Sympathetic Nerve.

viscera. The greater number, however, of these branches pass to a series of ganglia, or ganglionic masses, of variable size, situated in the large cavities of the trunk, the thorax, and abdomen; and are connected with the roots of the great arteries of the viscera. These ganglia are single and unsymmetrical, and are called the cardiac and semilunar. From these visceral ganglia numerous plexuses are derived, which entwine round the blood-vessels, and are conducted by them to the viscera.

The *Cephalic portion* of the sympathetic, consists of four ganglia. 1. The ophthalmic ganglion. 2. The spheno-palatine, or Meckel's ganglion. 3. The otic, or Arnold's ganglion. 4. The submaxillary ganglion.

These have been already described in connection with each of the three divisions of the fifth nerve.

Cervical Portion of the Sympathetic

The cervical portion of the sympathetic consists of three ganglia on each side, which are distinguished according to their position, as the superior, middle, and inferior cervical.

The *Superior Cervical Ganglion*, the largest of the three, is placed opposite the second and third cervical vertebrae, and sometimes as low as the fourth or fifth. It is of a reddish-grey colour, and usually fusiform in shape: sometimes broad, and occasionally constricted at intervals, so as to give rise to the opinion, that it consists of the coalescence of several smaller ganglia. It is in relation in front with the sheath of the internal carotid artery, and internal jugular vein; and behind, it lies on the Rectus capitis anticus major muscle.

Its branches may be divided into superior, inferior, external, internal, and anterior.

The *superior branch* appears to be a direct continuation of the ganglion. It is soft in texture, and of a reddish colour. It ascends by the side of the internal carotid artery, and, entering the carotid canal in the temporal bone, divides into two branches, which lie, one on the outer, and the other on the inner side, of that vessel.

The *outer branch*, the larger of the two, distributes filaments to the internal carotid artery, and forms the carotid plexus.

The *inner branch* also distributes filaments to the internal carotid, and, continuing onwards, forms the cavernous plexus.

Carotid Plexus

The carotid plexus is situated on the outer side of the internal carotid. Filaments from this plexus occasionally form a small gangliform swelling on the under surface of the artery, which is called the carotid ganglion. The carotid plexus communicates with the Gasserian ganglion of the fifth, with the sixth nerve, and spheno-palatine ganglion, and distributes filaments to the wall of the carotid artery, and to the dura mater (Valentin).

The *communicating branches* with the sixth nerve consist of one or two filaments, which join that nerve as it lies upon the outer side of the internal carotid. Other filaments are also connected with the Gasserian ganglion of the fifth nerve. The communication with the spheno-palatine ganglion is effected by the carotid portion of the Vidian nerve, which passes forwards, through the cartilaginous substance filling in the foramen lacerum medium, along the pterygoid canal, to the spheno-palatine ganglion. In this canal it joins the petrosal branch of the Vidian.

Cavernous Plexus

The cavernous plexus is situated below, and to the inner side of that part of the internal carotid, which is placed by the side of the sella Turcica, in the cavernous sinus, and is formed chiefly by the internal division of the ascending branch from the superior cervical ganglion. It communicates with the third, fourth, fifth, and sixth nerves, and with the ophthalmic ganglion, and distributes filaments to the wall of the internal carotid. The branch of communication with the third nerve joins it at its point of division; the branch to the fourth nerve joins it as it lies on the outer wall of the cavernous sinus; other filaments are connected with the under surface of the trunk of the ophthalmic nerve; and a second filament of communication joins the sixth nerve.

The filament of connection with the ophthalmic ganglion arises from the anterior part of the cavernous plexus; it accompanies the nasal nerve, or continues forwards as a separate branch.

The terminal filaments from the carotid and cavernous plexuses are prolonged along the internal carotid, forming plexuses which entwine around the cerebral and ophthalmic arteries; along the former vessel they may be traced on to the pia mater; along the latter, into the orbit, where they accompany each of the subdivisions of the vessel, a separate plexus passing with the arteria centralis retinae into the interior of the eyeball.

The *inferior*, or *descending branch* of the superior cervical ganglion communicates with the middle cervical ganglion.

The *external branches* are numerous, and communicate with the cranial nerves, and with the first four cervical nerves. The branches of communication with the cranial nerves consist of delicate filaments, which pass from the superior cervical ganglion to the ganglion of the trunk of the pneumogastric, and to the ninth nerve. A separate filament from the cervical ganglion subdivides and joins the petrosal ganglion of the glosso-pharyngeal, and the ganglion of the root of the pneumogastric in the jugular foramen.

The *internal branches* are three in number; pharyngeal, laryngeal, and the superior cardiac nerve. The pharyngeal branches pass obliquely inwards to the side of the pharynx, where they communicate with branches from the pneumo-gastric, glosso-pharyngeal, and external laryngeal nerves, and assist in forming the *pharyngeal plexus*. The laryngeal branches unite with the superior laryngeal nerve and its branches.

The superior cardiac nerve will be described in connection with the other cardiac nerves.

The *anterior branches* ramify upon the external carotid artery and its branches, forming around each a delicate plexus, on the nerves composing which small ganglia are occasionally found. These ganglia have been named, according to their position, intercarotid (one placed at the angle of bifurcation of the common carotid), lingual, temporal, and pharyngeal. The plexuses accompanying some of these arteries have important communications with other nerves. That surrounding the external carotid, is connected with the digastric branch of the facial; that surrounding the facial, communicates with the submaxillary ganglion by one or two filaments; and that accompanying the middle meningeal artery, sends offsets which pass to the otic ganglion and to the intumescentia gangvioformis of the facial nerve.

The MIDDLE CERVICAL GANGLION (thyroid ganglion) is the smallest of the three cervical ganglia, and is occasionally altogether wanting. It is placed opposite the fifth cervical vertebra, usually upon the inferior thyroid artery; hence the name "thyroid ganglion," assigned to it by Haller.

Its *superior branches* ascend to communicate with the superior cervical ganglion.

Its *inferior branches* descend to communicate with the inferior cervical ganglion.

Its *external branches* pass outwards to join the fifth and sixth cervical nerves. These branches are not constantly found.

Its *internal branches* are the thyroid, and the middle cardiac nerve.

The *thyroid branches* are small filaments, which accompany the inferior thyroid artery to the thyroid gland; they communicate, on the artery, with the superior cardiac nerve, and in the gland with branches from the recurrent and external laryngeal nerves.

The middle cardiac nerve is described with the other cardiac nerves.

The INFERIOR CERVICAL GANGLION is situated between the base of the transverse process of the last cervical vertebra and the neck of the first rib, on the inner side of the superior intercostal artery. Its form is irregular; it is larger in size than the preceding, and frequently joined with the first thoracic ganglion.

Its *superior branches* communicate with the middle cervical ganglion.

Its *inferior branches* descend, some in front, others behind the subclavian artery, to join the first thoracic ganglion. The most important of these branches constitutes the inferior cardiac nerve, to be presently described.

The *external branches* consist of several filaments, some of which communicate with the seventh and eighth cervical nerves; others accompany the vertebral artery along the vertebral canal, forming a plexus around this vessel, supplying it with filaments, and communicating with the cervical spinal nerves as high as the fourth.

CARDIAC NERVES

The cardiac nerves are three in number; superior, middle, and inferior, one being derived from each of the cervical ganglia.

The *Superior Cardiac Nerve* (nervus superficialis cordis) arises by two or more branches from the superior cervical ganglion, and occasionally receives a filament from the cord of communication between the first and second cervical ganglia. It runs down the neck behind the common carotid artery, lying upon the Longus colli muscle; and crosses in front of the inferior thyroid artery, and the recurrent laryngeal nerve.

The *right superior cardiac nerve*, at the root of the neck, passes either in front or behind the subclavian artery, and along the arteria innominata, to the back part of the arch of the aorta, to the deep cardiac plexus. This nerve, in its course, is connected with other branches of the sympathetic; about the middle of the neck it receives filaments from the external laryngeal nerve; lower down, one or two twigs from the pneumogastric; and as it enters the thorax, it joins with the recurrent laryngeal. Filaments from this nerve accompany the inferior thyroid artery to the thyroid gland.

The *left superior cardiac nerve* runs by the side of the left carotid artery, and in front of the arch of the aorta, to the superficial cardiac plexus; it occasionally passes behind this vessel, and terminates in the deep cardiac plexus.

The *Middle Cardiac Nerve* (nervus cardiacus magnus), the largest of the three, arises from the middle cervical ganglion, or from the interganglionic cord between the middle and inferior ganglia. On the right side, it descends behind the common carotid artery; and at the root of the neck passes either in front or behind the subclavian artery; it then descends on the trachea, receives a few filaments from the recurrent laryngeal nerve, and joins the deep cardiac plexus. In the neck, it communicates with the superior cardiac and recurrent laryngeal nerves. On the left side, the middle cardiac nerve enters the chest between the left carotid and subclavian arteries, and joins the left side of the deep cardiac plexus.

The *Inferior Cardiac Nerve* (nervus cardiacus minor) arises from the inferior cervical or first thoracic ganglion. It passes down behind the subclavian artery, and along the front of the trachea, to join the deep cardiac plexus. It communicates freely behind the subclavian artery with the recurrent laryngeal and middle cardiac nerves.

The *Great* or *Deep Cardiac Plexus* (*Plexus Magnus Profundus*—Scarpa) is situated in front of the trachea at its bifurcation, above the point of division of the pulmonary artery, and behind the arch of the aorta. It is formed by the cardiac nerves derived from the cervical ganglia of the sympathetic, and the cardiac branches of the recurrent laryngeal and pneumogastric. The only cardiac nerves which do not enter into the formation of this plexus are the left superior cardiac nerve and the left inferior cardiac branch from the pneumogastric. The branches derived from the great cardiac plexus form the posterior coronary plexus and part of the anterior coronary plexus, whilst a few filaments proceed to the pulmonary plexuses, and to the auricles of the heart.

The branches from the *right side* of this plexus pass some in front and others behind the right pulmonary artery; the former, the more numerous, transmit a few filaments to the anterior pulmonary plexus, and are continued along the trunk of the pulmonary artery, to form part of the anterior coronary plexus; those behind the pulmonary artery distribute a few filaments to the right auricle, and form part of the posterior coronary plexus.

The branches from the *left side* of the cardiac plexus distribute a few filaments to the left auricle of the heart and the anterior pulmonary plexus, and then pass on to form the greater part of the posterior coronary plexus, a few branches passing to the superficial cardiac plexus.

The *Superficial* or *Anterior Cardiac Plexus* lies beneath the arch of the aorta, in front of the right pulmonary artery. It is formed by the left superior cardiac nerve, the left (and occasionally the right) inferior cardiac branches of the pneumogastric, and by filaments from the deep cardiac plexus. A small ganglion (cardiac ganglion of Wrisberg) is occasionally found connected with these nerves at their point of junction. This ganglion, when present, is situated immediately beneath the arch of the aorta, on the right side of the ductus arteriosus. The superficial cardiac plexus forms the anterior part of the great coronary plexus, and several filaments pass along the pulmonary artery to the left anterior pulmonary plexus.

The *Posterior Coronary Plexus* is formed chiefly by filaments from the left side of the deep cardiac plexus, and by a few from the right side. It surrounds the branches of the coronary artery at the back of the heart, and its filaments are distributed with those vessels to the muscular substance of the ventricles.

The *Anterior Coronary Plexus* is prolonged chiefly from the superficial cardiac plexus, but receives filaments from the deep cardiac plexus. Passing forwards between the aorta and pulmonary artery, it accompanies the right coronary artery on the anterior surface of the heart.

Valentin has described nervous filaments ramifying under the endocardium; but they are less distinct in man than in mammalia; and Remak and Lee have found, in several mammalia (the latter in man), numerous small ganglia on the branches of these nerves, both on the surface of the heart and in its muscular substance.

Thoracic Part of the Sympathetic

The thoracic portion of the sympathetic consists of a series of ganglia, which usually correspond in number to that of the vertebrae; but, from the occasional coalescence of two, their number is uncertain. These ganglia are placed on each side of the spine, resting against the heads of the ribs, and covered by the pleura costalis: the last two are, however, anterior to the rest, being placed on the side of the bodies of the vertebrae. The ganglia are small in size, and of a greyish colour. The first, larger than the rest, is of an elongated form, and usually blended with the last cervical. They are connected together by cord-like prolongations from their substance.

The *external branches* from each ganglion, usually two in number, communicate with each of the dorsal spinal nerves.

The *internal branches from the six upper ganglia* are very small, and distribute filaments to the thoracic aorta and its branches, besides small branches to the bodies of the vertebrae and their ligaments.

Branches from the third and fourth ganglia form part of the posterior pulmonary plexus.

The *branches of the six lower ganglia* are large and white in colour; they distribute filaments to the aorta, and unite to form the three splanchnic nerves. These are named, the *great*, the *lesser*, and the *smallest* or *renal splanchnic*.

The *Great Splanchnic Nerve* is of a white colour, firm in texture, and bears a marked contrast to the ganglionic nerves. It is formed by branches from all the thoracic ganglia from the sixth to the tenth, receiving filaments (according to Mr. Beck) from all the thoracic ganglia above the sixth. These roots unite to form a large round cord of considerable size. It descends obliquely inwards in front of the bodies of the vertebrae along the posterior mediastinum, perforates the crus of the Diaphragm, and terminates in the semilunar ganglion, distributing filaments to the renal plexus and supra-renal gland.

The *Lesser Splanchnic Nerve* is formed by filaments from the tenth and eleventh ganglia, and from the cord between them. It pierces the Diaphragm with the preceding nerve, and joins the coeliac plexus. It communicates in the chest with the great splanchnic nerve, and occasionally sends filaments to the renal plexus.

The *Smallest* or *Renal Splanchnic Nerve* arises from the last ganglion, and piercing the Diaphragm, terminates in the renal plexus and lower part of the coeliac plexus. It occasionally communicates with the preceding nerve.

A striking analogy appears to exist between the splanchnic and the cardiac nerves. The cardiac nerves are three in number; they arise from the three cervical ganglia, and are distributed to a large and important organ in the thoracic cavity. The splanchnic nerves, also three in number, are connected with all the dorsal ganglia, and are distributed to important organs in the abdominal cavity.

The *Epigastric* or *Solar Plexus* supplies all the viscera in the abdominal cavity. It consists of a dense network of nerves and ganglia, situated behind the stomach and in front of the aorta and crura of the Diaphragm. It surrounds the coeliac axis and root of the superior mesenteric artery, extending downwards as low as the pancreas, and outwards to the supra-renal capsules. This plexus, and the ganglia connected with it, receive the great splanchnic nerve of both sides, part of the lesser splanchnic nerves, and the termination of the right pneumogastric. It distributes filaments, which accompany, under the name of plexuses, all the branches from the front of the abdominal aorta.

The *semilunar ganglia*, two in number, one on each side, are the largest ganglia in the body. They are large irregular gangliform masses, formed by the aggregation of smaller ganglia, having interspaces between them. They are situated by the side of the coeliac axis and superior mesenteric artery, close to the supra-renal glands: the one on the right side lies beneath the vena cava; the upper part of each ganglion is joined by the greater and lesser splanchnic nerves, and to the inner side of each the branches of the solar plexus are connected.

From the solar plexus are derived the following:

Phrenic or Diaphragmatic plexus.	Renal plexus.
Gastric plexus.	Superior mesenteric plexus.
Hepatic plexus.	Spermatic plexus.
Splenic plexus.	Inferior mesenteric plexus.
Supra-renal plexus.	

The *Phrenic Plexus* accompanies the phrenic artery to the Diaphragm, which it supplies, some filaments passing to the supra-renal gland. It arises from the upper part of the semilunar ganglion, and is larger on the right than on the left side. In connexion with this plexus, on the right side, at its point of junction with the phrenic nerve, is a small ganglion (ganglion diaphragmaticum). This ganglion is placed on the under surface of the Diaphragm, near the supra-renal gland. Its branches are distributed to the vena cava, supra-renal gland, and the hepatic plexus. The ganglion is absent on the left side.

The *Supra-Renal Plexus* is formed by branches from the solar plexus, from the semilunar ganglion, and from the splanchnic and phrenic nerves, a ganglion being formed at the point of junction of the latter nerve. It supplies the supra-renal gland.

The branches of this plexus are remarkable for their large size, in comparison with the size of the organ they supply.

The *Renal Plexus* is formed by filaments from the solar plexus, the outer part of the semilunar ganglion, and the aortic plexus. It is also joined by filaments from the lesser and smallest splanchnic nerves. The nerves from these sources, fifteen or twenty in number, have numerous ganglia developed upon them. They accompany the branches of the renal artery into the kidney; some filaments on the right side being distributed to the vena cava, and others to the spermatic plexus, on both sides.

The *Spermatic Plexus* is derived from the renal plexus, receiving branches from the aortic plexus. It accompanies the spermatic vessels to the testes.

In the female, the *ovarian plexus* is distributed to the ovaries and fundus of the uterus.

The *Coeliac Plexus*, of large size, is a direct continuation from the solar plexus: it surrounds the caeliac artery, and subdivides into the gastric, hepatic, and splenic plexuses. It receives branches from one or more of the splanchnic nerves, and, on the left side, a filament from the pneumogastric.

The *Gastric Plexus* accompanies the gastric artery along the lesser curvature of the stomach, and joins with branches from the left pneumogastric nerve. It is distributed to the stomach.

The *Hepatic Plexus*, the largest offset from the caeliac plexus, receives filaments from the left pneumogastric and right phrenic nerves. It accompanies the hepatic artery, ramifying in the substance of the liver, upon its branches, and upon those of the vena portae.

Branches from this plexus accompany all the divisions of the hepatic artery. Thus there is a pyloric plexus accompanying the pyloric branch of the hepatic, which joins with the gastric plexus, and pneumogastric nerves. There is also a gastro-duodenal plexus, which subdivides into the pancreatico-duodenal plexus, which accompanies the pancreatico-duodenal artery, to supply the pancreas and duodenum, joining with branches from the mesenteric plexus; and a gastro-epiploic plexus, which accompanies the right gastro-epiploic artery along the greater curvature of the stomach, and anastomoses with branches from the splenic plexus. A cystic plexus, which supplies the gall-bladder, also arises from the hepatic plexus, near the liver.

The *Splenic Plexus* is formed by branches from the right and left semilunar ganglia, and from the right pneumogastric nerve. It accompanies the splenic artery and its branches to the substance of the spleen, giving off, in its course, filaments to the pancreas (pancreatic plexus), and the left gastro-epiploic plexus, which accompanies the gastro-epiploica sinistra artery along the convex border of the stomach.

The *Superior Mesenteric Plexus* is a continuation of the lower part of the great solar plexus, receiving a branch from the junction of the right pneumogastric nerve with the caeliac plexus. It surrounds the superior mesenteric artery, which it accompanies into the mesentery, and divides into a number of secondary plexuses, which are distributed to all the parts supplied by the artery, viz., pancreatic branches to the pancreas; intestinal branches, which supply the whole of the small intestine; and ileo-colic, right colic, and middle colic branches, which supply the corresponding parts of the great intestine. The nerves composing this plexus are white in colour, and firm in texture, and have numerous ganglia developed upon them near their origin.

The *Aortic Plexus* is formed by branches on each side, from the semilunar ganglia and renal plexuses, receiving filaments from some of the lumbar ganglia. It is situated upon the sides and front of the aorta, between the origins of the superior and inferior mesenteric arteries. From this plexus arises the inferior mesenteric, part of the spermatic, and the hypogastric plexuses; and it distributes filaments to the inferior cava.

The *Inferior Mesenteric Plexus* is derived chiefly from the left side of the aortic plexus. It surrounds the inferior mesenteric artery, and divides into a number of secondary plexuses, which are distributed to all the parts supplied by the artery, viz., the left colic and sigmoid plexuses, to the descending and sigmoid flexure of the colon; and the superior haemorrhoidal plexus, which supplies the upper part of the rectum, and joins in the pelvis with branches of the left hypogastric plexus.

The Lumbar Portion of the Sympathetic

The lumbar portion of the sympathetic is situated in front of the vertebral column, along the inner margin of the Psoas muscle. It consists usually of four ganglia, connected together by interganglionic cords. The ganglia are of small size, of a greyish colour, hordeiform in shape, and placed much nearer the median line than the thoracic ganglia.

The *superior* and *inferior branches* of the lumbar ganglia, serve to communicate between the chain of ganglia in this region. They are usually single, and of a white colour.

The *external branches* communicate with the lumbar spinal nerves. From the situation of the lumbar ganglia, these branches are longer than in the other regions. They are usually two in number for each ganglion, and accompany the lumbar arteries around the sides of the bodies of the vertebrae, passing beneath the fibrous arches from which the fibres of the Psoas muscle partly arise.

The *internal branches* pass inwards, in front of the aorta, and form the lumbar aortic plexus (already described). Other branches descend in front of the common iliac arteries, and join, over the promontory of the sacrum, to form the hypogastric plexus. Numerous delicate filaments are also distributed to the bodies of the vertebrae, and the ligaments connecting them.

Pelvic Portion of the Sympathetic

The pelvic portion of the sympathetic is situated in front of the sacrum, along the inner side of the anterior sacral foramina. It consists of four or five small ganglia on each side, connected together by inter-ganglionic cords. Below, they converge and unite on the front of the coccyx, by means of a small ganglion (ganglion impar).

The *superior* and *inferior branches*, are the cords of communication between the ganglia above and below.

The *external branches*, exceedingly short, communicate with the sacral nerves. They are two in number to each ganglion. The coccygeal nerve communicates either with the last sacral, or coccygeal ganglion.

The *internal branches* communicate, on the front of the sacrum, with the corresponding branches from the opposite side; some, from the first two ganglia, pass to join the pelvic plexus, and others form a plexus, which accompanies the middle sacral artery.

The *Hypogastric Plexus* supplies the viscera of the pelvic cavity. It is situated in front of the promontory of the sacrum, between the two common iliac arteries, and is formed by the union of numerous filaments, which descend on each side from the aortic plexus, from the lumbar ganglia, and from the first two sacral ganglia. This plexus contains no ganglia, and bifurcates, below, into two lateral portions, which form the inferior hypogastric, or pelvic plexuses.

Inferior Hypogastric, or Pelvic Plexus

The inferior hypogastric, or pelvic plexus, is situated at the side of the rectum and bladder in the male, and at the side of the rectum, vagina, and bladder, in the female. It is formed by a continuation of the hypogastric plexus, by branches from the second, third, and fourth sacral nerves, and by a few filaments from the sacral ganglia. At the point of junction of these nerves, small ganglia are found. From this plexus numerous branches are distributed to all the viscera of the pelvis. They accompany the branches of the internal iliac artery.

The *Inferior Haemorrhoidal Plexus* arises from the back part of the pelvic plexus. It supplies the rectum, joining with branches of the superior haemorrhoidal plexus.

The *Vesical Plexus* arises from the fore part of the pelvic plexus. The nerves composing it are numerous, and contain a large proportion of spinal nerve-fibres. They accompany the vesical arteries, and are distributed to the side and base of the bladder. Numerous filaments also pass to the vesicular seminales, and vas deferens: those accompanying the vas deferens join, on the spermatic cord, with branches from the spermatic plexus.

The *Prostatic Plexus* is continued from the lower part of the pelvic plexus. The nerves composing it are of large size. They are distributed to the prostate gland, vesiculae seminales, and erectile structure of the penis. The nerves supplying the erectile structure of the penis, consist of two sets, the small and large cavernous nerves. They are slender filaments, which arise from the fore part of the prostatic plexus, and after joining with branches from the internal pudic nerve, pass forwards beneath the pubic arch.

The *small cavernous nerves* perforate the fibrous covering of the penis, near its root.

The *large cavernous nerve* passes forwards along the dorsum of the penis, joins with the dorsal branch of the pudic nerve, and is distributed to the corpus cavernosum and spongiosum.

The *Vaginal Plexus* arises from the lower part of the pelvic plexus. It is lost on the walls of the vagina, being distributed to the erectile tissue at its anterior part, and to the mucous membrane. The nerves composing this plexus, contain, like the vesical, a large proportion of spinal nerve-fibres.

The *Uterine Nerves* arise from the lower part of the hypogastric plexus, above the point where the branches from the sacral nerves join the pelvic plexus. They accompany the uterine arteries to the side of the organ between the layers of the broad ligament, and are distributed to the cervix and lower part of the body of the uterus, penetrating its substance.

Other filaments pass separately to the body of the uterus and Fallopian tube.

Branches from the hypogastric plexus accompany the uterine arteries into the substance of the uterus. Upon these filaments ganglionic enlargements are found.

ORGANS OF SENSE

The Organs of the Senses are the instruments by which the mind is brought into relation with external objects. These organs are five in number, viz. the organs of touch, of smell, of taste, of hearing, and of sight.

THE SKIN

The skin is the principal seat of the sense of touch, and may be regarded as a covering of protection to the exterior of the body. It consists of two layers, the derma or cutis vera, and the epidermis or cuticle.

The *Derma*, or *True Skin*, consists of fibro-areolar tissue, intermixed with numerous blood-vessels, lymphatics, and nerves. The fibro-areolar tissue forms the framework of the cutis; it is composed of firm interlacing bundles of white fibrous tissue, intermixed with a much smaller proportion of yellow elastic fibres, the amount of which varies in different parts. The fibro-areolar tissue is more abundant in the deeper layers of the cutis, where it is dense and firm, the meshes being large, and gradually becoming blended with the subcutaneous areolar tissue; towards the surface, the fibres become finer and more closely interlaced, the most superficial layer

274.—A Sectional View of the Skin (magnified).

being covered with numerous small, conical, vascular eminences, the papillae. From these differences in the structure of the cutis at different parts, it is usual to describe it as consisting of two layers; the deeper layer or corium, and the superficial or papillary layer.

The *Corium* consists of strong interlacing fibrous bands, composed chiefly of the white variety of fibrous tissue; but containing, also, some fibres of the yellow elastic tissue, which vary in amount in different parts. Towards the attached surface, the fasciculi are large and coarse; and the areolae left by their interlacing large and occupied by adipose tissue and the sudatory glands. This element of the skin becomes gradually blended with the subcutaneous areolar tissue. Towards the free surface, the fasciculi are much finer, and they have a closer interlacing, the most superficial layers consisting of a transparent, homogeneous matrix with imbedded nuclei.

The corium varies in thickness, from a quarter of a line to a line and a half, in different parts of the body. Thus, it is thicker in the more exposed regions, as the palm of the hand and sole of the foot; on the posterior aspect of the body, than the front; and on the outer, than the inner side of the limbs. In the eyelids, scrotum, and penis, it is exceedingly thin and delicate. The skin generally is thicker in the male than in the female.

The *areolae* are occupied by adipose tissue, hair follicles, and the sudatory and sebaceous glands; they are the channel by which the vessels and nerves are distributed to the more superficial strata of the corium, and to the papillary layer.

Plain muscular fibres are found in the superficial layers of the corium, wherever hairs are found; and in the subcutaneous areolar tissue of the scrotum, penis, perineum, and areolas of the nipple. In the latter situations, the fibres are arranged in bands, closely reticulated and disposed in super-imposed laminae.

The *Papillary Layer* is situated upon the free surface of the corium; it consists of numerous small, highly sensitive, and vascular eminences, the papillae, which rise perpendicularly from its surface, and form the essential element of the organ of touch. The papillae are conical-shaped eminences, having a round or blunted extremity, occasionally divided into two or more parts, and connected by their base with the free surface of the corium. On the general surface of the body, more especially in those parts which are endowed with slight sensibility, they are few in number, short, exceedingly minute, and irregularly scattered over the surface; but in other situations, as upon the palmar surface of the hands and fingers, upon the sole, and around the nipple, they are long, of large size, closely aggregated together, and arranged in curvilinear lines, forming the elevated ridges seen on the free surface of the epidermis. In these ridges, the larger papillae are arranged in a double row, with smaller papillae between them; and these rows are subdivided into small square-shaped masses by short transverse furrows regularly disposed, in the centre of each of which is the minute orifice of the duct of a sweat-gland. No papillae exist in the grooves between the ridges. In structure, the papillae resemble the superficial layer of the cutis; consisting of a homogeneous tissue, faintly fibrillated, and containing a few fine elastic fibres. The smaller papillae contain a single capillary loop; but in the larger the vessels are convoluted to a greater or less degree; each papilla also contains two or more nerve-fibres, which, after running in a waving manner through it, terminate in loops, or by a free extremity. No lymphatics have as yet been discovered in the papillae.

The *Epidermis*, or *Cuticle* (scarf-skin), is an epithelial structure, which forms a defensive covering to the surface of the true skin, being accurately moulded on the papillary layer of the derma. It varies in thickness in different parts. Where it is exposed to pressure and the influence of the atmosphere, as upon the palms of the hands and soles of the feet, it is thick, hard, and horny in texture; whilst that which lies in contact with the papillary layer, over the entire surface of the body, is soft and cellular in structure; hence the deeper layer has been called, the *rete mucosum*.

The free surface of the epidermis is marked by a network of linear furrows of variable size, marking out the surface into a number of spaces of polygonal or lozenge-shaped form. Some of these furrows are large, as opposite the flexures of the joints, and correspond to the folds in the derma produced by their movements. In other situations, as upon the back of the hand, they are exceedingly fine, and intersect one another at various angles: upon the palmar surface of the hand and fingers, and upon the sole, these lines are very distinct, and have a curvilinear arrangement, and depend upon the large size and peculiar arrangement of the papillae upon which the epidermis is placed. The deep surface of the epidermis is accurately moulded upon the papillary layer of the derma, each papilla being invested by its epidermic sheath; so that when this layer is removed by maceration, it presents a number of pits or depressions corresponding to the elevations of the papillae, as well as the furrows left in the interval between them. Fine tubular prolongations from this layer are continued into the ducts of the sudatory and sebaceous glands. In structure, the epidermis consists of flattened cells, agglutinated together, and having a laminated arrangement. In the deeper layers the cells are large, rounded or columnar, and filled with soft opaque contents. In the superficial layers the cells are flattened, transparent, dry, and firm, and their contents converted into a kind of horny matter. The difference in the structure of these layers is dependent upon the mode of growth of the epidermis. As the external layers desquamate, from their being constantly subjected to attrition, they are reproduced from beneath, successive layers gradually approaching towards the free surface, which, in their turn, die and are cast off.

These cells arise in the liquor sanguinis, which is poured out on the free surface of the derma; they contain nuclei, and form a thin stratum of closely-aggregated nucleated cells, which cover the entire extent of the papillary layer. The deepest layer of cells, according to Kölliker, are of a columnar form, and are arranged perpendicularly to the free surface of the derma, forming either a single or a double, or even triple, layer: the laminae succeeding these are composed of cells of a more rounded form, the contents of which are soft, opaque, granular, and soluble in acetic acid. As these cells successively approach the surface by the development of fresh layers from beneath, they assume a flattened form from the evaporation of their fluid contents, and finally form a transparent, dry, membranous scale, lose their nuclei, and apparently become changed in their chemical composition, as they are unaffected now by acetic acid.

The black colour of the skin in the negro, and the tawny colour among some of the white races, is due to the presence of pigment in the cells of the cuticle. This pigment is more especially distinct in the cells of the deeper layer, or rete mucosum, and is caused by the presence of pigment cells, similar to those found in the choroid. As these approach the surface and desiccate, the colour becomes partially lost.

The *arteries* which supply the skin divide into numerous branches in the subcutaneous tissue; they then pass through the areolae of the corium, and divide into a dense capillary plexus, which supplies the sudatory and sebiparous glands and the hair follicles, terminating in the superficial layers of the corium, by forming a capillary network, from which numerous fine branches ascend to the papillae.

The *lymphatic vessels* are arranged in a minute plexiform network in the superficial layers of the corium, where they become interwoven with the capillary and nervous plexuses; they are especially abundant in the scrotum and around the nipple.

The *nerves* which supply the skin ascend with the vessels through the areolae of the deep layers of the corium to the more superficial layers, where they form a minute plexiform mesh. From this plexus, the primitive nerve-fibres pass to be distributed to the papilla. The nerves are most numerous in those parts which are provided with the greatest sensibility.

Appendages of the Skin

The appendages of the skin are the nails, the hairs, the sudoriferous and sebaceous glands, and their ducts.

The nails and hairs are peculiar modifications of the epidermis, consisting essentially of the same cellular structure as that membrane.

The *Nails* are flattened elastic structures of a horny texture, placed upon the dorsal surface of the terminal phalanges of the fingers and toes. Each nail is convex on its outer surface, concave within, and is implanted by a portion called *the root* into a groove of the skin; the exposed portion is called *the body*, and the anterior extremity *the free edge*. The nail has a very firm adhesion to the cutis, being accurately moulded upon its surface, as the epidermis is in other parts. The part of the cutis beneath the body and root of the nail is called *the matrix*, from its being the part from which the nail is produced. Corresponding to the body, the matrix is thick, and covered with large highly vascular papillae, arranged in longitudinal rows, the colour of which is seen through the transparent tissue. Behind this, near the root of the nail, the papillae are small, less vascular, and have no regular arrangement; hence the portion of the nail corresponding to this part is of a whiter colour, and called *lunula*, from its form.

The cuticle, as it passes forwards on the dorsal surface of the finger, is attached to the surface of the nail a little in advance of its root: at the extremity of the finger, it is connected with the under surface of the nail, a little behind its free edge. The cuticle and horny structure of the nail (both epidermic structures) are thus seen to be directly continuous with each other. The nails, in structure, consist of cells having a laminated arrangement, and these are almost essentially similar to those composing the epidermis. The deepest layer of cells which lie in contact with the papillae at the root and under surface of the nail are of elongated form, arranged perpendicularly to the surface, and provided with nuclei; those which succeed these are of a rounded or polygonal form, the more superficial ones becoming broad, thin, and flattened, and so closely compacted together as to make the limits of each cell very indistinct.

It is by the successive growth of new cells at the root and under surface of the body of the nail, that it advances forwards, and maintains a due thickness, whilst, at the same time, the growth of the nail in the proper direction is secured. As these cells in their turn become displaced by the growth of new cells, they assume a flattened form, lose their nuclei, and finally become closely compacted together into a firm dense horny texture. In *chemical composition*, the nails resemble the epidermis. According to Mulder, they contain a somewhat larger proportion of carbon and sulphur.

Hairs are peculiar modifications of the epidermis, and consist essentially of the same structure as that membrane. They are found on nearly every part of the surface of the body, excepting the palms of the hands and soles of the feet, and vary much in length, thickness, and colour in different parts of the body, and in the different races of mankind. In some parts they are so short as not to project beyond the follicle containing them; in other parts, as upon the scalp, they are of considerable length; along the margin of the eyelids and upon the face they are remarkable for their thickness. A hair consists of a *root*, the part implanted in the skin; the *shaft*, the portion projecting from its surface; and the *point*. They generally present a cylindrical or more or less flattened form, and a reniform outline upon a transverse section.

The *root of the hair* presents a bulbous enlargement at its extremity; it is whiter in colour, and softer in texture, than the stem, and is lodged in a follicular involution of the epidermis, called the *hair follicle*. When the hair is of considerable length, the follicle extends into the subcutaneous cellular tissue. The hair follicle is bulbous at its deep extremity, like the hair which it contains, and has opening into it, near its free extremity, the orifices of the ducts of one or more sebaceous glands. In structure, the hair follicle consists of two coats; an outer or dermic, and an inner or cuticular. The outer coat is formed mainly of areolar tissue; it is continuous with the corium, is highly vascular, and supplied by numerous minute nervous filaments. The inner or cuticular lining is continuous with the epidermis, and, at the bottom of the hair follicle, with the root of the hair; this cuticular lining resembles the epidermis in the peculiar rounded form and soft character of those cells which lie in contact with the outer coat of the hair follicle, and the thin, dry, and scaly character of those which lie near the surface of the hair, to which they are closely adherent. When the hair is plucked from its follicle, this cuticular lining most commonly adheres to it, and forms what is called the *root sheath*. At the bottom of each hair follicle is a small conical-shaped vascular eminence or papilla, similar in every respect to those found upon the surface of the skin; it is continuous with the dermic layer of the follicle, is highly vascular, and probably supplied with nervous fibrils: this is the part through which material is supplied for the production and constant growth of the hair. The root of the hair rests upon this conical-shaped eminence, and is continuous with the cuticular lining of the follicle at this part. It consists of nucleated cells, similar in every respect to those which in other situations form the epidermis. These cells gradually enlarge as they are pushed upwards into the soft bulb, and some of them contain pigment granules, which exist either in separate cells, or the granules are separate but aggregated around the nucleus; it is these granules which give rise to the colour of the hair. It occasionally happens that these pigment granules completely fill the cells in the centre of the bulb, which gives rise to the dark tract of pigment often found, of greater or less length, in the axis of the hair.

The *shaft of the hair* consists of a central part, or medulla; surrounding this is the fibrous part of the hair, covered externally by the cortex. The medulla occupies the centre of the shaft, and ceases towards the point of the hair. It is usually wanting in the fine hairs covering the surface of the body, and commonly in those of the head. It is more opaque and deeper coloured than the fibrous part, and consists of cells containing pigment or fat granules. The fibrous portion of the hair constitutes the chief part of the stem; its cells are elongated, and unite to form flattened fusiform fibres. These also contain pigment granules, which assume a linear arrangement. The cells which form the cortex of the hair consist of a single layer, which surrounds those about to form the fibrous layer; they are converted into thin flat scales, having an imbricated arrangement.

The *Sebaceous Glands* are small sacculated glandular organs, lodged in the substance of the corium, or sub-dermoid tissue. They are found in all parts of the skin, but are most abundant in the face, and in those parts which are exposed to friction. Their orifices open most frequently into the hair follicles, but occasionally upon the general surface. Each gland consists of a single duct, which is more or less capacious, and terminates in a lobulated pouch-like extremity. The basement membrane forming the wall of the sac, as well as the duct, is lined by epithelium, which is filled with particles of sebaceous matter; and this becoming detached into the cavity of the sac, as its growth is renewed, constitutes the secretion. The number of sacculi connected with the duct vary from two to five, or even twenty, in number. On the nose and face the glands are of large size, distinctly lobulated, and often become much enlarged from the accumulation of pent-up secretion. The largest sebaceous glands are those found in the eyelids, the Meibomian glands.

The *Sudoriferous* or *Sweat-Glands* are found in almost every part of the skin, and are situated in small pits in the deep parts of the corium, or in the subcutaneous areolar tissue, surrounded by a quantity of adipose tissue. They are small, round, reddish bodies, consisting of one or more convoluted tubuli, from which the efferent duct proceeds upwards through the corium and cuticle, and opens upon the surface by a slightly enlarged orifice. The efferent duct, as it passes through the corium, pursues a curved and slightly serpentine course, and if the epidermis is thin, opens directly upon the surface; but where the epidermis is thicker, it assumes a spiral arrangement, the separate windings of the tube being as close and as regular as those of a common screw. The spiral course of these ducts is especially distinct in the thick cuticle of the palm of the hand and sole of the foot. The size of these glands varies. They are especially large in those regions where the amount of perspiration is great, as in the axillae, where they form a thin mammillated layer of a reddish colour, which corresponds exactly to the situation of the hair in this region; they are large, also, in the groin. Their number varies. They are most numerous on the palm of the hand, and present, according to Krause, 2,800 orifices on a square inch of the integument, and a rather less number on the sole of the foot. In both of these situations, the orifices of the ducts are exceedingly regular, and correspond to the small transverse grooves which intersect the ridges of papillae. In other situations they are more irregularly scattered; but in nearly equal numbers over parts including the same extent of surface. Each gland consists of a single tube intricately convoluted, at one end terminating by a blind extremity; at the other end opening upon the surface of the skin. In the larger glands, this single duct usually divides and subdivides dichotomously; and these smaller ducts ultimately terminate in short caecal pouches, rarely anastomosing. The wall of the duct is thick; the width of the canal rarely exceeding one-third of its diameter. The tube, both in the gland and where it forms the excretory duct, consists of two layers: an outer, formed by fine areolar tissue; and an epithelium lining its interior. The external, or fibro-cellular coat, is thin, continuous with the superficial layer of the corium, and extends only as high as the surface of the true skin. The epithelial lining is much thicker, continuous with the epidermis, and alone forms the spiral portion of the tube. When the cuticle is carefully removed from the surface of the cutis, these convoluted tubes of epidermis may be drawn out, and form nipple-shaped projections on its under surface. According to Kölliker, a layer of non-striated muscular fibres, arranged longitudinally, is found between the areolar and epithelial coats of the ducts of the larger sweat-glands, as in the axilla, root of the penis, on the labia majora, and around the anus.

The contents of the smaller sweat-glands are quite fluid; but in the larger glands, the contents are semi-fluid and opaque, and contain a number of coloured granules, and cells which appear analogous to epithelial cells.

The Tongue

The tongue is the organ of the special sense of taste. It is situated in the floor of the mouth, in the interval between the two lateral portions of the body of the lower jaw. Its base, or root, is directed backwards, and connected with the os hyoides by numerous muscles, to the epiglottis by three folds of mucous membrane, which form the glosso-epiglottic ligaments, and to the soft palate and pharynx by means of the anterior and posterior pillars of the fauces. Its apex or tip, thin and narrow, is directed forwards against the inner surface of the lower incisor teeth. The under surface of the tongue, at its back part, is connected with the lower jaw by the Genio-hyo-glossi muscles; from its sides, the mucous membrane is reflected to the inner surface of the gums; and, in front, a distinct fold of that membrane, the fraenum linguae, is formed beneath its under surface.

The *tip* of the tongue, part of its under surface, its sides, and dorsum, are free.

The *dorsum* of the tongue is convex, marked along the middle line by a raphe, which divides it into two symmetrical halves; and this raphe terminates behind, about half an inch from the base of the organ, a little in front of a deep mucous follicle, the *foramen caecum*. The anterior two-thirds of this surface are rough, and covered with papillae; the posterior third is more smooth, and covered by the projecting orifices of numerous muciparous glands.

The *mucous membrane* invests the entire extent of the free surface of the tongue. On the under surface of the organ it is thin and smooth, and may be traced on either side of the fraenum, through the ducts of the submaxillary glands; and between the sides of the tongue and the lower jaw, through the ducts of the sublingual glands. On being traced over the borders of the organ, it gradually assumes its papillary character; and on the dorsum becomes exceedingly dense, and gives support to numerous papillae.

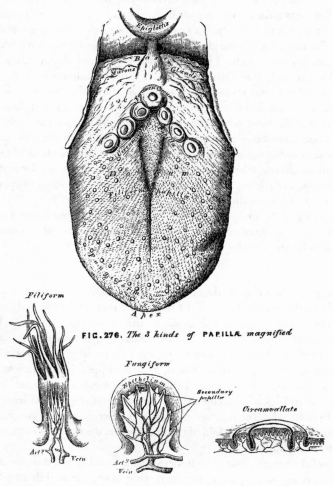

275.—Upper Surface of the Tongue.

FIG. 276. The 3 kinds of PAPILLÆ magnified

The papillae of the tongue are the papillae maximae (circumvallatae), papillae mediae (fungiformes), and papillae minimae (conicae and filiformes).

The *Papillae maximae* (circumvallatae) are of large size, and vary from eight to fifteen in number. They are situated at the back part of the dorsum of the tongue, near its root, forming a row on each side, which, running backwards and inwards, meet in the middle line at the foramen caecum. Each papilla resembles an inverted cone, the apex of which is attached to the bottom of a cup-shaped depression of the mucous membrane; the broad base being exposed upon the surface, and covered with numerous small papillae. This cup-shaped depression forms a kind of fossa around the papilla, having a circular elevated margin covered with smaller papillae.

The *Papillae mediae* (fungiformes), more numerous than the preceding, are scattered irregularly over the dorsum of the tongue; but are found chiefly at its sides and apex. They are easily recognised, among the other papillae, by their large size, rounded eminences, and deep red colour. They are narrow at their attachment to the tongue, but broad and rounded at their free extremities, and covered with secondary papillae.

The *Papillae minimae* (conicae et filiformes) cover the anterior two-thirds of the dorsum of the tongue. They are very minute, more or less conical or filiform in shape, and arranged in lines corresponding in direction with the two rows of the papillae circumvallatae; excepting at the apex of the organ, where their direction is transverse. The filiform papillae are of a whitish tint, owing to the thickness and density of their epithelium; they are covered with numerous secondary papillae, are firmer and more elastic than the papillae of mucous membrane generally, and often enclose minute hairs.

Structure of the Papillae. The papillae, in structure, resemble those of the cutis, consisting of a cone-shaped projection of homogeneous tissue, covered with a thick layer of squamous epithelium, and contain one or more capillary loops, amongst which nerves are distributed in great abundance. In the papillae circumvallatae, the nerves are numerous and of large size; in the papillae fungiformes they are also numerous, and terminate in a plexiform network, from which brush-like branches proceed; in the papillae filiformes, their mode of termination is uncertain.

Besides the papillae, the mucous membrane of the tongue is provided with numerous follicles and glands.

The *Follicles* are found scattered over its entire surface, but are especially numerous between the papillae circumvallatae and the epiglottis.

The *Mucous Glands* (lingual), similar in structure to the labial and buccal, are found chiefly beneath the mucous membrane of the posterior third of the dorsum of the tongue. There is a small group of these glands beneath the tip of the tongue, a few along the borders of the organ, and some in front of the circumvallate papillae projecting into the muscular substance. Their ducts open either upon the surface, or into the depressions around the large papillae.

The tongue consists of two symmetrical halves, separated from each other, in the middle line, by a fibrous septum. Each half is composed of muscular fibres arranged in various directions, containing much interposed fat, and supplied by vessels and nerves: the entire organ is invested by mucous membrane, and a submucous fibrous stratum. The latter membrane invests the greater part of the surface of the tongue, and into it the muscular fibres are inserted that pass to the surface. It is thicker behind than in front, and is continuous with the sheaths of the muscles attached to it.

The *Fibrous Septum* consists of a vertical layer of fibrous tissue, extending throughout the entire length of the middle line of the tongue, from the base to the apex. It is thicker behind than in front, and occasionally contains a small fibro-cartilage, about a quarter of an inch in length. It is well displayed by making a vertical section through the organ. Another strong fibrous lamina, termed the *hyo-glossal membrane*, connects the under surface of the base of the tongue to the body of the hyoid bone. This membrane receives, in front, some of the fibres of the Genio-hyo-glossi.

Each half of the tongue consists of extrinsic and intrinsic muscles. The former have been already described; they are the Hyo-glossus, Genio-hyo-glossus, Stylo-glossus, Palato-glossus, and part of the Superior constrictor. The intrinsic muscles are the Superior longitudinal, Inferior longitudinal, and Transverse.

The *Superior Longitudinal Fibres* (*lingualis superficialis*) form a superficial stratum of oblique and longitudinal fibres on the upper surface of the organ, beneath the mucous membrane, and extend from the apex backwards to the hyoid bone, the individual fibres being attached in their course to the submucous and glandular structures.

The *Inferior Longitudinal Fibres* are formed by the Lingualis muscle, already described (p. 238).

The *Transverse Fibres* are placed between the two preceding layers; they are intermixed with a considerable quantity of adipose substance, and form the chief part of the substance of the organ. They are attached internally to the median fibrous septuni; and, passing outwards, the posterior ones taking an arched course, are inserted into the dorsum and margins of the organ, intersecting the other muscular fibres.

The *Arteries* of the tongue are supplied from the lingual, the facial, and ascending pharyngeal.

The *Nerves* of the tongue are three in number in each half: the gustatory branch of the fifth, which is distributed to the papillae at the fore part and sides of the tongue; the lingual branch of the glosso-pharyngeal, which is distributed to the mucous membrane at the base and side of the tongue, and to the papillae circumvallatae; and the hypo-glossal nerve, which is distributed to the muscular substance of the tongue. The two former are nerves of common sensation and of taste; the latter is the motor nerve of the tongue.

The Nose

The Nose is the special organ of the sense of smell: by means of the peculiar properties of its nerves, it protects the lungs from the inhalation of deleterious gases, and assists the organ of taste in discriminating the properties of food.

277.—Cartilages of the Nose.

The organ of smell consists of two parts, one external, the nose; the other internal, the nasal fossae.

The *Nose* is the most anterior and prominent part of the organ of smell. It is of a triangular form, directed vertically downwards, and projects from the centre of the face, immediately above the upper lip. Its summit, or root, is connected directly with the forehead. Its inferior part, the base of the nose, presents two elliptical orifices, the nostrils, separated from each other by an antero-posterior septum, the columna. The margins of these orifices are provided with a number of stiff hairs, or *vibrissae*, which arrest the passage of foreign substances carried with the current of air intended for respiration. The lateral surfaces of the nose, form, by their union, the dorsum, the direction of which varies considerably in different individuals. The dorsum terminates below in a rounded eminence, the lobe of the nose.

The nose is composed of a framework of bones and cartilages, the latter being slightly acted upon by certain muscles. It is covered externally by the integument, internally by mucous membrane, and supplied with vessels and nerves.

The *bony framework* occupies the upper part of the organ: it consists of the nasal bones, and the nasal processes of the superior maxillary.

The *cartilaginous framework* consists of five pieces, the two upper, and the two lower lateral cartilages, and the cartilage of the septum.

The *upper lateral cartilages* are situated below the free margin of the nasal bones: each cartilage is flattened, and triangular in shape. Its anterior margin is thicker than the posterior, and connected with the fibro-cartilage of the septum. Its posterior margin is attached to the nasal process of the superior maxillary and nasal bones. Its inferior margin is connected by fibrous tissue with the lower lateral cartilage: one surface is turned outwards, the other inwards towards the nasal cavity.

278.—Bones and Cartilages of Septum of Nose. Right Side.

The *lower lateral cartilages* are two thin, flexible plates, situated immediately below the preceding, and curved in such a manner as to form the inner and outer walls of each orifice of the nostril. The portion which forms the inner wall, thicker than the rest, is loosely connected with the same part of the opposite cartilage, and forms a small part of the columna. Its outer extremity, free, rounded, and projecting, forms, with the thickened integument and subjacent tissue, the lobe of the nose. The part which forms the outer wall is curved to correspond with the ala of the nose: it is oval and flattened, narrow behind, where it is connected with the nasal process of the superior maxilla by a tough fibrous membrane, in which is found three or four small cartilaginous plates (sesamoid cartilages), *cartilagines minores*. Above, it is connected to the lateral fibro-cartilage and front part of the cartilage of the septum; below, it is separated from the margin of the nostril by dense cellular tissue; and in front, it forms, with its fellow, the prominence of the tip of the nose.

The *cartilage of the septum* is somewhat triangular in form, thicker at its margins than at its centre, and completes the separation between the nasal fossae in front. Its anterior margin, thickest above, is connected from above downwards with the nasal bones, the front part of the two upper lateral cartilages, and the inner portion of the lower lateral cartilages. Its posterior margin is connected with the perpendicular lamella of the ethmoid; its inferior margin with the vomer and palate processes of the superior maxillary bones.

These various cartilages are connected to each other, and to the bones, by a tough fibrous membrane, the perichondrium, which allows the utmost facility of movement between them.

The *Muscles of the Nose* are situated immediately beneath the integument: they are (on each side) the Pyramidalis nasi, the Levator labii superioris alaeque nasi, the Dilator naris, anterior and posterior, the Compressor nasi, the Compressor narium minor, and the Depressor alae nasi. (See p. 221).

The *Integument* covering the dorsum and sides of the nose is thin, and loosely connected with the subjacent parts; but where it forms the tip, or lobe, and the alae of the nose, it is thicker, and more firmly adherent. It is furnished with a large number of sebaceous follicles, the orifices of which are usually very distinct.

The *Mucous Membrane*, lining the interior of the nose, is continuous with the skin externally, and with that which lines the nasal fossae within.

The *Arteries of the Nose* are the lateralis nasi, from the facial, and the nasal artery of the septum, from the superior coronary, which supplies the alae and septum; the sides and dorsum being supplied from the nasal branch of the ophthalmic and infra-orbital.

The *Veins of the Nose* terminate in the facial and ophthalmic.

The *Nerves of the Nose* are branches from the facial, infra-orbital, and infra-trochlear, and a filament from the nasal branch of the ophthalmic.

Nasal Fossae

The nasal fossae are two irregular cavities, situated in the middle of the face, and extending from before backwards. They open in front by the two anterior nares, and terminate in the pharynx, behind, by the posterior nares. The boundaries of these cavities, and the openings which are connected with them, as they exist in the skeleton, have been already described (p. 73).

The *Mucous Membrane* lining the nasal fossae is called *pituitary*, from the nature of its secretion; or Schneiderian, from Schneider, the first anatomist who showed that the secretion proceeded from the mucous membrane, and not, as was formerly imagined, from the brain. It is intimately adherent to the periosteum, or perichondrium, over which it lies. It is continuous externally with the skin, through the anterior nares, and with the mucous membrane of the pharynx, through the posterior nares. From the nasal fossae its continuity may be traced with the conjunctiva, through the nasal duct and lachrymal canals; with the lining membrane of the tympanum and mastoid cells, through the Eustachian tube; and with the frontal, ethmoidal, and sphenoidal sinuses, and the antrum maxillare, through the several openings in the meatuses. The mucous membrane is thickest, and most vascular, over the turbinated bones, especially the inferior; from which circumstance, these bones in their recent state appear longer and more prominent than in the skeleton. It is also thick over the septum; but, in the intervals between the spongy bones, and on the floor of the nasal fossae, it is very thin. Where it lines the various sinuses and the antrum maxillare, it is thin and pale.

The surface of the membrane is covered with a layer of epithelium, thicker and non-ciliated at the upper part of the nasal fossae, corresponding with the distribution of the olfactory nerve, but ciliated throughout the rest of its extent, excepting near the aperture of the nares.

This membrane is also provided with a nearly continuous layer of branched mucous glands, the ducts of which open upon its surface. They are most numerous at the middle and back parts of the nasal fossae, and largest at the lower and back part of the septum.

Owing to the great thickness of this membrane, the nasal fossae are much narrower, and the turbinated bones, especially the lower ones, appear larger, and more prominent, than in the skeleton. From the same circumstance, also, the various apertures communicating with the meatus, are either narrowed or completely closed.

In the *superior meatus*, the aperture of communication with the posterior ethmoidal cells is considerably diminished in size, and the spheno-palatine foramen completely covered in.

In the *middle meatus*, the opening of the infundibulum is partially hidden by a projecting fold of mucous membrane, and the orifice of the antrum is contracted to a small circular aperture, much narrower than in the skeleton.

In the *inferior meatus*, the orifice of the nasal duct is partially hidden by either a single or double valvular mucous fold, and the anterior palatine canal either completely closed in, or a tubular *cul-de-sac* of mucous membrane is continued a short distance into it.

In the *roof*, the opening leading to the sphenoidal sinuses is narrowed, and the apertures in the cribriform plate of the ethmoid completely closed in.

279.—Nerves of Septum of Nose. Right Side.

The *Arteries of the Nasal Fossae*, are the anterior and posterior ethmoidal, from the ophthalmic, which supply the ethmoidal cells, frontal sinuses and roof of the nose; the spheno-palatine, from the internal maxillary, which supplies the mucous membrane covering the spongy bone, the meatuses, and septum; and the alveolar branch of the internal maxillary, which supplies the lining membrane of the antrum. The ramifications of these vessels form a close, plexiform network, beneath and in the substance of the mucous membrane.

The *Veins of the Nasal Fossae* form a close network beneath the mucous membrane. They pass, some with the veins accompanying the spheno-palatine artery through the spheno-palatine foramen; and others, through the alveolar branch, join the facial vein; some accompany the ethmoidal arteries, and terminate in the ophthalmic vein; and lastly, a few communicate with the veins in the interior of the skull, through the foramina in the cribriform plate of the ethmoid bone.

The *Nerves* are the olfactory, the nasal branch of the ophthalmic, filaments from the anterior dental branch of the superior maxillary, the Vidian, naso-palatine, descending anterior palatine, and spheno-palatine branches of Meckel's ganglion.

The *Olfactory*, the special nerve of the sense of smell, is distributed over the upper third of the septum, and over the surface of the superior and middle spongy bones.

The *Nasal Branch of the Ophthalmic* distributes filaments to the upper and anterior part of the septum, and outer wall of the nasal fossae.

Filaments from the Anterior Dental Branch of the Superior Maxillary supply the inferior meatus and inferior turbinated bone.

The *Vidian Nerve* supplies the upper and back part of the septum, and superior spongy bone: and the upper anterior nasal branches from the spheno-palatine ganglion, have a similar distribution.

The *Naso-Palatine Nerve* supplies the middle of the septum.

The *Larger*, or *Anterior Palatine Nerve*, supplies the middle and lower spongy bones.

The Eye

The eyeball is contained in the cavity of the orbit. In this situation it is securely protected from injury, whilst its position is such as to ensure the most extensive range of sight. It is acted upon by numerous muscles, by which it is capable of being directed to any part, supplied by numerous vessels and nerves, and is additionally protected in front by several appendages, such as the eyebrow, eyelids, etc.

The eyeball is spherical in form, having the segment of a smaller and more prominent sphere engrafted upon its anterior part. It is from this circumstance, that the antero-posterior diameter of the eyeball, which measures about an inch, exceeds the transverse diameter by about a line. The segment of the larger sphere, which forms about five-sixths of the globe, is opaque, and formed by the sclerotic, the tunic of protection to the eyeball; the smaller sphere, which forms the remaining sixth, is transparent, and formed by the cornea. The axes of the eyeballs are nearly parallel, and do not correspond to the axes of the orbits, which are directed outwards. The optic nerves follow the direction of the axes of the orbits, and enter the eyeball a little to their inner or nasal side. The eyeball is composed of several investing tunics, and of fluid and solid refracting media, called *humours*.

The tunics are three in number:
    1. Sclerotic and Cornea.
    2. Choroid, Iris, and Ciliary Processes.
    3. Retina.

The refracting media, or humours, are also three:

    Aqueous.   Crystalline (lens) and Capsule.   Vitreous.

The sclerotic and cornea form the most external tunic of the eyeball; they are essentially fibrous in structure, the sclerotic being opaque and forming the posterior five-sixths of the globe, the cornea, which forms the remaining sixth, being transparent.

280.—A Vertical Section of the Eyeball (Enlarged).

The *Sclerotic* (σκληρος, hard) (fig. 280) has received its name from its extreme density and hardness; it is a firm unyielding fibrous membrane, serving to maintain the peculiar form of the globe. It is much thicker behind than in front. The *external surface* is of a white colour, and quite smooth, except at the points where the Recti and Obliqui muscles are inserted into it, and covered, for part of its extent, by the conjunctival membrane; hence the whiteness and brilliancy of the front of the eyeball. Its *inner surface* is stained of a brown colour, marked by grooves, in which are lodged the ciliary nerves, and connected by an exceedingly fine cellular tissue (*lamina fusca*) with the outer surface of the choroid. Behind, it is pierced by the optic nerve a little to its inner or nasal side, and is continuous with its fibrous sheath, which is derived from the dura mater. At the point where the optic nerve passes through the sclerotic, this membrane forms a thin cribriform lamina (the *lamina cribrosa*); the minute orifices in this part serve for the transmission of the nervous filaments, and the fibrous septa dividing them from one another are continuous with the membranous processes which separate the bundles of nerve-fibres. One of these openings, larger than the rest, occupies the centre of this lamella; it is called the *porus opticus*, and transmits the arteria centralis retinae to the interior of the eyeball. Around the cribriform lamella are numerous smaller apertures for the transmission of the ciliary vessels and nerves. In front, this membrane is continuous with the cornea by direct continuity of tissue; but the opaque sclerotic overlaps it rather more on the outer than upon its inner surface.

Structure. The sclerotic is formed of white fibrous tissue intermixed with fine elastic fibres, and fusiform nucleated cells. These are aggregated into bundles, which are arranged chiefly in a longitudinal direction. It yields gelatin on boiling. Its vessels are not numerous, the capillaries being of small size, uniting at long and wide intervals. The existence of nerves in it is doubtful.

The *Cornea* is the projecting transparent part of the external tunic of the eyeball, and forms the anterior sixth of the globe. Its form is not quite circular, being a little broader in the transverse than in the vertical direction, in consequence of the sclerotic overlapping the margin above and below. It is concavo-convex, and projects forwards from the sclerotic in the same manner that a watch-glass does from its case. Its degree of curvature varies in different individuals, and in the same individual at different periods of life, being more prominent in youth than in advanced life, when it becomes flattened. This difference in the greater or smaller convexity of the cornea influences considerably the refractive power of the eye, and is the chief cause of the long or short sight peculiar to different individuals. It is of uniform thickness throughout, but its posterior surface is perfectly circular in outline, and exceeds the anterior surface slightly in extent, from the latter being overlapped by the sclerotic.

Structure. The cornea consists of five layers: namely, of a thick central fibrous part, the cornea proper; in front of this is the anterior elastic lamina, covered by the conjunctiva; behind, the posterior elastic lamina, covered by the lining membrane of the anterior chamber of the eyeball.

The *proper substance* of the cornea is fibrous in structure, tough, unyielding, perfectly transparent, and continuous with the sclerotic, with which it is in structure identical. The anastomosing fusiform cells of which it is composed are arranged in superimposed flattened laminae, at least sixty in number, all of which have the same direction, the contiguous laminae becoming united at frequent intervals. If the relative position of the component parts of this tissue is in any way altered, either by pressure or by an increase of its natural tension, it immediately presents an opaque milky appearance. The interstices between the laminae are tubular, and usually contain a small amount of perfectly clear transparent fluid.

The *anterior* and *posterior elastic laminae*, which invest the proper structure of the cornea behind and in front, present an analogous structure. They consist of a hard, elastic, and perfectly transparent homogeneous membrane, of extreme thinness, which is not rendered opaque by either water, alcohol, or acids. This membrane is intimately connected by means of a fine cellular web to the proper substance of the cornea both in front and behind. Its most remarkable property is its extreme elasticity, and the tendency which it presents to curl up, or roll upon itself, with the attached surface innermost, when separated from the proper substance of the cornea. Its use appears to be (as suggested by Dr. Jacob), "to preserve the requisite permanent correct curvature of the flaccid cornea proper."

The *conjunctival epithelium*, which covers the front of the anterior elastic lamina, consists of two or three layers of transparent nucleated cells, the deepest being of an oblong form and placed perpendicular to the surface, the superficial ones more flattened.

The *epithelial lining* of the aqueous chamber covers the posterior surface of the posterior elastic lamina. It consists of a single layer of polygonal transparent nucleated cells, similar to those found lining other serous cavities.

Arteries and Nerves. The cornea is a non-vascular structure, the capillary vessels terminating in loops at its circumference. Lymphatic vessels have not as yet been demonstrated in it. The nerves are numerous, twenty or thirty in number: they are derived from the ciliary nerves, and enter the laminated substance of the cornea. They ramify throughout its substance in a delicate network.

Dissection. In order to separate the sclerotic and cornea, so as to expose the second tunic, the eyeball should be immersed in water contained in a small vessel. A fold of the sclerotic near its anterior part being then pinched up, an operation not easily performed from the extreme tension of the membrane, it should be divided with a pair of blunt-pointed scissors. As soon as the choroid is exposed, the end of a blow-pipe should be introduced into the orifice, and a stream of air forced into it, so as to separate the slight cellular connection between the sclerotic and choroid. The sclerotic should now be divided around its entire circumference, and may be removed in separate portions. The front segment being now drawn forwards, the handle of the scalpel should be pressed gently against it at its connection with the iris, and these being separated, a quantity of perfectly transparent fluid will escape; this is the aqueous humour. In the course of this dissection, the ciliary nerves may be seen lying in the loose cellular tissue between the choroid and sclerotic, or contained in delicate grooves on the inner surface of the latter membrane.

281.—The Choroid and Iris (enlarged).

Second Tunic. This is formed by the choroid behind; the iris and ciliary processes in front; and by the ciliary ligament, and Ciliary muscle, at the point of junction of the sclerotic and cornea.

The choroid is the vascular and pigmentary tunic of the eyeball, investing the posterior five-sixths of the globe, and extending as far forwards as the cornea; the ciliary processes being appendages of the choroid developed from its inner surface in front. The iris is the circular-shaped muscular septum, which hangs vertically behind the cornea, presenting in its centre a large circular aperture, the pupil. The ciliary ligament and Ciliary muscle form the white ring observed at the point where the choroid and iris join with each other, and with the sclerotic and cornea.

The *Choroid* is an extremely thin membrane, highly vascular, of a dark brown, or chocolate colour, which invests the posterior five-sixths of the central part of the globe. It is pierced behind by the optic nerve, and terminates in front at the ciliary ligament, where it bends inwards, and forms on its inner surface a series of folds, or plaitings, the ciliary processes. It is thicker behind than in front. Externally, it is connected by a fine cellular web (*membrana fusca*) with the inner surface of the sclerotic. Its inner surface is smooth, and lies in contact with the the retina. The choroid is composed of three layers, external, middle, and internal.

282.—The Veins of the Choroid (enlarged).

The *external layer* consists of the larger branches of the short ciliary arteries, which run forwards between the veins before they bend downwards to terminate on the inner surface. This coat consists, however, principally of veins, which are disposed in curves, hence their name, *venae vorticosae*. They converge to four or five equidistant trunks, which pierce the sclerotic midway between the margin of the cornea and the entrance of the optic nerve. Interspersed between the vessels, are lodged dark star-shaped pigment cells, the fibrous offsets from which, communicating with similar branchings from neighbouring cells, form a delicate network, which, towards the inner surface of the choroid, loses its pigmentary character.

The *middle layer* consists of an exceedingly fine capillary plexus, formed by the short ciliary vessels, and is known as the *tunica Ruyschiana*. The network is exceedingly close, and finer at the hinder part of the choroid than in front. About half an inch behind the cornea, its meshes become larger, and are continuous with those of the ciliary processes.

The *internal*, or *pigmentary layer*, is an exceedingly delicate membrane, consisting of a single layer of hexagonal, nucleated cells, loaded with pigment granules, and applied to each other so as to resemble a tesselated pavement. Each cell contains a nucleus, and is filled with grains of pigment, which are in greater abundance at the circumference of the cell. In perfect albinos this epithelium contains no pigment, and none is present in the star-shaped cells found in the other layers of the choroid.

The ciliary processes should be next examined: they may be exposed, either by detaching the iris from its connection with the ciliary ligament, or by making a transverse section of the globe, and examining them from behind.

283.—The Arteries of the Choroid and Iris.
The Sclerotic has been mostly removed. (Enlarged)

The Ciliary processes are formed by the plaiting or folding inwards of the middle and internal layers of the choroid, at its anterior margin, and are received between corresponding foldings of the suspensory ligament of the lens, thus establishing a communication between the choroid and inner tunic of the eye. They are arranged in a circle, behind the iris, around the margin of the lens. They vary between sixty and eighty in number, lie side by side, and may be divided into large and small; the latter, consisting of about one-third of the entire number, are situated in the spaces between the former, but without regular alternation. The larger processes are each about one-tenth of an inch in length, and hemispherical in shape, their periphery being attached to the ciliary ligament, and continuous with the middle and inner layers of the choroid: the opposite margin is free, and rests upon the circumference of the lens. Their anterior surface is turned towards the back of the iris, with the circumference of which it is continuous. The posterior surface is closely connected with the suspensory ligament of the lens.

Structure. The ciliary processes are similar in structure to the choroid: the vessels are larger, having chiefly a longitudinal direction. Externally they are covered with several layers of pigment cells; the component cells are small, rounded, and full of pigment granules.

The *Iris* (*iris*, a rainbow) has received its name from the varied colour it presents in different individuals. It is a thin, circular-shaped, contractile curtain, suspended in the aqueous humour behind the cornea, and in front of the lens, being perforated at the nasal side of its centre by a circular aperture for the transmission of the light, the pupil. By its circumference it is intimately connected with the choroid; externally to this is the ciliary ligament, by which it is connected to the sclerotic and cornea; its inner edge forms the margin of the pupil; its surfaces are flattened, and look forwards and backwards, the anterior surface towards the cornea, the posterior towards the ciliary processes and lens. The anterior surface is variously coloured in different individuals, and marked by lines which converge towards the pupil. The posterior surface is of a deep purple tint, from being covered by dark pigment; it is hence named *uvea*, from its resemblance in colour to a ripe grape.

Structure. The iris is composed of a fibrous stroma, muscular fibres, and pigment cells.

The *fibrous stroma* consists of fine, delicate bundles of fibrous tissue, which have a circular direction at the circumference: but the chief mass radiate towards the pupil. They form, by their interlacement, a delicate mesh, in which the pigment cells, vessels, and nerves, are contained.

The *muscular fibre* is involuntary, and consists of circular and radiating fibres. The *circular fibres* (sphincter of the pupil), surround the margin of the pupil on the posterior surface of the iris, like a sphincter, forming a narrow band, about one-thirtieth of an inch in width; those near the free margin being closely aggregated; those more external are separated somewhat, and form less complete circles. The *radiating fibres* (dilator of the pupil), converge from the circumference towards the centre, and blend with the circular fibres near the margin of the pupil. The circular fibres contract the pupil, the radiating fibres dilate it.

The *pigment cells* are found in the stroma of the iris, and also as a distinct layer on its anterior and posterior surfaces. In the stroma, the cells are ramified, and contain yellow or brown pigment, according to the colour of the eye. On the front of the iris, there is a single layer of oval or rounded cells, with branching offsets. On the back of the iris, there are several layers of small, round cells, filled with dark pigment. This layer is continuous with the pigmentary covering of the ciliary processes.

The *arteries of the iris* are derived from the long and anterior ciliary, and from the vessels of the ciliary processes.

Membrana Pupillaris. In *the foetus*, the pupil is closed by a delicate, transparent vascular membrane, the membrana pupillaris, which divides the space in which the iris is suspended into two distinct chambers. This membrane contains numerous minute vessels, continued from the margin of the iris to those on the front part of the capsule of the lens. These vessels have a looped arrangement, converging towards each other without anastomosing. Between the seventh and eighth month, this membrane commences to disappear, by its gradual absorption from the centre towards the circumference, and, at birth, only a few fragments remain. Sometimes it remains permanent, and produces blindness.

The *Ciliary Ligament* is a narrow ring of circular fibres, about one-fortieth of an inch thick, and of a whitish colour, which serves to connect the external and middle tunics of the eye. It is placed around the circumference of the iris, at its point of connection with the external layer of the choroid, the cornea, and sclerotic. Its component fibres are delicate, and resemble those of elastic tissue. At its point of connection with the sclerotic, a minute canal is situated between the two, called the *sinus circularis iridis*.

The *Ciliary Muscle* (Bowman) consists of unstriped fibres: it forms a greyish, semitransparent, circular band, about one-eighth of an inch broad, on the outer surface of the fore part of the choroid. It is thickest in front, and gradually becomes thinner behind. Its fibres are soft, of a yellowish-white colour, longitudinal in direction, and arise at the point of junction of the cornea and sclerotic. Passing backwards, they are attached to the choroid, in front of the retina, and correspond by their inner surface to the plicated part of the former membrane. Mr. Bowman supposes that this muscle is so placed as to advance the lens, by exercising compression on the vitreous body, and by drawing the ciliary processes towards the line of junction of the sclerotic and cornea, and by this means to adjust the eye to the vision of near objects.

The *Retina* may be easily exposed by carefully removing the choroid from its external surface. The retina is the delicate nervous membrane, upon the surface of which the images of external objects are received. Its outer surface is in contact with the pigmentary layer of the choroid; its inner surface with the vitreous body. Behind, it is continuous with the optic nerve; and in front extends nearly as far forwards as the ciliary ligament, where it terminates by a jagged margin, the *ora serrata*. It is soft,

284.—The Arteria Centralis Retinæ, Yellow Spot, &c. the Anterior half of the Eyeball being removed. (Enlarged).

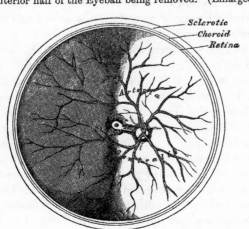

transparent, of a pinkish-grey colour in the fresh state, and gradually diminishes in thickness from behind forwards. In the centre of the posterior part of the globe, the retina presents a yellow spot, *limbus luteus* (*Soemmering*), of a circular form, and having a central depression, *fovea centralis*. The retina in this situation is exceedingly thin; so much so, that the dark colour of the choroid is distinctly seen through it; so that it presents more the appearance of a foramen, and hence the name "foramen of Soemmering" at first given to it. It exists only in man, the quadrumana, and some saurian reptiles. About $1/10$ inch to the inner side of the yellow spot, is the entrance of the optic nerve; and the arteria centralis retinae piercing its centre.

Structure. The retina is composed of three layers, together with blood-vessels.

   External or columnar layer (Jacob's membrane).
   Middle or granular layer.
   Internal or nervous layer.

The blood-vessels do not form a distinct layer; they ramify in the substance of the internal layer.

The *External*, or *Jacob's Membrane*, is exceedingly thin, and can be detached from the external surface of the retina by the handle of the scalpel, in the form of a flocculent film. It is thicker behind than in front, and consists of rod-like bodies of two kinds: 1. *Columnar rods*, solid, nearly of uniform size, and arranged perpendicular to the surface. 2. *Bulbous particles*, or *cones*, which are interspersed at regular intervals among the former; these are conical or flask-shaped, their broad ends resting upon the granular layer, the narrow pointed extremity being turned towards the choroid; they are not solid, like the columnar rods, but consist of an external membrane with fluid contents. By their deep ends, both kinds are joined to the fibres of Müller.

The *Middle*, or *Granular Layer*, forms about one-third of the entire thickness of the retina. It consists of two laminae of rounded or oval nuclear particles, separated from each other by an intermediate layer, which is transparent, finely fibrillated, and contains no blood-vessels. The outermost layer is the thickest, and its constituent particles are globular. The innermost layer is the thinnest; its component particles are flattened, looking like pieces of money seen edgeways; hence it has been called by Bowman, the *nummular layer*.

The *Internal*, or *Nervous Layer*, consists of nerve-fibres and nerve-cells. The expansion of the optic nerve forms a thin, semi-transparent, fibrous membrane, thicker behind than in front. It is composed of nerve-fibres collected into bundles, which communicate, forming a delicate net, with flattened elongated meshes. The nerve-fibres which form this layer, differ from the fibres of the optic nerve in this respect: they lose their dark outline, and their tendency to become varicose; and consist only of the central part, or axis, of the nerve tubes. The mode of termination of the nerve-fibres is unknown. Recent observers have stated, that some of the nerve fibres are continuous with the caudate prolongations of the nerve-cells external to the fibrous layer. The nerve-cells are placed on the outer side of the fibrous layer; they are round or pear-shaped transparent cells, nucleated, with granular contents, furnished with caudate prolongations, some of which join the fibres of the optic nerve, whilst others are directed externally towards the granular layer.

An extremely thin and delicate structureless membrane lines the inner surface of the retina, and separates it from the vitreous body; it is called the *membrana limitans*.

The *Radiating Fibres* of the retina, described by Heinrich Müller, consist of extremely fine fibrillated threads, which are connected externally with each of the rods of the columnar layer, of which they appear to be direct continuations; and, passing through the entire substance of the retina, are united to the outer surface of the membrana limitans. In their course through the retina, they become connected with the nuclear particles of the granular layer, and give off branching processes opposite its innermost lamina; as they approach the fibrous expansion of the optic nerve, they are collected into bundles, which pass through the areolae between its fibres, and are finally attached to the inner surface of the membrana limitans, where each fibre terminates in a triangular enlargement.

The *Arteria Centralis Retinae* and its accompanying vein pierce the optic nerve, and enter the globe of the eye through the porus opticus. It immediately divides into four or five branches, which at first run between the hyaloid membrane and the nervous layer; but they soon enter the latter membrane, and form a close capillary network in its substance. At the ora serrata they terminate in a single vessel, which bounds the terminal margin of the retina.

The structure of the retina at the yellow spot, presents some modifications. Jacob's membrane is thinner, and of its constituents only the cones are present; but they are small, and more closely aggregated than in any other part. The granular layer is absent over the fovea centralis. Of the two elements of the nervous layer, the nerve-fibres extend only to the circumference of the spot; but the nerve-cells cover its entire surface. The radiating fibres are found at the circumference, and here only extend to the inner strata of the granular layer. Of the capillary vessels, the larger branches pass around the spot; but the smaller capillaries meander through it. The colour of the spot appears to imbue all the layers except Jacob's membrane; it is of a rich yellow, deepest towards the centre, and does not appear to consist of pigment-cells, but resembles more a staining of the constituent parts.

Humours of the Eye

The *Aqueous Humour* completely fills the anterior and posterior chambers of the eyeball. It is small in quantity (scarcely exceeding, according to Petit, four or five grains in weight), has an alkaline reaction, in composition is little more than water, less than one-fiftieth of its weight being solid matter, chiefly chloride of sodium.

The *anterior chamber* is the space bounded in front by the cornea; behind, by the front of the iris and ciliary ligament.

The *posterior chamber*, smaller than the anterior, is bounded in front by the iris; behind, by the capsule of the lens and its suspensory ligament, and the ciliary processes.

In the adult, these two chambers communicate through the pupil; but in the foetus before the seventh month, when the pupil is closed by the membrana pupillaris, the two chambers are quite separate.

It has been generally supposed that the two chambers are lined by a distinct membrane, the *secreting membrane of the aqueous humour*, analogous in structure to that of a serous sac. Only an epithelial covering can, however, be found on the posterior surface of the cornea. That the two chambers do, however, secrete this fluid separately, is shown from its being found in both spaces before the removal of the membrana pupillaris. It is probable that the parts concerned in the secretion of the fluid, are the posterior surface of the cornea, both surfaces of the iris, and the ciliary processes.

Vitreous Body

The *Vitreous Body* forms about four-fifths of the entire globe. It fills the concavity of the retina, and is hollowed in front for the reception of the lens and its capsule. It is perfectly transparent, of the consistence of thin jelly, and consists of an albuminous fluid enclosed in a delicate, transparent membrane, the *hyaloid*. This membrane invests the outer surface of the vitreous body; it is intimately connected in front with the suspensory ligament of the lens; and is continued into the back part of the capsule of the lens. It has been supposed, by Hannover, that from its inner surface numerous thin lamellae are prolonged inwards in a radiating manner, forming spaces in which the fluid is contained. In the adult, these lamellae cannot be detected even after careful microscopic examination; but in the foetus a peculiar fibrous texture pervades the mass, the fibres joining at numerous points, presenting minute nuclear granules at their point of junction. The fluid from the vitreous body resembles nearly pure water; it contains, however, some salts, and a little albumen.

In the *foetus*, the centre of the vitreous humour presents a tubular canal, through which a minute artery passes along the vitreous body to the capsule of the lens. In the *adult*, no vessels penetrate its substance; so that its nutrition must be carried on by the vessels of the retina and ciliary processes, situated upon its exterior.

Crystalline Lens and its Capsule

The *Crystalline Lens*, enclosed in its *Capsule*, is situated immediately behind the pupil, in front of the vitreous body, and surrounded by the ciliary processes, which slightly overlap its margin.

The *Capsule of the Lens* is a transparent, highly elastic, and brittle membrane, which closely surrounds the lens. It rests, behind, in a depression in front of the vitreous body: in front, it forms part of the posterior chamber of the eye; and it is retained in its position chiefly by the suspensory ligament of the lens. The capsule is much thicker in front than behind, structureless in texture; and when ruptured, the edges roll up with the outer surface innermost, like the elastic laminae of the cornea. The lens is connected to the inner surface of the capsule by a single layer of transparent, polygonal, nucleated cells. These, after death, absorb moisture from the fluids of the eye; and, breaking down, form the *liquor Morgagni*.

In *the foetus*, a small branch from the arteria centralis retinae runs forwards, as already mentioned, through the vitreous humour to the posterior part of the capsule of the lens, where its branches radiate and form a plexiform network, which covers its surface, and are continuous around the margin of the capsule, with the vessels of the pupillary membrane, and with those of the iris. In the adult, no vessels enter its substance.

The *Lens* is a transparent, double convex body, the convexity being greater on the posterior, than on the anterior, surface. It measures about a third of an inch in the transverse diameter, and about one-fourth in the antero-posterior. It consists of concentric layers, of which the external, in the fresh state, are soft and easily detached; those beneath are firmer, the central ones forming a hardened nucleus. These laminae are best demonstrated by boiling, or immersion in alcohol. The same re-agents demonstrate that the lens consists of three triangular segments, the sharp edges of which are directed towards the centre, the bases towards the circumference.

285.—The Crystalline Lens, hardened and divided. (Enlarged.)

The laminae consist of minute parallel fibres, which are united to each other by means of wavy margins, the convexities upon one fibre fitting accurately into the concavities of the adjoining fibre.

The *changes produced in the lens by age*, are the following:

In *the foetus*, its form is nearly spherical, its colour of a slightly reddish tint, not perfectly transparent, and so soft as to readily break down on the slightest pressure.

In *the adult*, the posterior surface is more convex than the anterior, it is colourless, transparent, and firm in texture.

In *old age*, it becomes flattened on both surfaces, slightly opaque, of an amber tint, and increases in density.

The *suspensory ligament of the lens* is a thin, transparent membranous structure, placed between the vitreous body and the ciliary processes of the choroid: it connects the anterior margin of the retina with the anterior surface of the lens, near its circumference. It assists in retaining the lens in its position. Its outer surface presents a number of folds or plaitings, in which the corresponding folds of the ciliary processes are received. These plaitings are arranged round the lens in a radiating form, and are stained by the pigment of the ciliary processes. The suspensory ligament consists of two layers, which commence behind, at the *ora serrata*. The external, a tough, milky, granular membrane, covers the inner surface of the ciliary processes, and extends as far forwards as their anterior free extremities. The *inner layer*, an elastic transparent, fibro-membranous structure, extends as far forwards as the anterior surface of the capsule of the lens, near its circumference. That portion of this membrane which intervenes between the ciliary processes and the capsule of the lens, forms part of the boundary of the posterior chamber of the eye. The posterior surface of this layer is turned towards the hyaloid membrane, being separated from it at the circumference of the lens by a space called the canal of Petit.

The *canal of Petit* is about one-tenth of an inch wide. It is bounded in front by the suspensory ligament; behind, by the hyaloid membrane, its base being formed by the capsule of the lens. When inflated with air, it is sacculated at intervals, owing to the foldings on its anterior surface.

The *vessels of the globe of the eye* are the short, long, and anterior ciliary arteries, and the arteria centralis retinae.

The *Short Ciliary Arteries* pierce the back part of the sclerotic, around the entrance of the optic nerve, and divide into branches which run parallel with the axis of the eyeball: they are distributed to the middle layer of the choroid, and ciliary processes.

The *Long Ciliary Arteries*, two in number, pierce the back part of the sclerotic, and run forward, between this membrane and the choroid, to the Ciliary muscle, where they each divide into an upper and lower branch; these anastomose, and form a vascular circle around the outer circumference of the iris: from this circle branches are given off which unite, near the margin of the pupil, in a smaller vascular circle. These branches, in their course, supply the muscular structure.

The *Anterior Ciliary Arteries*, five or six in number, are branches of the muscular and lachrymal branches of the ophthalmic. They pierce the eyeball, at the anterior part of the sclerotic, immediately behind the margin of the cornea, and are distributed to the ciliary processes, some branches joining the greater vascular circle of the iris.

The *Arteria Centralis Retina* has been already described.

The *Veins*, usually four in number, are formed mainly by branches from the surface of the choroid. They perforate the sclerotic, midway between the cornea and the optic nerve, and end in the ophthalmic vein.

The *Nerves of the Eyeball*, are the optic, the long ciliary nerves from the nasal branch of the ophthalmic, and the short ciliary nerves from the ciliary ganglion.

Appendages of the Eye

The appendages of the eye (*tutamina oculi*), include the eyebrows, the eyelids, the conjunctiva, and the lachrymal apparatus, viz., the lachrymal gland, the lachrymal sac, and the nasal duct.

The *Eyebrows* (*supercilia*) are two arched eminences of integument, which surmount the upper circumference of the orbit on each side, and support numerous short, thick hairs, directed obliquely on the surface. In structure, they consist of thickened integument, connected beneath with the Orbicularis palpebrarum, Corrugator supercilii, and Occipito-frontalis muscles. These muscles serve, by their action on this part, to control to a certain extent the amount of light admitted into the eye.

The *Eyelids* (*palpebrae*) are two thin, moveable folds, placed in front of the eye, protecting it from injury by their closure. The upper lid is the larger, the more moveable of the two, and supplied by a separate elevator muscle, the *Levator palpebrae superioris*. When the eyelids are opened, an elliptical space (*fissura palpebrarum*), is left between their margins, the angles of which correspond to the junction of the upper and lower lids, and are called *canthi*.

The *outer canthus* is more acute than the inner, and the lids here lie in close contact with the globe: but the *inner canthus* is prolonged for a short distance inwards, towards the nose, and the two lids are separated by a triangular space, the *lacus lachrymalis*. At the commencement of the lacus lachrymalis, on the margin of each eyelid, is a small conical elevation, the *lachrymal papilla*, or tubercle, the apex of which is pierced by a small orifice, the *punctum lachrymale*, the commencement of the lachrymal canal.

Structure of the Eyelids. The eyelids are composed of the following structures, taken in their order from without inwards:

Integument, areolar tissue, fibres of the Orbicularis muscle, tarsal cartilage, fibrous membrane, Meibomian glands, and conjunctiva. The upper lid has, in addition, the aponeurosis of the Levator palpebrae.

The *Integument* is extremely thin, and continuous at the margin of the lids with the conjunctiva.

The *Subcutaneous Areolar Tissue* is very lax and delicate, seldom contains any fat, and is extremely liable to serous infiltration.

The *Fibres of the Orbicularis Muscle*, where they cover the palpebrae, are thin, pale in colour, and possess an involuntary action.

The *Tarsal Cartilages* are two thin, elongated plates of fibro-cartilage, about an inch in length. They are placed one in each lid, contributing to their form and support.

The *superior*, the larger, is of a semilunar form, about one-third of an inch in breadth at the centre, and becoming gradually narrowed at each extremity. Into the fore part of this cartilage the aponeurosis of the Levator palpebrae is attached.

The *inferior* tarsal cartilage, the smaller, is thinner, and of an elliptical form.

The *free*, or *ciliary margin* of the cartilages is thick, and presents a perfectly straight edge. The *attached*, or *orbital margin*, is connected to the circumference of the orbit by the fibrous membrane of the lids. The outer angle of each cartilage is attached to the malar bone, by the external palpebral or tarsal ligament. The inner angles of the two cartilages terminate at the commencement of the lacus lachrymalis, being fixed to the margins of the orbit by the tendo oculi.

The *Fibrous Membrane of the Lids*, or tarsal ligament, is a layer of fibrous membrane, beneath the Orbicularis, attached, externally, to the margin of the orbit, and internally to the orbital margin of the lids. It is thick and dense at the outer part of the orbit, but becomes thinner as it approaches the cartilages. This membrane serves to support the eyelids, and retains the tarsal cartilages in their position.

The *Meibomian Glands* (fig. 286) are situated upon the inner surface of the eyelids, between the tarsal cartilages and conjunctiva, and may be distinctly seen through the mucous membrane on everting the eyelids, presenting the appearance of parallel strings of pearls. They are about thirty in number in the upper cartilage,

286.—The Meibomian Glands, etc., seen from the Inner Surface of the Eyelids.

and somewhat fewer in the lower. They are imbedded in grooves in the inner surface of the cartilages, and correspond in length with the breadth of each cartilage; they are, consequently, longer in the upper than in the lower eyelid. Their ducts open on the free margin of the lids by minute foramina, which correspond in number to the follicles. These glands are a variety of the cutaneous sebaceous glands, each consisting of a single straight tube or follicle, having a coecal termination, into which open a number of small secondary follicles. The tubes consist of a basement membrane, covered by a layer of scaly epithelium; the cells are charged with sebaceous matter, which constitutes the secretion. The peculiar parallel arrangement of these glands side by side, forms a smooth layer, admirably adapted to the surface of the globe, over which they constantly glide. The use of their secretion is to prevent adhesion of the lids.

The *Eyelashes* (*Cilia*) are attached to the free edges of the eyelids; they are short, thick, curved hairs, arranged in a double or triple row at the margin of the lids: those of the upper lid, more numerous and longer than the lower, curve upwards; those of the lower lid curve downwards, by which means they do not interlace in closing the lids.

The *Conjunctiva* is the mucous membrane of the eye. It lines the inner surface of the eyelids, and is reflected over the fore part of the sclerotic and cornea. In each of these situations, its structure presents some peculiarities.

The *palpebral portion of the conjunctiva* is thick, opaque, highly vascular, and covered with numerous papillae, which, in the disease called *granular lids*, become greatly hypertrophied. At the margin of the lids, it becomes continuous with the lining membrane of the ducts of the Meibomian glands, and, through the lachrymal canals, with the lining membrane of the lachrymal sac and nasal duct. At the outer angle of the upper lid, it may be traced along the lachrymal ducts into the lachrymal gland; and at the inner angle of the eye, it forms a semilunar fold, the *plica semilunaris*. The folds formed by the reflection of the conjunctiva from the lids on to the eye are called the *superior* and *inferior palpebral folds*, the former being the deeper of the two. Upon the *sclerotic*, the conjunctiva is loosely connected to the globe; it becomes thinner, loses its papillary structure, is transparent, and only slightly vascular in health. Upon the *cornea*, the conjunctiva is extremely thin and closely adherent, and no vessels can be traced into it in the adult in a healthy state. In the *foetus*, fine capillary loops extend, for some little distance forwards, into this membrane; but in the *adult*, they pass only to the circumference of the cornea.

The *Caruncula Lachrymalis* is a small, reddish, conical-shaped body, situated at the inner canthus of the eye, and filling up the small triangular space in this situation, the *lacus lachrymalis*. It consists of a cluster of follicles similar in structure to the Meibomian, covered with mucous membrane, and is the source of the whitish secretion which constantly collects at the inner angle of the eye. A few slender hairs are attached to its surface. On the outer side of the caruncula is a slight semilunar fold of mucous membrane, the concavity of which is directed towards the cornea; it is called the *plica semilunaris*. Between its two layers is found a thin plate of cartilage. This structure is considered to be the rudiment of the third eyelid in birds, the *membrana nictitans*.

Lachrymal Apparatus (fig. 287)

The lachrymal apparatus consists of the lachrymal gland, which secretes the tears, and its excretory ducts, which convey the fluid to the surface of the eye. This fluid is carried away by the lachrymal canals into the lachrymal sac, and along the nasal duct into the cavity of the nose.

287.—The Lachrymal Apparatus. Right Side.

The *Lachrymal Gland* is lodged in a depression at the outer angle of the orbit, on the inner side of the external angular process of the frontal bone. It is of an oval form, about the size and shape of an almond. Its upper convex surface is in contact with the periosteum of the orbit, to which it is connected by a few fibrous bands. Its under concave surface rests upon the convexity of the eyeball, and upon the Superior and External recti muscles. Its vessels and nerves enter its posterior border, whilst its anterior margin is closely adherent to the back part of the upper eyelid, and is covered, on its inner surface, by a reflection of the conjunctiva. This margin is separated from the rest of the gland by a slight depression, and is hence sometimes described as a separate lobe, called the *palpebral portion of the gland*. In structure and general appearance, it resembles the salivary glands. Its ducts, about seven in number, run obliquely beneath the mucous membrane for a short distance, and separating from each other, open by a series of minute orifices on the upper and outer half of the conjunctiva, near its reflexion on to the globe. These orifices are arranged in a row, so as to disperse the secretion over the surface of the membrane.

The *Lachrymal Canals* commence at the minute orifices, *puncta lachrymalia*, seen on the margin of the lids, at the outer extremity of the lacus lachrymalis. They commence on the summit of a slightly elevated papilla, the *papilla lachrymalis*, and lead into minute canals, the *canaliculi*, which proceed inwards to terminate in the lachrymal sac. The *superior canal*, the smaller and longer of the two, at first ascends, and then bends at an acute angle, and passes inwards and downwards to the lachrymal sac. The *inferior canal* at first descends, and then abruptly changing its course, passes almost horizontally inwards. They are dense and elastic in structure, and somewhat dilated at their angle.

The *Lachrymal Sac* is the upper dilated extremity of the nasal duct, and is lodged in a deep groove formed by the lachrymal and nasal process of the superior maxillary bone. It is oval in form, its upper extremity being closed in and rounded, whilst below it is continued into the nasal duct. It is covered by a fibrous expansion derived from the tendo oculi, which is attached to the ridge on the lachrymal bone, and also by the Tensor tarsi muscle. In structure, it consists of a fibrous elastic coat, lined internally by mucous membrane, which is continuous, through the canaliculi, with the mucous lining of the conjunctiva, and through the nasal duct with the pituitary membrane of the nose.

The *Nasal Duct* is a membranous canal, about three quarters of an inch in length, which extends from the lower part of the lachrymal sac to the inferior meatus of the nose, where it terminates by a somewhat expanded orifice, provided with an imperfect valve formed by the mucous membrane. It is contained in an osseous canal, formed by the superior maxillary, the lachrymal, and the inferior turbinated bones, is narrower in the middle than at each extremity, and takes a direction downwards, backwards, and a little outwards. It is lined by mucous membrane, which is continuous below with the pituitary lining of the nose. In the canaliculi, this membrane is provided with scaly epithelium, but in the lachrymal sac and nasal duct it is ciliated as in the nose.

The Ear

The organ of hearing consists of three parts; the external ear, the middle ear or tympanum, and the internal ear or labyrinth.

The *External Ear* consists of an expanded portion or pinna, and the auditory canal or meatus. The former serves to collect the vibrations of the air constituting sound, and the latter conducts those vibrations to the tympanum.

The *Pinna*, or *Auricle* (fig. 288), consists of a layer of cartilage, covered by integument, and connected to the commencement of the auditory canal; it is of an ovoid form, its surface uneven, with its larger end directed upwards. Its outer surface is irregularly concave, directed slightly forwards, and presents numerous eminences and depressions, which result from the foldings of its fibro-cartilaginous element. To each of these, names have been assigned. Thus, the external prominent rim of the auricle is called the *helix*. Another curved prominence parallel with, and in front of the helix, is called the *antihelix*; this bifurcates above into two parts, so as to enclose a triangular depression, the fossa of the antihelix. The narrow curved depression between the helix and antihelix, is called the *fossa of the helix* (*fossa innominata, scaphoidea*). The antihelix describes a circuit around a deep, capacious cavity, the *concha*, which is partially divided into two parts by

288.—The Pinna or Auricle. Outer Surface.

the commencement of the helix. In front of the concha, and projecting backwards over the meatus, is a small pointed eminence, the *tragus*; so called from its being generally covered, on its under surface, with a tuft of hairs, resembling a goat's beard. Opposite the tragus, and separated from it by a deep notch (*incisura intertragica*), is a small tubercle, the *antitragus*. Below this is the *lobule*, composed of tough areolar and adipose tissues, wanting the firmness and elasticity of the rest of the pinna.

Structure of the Pinna. The pinna is composed of a thin plate of cartilage, covered with integument, and connected to the surrounding parts by ligaments, and a few muscular fibres.

The *Integument* is thin, closely adherent to the cartilage, and furnished with sebaceous glands, which are most numerous in the concha and scaphoid fossa.

The *Cartilage of the Pinna* consists of one single piece; it gives form to this part of the ear, and upon its surface are found all the eminences and depressions above described. It does not enter into the construction of all parts of the auricle, and presents several intervals or fissures in its substance, which partially separate the different parts. Thus, it does not form a constituent part of the lobule; it is deficient, also, between the tragus and beginning of the helix, the notch between them being filled up by dense fibrous tissue. The fissures in the cartilage are the fissure of the helix, a short, vertical slit, situated at the fore part of the pinna, immediately behind a small conical projection of cartilage, opposite the first curve of the helix (process of the helix); another fissure, the fissure of the tragus, is seen upon the anterior surface of the tragus. The antihelix is divided below, by a deep fissure, into two parts; one part terminates by a pointed, tail-like extremity (*processus caudatus*); the other is continuous with the antitragus. The cartilage of the pinna is very pliable, elastic, of a yellowish colour, and is similar in structure to the cartilages of the alae nasi.

The *Ligaments of the Pinna* consist of two sets. 1. Those connecting it to the side of the head. 2. Those connecting the various parts of its cartilage together.

The former, the most important, are two in number, anterior and posterior. The *anterior ligament* extends from the process of the helix, to the root of the zygoma. The *posterior ligament* passes from the posterior surface of the concha, to the outer surface of the mastoid process of the temporal bone. A few fibres connect the tragus to the root of the zygoma.

Those connecting the various parts of the cartilage together, are also two in number. Of these, one is a strong fibrous band, stretching across from the tragus to the commencement of the helix, completing the meatus in front, and partly encircling the boundary of the concha; the other extends between the concha and the processus caudatus.

The *Muscles of the Pinna* (fig. 289), like the ligaments, also consist of two sets. 1. Those which connect it with the side of the head, moving the pinna as a whole, viz. the Attollens, Attrahens and Retrahens aurem, already described; and the proper muscles of the pinna, which extend from one part of the auricle to another. These are the

Helicis major.	Antitragicus.
Helicis minor.	Transversus auriculae.
Tragicus.	Obliquus auris.

289.—The Muscles of the Pinna.

The *Helicis major* is a narrow, vertical band of muscular fibres, situated upon the anterior margin of the helix. It arises, below, from the tubercle of the helix, and is inserted into the anterior border of the helix, just where it is about to curve backwards. It is pretty constant in its existence.

The *Helicis minor* is an oblique fasciculus, firmly attached to that part of the helix which commences from the bottom of the concha.

The *Tragicus* is a short, flattened band of muscular fibres, situated upon the outer surface of the tragus; the direction of its fibres being vertical.

The *Antitragicus* arises from the outer part of the antitragus; its fibres are inserted into the processus caudatus of the helix. This muscle is usually very distinct.

The *Transversus auriculae* is placed on the cranial surface of the pinna. It consists of radiating fibres, partly tendinous and partly muscular, extending from the convexity of the concha, to the prominence corresponding with the groove of the helix.

The *Obliquus auris* (Tod) consists of a few fibres extending from the upper and back part of the concha, to the convexity immediately above it.

The *Arteries of the Pinna* are the posterior auricular, from the external carotid; the anterior auricular, from the temporal; and an auricular branch from the occipital artery.

The *Veins* accompany the corresponding arteries.

The *Nerves* are the auricularis magnus, from the cervical plexus; the posterior auricular, from the facial; the auricular branch of the pneumogastric; and the auriculo-temporal branch of the inferior maxillary nerve.

The AUDITORY CANAL (fig. 290), (*meatus auditorius externus*), conveys the vibrations of the air to the tympanum, and extends from the bottom of the concha to the membrana tympani. It is about an inch and a quarter in length, its direction obliquely forwards and inwards, and it is slightly curved upon itself, so as to be higher in the middle than at either extremity. It forms an oval cylindrical canal, somewhat flattened from before backwards, the greatest diameter being in the vertical direction at the external orifice; but, in the transverse direction, at the tympanic end. The membrana tympani, which occupies the termination of the meatus, is obliquely directed, in consequence of the floor of this canal being longer than the roof, and the anterior wall longer than the posterior. The auditory canal is formed partly by cartilage and fibrous tissue, and partly by bone.

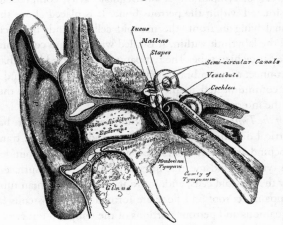

290.—A Front View of the Organ of Hearing. Right Side.

The *cartilaginous portion* is about half an inch in length, being rather less than half the canal; it is formed by the cartilage of the concha and tragus being prolonged inwards to the circumference of the auditory process, to which it is firmly attached. This tube is deficient at the upper and back part, its place being supplied by fibrous membrane. This part of the canal is rendered extremely moveable, by two or three deep fissures (*incisurae Santorini*) which extend through the cartilage in a vertical direction.

The *osseous portion* of the meatus is about three-quarters of an inch in length, and is longer and narrower than the cartilaginous portion. It is directed inwards and a little forwards, forming a slight curve in its course, the convexity of which is upwards and backwards, and it terminates in the external orifice of the tympanum. Its inner end is smaller than the outer, and it is slightly contracted in the middle. Its vertical transverse section is oval, the greatest diameter being from above downwards. The front and lower parts of this canal are formed by a curved plate of bone, presenting a rough margin externally, to which the cartilage of the pinna is attached. This osseous plate, in the foetus, exists as a separate ring of bone (tympanic bone), incomplete at its upper part.

The *skin* lining the meatus is very thin, closely adherent to the cartilaginous and osseous portions of the tube, and covers the surface of the membrana tympani, forming its outer layer. After maceration, the thin pouch of epidermis, when withdrawn, preserves the form of the meatus. The skin near the orifice is supplied with hairs, sebaceous glands, and numerous ceruminous glands, which furnish an abundant secretion, calculated to prevent the ingress of insects and particles of dust.

The *Arteries* supplying the meatus are branches from the posterior auricular, internal maxillary, and temporal.

The *Nerves* are chiefly derived from the temporo-auricular branch of the inferior maxillary nerve.

Middle Ear, or Tympanum

The middle ear, or tympanum, is an irregular cavity, compressed from without inwards, and situated within the petrous bone. It is placed above the jugular fossa, the carotid canal lying in front, the mastoid cells behind, the meatus auditorius externally, and the labyrinth within. It is filled with air, and communicates with the pharynx by the Eustachian tube. The tympanum is traversed by a chain of moveable bones, which connect the membrana tympani with the labyrinth, and serve to convey the vibrations communicated to the membrana tympani across the cavity of the tympanum to the internal ear.

The *Cavity of the Tympanum* measures about five lines from before backwards, three lines in the vertical direction, and between two and three in the transverse, being a little broader behind and above than below and in front. It is bounded externally by the membrana tympani and meatus: internally, by the outer surface of the internal ear; behind, by the mastoid cells; and, in front, by the Eustachian tube and canal for the Tensor tympani. Its roof and floor are formed by thin osseous lamellae, which connect the squamous and petrous portions of the temporal bone.

The *roof* is broad, flattened, and formed of a thin plate of bone, which separates the cranial and tympanic cavities.

The *floor* is narrow, and corresponds to the jugular fossa, which lies beneath.

The *outer wall* is formed by the membrana tympani, a small portion of bone being seen above and below this membrane. It presents three small apertures, the iter chordae posterius, the Glasserian fissure, and the iter chordae anterius.

The *Aperture of the Iter Chordae Posterius* is behind the aperture for the membrana tympani, close to its margin, at a level with its centre; it leads into a minute canal, which descends in front of the aqueductus Fallopii, and terminates in this canal near the stylo-mastoid foramen. It transmits the chorda tympani nerve.

The *Glasserian Fissure* opens just above and in front of the orifice of the membrana tympani; in this situation it is a mere slit, about a line in length. It gives passage to the long process of the malleus, the Laxator tympani muscle, and some tympanic vessels.

The *Aperture of the Iter Chordae Anterius* is seen just above the preceding fissure; it leads into a canal which runs parallel with the Glasserian fissure, and transmits the chorda tympani nerve.

The *internal wall of the tympanum* is vertical in direction, and looks directly outwards. It presents for examination the following parts:

        Fenestra ovalis.   Ridge of the Aqueductus Fallopii.
        Fenestra rotunda.   Pyramid.
        Promontory.   Opening for the Stapedius.

291.—View of Inner Wall of Tympanum (Enlarged).

The *Fenestra Ovalis* is a reniform opening, leading from the tympanum into the vestibule; its long diameter is directed horizontally, and its convex border is upwards. The opening in the recent state is closed by the lining membrane common to both cavities, and is occupied by the base of the stapes. This membrane is placed opposite the membrana tympani, and is connected with it by the ossicula auditus.

The *Fenestra Rotunda* is an oval aperture, placed at the bottom of a funnel-shaped depression, leading into the cochlea. It is placed below and rather behind the fenestra ovalis, from which it is separated by a rounded elevation, the promontory; it is closed in the recent state by a membrane (*membrana tympani secundaria*, Scarpa). This membrane is concave towards the tympanum, convex towards the cochlea. It consists of three layers: the external, or mucous, is derived from the mucous lining of the tympanum; the internal, or serous, from the lining membrane of the cochlea; and an intermediate, or fibrous layer.

The *Promontory* is a rounded hollow prominence, formed by the projection outwards of the first turn of the cochlea; it is placed between the fenestrae, and furrowed on its surface by three small grooves, which lodge branches of the tympanic plexus.

The *Rounded Eminence of the Aqueductus Fallopii* is placed between the fenestra ovalis and roof of the tympanum; it is the prominence of the bony canal in which the portio dura is contained. It traverses the inner wall of the tympanum above the fenestra ovalis, and behind that opening, curves nearly vertically downwards along the posterior wall.

The *Pyramid* is a conical eminence, situated immediately behind the fenestra ovalis, and in front of the vertical portion of the eminence above described; it is hollow in the interior, and contains the Stapedius muscle; its summit projects forwards towards the vestibular fenestra, and presents a small aperture, which transmits the tendon of this muscle. The cavity in the pyramid is prolonged into a canal, which descends vertically, parallel with the aqueductus Fallopii, and terminates at the base of the petrous portion of the temporal bone, in front and to the inner side of the stylo-mastoid foramen. This canal communicates with the aqueductus Fallopii.

The *posterior wall of the tympanum* is wider above than below, and presents for examination the

Openings of the Mastoid Cells.

They consist of one large irregular aperture, and several smaller openings, situated at the upper part of the posterior wall; they lead into canals, which communicate with large irregular cavities contained in the interior of the mastoid process. These cavities vary considerably in number, size, and form; they are lined by mucous membrane, continuous with that covering the cavity of the tympanum.

The *anterior wall of the tympanum* is wider above than below; it corresponds with the carotid canal, from which it is separated by a thin plate of bone; it presents for examination the

Canal for the Tensor tympani.   Orifice of the Eustachian Tube.
The Processus Cochleariformis.

The orifice of the *canal for the tensor Tympani*, and the *orifice of the Eustachian tube*, are situated at the upper part of the anterior wall, being separated from each other by a thin, delicate, horizontal plate of bone, the processus cochleariformis. These canals run from the tympanum forward, inward, and a little downward, to the retiring angle between the squamous and petrous portions of the temporal bone.

The *canal for the tensor Tympani* is the superior and the smaller of the two; it is rounded, and lies beneath the upper surface of the petrous bone, close to the hiatus Fallopii. The tympanic end of this canal forms a conical eminence, which is prolonged backwards into the cavity of the tympanum, and is perforated at its summit by an aperture, which transmits the tendon of the muscle contained in it. This eminence is sometimes called the *anterior pyramid*. The canal contains the Tensor tympani muscle.

The *Eustachian tube* is the channel through which the tympanum communicates with the pharynx. Its length is from an inch and a half to two inches, and its direction downwards, forwards, and inwards. It is formed partly of bone, partly of cartilage and fibrous tissue.

The *osseous portion* is about half an inch in length. It commences in the lower part of the anterior wall of the tympanum, below the processus cochleariformis, and gradually narrowing, terminates in an oval dilated opening, at the angle of junction of the petrous and squamous portions, its extremity presenting a jagged margin, which serves for the attachment of the cartilaginous portion.

The *cartilaginous portion*, about an inch in length, is formed of a triangular plate of cartilage, curled upon itself, an interval being left below, between the non-approximated margins of the cartilage, which is completed by fibrous tissue. Its canal is narrow behind, wide, expanded, and somewhat trumpet-shaped in front, terminating by an oval orifice, placed at the upper part and side of the pharynx, behind the back part of the inferior meatus. Through this canal the mucous membrane of the pharynx is continuous with that which lines the tympanum.

The *Membrana Tympani* separates the cavity of the tympanum from the bottom of the external meatus. It is a thin semitransparent membrane, nearly oval in form, somewhat broader above than below, and directed very obliquely downwards and inwards. Its circumference is contained in a groove at the inner end of the meatus, which skirts the circumference of this part excepting above. The handle of the malleus descends vertically between the inner and middle layers of this membrane as far down as its centre, where it is firmly attached, drawing the membrane inwards, so that its outer surface is concave, its inner convex.

Structure. This membrane is composed of three layers, an external (cuticular), a middle (fibrous), and an internal (mucous). The *cuticular lining* is derived from the integument lining the meatus. The *fibrous layer* consists of fibrous and elastic tissues; some of the fibres radiate from near the centre to the circumference; others are arranged, in the form of a dense circular ring, around the attached margin of the membrane. The *mucous lining* is derived from the mucous lining of the tympanum. The vessels pass to the membrana tympani along the handle of the malleus, and are distributed between its layers.

Ossicles of the Tympanum

The tympanum is traversed by a chain of moveable bones, three in number, the malleus, incus, and stapes. The former is attached to the membrana tympani, the latter to the fenestra ovalis, the incus being placed between the two, to both of which it is connected by delicate articulations.

The *Malleus*, so named from its fancied resemblance to a hammer, consists of a head, neck, handle or manubrium, and two processes, viz., the processus gracilis and the processus brevis.

The *head* is the large upper extremity of the bone; it is oval in shape, and articulates posteriorly with the incus, being free in the rest of its extent.

The *neck* is the narrow contracted part just beneath the head; and below this is a prominence, to which the various processes are attached.

The *manubrium* is a vertical portion of bone, which is connected by its outer margin with the membrana tympani. It decreases in size towards its extremity, where it is curved slightly forwards, and flattened from within outwards.

The *processus gracilis* is a long and very delicate process, which passes from the eminence below the neck forwards and outwards to the Glasserian fissure, to which it is connected by bone and ligamentous fibres. It gives attachment to the Laxator tympani.

The *processus brevis* is a slight conical projection, which springs from the root of the manubrium, and lies in contact with the membrana tympani. Its summit gives attachment to the Tensor tympani.

The *Incus* has received its name from its resemblance to an anvil, but it does not look unlike a bicuspid tooth, with two roots, which differ in length, and are widely separated from each other. It consists of a body and two processes.

292.—The Small Bones of the Ear, seen from the Outside (Enlarged).

The *body* is somewhat quadrilateral, but compressed laterally. Its summit is deeply concave, and articulates with the malleus; in the fresh state, it is covered with cartilage and lined with synovial membrane.

The two processes diverge from one another nearly at right angles.

The *short process*, somewhat conical in shape, projects nearly horizontally backwards, and is attached to the margin of the opening leading into the mastoid cells by ligamentous fibres.

The *long process*, longer and more slender than the preceding, descends nearly vertically behind the handle of the malleus, and bending inwards, terminates in a rounded globular projection, the *os orbiculare*, tipped with cartilage, and articulating with the head of the stapes. In the foetus the os orbiculare exists as a separate bone, but becomes united to the long process of the incus in the adult.

The *Stapes*, so called from its close resemblance to a stirrup, consists of a head, neck, two branches, and a base.

The *head* presents a depression, tipped with cartilage, which articulates with the os orbiculare.

The *neck*, the constricted part of the bone below the head, receives the insertion of the Stapedius muscle.

The *two branches* (*crura*), diverge from the neck, and are connected at their extremities by a flattened, oval-shaped plate (the *base*), which forms the foot of the stirrup, and is fixed to the margin of the fenestra ovalis by ligamentous fibres.

Ligaments of the Ossicula. These small bones are connected with each other, and with the tympanum, by ligaments, and moved by small muscles. The articular surfaces of the malleus and incus, the orbicular process of the incus and head of the stapes, are covered with cartilage, connected together by delicate capsular ligaments, and lined by synovial membrane. The ligaments connecting the ossicula with the walls of the tympanum, are three in number, one for each bone.

The *Suspensory Ligament of the Malleus* is a delicate, round bundle of fibres, which descends perpendicularly from the roof of the tympanum to the head of the malleus.

The *Posterior Ligament of the Incus* is a short, thick ligamentous band, which connects the extremity of the short process of the incus to the posterior wall of the tympanum, near the margin of the opening of the mastoid cells.

The *Annular Ligament of the Stapes* connects the circumference of the base of this bone to the margin of the fenestra ovalis.

A Suspensory Ligament of the Incus has been described by Arnold, descending from the roof of the tympanum to the upper part of the incus, near its articulation with the malleus.

The muscles of the tympanum are three:

    Tensor tympani.   Laxator tympani.   Stapedius.

The *Tensor Tympani*, the largest, is contained in a bony canal, above the osseous portion of the Eustachian tube, from which it is separated by the processus cochleariformis. It arises from the under surface of the petrous bone, from the cartilaginous portion of the Eustachian tube, and from the osseous canal in which it is contained. Passing backwards, it terminates in a slender tendon, which is reflected outwards over the processus cochleariformis, and is inserted into the handle of the malleus, near its root. It is supplied by a branch from the otic ganglion.

The *Laxator Tympani major* (Soemmering), arises from the spinous process of the sphenoid bone, and from the cartilaginous portion of the Eustachian tube; passing backwards through the Glasserian fissure, it is inserted into the neck of the malleus, just above the processus gracilis. It is supplied by a branch from the chorda tympani.

The *Laxator Tympani minor* (Soemmering), arises from the upper and back part of the external meatus, passing forwards and inwards between the middle and inner layers of the membrana tympani; it is inserted into the handle of the malleus, and processus brevis. This is considered as a ligament by some anatomists.

The *Stapedius* is lodged in a cavity hollowed out of the interior of the pyramid. It arises from the sides of this conical cavity: its tendon emerges from the orifice at its apex, and, passing forwards, is inserted into the neck of the stapes. Its surface is aponeurotic, its interior fleshy, and its tendon occasionally contains a slender bony spine, which is constant in some mammalia. It is supplied by a filament from the facial nerve.

Actions. The Tensor tympani draws the membrana tympani inwards, and thus heightens its tension. The Laxator tympani draws the malleus outwards, and thus the tympanic membrane, especially at its fore part, is relaxed. The Stapedius depresses the back part of the base of the stapes, and raises its fore part. It probably compresses the contents of the vestibule.

The *Mucous Membrane of the Tympanum* is thin, vascular, and continuous with the mucous membrane of the pharynx, through the Eustachian tube. It invests the ossicula, and the muscles and nerves contained in the tympanic cavity; forms the internal layer of the membrana tympani; covers the foramen rotundum; and is reflected into the mastoid cells, which it lines throughout. In the tympanum and mastoid cells, this membrane is pale, thin, slightly vascular, and covered with ciliated epithelium. In the osseous portion of the Eustachian tube, the membrane is thin; but, in the cartilaginous portion, it is very thick, highly vascular, covered with ciliated laminar epithelium, and provided with numerous mucous glands.

The *Arteries supplying the Tympanum* are five in number, viz., the tympanic branch of the internal maxillary, which supplies the membrana tympani; the stylo-mastoid branch of the posterior auricular, which supplies the back part of the tympanum and mastoid cells. The smaller branches are, the petrosal branch of the middle meningeal, and branches from the ascending pharyngeal and internal carotid.

The *Veins of the Tympanum* terminate in the middle meningeal and pharyngeal veins, and, through these, in the internal jugular.

The *Nerves of the Tympanum* may be divided into: 1, Those supplying the muscles; 2, Those distributed to the lining membrane; 3, Branches communicating with other nerves.

Nerves to Muscles. The Tensor tympani is supplied by a branch from the otic ganglion; the Laxator tympani, by the chorda tympani; and the Stapedius, by a filament from the facial (Soemmering).

The *Nerves distributed to the Lining Membrane* are derived from the tympanic plexus.

Communications between the following nerves take place in the tympanum: the Tympanic branch, from the petrous ganglion of the glosso-pharyngeal; a filament from the carotid plexus; a branch which joins the great superficial petrosal nerve from the Vidian; and a branch to the otic ganglion (small superficial petrosal nerve).

The *Tympanic Branch of the Glosso-Pharyngeal* (Jacobson's nerve), enters the tympanum by an aperture in its floor, close to the inner wall, and ascends on to the promontory. It distributes filaments to the lining membrane of the tympanum, and divides into three branches, which are contained in grooves on the promontory, and serve to connect this with other nerves. One branch runs in a groove, forwards and downwards, to an aperture situated at the junction of the anterior and inner walls, just above the floor, and enters the carotid canal, to communicate with the carotid plexus of the sympathetic. The second branch is contained in a groove which runs vertically upwards to an aperture on the inner wall of the tympanum, just beneath the anterior pyramid, and in front of the fenestra ovalis. The canal leading from this opens into the hiatus Fallopii, where it joins the great petrosal nerve. The third branch ascends towards the anterior surface of the petrous bone; it then passes through a small aperture in the sphenoid and temporal bones to the exterior of the skull, and joins the otic ganglion. As this nerve passes by the gangliform enlargement of the facial, it has a connecting filament with it.

The *Chorda Tympani* quits the facial near the stylo-mastoid foramen, enters the tympanum at the base of the pyramid, and arches forwards across its cavity, between the handle of the malleus and long process of the incus, to an opening internal to the fissura Glasseri. It is invested by a reflection of the lining membrane of the tympanum.

Internal Ear, or Labyrinth

The internal ear is the essential part of the acoustic organ, receiving the ultimate distribution of the auditory nerve. It is called the *labyrinth*, from the complexity of its communications, and consists of three parts, the vestibule, semicircular canals, and cochlea. It consists of a series of cavities, channelled out of the substance of the petrous bone, communicating externally with the cavity of the tympanum, through the fenestra ovalis and rotunda; and internally, with the meatus auditorius internus, which contains the auditory nerve. Within the osseous labyrinth, is contained the membranous labyrinth, upon which the ramifications of the auditory nerve are distributed.

The *Vestibule* is the common central cavity of communication between the parts of the internal ear. It is situated on the inner side of the tympanum, behind the cochlea, and in front of the semicircular canals. It is somewhat ovoidal in shape from before backwards, flattened from side to side, and measures about one-fifth of an inch from before backwards, as well as from above downwards, being narrower from without inwards. On its *outer*, or *tympanic wall*, is the fenestra ovalis, closed, in the

293.—The Osseous Labyrinth, laid open. (Enlarged).

recent state, by the base of the stapes, and its annular ligament. On its *inner wall*, at its fore part, is a small circular depression, *fovea hemispherica*; it is perforated at its anterior and inferior part by several minute holes (*macula cribrosa*), for the passage of filaments of the auditory nerve; and behind it is a vertical ridge, the pyramidal eminence. At the back part of the inner wall is the orifice of the *aqueductus vestibuli*, which extends to the posterior surface of the petrous portion of the temporal bone. It transmits a small vein, and, according to some, contains a tubular prolongation of the lining membrane of the vestibule which ends in a *cul-de-sac*, between the layers of the dura mater within the cranial cavity. On the *upper wall*, or *roof*, is a transversely-oval depression, *fovea hemi-elliptica*, separated from the fovea hemispherica by the pyramidal eminence, already mentioned. Behind, the semicircular canals open into the vestibule by five orifices. In front, is a larger oval opening which communicates with the scala vestibuli of the cochlea by a single orifice, *apertura scalae vestibuli cochleae*.

The *Semicircular Canals* are three bony canals, situated above and behind the vestibule. They are of unequal length, compressed from side to side, and describe the greater part of a circle. They measure about one-twentieth of an inch in diameter, and each presents a dilatation at one end, called the *ampulla*, which measures more than twice the diameter of the tube. These canals open into the vestibule by five orifices, one of the apertures being common to two of the canals.

The *superior semicircular canal* is vertical in direction, its arch forming a round projection on the anterior surface of the petrous bone. It forms about two-thirds of a circle. Its outer extremity, which is ampullated, commences by a distinct orifice in the upper part of the vestibule; the opposite end of the canal, which is not dilated, joins with the corresponding part of the posterior canal, and opens by a common orifice with it in the back part of the vestibule.

The *posterior semicircular canal*, also vertical in direction, is directed backwards to the posterior surface of the petrous bone: it is the longest of the three, its ampullated end commencing at the lower and back part of the vestibule, its opposite end joining the common canal already mentioned.

The *external*, or *horizontal canal*, is the shortest of the three, its arch being directed outwards and backwards. Its ampullated end corresponds to the upper and outer angle of the vestibule, just above the fenestra ovalis; its opposite end opens by a distinct orifice at the upper and back part of the vestibule.

The *Cochlea* bears some resemblance to a common snail-shell: it forms the anterior part of the labyrinth, is conical in form, and placed almost horizontally in front of the vestibule; its apex is directed forwards and outwards towards the upper and front part of the inner wall of the tympanum; its base corresponds with the anterior depression at the bottom of the internal auditory meatus; and is perforated by numerous apertures, for the passage of the cochlear branch of the auditory nerve. It measures about a quarter of an inch in length, and its breadth towards the base is about the same. It consists of a conical-shaped central axis, the modiolus or columella; of a canal wound spirally round the axis for two turns and a half; and of a delicate lamina (the *lamina spiralis*) contained within the canal, which follows its windings, and subdivides it into two.

294.—The Cochlea laid open (enlarged).

The *central axis*, or *modiolus*, is conical in form, and extends from the base to the apex of the cochlea. Its base is broad, corresponds with the first turn of the cochlea, and is perforated by numerous orifices, which transmit filaments of the cochlear branch of the auditory nerve; the axis diminishes rapidly in size in the second coil, and terminates within the last half coil, or cupola, in an expanded, delicate, bony lamella, which resembles the half of a funnel, divided longitudinally, and called the *infundibulum*; the broad part of this funnel is directed towards the summit of the cochlea, and blends with the last half-turn of the spiral canal of the cochlea, the cupola. The outer surface of the modiolus is formed of the wall of the spiral canal, and is dense in structure; but its centre is channelled, as far as the last half-coil, by numerous branching canals, which transmit nervous filaments in regular succession into the canal of the cochlea, or upon the surface of the lamina spiralis. One of these, larger than the rest, occupies the centre of the modiolus, and is named the *tubulus centralis modioli*; it extends from the base to the extremity of the modiolus, and transmits a small nerve and artery (*arteria centralis modioli*).

The *spiral canal* (fig. 294) takes two turns and a half round the modiolus. It is about an inch and a half in length, measured along its outer wall; and diminishes gradually in size from the base to the summit, where it terminates in a *cul-de-sac*, the cupola, which forms the apex of the cochlea. The commencement of this canal is about the tenth of an inch in diameter: it diverges from the modiolus towards the tympanum and vestibule, and presents three openings. One, the *fenestra rotunda*, communicates with the tympanum: in the recent state, this aperture is closed by a membrane, the *membrana tympani secundaria*. Another aperture, of an oval form, enters the vestibule. The third is the aperture of the *aqueductus cochleae*, which leads to a minute funnel-shaped canal, which opens on the basilar surface of the petrous bone, and transmits a small vein.

The interior of the spiral canal is divided into two passages (*scalae*) by a thin, osseous, and membranous lamina, which winds spirally round the modiolus. This is the *lamina spiralis*, the essential part of the cochlea upon which the nerve tubules are distributed. The osseous part of the spiral lamina extends about half way across the diameter of the spiral canal; it is called the *osseous zone*. It commences in the vestibule between the tympanic and vestibular openings of the cochlea, and, gradually becoming narrower in its course, terminates in a projecting hook, the hamular process, just where the expansion of the infundibulum commences. The lamina spiralis consists of two thin lamellae of bone, between which are numerous canals for the passage of nervous filaments, which open chiefly on the lower or tympanic surface. At the point where the osseous lamina is attached to the modiolus, and following its windings, is a small canal, called by Rosenthal the *canalis spiralis modioli*. In the recent state, the osseous zone is continued to the opposite wall of the canal by a membranous and muscular layer (membranous zone), so as to form a complete partition in the tube of the cochlea. Two passages, or scalae, are thus formed, by a division of the canal of the cochlea into two. One, the *scala tympani*, is closed below by the membrane of the fenestra rotunda; the other, the *scala vestibuli*, communicates, by an oval aperture, with the vestibule. Near the termination of the scala vestibuli, close by the fenestra rotunda, is the orifice of the aqueductus cochleae. The scalae communicate, at the apex of the cochlea, by an opening common to both, the *helicotrema*, which exists in consequence of the deficiency of the lamina spiralis in the last half coil of the canal.

In *structure*, the membranous zone is a transparent glassy lamina, presenting near its centre a number of minute transverse lines, which radiate outwards, and give it a fibrous appearance; and at its circumference, where connected with the outer wall of the spiral canal, it is composed of a semi-transparent structure, the Cochlearis muscle (Todd and Bowman), connective tissue (Kölliker).

The vestibular surface of the osseous portion of the lamina spiralis is covered, for about the outer fifth of its surface, with a thin layer, resembling cartilage in texture. It is described as the *denticulate lamina* (Todd and Bowman), from its presenting a series of wedge-shaped teeth which form its free margin, and which project into the vestibular scalae.

The *Inner surface* of the osseous labyrinth is lined by an exceedingly thin fibroserous membrane, analogous to a periosteum, from its close adhesion to the inner surface of these cavities, and performing the office of a serous membrane by its free surface. It lines the vestibule, and from this cavity is continued into the semicircular canals and the scala vestibuli of the cochlea, and through the helicotrema into the scala tympani. Two delicate tubular processes are prolonged along the aqueducts of the vestibule and cochlea, to the inner surface of the dura mater. This membrane is continued across the fenestra ovalis and rotunda, and consequently has no communication with the lining membrane of the tympanum. Its attached surface is rough and fibrous, and closely adherent to the bone; its free surface is smooth and pale, covered with a layer of epithelium, and secretes a thin, limpid fluid, the *aqua labyrinthi* (*perilymph* (Blainville), *liquor Cotunnii*). In the vestibule and semicircular canals, it separates the osseous from the membranous labyrinth; but in the cochlea it lines the two surfaces of the bony lamina spiralis; and being continued from its free margin across the canal to its outer wall, forms the lamina spiralis membranacea, serving to complete the separation between the two scalae.

The Membranous Labyrinth

The membranous labyrinth (fig. 295) is a closed membranous sac, containing fluid, upon which the ultimate ramifications of the auditory nerve are distributed. It has the same general form as the vestibule and semicircular canals, in which it is enclosed; but is considerably smaller, and separated from their lining membrane by the perilymph.

The *Vestibular portion* consists of two sacs, the utricle and the saccule.

The *Utricle* is the larger of the two, of an oblong form, compressed laterally, and occupies the upper and back part of the vestibule, lying in contact with the fovea semi-elliptica. Numerous filaments of the auditory nerve are distributed to the wall of this sac; and its cavity communicates, behind, with the membranous semicircular canals by five orifices.

The *Sacculus* is the smaller of the two vestibular sacs: it is globular in form, lies in the fovea hemispherica, near the opening of the vestibular scala of the cochlea, and receives numerous nervous filaments, which enter from the bottom of the depression in which it is contained. Its cavity is apparently distinct from that of the utricle.

The *Membranous Semicircular Canals* are about one-third the diameter of the osseous canals, but in number, shape, and general form they are precisely similar; they are hollow, and open by five orifices into the utriculus, one being common to two canals. Their ampullae are thicker than the rest of the tubes, and nearly fill the cavities in which they are contained.

The membranous labyrinth is held in its position by the numerous nervous filaments distributed to the utriculus, the sacculus, and to the ampulla of each canal. These nerves enter the vestibule through the minute apertures on its inner wall.

295.—The Membranous Labyrinth detached (enlarged).

Structure. The wall of the membranous labyrinth is semi-transparent, and consists of three layers. The *outer* layer is a loose and flocculent tissue, containing blood-vessels and numerous pigment cells, analogous to those in the choroid. The *middle layer*, thicker and more transparent, bears some resemblance to the hyaloid membrane, but it presents in parts marks of longitudinal fibrillation and elongated nuclei on the addition of acetic acid. The *inner layer* is formed of polygonal nucleated epithelial cells, which secrete the fluid contained in its interior.

The *Endolymph* (*liquor Scarpae*) is a limpid serous fluid, which fills the membranous labyrinth; in composition, it closely resembles the perilymph.

The *Otoliths* are two small rounded bodies, consisting of a mass of minute crystalline grains of carbonate of lime, held together in a mesh of delicate fibrous tissue, and contained in the wall of the utriculus and sacculus, opposite the distribution of the nerves. A calcareous material is also, according to Bowman, sparingly scattered in the cells lining the ampulla of each semicircular canal.

The *Arteries of the Labyrinth* are the internal auditory, from the basilar or superior cerebellar, the stylo-mastoid, from the posterior auricular, and, occasionally, branches from the occipital. The internal auditory divides at the bottom of the internal meatus into two branches, cochlear and vestibular.

The *cochlear branch* subdivides into from twelve to fourteen twigs, which traverse the canals in the modiolus, and are distributed, in the form of a capillary network, in the substance of the lamina spiralis.

The *vestibular branches* accompany the nerves, and are distributed, in the form of a minute capillary network, in the substance of the membranous labyrinth.

The *Veins* of the vestibule and semicircular canals accompany the arteries, and receiving those of the cochlea at the base of the modiolus, terminate in the superior petrosal sinus.

The *Auditory Nerve*, the special nerve of the sense of hearing, divides, at the bottom of the internal auditory meatus, into two branches, the cochlear and vestibular. The trunk of the nerve, as well as the branches, contains numerous ganglion cells with caudate prolongations.

The *Vestibular Nerve*, the most posterior of the two, divides into three branches, superior, middle, and inferior.

The *superior vestibular branch*, the largest, divides into numerous filaments, which pass through minute openings at the upper and back part of the *cul-de-sac* at the bottom of the meatus, and entering the vestibule, are distributed to the utricle, and to the ampulla of the external and superior semicircular canals.

The *middle vestibular branch* consists of numerous filaments, which enter the vestibule by a smaller cluster of foramina, placed below those above mentioned, and which correspond to the bottom of the fovea hemispherica; they are distributed to the sacculus.

The *inferior* and smallest branch passes backwards in a canal behind the foramina for the nerves of the sacculus, and is distributed to the ampulla of the posterior semicircular canal.

The nervous filaments enter the ampullary enlargement at a deep depression seen on their external surface, and a corresponding elevation is seen within, the nerve fibres ending in loops and in free extremities. In the utricle and saccule the nerve fibres spread out, some blending with the calcareous matter, others radiating on the inner surface of the wall of each cavity, becoming blended with a layer of nucleated cells, and terminating in a thin fibrous film.

The *Cochlear Nerve* divides into numerous filaments at the base of the modiolus, which ascend along its canals, and then bending outwards at right angles, pass between the plates of the bony lamina spiralis close to its tympanic surface. Between the plates of the spiral lamina, the nerves form a plexus, which contains ganglion cells; and from the margin of the osseous zone, branches from this plexus are distributed to the membranous part of the septum, where they are arranged in small conical-shaped bundles, parallel with one another. The filaments which supply the apical portion of the lamina spiralis are conducted to this part through the tubulus centralis modioli.

ORGANS OF DIGESTION

The Apparatus for the digestion of the food consists of the alimentary canal, and of certain accessory organs.

The *Alimentary Canal* is a musculo-membranous tube, about thirty feet in length, extending from the mouth to the anus, and lined throughout its entire extent by mucous membrane. It has received different names in the various parts of its course: at its commencement, which comprises the mouth, we find every provision for the mechanical division of the food (mastication), and for its admixture with a peculiar fluid secreted by the salivary glands (insalivation); beyond this is the pharynx and the oesophagus, the organs of deglutition, which convey the food into that part of the alimentary canal (the stomach) in which the principal chemical changes occur; in the stomach the reduction and solution of the food takes place; in the small intestines, the nutritive principles of the food (the chyle), by its admixture with the bile and pancreatic fluid, are separated from that portion which passes into the large intestine, and which is expelled from the system.

Alimentary Canal.

Mouth.
Pharynx.
Oesophagus.
Stomach.

Small intestine { Duodenum. Jejunum. Ileum.

Large intestine { Caecum. Colon. Rectum.

Accessory Organs.

Teeth.

Salivary glands { Parotid. Submaxillary. Sublingual.

Liver.
Pancreas.
Spleen.

The MOUTH (fig. 296) is placed at the commencement of the alimentary canal; it is a nearly oval-shaped cavity, in which the mastication of the food takes place. It is bounded in front by the lips; laterally, by the cheeks and alveolar process of the upper and lower jaw; above, by the hard palate and teeth of the upper jaw; below, by the tongue, the mucous membrane stretched between the under surface of this organ and the inner surface of the jaws, and by the teeth of the lower jaw; behind, by the soft palate and fauces.

The *mucous membrane* lining the mouth is continuous with the integument at the free margin of the lips, and with the mucous lining of the fauces behind; it is of a pink rose tinge during life, and very thick where it covers the hard parts bounding this cavity.

The LIPS are two fleshy folds, which surround the orifice of the mouth, formed externally of integument, internally of mucous membrane, between which is found the Orbicularis oris muscle, the coronary vessels, some nerves, areolar tissue, fat, and numerous small labial glands. The inner surface of each lip is connected in the middle line to the gum of the corresponding jaw by a fold of mucous membrane, the *fraenum labii superioris* and *inferioris*, the former being the larger of the two.

The *labial glands* are situated between the mucous membrane and the Orbicularis oris, around the orifice of the mouth. They are rounded in form, about the size of a small pea, their ducts opening by small orifices upon the mucous membrane. In structure, they resemble the other salivary glands.

The CHEEKS form the sides of the face, and are continuous in front with the lips. They are composed, externally, of integument; internally, of mucous membrane; and, between the two, is a muscular stratum, besides a large quantity of fat, areolar tissue, vessels, nerves, and buccal glands.

296.—Sectional View of the Nose, Mouth, Pharynx, etc.

The *mucous membrane* lining the cheek, is reflected above and below upon the gums, and is continuous behind with the lining membrane of the soft palate. Opposite the second molar tooth of the upper jaw is a papilla, the summit of which presents the minute aperture of the duct of the parotid gland. The principal muscle of the cheek is the Buccinator; but numerous others enter into its formation, viz., the Zygomatici, Masseter, and the Platysma myoides.

The *buccal glands* are placed between the mucous membrane and Buccinator muscle: they are similar in structure, but smaller, than the labial glands. Two or three, of larger size than the rest, are placed between the Masseter and Buccinator muscles: their ducts open into the mouth, opposite the last molar tooth. They are called *molar glands*.

The GUMS are composed of a dense fibrous tissue, closely connected to the periosteum of the alveolar processes, and surrounding the necks of the teeth. They are covered by smooth and vascular mucous membrane, which is remarkable for its limited sensibility. Around the necks of the teeth, this membrane presents numerous fine papillae; and from this point it is reflected into the alveolus, where it is continuous with the periosteal membrane lining that cavity.

THE TEETH

The human subject is provided with two distinct sets of teeth, which make their appearance at different periods of life. The first set appear in childhood, and are called the *temporary*, *deciduous*, or *milk teeth*. The second set, which also appear at an early period, continue until old age, and are named *permanent*.

The *Temporary Teeth* are twenty in number; four incisors, two canine, and four molars, in each jaw.

The *Permanent Teeth* are thirty-two in number; four incisors, two central and two lateral, two canine, four bicuspids, and six molars, in each jaw.

General characters. Each tooth consists of three portions: the crown, or body, which projects above the gum; the root, or fang, entirely concealed within the alveolus; and the neck, the constricted portion between the other two.

297.—The Permanent Teeth. External View.

The *roots of the teeth* are firmly implanted within the alveoli: these depressions are lined with periosteum, which is reflected on to the tooth at the point of the fang, and covers it as far as the neck. At the margin of the alveolus, the periosteum becomes continuous with the fibrous structure of the gums.

Permanent Teeth

The Incisors, or cutting teeth, are so named from their presenting a sharp, cutting edge, adapted for cutting the food. They are eight in number, and form the four front teeth in each jaw.

The *crown* is directed vertically, is wedge-like in form, being bevelled at the expense of its posterior surface, so as to terminate in a sharp, horizontal cutting edge, which, before being subject to attrition, presents three small prominent points. It is convex, smooth, and highly polished in front; slightly concave behind, where it is frequently marked by slight longitudinal furrows.

The *neck* is constricted.

The *fang* is long, single, conical, transversely flattened, thicker before than behind, and slightly grooved on each side, in the longitudinal direction.

The *Incisors of the Upper Jaw* are altogether larger and stronger than those of the lower jaw. They are directed obliquely downwards and forwards. The two central ones are larger than the two lateral, and their free edges sharp and chisel-like, being bevelled at the expense of their posterior edge: the root is more rounded.

The *Incisors of the Lower Jaw* are smaller than the upper: the two central ones are smaller than the two lateral, and are the smallest of all the incisor teeth.

The Canine Teeth (*cuspidati*) are four in number, two in the upper, and two in the lower jaw; one being placed behind each lateral incisor. They are larger and stronger than the incisors, especially the root, which sinks deeply into the jaw, and causes a well marked prominence upon its surface.

The *crown* is large and conical, very convex in front, a little hollowed and uneven posteriorly, and tapering to a blunted point, or cusp, which rises above the level of the other teeth.

The *root* is single, but longer and thicker than that of the incisors, conical in form, compressed laterally, and marked by a slight groove on each side.

The *Upper Canine Teeth* (vulgarly called eye-teeth), are larger and longer than the two lower, and situated a little behind them.

The *Lower Canine Teeth* are placed in front of the upper, so that their summits correspond to the interval between the upper canine tooth and the neighbouring incisor.

The Bicuspid Teeth (small, or false molars), are eight in number, four in each jaw, two being placed immediately behind each of the canine teeth. They are smaller and shorter than the canine.

The *crown* is compressed from without inwards, and surmounted by two pyramidal eminences, or cusps, separated by a groove, hence their name, *bicuspidate*. The outer of these cusps is larger and more prominent than the inner.

The *neck* is oval.

The *root* is generally single, compressed, and presenting a tendency to become double, as seen from the deep groove on each side. The apex is generally bifid.

The *Upper Bicuspids* are larger, and present a greater tendency to the division of their roots than the lower: this is especially marked in the second upper bicuspid.

The Molar Teeth (*multicuspidati*, true, or large molars), are the largest of the permanent set, and are adapted, from the great breadth of their crowns, for grinding and pounding the food. They are twelve in number, six in each jaw, three being placed behind each of the posterior bicuspids.

The *crown* is nearly cubical in form, rounded on each of its lateral surfaces, flattened in front and behind; the upper surface being surmounted by four or five tubercles, or cusps, (four in the upper, five in the lower molars), separated from each other by a crucial depression, hence their name, *multicuspidati*.

The *neck* is distinct, large, and rounded.

The *root* is subdivided into from two to five fangs, each of which presents an aperture at its summit.

The *First Molar Tooth* is the largest and broadest of all: its crown has usually five cusps, three outer and two inner. In the upper jaw, the root consists of three fangs, widely separated from one another, two being external, the other internal. The latter is the largest and the longest, slightly grooved, and sometimes bifid. In the lower jaw, the root consists of two fangs, one being placed in front, the other behind: they are both compressed from before backwards, and grooved on their contiguous faces, indicating a tendency to division.

The *Second Molar* is a little smaller than the first.

The *crown* has four cusps in the upper, and five in the lower jaw.

The *root* has three fangs in the upper jaw, and two in the lower, the characters of which are similar to the preceding tooth.

The *Third Molar Tooth* is called the *wisdom tooth* (*dens sapientiae*), from its late appearance through the gum. It is smaller than the others, and its axis is directed inwards.

The *crown* is small and rounded, and furnished with three tubercles.

The *root* is generally single, short, conical, slightly curved, and grooved so as to present traces of a subdivision into three fangs in the upper, and two in the lower jaw.

Temporary Teeth

The temporary, or milk teeth, are smaller, but resemble in form those of the permanent set. The hinder of the two temporary molars is the largest of all the milk teeth, and is succeeded by the second permanent bicuspid. The first upper molar has only three cusps, two external, one internal: the second upper molar has four cusps. The first lower molar has four cusps: the second lower molar has five. The fangs of the temporary molar teeth are smaller, and more diverging than those of the permanent set; but, in other respects, bear a strong resemblance to them.

Structure. On making a vertical section of a tooth (fig. 299), a hollow cavity will be found in the interior. This cavity is situated at the base of the crown, and is continuous with a canal which traverses the centre of each fang, and opens by a minute orifice at its extremity. The shape of the cavity corresponds somewhat with that of the tooth: it forms what is called the *pulp cavity*, and contains a soft, highly vascular, and sensitive substance, the dental pulp. The pulp is richly supplied with

298.—The Temporary, or Milk Teeth.
External View.

vessels and nerves, which enter the cavity through the small aperture at the point of each fang.

The solid portion of the tooth consists of three distinct structures, viz. ivory (tooth-bone, or dentine), which forms the larger portion of the tooth; enamel, which covers the exposed part, or crown; and the cortical substance, or cement (*crusta petrosa*), which is disposed as a thin layer on the surface of the fang.

The IVORY, or dentine (fig. 300), forms the principal mass of a tooth; in its central part is the cavity enclosing the pulp. It is a modification of the osseous tissue, from which it differs, however, in structure and chemical composition. On examination with the microscope, it is seen to consist of a number of minute wavy and branching tubes, having distinct parietes. They are called the *dental tubuli*, and are imbedded in a dense homogeneous substance, the *intertubular tissue*.

The *Dental Tubuli* are placed parallel with one another, and open at their inner ends into the pulp cavity. They pursue a wavy and undulating course towards the periphery. The direction of these tubes varies; they are vertical in the upper portion of the crown, oblique in the neck and upper part of the root, and towards the lower part of the root they are inclined downwards.

299.—Vertical Section of a Molar Tooth.

The tubuli, at their commencement, are about $1/4500$ of an inch in diameter; in their course they divide and subdivide dichotomously, so as to give to the cut surface of the dentine a striated appearance. From the sides of the tubes, especially in the fang, ramifications of extreme minuteness are given off, which join together in loops in the intertubular substance, or terminate in small dilatations, from which branches are given off. Near the periphery of the dentine, the finer ramifications of the tubuli terminate in a somewhat similar manner. In the fang, these ramifications occasionally pass into the crusta petrosa. The dental tubuli have comparatively thick walls, and contain, according to Mr. Tomes, slender cylindrical prolongations of the pulp-tissue.

300.—Vertical Section of a Bicuspid Tooth (magnified).

The *Intertubular Substance* is translucent, finely granular, and contains the chief part of the earthy matter of the dentine. After the earthy matter has been removed, by steeping a tooth in weak acid, the animal basis remaining is described by Henle as consisting of bundles of pale, granular, flattened fibres, running parallel with the tubes; but by Mr. Nasmyth as consisting of a mass of brick-shaped cells surrounding the tubules. By Czermack and Mr. Salter it is supposed to consist of laminae which run parallel with the pulp cavity, across the direction of the tubes.

Chemical Composition. According to Berzelius and Bibra, dentine consists of 28 parts of animal, and 72 of earthy matter. The animal matter is resolvable by boiling into gelatin. The earthy matter consists of phosphate of lime, carbonate of lime, fluoride of calcium, a trace; and phosphate of magnesia and other salts.

The ENAMEL is the hardest and most compact part of a tooth, and forms a thin crust over the exposed part of the crown, as far as the commencement of the fang. It is thickest on the grinding surface of the crown, until worn away by attrition, and becomes thinner towards the neck. It consists of a congeries of minute hexagonal rods. They lie parallel with one another, resting by one extremity upon the dentine, which presents a number of minute depressions for their reception; the other extremity forming the free surface of the crown. These fibres are directed vertically on the summit of the crown, horizontally at the sides; they are about the $1/5500$ of an inch in diameter, and pursue a more or less wavy course, which gives to the cut surface of the enamel a series of concentric lines.

Numerous minute interstices intervene between the enamel-fibres near their dentinal surface, a provision calculated to allow of the permeation of fluids from the dentinal tubuli into its substance. The enamel-rods consist of solid hexagonal or four-sided prisms, connected by their surfaces and ends, and filled with calcareous matter. If the latter is removed, by weak acid, from newly-formed or growing enamel, it will be found to present a network of delicate prismatic cells of animal matter.

Chemical Composition. According to Bibra, enamel consists of 96.5 percent of earthy matter, and 3.5 percent of animal matter. The earthy matter consists of phosphate of lime, with traces of fluoride of calcium; carbonate of lime, phosphate of magnesia and other salts.

The CORTICAL SUBSTANCE, or cement (*crusta petrosa*), is disposed as a thin layer on the roots of the teeth, from the termination of the enamel, as far as the apex of the fang, where it is usually very thick. In structure and chemical composition, it resembles bone. It contains, sparingly, the lacunae and canaliculi which characterise true bone: those placed near the surface have the canaliculi radiating from the side of the lacunae towards the periodontal membrane; and those more deeply placed, join with the adjacent dental tubuli. In the thicker portions of the crusta petrosa, the lamellae and Haversian canals peculiar to bone are also found. As age advances, the cementum increases in thickness, and gives rise to those bony growths, or exostoses, so common in the teeth of the aged; the pulp cavity becomes also partially filled up by a hard substance, intermediate in structure between dentine and bone (*osteo-dentine*, Owen; secondary dentine, Tomes). It appears to be formed by a slow conversion of the dental pulp, which shrinks or even disappears.

Development of the Teeth (figs. 301 to 306)

According to the observations of Arnold and Goodsir, the teeth are developed from the mucous membrane covering the edges of the maxillary arches. About the sixth week of foetal life (fig. 301), the mucous membrane covering the edge of the upper jaw, presents a semicircular depression or groove: this is the primitive dental groove (Goodsir), from the floor of which the germs of the ten deciduous or milk-teeth are developed. The germ of each tooth is formed by a conical elevation or papilla of mucous membrane (fig. 302); it constitutes the rudimentary pulp of a milk-tooth. The germs of the milk-teeth make their appearance in the following order: at the seventh week, the germ of the first deciduous molar of the upper jaw appears; at the eighth week, that for the canine tooth is developed; the two incisor papillae appear about the ninth week (the central preceding the lateral); lastly, the second molar papilla is seen at the tenth week, behind the anterior molar. The teeth of the lower jaw appear rather later, the first molar papilla being only just visible at the seventh week; and the tenth papilla not being developed before the eleventh week. This completes the first or papillary stage of their development.

The dental groove now becomes contracted, its margins thickened and prominent, and the groove is converted into follicles, for the reception of the papillae, by the growth of membranous septa, which pass across the groove between its borders (fig. 303). The follicles by this means become the alveoli, lined by periosteum, from the bottom of which a process of the mucous membrane of the gum rises, which is the germ of the future tooth. The follicle for the first molar is complete about the tenth week, the canine follows next, succeeded by the follicles for the incisors, which are completed about the eleventh or twelfth week; and, lastly, the follicle of the posterior deciduous molar is completed about the fourteenth week. These changes constitute the second or follicular stage.

About the thirteenth week, the papillae begin to grow rapidly, project from the follicles, and assume a form corresponding with that of the future teeth; the follicles soon become deeper, and from their margins small membranous processes, or opercula, are developed, which, meeting, unite and form a lid to the now closed cavity (fig. 304). These processes correspond in shape to the form of the crown of the tooth, and in number to the tubercles on its surface. The follicles of the incisor teeth have two opercula, the canine three, and the molars four or five each. The follicles are thus converted into dental sacs, and the contained papillae become pulps. The lips of the dental groove gradually advance over the follicles from behind forwards, and, uniting, gradually obliterate it. This completes the third or saccular stage, which takes place about the end of the fifteenth week.

The deep portion of the primitive dental groove is now closed in; but the more superficial portion, near the surface of the gum, still remains open; it is called by Mr. Goodsir, the *secondary dental groove:* from it are developed the ten anterior permanent teeth. About the fourteenth week, certain lunated depressions are formed, one behind each of the sacs of the rudimentary milk-teeth. They are ten in number in each jaw, and are formed successively from before backwards; they are the rudimentary follicles of the four permanent incisors, the two canine, and the four bicuspids. As the secondary dental groove closes in, these follicles become closed cavities of reserve (fig. 304). The cavities soon elongate, and recede from the surface into the substance of the gum, behind the sacs of the deciduous teeth, and a papilla projects from the bottom of each, which is the germ of the permanent tooth: at the same time, one or more opercular folds are developed from the sides of the cavity; and these uniting, divide it into two portions; the lower portion containing the papilla of the permanent tooth, the upper narrower portion becoming gradually contracted in the same way that the primitive dental groove was obliterated over the sacs of the deciduous teeth.

The six posterior permanent teeth in each jaw, three on each side, arise from successive extensions backwards of the posterior part of the primitive dental groove. During the fourth month, that portion of the dental groove which lies behind the last temporary molar follicle remains open, and from it is developed the papilla, the rudiment of the first permanent molar. The follicle in which it is contained becomes closed

Development of Teeth.

by its operculum, and the upper part of the now-formed sac elongates backwards to form a cavity of reserve, in which the papilla of the second permanent molar appears at the seventh month after birth. After a considerable interval, during which the sacs of the first and second permanent molars have considerably increased in size, the remainder of the cavity of reserve presents for the last time a series of changes similar to the preceding, and gives rise to the sac and papilla of the wisdom-tooth, which appears at the sixth year.

Growth of the Teeth. As soon as the dental sacs are formed by the closing in of the follicles, they gradually enlarge, as well as their contained papillae. The sacs consist of two layers: an internal highly vascular layer, lined by epithelium; and an external or areolo-fibrous membrane, analogous to the corium of the mucous membrane.

The *Dental Pulps* soon become moulded to the form of the future teeth, and are adherent by their bases to the bottom of the dental sacs; in the case of the molars, the base of the pulp is divided into two or more portions, which form the future fangs. During the fourth or fifth month of foetal life, a thin lamina or cap of dentine is formed on the most prominent point of the pulp of all the milk-teeth. In the incisor and canine teeth, this newly-formed lamina has the form of a hollow cone; in the molar teeth, as many separate laminae are found, as there are eminences upon its crown. These laminae grow at the expense of the pulp-substance, increasing in breadth by a growth around their margins, and in thickness by a similar formation in its substance; the separate cones (if a molar tooth) ultimately coalesce, and the crown is completely formed. The pulp now becomes constricted, so as to form the cervix; and the remaining portion becomes narrow and elongated, to form the fang. The growth of dentine takes place from the surface towards the interior, until nothing but the small *cavitas pulpae* remains in the centre of the tooth, communicating by the aperture left at the point of each fang, with the dental vessels and nerves.

As soon as the formation of the dentine has commenced, there is developed from the inner wall of the dental sac, a soft pulpy mass, the *enamel organ*, which is intimately united to the surface of the dental pulp, or its cap of dentine. It consists of a mesh of fibres, elastic and spongy, containing within its reticulations fluid albumen; and at the point of junction of each fibre, a transparent nucleus is visible. The surface towards the dentinal pulp is covered by a layer of elongated nucleated cells, the *enamel membrane*. The deposition of the enamel takes place on the outer surface of the cap of dentine.

The *Cementum* appears to be formed, at a later period of life, by the periodontal membrane, extending from the margin of the enamel downwards.

Eruption. When the calcification of the different tissues of the tooth is sufficiently advanced to enable it to bear the pressure to which it will be afterwards subjected, its eruption takes place, the tooth making its way through the gum. The gum is absorbed by the pressure of the crown of the tooth against it, which is itself pressed up by the increasing size of the fang. Concurrent with this, the septa between the dental sacs, at first fibrous in structure, soon ossify, and constitute the alveoli; these firmly embrace the necks of the teeth, and afford them a solid basis of support.

The eruption of the temporary teeth commences at the seventh month, and is complete about the end of the second year, those of the lower jaw preceding the upper.

The periods for the eruption of the temporary set are:

  7th month, central incisors.
  7th to 10th month, lateral incisors.
  12th to 14th month, anterior molars.
  14th to 20th month, canine.
  18th to 36th month, posterior molars.

Calcification of the permanent teeth commences a little before birth, and proceeds in the following order in the upper jaw, in the lower jaw appearing a little earlier. First molar, five or six months; the central incisor a little later; lateral incisors and canine, about the eighth or ninth month; the bicuspids at the second year; second molar, five or six years; wisdom-tooth, about twelve years.

Previous to the permanent teeth penetrating the gum, the bony partitions separating their sacs from the deciduous teeth are absorbed, the fangs of the temporary teeth disappear, and the permanent teeth become now placed under the loose crowns of the deciduous teeth; the latter finally become detached, when the permanent teeth take their place in the mouth.

The eruption of the permanent teeth takes place at the following periods, the teeth of the lower jaw preceding those of the upper by a short interval:

  6½ years, first molars.
  7th year, two middle incisors.
  8th year, two lateral incisors.
  9th year, first bicuspid.
  10th year, second bicuspid.
  11th to 12th year, canine.
  12th to 13th year, second molars.
  17th to 21st year, wisdom-teeth.

The Palate

The PALATE forms the roof of the mouth; it consists of two portions, the hard palate in front, the soft palate behind.

The *Hard Palate* is bounded in front and at the sides by the alveolar arches and gums; behind, it is continuous with the soft palate. It is covered by a dense structure, formed by the periosteum and mucous membrane of the mouth, which are intimately adherent. Along the middle line is a linear ridge or raphe, which terminates anteriorly in a small papilla, corresponding with the inferior opening of the anterior palatine fossa. This papilla receives filaments from the naso-palatine and anterior palatine nerves. On either side and in front of the raphe, the mucous membrane is thick, pale in colour, and corrugated; behind, it is thin, smooth, and of a deeper colour: it is covered with squamous epithelium, and furnished with numerous glands (palatal glands), which lie between the mucous membrane and the surface of the bone.

The *Soft Palate* (*Velum pendulum palati*) is a moveable fold, suspended from the posterior border of the hard palate, forming an incomplete septum between the mouth and pharynx. It consists of a fold of mucous membrane, enclosing muscular fibres, an aponeurosis, vessels, nerves, and mucous glands. When occupying its usual position (a relaxed pendent state), its anterior surface is concave, continuous with the roof of the mouth, and marked by a median ridge or raphe, which indicates its original separation into two lateral halves. Its posterior surface is convex, and continuous with the mucous membrane covering the floor of the posterior nares. Its upper border is attached to the posterior margin of the hard palate, and its sides are blended with the pharynx. Its lower border is free.

Hanging from the middle of its lower border is a small conical-shaped pendulous process, the *uvula*; and arching outwards and downwards from the base of the uvula on each side, are two curved folds of mucous membrane, containing muscular fibres, called the *arches* or *pillars of the soft palate*.

The *anterior pillar* runs downwards and forwards to the side of the base of the tongue, and is formed by the projection of the Palato-glossus muscle, covered by mucous membrane.

The *posterior pillars* are more closely approximated and larger than the anterior; they run downwards and backwards to the sides of the pharynx, and are formed by the projection of the Palato-pharyngei muscles, covered by mucous membrane. The anterior and posterior pillars are separated below by a triangular interval, in which the tonsil is lodged.

The space left between the arches of the palate on the two sides is called the *isthmus of the fauces*. It is bounded above by the free margin of the palate; below, by the tongue; and on each side, by the pillars of the soft palate and tonsils.

The *mucous membrane* of the soft palate is thin, and covered with squamous epithelium on both surfaces, excepting near the orifice of the Eustachian tube, where it is columnar and ciliated. The palatine glands form a continuous layer on its posterior surface and around the uvula.

The *aponeurosis* of the soft palate is a thin but firm fibrous layer, attached above to the hard palate, and becoming thinner towards the free margin of the velum. It is blended with the aponeurotic tendon of the Tensor palati muscle.

The *muscles* of the soft palate are five on each side; the Levator palati, Tensor palati, Palato-glossus, Palato-pharyngeus, and the Azygos uvulae. (See p. 243.)

The *Tonsils* (*Amygdalae*) are two glandular organs, situated one on each side of the fauces, between the anterior and posterior pillars of the soft palate. The tonsils are of a rounded form, and vary considerably in size in different individuals. Externally the tonsil is in relation with the inner surface of the Superior constrictor, and with the internal carotid and ascending pharyngeal arteries, and corresponds to the angle of the lower jaw. Its *inner surface* presents from twelve to fifteen orifices, leading into small recesses, from which numerous follicles branch out into the substance of the gland. These follicles are lined by a continuation of the mucous membrane of the pharynx, covered with epithelium, their walls being formed by a layer of closed capsules imbedded in the submucous tissue. These capsules are analogous to those of Peyer's glands; they contain a thick greyish secretion.

The *arteries* supplying the tonsil are the dorsalis linguae from the lingual, the ascending palatine and tonsillar from the facial, the ascending pharyngeal from the external carotid, and the descending palatine branch of the internal maxillary.

The *veins* terminate in the tonsillar plexus, on the outer side of the tonsil.

The *nerves* are derived from the fifth, and from the glosso-pharyngeal.

The Salivary Glands

The principal salivary glands communicating with the mouth, and pouring their secretion into its cavity, are the parotid, submaxillary, and sublingual.

The *Parotid Gland* (παρὰ, *near*; ους, ωτος *the ear*) (fig. 307), so called from being placed near the ear, is the largest of the three salivary glands, varying in weight from half an ounce to an ounce. It lies upon the side of the face, immediately in front of the external ear. It is bounded above by the zygoma; below, by the angle of the jaw, and by an imaginary line drawn between it and the Sterno-mastoid muscle; anteriorly, it extends to a variable extent over the Masseter muscle; posteriorly, it is bounded by the external meatus, the mastoid process, and the insertion of the Sterno-mastoid muscle, which it slightly overlaps. Its outer surface is smooth, covered by the integument, fascia, and Platysma, and has one or two lymphatic glands resting on it. Its inner surface passes behind the ramus of the lower jaw, between it and the mastoid process. This portion of the gland is deeply seated, extending above into the back part of the glenoid cavity; behind and below, lying upon the styloid process and Styloid muscles; and, in front, advancing forwards beneath the ramus of the jaw, between the two Pterygoid muscles. Embedded in its substance is the external carotid, which ascends behind the ramus of the jaw; the posterior auricular artery emerges from it behind; the temporal artery above; the transverse facial in front; and the internal maxillary winds through it inwards, behind the neck of the jaw. Superficial to the carotid artery is the trunk formed by the union of the temporal and internal maxillary veins. It is traversed, from before backwards, by the facial nerve and its branches, which emerge at its anterior border; the great auricular nerve pierces the gland to join the facial, and

307.—The Salivary Glands.

the temporal branch of the inferior maxillary nerve lies above the upper part of the gland. The internal carotid artery and internal jugular vein lie close to its deep surface.

The *duct* of the parotid gland (Steno's) is about two inches and a half in length. It commences upon the inner surface of the cheek by a small orifice, opposite the second molar tooth of the upper jaw; running obliquely for a short distance beneath the mucous membrane, it pierces the Buccinator muscle, and crosses the Masseter to the anterior border of the gland, in the substance of which it subdivides into numerous branches. The direction of the duct corresponds to a line drawn across the face about a finger's breadth below the zygoma, from the lower part of the concha, to midway between the free margin of the upper lip and the ala of the nose. While crossing the Masseter, it receives the duct of a small detached portion of the gland, which occasionally exists as a separate lobe, the *socia parotidis*. The parotid duct is dense, of considerable thickness, and its canal about the size of a crow-quill; it consists of an external or fibrous coat, of considerable density, containing contractile fibres, and of an internal or mucous coat, lined with columnar epithelium.

The *arteries* supplying the parotid gland are derived from the external carotid, and from the branches of that vessel in or near its substance.

The *veins* follow a similar course.

The *lymphatics* terminate in the superficial and deep cervical glands, passing in their course through two or three lymphatic glands, placed on its surface and in its substance.

The *nerves* are derived from the carotid plexus of the sympathetic, the facial, superficial temporal, and great auricular nerves.

The *Submaxillary Gland* is situated below the jaw, in the anterior part of the submaxillary triangle of the neck. It is irregular in form, and weighs about two drachms. It is covered in by the body of the lower jaw, the integument, Platysma, and deep cervical fascia; and lies upon the Mylo-hyoid, Hyo-glossus, and Stylo-glossus muscles, a portion of the gland passing beneath the posterior border of the Mylo-hyoid. Behind, it is separated from the parotid gland by the stylo-maxillary ligament, and from the sublingual gland in front by the Mylo-hyoid muscle. The facial artery lies in a groove in its posterior and upper border.

The *duct* of the submaxillary gland (Wharton's) is about two inches in length, and its walls much thinner than those of the parotid duct. It commences by a narrow orifice on the summit of a small papilla, at the side of the fraenum linguae. Passing between the sublingual gland and the Genio-hyo-glossus muscle, it runs backwards and outwards between the Mylo-hyoid, and the Hyo-glossus and Genio-hyo-glossus muscles, to the deep portion of the gland, where it divides into numerous branches.

The *arteries* supplying the submaxillary gland are branches of the facial and lingual. Its *veins* follow the course of the arteries. The *nerves* are derived from the submaxillary ganglion, from the mylo-hyoid branch of the inferior dental, and from the sympathetic.

The *Sublingual Gland* is the smallest of the salivary glands. It is situated beneath the mucous membrane of the floor of the mouth, on either side of the fraenum linguae, in contact with the inner surface of the lower jaw, close to the symphysis. It is narrow, flattened, in shape somewhat like an almond, and weighs about a drachm. It is in relation, *above*, with the mucous membrane; *below*, with the Mylo-hyoid muscle; *in front*, with the depression on the side of the symphysis of the lower jaw, and with its fellow of the opposite side; *behind*, with the deep part of the submaxillary gland; and *internally*, with the Genio-hyo-glossus, from which it is separated by the lingual nerve and Wharton's duct. Its excretory ducts (*ductus Riviniani*), from eight to twenty in number, open separately into the mouth, on the elevated crest of mucous membrane, caused by the projection of the gland, on either side of the fraenum linguae. One or more join to form a tube which opens into, or near, the Whartonian duct: it is called the *duct of Bartholine*.

Vessels and Nerves. The sublingual gland is supplied with blood from the sublingual and submental arteries. Its nerves are derived from the gustatory.

Structure. The salivary are conglomerate glands, consisting of numerous lobes, which are made up of smaller lobules, connected together by dense areolar tissue, vessels, and ducts. Each lobule consists of numerous closed vesicles, which open into a common duct: the wall of each vesicle is formed of a delicate basement membrane, lined by epithelium, and covered on its outer surface with a dense capillary network. In the submaxillary and sublingual glands, the lobes are larger and more loosely united than in the parotid.

The Pharynx

The pharynx is that part of the alimentary canal which is placed behind the nose, mouth, and larynx. It is a musculo-membranous sac, somewhat conical in form, with the base upwards, and the apex downwards, extending from the base of the skull to the cricoid cartilage in front, and the fifth cervical vertebra behind.

The pharynx is about four inches and a half in length, and broader in the transverse than in the antero-posterior diameter. Its greatest breadth is opposite the cornua of the hyoid bone; its narrowest point at its termination in the oesophagus. It is limited, *above*, by the basilar process of the occipital bone; *below*, it is continuous with the oesophagus; *posteriorly*, it is connected by loose areolar tissue with the cervical portion of the vertebral column, and the Longi colli and Recti capitis antici muscles; *anteriorly*, it is incomplete, and is attached in succession to the internal pterygoid plate, the pterygo-maxillary ligament, the lower jaw, the tongue, hyoid bone, and larynx; *laterally*, it is connected to the styloid processes and their muscles, and is in contact with the common and internal carotid arteries, the internal jugular veins, and the eighth, ninth, and sympathetic nerves, and above, with a small part of the Internal pterygoid muscles.

It has seven openings communicating with it: the two posterior nares, the two Eustachian tubes, the mouth, larynx, and oesophagus.

The *posterior nares* are the two large apertures situated at the upper part of the anterior wall of the pharynx.

The *two Eustachian tubes* open one at each side of the upper part of the pharynx, at the back part of the inferior meatus. Below the nasal fossae is the posterior surface of the soft palate and uvula, the large aperture of the mouth, the base of the tongue, the epiglottis, and the cordiform opening of the larynx.

The *oesophageal opening* is the lower contracted portion of the pharynx.

Structure. The pharynx is composed of three coats: a mucous coat, a muscular layer, and a fibrous coat.

The *fibrous coat* is situated between the mucous and muscular layers, and is called the *pharyngeal aponeurosis*. It is thick above, where the muscular fibres are wanting, and firmly connected to the basilar process of the occipital and petrous portion of the temporal bones. As it descends, it diminishes in thickness, and is gradually lost.

The *mucous coat* is continuous with that lining the Eustachian tubes, the nares, the mouth, and the larynx. It is covered by columnar ciliated epithelium, as low down as a level with the floor of the nares; below that point, it is of the squamous variety.

The *muscular coat* has been already described (p. 240).

The *pharyngeal glands* are of two kinds, the simple, or compound follicular, which are found in considerable numbers beneath the mucous membrane, throughout the entire pharynx; and the racemose, which are especially numerous at the upper part of the pharynx, and form a thick layer, across the back of the fauces, between the two Eustachian tubes.

The Oesophagus

The oesophagus is a membranous canal, about nine inches in length, extending from the pharynx to the stomach. It commences at the lower border of the cricoid cartilage, opposite the fifth cervical vertebra, descends along the front of the spine, through the posterior mediastinum, passes through the Diaphragm, and, entering the abdomen, terminates at the cardiac orifice of the stomach, opposite the ninth dorsal vertebra. The general direction of the oesophagus is vertical; but it presents two or three slight curvatures in its course. At its commencement, it is placed in the median line; but it inclines to the left side at the root of the neck, gradually passes to the middle line again, and finally, again deviates to the left, as it passes forwards to the oesophageal opening of the Diaphragm. The oesophagus also presents an antero-posterior flexure, corresponding to the curvature of the cervical and thoracic portions of the spine. It is the narrowest part of the alimentary canal, being most contracted at its commencement, and at the point where it passes through the Diaphragm.

Relations. In the *neck*, the oesophagus is in relation, *in front*, with the trachea; and, at the lower part of the neck, where it projects to the left side, with the thyroid gland and thoracic duct; *behind*, it rests upon the vertebral column and Longus colli muscle; on *each side*, it is in relation with the common carotid artery (especially the left, as it inclines to that side), and part of the lateral lobes of the thyroid gland; the recurrent laryngeal nerves ascend between it and the trachea.

In the *thorax*, it is at first situated a little to the left of the median line: it passes across the left side of the transverse part of the aortic arch, descends in the posterior mediastinum, along the right side of the aorta, until near the Diaphragm, where it passes in front and a little to the left of this vessel, previous to entering the abdomen. In is in relation, *in front*, with the trachea, the arch of the aorta, the left bronchus, and the posterior surface of the pericardium; *behind*, it rests upon the vertebral column, the Longus colli, and the intercostal vessels; and *below*, near the Diaphragm, upon the front of the aorta; *laterally*, it is covered by the pleurae; the vena azygos major lies on the right, and the descending aorta on the left side. The pneumogastric nerves descend in close contact with it, the right nerve passing down behind, and the left nerve in front of it.

Surgical Anatomy. The relations of the oesophagus are of considerable practical interest to the surgeon, as he is frequently required, in cases of stricture of this tube, to dilate the canal by a bougie, when it becomes of importance that its direction, and relations to surrounding parts, should be remembered. In cases of malignant disease of the oesophagus, where its tissues have become softened from infiltration of the morbid deposit, the greatest care is requisite in directing the bougie through the strictured part, as a false passage may easily be made, and the instrument may pass into the mediastinum, or into one or the other pleural cavity, or even into the pericardium.

The student should also remember that contraction of the oesophagus, and consequent symptoms of stricture, are occasionally produced by an aneurism of some part of the aorta pressing upon this tube. In such a case, the passage of a bougie could only facilitate the fatal issue.

It occasionally happens that a foreign body becomes impacted in the oesophagus, which can neither be brought upwards nor moved downwards. When all ordinary means for its removal have failed, excision is the only resource. This of course can only be performed when it is not very low down. If the foreign body is allowed to remain, extensive inflammation and ulceration of the oesophagus may ensue. In one case with which I am acquainted, the foreign body ultimately penetrated the intervertebral substance, and destroyed life by inflammation of the membranes and substance of the cord.

The operation of oesophagotomy is thus performed. The patient being placed upon his back, with the head and shoulders slightly elevated, an incision, about four inches in length, should be made on the left side of the trachea, from the thyroid cartilage downwards, dividing the skin and Platysma. The edges of the wound being separated, the Omo-hyoid muscle, and the fibres of the Sterno-hyoid and Sterno-thyroid muscles, must be cut through; the sheath of the carotid vessels being exposed, should be drawn outwards, and retained in that position by retractors; the oesophagus will then be exposed, and should be divided over the foreign body, which should then be removed. Great care is necessary to avoid wounding the thyroid vessels, the thyroid gland, and the laryngeal nerves.

Structure. The oesophagus has three coats: an external, or muscular; a middle, or cellular; and an internal, or mucous coat.

The *muscular coat* is composed of two planes of fibres of considerable thickness, an external longitudinal, and an internal circular.

The *longitudinal fibres* are arranged, at the commencement of the tube, in three fasciculi; one in front, which is attached to the vertical ridge on the posterior surface of the cricoid cartilage; and one at each side, which are continuous with the fibres of the Inferior constrictor: as they descend they blend together, and form a uniform layer, which covers the outer surface of the tube.

The *circular fibres* are continuous above with the Inferior constrictor: their direction is transverse at the upper and lower parts of the tube, but oblique in the central part.

The muscular fibres in the upper part of the oesophagus are of a red colour, and consist chiefly of the striped variety; but below, they consist entirely of the involuntary muscular fibre.

The *cellular coat* connects loosely the mucous and muscular coats.

The *mucous coat* is thick, of a reddish colour above, and pale below. It is loosely connected with the muscular coat, and disposed in longitudinal plicae, which disappear on distension of the tube. Its surface is studded with minute papillae, and it is covered throughout with a thick layer of squamous epithelium.

The *Oesophageal Glands* are numerous small compound glands, scattered throughout the tube: they are lodged in the submucous tissue, and open upon the surface by a long excretory duct. They are most numerous at the lower part of the tube, where they form a ring round the cardiac orifice.

The Abdomen

The abdomen is the largest cavity of the trunk of the body, and is separated, below, from the pelvic cavity by the brim of the pelvis. It is of an oval form, the extremities of the oval being directed upwards and downwards; it is wider above than below, and measures more in the vertical than in the transverse diameter.

Boundaries. It is bounded in *front* and at the *sides* by the lower ribs, the Transversales muscles, and venter ilii; *behind*, by the vertebral column, and the Psoae and Quadrati lumborum muscles; *above*, by the Diaphragm; *below*, by the brim of the pelvis. The muscles forming the boundaries of this cavity are lined upon their inner surface by a layer of fascia, differently named according to the part to which it is attached.

This cavity contains the greater part of the alimentary canal, some of the accessory organs to digestion, the liver, pancreas, and spleen, and the kidneys and supra-renal capsules. Most of these structures, as well as the wall of the cavity in which they are contained, are covered by an extensive and complicated serous membrane, the *peritoneum*.

The *apertures* found in the walls of the abdomen, for the transmission of structures to or from it, are the *umbilicus*, for the transmission (in the foetus) of the umbilical vessels; the *caval opening* in the Diaphragm, for the transmission of the inferior vena cava; the *aortic opening*, for the passage of the aorta, vena azygos, and thoracic duct; and the *oesophageal opening*, for the oesophagus and pneumogastric nerves. *Below*, are two apertures on each side; one for the passage of the femoral vessels, and the other for the transmission of the spermatic cord in the male, and the round ligament in the female.

Regions. For convenience of description of the viscera, as well as of reference to the morbid condition of the contained parts, the abdomen is artificially divided into certain regions. Thus, if two circular lines are drawn round the body, the one parallel with the cartilages of the ninth ribs, and the other with the highest point of the crests of the ilia, the abdominal cavity is divided into three zones, an upper, a middle, and a lower. If two parallel lines are drawn from the cartilage of the eighth rib on each side, down to the centre of Poupart's ligament, each of these zones is subdivided into three parts, a middle and two lateral.

The middle region of the upper zone is called the *epigastric* (ἐπι, *over*, γαστηρ, *the stomach*); and the two lateral regions, the *right and left hypochondriac* (ὑπο, *under*, χονδροι, *the cartilages*). The central region of the middle zone is the *umbilical*; and the two lateral regions, the *right and left lumbar*. The middle region of the lower zone is the *hypogastric* or *pubic region*; and the lateral regions are the *right and left inguinal*. The parts contained in these different regions are the following (fig. 308):

308.—The Regions of the Abdomen and their Contents.
(Edge of Costal Cartilages in dotted outline.)

Right Hypochondriac.

The right lobe of the liver and the gall-bladder, the duodenum, hepatic flexure of the colon, upper part of the right kidney, and right supra-renal capsule.

Right Lumbar.

Ascending colon, lower part of the right kidney, and some convolutions of the small intestines.

Right Inguinal.

The caecum, appendix caeci, ureter, and spermatic vessels.

Epigastric Region.

The middle and pyloric end of the stomach, left lobe of the liver and lobus Spigelii, the hepatic vessels, caeliac axis, semilunar ganglia, pancreas, parts of the aorta, vena cava, vena azygos, and thoracic duct.

Umbilical Region.

The transverse colon, part of the great omentum and mesentery, transverse part of the duodenum, and some convolutions of the jejunum and ileum.

Hypogastric Region.

Convolutions of the small intestines, the bladder in children, and in adults if distended, and the uterus during pregnancy.

Left Hypochondriac.

The splenic end of the stomach, the spleen and extremity of the pancreas, the splenic flexure of the colon, upper half of the left kidney, and left supra-renal capsule.

Left Lumbar.

Descending colon, lower part of left kidney, and some convolutions of the small intestines.

Left Inguinal.

Sigmoid flexure of the colon, ureter, and spermatic vessels.

The Peritoneum

The peritoneum (περιτείνειν, *to extend around*) is a serous membrane; and, like all membranes of this class, a shut sac. In the female, however, it is not completely closed, the Fallopian tubes communicating with it by their free extremities; and thus the serous membrane is continuous with their mucous lining.

The peritoneum invests, more or less completely, all the viscera contained in the abdominal and pelvic cavities, forming the visceral layer of the membrane; it is then reflected upon the internal surface of the parietes of those cavities, forming the parietal layer. The *free* surface of the peritoneum is smooth, moist, and covered by a thin, squamous epithelium; its *attached* surface is rough, being connected to the viscera and inner surface of the parietes by means of areolar tissue, called the *sub-peritoneal areolar tissue*. The parietal portion is loosely connected with the fascia lining the abdomen and pelvis; but more closely to the under surface of the Diaphragm, and in the middle line of the abdomen.

In order to trace the reflections of this membrane (fig. 309) (the abdomen having been opened), the liver should be raised upwards and supported in that position, and the stomach should be depressed, when a thin membranous layer is seen passing from

309.—The Reflections of the Peritoneum, as seen in a vertical Section of the Abdomen.

the transverse fissure of the liver, to the upper border of the stomach: this is the *lesser*, or *gastro-hepatic omentum*. It consists of two thin, delicate layers of peritoneum, an anterior and a posterior, between which are contained the hepatic vessels and nerves. Of these two layers, the anterior should first be traced, and then the posterior.

The *anterior layer* descends to the lesser curvature of the stomach, and covers its anterior surface as far as the great curvature; it descends for some distance in front of the small intestines, and, returning upon itself to the transverse colon, forms the *external layer of the great omentum*; it then covers the under surface of the transverse colon, and, passing to the back part of the abdominal cavity, forms the *inferior layer of the transverse meso-colon*. It may then be traced, in front of the duodenum, the aorta, and vena cava, as far as the superior mesenteric artery, along which it passes to invest the small intestines, and, returning to the vertebral column, forms the *mesentery*; whilst on either side, it covers the ascending and descending colon, and is thus continuous with the peritoneum lining the walls of the abdomen. From the root of the mesentery, it descends along the front of the spine into the pelvis, and surrounds the upper part of the rectum, which it holds in its position by means of a distinct fold, the meso-rectum. Its course in the male and female now differs.

In the *male*, it forms a fold between the rectum and bladder, the *recto-vesical fold*, and ascends over the posterior surface of the latter organ as far as its summit.

In the *female*, it descends into the pelvis in front of the rectum, covers a small part of the posterior wall of the vagina, and is then reflected on to the uterus, the fundus and body of which it covers. From the sides of the uterus, it is reflected on each side to the wall of the pelvis, forming the broad ligaments; and from its anterior surface it ascends upon the posterior wall of the bladder, as far as its summit. From this point it may be traced, as in the male, ascending upon the anterior parietes of the abdomen, to the under surface of the Diaphragm; from which it is reflected upon the liver, forming the upper layer of the coronary, and the lateral and longitudinal ligaments. It then covers the upper and under surfaces of the liver, and at the transverse fissure becomes continuous with the anterior layer of the lesser omentum, the point from whence its reflection was originally traced.

The *posterior layer* of the lesser omentum descends to the lesser curvature of the stomach, and covers its posterior surface as far as the great curvature; it then descends for some distance in front of the small intestines, and, returning upon itself to the transverse colon, forms the internal layer of the great omentum; it covers the upper surface of the transverse colon, and, passing backwards to the spine, forms the upper layer of the transverse meso-colon. Ascending in front of the pancreas and crura of the Diaphragm, it lines the back part of the under surface of this muscle, from which it is reflected on to the posterior border of the liver, forming the inferior layer of the coronary ligament. From the under surface of the liver, it may be traced to the transverse fissure, where it is continuous with the posterior layer of the lesser omentum, the point from whence its reflection was originally traced.

The space included in the reflections of this layer of the peritoneum, is called the *lesser cavity of the peritoneum*, or *cavity of the great omentum*. It is bounded, in *front*, by the lesser omentum, the stomach, and the descending part of the great omentum; *behind*, by the ascending part of the great omentum, the transverse colon, transverse meso-colon, and its ascending layer; *above*, by the liver; and *below*, by the folding of the great omentum. This space communicates with the general peritoneal cavity through the foramen of Winslow, which is situated behind the right free border of the lesser omentum.

The *Foramen of Winslow* is bounded *in front* by the lesser omentum, enclosing the vena portae and the hepatic artery and duct; *behind*, by the inferior vena cava; *above*, by the lobus Spigelii; *below*, by the hepatic artery curving forwards from the caeliac axis.

This foramen is nothing more than a constriction of the general peritoneal cavity at this point, caused by the hepatic and gastric arteries passing forwards from the caeliac axis to reach their respective viscera.

The viscera thus shown to be almost entirely invested by peritoneum are the liver, stomach, spleen, first portion of duodenum, jejunum, and ileum, transverse colon, sigmoid flexure, upper end of rectum, uterus, and ovaries.

Those partially covered by it are the descending and transverse portions of the duodenum, the caecum, the ascending and descending colon, the middle portion of the rectum, and the upper part of the vagina and posterior wall of bladder.

Those receiving *no* investment from it, are the lower end of the rectum, the neck, base, and anterior surface of the bladder, and the lower part of the vagina.

Numerous folds are formed by the peritoneum, extending between the various organs. These serve to hold them in position, and, at the same time, enclose the vessels and nerves proceeding to each part. Some of the folds are called *ligaments*, from their serving to support the organs in position. Others constitute the *mesenteries*, which connect certain parts of the intestine with the abdominal wall; and lastly, those are called *omenta*, which proceed from one viscus to another.

The LIGAMENTS, formed by folds of the peritoneum, include those of the liver, spleen, bladder, and uterus. They are described with their respective organs.

The OMENTA are the lesser omentum, the great omentum, and the gastro-splenic omentum.

The *Lesser*, or *Gastro-hepatic Omentum*, is the duplicature which extends between the transverse fissure of the liver, and the lesser curvature of the stomach. It is extremely thin, and consists, as before said, of two layers of peritoneum. At the left border, its two layers pass on to the end of the oesophagus; but, at the right border, where it is free, they are continuous, and form a free rounded margin, which contains between its layers the hepatic artery, the ductus communis choledocus, the portal vein, lymphatics, and hepatic plexus of nerves; all these structures being enclosed in loose areolar tissue, called *Glisson's capsule*.

The *Great Omentum* (*gastro-colic*) is the largest peritoneal fold. It consists of four layers of peritoneum, two of which descend from the stomach, one from its anterior, the other from its posterior surface; these, uniting at its lower border, descend in front of the small intestines, as low down as the pelvis; and the same two ascend again as far as the transverse colon, where they separate and enclose that part of the intestine. These separate layers may be easily demonstrated in the young subject, but, in the adult, they are more or less inseparably blended. The left border of the great omentum is continuous with the gastro-splenic omentum: its right border extends as far only as the duodenum. The great omentum is usually thin, presents a cribriform appearance, and always contains some adipose tissue, which, in fat subjects, accumulates in considerable quantity. Its use appears to be to protect the intestines from cold, and to facilitate their movement upon each other during their vermicular action.

The *Gastro-Splenic Omentum* is the fold which connects the concave surface of the spleen to the *cul-de-sac* of the stomach, being continuous by its lower border with the great omentum. It contains the splenic vessels and vasa brevia.

The MESENTERIES are, the mesentery proper, the meso-caecum, the ascending, transverse, and descending meso-colon, and the meso-rectum.

The *Mesentery* (μεσον, εντερον), so called from being connected to the middle of the cylinder of the small intestine, is the broad fold of peritoneum which connects the convolutions of the jejunum and ileum with the posterior wall of the abdomen. Its *root*, the part connected with the vertebral column, is narrow, about six inches in length, and directed obliquely from the left side of the second lumbar vertebra, to the right sacro-iliac symphysis. Its intestinal border is much longer; and here its two layers separate, so as to enclose the intestine, and form its peritoneal coat. Its breadth, between its vertebral and intestinal border, is about four inches. Its *upper border* is continuous with the under surface of the transverse meso-colon; its *lower border*, with the peritoneum covering the caecum and ascending colon. It serves to retain the small intestines in their position, and contains between its layers the mesenteric vessels and nerves, the lacteal vessels, and mesenteric glands.

The *Meso-Caecum*, when it exists, serves to connect the back part of the caecum with the right iliac fossa; more frequently, the peritoneum passes merely in front of this portion of the large intestine.

The *Ascending Meso-Colon* is the fold which connects the back part of the ascending colon with the posterior wall of the abdomen.

The *Descending Meso-Colon* retains the descending colon in connection with the posterior abdominal wall; more frequently, the peritoneum merely covers the anterior surface and sides of these two portions of the intestine.

The *Transverse Meso-Colon* is a broad fold, which connects the transverse colon with the posterior wall of the abdomen. It is formed of the two ascending layers of the great omentum, which, after separating to surround the transverse colon, join behind it, and are continued backwards to the spine, where they diverge in front of the duodenum, as already mentioned. This fold contains between its layers the vessels which supply the transverse colon.

The *Sigmoid Meso-Colon* is the fold of peritoneum which retains the sigmoid flexure in connection with the left iliac fossa.

The *Meso-Rectum* is the narrow fold which connects the upper part of the rectum with the front of the sacrum. It contains the haemorrhoidal vessels.

The *Appendices Epiploicae* are small pouches of the peritoneum filled with fat, and situated along the colon and upper part of the rectum. They are chiefly appended to the transverse colon.

THE STOMACH

The stomach is the principal organ of digestion. It is the most dilated part of the alimentary canal, serving for the solution and reduction of the food, which

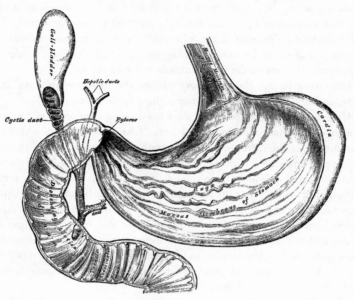

310.—The Mucous Membrane of the Stomach and Duodenum, with the Bile Ducts.

constitutes the process of chymification. It is situated in the left hypochondriac, the epigastric, and part of the right hypochondriac regions. Its form is irregularly conical, curved upon itself, and presenting a rounded base, turned to the left side. It is placed immediately behind the anterior wall of the abdomen, above the transverse colon, below the liver and Diaphragm. Its size varies considerably in different subjects, and also according to its state of distension. When moderately full, its transverse diameter is about twelve inches, its vertical diameter about four. Its weight, according to Clendenning, is about four ounces and a half. It presents for examination two extremities, two orifices, two borders, and two surfaces.

Its *left extremity* is called the *greater*, or *Splenic end*. It is the largest part of the stomach, and expands for two or three inches to the left of the point of entrance of the oesophagus. This expansion is called the great *cul-de-sac*, or *fundus*. It lies beneath the ribs, in contact with the spleen, to which it is connected by the gastro-splenic omentum.

The *lesser*, or *pyloric end*, is much smaller than the preceding, and situated on a plane anterior and inferior to it. It lies in contact with the wall of the abdomen, the under surface of the liver, and the neck of the gall-bladder.

The *oesophageal*, or *cardiac orifice*, communicates with the oesophagus: it is the highest part of the stomach, and somewhat funnel-shaped in form.

The *pyloric orifice* communicates with the duodenum, the aperture being guarded by a valve.

The *lesser curvature* extends between the oesophageal and cardiac orifices, along the upper border of the organ, and is connected to the under surface of the liver by the lesser omentum.

The *greater curvature* extends between the same points, along the lower border, and gives attachment to the great omentum. The surfaces of the organ are limited by these two curvatures.

The *anterior surface* is directed upwards and forwards, and is in relation with the Diaphragm, the under surface of the left lobe of the liver, and, in the epigastric region, with the abdominal parietes.

The *posterior surface* is directed downwards and backwards, and is in relation with the pancreas and great vessels of the abdomen, the crura of the Diaphragm, and the solar plexus.

The stomach is held in position by the lesser omentum, which extends from the transverse fissure of the liver to its lesser curvature, and by a fold of peritoneum, which passes from the Diaphragm on to the oesophageal end of the stomach, the gastro-phrenic ligament; this constitutes the most fixed point of the stomach, whilst the pyloric end and greater curvature are the most moveable parts; hence, when this organ becomes greatly distended, the greater curvature is directed forwards, whilst the anterior and posterior surfaces are directed, the former upwards, and the latter downwards.

Alterations in Position. There is no organ in the body the position and connexions of which present such frequent alterations as the stomach. *During inspiration* it is displaced downwards by the descent of the Diaphragm, and elevated by the pressure of the abdominal muscles during expiration. Its position to the surrounding viscera is also changed, according to the empty or distended state of the organ. *When empty*, it occupies only a small part of the left hypochondriac region, the spleen lying behind it; the left lobe of the liver covers it in front, and the under surface of the heart rests upon it above, and in front, being separated from it by the left lobe of the liver and pericardium. Hence it is, that, in gastralgia, the pain is generally referred to the heart, and is often accompanied by palpitation and intermission of the pulse. *When the stomach is distended* the Diaphragm is forced upwards, contracting the cavity of the chest; hence the dyspnoea complained of from inspiration being impeded. The heart is also displaced upwards; hence the oppression in this region, and the palpitation experienced in extreme distension of the stomach. *Pressure from without*, as in the pernicious practice of tight lacing, pushes the stomach down towards the pelvis. In disease, also, the position and connexions of the organ may be greatly changed, from the accumulation of fluid in the chest or abdomen, or when the size of any of the surrounding viscera undergoes alteration.

On looking into the pyloric end of the stomach, the mucous membrane is found projecting inwards in the form of a circular fold, the pylorus, leaving a narrow circular aperture, about half an inch in diameter, by which the stomach communicates with the duodenum.

The *pylorus* is formed by a reduplication of the mucous membrane of the stomach, containing its circular muscular fibres, which are here aggregated into a thick circular ring, the longitudinal fibres and serous membrane being continued over the fold without assisting in its formation. The aperture is occasionally oval. Sometimes the circular fold is replaced by two crescentic folds, placed one above and the other below the pyloric orifice; and, more rarely, there is only one crescentic fold.

Structure. The stomach consists of four coats, a serous, a muscular, a cellular, and a mucous coat, together with vessels and nerves.

The *serous coat* is derived from the peritoneum, and covers the entire surface of the organ, excepting along the greater and lesser curvatures, at the points of attachment of the greater and lesser omenta; here the two layers of peritoneum leave a small triangular space, along which the nutrient vessels and nerves pass.

The *muscular coat* (fig. 311) is situated immediately beneath the serous covering. It consists of three sets of fibres, longitudinal, circular, and oblique.

311.—The Muscular Coat of the Stomach. The Innermost Layer is not seen.

The *longitudinal fibres* are placed most externally; they are continuous with the longitudinal fibres of the oesophagus, radiating in a stellate manner from the cardiac orifice. They are most distinct along the curvatures, especially the lesser, but are very thinly distributed over the surfaces. At the pyloric end, they are more thickly distributed, and continuous with the longitudinal fibres of the small intestine.

The *circular fibres* form a uniform layer over the whole extent of the stomach, beneath the longitudinal fibres. At the pylorus, they become thicker, and are aggregated into a circular ring, which projects into the cavity, and forms, with the fold of mucous membrane covering its surface, the pyloric valve.

The *internal* or *oblique fibres* are limited chiefly to the great end of the stomach, where they are disposed as a thick uniform layer, and from this point they descend obliquely upon its anterior and posterior surfaces, upon which they become lost.

The *cellular coat* consists of a loose filamentous areolar tissue, connecting the mucous and muscular layers. It is sometimes called the *submucous coat*. It supports the blood-vessels previous to their distribution to the mucous membrane; hence it is sometimes called the *vascular coat*.

The *mucous membrane* is thick; its surface smooth, soft, and of a pulpy consistence.

During infancy, and immediately after death, it is of a pinkish tinge; but in adult life and in old age it becomes of a pale straw or ash-grey colour. It is thin at the cardiac extremity, but thicker towards the pylorus. During the contracted state of the organ it is thrown into plaits or rugae, which, for the most part, have a longitudinal direction, and are most numerous towards the lesser end of the stomach and along the greater curvature. These folds are entirely obliterated when the organ becomes distended.

Structure of the mucous membrane (fig. 312). The entire surface of the mucous membrane is covered with small shallow depressions or alveoli, of a polygonal or hexagonal form, which vary from 1-100th to 1-200th of an inch in diameter. At the bottom of the alveoli, as well as in the intervals between them, are seen the orifices of minute tubes, the *gastric follicles*, which are situated perpendicularly side by side, over the entire surface of the mucous membrane. They are short, and simply tubular in character towards the cardia, but at the pyloric end they are longer, more convoluted, and terminate in dilated saccular extremities, or subdivide into from two to six tubular branches. The gastric follicles are composed of a homogeneous basement membrane, lined upon its free surface by a layer of cells, which differ in their character in different parts of the stomach. Towards the pylorus, these tubes are lined throughout with columnar epithelium. They are termed the *mucous glands*, and are supposed to secrete the gastric mucus. In other parts of the organ, the deep part of each tube is filled with nuclei, and a mass of granules; above these are a mass of nucleated cells, the upper fourth of the tube being lined by columnar epithelium. These are called the *peptic glands*, the supposed agents in the secretion of the gastric juice.

312.—Minute Anatomy of Mucous Membrane of Stomach.

Simple follicles are found in greater or less numbers over the entire surface of the mucous membrane; they are most numerous near the pyloric end of the stomach, and especially distinct in early life. The epithelium lining the mucous membrane of the stomach and its alveoli is of the columnar variety.

Vessels and Nerves. The *arteries* supplying the stomach are the gastric, pyloric and right gastro-epiploic branches of the hepatic, the left gastro-epiploic and vasa brevia from the splenic. They supply the muscular coat, ramify in the submucous coat, and are finally distributed to the mucous membrane. The *veins* accompany the arteries, and terminate in the splenic and superior mesenteric veins. The *lymphatics* are numerous: they consist of a superficial and deep set, which pass through the lymphatic glands found along the two curvatures of the organ. The *nerves* are the terminal branches of the right and left pneumogastric, the former being distributed upon the back, and the latter upon the front part of the organ. Branches from the sympathetic also supply the organ.

The Small Intestines

The small intestine is that part of the alimentary canal in which the food is mixed with the bile and pancreatic juice, and where the separation of the nutritive principles of the food, the chyle, is effected: this constitutes chylification.

The small intestine is a convoluted tube, about twenty feet in length, which gradually diminishes in size from its commencement to its termination. It is contained in the central and lower parts of the abdominal and pelvic cavities, surrounded above and at the sides by the large intestine, in relation in front with the mesentery and abdominal parietes, and connected to the spine by a fold of peritoneum, the mesentery. The small intestine is divisible into three portions; the duodenum, jejunum, and ileum.

The *Duodenum* has received its name from being about equal in length to the breadth of twelve fingers (eight or ten inches). It is the shortest, the widest, and the most fixed part of the small intestine; it has no mesentery, and is only partially covered by the peritoneum. Its course presents a remarkable curve, somewhat like a horse-shoe in form; the convexity being directed towards the right, and the concavity to the left, embracing the head of the pancreas. Commencing at the pylorus, it ascends obliquely to the under surface of the liver; it then descends in front of the right kidney, and passes nearly transversely across the front of the second lumbar vertebra, terminating in the jejunum on the left side of this bone. Hence the duodenum has been divided into three portions; ascending, descending, and transverse.

The first, or *ascending portion*, about two inches in length, is free and moveable, and completely invested by the peritoneum. It commences at the pylorus, and passes obliquely upwards and backwards to the under surface of the liver. It is in relation, above and in front, with the liver and neck of the gall-bladder; behind, with the right border of the lesser omentum, the hepatic artery and duct, and vena portae. This portion of the intestine is usually found stained with bile, especially on its anterior surface.

The second, or *descending portion*, about three inches in length, is firmly fixed by the peritoneum and pancreas. It passes from the neck of the gall-bladder vertically downwards, in front of the right kidney, as far as the third lumbar vertebra. It is covered by peritoneum only on its anterior surface. It is in relation, by its anterior surface, with the right arch of the colon and meso-colon; by its posterior surface, it is connected to the front of the right kidney by loose areolar tissue; at its inner side is the head of the pancreas, and the common choledic duct. The common bile and pancreatic ducts perforate this side of the intestine obliquely, a little below its middle.

The third, or *transverse portion*, the longest and narrowest part of the duodenum, passes across the front of the spine, ascending from the third to the second lumbar vertebra, terminating in the jejunum on the left side of the second lumbar vertebra. In front, it is covered by the ascending and descending layers of the transverse meso-colon, and crossed by the superior mesenteric vessels; behind, it rests upon the aorta, the vena cava, and the crura of the Diaphragm; above it, is the lower border of the pancreas, the superior mesenteric vessels passing forwards between the two.

The *arteries* supplying the duodenum are the pyloric and superior pancreatico-duodenal branches of the hepatic, and the inferior pancreatico-duodenal branch of the superior mesenteric.

The *veins* terminate in the gastro-duodenal and superior mesenteric.

Its *nerves* are derived from the solar plexus.

The *Jejunum* (*jejunus*, empty), so called from being usually found empty after death, includes the upper two-fifths of the rest of the small intestine. It commences at the duodenum on the left side of the second lumbar vertebra, and terminates in the ileum; its convolutions being chiefly confined to the umbilical and left iliac regions. The jejunum is wider, its coats thicker, more vascular, and of a deeper colour than those of the ileum; but there is no characteristic mark to distinguish the termination of the one, or the commencement of the other.

The *Ileum* (ειλειν, *to twist*), so called from its numerous coils or convolutions, includes the remaining three-fifths of the small intestine, occupying chiefly the umbilical, hypogastric, right iliac, and occasionally the pelvic, regions, and terminating in the right iliac fossa by opening into the inner side of the commencement of the large intestine. The ileum is narrower, its coats thinner and less vascular than those of the jejunum; a given length of it weighing less than the same length of jejunum.

Structure. The wall of the small intestine is composed of four coats; serous, muscular, cellular, and mucous.

The *serous coat* is derived from the peritoneum. The first, or ascending portion of the duodenum, is completely surrounded by this membrane; the second, or descending portion, is covered by it only in front; and the third, or transverse, portion lies behind the ascending and descending layers of the transverse meso-colon, by which it is covered in front. The remaining portion of the small intestine is completely surrounded by the peritoneum, excepting along the attached or mesenteric border of the intestine; here a space is left for the vessels and nerves to pass to the intestine.

The *muscular coat* consists of two layers of fibres, an external or longitudinal, and an internal or circular layer. The *longitudinal* fibres are thinly scattered over the surface of the intestine, and are most distinct along its free border. The *circular* fibres form a thick, uniform layer; they surround the cylinder of the intestine in the greater part of its circumference, but do not form complete rings. The muscular coat is thicker at the upper, than at the lower part of the small intestine.

The *cellular*, or *submucous coat*, connects together the mucous and muscular layers. It consists of a loose, filamentous, areolar tissue, which forms a nidus for the subdivision of the nutrient vessels, previous to their distribution to the mucous surface.

The *mucous membrane* is thick and highly vascular at the upper part of the small intestine, but somewhat paler and thinner below. It presents for examination the following constituents:

Epithelium.
Valvulae conniventes.
Villi.
Glands. { Simple follicles.
Duodenal glands.
Solitary glands.
Agminate or Peyer's glands.

The *Epithelium*, covering the mucous membrane of the small intestines, is of the columnar variety throughout.

The *Valvulae Conniventes* (valves of Kerkring) are reduplications or foldings of the mucous membrane, containing no muscular fibres. They extend transversely across the cylinder of the intestine for about one-half or three-fourths of its circumference. The larger folds are about two inches in length, and two-thirds of an inch wide at their broadest part; but the greater number are of smaller size. The larger and smaller folds alternate with each other. They are not found at the commencement of the duodenum, but begin to appear about one or two inches beyond the pylorus. In the lower part of the descending portion, below the point where the common choledic and pancreatic ducts enter the intestine; they are very large and closely approximated. In the transverse portion of the duodenum and upper half of the jejunum, they are large and numerous; and from this point, as far as the middle of the ileum, where they terminate, they diminish considerably in size. In the lower half of the ileum, the mucous lining is without folds; hence the comparative thinness of this portion of the intestine, as compared with the duodenum and jejunum. The valvulae conniventes retard the passage of the food along the intestines, and afford a more extensive surface for absorption.

The *Villi* are minute, highly vascular, projecting processes, covering the mucous membrane of the small intestine throughout its whole extent, and giving to its entire surface a beautiful velvety appearance. In shape, some are triangular and laminated, others conical or cylindrical, with clubbed, or filiform extremities. They are largest and most numerous in the duodenum and jejunum, and become fewer and smaller in the ileum. Krause estimates their number in the upper part of the small intestine, at from fifty to ninety in a square line; and in the lower part, from forty to seventy; the total number for the whole length of the intestine being four millions.

313.—Two Villi magnified.

In *structure* each villus consists of a network of blood and lacteal vessels, with granular corpuscles and fat globules in their interstices, enclosed in a thin prolongation from the mucous surface, covered by columnar epithelium, which is arranged perpendicular to the surface. The mode of origin of the lacteals within the villi is unknown.

The *Simple Follicles*, or crypts of Lieberkühn, are found in considerable numbers over every part of the mucous membrane of the small intestine. They consist of minute caecal pouches of the mucous membrane, arranged perpendicularly to the surface, upon which they open by small circular apertures. Their walls are thin, and lined with columnar epithelium.

The *Duodenal*, or Brunner's glands, are limited to the duodenum and commencement of the jejunum (Huschke). They are small, flattened, granular bodies, imbedded in the submucous areolar tissue, and open upon the surface of the mucous membrane by minute excretory ducts. They are most numerous near the pylorus. In structure, they are analogous to the salivary glands and pancreas.

The *Solitary Glands* (*glandulae solitariae*), are found scattered throughout the mucous membrane of the small intestine, but are most numerous in the lower part of the ileum. They are small, round, whitish, prominent bodies, about the size of a millet-seed, consisting of a closed saccular cavity, having no excretory duct, and containing an opaque white secretion. Their free surface is covered with villi, and around them is scattered irregularly the orifices of the follicles of Lieberkühn. Their use is not known.

314.—Patch of Peyer's Glands. From the lower part of the Ileum.

The *Aggregate*, or Peyer's glands, are largest and most numerous in the ileum. In the lower part of the jejunum they are small, of a circular form, and few in number, and they are occasionally seen in the duodenum. They consist of circular or oval patches, from twenty to thirty in number, and varying in length from half an inch to four inches. They are placed lengthwise in the intestine, covering the portion of the tube most distant from the attachment of the mesentery. Each patch is formed of a group of small, round, whitish vesicles, covered with mucous membrane. Each vesicle consists of a moderately thick external capsule, having no excretory duct, and containing an opaque white secretion. Each is surrounded by a zone, or wreath of simple follicles, and the

315.—A portion of the above magnified. *Orifices of Lieberkühn's Follicles.*

interspaces between them covered with villi. These vesicles are usually closed; but it is supposed they open at intervals to discharge the secretion contained within them. The mucous and sub-mucous coats of the intestine are intimately adherent, and highly vascular, opposite the Peyerian glands. Their use is not known.

The Large Intestine

The large intestine extends from the termination of the ileum to the anus. It is about five feet in length; being one-fifth of the whole extent of the intestinal canal. It is largest at its commencement at the caecum, and gradually diminishes as far as the rectum, where there is a dilatation of considerable size, just above the anus. It differs from the small intestine in its greater size, its more fixed position, and its sacculated form. The large intestine, in its course, describes an arch, which surrounds the convolutions of the small intestine. It commences in the right iliac fossa, in a dilatation of considerable size, the caecum. It ascends through the right lumbar and hypochondriac regions, to the under surface of the liver; passes transversely across the abdomen, on the confines of the epigastric and umbilical regions, to the left hypochondriac region; descends through the left lumbar region to the left iliac fossa, where it becomes convoluted, and forms the sigmoid flexure; finally, it enters the pelvis, and descends along its posterior surface to the anus. The large intestine is divided into the caecum, colon, and rectum.

The *Caecum* (*caecus*, blind), is the large blind pouch, or *cul-de-sac*, extending downwards from the commencement of the large intestine. It is the most dilated part of this tube, measuring about two and a half inches, both in its vertical and transverse diameters. It is situated in the right iliac fossa, immediately behind the anterior abdominal wall, being retained in its place by the peritoneum, which passes over its anterior surface and sides; its posterior surface being connected by loose areolar tissue with the iliac fascia. Occasionally, it is almost completely surrounded by peritoneum, which forms a distinct fold, the meso-caecum, by which it is held in connection with the iliac fossa. This fold allows the caecum considerable freedom of movement. Attached to its lower and back part, is the appendix vermiformis, a long, narrow, worm-shaped tube, the rudiment of the lengthened caecum found in all the mammalia, except the ourang-outang and wombat. The appendix varies from three to six inches in length, its average diameter being about equal to that of a goose-quill. It is usually directed upwards and inwards behind the caecum, coiled upon itself, and terminates in a blunt point, being retained in its position by a fold of peritoneum, which sometimes forms a mesentery for it. Its canal is small, and communicates with the caecum by an orifice which is sometimes guarded with an incomplete valve. Its coats are thick, and its mucous lining furnished with a large number of solitary glands.

316.—The Cæcum and Colon laid open to shew the Ilio-Cæcal Valve.

Ilio-Cecal Valve. The lower end of the ileum terminates at the inner and back part of the large intestine, opposite the junction of the caecum with the colon. At this point, the mucous membrane forms two valvular folds, which project into the large intestine, and are separated from each other by a narrow elongate aperture. This is the ilio-caecal valve (*valvula Bauhini*). Each fold is semilunar in form, and projects vertically inwards, the upper one being attached by its convex border to the point of junction of the ileum with the colon; the lower segment, the larger, being connected at the point of junction of the ileum with the caecum. Their concave margins are free, project into the intestine, separated from one another by a narrow elongated aperture, which has a transverse direction. At each end of this aperture, the two segments of the valve coalesce, and are continued, as a narrow membranous ridge, around the canal of the intestine for a short distance, forming the fraena, or retinacula of the valve. The left end of this aperture, is rounded: the right end is narrow and pointed.

Each segment of the valve is formed of a reduplication of the mucous membrane, and of the circular muscular fibres of the intestine, the longitudinal fibres and peritoneum being continued uninterruptedly across from one intestine to the other. When these are divided or removed, the ileum may be drawn outwards, and all traces of the valve will be lost, the ileum appearing to open into the large intestine by a funnel-shaped orifice of large size.

The surface of each segment of the valve directed towards the ileum is covered with villi, and presents the characteristic structure of the mucous membrane of the small intestine; whilst that turned towards the large intestine is destitute of villi, and marked with the orifices of the numerous tubuli peculiar to this membrane. These differences in structure continue as far as the free margin of the valve.

When the caecum is distended, the margins of the opening are approximated, so as to prevent any reflux into the ileum.

The *Colon* is divided into four parts, the ascending, transverse, descending, and the sigmoid flexure.

The *ascending colon* is smaller than the caecum. It passes upwards from the right iliac fossa, to the under surface of the liver, on the right of the gall-bladder, where it bends abruptly inwards to the left, forming the hepatic flexure. It is retained in position to the wall of the abdomen by the peritoneum, which covers its anterior surface and sides, its posterior surface being connected by loose areolar tissue with the Quadratus lumborum and right kidney; sometimes the peritoneum almost completely invests it, and forms a distinct but narrow meso-colon. It is in relation, in front, with the convolutions of the ileum and the abdominal parietes; behind, it lies on the Quadratus lumborum muscle, and right kidney.

The *transverse colon*, the longest part of the large intestine, passes transversely from right to left across the abdomen, opposite the confines of the epigastric and umbilical zones, into the left hypochondriac region, where it curves downwards beneath the lower end of the spleen, forming its splenic flexure. In its course it describes an arch, the concavity of which is directed backwards towards the vertebral column; hence the name, *transverse arch of the colon*. This is the most moveable part of the colon, being completely invested by peritoneum and connected to the spine behind by a large and wide duplicature of this membrane, the *transverse meso-colon*. It is in relation, by its upper surface, with the liver and gall-bladder, the great curvature of the stomach, and the lower end of the spleen; by its under surface, with the small intestines; by its anterior surface, with the anterior layers of the great omentum and the abdominal parietes; by its posterior surface, with the transverse meso-colon.

The *descending colon* passes almost vertically downwards through the left hypochondriac and lumbar regions to the left iliac fossa, where it terminates in the sigmoid flexure. It is retained in position by the peritoneum, which covers its anterior surface and sides, its posterior surface being connected by areolar tissue with the left crus of the Diaphragm, the left kidney, and the Quadratus lumborum. It is smaller in calibre and more deeply placed than the ascending colon.

The *sigmoid flexure* is the narrowest part of the colon; it is situated in the left iliac fossa, commencing at the termination of the descending colon, at the margin of the crest of the ilium, and terminating in the rectum, opposite the left sacro-iliac symphysis. It curves in the first place upwards, and then descends vertically, and to one or the other side like the letter *f*, hence its name; and is retained in its place by a loose fold of peritoneum, the *iliac meso-colon*. It is in relation, in front, with the small intestines and abdominal parietes; behind, with the iliac fossa.

The *Rectum* is the terminal part of the large intestine, and extends from the sigmoid flexure to the anus; it varies in length from six to eight inches, and has received its name from being somewhat less flexuous than any other part of the intestinal canal. It commences opposite the left sacro-iliac symphysis, passes obliquely downwards from left to right to the middle of the sacrum, forming a gentle curve to the right side. Regaining the middle line, it descends in front of the lower part of the sacrum and coccyx, and near the extremity of the latter bone inclines backwards to terminate at the anus, being curved both in the lateral and antero-posterior directions. The rectum is, therefore, not straight, the upper part being directed obliquely from the left side to the median line, the middle portion being curved in the direction of the hollow of the sacrum and coccyx, the lower portion presenting a short curve in the opposite direction. The rectum is cylindrical, not sacculated like the rest of the large intestine; it is narrower at its upper part than the sigmoid flexure, gradually increases in size as it descends, and immediately above the anus presents a considerable dilatation, capable of acquiring an enormous size. The rectum is divided into three portions, upper, middle, and lower.

The *upper portion*, which includes about half the length of the tube, extends obliquely from the left sacro-iliac symphysis to the centre of the third piece of the sacrum. It is almost completely surrounded by peritoneum, and connected to the sacrum behind by a duplicature of this membrane, the *meso-rectum*. It is in relation behind with the Pyriformis muscle, the sacral plexus of nerves, and the branches of the internal iliac artery of the left side, which separate it from the sacrum and sacro-iliac symphysis; in front, it is separated, in the male, from the posterior surface of the bladder; in the female, from the posterior surface of the uterus and its appendages, by some convolutions of the small intestine.

The *middle portion* of the rectum is about three inches in length, and extends as far as the tip of the coccyx. It is closely connected to the concavity of the sacrum, and covered by peritoneum only on the upper part of its anterior surface. It is in relation, in front, with the triangular portion of the base of the bladder, the vesiculae seminales, and vasa deferentia; more anteriorly, with the under surface of the prostate. In the female, it is adherent to the posterior wall of the vagina.

The *third portion* is about an inch in length; it curves backwards at the fore part of the prostate gland, and terminates at the anus. This portion of the intestine receives no peritoneal covering. It is invested by the Internal sphincter, supported by the Levatores ani muscles, and surrounded at its termination by the External sphincter. In the male, it is separated from the membranous portion and bulb of the urethra by a triangular space; and, in the female, a similar space intervenes between it and the vagina. This space forms by its base the perinaeum.

Structure. The large intestine has four coats, serous, muscular, cellular, and mucous.

The *serous coat* is derived from the peritoneum, and invests the different portions of the large intestine to a variable extent. The caecum is covered only on its anterior surface and sides; more rarely it receives a complete investment, and is held in its position by a duplicature, the meso-caecum. The ascending and descending colon are usually covered only in front. The transverse colon is completely invested, excepting at the points corresponding to the attachment of the great omentum and transverse meso-colon. The sigmoid flexure is also completely surrounded, excepting at the point corresponding to the attachment of the iliac meso-colon. The upper part of the rectum is completely invested by the peritoneum; the upper portion is covered only on its anterior surface; and the lower third is entirely devoid of any serous investment.

The *muscular coat* consists of an external longitudinal and an internal circular layer of muscular fibres.

The *longitudinal fibres* are found as a uniform layer over the whole surface of the large intestine. In the caecum and colon, they are especially collected into three flat longitudinal bands, each being about half an inch in width. These bands commence at the attachment of the appendix vermiformis to the caecum: one, the posterior, is placed along the attached border of the intestine; the anterior band, the largest, becomes inferior along the arch of the colon, where it corresponds to the attachment of the great omentum, but is in front in the ascending and descending colon and sigmoid flexure; the third, or lateral band, is found on the inner side of the ascending and descending colon, and on the under border of the transverse colon. These bands are nearly one-half shorter than the other parts of the intestine, and serve to produce those sacculi characteristic of the caecum and colon; accordingly, when they are dissected off, the tube can be lengthened, and its sacculated character becomes lost. In the sigmoid flexure, the longitudinal fibres become more scattered, and upon its lower part, as well as around the rectum, they spread out and form a thick uniform layer.

The *circular fibres* form a thin layer over the caecum and colon, being especially accumulated in the intervals between the sacculi; in the rectum, they form a thick layer, especially at its lower end, where they become numerous, and form the Internal sphincter.

The *cellular coat* connects closely together the muscular and mucous layers.

The *mucous membrane*, in the caecum and colon, is pale, and of a greyish or pale yellow colour. It is quite smooth, destitute of villi, and raised into numerous crescentic folds, which correspond to the intervals between the sacculi. In the rectum, it is thicker, of a darker colour, more vascular, and connected loosely to the muscular coat, as in the oesophagus. In its contracted state, the lower part of the rectum is thrown into a number of longitudinal folds, similar to those in the oesophagus, the larger of which have been called by Morgagni, the *columns of the rectum*. Besides these, the mucous membrane forms three prominent valvular folds, specially described by Mr. Houston. They are all directed obliquely; one is situated near the commencement of the rectum, on the right side; another extends inwards from the side of the tube opposite the middle of the sacrum; and the third projects backwards from the fore part of the rectum, opposite the prostate gland; these folds become effaced when the canal is distended. The mucous membrane of the large intestine presents for examination: epithelium, simple follicles, and solitary glands.

317.—Minute Structure of Large Intestine.

The *epithelium* is of the columnar kind.

The *simple follicles* are minute tubular prolongations of the mucous membrane, arranged perpendicularly, side by side, over its entire surface; they are longer, more numerous, and placed in much closer apposition than those of the small intestine; and they open by minute rounded orifices upon the surface, giving it a cribriform appearance.

The *solitary glands* in the large intestine, are most abundant in the caecum and appendix vermiformis, being irregularly scattered over the rest of the intestine; they are small, prominent, circular bodies, of a whitish colour, perforated upon the central part of their free surface by a minute orifice.

The Liver

The liver is a glandular organ of large size, intended for the secretion of the bile. It is situated in the right hypochondriac region, and extends across the epigastrium into the left hypochondrium. It is the largest gland in the body, weighing from three to four pounds (from fifty to sixty ounces avoirdupois). It measures, in its transverse diameter, from ten to twelve inches; from six to seven in its antero-posterior; and is about three inches thick at the back part of the right lobe, which is the thickest part.

Its *upper surface* is convex, directed upwards and forwards, smooth, covered by peritoneum, and is in relation with the under surface of the Diaphragm; and below, to a small extent, with the abdominal parietes. This surface is divided into two unequal lobes, the right and left, by a fold of peritoneum, the *suspensory* or *broad ligament*.

Its *under surface* is concave, directed downwards and backwards, and in relation with the stomach and duodenum, the hepatic flexure of the colon, and the right kidney and supra-renal capsule. This surface is divided, by a longitudinal fissure, into a right and left lobe.

The *posterior border* is rounded and broad, and connected to the Diaphragm by the coronary ligament; it is in relation with the aorta, the vena cava, and the crura of the Diaphragm.

The *anterior border* is thin and sharp, and marked, opposite the attachment of the broad ligament, by a deep notch. In adult males, this border usually corresponds with the margin of the ribs; but in women and children, it projects usually below this point.

The *right extremity* of the liver is thick and rounded; whilst the *left* is thin and flattened.

Changes of Position. The student should make himself acquainted with the different circumstances under which the liver changes its position, as they are of importance as a guide in determining the existence of enlargement, or other disease of that organ.

Its position varies according to the posture of the body; in the upright and sitting postures, its lower border may be felt beneath the edges of the ribs; in the recumbent posture, it usually recedes beneath the ribs.

Its position varies with the ascent or descent of the Diaphragm. In a deep inspiration, the liver descends below the ribs; in expiration, it is raised to its ordinary level. Again, in emphysema, where the lungs are distended, and the Diaphragm descends very low, the liver is pushed down; but in some other diseases, as phthisis, where the Diaphragm is much arched, the liver rises very high up.

Pressure from without, as in tight lacing, by compressing the lower part of the chest, displaces the liver considerably, its anterior edge often extending as low as the crest of the ilium; and its convex surface is often, at the same time, deeply indented from pressure of the ribs.

Its position varies greatly, according to the greater or less distension of the stomach and intestines. When the intestines are empty, the liver descends in the abdomen; but when they are distended, it is pushed upwards. Its relations with surrounding organs may also be changed by the growth of tumours, or from collections of fluid in the thoracic or abdominal cavities.

LIGAMENTS. The ligaments of the liver (fig. 318) are five in number: four being formed of folds of peritoneum; the fifth, the *ligamentum teres*, is a round, fibrous cord, resulting from the obliteration of the umbilical vein. The ligaments are the longitudinal, two lateral, coronary, and round.

The *Longitudinal Ligament* (broad, falciform, or suspensory ligament) is a broad and thin antero-posterior peritoneal fold, falciform in shape, its base being directed forwards, its apex backwards. It is attached by one margin to the under surface of the Diaphragm, and the posterior surface of the sheath of the right Rectus muscle as low down as the umbilicus; by its hepatic margin, it extends from the notch on the anterior margin of the liver, as far back as its posterior border. It consists of two layers of peritoneum closely united together. Its anterior free margin contains between its layers the round ligament.

The *Lateral Ligaments*, two in number, right and left, are triangular in shape. They are formed of two layers of peritoneum united, and extend from the sides of the

318.—The Liver. Upper Surface.

Diaphragm to the adjacent margins of the posterior border of the liver. The left is the longer of the two, and lies in front of the oesophageal opening in the Diaphragm; the right lies in front of the inferior vena cava.

The *Coronary Ligament* connects the posterior border of the liver to the Diaphragm. It is formed by the reflection of the peritoneum from the Diaphragm on to the upper and lower margins of the posterior border of the organ. The coronary ligament consists of two layers, which are continuous on each side with the lateral ligaments; and in front, with the longitudinal ligament. Between the layers, a large oval interspace is left uncovered by peritoneum, and connected to the Diaphragm by firm areolar tissue. This space is subdivided, near its left extremity, into two parts by a deep notch (sometimes a canal), which lodges the inferior vena cava, and into which open the hepatic veins.

The *Round Ligament* is a fibrous cord, resulting from the obliteration of the umbilical vein. It ascends from the umbilicus in the anterior free margin of the longitudinal ligament, to the notch in the anterior border of the liver, from which it may be traced along the longitudinal fissure on the under surface of the liver, as far back as the inferior vena cava.

FISSURES. Five fissures are seen upon the under surface of the liver, which serve to divide it into five lobes. They are the longitudinal fissure, the fissure of the ductus venosus, the transverse fissure, the fissure for the gall-bladder, and the fissure for the vena cava.

The *Longitudinal Fissure* is a deep groove, which extends from the notch on the anterior margin of the liver, to the posterior border of the organ. It separates the right and left lobes; the transverse fissure joins it, at right angles, about one-third from its posterior extremity, and divides it into two parts. The anterior half is called the *umbilical fissure:* it is deeper than the posterior part, and lodges the umbilical vein in the foetus, or its fibrous cord (the round ligament) in the adult. This fissure is often partially bridged over by a prolongation of the hepatic substance, the *pons hepatis.*

319.—The Liver. Under Surface.

The *Fissure of the Ductus Venosus* is the back part of the longitudinal fissure; it is shorter and shallower than the anterior portion. It lodges in the foetus the ductus venosus, and in the adult a slender fibrous cord, the obliterated remains of that vessel.

The *Transverse*, or *Portal Fissure*, is a short but deep fissure, about two inches in length, extending transversely across the under surface of the right lobe, nearer to its posterior than its anterior border. It joins, nearly at right angles, with the longitudinal fissure. By the older anatomists, this fissure was considered the gateway (*porta*) of the liver; hence the large vein which enters at this point was called the *portal vein*. Besides this vein, the fissure transmits the hepatic artery and nerves, and the hepatic duct and lymphatics. At their entrance into the fissure, the hepatic duct lies to the right, the hepatic artery to the left, and most posteriorly the portal vein.

The *Fissure for the Gall-bladder* (*fossa cystis felleae*) is a shallow, oblong fossa, placed on the under surface of the right lobe, parallel with the longitudinal fissure. It extends from the anterior free margin of the liver, which is occasionally notched for its reception, to near the right extremity of the transverse fissure.

The *Fissure for the Vena Cava* is a short deep fissure, occasionally a complete canal, which extends obliquely upwards from a little behind the right extremity of the transverse fissure, to the posterior border of the organ, where it joins the fissure for the ductus venosus. On slitting open the inferior vena cava which is contained in it, a deep fossa is seen, at the bottom of which the hepatic veins communicate with this vessel. This fissure is separated from the transverse fissure by the lobus caudatus; and from the longitudinal fissure by the lobus Spigelii.

LOBES. The lobes of the liver, like the ligaments and fissures, are also five in number: the right lobe, the left lobe, the lobus quadratus, the lobus Spigelii, and the lobus caudatus.

The *right lobe* is much larger than the left; the proportion between them being as six to one. It occupies the right hypochondrium, and is separated from the left lobe, on its upper surface, by the longitudinal ligament; on its under surface, by the longitudinal fissure; and in front, by a deep notch. It is of a quadrilateral form, its under surface being marked by three fissures: the transverse fissure, the fissure for the gall-bladder, and the fissure for the inferior vena cava; and by two shallow impressions, one in front (*impressio colica*), for the hepatic flexure of the colon; and one behind (*impressio renalis*), for the right kidney and supra-renal capsule.

The *left lobe* is smaller and more flattened than the right. It is situated in the epigastric and left hypochondriac regions, sometimes extending as far as the upper border of the spleen. It upper surface is convex; its under concave surface rests upon the front of the stomach; and its posterior border is in relation with the cardiac orifice of the stomach.

The *Lobus Quadratus*, or square lobe, is situated on the under surface of the right lobe, bounded in front by the free margin of the liver; behind, by the transverse fissure; on the right, by the fissure for the gall-bladder; and, on the left, by the umbilical fissure.

The *Lobus Spigelii* projects from the back part of the under surface of the right lobe. It is bounded, in front, by the transverse fissure; on the right, by the fissure for the vena cava; and, on the left, by the fissure for the ductus venosus.

The *Lobus Caudatus*, or tailed lobe, is a small elevation of the hepatic substance, extending obliquely outwards, from the base of the lobus Spigelii, to the under surface of the right lobe. It separates the right extremity of the transverse fissure from the commencement of the fissure for the inferior cava.

Vessels. The vessels connected with the liver are also five in number: they are the hepatic artery, the portal vein, the hepatic vein, the hepatic duct, and lymphatics.

The *Hepatic Artery*, *Portal Vein*, and *Hepatic Duct*, accompanied by numerous lymphatic vessels and nerves, ascend to the transverse fissure, between the layers of the gastro-hepatic omentum; the hepatic duct lying to the right, the hepatic artery to the left, and the portal vein behind and between the other two. They are enveloped in a loose areolar tissue, the capsule of Glisson, which accompanies the vessels in their course through the portal canals, which are hollowed out of the interior of the organ.

The *Hepatic Veins* convey the blood from the liver. They commence at the circumference of the organ, and proceed towards the deep fossa in its posterior border, where they terminate by two large, and several smaller branches, in the inferior vena cava.

The hepatic veins have no cellular investment; consequently their parietes are adherent to the walls of the canals through which they run: so that, on a section of the organ, these veins remain widely open and solitary, and may be easily distinguished from the branches of the portal vein, which are more or less collapsed, and always accompanied by an artery and duct.

The *Lymphatics* are large and numerous, consisting of a deep and superficial set. They have been already described.

Nerves. The nerves of the liver are derived from the hepatic plexus of the sympathetic, from the pneumogastric nerves, especially the left, and from the right phrenic.

Structure. The substance of the liver is composed of lobules, held together by an extremely fine areolar tissue, of the ramifications of the portal vein, hepatic duct, hepatic artery, hepatic veins, lymphatics, and nerves; the whole being invested by a fibrous and a serous coat.

The *serous coat* is derived from the peritoneum, and invests the entire surface of the organ, excepting at the point corresponding to the attachment of its various ligaments, and at the bottom of the different fissures, where it is deficient. It is intimately adherent to the fibrous coat.

The *fibrous coat* lies beneath the serous investment, and covers the entire surface of the organ. It is difficult of demonstration, excepting where the serous coat is deficient. At the transverse fissure, it is continuous with the capsule of Glisson; and, on the surface of the organ, with the areolar tissue separating the lobules.

The *Lobules* form the chief mass of the hepatic substance: they may be seen either on the surface of the organ, or by making a section through the gland. They are small granular bodies, about the size of a millet-seed, measuring from one-twentieth to one-tenth of an inch in diameter. When divided longitudinally, they have a foliated margin, and, if transversely, a polygonal outline. The bases of the lobules are clustered round the smallest branches (sub-lobular) of the hepatic veins, to which each is connected by means of a small branch, which issues from the centre of each lobule (intra-lobular). The remaining part of the surface of each lobule is imperfectly isolated from the surrounding lobules, by a thin stratum of areolar tissue, or by the smaller vessels and ducts.

Fig. 320.

If one of the hepatic veins be laid open, the bases of the lobules may be seen through the thin wall of the vein, on which they rest, arranged in the form of a tessellated pavement, the centre of each polygonal space presenting a minute aperture, the mouth of a sublobular vein.

Each lobule is composed of a mass of cells; of a plexus of biliary ducts; of a venous plexus, formed by branches of the portal vein; of a branch of an hepatic vein (intralobular); of minute arteries; and probably, of nerves and lymphatics.

The hepatic cells form the chief mass of the substance of a lobule, and lie in the interspaces of the capillary plexus, being probably contained in a tubular network, which forms the origin of the biliary ducts. The smallest branches of the vena

H. Longitudinal section of an hepatic vein; *a.* portion of the canal, from which the vein has been removed; *b.* orifices of ultimate twigs of the vein (sub-lobular), situated in the centre of the lobules. After Kiernan.

portae pass between the lobules, around which they form a plexus, the interlobular. Branches from this plexus enter the lobules, and form a network in its circumference. The radicles of the portal vein communicate with those of the hepatic vein, which occupy the centre of the lobule; and the latter converge to form the intralobular vein, which issues from the base of the lobule, and joins the hepatic vein. The portal vein carries the blood to the liver, from which the bile is secreted; the hepatic vein carries from the liver the superfluous blood; and the bile duct carries from the liver the bile secreted by the hepatic cells.

The *Hepatic Cells* form the chief mass of each lobule: they are of a more or less spheroidal form; but may be rounded, flattened, or many-sided, from mutual compression. They vary in size from the $1/1000$th to the $1/2000$th of an inch in diameter, and contain a distinct nucleus in the interior, or even sometimes two. In the nucleus is a highly refracting nucleolus, with granules. The cell-contents are viscid, and contain yellow particles, the colouring matter of the bile, and oil globules. The cells adhere together by their surfaces, so as to form rows, which radiate from the centre towards the circumference of the lobule. These cells are the chief agents in the secretion of the bile; and, according to the extensive researches of Dr. Beale, lie in a tubular network, which forms the origin of the ducts.

Biliary Ducts. The precise mode of origin of the biliary ducts is uncertain. Mr. Kiernan's original view, confirmed as it is by the researches of Dr. Beale, shows that the ducts commence within the lobules, in a plexiform network (lobular biliary plexus), in which the hepatic cells lie. The lobular ducts, on leaving the lobules, form a plexus (interlobular) between the lobules; and the interlobular branches unite into vaginal branches, which lie in the portal canals, with branches of the portal vein and hepatic duct. The ducts finally join into two large trunks which leave the liver at the transverse fissure.

The *Portal Vein*, on entering the liver at the transverse fissure, divides into primary branches, which are contained in the portal canals, together with branches of the hepatic artery and duct, and the nerves and lymphatics. In the larger portal canals, the vessels are separated from the parietes, and joined to each other, by a loose cellular web, the capsule of Glisson. The veins, as they lie in the portal canals, give off vaginal branches, which form a plexus (vaginal plexus) in Glisson's capsule. From this plexus, and from the portal vein itself, small branches are given off, which pass between the lobules (interlobular veins); these cover the entire surface of the lobules, excepting their bases. The lobular branches are derived from the interlobular veins: they penetrate into the lobule, and form a capillary plexus within them. From this plexus the intralobular vein arises.

Longitudinal section of a small portal vein and canal, after Kiernan. *a.* Portions of the canal, from which the vein has been removed; *b.* side of the portal vein in contact with the canal; *c.* the side of the vein which is separated from the canal by the hepatic artery and duct, with areolar tissue (Glisson's capsule); *d.* internal surface of the portal vein, through which is seen the outlines of the lobules and the openings of the interlobular veins; *f.* vaginal veins of Kiernan; *g.* hepatic artery; *h.* hepatic duct.

The *Hepatic artery* enters the liver at the transverse fissure, with the portal vein and hepatic duct, and ramifies with these vessels through the portal canals. It gives off vaginal branches, which ramify in the capsule of Glisson; and other branches which are distributed to the coats of the vena portae and hepatic duct. From the vaginal plexus, inter-lobular branches are given off, which ramify through the interlobular fissures, a few branches being distributed to the lobules. Kiernan supposes that the branches of the hepatic artery terminate in a capillary plexus which communicates with the branches of the vena portae.

The *Hepatic Veins* commence in the interior of each lobule by a plexus, the branches of which converge to form the intralobular vein.

The *intralobular vein* passes through the centre of the lobule, and leaves it at its base to terminate in a sublobular vein.

The *sublobular veins* unite with neighbouring branches to form larger veins, and they join to form the large hepatic trunks, which terminate in the vena cava.

Fig. 322.

A transverse section of a small portal canal and its vessels, after Kiernan. 1. Portal vein; 2. interlobular branches; 3. branches of the vein, termed by Mr. Kiernan, vaginal, also giving off interlobular branches; 4. hepatic duct; 5. hepatic artery.

GALL BLADDER

The gall bladder is the reservoir for the bile; it is a conical or pear-shaped membranous sac, lodged in a fossa on the under surface of the right lobe of the liver, and extending from near the right extremity of the transverse fissure to the anterior free margin of the organ. It is about four inches in length, one inch in breadth at its widest part, and holds from eight to ten drachms. It is divided into a fundus, body, and neck. The *fundus*, or broad extremity, is directed downwards, forwards, and to the right, and occasionally projects from the anterior border of the liver: the *body* and *neck* are directed upwards and backwards to the left. The gall bladder is held in its position by the peritoneum, which, in the majority of cases, passes over its under surface, but it occasionally invests it, and is connected to the liver by a kind of mesentery.

Relations. The *body* of the gall bladder is in relation by its upper surface with the liver, to which it is connected by areolar tissue and vessels; by its under surface, with the first portion of the duodenum, occasionally the pyloric end of the stomach, and the hepatic flexure of the colon. The *fundus* is completely invested by peritoneum; it is in relation, in front, with the abdominal parietes, immediately below the tenth costal cartilage; behind, with the transverse arch of the colon. The *neck* is narrow, and curved upon itself like the italic letter ʃ; at its point of connection with the body and with the cystic duct, it presents a well marked constriction.

When the gall bladder is distended with bile or calculi, the fundus may be felt through the abdominal parietes, especially in an emaciated subject; the relations of this sac will also serve to explain the occasional occurrence of abdominal biliary fistulae, through which biliary calculi may pass out, and of the passage of calculi from the gall bladder into the stomach, duodenum, or colon, which occasionally happens.

Structure. The gall bladder consists of three coats, serous, fibrous and muscular, and mucous.

The *external* or *serous coat* is derived from the peritoneum; it completely invests the fundus, but covers the body and neck only on their under surface.

The *middle* or *fibrous coat* is a thin but strong fibrous layer, which forms the framework of the sac, consisting of dense fibres which interlace in all directions. Plain muscular fibres are also found in this coat, disposed chiefly in a longitudinal direction, a few running transversely.

The *internal* or *mucous coat* is loosely connected with the fibrous layer. It is generally tinged of a yellowish brown colour, and is everywhere elevated into minute rugae, by the union of which numerous meshes are formed; the depressed intervening spaces having a polygonal outline. The meshes are smaller at the fundus and neck, being most developed about the centre of the sac.

Opposite each curve of the neck of the gall bladder, the mucous membrane projects inwards so as to form a large valvular fold. The mucous membrane is covered by columnar epithelium, and secretes an abundance of thick viscid mucus; it is continuous through the hepatic duct with the mucous membrane lining the ducts of the liver, and through the ductus communis choledocus with the mucous membrane of the alimentary canal.

The *Biliary Ducts* are the hepatic, the cystic, and the ductus communis choledocus.

The *Hepatic Duct* is formed of two trunks of nearly equal size, which issue from the liver at the transverse fissure, one from the right and one from the left lobe; these unite, and pass downwards and to the right for about an inch and a half to join at an acute angle with the cystic duct, to form the common choledic duct.

The *Cystic Duct*, the smallest of the three biliary ducts, is about an inch in length. It passes obliquely downwards and to the left from the neck of the gall bladder, and joins the hepatic duct to form the common duct. It lies in the gastro-hepatic omentum in front of the vena cava, the cystic artery lying to its left side. The mucous membrane lining its interior is thrown into a series of crescentic folds, from five to twelve in number, which project into the duct in regular succession, and are directed obliquely round the tube, presenting much the appearance of a continuous spiral valve. They exist only in the human subject. When the duct has been distended, the interspaces between the folds are dilated, so as to give to its exterior a sacculated appearance.

The *Ductus communis choledocus*, the largest of the three, is the common excretory duct of the liver and gall bladder. It is about three inches in length, of the diameter of a goose-quill, and formed by the junction of the cystic and hepatic ducts. It descends along the right border of the lesser omentum, behind the first portion of the duodenum, in front of the vena portae, and to the right of the hepatic artery; it then passes between the pancreas and descending portion of the duodenum, and running for a short distance along the right side of the pancreatic duct, near its termination, passes with it obliquely between the mucous and muscular coats, the two opening by a common orifice upon the summit of a papilla, situated at the inner side of the descending portion of the duodenum, a little below its middle.

Structure. The coats of the biliary ducts are composed of an external or fibrous and an internal or mucous layer. The fibrous coat is composed of a strong areolar fibrous tissue. The mucous coat is continuous with the lining membrane of the hepatic ducts and gall bladder, and also with that of the duodenum. It is provided with numerous glands, the orifices of which are scattered irregularly in the larger ducts, but in the smaller hepatic ducts are disposed in two longitudinal rows, one on each side of the vessel. These glands are of two kinds. Some are ramified tubes, which occasionally anastomose, and from the sides of which saccular dilatations are given off; others are small clustered cellular glands, which open either separately into the hepatic duct, or into the ducts of the tubular glands.

The Pancreas

Dissection. The pancreas may be exposed for dissection in three different ways: 1. By raising the liver, drawing down the stomach, and tearing through the gastro-hepatic omentum. 2. By raising the stomach, the arch of the colon, and great omentum upwards, and then dividing the inferior layer of the transverse meso-colon. 3. By dividing the two layers of peritoneum, which descend from the great curvature of the stomach to form the great omentum; turning this organ upwards, and then cutting through the ascending layer of the transverse meso-colon.

The *Pancreas* (παν-κρέας, *all flesh*) is a conglomerate gland, analogous in its structure to the salivary glands. In shape, it is transversely oblong, flattened from before backwards, and bears some slight resemblance to a hammer, its right extremity being broad, and presenting a sort of angular bend from above downwards, called the *head*; its left extremity gradually tapers to form the tail, the intermediate portion being called the *body*. It is situated transversely across the posterior wall of the abdomen, at the back of the epigastric and both hypochondriac regions. Its length varies from six to eight inches, its breadth an inch and a half, and its thickness from half an inch to an inch, being thicker at its right extremity and along its upper border. Its weight varies from two to three and a half ounces, but it may reach six ounces.

The *right extremity* or *head* of the pancreas (fig. 323) is curved upon itself from above downwards, and is embraced by the concavity of the duodenum. The common bile duct descends behind, between the duodenum and pancreas; and the pancreatico-duodenal artery descends in front between the same parts. Upon its posterior part is a lobular fold of the gland, which passes transversely to the left, behind the superior

323.—The Pancreas and its Relations.

mesenteric vessels, forming the posterior part of the canal, in which they are contained. It is sometimes detached from the rest of the gland, and is called the *lesser pancreas*.

The *lesser end* or *tail* of the pancreas is narrow; it extends to the left as far as the spleen, and is placed over the left kidney and supra-renal capsule.

The *body* of the pancreas is convex in front, and covered by the ascending layer of the transverse meso-colon and the posterior surface of the stomach.

The *posterior surface* is concave, and has the following structures interposed between it and the first lumbar vertebra. The superior mesenteric artery and vein, and commencement of the vena portae, the vena cava, the aorta, the left kidney, supra-renal capsule, and corresponding renal vessels.

The *upper border* is thick, and has resting upon it, near its centre, the coeliac axis; the splenic artery and vein are lodged in a deep groove or canal in this border; and to the right, the first part of the duodenum and the hepatic artery are in relation with it.

The *lower border*, thinner than the upper, is separated from the transverse portion of the duodenum by the superior mesenteric artery and vein; to the left of this the inferior mesenteric vein ascends behind it to join the splenic vein.

The *Pancreatic Duct*, called the *canal of Wirsung* from its discoverer, extends transversely from left to right through the substance of the pancreas, nearer to its lower than its upper border, and lying nearer its anterior than its posterior surface. In order to expose it, the superficial portion of the gland must be removed. It commences by an orifice common to it and the ductus communis choledocus upon the summit of an elevated papilla, situated at the inner side of the descending portion of the duodenum, a little below its middle. Passing very obliquely through the mucous and muscular coats, it separates itself from the common choledic duct, and, ascending slightly, runs from right to left through the middle of the gland, giving off numerous branches, which pass to be distributed to its separate lobules.

Sometimes the pancreatic and common choledic ducts open separately into the duodenum. The excretory duct of the lesser pancreas is called the *ductus pancreaticus minor*; it opens into the main duct near the duodenum, and sometimes separately into that intestine, at a distance of an inch or more from the termination of the principal duct.

The pancreatic duct, near the duodenum, is about the size of an ordinary quill; its walls are thin, consisting of two coats, an external fibrous and an internal mucous; the latter is thin, smooth, and furnished, near its termination, with a few scattered follicles.

Sometimes the pancreatic duct is double, up to its point of entrance into the duodenum.

In *structure*, the pancreas resembles that of the salivary glands; but it is looser and softer in its texture.

The *arteries* of the pancreas are derived from the splenic, the pancreatico-duodenal branch of the hepatic, and the superior mesenteric.

Its *veins* open into the splenic and superior mesenteric veins.

Its *lymphatics* terminate in the lumbar glands.

The *nerves* are filaments from the splenic plexus.

The Spleen

The spleen is usually classified together with the thyroid, supra-renal glands, and thymus, as one of the ductless glands, from its possessing no excretory duct. It is of an oblong flattened form, soft, of very brittle consistence, highly vascular, of a dark bluish-red colour, and situated in the left hypochondriac region, embracing the cardiac end of the stomach. It is invested by peritoneum, and connected with the stomach by the gastro-splenic omentum.

Relations. Its *external surface* is convex, smooth, and in relation with the under surface of the Diaphragm, which separates it from the ninth, tenth, and eleventh ribs of the left side.

The *internal surface* is slightly concave, and divided by a vertical fissure, the *hilus*, into an anterior or larger, and a posterior or smaller portion. The hilus is pierced by several large, irregular apertures, for the entrance and exit of vessels and nerves. At the margins of the hilus, the two layers of peritoneum are reflected from the surface of the organ on to the cardiac end of the stomach, forming the gastro-splenic omentum, which contains between its layers the splenic vessels and nerves, and the vasa brevia. The internal surface is in relation, in front, with the great end of the stomach; below, with the tail of the pancreas; and behind, with the left crus of the Diaphragm and corresponding supra-renal capsule.

Its *upper end*, thick and rounded, is in relation with the Diaphragm, to which it is connected by a fold of peritoneum, the suspensory ligament.

Its *lower end* is pointed; it is in relation with the left extremity of the transverse arch of the colon.

Its *anterior margin* is free, rounded, and often notched, especially below.

Its *posterior margin* is rounded, and lies in relation with the left kidney, to which it is connected by loose areolar tissue.

The spleen is held in its position by two folds of peritoneum; one, the *gastro-splenic omentum*, connects it with the stomach; and the other, the *suspensory ligament*, with the under surface of the Diaphragm.

The size and weight of the spleen are liable to very extreme variations at different periods of life, in different individuals, and in the same individual under different conditions. In the *adult*, in whom it attains its greatest size, it is usually about five inches in length, three or four inches in breadth, and an inch or an inch and a half in thickness, and weighs about seven ounces. At *birth*, its weight, in proportion to the entire body, is almost equal to what is observed in the adult, being as 1 to 350; whilst in the adult it varies from 1 to 320 and 400. In *old age*, the organ not only decreases in weight, but decreases considerably in proportion to the entire body, being as 1 to 700. The size of the spleen is increased during and after digestion, and varies considerably, according to the state of nutrition of the body, being large in highly fed, and small in starved animals. In intermittent and other fevers, it becomes much enlarged, weighing occasionally from 18 to 20 pounds.

Structure. The spleen is invested by two coats; an external serous, and an internal fibrous elastic coat.

The *external*, or *serous coat*, is derived from the peritoneum; it is thin, smooth, and in the human subject intimately adherent to the fibrous elastic coat. It invests almost the entire organ; being reflected from it, at the hilus, on to the great end of the stomach, and at the upper end of the organ on to the Diaphragm.

The *fibrous elastic coat* forms the framework of the spleen. It invests the exterior of the organ, and at the hilus is reflected inwards upon the vessels in the form of vaginae or sheaths. From these sheaths, as well as from the inner surface of the fibro-elastic coat, numerous small fibrous bands, *trabeculae* (fig. 324), are given off in all directions; these uniting, constitute the areolar framework of the spleen. The proper coat, the sheaths of the vessels, and the trabeculae, consist of a dense mesh of the white and yellow elastic fibrous tissues, the latter considerably predominating. It is owing to the presence of this tissue, that the spleen possesses a considerable amount of elasticity, admirably adapted for the very considerable variations in size that it presents under certain circumstances. In some of the mammalia, in addition to the usual constituents of this tunic, are found numerous pale, flattened, spindle-shaped, nucleated fibres, like unstriped muscular fibre. It is probably owing to this structure, that the spleen possesses, when acted upon by the galvanic current, faint traces of contractility.

324.—Transverse Section of the Spleen, showing the Trabecular Tissue, and the Splenic Vein and its Branches.

The *proper substance* of the spleen occupies the interspaces of the areolar framework of the organ; it is a soft, pulpy mass, of a dark reddish-brown colour, consisting of colourless and coloured elements.

The *colourless elements* consist of granular matter; nuclei, about the size of the red blood-discs, homogeneous or granular in structure; and nucleated vesicles in small numbers. These elements form, probably, one-half or two-thirds of the whole substance of the pulp, filling up the interspaces formed by the partitions of the spleen, and lying in close contact with the walls of the capillary vessels, so as to be readily acted upon by the nutrient fluid which permeates them. These elements form a large part of the entire bulk of the spleen in well-nourished animals; whilst they diminish in number, and occasionally are not found at all, in starved animals. The application of chemical tests shows that they are essentially a proteine compound.

The *coloured elements* of the pulp consist of red blood-globules and of coloured corpuscles, either free, or included in cells. Sometimes, unchanged blood-discs are seen included in a cell; but more frequently the included blood-discs are altered both in form and colour. Besides these, numerous deep-red, or reddish-yellow, or black corpuscles and crystals, either single or aggregated in masses, are seen diffused throughout the pulp-substance: these, in chemical composition, are closely allied to the haematin of the blood.

Malpighian Corpuscles. On examining the cut surface of a healthy spleen, a number of small semi-opaque bodies, of gelatinous consistence, are seen disseminated throughout its substance; these are the splenic or Malpighian corpuscles (fig. 325). They may be seen at all periods of life; but they are more distinct in early, than in adult life or old age; and they are much smaller in man, than in most mammalia.

325.—The Malpighian Corpuscles, and their Relation with the Splenic Artery and its Branches.

They are of a spherical or ovoid form, vary considerably in size and number, and are of a semi-opaque whitish colour. They are appended to the sheaths of the smaller arteries and their branches, presenting a resemblance to the buds of the moss rose. Each consists of a membranous capsule, composed of fine pale fibres, which interlace in all directions. In man, the capsule is homogeneous in structure, and formed by a prolongation from the sheaths of the small arteries to which the corpuscles are attached. The blood-vessels ramifying on the surface of the corpuscles, consist of the larger ramifications of the arteries to which the sacculus is connected; and also of a delicate capillary plexus, similar to that surrounding the vesicles of other glands.

These vesicles have also a close relation with the veins (fig. 326). These vessels, which are of considerable size even at their origin, commence on the surface of each vesicle throughout the whole of its circumference, forming a dense venous mesh, in which each of these bodies is enclosed. It is probable, that from the blood contained in the capillary network, the material is separated which is occasionally stored up in their cavity; the veins being so placed as to carry off, under certain conditions, those contents that are again to be discharged into the circulation. Each capsule contains a soft, white, semi-fluid substance, consisting of granular matter, nuclei similar to those found in the pulp, and a few nucleated cells, the composition of which is apparently albuminous. These bodies are very large after the early periods of digestion, in well-fed animals, and especially in those fed upon albuminous diet. In starved animals, they disappear altogether.

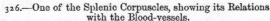

326.—One of the Splenic Corpuscles, showing its Relations with the Blood-vessels.

The *Splenic Artery* is remarkable for its large size, in proportion to the size of the organ; and also for its tortuous course. It divides into from four to six branches, which enter the hilus of the organ, and ramify throughout its substance (fig. 327), receiving sheaths from an involution of the external fibrous tunic, the same sheaths also investing the nerves and veins. Each branch runs in the transverse axis of the organ, from within outwards, diminishing in size during its transit, and giving off, in its passage, smaller branches, some of which pass to the anterior, others to the posterior part: these ultimately terminate in the proper substance of the spleen, in small tufts or pencils of capillary vessels, which lie in direct contact with the pulp. Each of the larger branches of the artery supplies chiefly that region of the organ in which the branch ramifies, having no anastomosis with the majority of the other branches.

327.—Transverse Section of the Human Spleen, showing the Distribution of the Splenic Artery and its Branches.

The *Capillaries*, supported by the minute trabeculae, traverse the pulp in all directions, and terminate either directly in the veins, or open into lacunar spaces, from which the veins originate.

The *Veins* are of large size, as compared with the size of the organ; and their distribution is limited, like that of the arteries, to the supply of a particular part of the gland; they are much larger and more numerous than the arteries. They originate, 1st, as continuations of the capillaries of the arteries; 2nd, by intercellular spaces communicating with each other; 3rd, by distinct caecal pouches. By their junction they form from four to six branches, which emerge from the hilus; and these uniting, form the splenic vein, the largest branch of the vena portae.

The *Lymphatics* form a deep and superficial set; they pass through the lymphatic glands at the hilus, and terminate in the thoracic duct.

The *Nerves* are derived from branches of the right and left semilunar ganglia, and right pneumogastric nerve.

THE THORAX

The Thorax is a conical, osseous framework, connected with the middle region of the spine. It is the largest of the three cavities of the trunk, narrow above, broad below, flattened before and behind, and somewhat cordiform on a transverse section.

Boundaries. It is bounded in front by the sternum, the six upper costal cartilages, the ribs, and intercostal muscles; at the sides, by the ribs and intercostal muscles; and behind, by the same structures and the dorsal portion of the vertebral column.

The *Superior Opening* of the thorax is bounded on each side by the first rib; in front, by the upper border of the sternum; and behind, by the first dorsal vertebra. It is broader from side to side, than from before backwards; and its direction is backwards and upwards.

The *Lower Opening*, or *Base*, is bounded in front by the ensiform cartilage; behind, by the last dorsal vertebra; and on each side by the last rib, the Diaphragm filling in the intervening space. Its direction is obliquely downwards and backwards; so that the cavity of the thorax is much deeper on the posterior, than on the anterior wall. It is wider transversely than from before backwards, and its general direction is convex towards the chest; but it is more flattened at the centre than at the sides, and rises higher on the right than on the left side, corresponding in the dead body to the upper border of the fifth rib, near the sternum, on the right side; and to the corresponding part of the sixth rib on the left side.

The parts which pass through the upper opening of the thorax are, from before backwards, the Sterno-hyoid and Sterno-thyroid muscles, the remains of the thymus gland, the trachea, oesophagus, thoracic duct, and the Longi colli muscles; on the sides, the arteria innominata on the right, the left carotid and left sub-clavian arteries, the internal mammary and superior intercostal arteries, the right and left venae innominatae, the pneumogastric, sympathetic, phrenic, and cardiac nerves, and the recurrent laryngeal nerve of the left side. The apex of each lung, covered by the pleura, also projects through this aperture, a little above the margin of the first rib.

The viscera contained in the thoracic cavity are, the great central organ of circulation, the heart, enclosed in its membranous bag, the pericardium; and the organs of respiration, the lungs, invested by the pleurae.

The Pericardium

The pericardium is a conical membranous sac, in which the heart and the commencement of the great vessels are contained. It is placed behind the sternum, and the cartilages of the fourth, fifth, sixth, and seventh ribs of the left side, in the interval between the pleurae.

Its *apex* is directed upwards, and surrounds the great vessels about two inches above their origin from the base of the heart. Its *base* is attached to the central tendon of the Diaphragm, extending a little farther to the left than the right side. In *front*, it is separated from the sternum by the remains of the thymus gland above, and a little loose areolar tissue below; and is covered by the margins of the lungs, especially the left. *Behind*, it rests upon the bronchi, the oesophagus, and the descending aorta. *Laterally*, it is covered by the pleurae; the phrenic vessels and nerve descending between the two membranes on either side.

The *Pericardium* is a fibro-serous membrane, and consists, therefore, of two layers: an external fibrous, and an internal serous.

The *fibrous layer* is a strong, dense fibrous membrane. Above, it surrounds the great vessels arising from the base of the heart, on which it is continued in the form of tubular prolongations, which are gradually lost upon their external coats; the strongest being that which encloses the aorta. Below, it is attached to the central tendon of the Diaphragm, and, on the left side, to its muscular fibres.

The vessels receiving fibrous prolongations from this membrane are the aorta, the superior vena cava, and the pulmonary arteries and veins. As the inferior cava enters the pericardium, through the central tendon of the Diaphragm, it receives no covering from the fibrous layer.

The *serous layer* invests the heart, and is then reflected on the inner surface of the pericardium. It consists, therefore, of a visceral and a parietal portion. The former invests the surface of the heart, and the commencement of the great vessels, to the extent of two inches from their origin; from these it is reflected upon the inner surface of the fibrous layer, lining, below, the upper surface of the central tendon of the Diaphragm. The serous membrane encloses the aorta and pulmonary artery in a single tube; but it only partially covers the superior and inferior vena cavae, and the four pulmonary veins. Its inner surface is smooth and glistening, and secretes a thin fluid, which serves to facilitate the movements of the contained organ.

The arteries of the pericardium are derived from the internal mammary, the bronchial, the oesophageal, and the phrenic.

The Heart

The heart is a hollow muscular organ, of a conical form, placed between the lungs, and enclosed in the cavity of the pericardium.

Position. The heart is placed obliquely in the chest: the broad attached end, or base, is directed upwards and backwards to the right, and corresponds to the interval between the fifth and eighth dorsal vertebrae: the apex is directed downwards and forwards to the left, and corresponds to the interspace between the cartilages of the fifth and sixth ribs, one inch to the inner side, and two inches below the left nipple. The heart projects further into the left than into the right cavity of the chest, extending from the median line about three inches in the former direction, and only one and a half in the latter. Its upper border would correspond to a line drawn across the sternum, on a level with the upper border of the third costal cartilage; and its lower border, to a line drawn across the lower end of the same bone, from the costo-xiphoid articulations of the right side, to the part corresponding to the apex. Its upper surface is rounded and convex, directed upwards and forwards, and formed chiefly by the right ventricle and part of the left. Its under surface is flattened, and rests upon the Diaphragm. Of its two borders, the right is the longest and thinnest, the left shorter, but thick and round.

Size. The heart, in the adult, measures about five inches in length, three inches

328.—The Right Auricle and Ventricle laid open, the Anterior Walls of both being removed.

and a half in the broadest part of its transverse diameter, and two inches and a half in its antero-posterior. The prevalent weight, in the male, varies from ten to twelve ounces; in the female, from eight to ten: its proportion to the body being as 1 to 169, in males; 1 to 149, in females. The heart continues increasing in weight, and also in length, breadth, and thickness, up to an advanced period of life; and this is more marked in men than in women.

The heart is subdivided by a longitudinal muscular septum, into two lateral halves, which are named respectively, from their position, right and left; and a transverse constriction divides each half of the organ into two cavities, the upper cavity on each side being called the *auricle*, the lower the *ventricle*. The right is the venous side of the heart, receiving into its auricle the dark venous blood from the entire body, by the superior and inferior cavae, and coronary sinus. From the auricle, the blood passes into the right ventricle; and from the right ventricle, through the pulmonary artery, into the lungs. The blood, arterialized by its passage through the lungs, is returned to the left side of the heart by the pulmonary veins, which open into the left auricle; from the left auricle the blood passes into the left ventricle, and from the left ventricle is distributed, by the aorta and its subdivisions, through the entire body. This constitutes the circulation of the blood in the adult.

This division of the heart into four cavities, is indicated upon its surface in the form of grooves. Thus, the great transverse groove separating the auricles from the ventricles, is called the *auriculo-ventricular groove*. It is deficient, in front, from being crossed by the root of the pulmonary artery, and contains the trunk of the nutrient vessels of the heart. The auricular portion occupies the base of the heart, and is subdivided into two cavities by a median septum. The two ventricles are also separated into a right and left, by two longitudinal furrows, which are situated, one on its anterior, the other on its posterior surface: these extend from the base to the apex of the organ: the former being situated nearer to the left border of the heart, and the latter to the right. It follows, therefore, that the right ventricle forms the greater portion of the anterior surface of the heart, and the left ventricle more of its posterior surface.

Each of these cavities should now be separately examined.

The RIGHT AURICLE is a little larger than the left, its walls somewhat thinner, measuring about one line; and its cavity is capable of containing about two ounces. It consists of two parts, a principal cavity, or sinus, and an appendix auriculae.

The *sinus* is the large quadrangular-shaped cavity, placed between the two venae cavae: its walls are extremely thin, and it is connected below with the right ventricle, and internally, with the left auricle, being free in the rest of its extent.

The *appendix auriculae*, so called from its fancied resemblance to a dog's ear, is a small conical muscular pouch, the margins of which present a dentated edge. It projects from the sinus forwards and to the left side, overlapping the root of the pulmonary artery.

To examine the interior of the auricle, a transverse incision should be made along its ventricular margin, from its right border to the appendix; and, from the middle of this, a second incision should be carried upwards, along the inner side of the two venae cavae.

The following parts present themselves for examination:

Openings. { Superior cava. Inferior cava. Coronary sinus. Foramina Thebesii. Auriculo-ventricular.

Valves. { Eustachian. Coronary.

Relics of Foetal structure. { Annulus ovalis. Fossa ovalis.

Musculi pectinati.

OPENINGS. The *Superior Vena Cava* returns the blood from the upper half of the body, and opens into the upper and front part of the auricle, the direction of its orifice being downwards and forwards.

The *Inferior Vena Cava*, larger than the superior, returns the blood from the lower half of the body, and opens into the lowest part of the auricle, near the septum, the direction of its orifice being upwards and inwards. The direction of a current of blood through the superior vena cava, would consequently be towards the auriculo-ventricular orifice; whilst the direction of the blood through the inferior cava, would be towards the auricular septum. This is the normal direction of the two currents in foetal life.

The *Tuberculum Loweri* is a small projection on the right wall of the auricle, between the two cavae. This is most distinct in the hearts of quadrupeds: in man, it is scarcely visible. It was supposed by Lower, to direct the blood from the superior cava towards the auriculo-ventricular opening.

The *Coronary Sinus* opens into the auricle, between the inferior vena cava, and the auriculo-ventricular opening. It returns the blood from the substance of the heart, and is protected by a semicircular fold of the lining membrane of the auricle, the coronary valve. The sinus, before entering the auricle, is considerably dilated. Its wall is partly muscular, and, at its junction with the great coronary vein, is somewhat constricted, and furnished with a valve, consisting of two unequal segments.

The *Foramina Thebesii* are numerous minute apertures, the mouths of small veins (*venae cordis minimae*), which open on various parts of the surface of the auricle. They return the blood directly from the muscular substance of the heart. Some of these foramina are minute depressions in the walls of the heart, presenting a closed extremity.

The *Auriculo-Ventricular Opening* is the large oval aperture of communication between the auricle and ventricle, to be presently described.

VALVES. The *Eustachian Valve* is situated between the anterior margin of the inferior cava, and the auriculo-ventricular orifice. It is semilunar in form, its convex margin attached to the wall of the vein; its concave margin, which is free, terminating in two cornua, of which the left is attached to the anterior edge of the annulus ovalis; the right being lost on the wall of the auricle.

In the *foetus*, this valve is of large size, and serves to direct the blood from the inferior cava, through the foramen ovale, into the left auricle.

In the *adult*, it is occasionally persistent, and may assist in preventing the reflux of blood into the inferior cava: more commonly, it is small, and its free margin presents a cribriform, or filamentous appearance: occasionally, it is altogether wanting. It is formed by a duplicature of the lining membrane of the auricle, containing a few muscular fibres.

The *Coronary Valve* is a semicircular fold of the lining membrane of the auricle, protecting the orifice of the coronary sinus. It prevents the regurgitation of blood into the sinus during the contraction of the auricle. This valve is occasionally double.

The *Fossa Ovalis* is an oval depression, corresponding to the situation of the foramen ovale in the foetus. It is situated at the lower part of the septum auricularum, above the orifice of the inferior vena cava.

The *Annulus Ovalis* is the prominent oval margin of the foramen ovale. It is most distinct above, and at the sides; below, it is deficient. A small slit-like valvular opening is occasionally found, at the upper margin of the fossa ovalis, which leads upwards, beneath the annulus, into the left auricle, it is the remains of the aperture between the two auricles in the foetus.

The *Musculi Pectinati* are small, prominent muscular columns, running transversely across the inner surface of the appendix auriculae, and adjoining portion of the wall of the sinus. They have received the name, *pectinati*, from the fancied resemblance they bear to the teeth of a comb.

The RIGHT or ANTERIOR VENTRICLE is somewhat triangular in form, and extends from the right auricle to near the apex. Its anterior or upper surface is rounded and convex, and forms the larger part of the front of the heart. Its posterior or under surface is flattened, rests upon the Diaphragm, and forms only a small part of this surface. Its inner wall is formed by the partition between the two ventricles, the *septum ventriculorum*, the surface of which is convex, and bulges into the cavity of the right ventricle. Superiorly, the ventricle forms a conical prolongation, the *infundibulum*, or *conus arteriosus*, from which the pulmonary artery arises. The walls of the right ventricle are thinner than those of the left, the proportion between them being as 1 to 2 (Bizot). The thickest part of the wall is at the base, and it gradually becomes thinner towards the apex. Its cavity, which equals that of the left ventricle, is capable of containing about two fluid ounces.

To examine its interior, an incision should be made a little to the right of the anterior ventricular groove from the pulmonary artery to the apex of the heart, and from thence carried up along the right border of the ventricle, as far as the auriculo-ventricular opening.

The following parts present themselves for examination:

Openings. { Auriculo-ventricular. Opening of the Pulmonary Artery.
Valves. { Tricuspid. Semilunar.

And a muscular and tendinous apparatus connected with the tricuspid valves.

Columnae carneae. Chordae tendineae.

The *Auriculo-Ventricular Orifice* is the large oval aperture of communication between the auricle and ventricle. It is situated at the base of the right ventricle, near the right border of the heart, and corresponds to the sternum between the third costal cartilages. The opening is about an inch in diameter, oval from side to side, surrounded by a fibrous ring, covered by the lining membrane of the heart, and is rather larger than the corresponding aperture on the left side, being sufficiently large to admit the ends of three fingers. It is guarded by the tricuspid valve.

The *Opening of the Pulmonary Artery* is circular in form, and situated at the summit of the conus arteriosus, close to the septum ventriculorum. It is placed on the left side, and in front of the auriculo-ventricular opening, upon the anterior aspect of the heart, and corresponds to the upper border of the third costal cartilage of the left side, close to the sternum. Its orifice is guarded by the semilunar valves.

The *Tricuspid Valve* consists of three segments of a triangular or trapezoidal shape, formed by a duplicature of the lining membrane of the heart, strengthened by a layer of fibrous tissue, and containing, according to Kürschner and Senac, muscular fibres. These segments are connected by their bases to the auriculo-ventricular orifice, and by their sides with one another, so as to form a continuous annular membrane, which is attached around the margin of the auriculo-ventricular opening, their free margin and ventricular surfaces affording attachment to a number of delicate tendinous cords, the *chordae tendineae*. The largest of these three segments is placed towards the left side of the auriculo-ventricular opening interposed between it and the pulmonary artery, so as to prevent the filling of that tube during the distension of the ventricle. Another segment corresponds to the front of the ventricle; and a third to its posterior wall. The central part of each segment is thick and strong; and the lateral margins thin and indented. The chordae tendineae are connected with the adjacent margins of the principal segments of the valve, and are further attached to each segment in the following manner: 1. Three or four reach the attached margin of each segment, where they are continuous with the auriculo-ventricular tendinous ring. 2. Others, four to six in number, are attached to the central thickened part of each segment. 3. The most numerous and finest are connected with the marginal portion of each segment.

The *Columnae Carneae* are the rounded muscular columns which are observed projecting from the whole of the inner surface of the ventricle, excepting the infundibulum, and interlacing in every direction. They may be classified according to their mode of connection with the ventricle, into three sets. The first set merely form prominent ridges on the inner surface of the ventricle, being attached their entire length on one side, as well as by their extremities. The second set are attached by their two extremities, but are free in the rest of their extent; whilst the third set (*columnae papillares*), three or four in number, are attached by one extremity to the wall of the heart, the opposite extremity giving insertion to the *chordae tendineae*.

The *Semilunar Valves*, three in number, guard the orifice of the pulmonary artery. They consist of three semicircular folds, formed by a duplicature of the lining membrane, strengthened by fibrous tissue. They are attached, by their convex margins, to the wall of the artery, at its junction with the ventricle, the straight border being free, and directed upwards in the course of the vessel, against the sides of which they are pressed during the passage of the blood along its canal. The free margin of each valve is somewhat thicker than the rest, and strengthened by a bundle of tendinous fibres; it presents, at its middle, a small projecting fibro-cartilaginous nodule, called *corpus Arantii*. From this nodule tendinous fibres radiate through the valve to its attached margin, and these fibres form a constituent part of its substance throughout its whole extent, excepting two narrow lunated portions, placed one on either side of the nodule, immediately behind the free margin; here the valve is thin, and formed merely by the lining membrane. During the passage of the blood along the pulmonary artery, these valves are pressed against the sides of its cylinder, and the course of the blood along the tube is uninterrupted; but during the ventricular diastole, when the current of blood along the pulmonary artery is checked, and partly thrown back by its elastic walls, these valves become immediately expanded, and effectually close the entrance of the tube. When the valves are closed, the lunated portions of each are brought into contact with one another by their opposed surfaces, the three fibro-cartilaginous nodules filling up the small triangular space that would be otherwise left by the approximation of three semilunar folds.

Between the semilunar valves and the commencement of the pulmonary artery are three pouches or dilatations, one behind each valve. These are the pulmonary sinuses (*sinuses of Valsalva*). Similar sinuses exist between the semilunar valves and the commencement of the aorta; they are larger than the pulmonary sinuses.

The LEFT AURICLE is rather smaller but thicker than the right, measuring about one line and a half; it consists of two parts, a principal cavity or sinus, and an appendix auriculae.

The *sinus* is cuboidal in form, and concealed in front by the pulmonary artery and aorta; internally, it is separated from the right auricle by the septum auricularum; and behind, it receives on each side the pulmonary veins, being free in the rest of its extent.

The *appendix auriculae* is somewhat constricted at its junction with the auricle; it is longer, narrower, and more curved than that of the right side, and its margins more deeply indented, presenting a kind of foliated appearance. Its direction is forwards towards the right side, overlapping the root of the pulmonary artery.

In order to examine its interior, a horizontal incision should be made along the attached border of the auricle to the ventricle, and from the middle of this a second incision should be carried upwards.

The following parts then present themselves for examination:

>The Openings of the four Pulmonary Veins.
>Auriculo-Ventricular Opening.
>Musculi Pectinati.

The *Pulmonary Veins*, four in number, open two into the right, and two into the left side of the auricle. The two left veins frequently terminate by a common opening. They are not provided with valves.

The *Auriculo-Ventricular Opening* is the large oval aperture of communication between the auricle and ventricle. It is rather smaller than the corresponding opening on the opposite side.

329.—The Left Auricle and Ventricle laid open, the Anterior Walls of both being removed.

The *Musculi Pectinati* are fewer in number and smaller than on the right side; they are confined to the inner surface of the appendix.

On the inner surface of the septum auricularum may be seen a lunated impression, bounded below by a crescentic ridge, the concavity of which is turned upwards. It corresponds to the foramen ovale in the foetus. This depression is just above the fossa ovalis in the right auricle.

The LEFT VENTRICLE is longer and more conical in shape than the right ventricle. It forms a small part of the left side of the anterior surface of the heart, and a considerable part of its posterior surface. It also forms the apex of the heart by its projection beyond the right ventricle. Its walls are much thicker than those of the right ventricle, the proportion being as 2 to 1 (Bizot). They are also thickest in the broadest part of the ventricle, becoming gradually thinner towards the base, and also towards the apex, which is the thinnest part.

Its cavity should be opened by making an incision through its anterior wall along the left side of the ventricular septum, and carrying it round the apex and along its posterior surface to the auriculo-ventricular opening.

The following parts present themselves for examination:

Openings. { Auriculo-ventricular. Aortic. }   Valves. { Mitral. Semilunar. }

Chordae tendineae.   Columnae carneae.

The *Auriculo-ventricular Opening* is placed behind, and to the left side of, the aortic orifice. This opening is a little smaller than the corresponding aperture of the opposite side; and, like it, is broader in the transverse, than in the antero-posterior diameter. Its position corresponds to the centre of the sternum. It is surrounded by a dense fibrous ring, covered by the lining membrane of the heart, and is guarded by the mitral valve.

The *Aortic Opening* is a small circular aperture, in front and to the right side of the auriculo-ventricular, from which it is separated by one of the segments of the mitral valve. Its orifice is guarded by the semilunar valves. Its position corresponds to the junction of the third left costal cartilage with the sternum.

The *Mitral Valve* is attached to the circumference of the auriculo-ventricular orifice, in the same way that the tricuspid valve is on the opposite side. It is formed by a duplicature of the lining membrane, strengthened by fibrous tissue, and contains a few muscular fibres. It is larger in size, thicker, and altogether stronger than the tricuspid, and consists of two segments of unequal size. The larger segment is placed in front, between the auriculo-ventricular and aortic orifices, and is said to prevent the filling of the aorta during the distension of the ventricle. Two smaller segments are usually found at the angle of junction of the larger. The mitral valves are furnished with numerous chordae tendineae; their mode of attachment is precisely similar to those on the right side; but they are thicker, stronger, and less numerous.

The *Semilunar Valves* surround the orifice of the aorta; they are similar in structure, and in their mode of attachment, to those of the pulmonary artery. They are, however, larger, thicker, and stronger than those of the right side; the lunulae are more distinct, and the corpora Arantii larger and more prominent. Between each valve and the cylinder of the aorta is a deep depression, the *sinus aortici* (sinuses of Valsalva); they are larger than those at the root of the pulmonary artery.

The *Columnae Carneae* admit of a subdivision into three sets, like those upon the right side; but they are smaller, more numerous, and present a dense interlacement, especially at the apex, and upon the posterior wall. Those attached by one extremity only, the *musculi papillares*, are two in number, being connected one to the anterior, the other to the posterior wall; they are of large size, and terminate by free rounded extremities, from which the chordae tendineae arise.

The *Endocardium* is the serous membrane which lines the internal surface of the heart; it assists in forming, by its reduplications, the valves contained in this organ, and is continuous with the lining membrane of the great blood-vessels. It is a thin, smooth, transparent membrane, giving to the inner surface of the heart its glistening appearance. It is more opaque on the left, than on the right side of the heart, thicker in the auricles than in the ventricles, and thickest in the left auricle. It is thin on the musculi pectinati, and on the columnae carneae; but thicker on the smooth part of the auricular and ventricular walls, and on the tips of the musculi papillares.

Structure. The heart consists of muscular fibres, and of fibrous rings which serve for their attachment.

The *Fibrous Rings* surround the auriculo-ventricular and arterial orifices: they are stronger upon the left, than on the right side of the heart. The auriculo-ventricular rings serve for the attachment of the muscular fibres of the auricles and ventricles, and also for the mitral and tricuspid valves; the left one is closely connected, by its right margin, with the aortic arterial ring. Between these and the right auriculo-ventricular ring, is a fibro-cartilaginous mass; and in some of the larger animals, as the ox and elephant, a portion of bone.

The fibrous rings surrounding the arterial orifices, serve for the attachment of the great vessels and semilunar valves. Each ring receives, by its ventricular margin, the attachment of the muscular fibres of the ventricles; its opposite margin presents three deep semicircular notches, within which the middle coat of the artery (which presents three convex semicircular segments) is firmly fixed; the attachment of the artery to its fibrous ring being strengthened by the thin cellular coat and serous membrane externally, and by the endocardium within. It is opposite the margins of these semicircular notches, in the arterial rings, that the endocardium, by its reduplication, forms the semilunar valves, the fibrous structure of the ring being continued into each of the segments of the valve at this part. The middle coat of the artery in this situation is thin, and the sides of the vessel dilated to form the sinuses of Valsalva.

The *Muscular Structure* of the heart consists of bands of fibres, which present an exceedingly intricate interlacement. They are of a deep red colour, involuntary, but marked with transverse striae.

The muscular fibres of the heart admit of a subdivision into two kinds; those of the auricles, and those of the ventricles; which are quite independent of one another.

Fibres of the Auricles. These are disposed in two layers; a superficial layer common to both cavities, and a deep layer proper to each. The *superficial fibres* are most distinct on the anterior surface of the auricles, across the bases of which they run in a transverse direction, forming a thin, but incomplete, layer. Some of these fibres pass into the septum auricularum. The *internal* or *deep fibres* proper to each auricle consist of two sets, looped, and annular fibres. The *looped fibres* pass upwards over each auricle, being attached by both extremities to the corresponding auriculo-ventricular rings, in front and behind. The *annular fibres* surround the whole extent of the appendices auriculae, and are continued upon the walls of the venae cavae and coronary sinus on the right side, and upon the pulmonary veins on the left side, at their connection with the heart. In the appendices, they interlace with the longitudinal fibres.

Fibres of the Ventricles. These, as in the auricles, are disposed in layers, some of which are common to both ventricular cavities, whilst others belong exclusively to one ventricle, the latter being chiefly found towards the base of the heart. The greater majority of these fibres are connected by both ends with the auriculo-ventricular fibrous rings, either directly or indirectly through the chordae tendineae; some, however, are attached to the fibrous rings surrounding the arterial orifices.

The *superficial fibres* are either longitudinal, or more commonly oblique or spiral in their direction, and towards the apex are arranged in the form of twisted loops; the deeper fibres are circular.

The *Spiral fibres* are disposed in layers of various degrees of thickness: the most superficial, on the front of the ventricles, run obliquely from right to left, and from above downwards. On the back of the ventricles they are directed more vertically, and pass from left to right.

The *superficial fibres* coil inwards at the apex of the heart, around which they are arranged in a whorl-like form, called the *vortex*, dipping beneath the edge of the deeper and shorter layers. If these fibres are carefully uncoiled, in a heart previously boiled, the cavity of the left, and then that of the right ventricle, will be exposed at this point. The layers of fibres successively met with, have a similar arrangement; the more superficial and longer turning inwards, and including the deeper and shorter bands. All these fibres ascend and spread out upon the inner surface of the ventricles, forming the walls, the septum, and the musculi papillares, which project from these cavities; and they are finally inserted into the auriculo-ventricular fibrous rings, or, indirectly, through the chordae tendineae. Of these spiral fibres, some enter at the interventricular furrows, and surround either ventricle singly; others pass across the furrows and embrace both cavities. On tracing those which form the vortex, back into the interventricular septum, they become interlaced with similar fibres from the right ventricle, and ascend vertically upon the right side of the septum, as far as its base, in the form of a long and broad band.

Circular Fibres. The circular fibres are situated deeply in the substance of the heart; towards the base they enter the anterior and posterior longitudinal furrows, so as to include each cavity singly, or, passing across them, surround both ventricles, more fibres passing across the posterior than the anterior furrow. They finally ascend in the substance of the ventricle, to be inserted into the fibrous rings at its base.

Vessels and Nerves. The *arteries* supplying the heart, are the anterior and posterior coronary.

The *veins* accompany the arteries, and terminate in the right auricle. They are the great cardiac vein, the small, or anterior cardiac veins, and the venae cordis minimae (*venae Thebesii*).

The *lymphatics* terminate in the thoracic and right lymphatic ducts.

The *nerves* are derived from the cardiac plexuses, which are formed partly from the spinal, and partly from the sympathetic system. In their distribution throughout its substance, they are furnished with small ganglia, which have been figured both by Remak and Lee.

Peculiarities in the Vascular System of the Foetus

The chief peculiarities in the heart of the foetus, are the direct communication between the two auricles by the foramen ovale, and the large size of the Eustachian valve. There are also several minor peculiarities. Thus, the position of the heart is vertical until the fourth month, when it commences to assume an oblique direction. Its size is also very considerable, as compared with the body, the proportion at the second month being as 1 to 50: at birth, it is as 1 to 120: whilst, in the adult, the average is about 1 to 160. At an early period of foetal life, the auricular portion of the heart is larger than the ventricular, the right auricle being more capacious than the left; but, towards birth, the ventricular portion becomes the larger. The thickness of both ventricles is, at first, about equal; but, towards birth, the left becomes much the thicker of the two.

The *Foramen Ovale* is situated at the lower and back part of the septum auricularum. It attains its greatest size at the sixth month.

The *Eustachian Valve* is developed from the anterior border of the inferior vena cava, at its entrance in the auricle, and, rising up on the left side of the opening of this vein, serves to direct the blood from the inferior vena cava through the foramen ovale into the left auricle.

The peculiarities in the arterial system of the foetus are the communication between the pulmonary artery and descending aorta, by means of the ductus arteriosus, and the communication between the common iliac arteries and the placenta, by the umbilical arteries.

The *Ductus Arteriosus* is a short tube, about half an inch in length at birth, and of the diameter of a goose-quill. It connects the left branch of the pulmonary artery with the termination of the arch of the aorta, just beyond the origin of the left subclavian artery. It serves to conduct the chief part of the blood of the right ventricle into the descending aorta.

The *Umbilical*, or *Hypogastric Arteries*, arise from the internal iliacs, in addition to the usual branches given off from these vessels in the adult. Ascending along the sides of the bladder to its fundus, they pass out of the abdomen at the umbilicus, and are continued along the umbilical cord to the placenta, coiling round the umbilical vein. They return the blood to the placenta which has been circulated in the system of the foetus.

The peculiarity in the venous system of the foetus is the communication established between the placenta, and the liver and portal vein, through the umbilical vein, and with the inferior vena cava by the ductus venosus.

Fœtal Circulation

The arterial blood destined for the nutrition of the foetus, is carried from the placenta to the foetus, along the umbilical cord, by the umbilical vein. The umbilical vein enters the abdomen at the umbilicus, and passes upwards along the free margin of the suspensory ligament of the liver, to the under surface of this organ, where it gives off two or three branches to the left lobe; and others to the lobus quadratus and

330.—Plan of the Fœtal Circulation.

In this plan the figured arrows represent the kind of blood, as well as the direction which it takes in the vessels. Thus—arterial blood is figured ⟫⋯⋯⋯⟩; venous blood, ⟫------⟩; mixed (arterial and venous blood), ⟫-·-·-·⟩.

lobus Spigelii. At the transverse fissure, it divides into two branches; of these, the larger joins the portal vein, and enters the right lobe: the smaller branch continues onwards, under the name of the ductus venosus, and joins the inferior vena cava. The blood, therefore, which traverses the umbilical vein, reaches the inferior cava in three different ways. The greater quantity circulates through the liver with the portal venous blood, before entering the cava by the hepatic veins: some enters the liver directly, and is also returned to the inferior cava by the hepatic veins: the smaller quantity passes directly into the vena cava, by the ductus venosus.

In the inferior cava, the blood carried by the ductus venosus becomes mixed with that returning from the lower extremities and viscera of the abdomen. It enters the right auricle, and, guided by the Eustachian valve, passes through the foramen ovale into the left auricle, where it becomes mixed with a small quantity of blood returned from the lung by the pulmonary veins. From the left auricle it passes into the left ventricle; and, from the left ventricle, into the aorta, from whence it is distributed almost entirely to the head and upper extremities, a small quantity being probably carried into the descending aorta. From the head and upper extremities, the blood is returned by the branches of the superior vena cava to the right auricle, where it becomes mixed with a small portion of the blood from the inferior cava. From the right auricle, it descends over the Eustachian valve into the right ventricle; and, from the right ventricle, into the pulmonary artery. The lungs of the foetus being solid, and almost impervious, only a small quantity of blood is distributed to them by the right and left pulmonary arteries, which is returned by the pulmonary veins to the left auricle: the greater part passes through the ductus arteriosus into the commencement of the descending aorta, where it becomes mixed with a small quantity of blood transmitted by the left ventricle into the aorta. Along this vessel it descends to supply the lower extremities and viscera of the abdomen and pelvis, the chief portion being, however, conveyed by the umbilical arteries to the placenta.

From the preceding account of the circulation of the blood in the foetus, it will be seen:

1. That the placenta serves the double purpose of a respiratory and nutritive organ, receiving the venous blood from the foetus, and returning it again re-oxygenated, and charged with additional nutritive material.

2. That nearly the whole of the blood of the umbilical vein traverses the liver before entering the inferior cava; hence the enormous size of this organ, especially at an early period of foetal life.

3. That the right auricle is the point of meeting of a double current, the blood in the inferior cava being guided by the Eustachian valve into the left auricle, whilst that in the superior cava descends into the right ventricle. At an early period of foetal life, it is highly probable that the two streams are quite distinct; for the inferior cava opens almost directly into the left auricle, and the Eustachian valve would exclude the current along the tube from entering the right ventricle. At a later period, as the separation between the two auricles becomes more distinct, it seems probable that some slight mixture of the two streams must take place.

4. The blood carried from the placenta to the foetus by the umbilical vein, mixed with the blood from the inferior cava, passes almost directly to the arch of the aorta,

and is distributed by the branches of this vessel to the head and upper extremities; hence the large size and perfect development of these parts at birth.

5. The blood contained in the descending aorta, chiefly derived from that which has already circulated through the head and limbs, together with a small quantity from the left ventricle, is distributed to the lower extremities; hence the small size and imperfect development of these parts at birth.

Changes in the Vascular System at Birth

At birth, when respiration is established, an increased amount of blood from the pulmonary artery passes through the lungs, which now perform their office as respiratory organs, and, at the same time, the placental circulation is cut off. The foramen ovale becomes gradually closed in by about the tenth day after birth, a valvular fold rises up on the left side of its margin, and ultimately above its upper part; this valve becomes adherent to the margins of the foramen for the greater part of its circumference, but above a valvular opening is left between the two auricles, which sometimes remains persistent.

The *ductus arteriosus* begins to contract immediately after respiration is established, becomes completely closed from the fourth to the tenth day, and ultimately degenerates into an impervious cord, which serves to connect the left pulmonary artery to the concavity of the arch of the aorta.

Of the *umbilical* or *hypogastric arteries*, their commencement forms the trunk of the corresponding internal iliac; the portion continued on to the bladder remains pervious, as the superior vesical artery; and the part between the fundus of the bladder and the umbilicus becomes obliterated between the second and fifth days after birth, and forms the anterior true ligament of this viscus.

The *umbilical vein* and *ductus venosus* become completely obliterated between the second and fifth days after birth, and ultimately dwindle to fibrous cords; the former becoming the round ligament of the liver, the latter, the fibrous cord, which, in the adult, may be traced along the fissure of the ductus venosus to the inferior cava.

ORGANS OF VOICE AND RESPIRATION

The Larynx

The Larynx is the organ of voice, placed at the upper part of the air passage.

It is situated between the trachea and base of the tongue, at the upper and fore part of the neck, where it forms a considerable projection in the middle line. On either side of it lie the great vessels of the neck; behind, it forms part of the boundary of the pharynx, and is covered by the mucous membrane lining this cavity.

The larynx is narrow and cylindrical below, but broad above, where it presents the form of a triangular box, being flattened behind and at the sides, whilst in front it is bounded by a prominent vertical ridge. It is composed of cartilages, connected together by ligaments, moved by numerous muscles, lined by mucous membrane, and supplied with vessels and nerves.

The cartilages of the larynx are nine in number, three single and three pairs:

Thyroid.	Two Arytenoid.
Cricoid.	Two Cornicula Laryngis.
Epiglottis.	Two Cuneiform.

The *Thyroid* (θυρεος ειδος, *like a shield*) is the largest cartilage of the larynx. It consists of two lateral lamellae or alae, united at an acute angle in front, forming a vertical projection in the middle line, which is prominent above, and called the *pomum Adami*. This projection is subcutaneous, more distinct in the male than in the female, and occasionally separated from the integument by a bursa mucosa.

331.—Side View of the Thyroid and Cricoid Cartilages.

Each lamella is quadrilateral in form. Its *outer surface* presents an oblique ridge, which passes downwards and forwards from a tubercle, situated near the root of the superior cornu. This ridge gives attachment to the Sterno-thyroid and Thyro-hyoid muscles; the portion of cartilage included between it and the posterior border, to part of the Inferior constrictor muscle.

The *inner surface* of each ala is smooth, concave, and covered by mucous membrane above and behind; but in front, in the receding angle formed by their junction, is attached the epiglottis, the true and false chordae vocales, the Thyro-arytenoid, and Thyro-epiglottidean muscles.

The *upper border* of the thyroid cartilage is deeply notched in the middle line, immediately above the pomum Adami, whilst on either side it is slightly concave. This border gives attachment throughout its whole extent to the thyro-hyoid membrane.

The *lower border* is connected to the cricoid cartilage, in the median line by the crico-thyroid membrane, and on each side by the Crico-thyroid muscle.

The *posterior borders*, thick and rounded, terminate above in the superior cornua, and below in the interior cornua. The two superior cornua are long and narrow, directed backwards, upwards, and inwards, and terminate in a conical extremity, which gives attachment to the thyro-hyoid ligament. The two inferior cornua are shorter and thicker; they pass forwards and inwards, and present, on their inner surfaces, a small oval articular facet for articulation with the side of the cricoid cartilage. The posterior border receives the insertion of the Stylo-pharyngeus and Palato-pharyngeus muscles on each side.

332.—The Cartilages of the Larynx. Posterior View.

The *Cricoid Cartilage* is so called from its resemblance to a signet ring (κρικος-εἶδος, *like a ring*). It is smaller but thicker and stronger than the thyroid cartilage, and forms the lower and back part of the cavity of the larynx.

Its *anterior half* is narrow, convex, affording attachment in front and at the sides to the Crico-thyroid muscles, and behind those to part of the Inferior constrictor.

Its *posterior half* is very broad, both from side to side and from above downwards; it presents in the middle line a vertical ridge for the attachment of the longitudinal fibres of the oesophagus; and on either side of this is a broad depression for the Crico-arytenoideus posticus muscle.

At the point of junction of the two halves of the cartilage on either side, is a small round elevation, for articulation with the inferior cornu of the thyroid cartilage.

The *lower border* of the cricoid cartilage is horizontal, and connected to the upper ring of the trachea by fibrous membrane.

Its *upper border* is directed obliquely upwards and backwards, owing to the great depth of its posterior surface. It gives attachment, in front, to the crico-thyroid membrane; at the sides, to part of the same membrane and to the lateral Crico-arytenoid muscle; behind, the highest point of the upper border is surmounted on each side by a smooth oval surface, for articulation with the arytenoid cartilage. Between the articular surfaces is a slight notch, for the attachment of part of the Arytenoideus muscle.

The *inner surface* of the cricoid cartilage is smooth, and lined by mucous membrane.

The *Arytenoid Cartilages* are so called from the resemblance they bear, when approximated, to the mouth of a pitcher (αρυταινα-ειδος, *like a pitcher*). They are two in number, and situated at the upper border of the cricoid cartilage, at the back of the larynx. Each cartilage is pyramidal in form, and presents for examination three surfaces, a base, and an apex.

The *posterior surface* is triangular, smooth, concave, and lodges part of the Arytenoideus muscle.

The *anterior surface*, somewhat convex and rough, gives attachment to the Thyro-arytenoid muscle, and to the false vocal cord.

The *internal surface* is narrow, smooth, and flattened, covered by mucous membrane, and lies almost in apposition with the cartilage of the opposite side.

The *base* of each cartilage is broad, and presents a concave smooth surface, for articulation with the cricoid cartilage. Of its three angles, the external one is short, rounded, and prominent, receiving the insertion of the posterior and lateral Crico-arytenoid muscles. The anterior one, also prominent, but more pointed, gives attachment to the true vocal cord.

The *apex* of each cartilage is pointed, curved backwards and inwards, and surmounted by a small conical-shaped, cartilaginous nodule, *corniculum laryngis* (cartilage of Santorini). This cartilage is sometimes united to the arytenoid, and serves to prolong it backwards and inwards. To it is attached the aryteno-epiglottidean fold.

The *Cuneiform Cartilages* (cartilages of Wrisberg) are two small, elongated, cartilaginous bodies, placed one on each side, in the fold of mucous membrane which extends from the apex of the arytenoid cartilage to the side of the epiglottis (*aryteno-epiglottidean fold*); they give rise to the small whitish elevations on the inner surface of the mucous membrane, just in front of the arytenoid cartilages.

The *Epiglottis* is a thin lamella of fibro-cartilage, of a yellowish colour, shaped like a leaf, and placed behind the tongue and in front of the superior opening of the larynx. During respiration, its direction is vertically upwards, its free extremity curving forwards towards the base of the tongue; but when the larynx is drawn up beneath the base of the tongue during deglutition, it is carried downwards and backwards, so as to completely close the opening of the larynx. Its free extremity is broad and rounded; its attached end is long and narrow, and connected to the receding angle between the two alae of the thyroid cartilage, just below the median notch, by a long, narrow, ligamentous band, the *thyro-epiglottic ligament*. It is also connected to the posterior surface of the body of the hyoid bone, by an elastic ligamentous band, the *hyo-epiglottic ligament*.

Its *anterior* or *lingual surface* is curved forwards towards the tongue, and covered by mucous membrane, which is reflected on to the sides and base of this organ, forming a median and two lateral folds, the *glosso-epiglottidean ligaments*.

Its *posterior* or *laryngeal surface* is smooth, concave from side to side, convex from above downwards, and covered by mucous membrane; when this is removed, the surface of the cartilage is seen to be studded with a number of small mucous glands, which are lodged in little pits upon its surface. To its sides the aryteno-epiglottidean folds are attached.

Structure. The epiglottis, cuneiform cartilages, and cornicula laryngis are composed of yellow cartilage, which shows little tendency to ossification; but the other cartilages resemble in structure the costal cartilages, and they become more or less ossified in old age.

Ligaments. The ligaments of the larynx are *extrinsic*, as those connecting the thyroid cartilage with the os hyoides; and *intrinsic*, as those connecting the several cartilaginous segments to each other.

The ligaments connecting the thyroid cartilage with the os hyoides, are three in number; the thyro-hyoidean membrane, and the two lateral thyro-hyoidean ligaments.

The *Thyro-hyoidean Membrane* is a broad, fibro-elastic, membranous layer, attached below to the upper border of the thyroid cartilage, and above to the upper border of the inner surface of the hyoid bone; being separated from the posterior surface of the hyoid bone by a synovial bursa. It is thicker in the middle line than at either side, in which situation it is pierced by the superior laryngeal nerve and artery.

The *two lateral Thyro-hyoidean Ligaments* are rounded, elastic cords, which pass between the superior cornua of the thyroid cartilage, to the extremities of the greater cornua of the hyoid bone. A small cartilaginous nodule (*cartilago triticea*), sometimes bony, is found in each.

The ligaments connecting the thyroid cartilage to the cricoid are also three in number; the crico-thyroidean membrane, and the capsular ligaments and synovial membrane.

The *Crico-thyroid Membrane* is composed mainly of yellow elastic tissue. It is of triangular shape; thick in front, where it connects together the contiguous margins of the thyroid and cricoid cartilages; thinner at each side, where it extends from the superior border of the cricoid cartilage, to the inferior margin of the true vocal cords, with which it is closely united in front.

The *anterior portion* of the crico-thyroid membrane is convex, concealed on each side by the Crico-thyroid muscle, subcutaneous in the middle line, and crossed horizontally by a small anastomotic arterial arch, formed by the junction of the crico-thyroid branches on either side.

The *lateral portions* are lined internally by mucous membrane, and covered by the lateral Crico-arytenoid and Thyro-arytenoid muscles.

A *Capsular Ligament* encloses the articulation of the inferior cornu of the thyroid with the side of the cricoid, on each side. The articulation is lined internally by synovial membrane.

The ligaments connecting the arytenoid cartilages to the cricoid, are two thin and loose capsular ligaments connecting together the articulating surfaces, lined internally by synovial membrane, and strengthened behind by a strong posterior crico-arytenoid ligament, which extends from the cricoid to the inner and back part of the base of the arytenoid cartilage.

The ligaments of the epiglottis are the hyo-epiglottic, the thyro-epiglottic, and the three glosso-epiglottic folds of mucous membrane which connect the epiglottis to the sides and base of the tongue. The latter have been already described.

The *Hyo-epiglottic Ligament* is an elastic fibrous band, which extends from the anterior surface of the epiglottis, near its apex, to the posterior surface of the body of the hyoid bone.

The *Thyro-epiglottic Ligament* is a long, slender, elastic cord, which connects the apex of the epiglottis with the receding angle of the thyroid cartilage, immediately beneath the median notch, above the attachment of the vocal cords.

Interior of the Larynx. The *superior aperture of the larynx* is a triangular or cordiform opening, wide in front, narrow behind, and sloping obliquely downwards and backwards. It is bounded in front by the epiglottis; behind, by the apices of the arytenoid cartilages, and the cornicula laryngis; and laterally, by a fold of mucous membrane, enclosing ligamentous and muscular fibres, stretched between the sides of the epiglottis and the apex of the arytenoid cartilage: these are the aryteno-epiglottidean folds, on the margins of which the cuneiform cartilages form a more or less distinct whitish prominence.

The *cavity of the larynx* is divided into two parts by the projection inwards of the chordae vocales, and Thyro-arytenoid muscles; between the two cords is a long and narrow triangular fissure or chink, the *glottis*, or *rima glottidis*. The portion of the cavity of the larynx above the glottis, is broad and triangular in shape above, and corresponds to the interval between the alae of the thyroid cartilage; the portion below the glottis is at first elliptical, and lower down of a circular form.

The *glottis* is the interval between the inferior vocal cords. The two superior or false vocal cords are placed above the latter, and formed almost entirely by a folding inwards of the mucous membrane; whilst the two inferior or true vocal cords are

333.—Interior of the Larynx, seen from above (enlarged).

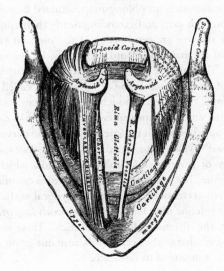

thick, strong, and formed partly by mucous membrane, and partly by ligamentous fibres. Between the true and false vocal cords, on each side, is an oval depression, the sinus or *ventricle of the larynx*, which leads upwards, on the outer side of the superior vocal cord, into a caecal pouch of variable size, the *sacculus laryngis*.

The *Rima Glottidis* is the narrow triangular fissure or chink between the inferior or true chordae vocales. Its length, in the male, measures rather less than an inch, its breadth varying at its widest part from a third to half an inch. In the female, these measurements are less by two or three lines. The form of the aperture varies; in ordinary respiration it is a narrow interval, being somewhat enlarged and rounded behind; but when dilated, it is triangular in form, narrow in front and broad behind.

The *Superior* or *False Vocal Cords*, so called because they are not concerned in the production of the voice, are two thick folds of mucous membrane, enclosing a delicate narrow fibrous band, the *superior thyro-arytenoid ligament*. This ligament consists of a thin band of elastic tissue, attached in front to the receding angle of the thyroid cartilage below the epiglottis, and behind to the tubercle on the anterior and inner surface of the arytenoid cartilage. The lower border of this ligament, enclosed in mucous membrane, forms a free crescentic margin, which constitutes the upper boundary of the ventricle of the larynx.

The *Inferior* or *True Vocal Cords*, so called from their being concerned in the production of sound, are two thick and strong fibrous bands (*inferior thyro-arytenoid ligaments*), covered externally by a thin and delicate mucous membrane. Each ligament consists of a compact band of parallel fibres of yellow elastic tissue, attached in front to the centre of the depression between the two alae of the thyroid cartilage, and behind to the anterior angle of the base of the arytenoid. Its inferior border is continuous with the thin lateral part of the crico-thyroid membrane. Its superior border forms the lower boundary of the ventricle of the larynx. Externally, the Thyro-arytenoideus muscle lies parallel with it. It is covered internally by mucous membrane, which is extremely thin, and closely adherent to its surface.

The *Ventricle of the Larynx* is an oblong fossa, situated between the superior and inferior vocal cords on each side, and extending nearly their entire length. This fossa is bounded above by the free crescentic edge of the superior vocal cord; below, by the straight margin of the true vocal cord; externally, by the corresponding Thyro-arytenoideus muscle. The anterior part of the ventricle leads up by a narrow opening into a caecal pouch of mucous membrane of variable size, called the *laryngeal pouch*.

The *Sacculus Laryngis*, or laryngeal pouch, is a membranous sac, placed between the superior vocal cord and the inner surface of the thyroid cartilage, occasionally extending as far as its upper border; it is conical in form, and curved slightly backwards, resembling in form a Phrygian cap. On the surface of its mucous membrane are the openings of sixty or seventy small follicular glands, which are lodged in the submucous areolar tissue. This sac is enclosed in a fibrous capsule, continuous below with the superior thyro-arytenoid ligament; its laryngeal surface is covered by the Aryteno-epiglottideus inferior muscle (*Compressor sacculi laryngis*, Hilton); whilst its exterior is covered by the Thyro-epiglottideus muscle. These muscles compress the sacculus laryngis, and discharge the secretion it contains upon the chordae vocales, the surface of which it is intended to lubricate.

Muscles. The intrinsic muscles of the larynx are eight in number; five of which are the muscles of the chordae vocales and rima glottidis; three are connected with the epiglottis.

The five muscles of the chordae vocales and rima glottidis are the

    Crico-thyroid.   Arytenoideus.
    Crico-arytenoideus posticus.   Thyro-arytenoideus.
    Crico-arytenoideus lateralis.

The *Crico-thyroid* is triangular in form and situated at the fore part and side of the cricoid cartilage. It arises from the front and lateral part of the cricoid cartilage; its fibres diverge, passing obliquely upwards and outwards, to be inserted into the lower and inner borders of the thyroid cartilage; from near the median line in front, as far back as the inferior cornu.

The inner borders of these two muscles are separated in the middle line by a triangular interval, occupied by the crico-thyroid membrane.

The *Crico-arytaenoideus posticus* arises from the broad depression occupying each lateral half of the posterior surface of the cricoid cartilage; its fibres pass upwards and outwards, and converge to be inserted into the outer angle of the base of the arytenoid cartilage. The upper fibres are nearly horizontal, the middle oblique, and the lower almost vertical.

The *Crico-arytaenoideus lateralis* is smaller than the preceding, and of an oblong form. It arises from the upper border of the side of the cricoid cartilage, and passing obliquely upwards and backwards, is inserted into the outer angle of the base of the arytenoid cartilage, in front of the preceding muscle.

The *Thyro-arytaenoideus* is a broad flat muscle, which lies parallel with the outer side of the true vocal cord. It arises in front from the lower half of the receding angle of the thyroid cartilage, and from the crico-thyroid ligament. Its fibres pass horizontally backwards and outwards, to be inserted into the base and anterior and outer surface of the arytenoid cartilage. This muscle consists of two fasciculi. The *inferior*, the thickest, is inserted into the anterior angle of the base of the arytenoid cartilage, and into the adjacent portion of its anterior surface; it lies parallel with the true vocal cord, to which it is occasionally adherent. The *superior* fasciculus, the thinnest, is inserted into the anterior surface and outer border of the arytenoid, above the preceding fibres; it lies on the outer side of the sacculus laryngis, immediately beneath its mucous lining.

The *Arytaenoideus* is a single muscle, filling up the posterior concave surface of the arytenoid cartilages. It arises from the posterior surface and outer border of one arytenoid cartilage, and is inserted into the corresponding parts of the opposite cartilage. It consists of three planes of fibres, two oblique and one transverse. The *oblique fibres*, the most superficial, form two fasciculi, which pass from the base of one cartilage to the apex of the opposite one. The *transverse fibres*, the deepest and most numerous, pass transversely across between the two cartilages; hence the arytenoideus was formerly considered as several muscles, under the names of *transversi* and *obliqui*. A few of the oblique fibres are occasionally continued round the outer margin of the cartilage, and blend with the Thyro-arytenoid or the Aryteno-epiglottidean muscle.

Fig. 334.—Muscles of Larynx, side View. Right Ala of Thyroid Cartilage removed.

The muscles of the epiglottis are the
Thyro-epiglottideus.
Arytaeno-epiglottideus superior.
Arytaeno-epiglottideus inferior.

The *Thyro-epiglottideus* is a delicate fasciculus, which arises from the inner surface of the thyroid cartilage, just external to the origin of the Thyro-arytenoid muscle, and spreading out upon the outer surface of the sacculus laryngis, some of its fibres are lost in the aryteno-epiglottidean fold, whilst others are continued forwards to the margin of the epiglottis (*depressor epiglottidis*).

The *Arytaeno-epiglottideus superior* consists of a few delicate muscular fasciculi, which arise from the apex of the arytenoid cartilage, and become lost in the fold of mucous membrane extending between the arytenoid cartilage and side of the epiglottis (*aryteno-epiglottidean folds*).

The *Arytaeno-epiglottideus inferior* (*Compressor sacculi laryngis*, Hilton) arises from the arytenoid cartilage, just above the attachment of the superior vocal cord, and passing forwards and upwards, spreads out upon the inner and upper part of the sacculus laryngis, and is inserted, by a broad attachment, into the margin of the epiglottis. This muscle is separated from the preceding by an indistinct areolar interval.

Actions. In considering the action of the muscles of the larynx, they may be conveniently divided into two groups, viz.: 1. Those which open and close the glottis. 2. Those which regulate the degree of tension of the vocal cords.

1. The muscles which open the glottis are the Crico-arytaenoidei postici, and those which close it are the Arytaenoideus, the Crico-arytaenoidei laterales, and the Thyro-arytaenoidei. 2. The muscles which regulate the tension of the vocal cords are the Crico-thyroidei, which tense and elongate them, and the Thyro-arytaenoidei, which relax and shorten them. The Thyro-epiglottideus is a depressor of the epiglottis, and the Arytaeno-epiglottidei constrict the superior aperture of the larynx, compress the sacculus laryngis, and empty it of its contents.

The *Crico-arytaenoidei postici* separate the chordae vocales, and consequently open the glottis, by rotating the base of the arytenoid cartilages outwards and backwards; so that their anterior angles, and the ligaments attached to them, become widely separated.

The *Crico-arytaenoidei laterales* close the glottis, by rotating the base of the arytenoid cartilages in the opposite direction to the last-named muscles, so as to approximate their anterior angles.

The *Arytaenoideus muscle* approximates the arytenoid cartilages, and thus closes the opening of the glottis.

The *Crico-thyroid muscles* effect the tension and elongation of the vocal cords.

The *Thyro-arytaenoidei muscles* approximate the anterior angles of the arytenoid cartilages, and thus constrict the glottis. According to Willis, they also draw the arytenoid cartilages, together with the part of the cricoid to which they are connected, forwards, and thus shorten and relax the vocal cords.

The *Thyro-epiglottidei* depress the epiglottis, and assist in compressing the sacculus laryngis. The Aryteno-epiglottideus superior, constricts the superior aperture of the larynx, when it is drawn upwards, during deglutition, and the opening closed by the epiglottis. The Arytaeno-epiglottideus inferior, together with some fibres of the Thyro-arytaenoidei, compress the sacculus laryngis.

The *Mucous Membrane of the Larynx* is continuous, above, with that lining the mouth and pharynx, and is prolonged through the trachea and bronchial tubes into the lungs. It lines both surfaces of the epiglottis, to which it is closely adherent, and forms the aryteno-epiglottidean folds, which encircle the superior aperture of the larynx. It lines the whole of the cavity of the larynx; forms, by its reduplication, the chief part of the superior, or false vocal cords; and from the ventricle is continued into the sacculus laryngis. It is then reflected over the true vocal cords, where it is thin, and very intimately adherent; covers the inner surface of the crico-thyroid membrane, and cricoid cartilage; and is ultimately continuous with the lining membrane of the trachea. It is covered with a columnar ciliated epithelium, below the superior vocal cord, but, above this point, the ciliae are found only in front, as high as the middle of the epiglottis. In the rest of its extent, the epithelium is of the squamous variety.

Glands. The mucous membrane of the larynx is furnished with numerous muciparous glands, the orifices of which are found in nearly every part: they are very numerous upon the epiglottis, being lodged in little pits in its substance: they are also found in large numbers along the posterior margin of the aryteno-epiglottidean fold, in front of the arytenoid cartilages, where they are termed the *arytenoid glands*. They exist also in large numbers upon the inner surface of the sacculus laryngis. None are found on the vocal cords.

The *Arteries* of the larynx are the laryngeal branches derived from the superior and inferior thyroid.

The *Veins* empty themselves into the superior, middle, and inferior thyroid veins.

The *Lymphatics* terminate in the deep cervical glands.

The *Nerves* are the superior laryngeal, and the inferior or recurrent laryngeal branches of the pneumogastric nerves, joined by filaments from the sympathetic. The superior laryngeal nerve supplies the mucous membrane of the larynx, and the Crico-thyroid muscle. The inferior laryngeal nerve supplies the remaining muscles. The Arytenoid muscle is supplied by both nerves.

The Trachea

The trachea, or air-tube, is a cartilaginous and membranous cylindrical tube, flattened posteriorly, extending from the lower part of the larynx, on a level with the fifth cervical vertebra, to opposite the third dorsal, where it divides into the two bronchi, one for each lung. The trachea measures about four inches and a half in length; its diameter, from side to side, is from three quarters of an inch to an inch; being always greater in the male than in the female.

335.—Front View of Cartilages of Larynx: the Trachea and Bronchi.

Relations. The anterior surface of the trachea is convex, and covered, *in the neck*, from above downwards, by the isthmus of the thyroid gland, the inferior thyroid veins, the arteria thyroidea ima, (when that vessel exists), the Sterno-hyoid and Sterno-thyroid muscles, the cervical fascia (in the interval between these muscles), and, more superficially, by the anastomosing branches between the anterior jugular veins: *in the thorax*, it is covered by the first piece of the sternum, the remains of the thymus gland, the left vena innominata, the arch of the aorta, the innominate and left carotid arteries, and the deep cardiac plexus. It lies upon the oesophagus, which is directed to the left, near the arch of the aorta; *laterally, in the neck*, it is in relation with the common carotid arteries, the lateral lobes of the thyroid gland, the inferior thyroid arteries, and recurrent laryngeal nerves; and, *in the thorax*, it lies in the interspace between the pleurae.

The *Right Bronchus*, wider, shorter, and more horizontal in direction than the left, is about an inch in length, and enters the right lung, opposite the fourth dorsal vertebra. The vena azygos arches over it, from behind; and the right pulmonary artery lies below, and then in front of it.

The *Left Bronchus* is smaller, more oblique, and longer than the right, being nearly two inches in length. It enters the root of the left lung, opposite the fifth dorsal vertebra, about an inch lower than the right bronchus. It crosses in front of the oesophagus, the thoracic duct, and the descending aorta; passes beneath the arch of the aorta, and has the left pulmonary artery lying at first above, and then in front of it.

The trachea is composed of imperfect cartilaginous rings, fibrous membrane, muscular fibres, longitudinal yellow elastic fibres, mucous membrane, and glands.

The *Cartilages* vary from sixteen to twenty in number: each forms an imperfect ring, which surrounds about two-thirds of the cylinder of the trachea, being imperfect behind, where the tube is completed by fibrous membrane. The cartilages are placed horizontally above each other, separated by narrow membranous intervals. They measure about two lines in depth, and half a line in thickness. Their outer surfaces are flattened, but, internally, they are convex, from being thicker in the middle than at the extremities. The cartilages are connected together, at their margins, by an elastic fibrous membrane, which covers both their surfaces; and in the space between their extremities, behind, forms a distinct layer. The peculiar cartilages are the first and the two last.

The *first cartilage* is broader than the rest, and sometimes divided at one end: it is connected by fibrous membrane with the lower border of the cricoid cartilage, with which, or with the succeeding cartilage, it is sometimes blended.

The *last cartilage* is thick and broad in the middle, in consequence of its lower border being prolonged downwards, and, at the same time, curved backwards, at the point of bifurcation of the trachea. It terminates on each side in an imperfect ring, which encloses the commencement of the bronchi. The cartilage above the last, is somewhat broader than the rest at its centre. Two or more of the cartilages often unite, partially or completely, and are sometimes bifurcated at their extremities. They are highly elastic, and seldom ossify, even in advanced life. In the right bronchus, the cartilages vary in number from six to eight; in the left, from nine to twelve. They are shorter and narrower than those of the trachea.

The *Muscular Fibres* are disposed in two layers, longitudinal and transverse. The longitudinal fibres are the most external, and arise by minute tendons from the termination of the tracheal cartilages, and from the fibrous membrane.

The *transverse fibres*, the most internal, form a thin layer, which extends transversely between the ends of the cartilages, at the posterior part of the trachea. The muscular fibres are of the involuntary class.

The *Elastic Fibres* are situated beneath the mucous membrane, enclosing the entire cylinder of the trachea; they are most abundant at its posterior part, where they are collected in longitudinal bundles.

The *Tracheal Glands* are found in great abundance at the posterior part of the trachea. They are small, flattened, ovoid bodies, placed between the fibrous and muscular coats, each furnished with an excretory duct, which opens on the surface of the mucous membrane. Some glands of smaller size are also found at the sides of the trachea, between the layers of fibrous tissue connecting the rings, and others immediately beneath the mucous coat. The secretion from these glands serves to lubricate the inner surface of the trachea.

The *Mucous Membrane* lining the tube is covered with columnar ciliated epithelium. It is continuous with that lining the larynx.

Vessels and Nerves. The trachea is supplied with blood by the inferior thyroid arteries.

The *Veins* terminate in the thyroid venous plexus.

The *Nerves* are derived from the pneumogastric and its recurrent branches, and from the sympathetic.

Surgical Anatomy. The air-passage may be opened in three different situations; through the crico-thyroid membrane (*laryngotomy*), through the cricoid cartilage and upper ring of the trachea (*laryngo-tracheotomy*), or through the trachea below the isthmus of the thyroid gland (*tracheotomy*). The student should, therefore, carefully consider the relative anatomy of the air-tube in each of these situations.

Beneath the integument of the laryngo-tracheal region, on either side of the median line, are the two anterior jugular veins. Their size and position vary; there is nearly always one, and frequently two: at the lower part of the neck they diverge, passing beneath the Sterno-mastoid muscles, and are frequently connected by a transverse communicating branch. These veins should, if possible, always be avoided in any operation on the larynx or trachea. If cut through, considerable haemorrhage is the result.

336.—Surgical Anatomy of Laryngo-Tracheal Region,
In the Infant.

Beneath the cervical fascia are the Sterno-hyoid and Sterno-thyroid muscles, the contiguous edges of the former being near the median line; and beneath these muscles the following parts are met with, from above downwards: the thyroid cartilage, the crico-thyroid membrane, the cricoid cartilage, the trachea, and the isthmus of the thyroid gland.

The crico-thyroid space is very superficial, and may be easily felt, beneath the skin, as a depressed spot, about an inch below the pomum Adami; it is crossed transversely by a small artery, the crico-thyroid, the division of which is seldom accompanied by any troublesome haemorrhage.

The isthmus of the thyroid gland usually crosses the second and third rings of the trachea; above it, is found a large transverse communicating branch between the superior thyroid veins, and the isthmus is covered by a venous plexus, formed between the thyroid veins of opposite sides. On the sides of the thyroid gland, and below it, the veins converge to a single median vessel, or to two trunks which descend along the median line of the front of the trachea, to open into the innominate vein by valved orifices. In the infant, the thymus gland ascends a variable distance along the front of the trachea; and the innominate artery crosses this tube obliquely at the root of the neck, from left to right. The arteria thyroidea ima, when that vessel exists, passes from below upwards along the front of the trachea. The upper part of the trachea lies comparatively superficial; but the lower part passes obliquely downwards and backwards, so as to be deeply placed between the converging Sterno-mastoid muscles. In the child, the trachea is smaller, more deeply placed, and more moveable than in the adult. In fat, or short-necked people, or in those in whom the muscles of the neck are prominently developed, the trachea is more deeply placed than in the opposite conditions.

From these observations, it must be evident that laryngotomy is anatomically the most simple operation, can most readily be performed, and should always be preferred when particular circumstances do not render the operation of tracheotomy absolutely necessary. The operation is performed thus: The head being thrown back and steadied by an assistant, the finger is passed over the front of the neck, and the crico-thyroid depression felt for. A vertical incision is then made through the skin, in the middle line over this spot, and the crico-thyroid membrane is divided to a sufficient extent to allow of the introduction of a large curved tube. The crico-thyroid artery is the only vessel of importance crossing this space. If it should be of large size, its division might produce troublesome haemorrhage.

Laryngo-tracheotomy, anatomically considered, is more dangerous than tracheotomy, on account of the small interspace between the cricoid cartilage and the isthmus of the thyroid gland: the communicating branches between the superior thyroid veins, which cover this spot, can hardly fail to be divided; and the greatest care will not, in some cases, prevent the division of part of the thyroid isthmus. If either of these structures is divided, the haemorrhage will be considerable.

Tracheotomy below the isthmus of the thyroid gland is performed thus: The head being thrown back and steadied by an assistant, an incision, an inch and a half or two inches in length, is made through the skin, in the median line of the neck, from a little below the cricoid cartilage, to the top of the sternum. The anterior jugular veins should be avoided, by keeping exactly in the median line; the deep fascia should then be divided, and the contiguous borders of the Sterno-hyoid muscles separated from each other. A quantity of loose areolar tissue, containing the inferior thyroid veins, must then be separated from the front of the trachea, with the handle of the scalpel; and when the trachea is well exposed, it should be opened by inserting the knife into it, dividing two or three of its rings from below upwards. It is a matter of the greatest importance to restrain, if possible, all haemorrhage before the tube is opened; otherwise, blood may pass into the trachea, and suffocate the patient.

The Pleurae

Each lung is invested, upon its external surface, by an exceedingly delicate serous membrane, the pleura, which encloses the organ as far as its root, and is then reflected upon the inner surface of the thorax. The portion of the serous membrane investing the surface of the lung is called the *pleura pulmonalis* (visceral layer of pleura); whilst that which lines the inner surface of the chest is called the *pleura costalis* (parietal layer of pleura). The interspace or cavity between these two layers is called the *cavity of the pleura*. Each pleura is therefore a shut sac, one occupying the right, the other the left half of the thorax; and they are perfectly separate, not communicating with each other. The two pleurae do not meet in the middle line of the chest, excepting at one point in front; an interspace being left between them, which contains all the viscera of the thorax, excepting the lungs: this is the *mediastinum*.

337.—A Transverse Section of the Thorax, showing the relative Position of the Viscera, and the Reflections of the Pleuræ.

Reflections of the Pleura (fig. 337). Commencing at the sternum, the pleura passes outwards, covers the costal cartilages, the inner surface of the ribs and Intercostal muscles, and at the back of the thorax passes over the thoracic ganglia and their branches, and is reflected upon the sides of the bodies of the vertebrae, where it is separated by a narrow interspace from the opposite pleura, the *posterior mediastinum*.

From the vertebral column, the pleura passes to the side of the pericardium, which it covers to a slight extent; it then covers the back part of the root of the lung, from the lower border of which a triangular fold descends vertically by the side of the posterior mediastinum to the Diaphragm. This fold is the broad ligament of the lung, the *ligamentum latum pulmonis*, and serves to retain the lower part of the organ in position. From the *root*, the pleura may be traced over the convex surface of the lung, the summit and base, and also over the sides of the fissures between the lobes. It covers its anterior surface, and front part of its root, and is reflected upon the side of the pericardium to the inner surface of the sternum.

Below, it covers the upper surface of the Diaphragm. *Above*, its apex projects, in the form of a *cul-de-sac*, through the superior opening of the thorax into the neck, extending about an inch above the margin of the first rib, and receives the summit of the corresponding lung: this sac is strengthened, according to Dr. Sibson, by a dome-like expansion of fascia, derived from the lower part of the Scaleni muscles.

A little above the middle of the sternum, the contiguous surfaces of the two pleurae are in contact for a slight extent; but above and below this point, the interval left between them by their non-approximation forms the anterior mediastinum.

The inner surface of the pleura is smooth, polished, and moistened by a serous fluid; its outer surface is intimately adherent to the surface of the lung, and to the pulmonary vessels as they emerge from the pericardium; it is also adherent to the upper surface of the Diaphragm; throughout the rest of its extent it is somewhat thicker, and may be separated from the adjacent parts with extreme facility.

The right pleural sac is shorter, wider, and reaches higher in the neck than the left.

The *arteries* of the pleura are derived from the intercostal, the internal mammary, the phrenic, inferior thyroid, thymic, pericardiac, and bronchial.

The *veins* correspond to the arteries.

The *lymphatics* are very numerous.

The *nerves* are derived from the phrenic and sympathetic (Luschka). Kölliker states that some accompany the ramifications of the bronchial arteries in the pleura pulmonalis.

Mediastinum

The mediastinum is the space left in the median line of the chest by the non-approximation of the two pleurae. It extends from the sternum in front to the spine behind, and contains all the viscera in the thorax, excepting the lungs. The mediastinum may be subdivided, for convenience of description, into the anterior, middle, and posterior.

The *Anterior Mediastinum* is bounded in front by the sternum, on each side by the pleura, and behind by the pericardium. Owing to the oblique position of the heart towards the left side, this space is not parallel with the sternum, but directed obliquely from above downwards, and to the left of the median line; it is broad below, narrow above, very narrow opposite the second piece of the sternum, the contiguous surfaces of the two pleurae being occasionally united over a small space. The anterior mediastinum contains the origins of the Sterno-hyoid and Sterno-thyroid muscles, the Triangularis sterni, the internal mammary vessels of the left side, the remains of the thymus gland, and a quantity of loose areolar tissue, containing some lymphatic vessels ascending from the convex surface of the liver.

The *Middle Mediastinum* is the broadest part of the interpleural space. It contains the heart enclosed in the pericardium, the ascending aorta, the superior vena cava, the bifurcation of the trachea, the pulmonary arteries and veins, and the phrenic nerves.

The *Posterior Mediastinum* is an irregular triangular space, running parallel with the vertebral column; it is bounded in front by the pericardium and roots of the lungs, behind by the vertebral column, and on either side by the pleura. It contains the descending aorta, the greater and lesser azygos veins and superior intercostal vein, the pneumogastric and splanchnic nerves, the oesophagus, thoracic duct, and some lymphatic glands.

The Lungs

The lungs are the essential organs of respiration; they are two in number, placed one in each of the lateral cavities of the chest, separated from each other by the heart and other contents of the mediastinum. Each lung is conical in shape, and presents for examination an apex, a base, two borders, and two surfaces.

The *apex* forms a tapering cone, which extends into the root of the neck, about an inch to an inch and a half above the level of the first rib.

The *base* is broad, concave, and rests upon the convex surface of the Diaphragm;

338.—Front View of the Heart and Lungs.

its circumference is thin, and fits into the space between the lower ribs and the costal attachment of the Diaphragm, extending lower down externally and behind than in front.

The *external* or *thoracic surface* is smooth, convex, of considerable extent, and corresponds to the form of the cavity of the chest, being deeper behind than in front.

The *inner surface* is concave. It presents, in front, a depression corresponding to the convex surface of the pericardium, and behind, a deep fissure (the hilum pulmonis), which gives attachment to the root of the lung.

The *posterior border* is rounded and broad, and is received in the deep concavity on either side of the spinal column. It is much longer than the anterior border, and projects below between the ribs and Diaphragm.

The *anterior border* is thin and sharp, and overlaps the front of the pericardium.

The anterior border of the right lung corresponds to the median line of the sternum, and is in contact with its fellow, the pleurae being interposed, as low as the fourth costal cartilage; below this, the contiguous borders are separated by a V-shaped interval, formed at the expense of the left lung, and corresponding to which the pericardium is exposed.

Each lung is divided into two lobes, an upper and lower, by a long and deep fissure, which extends from the upper part of the posterior border of the organ, about three inches from its apex, downwards and forwards to the lower part of its anterior border. This fissure penetrates nearly to the root. In the right lung the upper lobe is partially divided by a second and shorter fissure, which extends forwards and upwards from the middle of the preceding to the anterior margin of the organ, marking off a small triangular portion, the middle lobe. The left lung presents a deep V-shaped notch at the lower part of its anterior border, into which the apex of the heart is received.

The *right lung* is the largest; it is broader than the left, owing to the inclination of the heart to the left side; it is also shorter by an inch, in consequence of the Diaphragm rising higher on the right side to accommodate the liver. The right lung has also three lobes.

The *left lung* is smaller, narrower, and longer than the right, and has only two lobes.

A little above the middle of the inner surface of each lung, and nearer its posterior than its anterior border, is its root, by which the lung is connected to the heart and the trachea. It is formed by the bronchial tube, the pulmonary artery, the pulmonary veins, the bronchial arteries and veins, the pulmonary plexus of nerves, lymphatics, bronchial glands, and areolar tissue, all of which are enclosed by a reflection of the pleura. The root of the right lung lies behind the superior cava and upper part of the right auricle, and below the vena azygos. That of the left lung passes beneath the arch of the aorta, and in front of the descending aorta; the phrenic nerve and the anterior pulmonary plexus lie in front of each, and the pneumogastric and posterior pulmonary plexus behind each.

The structures composing the root of each lung are arranged in a similar manner from before backwards on both sides, viz.: the pulmonary veins most anterior; the pulmonary artery in the middle; and the bronchus, together with the bronchial vessels, behind. From above downwards, on the two sides, their arrangement differs, thus:

On the right side, their position is, bronchus, pulmonary artery, pulmonary veins; but on the left side, their position is, pulmonary artery, bronchus, pulmonary veins; which is accounted for by the bronchus being placed on a lower level on the left than on the right side.

The *weight* of both lungs together is about forty-two ounces, the right lung being two ounces heavier than the left, but much variation is met with according to the amount of blood or serous fluid they may contain. The lungs are heavier in the male than in the female, their proportion to the body being, in the former, as 1 to 37, in the latter, as 1 to 43. The specific gravity of the lung tissue varies from 345 to 746, water being 1000.

The *colour* of the lungs at birth is of a pinkish white; in adult life, mottled in patches, of a dark slate colour; and, as age advances, this mottling assumes a dark black colour. The colouring matter consists of granules of a carbonaceous substance, deposited in the areolar tissue near the surface of the organ. It increases in quantity as age advances, and is more abundant in males than in females. The posterior border of the lung is usually darker than the anterior. The surface of the lung is smooth, shining, and marked out into numerous polyhedral spaces, which represent the lobules of the organ, and the area of each of these spaces is crossed by numerous lighter lines.

The *substance* of the lung is of a light, porous, spongy texture; it floats in water, and crepitates when handled, owing to the presence of air in the tissue; it is also highly elastic; hence the collapsed state of these organs when they are removed from the closed cavity of the thorax.

Structure. The lungs are composed of an external serous coat, a subserous areolar tissue, containing a large proportion of elastic fibres, and the pulmonary substance or parenchyma.

The *serous coat* is derived from the pleura; it is thin, transparent, and invests the entire organ as far as the root.

The *subserous areolar tissue* contains a large proportion of elastic fibres; it invests the entire surface of the lung, and extends inwards between the lobules.

The *parenchyma* is composed of lobules, which, although closely connected together by an interlobular areolar tissue, are quite distinct from one another, being easily separable in the foetus. The lobules vary in size; those on the surface are large, of a pyramidal form, the base turned towards the surface; those in the interior are smaller, and of various forms. Each lobule is composed of one of the ramifications of the bronchial tube and its terminal air-cells, of the ramifications of the pulmonary and bronchial vessels, lymphatics, and nerves: all of these structures being connected together by areolar fibrous tissue.

The *bronchus* upon entering the substance of the lung, divides and subdivides dichotomously throughout the entire organ. Sometimes three branches arise together, and occasionally small lateral branches are given off from the sides of a main trunk. Each of the smaller subdivisions of the bronchi enters a pulmonary lobule (lobular bronchial tube), and again subdividing, ultimately terminates in the intercellular passages and air-cells of which the lobule is composed. Within the lungs the bronchial tubes are circular, not flattened, and their constituent elements present the following peculiarities of structure.

The *Cartilages* are not imperfect rings, but consist of laminated plates, of varied form and size, scattered irregularly along the sides of the tube, being most distinct at the points of division of the bronchi. They may be traced into tubes the diameter of which is only one-fourth of a line. Beyond this point, the tubes are wholly membranous. The fibrous coat, and longitudinal elastic fibres, are continued into the smallest ramifications of the bronchi. The muscular coat is disposed in the form of a continuous layer of annular fibres, which may be traced upon the smallest bronchial tubes: they consist of the unstriped variety of muscular fibre. The mucous membrane lines the bronchi and its ramifications throughout, and is covered with columnar ciliated epithelium.

According to the observations of Mr. Rainey,* the lobular bronchial tubes, on entering the substance of the lobules, divide and subdivide from four to nine times, according to the size of the lobule, continuing to diminish in size until they attain a diameter of $\frac{1}{50}$th to $\frac{1}{30}$th of an inch. They then become changed in structure, losing their cylindrical form, and are continued onwards as irregular passages (intercellular passages), through the substance of the lobule, their sides and extremities being closely covered by numerous saccular dilatations, the air-cells. This arrangement resembles most closely the naked eye appearances observed in the reticulated structure of the lung of the tortoise, and other reptilia.

The *air-cells* are small, polyhedral, alveolar recesses, separated from each other by thin septa, and communicating freely with the intercellular passages. They are well seen on the surface of the lung, and vary from $\frac{1}{200}$th to $\frac{1}{70}$th of an inch in diameter; being largest on the surface, at the thin borders, and at the apex; and smallest in the interior.

At the termination of the bronchial tubes, in the intercellular passages, their constituent elements become changed: their walls are formed by an interlacing of the longitudinal elastic bundles with fibrous tissue; the muscular fibres disappear, and the mucous membrane becomes thin and delicate, and lined with a layer of squamous epithelium. This membrane lines the air-cells, and forms by its reduplications the septa intervening between them.

The *Pulmonary Artery*, conveys the dark, impure venous blood to the lungs: it divides into branches which accompany the bronchial tubes and terminate in a dense capillary network upon the walls of the intercellular passages and air-cells. From this network, the radicles of the pulmonary veins arise, coalescing into large branches, they accompany the arteries, and return the blood, purified by its passage through the capillaries, to the left auricle of the heart. In the lung, the branches of the pulmonary artery are usually above and in front of a bronchial tube, the vein below.

The *Pulmonary Capillaries* form plexuses which lie immediately beneath the mucous membrane, on the walls and septa of the air-cells, and upon the walls of the intercellular passages. In the septa between the cells, the capillary network forms a single layer. The capillaries are very minute, the meshes being only slightly wider than the vessels: their walls are also exceedingly thin.

The *Bronchial Arteries* supply blood for the nutrition of the lung: they are derived

---

* Medico-Chirugical Transactions, vol. xxviii. 1845.

from the thoracic aorta, and, accompany the bronchial tubes, are distributed to the bronchial glands, and upon the walls of the larger bronchial tubes and pulmonary vessels, and terminate in the deep bronchial veins. Others are distributed in the interlobular areolar tissue, and terminate partly in the deep, partly in the superficial, bronchial veins. Lastly, some ramify upon the walls of the smallest bronchial tubes, and terminate in the pulmonary veins.

The *Superficial and Deep Bronchial Veins* unite at the root of the lung, and terminate on the right side in the vena azygos; on the left side, in the superior intercostal vein.

The *Lymphatics* consist of a superficial and deep set: they terminate at the root of the lung, in the bronchial glands.

Nerves. The lungs are supplied from the anterior and posterior pulmonary plexuses, formed chiefly by branches from the sympathetic and pneumogastric. The filaments from these plexuses accompany the bronchial tubes upon which they are lost. Small ganglia have been found by Remak upon the smaller branches of these nerves.

Thyroid Gland

The thyroid gland bears much resemblance in structure to other glandular organs, and is usually classified together with the thymus, supra-renal glands, and spleen, under the head of *ductless glands*, from its possessing no excretory duct. Its function is unknown, but, from its situation in connection with the trachea and larynx, is usually described with these, although taking no part in the function of respiration. It is situated at the upper part of the trachea, and consists of two lateral lobes, placed one on each side of this tube, connected together by a narrow transverse portion, the isthmus.

Its *anterior surface* is convex, and covered by the Sterno-hyoid, Sterno-thyroid, and Omo-hyoid muscles.

Its *lateral surfaces*, also convex, lie in contact with the sheath of the common carotid artery.

Its *posterior surface* is concave, and embraces the trachea and larynx. The posterior borders of the gland extend as far back as the lower part of the pharynx.

This gland is of a brownish red colour. Its weight varies from one to two ounces. It is larger in females than in males, and becomes slightly increased in size during menstruation. It occasionally becomes enormously hypertrophied, constituting the disease called bronchocele, or goitre. Each lobe is somewhat conical in shape, about two inches in length, and three-quarters of an inch in breadth, the right lobe being rather the larger of the two.

The *isthmus* connects the lower third of the two lateral lobes: it measures about half an inch in breadth, and the same in depth, and usually covers the second and third rings of the trachea. Its situation presents, however, many variations, a point of some importance in the operation of tracheotomy. Sometimes the isthmus is altogether wanting.

A third lobe, of conical shape, called the *pyramid*, occasionally arises from the left side of the upper part of the isthmus, or from the left lobe, and ascends as high as the hyoid bone. It is occasionally quite detached, or divided into two parts, or altogether wanting.

A few muscular bands are occasionally found attached, above, to the body of the hyoid bone, and connected, below, with the isthmus of the gland, or its pyramidal process, and named by Soemmering, the *Levator glandulae thyroidae*.

STRUCTURE. The thyroid consists of numerous minute closed vesicles, composed of a homogeneous membrane, enclosed in a dense capillary plexus, and connected together into imperfect lobules by areolar tissue. These vesicles are spherical or oblong, perfectly distinct, and contain a yellowish fluid, in which are found floating numerous "dotted corpuscles" and cells. The fluid coagulates by heat or alcohol, but preserves its transparency. In the foetus, and in young subjects, the corpuscles lie in a single layer, in contact with the inner surface of these cavities, and become detached during the process of growth.

The *Arteries* supplying the thyroid, are the superior and inferior thyroid, and sometimes an additional branch from the arteria innominata, which ascends from this vessel upon the front of the trachea. The arteries are remarkable for their large size and frequent anastomoses.

The *Veins* form a plexus on the surface of the gland, and on the front of the trachea, from which arise the superior, middle, and inferior thyroid veins; the two former terminating in the internal jugular, the latter in the vena innominata.

The *Lymphatics* are numerous, of large size, and terminate in the thoracic and right lymphatic ducts.

The *Nerves* are derived from the pneumogastric, and from the middle and inferior cervical ganglia of the sympathetic.

Chemical Composition. The thyroid gland consists of albumen, traces of gelatine, stearine, oleine, extractive matter, alkaline, and earthy salts, and water. The salts are chloride of sodium, alkaline sulphate, phosphate of potash, lime, magnesia, and a trace of oxide of iron.

THYMUS GLAND

The thymus gland presents much resemblance in structure to other glandular organs, and is classified under the head of the *ductless glands*, from its possessing no excretory duct.

The thymus gland is a temporary organ, attaining its full size at the end of the second year, when it ceases to grow, and gradually dwindles, until, at puberty, it has almost disappeared. If examined when its growth is most active, it will be found to consist of two lateral lobes, placed in close contact along the middle line, situated partly in the anterior mediastinum, partly in the neck, and extending from the fourth costal cartilage upwards, as high as the lower border of the thyroid gland. It is covered by the sternum, and by the origins of the Sterno-hyoid and Sterno-thyroid muscles. In the mediastinum, it rests upon the pericardium, being separated from the arch of the aorta and great vessels, by the thoracic fascia. In the neck, it lies on the front and sides of the trachea, behind the Sterno-hyoid and Sterno-thyroid muscles. The two lobes generally differ in size: they are occasionally united, so as to form a single mass; and sometimes separated by an intermediate lobe. The thymus is of a pinkish-grey colour, soft, and lobulated on its surfaces. It is about two inches in length, one and half in width, below, and about three or four lines in thickness. At birth, it weighs about half an ounce.

Structure. Each lateral lobe is composed of numerous lobules, held together by delicate areolar tissue; the entire gland being enclosed in an investing capsule of a similar, but denser structure. The primary lobules vary in size from a pin's head to a small pea. Each lobule contains in its interior a small cavity, which is surrounded with smaller or secondary lobules, which are also hollow within. The cavities of the secondary and primary lobules communicate; those of the latter opening into the great central cavity, or *reservoir of the thymus,* which extends through the entire length of each lateral half of the gland. The central cavity is lined by a vascular membrane, which is prolonged into all the subordinate cavities, and contains a milk-white fluid resembling chyle.

If the investing capsule and vessels, as well as the areolar tissue connecting the lobules, are removed from the surface of either lateral lobe, it will be seen that the central cavity is folded upon itself, and admits of being drawn out into a lengthened tubular cord, around which the primary lobules are attached in a spiral manner, like knots upon a rope. Such is the condition of the organ at an early period of its development; for Mr. Simon has shown, that the primitive form of the thymus is a linear tube, from which, as its development proceeds, lateral diverticula lead outwards, the tube ultimately becoming obscure, from its surface being covered with numerous lobules.

According to Oesterlen and Mr. Simon, the cavities in the secondary lobules are surrounded by rounded saccular dilatations or vesicles, which open into it. These vesicles are formed of a homogeneous membrane, enclosed in a dense capillary plexus.

The whitish fluid contained in the vesicles and central cavity of the thymus, contains numerous dotted corpuscles, similar to those found in the chyle. The corpuscles are flattened circular discs, measuring about $1/1000$ of an inch in diameter.

Chemical Composition. The solid animal constituents of the thymus are albumen and fibrine in large quantities, gelatine, and other animal matters. The salts are alkaline and earthy phosphates, with chloride of potassium. It contains about 80 percent of water.

The *arteries* supplying the thymus are derived from the internal mammary, and from the superior and inferior thyroid.

The *veins* terminate in the left vena innominata, and in the thyroid veins.

The *lymphatics* are of large size, arise in the substance of the gland, and are said to terminate in the internal jugular vein. Sir A. Cooper considered that these vessels carried into the blood the secretion formed in the substance of the thymus.

The *nerves* are exceedingly minute; they are derived from the pneumogastric and sympathetic. Branches from the descendens noni and phrenic reach the investing capsule, but do not penetrate into the substance of the gland.

THE URINARY ORGANS

The Kidneys

The Kidneys are two glandular organs, intended for the secretion of the urine. They are situated at the back part of the abdominal cavity, behind the peritoneum, one in each lumbar region, extending from the eleventh rib to near the crest of the ilium; the right one being lower than the left, from its vicinity to the liver. They are usually surrounded by a considerable quantity of fat, and are retained in their position by the vessels which pass to and from them.

Relations. The *anterior surface* of the kidney is convex, partially covered by the peritoneum above, and is in relation, on the right side, with the liver, descending portion of the duodenum, and ascending colon; and on the left side with the great end of the stomach, the lower end of the spleen, and descending colon.

The *posterior surface* is flattened, and rests upon the corresponding crus of the Diaphragm, in front of the eleventh and twelfth ribs, on the anterior lamella of the aponeurosis of the transversalis which separates it from the Quadratus lumborum muscle, and on the Psoas magnus.

The *superior extremity*, directed inwards, is thick and rounded, and embraced by the supra-renal capsule. It corresponds, on the left side, to the upper border of the eleventh rib, and on the right side to the lower border.

The *inferior extremity*, small and flattened, extends nearly as low as the crest of the ilium.

The *external border* is convex, and directed outwards towards the parietes of the abdomen.

The *internal border* is concave, directed downwards and forwards, and presents a deep notch, the *hilus of the kidney*, more marked behind than in front. At the hilus, the vessels, excretory duct, and nerves pass into or from the organ; the branches of the renal vein lying in front, the artery and its branches next, the excretory duct or ureter being behind and below. The hilus leads into a hollow space, the *sinus*, which occupies the interior of the gland.

Each kidney is about four inches in length, two inches in breadth, and about one inch in thickness; the left one being somewhat longer and thinner than the right. The weight of the kidney in the adult male varies from $4\frac{1}{2}$ oz. to 6 oz.; in the female, from 4 oz. to $5\frac{1}{2}$ oz.; the difference between the two being about half an ounce. The left is nearly always heavier than the right, by about two drachms. Their weight in proportion to the body, is about 1 to 240. The renal substance is dense, firm, extremely fragile, and of a deep red colour.

The kidney is invested by a proper fibrous capsule, formed of dense fibro-areolar tissue. It is thin, smooth, and easily removed from its surface, to which it is connected by fine fibrous processes and vessels; and at the hilus is continued inwards, lining the sides of the sinus, and at the bottom of that cavity forms sheaths around the blood-vessels, and the subdivisions of the excretory duct.

On making a vertical section through the organ, from its convex to its concave border, it appears to consist of two different substances, named, from their position, external or cortical, and internal or medullary, substance.

The *cortical substance* forms about three-fourths of the substance of the gland. It occupies the surface of the kidney, forming a layer about two lines in thickness, and sends numerous prolongations inwards, towards the sinus between the pyramids of the medullary substance.

339.—Vertical Section of Kidney.

The *medullary substance* consists of pale, reddish-coloured, conical masses, the *pyramids of Malpighi*; they vary in number from eight to eighteen; their bases are directed towards the cortical substance; whilst their apices, which are free, converge towards the sinus, and are named the *papillae* (mammillae) of the kidney. The kidney is thus seen to consist of a number of conical-shaped masses, surrounded by an investment of the cortical substance: these represent the separate lobules of which the human kidney in the foetus consists, a condition observed permanently in the kidneys of many of the lower animals. As the human kidney becomes developed, the adjacent lobules coalesce, so as to form a single gland, the surface of which, even in the adult, occasionally presents faint traces of a lobular subdivision.

The medullary portion is denser in structure than the cortical, and presents a striated appearance, from being composed of a number of minute diverging tubes (*tubuli uriniferi*). The tubuli uriniferi commence at the apices of the cones by small orifices, which vary from $1/300$ to $1/200$ of an inch; as they pass up in the medullary substance, towards the periphery, they pursue a diverging course, dividing and subdividing at very acute angles, until they reach the cortical substance, when they become convoluted, anastomose freely with each other, and retain the same diameter. The number of orifices on the entire surface of a single papilla is, according to Huschke, about a thousand; from four to five hundred large, and as many smaller ones. The tubuli uriniferi are formed of a transparent homogeneous basement membrane, lined by spheroidal epithelium, which occupies about two-thirds of the diameter of the tube. The tubes are separated from one another, in the medullary cones, by capillary vessels, which form oblong meshes parallel with the tubuli, and by an intermediate parenchymatous substance composed of cells.

340.—Minute Structure of Kidney.

The *cortical substance* is soft, reddish, granular, easily lacerated, and contains numerous small, red, globular bodies disseminated through it in every part, excepting towards the free surface. These are the Malpighian bodies. The cortical portion, like the tubular, is composed of the tubuli uriniferi, blood-vessels, lymphatics, and nerves, together with an intermediate parenchymatous substance.

As soon as the tubuli uriniferi enter the cortical substance (fig. 340), they become convoluted, and anastomose freely with each other; they are sometimes called the *tubes of Ferrein*. At the bases of the pyramids, the straight tubes are described as being collected into small conical fasciculi, the tortuous tubuli corresponding to which are prolonged upwards into the cortical portion of the kidney as far as the surface, forming a number of small conical masses, which are named the *pyramids of Ferrein*, several of which correspond to each medullary cone and its corresponding portion of cortical substance. According to Mr. Bowman, the tubuli uriniferi commence in the cortical substance as small, dilated, membranous capsules, the capsules of the Malpighian bodies; they also form loops, either by the junction of adjacent tubes, or, according to Toynbee, by the union of two branches proceeding from the same tube; they have also been seen to arise by free closed extremities.

The *Malpighian bodies* are found only in the cortical substance of the kidney. They are small rounded bodies, of a deep red colour, and of the average diameter of the $1/120$ of an inch. Each body is composed of a vascular tuft enclosed in a thin membranous capsule, the dilated commencement of a uriniferous tubule. The vascular tuft consists of the ramifications of a minute artery, the *afferent vessel*, which, after piercing the capsule, divides, in a radiated manner, into several branches, which ultimately terminate in a finer set of capillary vessels. From these a small vein, the *efferent vessel*, proceeds; this pierces the capsule near the artery, and forms a close venous plexus, with the efferent vessels from other Malpighian bodies, around the adjacent tubuli.

The capsular dilatation of the Malpighian body is not always placed at the commencement of the tube; it may occupy one side (Gerlach): hence their subdivision into lateral or terminal. The membrane composing it is thicker than that of the tubule; the epithelium lining its inner surface is thin, and in the frog provided with ciliae at the neck of the dilated portion; but in the human subject, ciliae have not been detected. According to Mr. Bowman, the surface of the vascular tuft lies free and uncovered in the interior of its capsule; but according to Gerlach, it is covered with a thick layer of nucleated cells, similar to those lining the inner surface of the capsule.

Ducts. The ureter, as it approaches the hilus, becomes dilated into a funnel-shaped membranous sac, the *pelvis*. It then enters the sinus, and subdivides usually into three prolongations, the *infundibula*; one placed at each extremity, and one in the middle of the organ: these subdivide into smaller tubes, the *calyces*, which surround, like a cup-like pouch or calyx, the bases of one or more of the papillae. The ureter, the pelvis, and the calyces consist of three coats, fibrous, muscular, and mucous.

The *external* or *fibro-elastic coat* is continuous, around the bases of the papillae, with the tunica propria investing the surface of the organ.

The *muscular coat* consists of a double layer of fibres placed between the fibrous and mucous coats.

The *internal* or *mucous coat* invests the papillae of the kidney, and is continued into the orifices upon their surfaces.

The *Renal artery* is large in proportion to the size of the organ which it supplies. Each vessel divides into four or five branches, which enter the hilus, and are invested by sheaths derived from the fibrous capsule; they penetrate the substance of the organ between the papillae, and enter the cortical substance in the intervals between the medullary cones; dividing and subdividing in their course towards the bases of the pyramids, where they form arches by their anastomoses: from these arches numerous vessels are distributed to the cortical substance, some of which enter the Malpighian corpuscles; whilst others form a capillary network around the uriniferous tubes.

The *Veins* of the kidney commence upon the surface of the organ, where they have a stellate arrangement; they pass inwards, and open into larger veins, which unite into arches around the bases of the medullary cones. After receiving the venous plexus from the tubular portion, they accompany the branches of the arteries to the sinus of the kidney, where they finally unite to form a single vein, which terminates in the inferior vena cava.

The *lymphatics* of the kidney consist of a superficial and deep set; they accompany the blood-vessels, and terminate in the lumbar glands.

The *nerves* are derived from the renal plexus, which is formed by filaments from the solar plexus and lesser splanchnic nerve; they accompany the branches of the arteries. From the renal plexus, some filaments pass to the spermatic plexus and ureter.

The Ureters

The URETER is the excretory duct of the kidney. It is a cylindrical membranous tube, from sixteen to eighteen inches in length, and of the diameter of a goose-quill. It is placed at the back part of the abdomen, behind the peritoneum; and extends obliquely downwards and inwards, from the lower part of the pelvis of the kidney, enters the cavity of the pelvis, and then passes downwards, forwards, and inwards, to the base of the bladder, into which it opens by a constricted orifice, after passing obliquely, for nearly an inch, between its muscular and mucous coats.

Relations. In its course from above downwards, it rests upon the Psoas muscle, being covered by the peritoneum, and crossed in front very obliquely by the spermatic vessels; the *right* ureter lying close to the outer side of the inferior vena cava. Opposite the sacrum, it crosses the common iliac artery, or the external iliac vessels, behind the ileum on the right side, and the sigmoid flexure of the colon on the left. In the pelvis, it enters the posterior false ligament of the bladder, runs below the obliterated hypogastric artery, the vas deferens, in the male, passing between it and the bladder. In the female, the ureter passes along the sides of the cervix uteri and upper part of the vagina. At the base of the bladder, it is situated about two inches from its fellow; lying, in the male, about an inch and a half behind the base of the prostate, at the posterior angle of the trigone vesicae.

Structure. The ureter is composed of two coats, an external or muscular, and an internal mucous coat.

The *muscular coat* consists of two layers of longitudinal fibres, and an intermediate transverse layer.

The *mucous coat* is smooth, and presents a few longitudinal folds, which become effaced by distension. It is continuous with the mucous membrane of the bladder below; whilst, above, it is prolonged over the papillae into the tubuli uriniferi. The epithelial cells lining it are spheroidal.

The *arteries* supplying the ureter are branches of the renal, spermatic, internal iliac, and inferior vesical.

The *nerves* are derived from the inferior mesenteric, spermatic, and hypo-gastric plexuses.

Supra-Renal Capsules

The supra-renal capsules are usually classified, together with the spleen, thymus, and thyroid, under the head of "ductless glands," as they have no excretory duct. They are two small flattened glandular bodies, of a yellowish colour, situated at the back part of the abdomen, behind the peritoneum, immediately in front of the upper end of either kidney; hence their name. The right one is somewhat triangular in shape, bearing a resemblance to a cocked hat; the left is more semilunar, and usually larger and higher than the right. They vary in size in different individuals, being sometimes so small as to be scarcely detected; at other times large. They measure from an inch and a quarter to nearly two inches in length, about an inch and a quarter in breadth, and from two to three lines in thickness. In weight, they vary from one to two drachms.

Relations. The *anterior surface* of the right supra-renal capsule is in relation with the under surface of the liver; that of the left with the pancreas and spleen. The *posterior surface* rests upon the crus of the Diaphragm, opposite the tenth dorsal vertebra. Their *upper thin convex border* is directed upwards and inwards. Their *lower thick concave border* rests upon the upper end of the kidneys, to which they are connected by the common investing areolar tissue. Their *inner borders* are in relation with the great splanchnic nerves and semi-lunar ganglia, and lie in contact on the right side with the inferior vena cava, and on the left side with the aorta. The surface of the supra-renal gland is surrounded by areolar tissue containing much fat, and closely invested by a thin fibrous coat, which is difficult to remove, on account of the numerous fibrous and vascular processes which enter the organ through the furrows on its anterior surface and base.

Structure. On making a perpendicular section, the gland is seen to consist of two substances, external or cortical, and internal medullary.

The *cortical substance* forms the chief part of the organ; it is of a deep yellow colour, and consists of narrow columnar masses placed perpendicularly to the surface.

The *medullary substance* is soft, pulpy, and of a dark brown or black colour; hence the name, *atrabiliary capsules*, given to these organs. In its centre is often seen a space formed by the breaking down of its component parts.

According to the researches of Oesterlen and Mr. Simon, the narrow columnar masses of which the cortical substance is composed measure about $1/700$th of an inch in diameter, and consist of small closed parallel tubes of limitary membrane containing dotted nuclei, together with much granular matter, oil globules, and nucleated cells. According to Ecker, the apparent tubular canals consist of rows of closed vesicles placed endwise, so as to resemble tubes; whilst Kölliker states, that these vesicles are merely loculi or spaces in the stroma of the organ, having no limitary membrane, and, from being situated endwise, present the appearance of linear tubes. Nucleated cells exist in large numbers in the supra-renal glands of ruminants, but more sparingly in man and other animals, but the granular matter appears to form their chief constituent; their size varies, and they present the singular peculiarity of undergoing no change when acted upon by most chemical reagents. The columnar masses are surrounded by a close capillary network, which runs parallel with them.

The medullary substance consists of nuclei and granular matter, uniformly scattered throughout a plexus of minute veins.

The *arteries* supplying the supra-renal glands are numerous and of large size, they are derived from the aorta, the phrenic, and the renal; they subdivide into numerous minute branches previous to entering the substance of the gland.

The *supra-renal vein* returns the blood from the medullary venous plexus, and receives several branches from the cortical substance; it opens on the right side into the inferior vena cava, on the left side into the left renal vein.

The *lymphatics* terminate in the lumbar glands.

The *nerves* are exceedingly numerous; they are derived from the solar and renal plexuses, and, according to Bergmann, from the phrenic and pneumo-gastric nerves. They have numerous small ganglia developed upon them.

The Pelvis

The cavity of the pelvis is that part of the general abdominal cavity which is below the level of the linea ileo-pectinea and the promontory of the sacrum.

Boundaries. It is bounded, behind, by the sacrum, the coccyx, and the great sacro-sciatic ligaments; in front and at the sides, by the pubes and ischia, covered by the Obturator muscles; above, it communicates with the cavity of the abdomen; and below, it is limited by the Levatores ani and Coccygei muscles, and the visceral layer of the pelvic fascia, which is reflected from the wall of the pelvis on to the viscera.

Contents. The viscera contained in this cavity are the urinary bladder, the lower end of the rectum, and the generative organs peculiar to each sex; they are covered by the peritoneum, and supplied with blood and lymphatic vessels and nerves.

The Bladder

The bladder is the reservoir for the urine. It is a musculo-membranous sac, situated in the pelvic cavity, behind the pubes, and in front of the rectum in the male, the uterus and vagina intervening between it and that intestine in the female. The shape, position, and relations of the bladder are greatly influenced by age, sex, and the degree of distension of the organ. *During infancy*, it is conical in shape, and projects above the upper border of the pubes into the hypo-gastric region. *In the adult*, when quite empty and contracted, it is a small triangular sac, placed deeply in the pelvis, flattened from before backwards, its apex reaching as high as the upper border of the symphysis pubis. When slightly distended, it has a rounded form, and partially fills the pelvic cavity; and when greatly distended, is ovoid in shape, rising into the abdominal cavity, often extending upwards nearly as high as the umbilicus. It is larger in its vertical diameter than from side to side, and its long axis is directed from above obliquely downwards and backwards in a line directed from some point between the pubes and umbilicus (according to its distension) to the coccyx. The bladder, when distended, is slightly curved forwards towards the anterior wall of the abdomen, so as to be more convex behind than in front. In the female, it is larger in the transverse than in the vertical diameter, and its capacity is said to be greater than in the male. The ordinary amount which it contains is about a pint.

The bladder is divided into a summit, body, base, and neck.

341.—Vertical Section of Bladder, Penis, and Urethra.

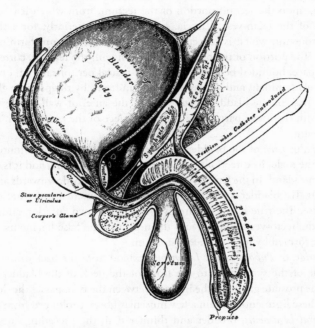

The *summit*, or apex, of the bladder is rounded and directed forwards and upwards; it is connected to the umbilicus by a fibro-muscular cord, the urachus, and also by means of two rounded fibrous cords, the obliterated portions of the hypo-gastric arteries, which are placed one on each side of the urachus.

The *urachus* is the obliterated remains of a tubular canal existing in the embryo, which connects the cavity of the bladder with a membranous sac placed external to the abdomen, opposite the umbilicus, called the *allantois*. In the infant at birth, it is occasionally found pervious, so that the urine escapes at the umbilicus, and calculi have been found in its canal. The summit of the bladder behind the urachus is covered by peritoneum, whilst the portion in front is uncovered by it, and rests upon the abdominal wall.

The *body* of the bladder in front is not covered by peritoneum, and is in relation with the triangular ligament of the urethra, the posterior surface of the symphysis pubis, the Internal obturator muscles, and, when distended, with the abdominal parietes.

The *posterior surface* is covered by peritoneum throughout. It corresponds, in the male, with the rectum; in the female, with the uterus, some convolutions of the small intestines being interposed.

The *side* of the bladder is crossed obliquely from below, upwards and forwards, by the obliterated hypogastric artery: above and behind this cord, the bladder is covered by peritoneum; but, below and in front of it, the serous covering is wanting, and it is connected to the pelvic fascia. The vas deferens passes, in an arched direction, from before backwards, along the side of the bladder, towards its base, crossing in its course the obliterated hypogastric artery, and passing along the inner side of the ureter.

The *base* of the bladder is broad, directed downwards and backwards, and rests, in the male, upon the second portion of the rectum, from which it is separated by a reflection of the recto-vesical fascia. It is covered posteriorly, for a slight extent, by the peritoneum, which is reflected from it upon the rectum, forming the recto-vesical fold. The portion of the bladder in relation with the rectum, corresponds to a triangular space, bounded behind by the recto-vesical peritoneal fold; on either side, by the vesicula seminalis and vas deferens; and in front, by the prostate gland. In the female, the base of the bladder is adherent to the anterior wall of the vagina, and separated from the lower part of the anterior surface of the cervix uteri, by a fold of the peritoneum.

The *cervix*, or *neck* of the bladder, is the constricted portion continuous with the urethra. In the male, its direction is oblique in the erect posture, and it is surrounded by the prostate gland. In the female, its direction is obliquely downwards and forwards.

Ligaments. The bladder is retained in its place by ligaments, which are divided into true and false. The true ligaments are five in number, two anterior, and two lateral, formed by the recto-vesical fascia, and the urachus. The false ligaments, also five in number, are formed by folds of the peritoneum.

The *Anterior*, or *Pubo-prostatic Ligaments*, extend from the back of the pubes, one on each side of the symphysis, to the front of the neck of the bladder, and upper surface of the prostate gland. In the interval between the two, passes the dorsal vein of the penis. These ligaments contain a few muscular fibres, prolonged from the bladder.

The *Lateral Ligaments*, broader and thinner than the preceding, are formed by reflected *portions* of the visceral layer of the pelvic fascia, which are attached to the lateral parts of the prostate, and to the sides of the base of the bladder.

The *Urachus* is the fibro-muscular cord already mentioned, extending between the apex of the bladder and the umbilicus. It is broad below, at its attachment to the bladder, and becomes narrower as it ascends.

The *False Ligaments* of the bladder are formed by peritoneum: they are five in number, two posterior, two lateral, and one superior.

The *two posterior* pass forwards, in the male, from the sides of the rectum; in the female, from the sides of the uterus, to the posterior and lateral aspect of the bladder: they form the lateral boundaries of the recto-vesical fold of peritoneum, and contain the obliterated hypogastric arteries, the ureters, and vessels and nerves.

The *two lateral* ligaments are reflections of the peritoneum, from the iliac fossae to the sides of the bladder.

The *superior* ligament is the prominent fold of peritoneum extending from the summit of the bladder to the umbilicus. It covers the urachus, and the obliterated hypogastric arteries.

Structure. The bladder is composed of four coats: a serous, a muscular, a cellular, and a mucous coat.

The *serous coat* is partial, and derived from the peritoneum. It invests the posterior surface, from opposite the termination of the two ureters to its summit, and is reflected from this point and from the sides, on to the abdominal and pelvic walls.

The *muscular coat* consists of two layers of unstriped muscular fibre, an external layer, composed of longitudinal fibres, and an internal layer of circular fibres.

The *longitudinal fibres* are most distinct on the anterior and posterior surfaces of the organ. They arise, in front, from the anterior ligaments of the bladder from the neck of the bladder, and, in the male, from the adjacent portion of the prostate gland. They spread out, and form a plexiform mesh, on the anterior surface of the bladder, being continued over the posterior surface and base of the organ to the neck, where they are inserted into the prostate, in the male, and into the vagina in the female.

Other longitudinal fibres arise, in the male, from the sides of the prostate, and spread out upon the sides of the bladder, intersecting with one another.

The *circular fibres* are very thinly and irregularly scattered on the body of the organ; but, towards its lower part, around the cervix and commencement of the urethra, they are disposed in a thick circular layer, forming the sphincter vesicae.

Two bands of oblique fibres, originating behind the orifices of the ureters, converge to the back part of the prostate gland, and are inserted, by means of a fibrous process, into the middle lobe of this organ. They are the *muscles of the ureters*, described by Sir C. Bell, who supposed that, during the contraction of the bladder, they served to retain the oblique direction of the ureters, and so prevent the reflux of urine into them.

The *cellular coat* consists of a layer of areolar tissue, connecting together the muscular and mucous coats, being intimately connected with the latter.

The *mucous coat* is thin, smooth, and of a pale rose colour. It is continuous through the ureters with the lining membrane of the uriniferous tubes, and below, with the urethra. It is connected loosely to the muscular coat, by a layer of areolar tissue, excepting at the trigone, where its adhesion is more close. It is provided with a few mucous follicles; and numerous small racemose glands, lined with columnar epithelium, exist near the neck of the organ. The epithelium covering it is intermediate in form between the columnar and squamous varieties.

Interior of the bladder. Upon the internal surface of the base of the bladder, immediately behind the urethral orifice, is a triangular, smooth surface, the apex of which is directed forwards: this is the *trigonum vesicae*, or *trigone vesicale*. It is paler in colour than the rest of the mucous membrane, and never presents any rugae, even in the collapsed condition of the organ, owing to its intimate adhesion to the subjacent tissues. It is bounded on each side by two slight ridges, which pass backwards and outwards to the orifices of the ureters, and correspond with the muscles of these tubes; at each posterior angle, are the orifices of the ureters, which are placed nearly two inches from each other, and about an inch and a half behind the orifice of the urethra. Projecting from the lower and anterior part of the bladder, into the orifice of the urethra, is a slight elevation of mucous membrane, called the *uvula vesicae*. It is formed by a thickening of the submucous areolar tissue, and lies just in front of the middle lobe of the prostate.

The *Arteries* supplying the bladder are the superior, middle, and inferior vesical, in the male, with additional branches from the uterine, in the female. They are all derived from the anterior trunk of the internal iliac.

The *Veins* form a complicated plexus around the neck, sides, and base of the bladder, and terminate in the internal iliac vein.

The *Lymphatics* accompany the blood-vessels, passing through the glands surrounding them.

The *Nerves* are derived from the hypogastric and sacral plexuses; the former supplying the upper part of the organ, the latter its base and neck.

342.—The Bladder and Urethra laid open. Seen from above.

Male Urethra

The urethra extends from the neck of the bladder to the meatus urinarius. It is curved in its course, so as to resemble, in its flaccid state, the italic letter ∫; but in the erect state it forms only a single curve, the concavity of which is directed upwards. Its length varies from eight to nine inches; and it is divided into three portions, the prostatic, membranous, and spongy, the structure and relations of which are essentially different.

The *Prostatic portion* is the widest and most dilatable part of the canal. It passes through the prostate gland, from its base to its apex, lying nearer its upper than its lower surface. It is about an inch and a quarter in length; and the form of the canal is spindle-shaped, being wider in the middle than at either extremity, and narrowest in front, where it joins the membranous portion. A transverse section of the canal in this situation is triangular, the apex directed downwards.

Upon the floor of the canal is a narrow longitudinal ridge, the *veru montanum*, or *caput gallinaginis*, formed by an elevation of the mucous membrane and its subjacent tissue. It is eight or nine lines in length, and a line and a half in height; and contains, according to Kobelt, muscular and erectile tissues. When distended, it may serve to prevent the passage of the semen backwards into the bladder. On each side of the ridge is a slightly depressed fossa, the *prostatic sinus*, the floor of which is perforated by numerous apertures, the *orifices of the prostatic ducts*, the ducts of the middle lobe opening behind the crest. At the fore part of the verumontanum, in the middle line, is a depression, the *sinus pocularis* (*vesicula prostatica*); and upon or within its margin are the slit-like openings of the ejaculatory ducts. The sinus pocularis forms a *cul-de-sac* about a quarter of an inch in length, which runs upwards and backwards in the substance of the prostate, beneath the middle lobe; its prominent upper wall partly forms the veru montanum. Its walls are composed of fibrous tissue, muscular fibres, and mucous membrane; and numerous small glands open on its inner surface. It has been called by Weber, who discovered it, the *uterus masculinus*, from its supposed homology with the female organ.

The *Membranous portion* of the urethra extends between the apex of the prostate, and the bulb of the corpus spongiosum. It is the narrowest part of the canal (excepting the orifice), and measures three-quarters of an inch along its upper, and half an inch along its lower surface, in consequence of the bulb projecting backwards beneath it below. Its upper concave surface is placed about an inch beneath the pubic arch, from which it is separated by the dorsal vessels and nerves of the penis, and some muscular fibres. Its lower convex surface is separated from the rectum by a triangular space, which constitutes the perinaeum. The membranous portion of the urethra perforates the deep perineal fascia; and two layers from this membrane are prolonged around it, the one forwards, the other backwards; it is also surrounded by the Compressor urethrae muscle. Its coverings are mucous membrane, elastic fibrous tissue, a thin layer of erectile tissue, muscular fibres, and a prolongation from the deep perineal fascia.

The *Spongy portion* is the longest part of the urethra, and is contained in the corpus spongiosum. It is about six inches in length, and extends from the termination of the membranous portion to the meatus urinarius. Commencing below the symphysis pubis, it ascends for a short distance, and then curves downwards. It is narrow and of uniform size in the body of the penis, measuring about a quarter of an inch in diameter; being dilated behind, within the bulb, where it forms the bulbous portion; and again anteriorly, within the glans penis, forming the fossa navicularis. A cross section of this canal in the body has its diameter transverse; but in the glans the diameter is directed vertically.

The *meatus urinarius* is the most contracted part of the urethra; it is a vertical slit, about three lines in length, bounded on each side by two small labia. The inner surface of the lining membrane of the urethra, especially on the floor of the spongy portion, presents the orifices of numerous mucous glands and follicles, situated in the submucous tissue, and named the *glands of Littré*. They vary in size, and their orifices are directed forwards; so that they may easily intercept the point of a catheter in its passage along the canal. One of these lacunae, larger than the rest, is situated on the upper surface of the fossa navicularis, about an inch and a half from the orifice; it is called the *lacuna magna*. Into the bulbous portion are found opening the ducts of Cowper's glands.

Structure. The urethra is composed of three coats, a mucous, muscular, and erectile.

The *mucous coat* forms part of the genito-urinary mucous membrane. It is continuous with the mucous membrane of the bladder, ureters, and kidneys; externally with the integument covering the glans penis; and is prolonged into the ducts of the numerous glands which open into the urethra, viz. Cowper's glands, the prostate gland, and through the ejaculatory ducts is continued into the vasa deferentia and vesiculae seminales. In the spongy and membranous portions, the mucous membrane is arranged in longitudinal folds when the organ is contracted. Small papillae are found upon it, near the orifice; and its epithelial lining is of the columnar variety, excepting near the meatus, where it is laminated.

The *muscular coat* consists of two layers of plain muscular fibres, an external longitudinal layer, and an internal circular. The muscular tissue is most abundant in the prostatic portion of the canal.

A thin layer of *erectile tissue* is continued from the corpus spongiosum around the membranous and prostatic portions of the urethra to the neck of the bladder.

MALE GENERATIVE ORGANS

Prostate Gland

The Prostate Gland (προϊστημι, *to stand before*) is a pale, firm, glandular body, which surrounds the neck of the bladder and commencement of the urethra. It is placed in the pelvic cavity, behind and below the symphysis pubis, posterior to the deep perinaeal fascia, and upon the rectum, through which it may be distinctly felt, especially when enlarged. In shape and size it resembles a horse-chestnut.

Its *base* is directed backwards towards the neck of the bladder.

The *apex* is directed forwards to the deep perinaeal fascia, which it touches.

Its *under surface* is smooth, and rests on the rectum, to which it is connected by a dense areolar fibrous tissue.

Its *upper surface* is flattened, marked by a slight longitudinal furrow, and placed about three-quarters of an inch below the pubic symphysis.

It measures about an inch and a half in its transverse diameter at the base, an inch in its anterior posterior, and three-quarters of an inch in depth; and its weight is about six drachms. It is held in its position by the anterior ligaments of the bladder (*pubo prostatic*); by the posterior layer of the deep perinaeal fascia, which invests the commencement of the membranous portion of the urethra and prostate gland; and by the anterior portion of the Levator ani muscle (*levator prostatae*), which passes down on each side from the symphysis pubis and anterior ligament of the bladder to the convex sides of the prostate.

The prostate consists of three lobes; two lateral and a middle lobe.

The *two lateral lobes* are of equal size, separated behind by a deep notch, and marked by a slight furrow upon their upper and lower surface, which indicates the bi-lobed condition of the organ in some animals.

The third or *middle lobe* is a small transverse band, occasionally a rounded or triangular prominence, placed between the two lateral lobes at the under and posterior part of the organ. It lies immediately beneath the neck of the bladder, behind the commencement of the urethra, and above the ejaculatory ducts. Its existence is not constant, but it is occasionally found at an early period of life, as well as in the adults, and in old age. In advanced life it often becomes enlarged and projects into the bladder, so as to impede the evacuation of the urine.

The prostate gland is perforated by the urethra and common seminal ducts. The urethra usually lies about one-third nearer its upper than its lower surface; occasionally the prostate surrounds only the lower three-fourths of this tube, and it more rarely runs through the lower than the upper part of the gland. The ejaculatory ducts pass forwards obliquely through a conical canal, situated in the lower part of the prostate, and open into the prostatic portion of the urethra.

Structure. The prostate is enclosed in a thin but firm fibrous capsule, distinct from that derived from the posterior layer of the deep perinaeal fascia, and separated from it by a plexus of veins. Its substance is of a pale reddish grey colour, very friable, but of great density. It consists of glandular substance and muscular tissue.

The *glandular substance* is composed of numerous follicular pouches, opening into elongated canals, which join to form from twelve to twenty small excretory ducts. The follicles are connected together by areolar tissue, supported by prolongations from the fibrous capsule, and enclosed in a delicate capillary plexus. The epithelium lining the canals is columnar, whilst that in the terminal vesicles is of the squamous variety.

The *muscular tissue* of the prostate is arranged in the form of circular bands around the urethra; it is continuous behind with the circular fibres of the sphincter vesicae, and in front with the circular fibres of the urethra. The muscular fibres are of the involuntary kind. The prostatic ducts open into the floor of the prostatic portion of the urethra.

Vessels and Nerves. The *arteries* supplying the prostate are derived from the internal pudic, vesical, and haemorrhoidal. Its *veins* form a plexus around the sides and base of the gland; they communicate in front with the dorsal vein of the penis, and terminate in the internal iliac vein. The *nerves* are derived from the hypogastric plexus.

The *Prostatic Secretion* is a milky fluid, having an acid reaction, and presenting on microscopic examination, molecular matter, the squamous and columnar forms of epithelium, and granular nuclei. In old age, this gland is liable to be enlarged, and its ducts are often filled with innumerable small concretions, of a brownish red colour, and of the size of a millet seed, composed of carbonate of lime and animal matter.

COWPER'S GLANDS

Cowper's Glands are two small rounded and somewhat lobulated bodies, of a yellowish colour, about the size of peas, placed beneath the fore-part of the membranous portion of the urethra, between the two layers of the deep perinaeal fascia. They lie close behind the bulb, and are enclosed by the transverse fibres of the Compressor urethrae muscle. Each gland consists of several lobules, held together by a fibrous investment. The excretory duct of each gland, nearly an inch in length, passes obliquely forwards beneath the mucous membrane, and opens by a minute orifice on the floor of the bulbous portion of the urethra. Their existence is said to be constant; they gradually diminish in size as age advances.

THE PENIS

The penis is the organ of copulation, and contains in its interior the larger portion of the urethra. It consists of a root, body, and the extremity or glans penis.

The *root* is broad and firmly connected to the rami of the pubes by two strong tapering fibrous processes, the crura, and to the front of the symphysis pubis by a fibrous membrane, the suspensory ligament.

The *extremity* or *glans penis* presents the form of an obtuse cone, flattened from above downwards. At its summit is a vertical fissure, the orifice of the urethra (meatus urinarius); and at the back part of this orifice a fold of mucous membrane passes backwards to the bottom of a depressed raphe, where it is continuous with the prepuce; this fold is termed the *fraenum preputii*. The base of the glans forms a rounded projecting border, the *corona glandis*; and behind the corona is a deep constriction, the *cervix*. Upon both of these parts numerous small lenticular sebaceous glands are found, the *glandulae Tysonii, odoriferae*. They secrete a sebaceous matter of very peculiar odour, which probably contains caseine, and becomes easily decomposed.

The *body* of the penis is the part between the root and the extremity. In the flaccid condition of the organ it is cylindrical, but when erect has a triangular prismatic form with rounded angles, the broadest side being turned upwards, and called the *dorsum*. It is covered by integument remarkable for its thinness, its dark colour, its looseness of connection with the deeper parts of the organ, and for containing no adipose tissue. At the root of the penis it is continuous with that upon the pubes and scrotum; and at the neck of the glans it leaves the surface, and becomes folded upon itself to form the prepuce.

The internal layer of the prepuce, which also becomes attached to the cervix, approaches in character to a mucous membrane; it is reflected over the glans penis, and at the meatus urinarius is continuous with the mucous lining of the urethra.

The mucous membrane covering the glans penis contains no sebaceous glands, but upon it are a number of small papillary elevations, which are highly sensitive.

The penis is composed of a mass of erectile tissue, enclosed in three cylindrical fibrous compartments. Of these, two, the corpora cavernosa, are placed side by side along the upper part of the organ; the third, or corpus spongiosum, encloses the urethra, and is placed below.

The *Corpora Cavernosa* form the chief part of the body of the penis. They consist of two cylindrical fibrous portions, placed side by side, and intimately connected along the median line for their anterior three-fourths, their posterior fourth being separated to form the two crura, by which the penis is connected to the rami of the pubes. Each crus commences by a thick-pointed process in front of the tuberosity of the ischium; and near its junction with its fellow, presents a slight enlargement, named by Kobelt, the *bulb of the corpus cavernosum*. Just beyond this point they become constricted, and retain an equal diameter to their anterior extremity, where they form a single rounded extremity, which is received into a fossa in the base of the glans penis. A median groove on the upper surface lodges the dorsal vein of the penis, and the deep groove on the under surface receives the corpus spongiosum. The root of the penis is connected to the symphysis pubis by the suspensory ligament.

Structure. The corpora cavernosa consist of a strong fibrous envelope, enclosing a fibrous reticular structure, containing erectile tissue in its meshes, and divided into two separate portions by an incomplete fibrous septum.

The *fibrous investment* is extremely dense, of considerable thickness, and highly elastic; it not only invests the surface of the organ, but sends off numerous fibrous bands (*trabeculae*) from its inner surface, as well as from the surfaces of the septum, which cross its interior in all directions, subdividing it into a number of separate compartments, which present a spongy structure, in which the erectile tissue is contained.

The *trabecular* structure fills the interior of the corpora cavernosa. Its component fibres are larger and stronger around the circumference than at the centre of the corpora cavernosa; they are also thicker behind than in front. The interspaces, on the contrary, are larger at the centre than at the circumference, their long diameter being directed transversely, and they are largest anteriorly. They are lined by a layer of squamous epithelium.

The *fibrous septum* forms an imperfect partition between the two corpora cavernosa; it is thick and complete behind, but in front it is incomplete, and consists of a number of vertical bands of fibrous tissue, which are arranged like the teeth of a comb, extending between the dorsal and urethral surface of the corpus cavernosum; hence the name *septum pectiniforme*. The fibrous investment and septum consist of longitudinal bands of white fibrous tissue, with numerous elastic and muscular fibres.

The trabeculae also consist of white fibrous tissue, elastic fibres, and plain muscular fibres, and enclose arteries and nerves.

The *Corpus Spongiosum* encloses the urethra, and is situated in the deep groove on the under surface of the corpora cavernosa. It commences posteriorly in front of the deep perinaeal fascia, between the diverging crura of the corpora cavernosa, where it forms a rounded enlargement, the bulb, and terminates anteriorly in another expansion, the glans penis, which overlays the anterior rounded extremity of the corpus cavernosum; its central portion or body is cylindrical, and tapers slightly from behind forwards.

The *bulb* varies in size in different subjects; it receives a fibrous investment from the anterior layer of the deep perinaeal fascia, and is surrounded by the Accelerator urinae muscle. The urethra enters the bulb nearer its upper than its lower surface, being surrounded by a layer of erectile tissue, named by Kobelt, the *colliculi bulbi*, a thin prolongation of which is continued backwards around the membranous and prostatic portions of the canal to the neck of the bladder, lying immediately beneath the mucous membrane. The portion of the bulb below the urethra presents a partial division into two lobes, being marked externally by a linear raphe, whilst internally there projects inwards, for a short distance, a thin fibrous septum, most distinct in early life.

Structure. The corpus spongiosum consists of a trabecular structure, enclosed in a strong fibrous envelope, and containing in its meshes erectile tissue. The fibrous envelope is thinner, whiter in colour, and more elastic than that of the corpus cavernosum. The trabeculae are delicate, uniform in size, and the meshes between them small, their long diameter, for the most part, corresponding with that of the penis. A thin layer of muscular fibres, continuous behind with those of the bladder, forms part of the outer coat of the corpus spongiosum.

Erectile tissue consists essentially of an intricate venous plexus, lodged in the interspaces between the trabeculae. The veins forming this plexus are so numerous, and communicate so freely with one another, as to present a cellular appearance when examined by means of a section; their walls are extremely thin and lined by squamous epithelium. The veins are smaller in the glans penis, corpus spongiosum, and circumference of the corpus cavernosum, than in the central part of the latter, where they are of large size and much dilated. They return the blood by a series of vessels, some of which emerge in considerable numbers from the base of the glans penis, and converge on the dorsum of the organ to form the dorsal vein: others pass out on the upper surface of the corpus spongiosum and join the dorsal vein: some emerge from the under surface between the corpus cavernosum and spongiosum, and receiving branches from the latter, wind round the sides of the penis to terminate in the dorsal vein; but the greater number pass out at the root of the penis, and join the prostatic plexus and pudendal veins.

The *arteries of the penis* are derived from the internal pudic. Those supplying the corpus cavernosum are the arteries of the corpus cavernosum, and branches from the dorsal artery of the penis, which perforate the fibrous capsule near the fore part of the organ. Those to the corpus spongiosum are the arteries of the bulb. Additional branches are described by Kobelt as arising from the trunk of the internal pudic; they enter the bulbous enlargement on the corpora cavernosa and spongiosum. The arteries, on entering the cavernous structure, divide into branches, which are supported and enclosed by the trabeculae; according to Müller, some of these branches terminate in a capillary network, which communicates with the veins as in other parts, whilst others are more convoluted, and assume a tendril-like appearance; hence the name *helicine arteries*, which is given to these peculiar vessels. The helicine arteries are most abundant in the back part of the corpus cavernosum and spongiosum; they have not been seen in the glans penis. The existence of these vessels is denied by Valentin, who describes the smallest branches of the arteries as terminating by wide funnel-shaped orifices, which open directly into the venous cavities.

The *lymphatics* of the penis consist of a superficial and deep set; the former terminate in the inguinal glands; the latter emerge from the corpora cavernosa and spongiosum, and passing beneath the pubic arch, join the deep lymphatics of the pelvis.

The *nerves* are derived from the internal pudic nerve and the hypogastric plexus. On the glans and bulb some filaments of the cutaneous nerves have Pacinian bodies connected with them.

The Testes and their Coverings

The testes are two small glandular organs, which secrete the semen: they are situated in the scrotum, being suspended by the spermatic cords. At an early period of foetal life, the testes are contained in the abdominal cavity, behind the peritoneum. Before birth, they descend to the inguinal canal, along which they pass with the spermatic cord, and, emerging at the external abdominal ring, they descend into the scrotum, becoming invested in their course by numerous coverings, derived from the serous, muscular, and fibrous layers of the abdominal parietes, as well as by the scrotum. The coverings of the testis are, the

- Skin ⎫
- Dartos ⎭ Scrotum.
- Intercolumnar, or External spermatic fascia.
- Cremaster muscle.
- Infundibuliform, or Fascia propia (Internal spermatic fascia).
- Tunica vaginalis.

The Scrotum is a cutaneous pouch, which contains the testes and part of the spermatic cords. It is divided into two lateral halves, by a median line, or raphe, which is continued forwards along the under surface of the penis, and backwards along the middle line of the perinaeum to the anus. Of these two lateral portions, the left is larger than the right, and corresponds with the greater length of the spermatic cord on the left side. Its external aspect varies under different circumstances: thus, under the influence of warmth, and in old and debilitated persons, it becomes elongated and flaccid; but, under the influence of cold, and in the young and robust, it is short, corrugated, and closely applied to the testes.

The scrotum consists of two layers, the integument and the dartos.

The *integument* is very thin, of a brownish colour, and generally thrown into folds or rugae. It is provided with sebaceous follicles, the secretion of which has a peculiar odour, and beset with thinly scattered, crisp hairs, the roots of which are seen through the skin.

The *dartos* is a thin layer of loose reddish tissue, endowed with contractility: it forms the proper tunic of the scrotum, is continuous, around the base of the scrotum, with the superficial fascia of the groin, perinaeum, and inner side of the thighs, and sends inwards a distinct septum, *septum scroti*, which divides it into two cavities for the two testes, the septum extending between the raphe, and under surface of the penis, as far as its root.

The dartos is closely united to the skin externally, but connected with the subjacent parts by delicate areolar tissue, upon which it glides with the greatest facility. The dartos is very vascular, and consists of a loose areolar tissue, containing unstriped muscular fibre. Its contractility is slow, and excited by cold and mechanical stimuli, but not by electricity.

The *Intercolumnar Fascia* is a thin membrane, derived from the margin of the pillars of the external abdominal ring, during the descent of the testis in the foetus, being prolonged downwards around the surface of the cord and testis. It is separated from the dartos by loose areolar tissue, which allows of considerable movement of the latter upon it, but is intimately connected with the succeeding layer.

The *Cremasteric Fascia* consists of scattered bundles of muscular fibres (*Cremaster muscle*), derived from the lower border of the Internal oblique muscle, and carried down during the descent of the testes.

The *Fascia Propria* is a thin membranous layer, which loosely invests the surface of the cord. It is a continuation downwards of the infundibuliform process of the fascia transversalis, and is also derived during the descent of the testis in the foetus.

The *Tunica Vaginalis* is described with the proper coverings of the testis. A more detailed account of these coverings may be found in the description of the surgical anatomy of inguinal hernia.

The *Arteries* supplying the coverings of the testes are: the superficial and deep external pudic, from the femoral; the superficial perinaeal branch of the internal pudic; and the cremasteric branch from the epigastric.

The *Veins* follow the course of the corresponding arteries.

The *Lymphatics* terminate in the inguinal glands.

The *Nerves* are, the ilio-inguinal and ilio-hypogastric branches of the lumbar plexus, the two superficial perinaeal branches of the internal pudic nerve, the inferior pudendal branch of the small sciatic nerve, and the genital branch of the genito-crural nerve.

The SPERMATIC CORD extends from the internal abdominal ring, where the structures of which it is composed converge, to the back part of the testicle. It is composed of arteries, veins, lymphatics, nerves, and the excretory duct of the testicle, connected together by areolar tissue, and invested by its proper coverings. In the abdominal wall, it passes obliquely along the inguinal canal, lying at first beneath the Internal oblique, and upon the fascia transversalis, but, nearer the pubes, it rests on Poupart's ligament, having the aponeurosis of the External oblique in front of it, and the conjoined tendon behind it. It then escapes at the external ring, and descends nearly vertically into the scrotum. The left cord is rather longer than the right, consequently the left testis hangs somewhat lower than its fellow.

The *Arteries of the Cord* are the spermatic, from the aorta; the artery of the vas deferens, from the superior vesical; and the cremasteric, from the epigastric artery.

The *spermatic artery* supplies the testicle. On approaching this gland, some branches supply the epididymis, others perforate the tunica albuginea behind, and spread out on its inner surface, or pass through the fibrous septum in its interior, to be distributed on the membranous septa, between the separate lobes.

The *artery of the vas deferens* is a long slender vessel, which accompanies the vas deferens, ramifying upon the coats of this duct, and anastomosing with the spermatic artery near the testis.

The *cremasteric branch* from the epigastric supplies the Cremaster muscle, and other coverings of the cord.

The *Spermatic Veins* leave the back part of the testis, and receive branches from the epididymis: they unite to form a plexus (pampiniform plexus), which forms the chief mass of the cord. They pass up in front of the vas deferens, and unite to form a single trunk, which terminates, on the right side, in the inferior vena cava, on the left side, in the left renal vein.

The *Lymphatics* are of large size, accompany the blood-vessels, and terminate in the lumbar glands.

The *Nerves* are the spermatic plexus from the sympathetic. This plexus is derived from the renal and aortic plexuses, joined by filaments from the hypogastric plexus, which accompany the artery of the vas deferens.

Testes

The testes are two small glandular organs, suspended in the scrotum by the spermatic cords. Each gland is of an oval form, compressed laterally and behind, and having an oblique position in the scrotum; its upper extremity being directed forwards and a little outwards; the lower, backwards and a little inwards: the anterior convex border looks forwards and downwards, the posterior or straight border, to which the cord is attached, backwards and upwards.

The anterior and lateral surfaces, as well as both extremities of the organ, are convex, free, smooth, and invested by the tunica vaginalis. The posterior border, to which the cord is attached, receives only a partial investment from this membrane. Lying upon the outer edge of this border, is a long, narrow, flattened body, named, from its relation to the testis, the epididymis (ἐπι διδυμος, testis). It consists of a central portion, or body, an upper enlarged extremity, the globus major, or head; and a lower pointed extremity, the tail, or globus minor. The globus major is intimately connected with the upper end of the testicle by means of its efferent ducts; and the globus minor is connected with its lower end by cellular tissue, and a reflection of the tunica vaginalis. The outer surface and upper and lower ends of the epididymis are free and covered by serous membrane; the body is also completely invested by it, excepting along its posterior border, and connected to the back of the testis by a fold of the serous membrane. Attached to the upper end of the testis, or to the epididymis, is a small pedunculated body, the use of which is unknown.

343.—The Testis in Situ. The Tunica Vaginalis having been laid open.

Size and Weight. The average dimensions of this gland are from one and a half to two inches in length, one inch in breadth, and an inch and a quarter in the antero-posterior diameter; and the weight varies from six to eight drachms, the left testicle being a little the larger.

The testis is invested by three tunics, the tunica vaginalis, tunica albuginea, and tunica vasculosa.

The *Tunica Vaginalis* is the serous covering of the testis. It is a pouch of serous membrane, derived from the peritoneum during the descent of the testis in the foetus, from the abdomen into the scrotum. After its descent, that portion of the pouch which extends from the internal ring to near the upper part of the gland, becomes obliterated, the lower portion remaining as a shut sac, which invests the outer surface of the testis, and is reflected on the internal surface of the scrotum; hence it may be described as consisting of a visceral and parietal portion.

The *visceral portion* (*tunica vaginalis propria*), covers the outer surface of the testis, as well as the epididymis, connecting the latter to the testis by means of a distinct fold. From the posterior border of the gland, it is reflected on to the internal surface of the scrotum.

The *parietal portion* of the serous membrane (*tunica vaginalis reflexa*), is far more extensive than the visceral portion, extending upwards for some distance in front, and on the inner side of the cord, and reaching below the testis. The inner surface of the tunica vaginalis is free, smooth, and covered by a layer of squamous epithelium. The interval between the visceral and parietal layers of this membrane, constitutes the cavity of the tunica vaginalis.

The *Tunica Albuginea* is the fibrous covering of the testis. It is a dense fibrous membrane, of a blueish-white colour, composed of bundles of white fibrous tissue, which interlace in every direction. Its outer surface is covered by the tunica vaginalis, except along its posterior border, and at the points of attachment of the epididymis; hence the tunica albuginea is usually considered as a fibro-serous membrane, like the dura mater and pericardium. This membrane surrounds the glandular structure of the testicle, and, at its posterior and upper border, is reflected into the interior of the gland, forming an incomplete vertical septum, called the *mediastinum testis* (*corpus Highmorianum*).

The *mediastinum testis* extends from the upper, nearly to the lower border of the gland, and is wider above than below. From the front and sides of this septum, numerous slender fibrous cords (*trabeculae*) are given off, which pass to be attached to the inner surface of the tunica albuginea: they serve to maintain the form of the testis, and join, with similar laminae given off from the inner surface of the tunica albuginea, to form spaces which enclose the separate lobules of the organ. The mediastinum supports the vessels and ducts of the testis in their passage to and from the substance of the gland.

The *Tunica Vasculosa* (*pia mater testis*), is the vascular layer of the testis, consisting of a plexus of blood-vessels, held together by a delicate areolar tissue. It covers the whole of the internal surface of the tunica albuginea, sending off numerous processes between the lobules, which are supported by the fibrous prolongations from the mediastinum testis.

Structure. The glandular structure of the testis consists of numerous lobules (*lobuli testis*). Their number, in a single testis, is estimated by Berres at 250, and by Krause, at 400. They differ in size according to their position, those in the middle of the gland being larger and longer. Each lobule is conical in shape, the base being directed towards the circumference of the organ, the apex towards the mediastinum. Each lobule is contained in one of the intervals between the fibrous cords and vascular processes, which extend between the mediastinum testis and the tunica albuginea, and consists of from one to three, or more, minute convoluted tubes, the *tubuli seminiferi*. The tubes may be separately unravelled, by careful dissection under water, and may be seen to commence either by free caecal ends, or by anastomotic loops. The total number of tubes is considered by Monro to be about 300, and the length of each about sixteen feet: by Lauth, their number is estimated at 840, and their average length two feet and a quarter. Their diameter varies from $1/200$th to $1/150$ of an inch. The tubuli are pale in colour in early life, but, in old age, they acquire a deep yellow tinge, from containing much fatty matter. They consist of a basement membrane, lined by epithelium, consisting of nucleated granular corpuscles, and are enclosed in a delicate plexus of capillary vessels. In the apices of the lobules, the tubuli become less convoluted, assume a nearly straight course, and unite together to form from twenty to thirty larger ducts, of about $1/50$th of an inch in diameter, and these, from their straight course, are called *vasa recta*.

344.—Vertical Section of the Testicle, to show the arrangement of the Ducts

The *Vasa Recta* enter the fibrous tissue of the mediastinum, and pass upwards and backwards, forming, in their ascent, a close network of anastomosing tubes, with exceedingly thin parietes; this constitutes the *rete testis*. At the upper end of the mediastinum, the vessels of the rete testis terminate in from twelve to fifteen or twenty ducts, the *vasa efferentia:* they perforate the tunica albuginea, and carry the seminal fluid from the testis to the epididymis. Their course is at first straight; they then become enlarged, and exceedingly convoluted, and form a series of conical masses, the *coni vasculosi*, which, together, constitute the globus major of the epididymis. Each cone consists of a single convoluted duct, from six to eight inches in length, the diameter of which gradually decreases from the testis to the epididymis. Opposite the bases of the cones, the efferent vessels open at narrow intervals into a single duct, which constitutes, by its complex convolutions, the body and globus minor of the epididymis. When the convolutions of this tube are unravelled, it measures upwards of twenty feet in length, and increases in breadth and thickness as it approaches the vas deferens. The convolutions are held together by fine areolar tissue, and by bands of fibrous tissue. A long narrow tube, the *vasculum aberrans* of Haller, is occasionally found connected with the lower part of the canal of the epididymis, or with the commencement of the vas deferens, and extending up into the cord for about two or three inches, where it terminates by a blind extremity, which is occasionally bifurcated. Its length varies from an inch and a half to fourteen inches, and sometimes it becomes dilated towards its extremity: more commonly, it retains the same diameter throughout. Its structure in similar to that of the vas deferens. Occasionally, it is found unconnected with the epididymis.

The *Vas Deferens*, the excretory duct of the testis, is the continuation of the epididymis. Commencing at the lower part of the globus minor, it ascends along the posterior and inner side of the testis and epididymis, and along the back part of the spermatic cord, through the spermatic canal, to the internal abdominal ring. From the ring it descends into the pelvis, crossing the external iliac vessels, and hooks round the outer side of the epigastric artery; at the side of the bladder, it curves backwards and downwards to its base, crossing outside the obliterated hypogastric artery, and to the inner side of the ureter. At the base of the bladder, it lies between it and the rectum, running along the inner border of the vesicula seminalis. In this situation, it becomes enlarged and sacculated; and, becoming narrowed, at the base of the prostate, unites with the duct of the vesicula seminalis to form the ejaculatory duct. This tube presents a hard and cordy sensation to the fingers, is about two feet in length, of cylindrical form, and about a line and a quarter in diameter. Its walls are of extreme density and thickness, measuring one-third of a line; and its canal is extremely small, measuring about half a line.

In *structure*, the vas deferens consists of three coats: an external, or cellular coat; a muscular coat, which is thick, dense, elastic, and consists of two longitudinal, and an intermediate circular layer of muscular fibres.

The *internal*, or *mucous coat*, is pale, and arranged in longitudinal plicae: its epithelial covering is of the columnar variety.

VESICULAE SEMINALES

The *Seminal Vesicles* are two lobulated membranous pouches, placed between the base of the bladder and the rectum, and supposed to serve as reservoirs for the semen. Each sac is somewhat pyramidal in form, the broad end being directed backwards, and the narrow end forwards towards the prostate. They measure about two and a half inches in length, about five lines in breadth, and from two to three lines in thickness. They vary, however, in size, not only in different individuals, but also in the same individual on the two sides. Their *upper surface* is in contact with the base of the bladder, extending from near the termination of the ureters to the base of the prostate gland. Their *under surface* rests upon the rectum, from which they are separated by the recto-vesical fascia. Their *posterior extremities* diverge backwards and outwards from each other. Their *anterior extremities* are pointed, and converge towards the base of the prostate gland, where each joins with the corresponding vas deferens to form the ejaculatory duct. Along the inner margin of each vesicula runs the enlarged and somewhat convoluted vas deferens. The inner border of the vesicula, and the corresponding vas deferens, form the lateral boundary of a triangular space, limited behind by the recto-vesical peritoneal fold; the portion of the bladder included in this space rests on the rectum, and corresponds with the trigonum vesicae in its interior.

345.—Base of the Bladder, with the Vasa Deferentia and Vesiculæ Seminales.

Structure. Each vesicula consists of a single tube, coiled upon itself, and giving off several irregular caecal diverticula; the separate coils, as well as the diverticula, being connected together by fibrous tissue. When uncoiled, this tube is about the diameter of a quill, and varies in length from four to six inches; it terminates posteriorly in a *cul-de-sac*, but its anterior extremity becomes constricted into a narrow straight duct, which joins on its inner side with the corresponding vas deferens, and forms the ejaculatory duct.

The *Ejaculatory Ducts*, two in number, one on each side, are formed by the junction of the duct of the vesicula seminalis with the vas deferens. Each duct is about three-quarters of an inch in length; it commences at the base of the prostate, and runs forwards and upwards in a canal in its substance, and along the side of the utriculus, to terminate by a separate slit-like orifice upon or within the margins of the sinus pocularis. The ducts diminish in size, and converge towards their termination.

Structure. The vesiculae seminales are composed of three coats: *external* or *fibro-cellular*, derived from the recto-vesical fascia; *middle* or *fibrous coat*, which is firm, dense, fibrous in structure, somewhat elastic, and contains, according to E. H. Weber, muscular fibres; and an *internal* or *mucous coat*, which is pale, of a whitish-brown colour, and presents a delicate reticular structure, like that seen in the gall-bladder, but the meshes are finer. It is lined by squamous epithelium. The coats of the ejaculatory ducts are extremely thin, the outer fibrous layer being almost entirely lost after their entrance into the prostate, a thin layer of muscular fibres and the mucous membrane forming the only constituent parts of these tubes.

Vessels and Nerves. The *arteries* supplying the vesiculae seminales are derived from the inferior vesical and middle haemorrhoidal. The *veins* and *lymphatics* accompany the arteries. The *nerves* are derived from the hypogastric plexus.

The *Semen* is a thick whitish fluid, having a peculiar odour. It consists of a fluid, the liquor seminis, and solid particles, the seminal granules, and spermatozoa.

The *liquor seminis* is transparent, colourless, and of an albuminous composition, containing particles of squamous and columnar epithelium, with oil globules and granular matter floating in it, besides the above-mentioned solid elements.

The *seminal granules* are round finely-granular corpuscles, measuring $1/4000$th of an inch in diameter.

The *spermatozoa*, or spermatic filaments, are the essential agents in producing fecundation. They are minute elongated particles, consisting of a small flattened oval extremity or body, and a long slender caudal filament. A small circular spot is observed in the centre of the body, and at its point of connection with the tail there is frequently seen a projecting rim or collar. The movements of these bodies are remarkable, and consist of a lashing and undulatory motion of the tail.

Descent of the Testes

The testes, at an early period of foetal life, are placed at the back part of the abdominal cavity, behind the peritoneum, in front and a little below the kidneys. The anterior surface and sides are invested by peritoneum; the blood-vessels and efferent ducts are connected with their posterior surface; and attached to the lower end is a peculiar structure, the gubernaculum testis, which is said to assist in their descent.

The *Gubernaculum Testis* attains its full development between the fifth and sixth months; it is a conical-shaped cord, attached above to the lower end of the epididymis, and below to the bottom of the scrotum. It is placed behind the peritoneum, lying upon the front of the Psoas muscle, and completely filling the inguinal canal. It consists of a soft transparent areolar tissue within, which often appears partially hollow, surrounded by a layer of striped muscular fibres, the Cremaster, which ascend upon this body to be attached to the testis. According to Mr. Curling, the gubernaculum, as well as these muscular fibres, divides below into three processes: the external and broadest is connected with Poupart's ligament in the inguinal canal; the middle process descends along the inguinal canal to the bottom of the scrotum, where it joins the dartos; the internal one is firmly attached to the os pubis and sheath of the Rectus muscle; some fibres, moreover, are reflected from the Internal oblique on to the front of the gubernaculum. Up to the fifth month, the testis is situated in the lumbar region, covered in front and at the sides by peritoneum, and supported in its position by a fold of this membrane, the *mesorchium*: between the fifth and sixth months the testis descends to the iliac fossa, the gubernaculum at the same time becoming shortened: during the seventh month, it enters the internal abdominal ring, a small pouch of peritoneum (*processus vaginalis*) preceding the testis in its course through the canal. By the end of the eighth month, the testis has descended into the scrotum, carrying down with it a lengthened pouch of peritoneum, which communicates by its upper extremity with the peritoneal cavity. Just before birth, the upper part of this pouch becomes closed, and this obliteration extends gradually downwards to within a short distance of the testis. The process of peritoneum surrounding the testis, which is now entirely cut off from the general peritoneal cavity, is called the *tunica vaginalis*.

Mr. Curling considers that the descent of the testis is effected by means of the muscular fibres of the gubernaculum; those fibres which proceed from Poupart's ligament and the Obliquus internus are said to guide the organ into the inguinal canal; those attached to the pubis draw it below the external abdominal ring; and those attached to the bottom of the scrotum complete its descent. During the descent of the organ these muscular fibres become gradually everted, forming a muscular layer, which becomes placed external to the process of the peritoneum, surrounding the gland and spermatic cord, and constitutes the Cremaster. In the female, a small cord, corresponding to the gubernaculum in the male, descends to the inguinal region, and ultimately forms the round ligament of the uterus. A pouch of peritoneum accompanies it along the inguinal canal, analogous to the processus vaginalis in the male; it is called the *canal of Nuck*.

FEMALE ORGANS OF GENERATION

The external Organs of Generation in the female are the mons Veneris, the labia majora and minora, the clitoris, the meatus urinarius, and the orifice of the vagina. The term "vulva" or "pudendum," as generally applied, includes all these parts.

The *Mons Veneris* is the rounded eminence in front of the pubes, formed by a collection of fatty tissue beneath the integument. It surmounts the vulva, and is covered with hair at the time of puberty.

346.—The Vulva. External Female Organs of Generation.

The *Labia Majora* are two prominent longitudinal cutaneous folds, extending downwards from the mons Veneris to the anterior boundary of the perinaeum, and enclosing an elliptical fissure, the common urino-sexual opening. Each labium is formed externally of integument, covered with hair; internally, of mucous membrane, which is continuous with the genito-urinary mucous tract; and between the two, of a considerable quantity of areolar tissue, fat, and a tissue resembling the dartos of the scrotum, besides vessels, nerves, and glands. The labia are thicker in front than behind, and joined together at each extremity, forming the anterior and posterior commissures. The interval left between the posterior commissure and the margin of the anus is about an inch in length, and constitutes the perinaeum. Just within the posterior commissure is a small, transverse fold, the *fraenulum pudendi* or *fourchette*, which is commonly ruptured in the first parturition, and the space between it and the commissure is called the *fossa navicularis*. The labia are analogous to the scrotum in the male.

The *Labia Minora* or *Nymphae* are two small folds of mucous membrane, situated within the labia majora, and extending from the clitoris obliquely downwards and outwards for about an inch and a half on each side of the orifice of the vagina, on the sides of which they are lost. They are continuous externally with the labia majora, internally with the inner surface of the vagina. As they converge towards the clitoris in front, each labium divides into two folds, which surround the glans clitoridis, the superior folds uniting to form the praeputium clitoridis, the inferior folds being attached to the glans, and forming the fraenum. The nymphae are composed of mucous membrane, covered by a thin epithelial layer. They contain a plexus of vessels in their interior, and are provided with numerous large mucous crypts which secrete abundance of sebaceous matter.

The *Clitoris* is an erectile structure, analogous to the corpus cavernosum of the penis. It is situated beneath the anterior commissure, partially hidden between the anterior extremities of the labia minora. It is an elongated organ, connected to the rami of the pubes and ischium on each side by two crura; the body is short, and concealed beneath the labia; its free extremity, the glans clitoridis, is a small rounded tubercle, consisting of spongy erectile tissue, and highly sensitive. The clitoris consists of two corpora cavernosa, composed of erectile tissue enclosed in a dense layer of fibrous membrane, and united together along their inner surfaces by an incomplete fibrous pectiniform septum. The clitoris is provided, like the penis, with a suspensory ligament, and with two small muscles, the Erectores clitoridis, which are inserted into the crura of the corpus cavernosum.

Between the clitoris above, and the entrance of the vagina below, is a triangular smooth surface, bounded on each side by the nymphae: this is the vestibule.

The orifice of the urethra (*meatus urinarius*), is situated at the back part of the vestibule, about an inch below the clitoris, and near the margin of the vagina, surrounded by a prominent elevation of the mucous membrane. Below the meatus urinarius, is the orifice of the vagina, an elliptical aperture, more or less closed in the virgin, by a membranous fold, the hymen.

The *Hymen* is a thin semilunar fold of mucous membrane, stretched across the lower part of the orifice of the vagina; its concave margin being turned upwards towards the pubes. Sometimes this membrane forms a complete septum across the orifice of the vagina, which constitutes an imperforate hymen. Occasionally, it forms a circular septum, perforated in the centre by a round opening; sometimes it is cribriform, or its free margin forms a membranous fringe, or it may be entirely absent. The hymen cannot, consequently, be considered as a proof of virginity. Its rupture, or the rudimentary condition of the membrane above referred to, gives rise to those small rounded elevations which surround the opening of the vagina, the *carunculae myrtiformes*.

Glands of Bartholine. On each side of the commencement of the vagina, are two round, or oblong bodies, of a reddish-yellow colour, and of the size of an almond: they are analogous to Cowper's glands in the male, and are called the *glands of Bartholine*. Each gland opens by means of a long single duct, upon the inner side of the nymphae, external to the hymen. Extending from the clitoris, along either side of the vagina, and lying a little behind the nymphae, are two large oblong masses, about an inch in length, consisting of a plexus of veins, enclosed in a thin layer of fibrous membrane. These bodies are narrow in front, rounded below, and are connected with the crura of the clitoris and rami of the pubes: they are termed by Kobelt, the *bulbi vestibuli*, and he considers them analogous to the bulb of the corpus spongiosum in the male. Immediately in front of this body is a smaller venous plexus, continuous with the bulbus vestibuli behind, and the glans clitoridis in front: it is called by Kobelt, the *pars intermedia*, and is considered by him as analogous to that part of the body of the corpus spongiosum which immediately succeeds the bulb.

347.—Section of Female Pelvis, showing position of Viscera.

Bladder

The bladder is situated at the anterior part of the pelvis. It is in relation, *in front*, with the os pubis; *behind*, with the uterus, some convolutions of the small intestine being interposed; its *base* lies in contact with the neck of the uterus, and with the anterior wall of the vagina. The bladder is larger in the female than in the male, and very broad in its transverse diameter.

Urethra

The urethra is a narrow membranous canal, about an inch and a half in length, extending from the neck of the bladder to the meatus urinarius. It is placed beneath the symphysis pubis, imbedded in the anterior wall of the vagina; and its direction is obliquely downwards and forwards, its course being slightly curved, the concavity directed upwards. Its diameter, when undilated, is about a quarter of an inch. The urethra perforates the triangular ligament, precisely as in the male, and is surrounded by the muscular fibres of the Compressor urethrae.

Structure. The urethra consists of three coats: muscular, erectile, and mucous.

The *muscular coat* is continuous with that of the bladder; it extends the whole length of the tube, and consists of a thick stratum of circular fibres.

A thin layer of spongy, erectile tissue, intermixed with much elastic tissue, lies immediately beneath the mucous coat.

The *mucous coat* is pale, continuous, externally, with the vulva, and internally with that of the bladder. It is thrown into longitudinal folds, one of which, placed along the floor of the canal, resembles the veru-montanum in the male urethra. It is lined by laminated epithelium, which becomes spheroidal at the bladder. Its external orifice is surrounded by a few mucous follicles.

The urethra, from not being surrounded by dense resisting structures, as in the male, admits of considerable dilatation, which enables the surgeon to remove with considerable facility, calculi, or other foreign bodies, from the cavity of the bladder.

Rectum

The rectum is more capacious, and less curved in the female, than in the male.

The *first portion* extends from the left sacro-iliac symphysis to the middle of the sacrum. It is surrounded by peritoneum; and its connections are similar to those in the male.

The *second portion* extends to the tip of the coccyx. It is covered by the peritoneum in front, for a short distance, at its upper part, and is in relation with the posterior wall of the vagina.

The *third portion* curves backwards, from the vagina to the anus, leaving a space which corresponds on the surface of the body to the perinaeum. Its extremity is surrounded by the Sphincter muscles, and its sides are supported by the Levatores ani.

The Vagina

The vagina is a membranous canal, extending from the vulva to the uterus. It is situated in the cavity of the pelvis, behind the bladder, and in front of the rectum. Its direction is curved forwards and downwards, following at first the line of the axis of the cavity of the pelvis, and afterwards that of the outlet. It is cylindrical in shape, flattened from before backwards, and its walls are ordinarily in contact with each other. Its length is about four inches along its anterior wall, and between five or six along its posterior wall. It is constricted at its commencement, and becomes dilated near the uterine extremity; it surrounds the vaginal portion of the cervix uteri, at some distance from the os, and its attachment extends higher up on the posterior than on the anterior wall.

Relations. Its *anterior surface* is concave, and in relation with the base of the bladder, and with the urethra. Its *posterior surface* is convex, and connected to the anterior wall of the rectum, for the lower three-fourths of its extent, the upper fourth being separated from that tube by the recto-uterine fold of peritoneum, which forms a *cul-de-sac* between the vagina and rectum. Its sides give attachment superiorly to the broad ligaments, and inferiorly to the Levatores ani muscles and recto-vesical fascia.

Structure. The vagina consists of an external, or muscular coat, a layer of erectile tissue, and an internal mucous lining.

The *muscular coat* consists of longitudinal fibres, which surround the vagina, and are continuous with the superficial muscular fibres of the uterus. The strongest fasciculi are those attached to the recto-vesical fascia on each side.

The *erectile tissue* is enclosed between two layers of fibrous membrane: it is more abundant at the lower than at the upper part of the vagina.

The *mucous membrane* is continuous, above, with that lining the uterus, and below, with the integument covering the labia majora. Its inner surface presents, along the anterior and posterior walls, a longitudinal ridge, or raphe, called the *column of the vagina*, and numerous transverse ridges, or rugae, extend outwards from the raphe on each side. These rugae are most distinct near the orifice of the vagina, especially in females before parturition. They indicate its adaptation for dilatation, and are calculated to facilitate its enlargement during parturition. The mucous membrane is covered with conical and filiform papillae, and provided with mucous glands and follicles, which are especially numerous in its upper part, and around the cervix uteri.

The Uterus

The uterus is the organ of gestation, receiving the fecundated ovum in its cavity, retaining and supporting it during the development of the foetus, and the principal agent in its expulsion at the time of parturition.

In the virgin state it is pear-shaped, flattened from before backwards, and situated in the cavity of the pelvis, between the bladder and rectum; it is retained in its position by the round and broad ligaments on each side, and projects into the upper end of the vagina below. Its upper end, or base, is directed upwards and forwards; its lower end, or apex, downwards and backwards, in the line of the axis of the inlet of the pelvis, and forming an angle with the vagina, the direction of which corresponds to the cavity and outlet of the pelvis. The uterus measures about three inches in length, two in breadth, at its upper part, and an inch in thickness. It weighs, at puberty, from seven to twelve drachms, and, after parturition, from one ounce and a half to two ounces. The uterus is divided into fundus, body, cervix, and os uteri.

The *Fundus* is the upper broad extremity of the organ: it is convex, covered by peritoneum, and placed on a line below the level of the brim of the pelvis.

The *Body* gradually narrows from the fundus to the neck. Its *anterior surface* is flattened, covered by peritoneum in the upper three-fourths of its extent, and separated from the bladder by some convolutions of the small intestine: the lower fourth is connected with the bladder. Its *posterior surface* is convex, covered by peritoneum throughout, and separated from the rectum by some convolutions of the intestine. Its *lateral margins* are concave, and give attachment to the broad ligaments, the ligament of the ovary, the Fallopian tube, and round ligament.

The *Cervix* is the lower rounded and narrowed portion of the uterus: around its circumference is attached the upper end of the vagina, which extends upwards a greater distance behind than in front.

At the vaginal extremity of the uterus is an aperture, the os uteri, round in the virgin, and transverse after parturition. It is bounded by two lips, an anterior one, which is thick, a posterior narrow and long.

Ligaments. The ligaments of the uterus are six in number: two anterior, two posterior, and two lateral. They are formed by peritoneum.

The *two anterior ligaments* (*vesico-uterine*) are two semilunar folds, which pass between the neck of the uterus and the posterior surface of the bladder.

The *two posterior ligaments* (*recto-uterine*) pass between the sides of the uterus and rectum.

The *two lateral*, or *broad ligaments*, pass from the sides of the uterus to the lateral walls of the pelvis, forming a septum across the pelvis, which divides this cavity into two portions. In the anterior part are contained the bladder, urethra, and vagina; in the posterior part, the rectum.

The *Cavity of the Uterus* is small in comparison with the size of the organ: that portion corresponding to the body is triangular, flattened from before backwards, so that its walls are closely approximated, and having its base directed upwards towards the fundus. At each superior angle is a funnel-shaped cavity, which constitutes the remains of the division of the body of the uterus into two cornua; and at the bottom of each cavity is the minute orifice of the Fallopian tube. At the inferior angle of the uterine cavity, is a small constricted opening, the internal orifice (*ostium internum*) which leads into the cavity of the cervix. The cavity in the cervix is somewhat cylindrical, flattened from before backwards, broader at the centre than at either extremity, and communicates, below, with the vagina. The walls of the canal present two longitudinal columns, which run, one on its anterior, the other on its posterior surface; and from these proceed a number of smaller oblique columns, giving the appearance of branches from the stem of a tree, and hence the name, *arbor vitae uterinus*, applied to it. These folds usually become very indistinct after the first labour.

Structure. The uterus is composed of three coats, an external serous coat, a middle, or muscular layer, and an internal mucous coat.

The *serous coat* is derived from the peritoneum: it invests the fundus and the whole of the posterior surface of the body of the uterus, but only the upper three-fourths of its anterior surface.

The *muscular coat* forms the chief bulk of the substance of the uterus. In the unimpregnated state, it is dense, firm, of a greyish colour, and cuts almost like cartilage. It is thick opposite the middle of the body and fundus, and thin at the orifices of the Fallopian tubes. It consists of bundles of unstriped muscular fibres, disposed in layers, intermixed with areolar tissue, blood, and lymphatic vessels and nerves. In the impregnated state, the muscular tissue becomes more prominently developed, and is disposed in three layers, external, middle, and internal.

The *external layer* is placed beneath the peritoneum, disposed as a thin plane on the anterior and posterior surfaces. It consists of fibres, which pass transversely across the fundus, and, converging at each superior angle of the uterus, are continued on the Fallopian tubes, the round ligament, and ligament of the ovary; some passing at each side into the broad ligament, and others running backwards from the cervix into the recto-uterine ligaments.

The *middle layer* of fibres presents no regularity in its arrangement, being disposed longitudinally, obliquely, and transversely.

The *internal*, or *deep layer*, consists of circular fibres arranged in the form of two hollow cones, the apices of which surround the orifices of the Fallopian tubes, their bases intermingling with one another on the middle of the body of the uterus. At the cervix, these fibres are disposed transversely.

The *mucous membrane* is thin, smooth, and closely adherent to the subjacent tissue. It is continuous, through the fimbriated extremity of the Fallopian tubes, with the peritoneum; and through the os uteri, with the mucous lining of the vagina.

In the body of the uterus, it is smooth, soft, of a reddish colour, lined by columnar-ciliated epithelium, and presenting, when viewed with a lens, the orifices of numerous branched tubular glands, which are visible before impregnation, but become much enlarged in the pregnant state.

In the cervix, the mucous membrane between the rugae and around the os uteri, is provided with numerous mucous follicles and glands. The small, transparent, vesicular elevations, so often found within the os and cervix uteri, are due to closure of the mouths of these follicles, and their distension with its proper secretion. They were called the *ovula of Naboth*. The mucous membrane covering the lower half of the cervix, presents numerous papillae.

Vessels and Nerves. The *Arteries* of the uterus are the two uterine, from the internal iliac; and the two ovarian, from the aorta. They are remarkable for their tortuous course in the substance of the organ, and for their frequent anastomoses.

The *Veins* are of large size, and correspond with the arteries. In the impregnated uterus, these vessels are termed the *uterine sinuses*, consisting of the lining membrane of the veins adhering to the walls of canals channelled through the substance of the uterus. They terminate in the uterine plexuses.

The *Lymphatics* are of large size in the impregnated uterus, and terminate in the pelvic and lumbar glands.

The *Nerves* are derived from the inferior hypogastric and spermatic plexuses, and from the third and fourth sacral nerves.

For an elaborate account of the nerves of this organ, the student should consult Dr. Lee's "Memoir on the Ganglia and Nerves of the Uterus," 1849; and Dr. Snow Beck's paper "On the Nerves of the Uterus." Phil. Trans. 1846.

The form, size, and situation of the uterus, varies at different periods of life, and under different circumstances.

In the foetus, the uterus is contained in the abdominal cavity, projecting beyond the brim of the pelvis. The cervix is considerably larger than the body.

At puberty, the uterus is pyriform in shape, and weighs from eight to ten drachms. It has descended into the pelvis, the fundus being just below the level of the brim of this cavity. The arbor vitae is distinct, and extends to the upper part of the cavity of that organ.

During, and after menstruation, the organ becomes enlarged, and more vascular, its surfaces rounder; the os externum is rounded, its labia swollen, and the lining membrane of the body thickened, softer, and of a darker colour.

During pregnancy, the uterus increases in weight from one pound and a half to three pounds. It becomes enormously enlarged, and projects into the hypogastric and lower part of the umbilical regions. This enlargement, which continues up to the sixth month of gestation, is partially due to increased development of pre-existing and new formed muscular tissue. The round ligaments are enlarged, and the broad ligaments become encroached upon by the uterus making its way between their laminae. The mucous membrane becomes more vascular, its mucous follicles and glands enlarged, the rugae and folds in the canal of the cervix become obliterated; the blood and lymphatic vessels, as well as the nerves, according to the researches of Dr. Lee, become greatly enlarged.

After parturition, the uterus nearly regains its usual size, weighing from two to three ounces, but its cavity is larger than in the virgin state, the external orifice is more marked and assumes a transverse direction, its edges present a fissured surface, its vessels are tortuous, and its muscular layers are more defined.

In old age, the uterus becomes atrophied, and paler and denser in texture; a more distinct constriction separates the body and cervix. The ostium internum, and occasionally the vaginal orifice, often becomes obliterated, and its labia almost entirely disappear.

Appendages of the Uterus

The appendages of the uterus are the Fallopian tubes, the ovaries and their ligaments, and the round ligaments. These structures, together with their nutrient vessels and nerves, and some scattered muscular fibres, are enclosed between the two folds of peritoneum, which constitute the broad ligaments: they are placed in the following order: in front is the round ligament; the Fallopian tube occupies the free margin of the broad ligament: the ovary and its ligament are behind and below the latter.

The *Fallopian Tubes*, or oviducts, convey the ova from the ovaries to the cavity of the uterus. They are two in number, one on each side, and situated in the free margin of the broad ligament, extending from each superior angle of the uterus to the sides of the pelvis. Each tube varies from three to five inches in length; its canal is exceedingly minute, and commences at the superior angle of the uterus by a minute orifice, the *ostium internum*, which will hardly admit a fine bristle; it continues narrow along the inner half of the tube; it then gradually widens into a trumpet-shaped extremity, which becomes contracted at its termination. This orifice is called the *ostium abdominale*, and communicates with the peritoneal cavity. Its margins are surrounded by a series of fringe-like processes, termed *fimbriae*, and one of these processes is connected with the outer end of the ovary. To this part of the tube the name *fimbriated extremity* is applied; it is also called *morsus diaboli*, from the peculiar manner in which it embraces the surface of the ovary during sexual excitement.

348.—The Uterus and its Appendages. Anterior View.

Structure. The Fallopian tube consists of three coats, serous, muscular, and mucous. The *external* or *serous coat* is derived from the peritoneum.

The *middle* or *muscular coat* consists of an external longitudinal and an internal or circular layer of muscular fibres continuous with those of the uterus.

The *internal* or *mucous coat* is continuous with the mucous lining of the uterus, and at the free extremity of the tube with the peritoneum. It is thrown into longitudinal folds in the outer part of the tube, which indicate its adaptation for dilatation, and is covered by columnar ciliated epithelium. This form of epithelium is also found on the inner and outer surfaces of the fimbriae.

The *Ovaries* (*testes muliebres*, Galen) are analogous to the testes in the male. They are two oblong flattened oval bodies, situated one on each side of the uterus in the posterior part of the broad ligament, behind and below the Fallopian tubes. Each ovary is connected with the broad ligament by its anterior margin; by its inner extremity to the uterus by a proper ligament, the ligament of the ovary; and by its outer extremity to the fimbriated extremity of the Fallopian tube by a short ligamentous cord. Each ovary is of a whitish colour, and presents either a smooth or puckered uneven surface. It is about an inch and a half in length, three quarters of an inch in width, and nearly half an inch thick, and weighs from one to two drachms. Its surfaces and posterior convex border are free, its anterior straight border being attached to the broad ligament.

Structure. The ovary is completely invested by peritoneum, excepting along its anterior attached margin; beneath this is the proper fibrous coat of the organ, the *tunica albuginea*, which is extremely dense and firm in structure, and encloses a firm reddish-white vascular structure, the *stroma*, formed apparently by delicate prolongations continued inwards from the external coat. In the meshes of this tissue are numerous vesicles, the ovisacs of the future ova, and termed the *Graafian vesicles*. Before impregnation they vary in number, from ten to fifteen or twenty, and are from the size of a pin's head to a pea; but Dr. Martin Barry has shewn, that a large number of microscopic ovisacs exist in the parenchyma of the organ, few of which produce ova. These vesicles have thin transparent walls, and are filled with a clear, colourless, albuminous fluid, the largest vesicles appearing on the surface of the ovary as semi-transparent elevations. Each vesicle includes, besides the fluid contents above mentioned, a small round vesicular body, the *ovum*. The Graafian vesicles have two coats, an external vascular and an internal coat, called the *ovi capsule*, which is lined by a granular epithelial layer, the *membrana granulosa*. On rupture of the vesicle, the ovum escapes through the fimbriated extremity of the Fallopian tube to the cavity of the uterus. After conception, a yellow mass, the *corpus luteum*, is found in one or both ovaries. The corpus luteum consists of a circular brownish yellow mass, of moderately firm consistence, containing a small cavity in its centre, originally occupied by the ovum, and now more or less obliterated. The cavity is lined by a puckered membrane, the remains of the ovisac. In recent corpora lutea, the aperture through which the ovum escaped from the ovisac remains visible; but at a later period this becomes closed, a small puckered cicatrix on the surface of the ovary indicating the existence of the aperture. Similar appearances, but of smaller size and wanting the central cavity, are sometimes found in the unimpregnated ovaries; these are false corpora lutea.

In the foetus, the ovaries are situated, like the testes, in the lumbar region, near the kidneys. They may be distinguished from those bodies at an early period by their elongated and flattened form, and by their position, which is at first oblique and then nearly transverse. They gradually descend into the pelvis.

The *Ligament of the Ovary* is a rounded cord, which extends from each superior angle of the uterus to the inner extremity of the ovary; it consists of fibrous tissue, and a few muscular fibres derived from the uterus.

The *Round Ligaments* are two rounded cords, between four and five inches in length, situated between the layers of the broad ligament, in front of and below the Fallopian tube. Commencing on each side at the superior angle of the uterus, each ligament passes forwards and outwards through the internal abdominal ring, along the inguinal canal to the labia majora, in which it becomes lost. Each ligament consists of areolar tissue, vessels, and nerves, besides a dense bundle of fibrous tissue, and muscular fibres prolonged from the uterus, enclosed in a duplicature of peritoneum, which in the foetus is prolonged in the form of a tubular process for a short distance into the inguinal canal: this process is called the *canal of Nuck*. It is generally obliterated in the adult, but sometimes remains pervious even in advanced life. It is analogous to the peritoneal pouch which accompanies the descent of the testis.

Vessels and Nerves. The *Arteries* of the ovaries and Fallopian tubes are the ovarian from the aorta. They anastomose with the termination of the uterine arteries, and enter the attached border of the ovary.

The *Veins* follow the course of the arteries; they form a plexus near the ovary, the *pampiniform plexus*.

The *Nerves* are derived from the spermatic plexus, the Fallopian tube receiving a branch from one of the uterine nerves.

Mammary Glands

The *mammae*, or breasts, are accessory glands of the generative system, which secrete the milk. They exist in the male as well as in the female; but in the former only in a rudimentary state, unless their growth is excited by peculiar circumstances. They are two large hemispherical eminences, situated towards the lateral aspect of the pectoral region, corresponding to the interval between the third and sixth or seventh ribs, and extending from the side of the sternum to the axilla. Their weight and dimensions differ at different periods of life and in different individuals. Before puberty they are of small size, but enlarge as the generative organs become more completely developed. They increase during pregnancy, and especially after delivery, and become atrophied in old age. The left mamma is generally a little larger than the right. Their base is nearly circular, flattened or slightly concave, and having their long diameter directed upwards and outwards towards the axilla; they are separated from the Pectoral muscles by a thin layer of superficial fascia. The outer surface of the mamma is convex, and presents, just below the centre, a small conical prominence, the nipple (*mammilla*). The surface of the nipple is dark-coloured, and surrounded by an areola having a coloured tint. In the virgin the areola is of a delicate rosy hue; about the second month of impregnation it enlarges, and acquires a darker tinge, which increases as pregnancy advances, becoming, in some cases, a dark brown, or even black colour. This colour diminishes as soon as lactation is over, but is never entirely lost through life. These changes in the colour of the areola are of extreme importance in forming a conclusion in a case of suspected pregnancy.

The *nipple* is a cylindrical or conical eminence, capable of undergoing a sort of erection from mechanical excitement. It is of a pink or brownish hue, its surface wrinkled and provided with papillae, and its summit perforated by numerous orifices, the apertures of the lactiferous ducts. Near the base of the nipple and upon the surface of the areola are numerous sebaceous glands, which become much enlarged during lactation, and present the appearance of small tubercles beneath the skin. These glands secrete a peculiar fatty substance, which serves as a protection to the integument of the nipple in the act of sucking. The nipple consists of numerous vessels, which form a kind of erectile tissue, intermixed with plain muscular fibres.

Structure. The mamma consists of gland tissue; of fibrous tissue, connecting its lobes; and of fatty tissue in the intervals between the lobes. The mammary gland, freed from cellular tissue and fat, is of a pale reddish colour, firm in texture, circular in form, flattened from before backwards, thicker in the centre than at the circumference, and presenting several inequalities on its surface, especially in front. It consists of numerous lobes, and these are composed of lobules, connected together by areolar tissue, blood-vessels, and ducts. The smallest lobules consist of a cluster of rounded vesicles, which open into the smallest branches of the lactiferous ducts; these ducts uniting, form larger ducts, which terminate in a single canal, corresponding with one of the chief subdivisions of the gland. The number of excretory ducts varies from fifteen to twenty: they are termed the *tubuli lactiferi, galactophori*. They converge towards the areola, beneath which they form dilatations, or ampullae, which serve as reservoirs for the milk, and, at the base of the nipple, become contracted, and pursue a straight course to its summit, perforating it by separate orifices considerably narrower than the ducts themselves. The ducts are composed of areolar tissue, with longitudinal and transverse elastic fibres, and longitudinal muscular fibres: their mucous lining is continuous, at the point of the nipple, with the integument: its epithelial lining is of the tessellated or scaly variety.

The *fibrous tissue* of the mamma invests the entire surface of the breasts, and sends down septa between its lobes, connecting them together.

The *fatty tissue* surrounds the surface of the gland, and occupies the intervals between its lobes and lobules. It usually exists in considerable abundance, and determines the form and size of the gland. There is no fat immediately beneath the areola and nipple.

Vessels and Nerves. The *Arteries* supplying the mammae are derived from the thoracic branches of the axillary, the intercostals, and internal mammary.

The *Veins* describe an anastomotic circle round the base of the nipple, called by Haller, the *circulus venosus*. From this, large branches transmit the blood to the circumference of the gland, and end in the axillary and internal mammary veins.

The *Lymphatics* run along the lower border of the Pectoralis major to the axillary glands.

The *Nerves* are derived from the anterior and lateral cutaneous nerves of the thorax.

THE SURGICAL ANATOMY OF INGUINAL HERNIA

Dissection (fig. 145). For the dissection of the parts concerned in inguinal hernia, a male subject, free from fat, should always be selected. The body should be placed in the prone position, the abdomen and pelvis raised by means of blocks placed beneath them, and the lower extremities rotated outwards, so as to make the parts as tense as possible. If the abdominal walls are flaccid, the cavity of the abdomen should be inflated by an aperture through the umbilicus. An incision should be made along the middle line, from the umbilicus to the pubes, and continued along the front of the scrotum; and a second incision, from the anterior superior spine of the ilium to just below the umbilicus. These incisions should divide the integument; and the triangular-shaped flap included between them should be reflected downwards and outwards, when the superficial fascia will be exposed.

The *superficial fascia* in this region consists of two layers, between which are found the superficial vessels and nerves, and the inguinal lymphatic glands.

The *superficial layer* is thick, areolar in texture, containing adipose tissue in its meshes, the quantity of which varies in different subjects. Below, it passes over Poupart's ligament, and is continuous with the outer layer of the superficial fascia of the thigh. This fascia is continued as a tubular prolongation around the outer surface of the cord and testis. In this situation, it changes its character: it becomes thin, destitute of adipose tissue, and of a pale reddish colour, and assists in forming the dartos. From the scrotum, it may be traced backwards to be continuous with the superficial fascia of the perinaeum. This layer should be removed, by dividing it across in the same direction as the external incisions, and reflecting it downwards and outwards, when the following vessels and nerves will be exposed:

The superficial epigastric, circumflex iliac, and external pudic vessels; the terminal filaments of the ilio-hypogastric and ilio-inguinal nerves; and the upper chain of inguinal lymphatic glands.

The *superficial epigastric artery* crosses Poupart's ligament, and ascends obliquely towards the umbilicus, lying midway between the spine of the ilium and the pubes. It supplies the integument, and anastomoses with the deep epigastric. This vessel is a branch of the common femoral artery, and pierces the fascia lata, below Poupart's ligament. Its accompanying vein empties itself into the internal saphena, piercing previously the cribriform fascia.

The *superficial external pudic artery* passes transversely inwards across the spermatic cord, and supplies the integument of the hypogastric region, and of the penis and scrotum. This vessel is usually divided in the first incision made in the operation for inguinal hernia, and usually requires the application of a ligature to suppress the haemorrhage.

The *circumflex iliac artery* passes outwards towards the crest of the ilium. The veins accompanying the latter vessels are usually much larger than the arteries: they terminate in the internal saphena vein.

Lymphatic vessels are found, taking the same course as the blood-vessels: they return the lymph from the superficial structures in the lower part of the abdomen, the scrotum, penis, and external surface of the buttock, and terminate in a small chain of lymphatic glands, three or four in number, which lie on a level with Poupart's ligament (p. 500).

Nerves. The terminal branch of the ilio-inguinal nerve emerges at the external abdominal ring; and the hypogastric branch of the ilio-hypogastric nerve perforates the aponeurosis of the external oblique, above and to the outer side of the external ring (pp. 602, 603).

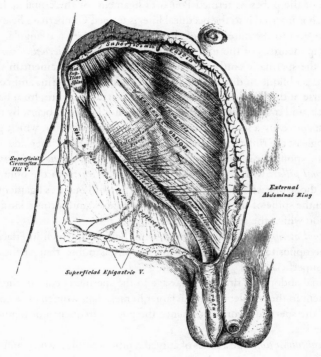

349.—Inguinal Hernia. Superficial Dissection.

The *deep layer of superficial fascia* should be divided across in the same direction as the external incisions, separated from the aponeurosis of the External oblique, to which it is connected by delicate areolar tissue, and reflected downwards and outwards. It is thin, aponeurotic in structure, and of considerable strength. It is intimately adherent, in the middle line, to the linea alba, and below, to the whole length of Poupart's ligament and upper part of the fascia lata. It forms a thin tubular prolongation around the outer surface of the cord, which blends with the superficial layer, and is continuous with the dartos of the scrotum. From the back of the scrotum, the conjoined layers may be traced into the perinaeum, where they are continuous with the deep layer of the superficial fascia in this region, which is attached, behind, to the triangular ligament, and on either side, to the rami of the pubes and ischia. The connections of this fascia serve to explain the course taken by the urine in extravasation of this fluid from rupture of the urethra; passing forwards from the perinaeum into the scrotum, it ascends on to the abdomen, but is prevented extending into the thighs by the attachment of the fascia to the rami of the pubes and ischia, on each side, and to the deep layer of Poupart's ligament in front.

The *aponeurosis of the External oblique muscle* is exposed on the removal of this fascia. It is a thin, strong, membranous aponeurosis, the fibres of which are directed obliquely downwards and inwards. It is attached to the anterior superior spinous process of the ilium, the spine of the pubes, the pectineal line, front of the pubes, and linea alba. That portion of the aponeurosis which extends from the anterior superior spine of the ilium, to the spine of the pubes, is termed Poupart's ligament, or the crural arch; and that portion which is inserted into the pectineal line, is termed Gimbernat's ligament.

Just above and to the outer side of the crest of the pubes, a triangular interval is seen in the aponeurosis of the External oblique, called the *external abdominal ring*; it transmits the spermatic cord in the male, and the round ligament in the female. This aperture is oblique in direction, somewhat triangular in form, and corresponds with the course of the fibres of the aponeurosis. It usually measures from base to apex about an inch, and transversely about half an inch. It is bounded below by the crest of the os pubis; above, by a series of curved fibres, the *intercolumnar*, which pass across the upper angle of the ring so as to increase its strength; and on either side, by the free borders of the aponeurosis, which are called the *columns* or *pillars of the ring*.

The *external pillar*, which, at the same time, is inferior from the obliquity of its direction, is the strongest; it is formed by that portion of Poupart's ligament, which is inserted into the spine of the pubes; it projects forwards, and forms a kind of groove or canal, upon which the spermatic cord rests.

The *internal* or *superior pillar* is a broad, thin, flat band, which interlaces with its fellow of the opposite side, in front of the symphysis pubis, that of the right side being most superficial.

The external abdominal ring gives passage to the spermatic cord in the male, and round ligament in the female; it is much larger in men than women on account of the large size of the spermatic cord, and hence the greater frequency of inguinal hernia in the former sex.

The *Intercolumnar fibres* are a series of curved tendinous fibres, which arch across the lower part of the aponeurosis of the External oblique. They have received their name from stretching across between the two pillars of the external ring; they increase the strength of the membrane which bounds the upper part of this aperture, and prevent the divergence of the pillars from one another. They are thickest below, where they are connected to the outer third of Poupart's ligament, and taking a curvilinear course, the convexity of which is directed downwards, are inserted into the linea alba. They are much thicker and stronger at the outer angle of the external ring than internally, and are more strongly developed in the male than in the female. These fibres are continuous with a thin fascia, which is closely connected to the margins of the external ring, and has received the name of the *intercolumnar* or *external spermatic fascia*; it forms a tubular prolongation around the outer surface of the cord and testis, and encloses them in a distinct sheath. The sac of an inguinal hernia, in passing through the external abdominal ring, receives an investment from the intercolumnar fascia.

The finger should be introduced a slight distance into the external ring, and if the limb is extended and rotated outwards, the aponeurosis of the External oblique, together with the iliac portion of the fascia lata, will be felt to become tense, and the external ring much contracted; if the limb is, on the contrary, flexed upon the pelvis and rotated inwards, this aponeurosis will become lax, and the external ring sufficiently enlarged to admit the finger with comparative ease; hence the latter position should always be assumed in cases where the taxis is applied for the reduction of an inguinal hernia, in order that the abdominal walls may be as much relaxed as possible.

The aponeurosis of the External oblique should be removed by dividing it across in the same direction as the external incisions, and reflecting it outwards; great care is requisite in separating it from the aponeurosis of the muscle beneath. The lower part of the Internal oblique and the Cremaster are then exposed, together with the inguinal canal, which contains the spermatic cord (fig. 350). The mode of insertion of Poupart's ligament and Gimbernat's ligament into the pubes should also be examined.

Poupart's Ligament, or the crural arch, extends from the anterior superior spine of the ilium to the spine of the pubes. It is also attached to the pectineal line to the extent of about an inch, forming Gimbernat's ligament. Its general direction is curved towards the thigh, where it is continuous with the fascia lata. Its outer half is rounded, oblique in its direction, and continuous with the iliac fascia. Its inner half gradually widens at its attachment to the pubes, is more horizontal in direction, and lies behind the spermatic cord.

Gimbernat's Ligament is that portion of the aponeurosis of the External oblique which is inserted into the pectineal line; it is thin, membranous in structure, triangular in shape, the base directed outwards, and passes upwards and backwards beneath the spermatic cord, from the spine of the os pubis to the pectineal line, to the extent of about half an inch.

The *Internal oblique Muscle* has been described (p. 264). The part which is now exposed is partly muscular and partly tendinous in structure. Those fibres which arise from the outer part of Poupart's ligament are thin, pale in colour, curve downwards, and terminate in an aponeurosis, which passes in front of the Rectus and Pyramidalis muscles, to be inserted into the crest of the os pubis and pectineal line, to the extent of half an inch, in common with that of the Transversalis muscle, forming by their junction the conjoined tendon. This tendon is placed immediately behind Gimbernat's ligament and the external abdominal ring, and serves to strengthen what would otherwise be a very weak point in the abdominal wall. When the pouch of inguinal hernia passes directly through the external ring, forming what is called *direct inguinal hernia*, the conjoined tendon usually forms one of its coverings.

The *Triangular Ligament* is a band of tendinous fibres, of a triangular shape, which is continued from Poupart's ligament at its attachment to the pectineal line upwards and inwards, beneath the inner pillar of the external ring to the linea alba.

The *Cremaster* is a slender muscular fasciculus, which arises from the middle of Poupart's ligament at the inner side of the Internal oblique, being connected with this muscle, and also occasionally with the Transversalis. It passes along the outer side of the spermatic cord, descends with it through the external ring upon the front

350.—Inguinal Hernia, showing the Internal Oblique, Cremaster, and Spermatic Canal.

and sides of the cord, and forms a series of loops, which differ in thickness and length in different subjects. Those at the upper part of the cord are exceedingly short, but they become in succession longer and longer, the longest reaching down as low as the testicle, where a few are inserted into the tunica vaginalis. These loops are united together by areolar tissue, and form a thin covering over the cord, the *fascia cremasterica*. The fibres ascend along the inner side of the cord, and are inserted, by a small pointed tendon, into the crest of the os pubis and front of the sheath of the Rectus muscle.

It will be observed, that the origin and insertion of the Cremaster is precisely similar with that of the lower fibres of the Internal oblique. This fact affords an easy explanation of the manner in which the testicle and cord are invested by this muscle. At an early period of foetal life, the testis is placed at the lower and back part of the abdominal cavity, but during its descent towards the scrotum, which takes place before birth, it passes beneath the arched border of the Internal oblique. In its passage beneath this muscle some fibres are derived from its lower part, which accompany the testicle and cord into the scrotum.

It occasionally happens that the loops of the Cremaster surround the cord, some lying behind as well as in front. It is probable, that under these circumstances the testis, in its descent, passed through instead of beneath the fibres of the Internal oblique.

In the descent of an oblique inguinal hernia, which takes the same course as the spermatic cord, the Cremaster muscle forms one of its coverings. This muscle becomes largely developed in cases of hydrocele and large old scrotal hernia. No such muscle exists in the female, but an analogous structure is developed in those cases where an oblique inguinal hernia descends beneath the margin of the Internal oblique.

The Internal oblique should be detached from Poupart's ligament, separated from the Transversalis to the same extent as in the previous incisions, and reflected inwards on to the sheath of the Rectus (fig. 354). The circumflex ilii vessels, which lie between these two muscles, form a valuable guide to their separation.

The *Transversalis Muscle* has been previously described (p. 265). Its lower part is partly fleshy and partly tendinous in structure; this portion arises from the outer third of Poupart's ligament, and arching downwards and inwards over the cord, terminates in an aponeurosis, which is inserted into the linea alba, the crest of the pubes, and into the pectineal line to the extent of an inch, forming, together with the Internal oblique, the conjoined tendon. Between the lower border of this muscle and Poupart's ligament, a space is left in which is seen the fascia transversalis.

The *Inguinal*, or *Spermatic Canal*, contains the spermatic cord in the male, and the round ligament in the female. It is an oblique canal, about an inch and a half in length, directed downwards and inwards, and placed parallel with, and a little above, Poupart's ligament. It communicates, above, with the cavity of the abdomen, by means of the internal abdominal ring, which is the point where the cord enters the spermatic canal; and terminates, below, at the external ring. It is bounded in front, in its whole length, by the aponeurosis of the External oblique, and by the Internal oblique for its outer third; behind, by the transversalis fascia, the conjoined tendon of the Internal oblique and Transversalis, and the triangular ligament; above, by the arched fibres of the Internal oblique and Transversalis; below, by the union of the fascia transversalis with Poupart's ligament. That form of protrusion in which the intestine follows the course of the spermatic cord along the spermatic canal, is called *oblique inguinal hernia*.

The *Fascia Transversalis* is a thin aponeurotic membrane, which lies between the inner surface of the Transversalis muscle and the peritoneum. It forms part of the general layer of fascia which lines the interior of the abdominal and pelvic cavities, and is directly continuous with the iliac and pelvic fasciae.

In the inguinal region, the transversalis fascia is thick and dense in structure, and joined by fibres from the aponeurosis of the Transversalis; but it becomes thin and cellular as it ascends to the Diaphragm. Below, it has the following attachments: external to the femoral vessels, it is connected to the posterior margin of Poupart's ligament, and is there continuous with the iliac fascia. Internal to these vessels, it is thin, and attached to the pubes and pectineal line, behind the conjoined tendon with which it is united; and, corresponding to the point where the femoral vessels pass into the thigh, this fascia descends in front of them, forming the anterior wall of the crural sheath.

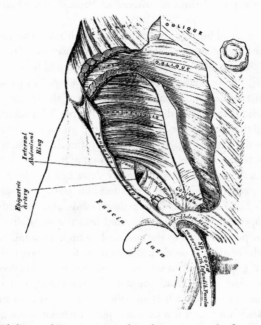

351.—Inguinal Hernia, showing the Transversalis Muscle, the Transversalis Fascia, and the Internal Abdominal Ring.

The *Internal Abdominal Ring* is situated in the transversalis fascia, midway between the anterior superior spine of the ilium and the spine of the pubes, and about half an inch above Poupart's ligament. It is of an oval form, the extremities of the oval directed upwards and downwards, varies in size in different subjects, and is much larger in the male than the female. It is bounded, above, by the arched fibres of the Transversalis muscle, and internally, by the epigastric vessels. It transmits the spermatic cord in the male, and the round ligament in the female; and from its circumference, a thin, funnel-shaped membrane, the infundibuliform, or transversalis fascia, is continued round the cord and testis, enclosing them in a distinct pouch. When the sac of an oblique inguinal hernia passes through the internal ring, the transversalis fascia forms one of its coverings.

Between the peritoneum and the transversalis fascia, is a quantity of loose areolar tissue. In some subjects it is of considerable thickness, and loaded with adipose tissue. Opposite the internal ring, it is continued round the surface of the cord, forming for it a loose sheath.

The *Epigastric Artery* bears a very important relation to the internal abdominal ring. This vessel lies between the transversalis fascia and peritoneum, and passes obliquely upwards and inwards, from its origin from the external iliac, to the margin of the sheath of the Rectus muscle. In this course, it lies along the lower and inner margin of the internal ring, and beneath the commencement of the spermatic cord, the vas deferens curving round it as it passes from the ring into the pelvis.

The *Peritoneum*, corresponding to the inner surface of the internal ring, presents a well-marked depression, the depth of which varies in different subjects. A thin fibrous band is continued from it along the front of the cord, for a variable distance, and becomes ultimately lost. This is the remains of the pouch of peritoneum which, in the foetus, accompanies the cord and testis into the scrotum, the obliteration of which commences soon after birth. In some cases, the fibrous band can only be traced a short distance; but occasionally, it may be followed, as a fine cord, as far as the upper end of the tunica vaginalis. Sometimes the tube of peritoneum is only closed at intervals, and presents a sacculated appearance; or a single pouch may extend along the whole length of the cord, which may be closed above; or the pouch may be directly continuous with the peritoneum by an opening at its upper part.

Inguinal Hernia

Inguinal hernia includes that form of protrusion which makes its way through the abdomen in the inguinal region.

There are two principal varieties of inguinal hernia: external, or oblique, and internal, or direct.

External, or *Oblique Inguinal Hernia*, the most frequent of the two, is that form of protrusion which takes the same course as the spermatic cord. It is called *external*, from the neck of the sac being on the outer or iliac side of the epigastric artery.

Internal, or *Direct Inguinal Hernia*, is that form of protrusion which does not follow the same course as the cord, but protrudes through the abdominal wall on the inner or pubic side of the epigastric artery.

Oblique Inguinal Hernia

In oblique inguinal hernia, the intestine escapes from the abdominal cavity at the internal ring, pushing before it a pouch of peritoneum, which forms the hernial sac. As it enters the inguinal canal, it receives an investment from the subserous areolar tissue, and is enclosed in the infundibuliform process of the transversalis fascia. In passing along the inguinal canal, it displaces upwards the arched fibres of the Transversalis and Internal oblique muscles, and is surrounded by the fibres of the Cremaster. It then passes along the front of the cord, and escapes from the inguinal canal at the external ring, receiving an investment from the intercolumnar fascia. Lastly, it descends into the scrotum, receiving coverings from the superficial fascia and the integument.

The various coverings of this form of hernia, after it has passed through the external ring, are, from without inwards, the integument, superficial fascia, intercolumnar fascia, Cremaster muscle, transversalis fascia, subserous cellular tissue, and peritoneum.

This form of hernia lies in front of the vessels of the spermatic cord, and seldom extends below the testis, on account of the intimate adhesion of the coverings of the cord to the tunica vaginalis.

The *seat of stricture* in oblique inguinal hernia, is either at the external ring, in the inguinal canal, caused by the fibres of the Internal oblique or Transversalis; or at the internal ring, more frequently in the latter situation. If it is situated at the external ring, the division of a few fibres at one point of its circumference, is all that is necessary for the replacement of the hernia. If in the inguinal canal, or at the internal ring, it will be necessary to divide the aponeurosis of the External oblique so as to lay open the inguinal canal. In dividing the stricture, the direction of the incision should be directly upwards.

When the intestine passes along the spermatic canal, and escapes from the external ring into the scrotum, it is called *common oblique inguinal*, or *scrotal hernia*. If the intestine does not escape from the external ring, but is retained in the inguinal canal, it is called *incomplete inguinal hernia*, or *bubonocele*. In each of these cases, the coverings which invest it will depend upon the extent to which it descends in the inguinal canal.

There are two other varieties of oblique inguinal hernia: the congenital, and infantile.

Congenital Hernia is liable to occur in those cases where the pouch of peritoneum which accompanies the cord and testis in its descent in the foetus remains unclosed, and communicates directly with the peritoneum. The intestine descends along this pouch into the cavity of the tunica vaginalis, and lies in contact with the testis. This form of hernia has no proper sac, being contained within the tunica vaginalis.

In Infantile Hernia, the hernial sac descends along the inguinal canal into the scrotum, behind the pouch of peritoneum which accompanies the cord and testis into the same part. The abdominal aperture of this pouch is closed, but the portion contained in the inguinal canal remains unobliterated. The hernial sac is consequently invested, more or less completely, by the posterior layer of the tunica vaginalis, from which it is separated by a little loose areolar tissue: so that in operating upon this variety of hernia, three layers of peritoneum would require division; the first and second being the layers of the tunica vaginalis, the third the anterior layer of the hernial sac.

Direct Inguinal Hernia

In direct inguinal hernia, the protrusion makes its way through some part of the abdominal wall internal to the epigastric artery, and passes directly through the abdominal parietes and external ring. At the lower part of the abdominal wall is a triangular space (Hesselbach's triangle), bounded, externally, by the epigastric artery; internally, by the margin of the Rectus muscle; below, by Poupart's ligament. The conjoined tendon is stretched across the inner two-thirds of this space, the remaining portion of the space being filled in by the transversalis fascia.

In some cases, the hernial protrusion escapes from the abdomen on the outer side of the conjoined tendon, pushing before it the peritoneum, the subserous cellular tissue, and the transversalis fascia. It then enters the inguinal canal, passing along nearly its whole length, and finally emerges from the external ring, receiving an investment from the intercolumnar fascia. The coverings of this form of hernia are precisely similar to those investing the oblique form of protrusion.

In other cases, and this is the more frequent variety, the intestine is either forced through the fibres of the conjoined tendon, or the tendon is gradually distended in front of it, so as to form a complete investment for it. The intestine then enters the lower end of the inguinal canal, escapes at the external ring, lying on the inner side of the cord, and receives additional coverings from the superficial fascia and the integument. This form of hernia has the same coverings as the oblique variety, excepting that the conjoined tendon is substituted for the Cremaster, and the infundibuliform fascia is replaced by a part of the general fascia transversalis.

The *seat of stricture* in both varieties of direct hernia is most frequently at the neck of the sac, or at the external ring. In that form of hernia which perforates the conjoined tendon, it not unfrequently occurs at the edges of the fissure through which the gut passes. In dividing the stricture, the incision should in all cases be directed upwards.

If the hernial protrusion passes into the inguinal canal, but does not escape from the external abdominal ring, it forms what is called *incomplete direct hernia*. This form of hernia is usually of small size, and, in corpulent persons, very difficult of detection.

Direct inguinal hernia is of much less frequent occurrence than the oblique, their comparative frequency being, according to Cloquet, as one to five. It occurs far more frequently in men than women, on account of the larger size of the external ring in the former sex. It differs from the oblique in its smaller size and globular form, dependent most probably on the resistance offered to its progress by the transversalis fascia and conjoined tendon. It differs also in its position, being placed over the pubes, and not in the course of the inguinal canal. The epigastric artery runs along the outer or iliac side of the neck of the sac, and the spermatic cord along its external and posterior side, not directly behind it, as in oblique inguinal hernia.

SURGICAL ANATOMY OF FEMORAL HERNIA

The dissection of the parts comprised in the anatomy of femoral hernia should be performed, if possible, upon a female subject free from fat. The subject should lie upon its back; a block is first placed under the pelvis, the thigh everted, and the knee slightly bent, and retained in this position. An incision should then be made from the anterior superior spinous process of the ilium along Poupart's ligament to the symphysis pubis; a second incision should be carried transversely across the thigh about six inches beneath the preceding; and these are to be connected together by a vertical one carried along the inner side of the thigh. These several incisions should divide merely the integument; this is to be reflected outwards, when the superficial fascia will be exposed.

The *Superficial Fascia* at the upper part of the thigh consists of two layers, between which are found the cutaneous vessels and nerves, and numerous lymphatic glands.

The *superficial layer* is a thick and dense cellulo-fibrous membrane, in the meshes of which is found a considerable amount of adipose tissue, which varies in quantity in different subjects; this layer may be traced upwards over Poupart's ligament to be continuous with the superficial fascia of the abdomen; whilst below, and on the inner and outer sides of the limb, it is continuous with the superficial fascia covering the rest of the thigh. This layer should be detached by dividing it across in the same direction as the external incisions; its removal will be facilitated by commencing at the lower and inner angle of the space, detaching it at first from the front of the internal saphena vein, and dissecting it off from the anterior surface of this vessel and its branches; it should then be reflected outwards, in the same manner as the integument. The cutaneous vessels and nerves, and superficial inguinal glands, are then exposed, lying upon the deep layer of superficial fascia. These are the internal saphenous vein, and the superficial epigastric, superficial circumflexa ilii, and superficial pudic vessels, as well as numerous lymphatics ascending with the saphena vein to the inguinal glands.

The *Internal Saphena Vein* is a vessel of considerable size, which ascends obliquely upwards along the inner side of the thigh, below Poupart's ligament. It passes through the saphenous opening in the fascia lata to terminate in the femoral vein. This vessel is accompanied by numerous lymphatics, which return the lymph from the dorsum of the foot and inner side of the leg and thigh; they terminate in the inguinal glands, which surround the saphenous opening. Converging towards the same point are the superficial epigastric vessels, which run across Poupart's ligament, obliquely upwards and inwards, to the lower part of the abdomen; the circumflexa ilii vessels pass obliquely outwards along Poupart's ligament to the crest of the ilium; and the superficial external pudic vessels, pass inwards to the perineal and scrotal regions. These vessels supply the subcutaneous areolar tissue and the integument, and are accompanied by numerous lymphatic vessels, which return the lymph from the same parts to the inguinal glands.

The *Superficial Inguinal Glands* are arranged in two groups, one of which is

352.—Femoral Hernia. Superficial Dissectiou.

disposed parallel with Poupart's ligament; the other is placed beneath this ligament, surrounding the termination of the saphena vein, and following (occasionally) the course of this vessel a short distance along the thigh. The upper chain receives the lymphatic vessels from the penis, scrotum, lower part of the abdomen, perinaeum, and buttock; the lower chain receives the lymphatic vessels from the lower extremity.

The *Nerves* supplying the integument of this region are derived from the ilio-inguinal, the genito-crural, and anterior crural. The ilio-inguinal nerve may be found on the inner side of the internal saphena vein, the terminal branch of the genito-crural nerve outside the vein, and the middle and external cutaneous nerves more external.

The deep layer of superficial fascia should be divided in the same direction as the external incisions, and separated from the fascia lata; this is easily effected, from its extreme thinness. It is a thin but dense membrane, placed beneath the subcutaneous vessels and nerves, and upon the surface of the fascia lata. It is intimately adherent above to the lower margin of Poupart's ligament, and about one inch below this ligament covers the saphenous opening in the fascia lata, is closely united to its circumference, and is connected to the sheath of the femoral vessels corresponding to its under surface. The portion of fascia covering this aperture is perforated by the internal saphena vein, and by numerous blood and lymphatic vessels; hence it has been termed, from its sieve-like appearance, the *cribriform fascia*. A femoral hernia, in passing through the saphenous opening, receives the cribriform fascia as one of its coverings.

The deep layer of superficial fascia, together with the cribriform fascia, having been removed, the fascia lata is exposed.

The *Fascia Lata*, already described (p. 315), is a dense fibrous aponeurosis, which forms an uniform investment for the whole of this region of the limb. At the upper and inner part of the thigh, a large oval-shaped aperture is observed in it; it transmits the internal saphenous vein and other small vessels, and is called the *saphenous opening*. In order the more correctly to consider the mode of formation of this aperture, the fascia lata in this part of the thigh is described as consisting of two portions, an iliac portion and a pubic portion.

The *iliac portion* of the fascia lata is situated on the outer side of the saphenous opening, covering the outer surface of the Sartorius, the Rectus, and the Psoas and Iliacus muscles. It is attached externally to the crest of the ilium and its anterior superior spine, to the whole length of Poupart's ligament as far internally as the spine of the pubes, and into the pectineal line in conjunction with Gimbernat's ligament, where it becomes continuous with the pubic portion. From the spine of the pubes, it is reflected downwards and outwards, forming an arched margin, the outer boundary (*superior cornu*) of the saphenous opening. This is sometimes called the *falciform process* of the fascia lata (femoral ligament of Hey); it overlies and is adherent to the sheath of the femoral vessels beneath; to its edge is attached the cribriform fascia, and it is continuous below with the pubic portion of the fascia lata by a well-defined curved margin.

The *pubic portion* of the fascia lata is situated at the inner side of the saphenous opening: at the lower margin of this aperture, it is continuous with the iliac portion:

353.—Femoral Hernia, shewing Fascia Lata and Saphenous Opening.

traced upwards, it covers the surface of the Pectineus, Adductor longus, and Gracilis muscles; and passing behind the sheath of the femoral vessels, to which it is closely united, is continuous with the sheath of the Psoas and Iliacus muscles, and is finally lost in the fibrous capsule of the hip-joint. This fascia is attached above to the pectineal line, and internally to the margin of the pubic arch. It may be observed from this description, that the iliac portion of the fascia lata passes in front of the femoral vessels, the pubic portion behind them; an apparent aperture consequently exists between the two, through which the internal saphena joins the femoral vein.

The *Saphenous Opening* is an oval-shaped aperture, measuring about an inch and a half in length, and half an inch in width. It is situated at the upper and inner part of the thigh, below Poupart's ligament, on the pubic side of its centre, and is directed obliquely downwards and outwards.

Its *outer margin* is of a semilunar form, thin, strong, sharply-defined, and lies on a plane considerably anterior to the inner margin. If this edge is traced upwards, it will be seen to form a curved elongated process or cornu (the superior cornu), or *falciform process of Burns*, which ascends in front of the femoral vessels, and curving inwards, is attached to Poupart's ligament and to the spine of the pubis and pectineal line, where it is continuous with the pubic portion. If traced downwards, it is found continuous with another curved margin, the concavity of which is directed upwards and inwards; this is the inferior cornu of the saphenous opening, and is blended with the pubic portion of the fascia lata covering the Pectineus muscle.

The *inner boundary* of the opening is on a plane posterior to the outer margin, and behind the level of the femoral vessels; it is much less prominent and defined than the outer, from being stretched over the subjacent Pectineus muscle. It is through this aperture that a femoral hernia passes after descending along the crural canal.

If the finger is introduced into the saphenous opening while the limb is moved in different directions, the aperture will be found to be greatly constricted on extending the limb, or rotating it outwards, and to be relaxed on flexing the limb and inverting it: hence the necessity of placing the limb in the latter position in employing the taxis for the reduction of a femoral hernia.

The iliac portion of the fascia lata, together with its falciform process, should now be removed, by detaching it from the lower margin of Poupart's ligament, carefully dissecting it from the subjacent structures, and turning it aside, when the sheath of the femoral vessels is exposed descending beneath Poupart's ligament (fig. 354).

The *Crural Arch*, or *Poupart's Ligament*, is the lower border of the aponeurosis of the External oblique muscle, which stretches across between the anterior superior spine of the ilium, to the spine of the os pubis and pectineal line; the portion corresponding to the latter insertion, is called *Gimbernat's ligament*. Its direction is curved downwards towards the thigh, its outer half being oblique, its inner half nearly horizontal. Nearly the whole of the space included between the crural arch and innominate bone is filled in by the parts which descend from the abdomen into the thigh. The outer half of the space is occupied by the Iliacus and Psoas muscles, together with the external cutaneous and anterior crural nerves. The pubic side of the space is occupied by the femoral vessels included in their sheath, a small oval-shaped interval existing between the femoral vein and the inner wall of the sheath, which is occupied merely by a little loose areolar tissue, and occasionally a small lymphatic gland; this is the crural canal, along which a portion of gut descends in femoral hernia.

Gimbernat's Ligament is that part of the aponeurosis of the External oblique muscle, which is reflected downwards and outwards to be inserted into the pectineal line of the os pubis. It is about an inch in length, larger in the male than in the female, almost horizontal in direction in the erect posture, and of a triangular form, the base directed outwards. Its *base*, or outer margin, is concave, thin and sharp, lies in contact with the crural sheath, and is blended with the pubic portion of the fascia lata. Its *apex* corresponds to the spine of the pubes. Its *posterior margin* is attached to the pectineal line. Its *anterior margin* is continuous with Poupart's ligament.

Crural Sheath. If Poupart's ligament is divided, the femoral or crural sheath may be demonstrated as a continuation downwards of the fasciae that line the abdomen, the transversalis fascia passing down in front of the femoral vessels, and the iliac fascia descending behind them; these fasciae are directly continuous on the iliac side of the femoral artery, but a small space exists between the femoral vein and the point where they are continuous on the pubic side of this vessel, which constitutes the femoral or crural canal. The femoral sheath is closely adherent to the contained vessels about an inch below the saphenous opening, becoming blended with the areolar sheath of the vessels, but opposite Poupart's ligament it is much larger than is required to contain them; hence the funnel-shaped form which it presents. The outer border of the sheath is perforated by the genito-crural nerve. Its inner border is pierced by the internal saphena vein, and numerous lymphatic vessels. In front it is covered by the iliac portion of the fascia lata; and behind it is the pubic portion of the same fascia.

354.—Femoral Hernia. Iliac Portion of Fascia Lata removed, and Sheath of Femoral Vessels and Femoral Canal exposed.

Deep Crural Arch. Passing across the front of the crural sheath, and closely connected with it, is a thickened band of fibres, called the *deep crural arch*. It is apparently a thickening of the fascia transversalis, joined externally to the centre of Poupart's ligament, and arching across the front of the crural sheath, to be inserted by a broad attachment into the pectineal line, behind the conjoined tendon. In some subjects, this structure is not very prominently marked, and not unfrequently it is altogether wanting.

If the anterior wall of the sheath is removed, the artery and vein are seen lying side by side, a thin septum separating the two vessels, and another septum separates the vein from the inner wall of the sheath, the septa stretching between the anterior and posterior wall of the sheath, so that each vessel is enclosed in a separate compartment. The interval left between the vein and the inner wall of the sheath is not filled up by any structure, excepting a little loose areolar tissue, a few lymphatic vessels, and occasionally a lymphatic gland; this is the femoral or crural canal, through which a portion of intestine descends in femoral hernia.

The *Crural Canal* is the narrow interval between the femoral vein and the inner wall of the crural sheath. Its length is from a quarter to half an inch, and it extends from Gimbernat's ligament to the upper part of the saphenous opening.

Its *anterior wall* is very narrow, and formed by the fascia transversalis, Poupart's ligament, and the falciform process of the fascia lata.

Its *posterior wall* is formed by the iliac fascia and the pubic portion of the fascia lata.

Its *external wall* is formed by the fibrous septum covering the inner side of the femoral vein.

Its *inner wall* is formed by the junction of the transversalis and iliac fasciae, which forms the inner side of the femoral sheath.

This canal has two orifices: a lower one, the *saphenous opening*, closed by the cribriform fascia; an upper one, the *femoral* or *crural ring*, closed by the septum crurale.

The *Femoral* or *Crural Ring* (fig. 355) is the upper opening of the femoral canal, and leads into the cavity of the abdomen. It is bounded in front by Poupart's ligament and the deep crural arch; behind, by the pubes, covered by the Pectineus muscle, and the pubic portion of the fascia lata; internally, by Gimbernat's ligament, the conjoined tendon, the transversalis fascia, and the deep crural arch; externally, by the femoral vein, covered by its sheath. The femoral ring is of an oval form, its long diameter, directed transversely, measures about half an inch, and it is larger in the female than in the male; hence one of the reasons of the greater frequency of femoral hernia in the former sex.

Position of Parts around the Ring. The spermatic cord in the male, and round ligament in the female, lie immediately above the anterior margin of the femoral ring, and may be divided in an operation for femoral hernia if the incisions for the relief of the stricture are not of limited extent. In the female this is of little importance, but in the male the spermatic artery may be divided.

The *femoral vein* lies on the outer side of the ring.

The *epigastric artery*, in its passage inwards from the external iliac to the umbilicus, passes across the upper and outer angle of the crural ring, and is consequently in great danger of being wounded if the stricture is divided in a direction upwards and outwards.

The *communicating branch* between the epigastric and obturator lies in front of the ring.

The circumference of the ring is thus seen to be bounded by vessels in every part excepting internally and behind. It is in the former position that the stricture is divided in cases of strangulated femoral hernia.

355.—Hernia. The Relations of the Femoral and Internal Abdominal Rings, seen from within the Abdomen. Right Side.

356.—Variations in Origin and Course of Obturator Artery.

The *obturator artery*, when it arises by a common trunk with the epigastric, which occurs once in every three subjects and a half, bears a very important relation to the crural ring (fig. 356). In some cases, it descends on the inner side of the external iliac vein to the obturator foramen, and will consequently lie on the outer side of the crural ring, where there is little danger of its being wounded in the operation for dividing the stricture in femoral hernia. Occasionally, however, this vessel curves along the free margin of Gimbernat's ligament in its passage to the obturator foramen; it would, consequently, skirt along the greater part of the circumference of the crural canal, and could hardly fail in being wounded in the operation.

Septum Crurale. The femoral ring is closed by a layer of condensed areolar tissue, called, by J. Cloquet, the *septum crurale*. This serves as a barrier to the protrusion of a hernia through this part. Its upper surface is slightly concave, and supports a small lymphatic gland, by which it is separated from the subserous areolar tissue and peritoneum. Its under surface is turned towards the femoral canal. The septum crurale is perforated by numerous apertures for the passage of lymphatic vessels, connecting the deep inguinal glands with those surrounding the external iliac artery.

The size of the femoral canal, the degree of tension of its orifices, and, consequently, the degree of constriction of a hernia, varies according to the position of the limb. If the leg and thigh are extended, abducted, or everted, the femoral canal and its orifices are extremely tense from the traction on these parts by Poupart's ligament and the fascia lata, as may be ascertained by passing the finger along it. If, on the contrary, the thigh is flexed upon the pelvis, and, at the same time, adducted and rotated inwards, the femoral canal and its orifices become considerably relaxed; for this reason, the limb should always be placed in the latter position when the application of the taxis is made in attempting the reduction of a femoral hernia.

The septum crurale is separated from the peritoneum by a quantity of loose subserous areolar tissue. In some subjects this tissue contains a considerable amount of adipose substance, which, when protruded forwards in front of the sac of a femoral hernia, may be mistaken for a portion of omentum.

Descent of the Hernia. From the preceding description, it follows, that the femoral ring must be a weak point in the abdominal wall; hence it is, that when violent or long-continued pressure is made upon the abdominal viscera, a portion of intestine may be forced into it, constituting a femoral hernia; and the larger size of this aperture in the female serves to explain the frequency of this form of hernia in women.

When a portion of intestine is forced through the femoral ring, it carries before it a pouch of peritoneum, which forms what is called the *hernial sac*; it receives an investment from the subserous areolar tissue, and from the septum crurale, and descends vertically along the crural canal in the inner compartment of the sheath of the femoral vessels as far as the saphenous opening: at this point, it changes its course, being prevented extending further down the sheath on account of its greater narrowness and close contact with the vessels, and also from the close attachment of the superficial fascia and crural sheath to the lower part of the circumference of the saphenous opening; it is, consequently, directed forwards, pushing before it the cribriform fascia, and curves upwards on to the falciform process of the fascia lata and lower part of the tendon of the External oblique, being covered by the superficial fascia and integument. While the hernia is contained in the femoral canal, it is usually of small size, owing to the resisting nature of the surrounding parts; but when it has escaped from the saphenous opening into the loose areolar tissue of the groin, it becomes considerably enlarged. The direction taken by a femoral hernia in its descent is at first downwards, then forwards and upwards; this should be borne in mind, as in the application of the taxis for the reduction of a femoral hernia, pressure should be directed precisely in the reverse order.

Coverings of the Hernia. The coverings of a femoral hernia from within outwards are peritoneum, subserous areolar tissue, the septum crurale, crural sheath, cribriform fascia, superficial fascia, and integument.

Varieties of Femoral Hernia. If the intestine descends along the femoral canal only as far as the saphenous opening, and does not escape from this aperture, it is called *incomplete femoral hernia*. The small size of the protrusion in this form of hernia, on account of the firm and resisting nature of the canal in which it is contained, renders it an exceedingly dangerous variety of this disease, from the extreme difficulty of detecting the existence of the swelling, especially in corpulent subjects. The coverings of an incomplete femoral hernia would be, from without inwards, integument, superficial fascia, falciform process of fascia lata, fascia propria, septum crurale, subserous cellular tissue, and peritoneum. When, however, the hernial tumour protrudes through the saphenous opening, and directs itself forwards and upwards, it forms a *complete femoral hernia*. Occasionally, the hernial sac descends on the iliac side of the femoral vessels, or in front of these vessels, or even behind them, the sac in the latter instance lying immediately upon the Pectineus muscle, separated from the femoral vessels by the pubic portion of the fascia lata.

The *Seat of Stricture* of a femoral hernia varies: it may be in the peritoneum at the neck of the hernial sac; in the greater number of cases it would appear to be at the point of junction of the falciform process of the fascia lata with the lunated edge of Gimbernat's ligament; or at the margin of the saphenous opening in the thigh. The stricture should in every case be divided in a direction upwards and inwards; and the extent necessary in the majority of cases is about two or three lines. By these means, all vessels or other structures of importance, in relation with the neck of the hernial sac, will be avoided.

SURGICAL ANATOMY OF THE PERINAEUM AND ISCHIO-RECTAL REGION

D*issection*. The student should select a well-developed muscular subject, free from fat; and the dissection should be commenced early, in order that the parts may be examined in as recent a state as possible. A staff having been introduced into the bladder, and the subject placed in the position shown in fig. 357, the scrotum should be raised upwards, and retained in that position, and the rectum moderately distended with tow.

The space which is now exposed, corresponds to the inferior aperture, or outlet of the pelvis. Its deep boundaries are, in front, the pubic arch and sub-pubic ligament; behind, the tip of the coccyx; and on each side, the rami of the pubes and ischia, the tuberosities of the ischia, and great sacro-sciatic ligaments. The space included by these boundaries is somewhat lozenge-shaped, and is limited on the surface of the body by the scrotum in front, by the buttocks behind, and on each side by the inner side of the thighs. It measures, from before backwards, about four inches, and about three in the broadest part of its transverse diameter, between the ischial tuberosities. A line drawn transversely between the anterior part of the tuberosity of the ischium, on either side, in front of the anus, subdivides this space into two portions. The anterior portion contains the penis and urethra, and is called the *perinaeum*. The posterior portion contains the termination of the rectum, and is called the *ischio-rectal region*.

ISCHIO-RECTAL REGION

The ischio-rectal region corresponds to the portion of the outlet of the pelvis situated immediately behind the perinaeum: it contains the termination of the rectum. A deep fossa, filled with fat, is situated on either side of the intestine, between it and the tuberosity of the ischium: this is called the *ischio-rectal fossa*.

The *Ischio-Rectal Region* presents, in the middle line, the aperture of the anus: around this orifice the integument is thrown into numerous folds, which are obliterated on distension of the intestine. The integument is of a dark colour, continuous with the mucous membrane of the rectum, and provided with numerous follicles, which occasionally inflame and suppurate, and may be mistaken for fistulae. The veins around the margin of the anus are occasionally much dilated, forming a number of hard, pendant masses, of a dark bluish colour, covered partly by mucous membrane, and partly by the integument. These tumours constitute the disease called *external piles*.

Dissection. Make an incision through the integument, along the median line, from the base of the scrotum to the anterior extremity of the anus; carry it around the margins of this aperture to its posterior extremity, and continue it backwards about an inch behind the tip of the coccyx. A transverse incision should now be carried across the base of the scrotum, joining the anterior extremity of the preceding; a second, carried in the same direction, should be made in front of the anus; and a third, at the posterior extremity of the gut. These incisions should be sufficiently extensive to enable the dissector to raise the integument from the inner side of the thighs. The flaps of skin corresponding to the ischio-rectal region (fig. 357—2), should now be removed. In dissecting the integument from this region, great care is required, otherwise the External sphincter will be removed, as it is intimately adherent to the skin.

The *Superficial Fascia* is exposed on the removal of the skin: it is very thick, areolar in texture, and contains much fat in its meshes. In it are found ramifying two or three cutaneous branches of the small sciatic nerve; these turn round the inferior border of the Gluteus maximus, and are distributed to the integument in this region.

357.—Dissection of Perinæum and Ischio-Rectal Region.

The *External Sphincter* is a thin flat plane of muscular fibres, elliptical in shape, and intimately adherent to the integument surrounding the margin of the anus. It measures about three or four inches in length, from its anterior to its posterior extremity, being about an inch in breadth, opposite the anus. It arises from the tip of the coccyx, by a narrow tendinous band; and from the superficial fascia in front of that bone; and is inserted into the tendinous centre of the perinaeum, joining with the Transversus perinaei, and the other muscles inserted into this part. Like other sphincter muscles, it consists of two planes of muscular fibre, which surround the margin of the anus, and join at the commissure before and behind.

Relations. By its *superficial surface*, with the integument; by its *deep surface* it is in contact with the Internal sphincter; and is separated from the Levator ani by loose areolar tissue.

The Sphincter ani is a voluntary muscle, supplied by the haemorrhoidal branch of the fourth sacral nerve. This muscle is divided in the operation for fistula in ano; and also in some cases of fissure of the rectum, especially if attended with much pain or spasm. The object of its division is to keep the parts at rest and in contact during the healing process.

The *Internal Sphincter* is a muscular ring, about half an inch in breadth, which surrounds the lower extremity of the rectum, about an inch from the margin of the anus. This muscle is about two lines in thickness, and is formed by an aggregation of the involuntary circular fibres of the intestine. It is paler in colour, and less coarse in texture, than the External sphincter.

The *Ischio-Rectal Fossa* is situated between the end of the rectum and the tuberosity of the ischium, on each side. It is triangular in shape, its base directed to the surface is formed by the integument of the ischio-rectal region; its *apex*, directed upwards, corresponds to the point of division of the obturator fascia, and the thin membrane given off from it, which covers the outer surface of the Levator ani (ischio-rectal fascia). Its dimensions are about an inch in breadth, at the base, and about two inches in depth, being deeper behind than in front. It is bounded, *internally*, by the Sphincter ani, Levator ani, and Coccygeus muscles; *externally*, by the tuberosity of the ischium, and the obturator fascia, which covers the inner surface of the Obturator internus muscle; *in front*, it is limited by the line of junction of the superficial and deep perinaeal fasciae; and *behind*, by the margin of the Gluteus maximus, and the great sacro-sciatic ligament. This space is filled with a large mass of adipose substance, which explains the frequency with which abscesses in the neighbourhood of the rectum burrow to a considerable depth.

If the subject has been injected, on placing the finger on the outer wall of this fossa, the internal pudic artery, with its accompanying veins and nerve, will be felt about an inch and a half above the margin of the ischial tuberosity, but approaching nearer the surface as they pass forwards along the inner margin of the pubic arch. These structures are enclosed in a sheath formed by the obturator fascia, the pudic nerve lying below the artery. Crossing the space transversely, about its centre, are the inferior haemorrhoidal vessels and nerves, branches of the pudic; they are distributed to the integument of the anus, and to the muscles of the lower end of the rectum. These vessels are occasionally of large size, and may give rise to troublesome haemorrhage, when divided in the operation of lithotomy, or for fistula in ano. At the back part of this space may be seen a branch of the fourth sacral nerve; and, at the fore part of the space, a cutaneous branch of the perinaeal nerve.

Perinaeum

The perinaeal space is of a triangular form: its deep boundaries are limited, laterally, by the rami of the pubes and ischia, meeting in front at the pubic arch; behind, by an imaginary transverse line, extending between the tuberosity of the ischium on either side. The lateral boundaries vary, in the adult, from three inches to three inches and a half in length; and the base from two to three inches and a half in breadth; the average diameter being two inches and three-quarters. The variations in the diameter of this space are of extreme interest in connection with the operation of lithotomy, and the extraction of a stone from the cavity of the bladder. In those cases where the tuberosities of the ischia are approximated, it would be necessary to make the incisions in the lateral operation of lithotomy much less oblique, than if the tuberosities were widely separated, and the perinaeal space consequently wider. The perinaeum is subdivided by the median raphe into two equal parts. Of these, the left is the one usually selected to commence the primary incisions in the operation of lithotomy.

In the middle line the perinaeum is convex, and corresponds to the bulb of the urethra. The skin covering it is of a dark colour, thin, freely moveable upon the subjacent parts, and covered with short crisp hairs, which should be removed before the dissection of the part is commenced. In front of the anus, a prominent line commences, the raphe, continuous in front with the raphe of the scrotum. The flaps of integument corresponding to this space having been removed, in the manner shown in fig. 357—1, the superficial fascia is exposed.

The *Superficial Fascia* consists of two layers, as in other regions of the body, superficial and deep.

The *superficial layer* is thick, loose, and areolar in texture, and contains much adipose tissue in its meshes, the amount of which varies in different subjects. In front, it is continuous with the dartos of the scrotum; behind, it is continuous with the subcutaneous areolar tissue surrounding the anus; and, on either side, with the same fascia on the inner side of the thighs. This layer should be carefully removed, after it has been examined, when the deep layer will be exposed.

The *deep layer of superficial fascia* (superficial perinaeal fascia) is thin, aponeurotic in structure, and of considerable strength, serving to bind down the muscles of the root of the penis. It is continuous, in front, with the dartos of the scrotum; on either

358.—The Perinæum. The Integument and Superficial Layer of Superficial Fascia reflected.

side, it is firmly attached to the margins of the rami of the pubes and ischia, external to the crus penis, and as far back as the tuberosity of the ischium; posteriorly, it curves down behind the Transversus perinaei muscle, to join the lower margin of the deep perinaeal fascia. This fascia not only covers the muscles in this region, but sends down a vertical septum from its under surface which separates the back part of the subjacent space into two, being incomplete in front.

In rupture of the anterior portion of the urethra, accompanied by extravasation of urine, the fluid makes it way forwards, beneath this fascia, into the areolar tissue of the scrotum, penis, and anterior and lateral portions of the abdomen: it rarely extends into the areolar tissue on the inner side of the thighs, or backwards around the anus. This limitation of the extravasated fluid to the parts above-named, is easy of explanation, when the attachments of the deep layer of the superficial fascia are considered. When this fascia is removed, the muscles connected with the penis and urethra will be exposed: these are, in the middle line, the Accelerator urinae; on each side, the Erector penis; and behind, the Transversus perinaei.

The *Accelerator urinae* is placed in the middle line of the perinaeum, immediately in front of the anus. It consists of two symmetrical halves, united along the median line by a tendinous raphe. It arises from the central tendon of the perinaeum, and from the median raphe in front. From this point, its fibres diverge like the plumes of a pen; the most posterior form a thin layer, which are lost on the anterior surface

359.—The Superficial Muscles and Vessels of the Perinæum

of the triangular ligament; the middle fibres encircle the bulb and adjacent part of the corpus spongiosum, and join with the muscle of the opposite side, on the upper part of this body, in a strong aponeurosis; the anterior fibres, the longest and most distinct, spread out over the sides of the corpus cavernosum, to be inserted partly into this body, anterior to the Erector penis; partly terminating in a tendinous expansion, which covers the dorsal vessels of the penis. The latter fibres are best seen by dividing the muscle longitudinally, and dissecting it outwards from the surface of the urethra.

Action. This muscle may serve to accelerate the flow of the urine or semen along the canal of the urethra. The middle fibres are supposed, by Krause, to assist in the erection of the corpus spongiosum, by compressing the erectile tissue of the bulb. The anterior fibres, according to Tyrrel, also contribute to the erection of the penis, as they are inserted into, and continuous with, the fascia penis, compressing the dorsal vein during the contraction of the muscle.

The *Erector Penis* covers the unattached part of the crus penis. It is an elongated muscle, broader in the middle than at either extremity, and situated on either side of the lateral boundary of the perinaeum. It arises by tendinous and fleshy fibres from the inner surface of the tuberosity of the ischium, behind the crus penis, from the surface of the crus, and from the adjacent portions of the ramus of the pubes. From these points, fleshy fibres succeed, which end in an aponeurosis which is inserted into the sides and under surface of the crus penis. This muscle compresses the crus penis, and thus serves to maintain this organ erect.

The *Erector Clitoridis* resembles the Erector penis in the male, but is smaller than that muscle.

The *Transversus Perinaei* is a narrow muscular slip, which passes more or less transversely across the back part of the perinaeal space. It arises by a small tendon from the inner side of the ascending ramus of the ischium, and, passing obliquely forwards and inwards, is inserted into the central tendinous point of the perinaeum, joining in this situation with the muscle of the opposite side, the Sphincter ani behind, and the Accelerator urinae in front.

Between the muscles just examined, a triangular space exists, bounded internally by the Accelerator urinae, externally by the Erector penis, the base corresponding to the Transversus perinaei. The floor of this space is formed by the triangular ligament of the urethra (deep perinaeal fascia), and, running from behind forwards in it, are the superficial perinaeal vessels and nerves, the transversus perinaei artery coursing along the posterior boundary of the space, resting upon the Transversus perinaei muscle.

In the lateral operation of lithotomy, the knife is carried obliquely across the back part of this space, downwards and outwards, into the ischio-rectal fossa, dividing the Transversus perinaei muscle and artery, the posterior fibres of the Accelerator urinae, the superficial perinaeal vessels and nerve, and, more posteriorly, the external haemorrhoidal vessels.

The superficial and transverse perinaeal arteries are described at p. 440; and the superficial perinaeal and inferior pudendal nerves, at pp. 610–12.

The Accelerator Urinae and Erector penis muscles, should now be removed, when the deep perinaeal fascia will be exposed, stretching across the front part of the outlet of the pelvis. The urethra is seen perforating its centre, just behind the bulb; and on either side is the crus penis, connecting the corpus spongiosum with the ramus of the ischium and pubes.

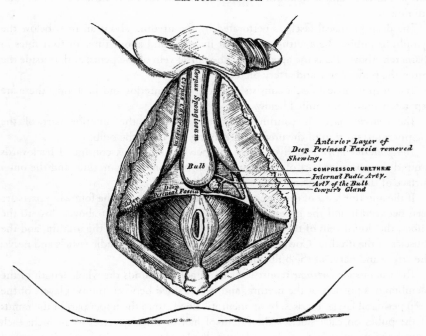

360.—Deep Perinæal Fascia. On the left side, the anterior layer has been removed.

The muscles of the perinaeum in the female are, the

Sphincter vaginae.	Compressor urethrae.
Erector clitoridis.	Sphincter ani.
Transversus perinaei.	Levator ani.
Coccygeus.	

The *Sphincter Vaginae* surrounds the orifice of the vagina, and is analogous to the Accelerator urinae in the male. It is attached, posteriorly, to the central tendon of the perinaeum, where it blends with the Sphincter ani. Its fibres pass forwards on each side of the vagina, to be inserted into the corpora cavernosa and body of the clitoris.

The *Erector Clitoridis* resembles the Erector penis in the male, but is smaller than it.

The *Transversus Perinaei* is inserted into the side of the Sphincter vaginae, and the Levator ani into the side of the vaginae. The other muscles are precisely similar to those in the male.

The *Deep Perinaeal Fascia* (triangular ligament), is a dense membranous lamina, which closes the front part of the outlet of the pelvis. It is triangular in shape, about an inch and a half in depth, attached above, by its apex, to the under surface of the symphysis pubis and sub-pubic ligament; and, on each side, to the rami of the ischia and pubes, beneath the crura penis. Its inferior margin, or base, is directed towards the rectum, and connected to the central tendinous point of the perinaeum. It is continuous with the deep layer of the superficial fascia, in front of the Transversus perinaei muscle, and with a thin fascia, which covers the outer surface of the Levator ani muscle.

The deep perinaeal fascia is perforated by the urethra, about an inch below the symphysis pubis. The aperture is circular in form, and about three or four lines in diameter. Above this is the aperture for the dorsal vein of the penis; and, outside the latter, the pudic nerve and artery pierce it.

The deep perinaeal fascia consists of two layers, anterior and posterior: these are separated above, but united below.

The *anterior layer* is continued forwards, around the anterior part of the membranous portion of the urethra, becoming lost upon the bulb.

The *posterior layer* is derived from the pelvic fascia: it is continued backwards around the posterior part of the membranous portion of the urethra, and the outer surface of the prostate gland.

If the anterior layer of this fascia is detached on either side, the following parts are seen between it and the posterior layer: the sub-pubic ligament above, close to the pubes; the dorsal vein of the penis; the membranous portion of the urethra, and the muscles of the urethra; Cowper's glands, and their ducts; the pudic vessels and nerve; the artery and nerve of the bulb, and a plexus of veins.

The *Compressor Urethrae* (constrictor urethrae), surrounds the whole length of the membranous portion of the urethra, and is contained between the two layers of the deep perinaeal fascia. It arises, by aponeurotic fibres, from the upper part of the ramus of the pubes on each side, to the extent of half or three quarters of an inch; each segment of the muscle passes inwards, and divides into two fasciculi, which surround the urethra from the prostate gland behind, to the bulbous portion of the urethra in front; and unite, at the upper and lower surfaces of this tube, with the muscle of the opposite side, by means of a tendinous raphe.

Circular Muscular Fibres surround the membranous portion of the urethra, from the bulb in front to the prostate gland behind; they are placed immediately beneath the transverse fibres already described, and are continuous with the circular fibres of the bladder. These fibres are involuntary.

Cowper's Glands are situated immediately below the membranous portion of the urethra, close behind the bulb, and below the artery of the bulb (p. 776).

The *Pudic Vessels and Nerves* are placed along the margin of the pubic arch (p. 439).

The *Artery of the Bulb* passes transversely inwards, from the internal pudic along the base of the triangular ligament, between the two layers of fascia, accompanied by a branch of the pudic nerve (p. 440).

If the posterior layer of the deep perinaeal fascia is removed, and the crus penis of one side detached from the bone, the under or perinaeal surface of the Levator ani is brought fully into view. This muscle, with the triangular ligament in front and the Coccygeus and Pyriformis behind, closes in the outlet of the pelvis.

The *Levator ani* is a broad thin muscle, situated on each side of the pelvis. It is attached to the inner surface of the sides of the true pelvis, and descending, unites with its fellow of the opposite side to form the floor of the pelvic cavity. It supports the viscera in this cavity, and surrounds the various structures which pass through it. It arises, in front, from the posterior surface of the body and ramus of the pubes, on the outer side of the symphysis; posteriorly, from the inner surface of the spine of the ischium: and between these two points, from the angle of division between the obturator and recto-vesical layers of the pelvic fascia at their under part: the fibres pass downwards to the middle line of the floor of the pelvis, and are inserted, the most posterior fibres into the sides of the apex of the coccyx; those placed more anteriorly unite with the muscle of the opposite side, in a median fibrous raphe, which extends between the coccyx and the margin of the anus. The middle fibres, which form the larger portion of the muscle, are inserted into the side of the rectum, blending with the fibres of the Sphincter muscles: lastly, the anterior fibres, the longest, descend upon the side of the prostate gland to unite beneath it with the muscle of the opposite side, blending with the fibres of the External sphincter and Transversus perinaei muscles, at the tendinous centre of the perinaeum.

The anterior portion is occasionally separated from the rest of this muscle by cellular tissue. From this circumstance, as well as from its peculiar relation with the prostate gland, descending by its side and surrounding it as in a sling, it has been described by Santorini and others as a distinct muscle, under the name of the *Levator prostatae*. In the female, the anterior fibres of the Levator ani descend upon the sides of the vagina.

Relations. By its *upper* or *pelvic surface* with the recto-vesical fascia, which separates it from the viscera of the pelvis and from the peritoneum. By its *outer* or *perinaeal surface*, it forms the inner boundary of the ischio-rectal fossa; is covered by a quantity of fat, and by a thin layer of fascia continued from the deep perinaeal fascia. Its *posterior border* is continuous with the Coccygeus muscle. Its *anterior border* is separated from the muscle of the opposite side by a triangular space, through which the urethra, and, in the female, the vagina passes from the pelvis.

Actions. This muscle supports the lower end of the rectum and vagina, and also the bladder during the efforts of expulsion.

The *Coccygeus* is situated behind and parallel with the preceding. It is a triangular plane of muscular and tendinous fibres, arising, by its apex, from the spine of the ischium and lesser sacro-sciatic ligament, and is inserted, by its base, into the margin of the coccyx and into the side of the lower piece of the sacrum. This muscle is continuous with the posterior border of the Levator ani, and closes in the back part of the outlet of the pelvis.

Relations. By its *inner* or *pelvic surface*, with the rectum. By its *external surface*, with the lesser sacro-sciatic ligament. By its *posterior border*, with the Pyriformis.

Action. The Coccygaei muscles raise and support the coccyx after it has been pressed backwards during defecation or parturition.

Position of the Viscera at the Outlet of the Pelvis. Divide the central tendinous point of the perinaeum, and separate the rectum from its connexions by dividing the fibres of the Levator ani, which descend upon the sides of the prostate gland and rectum;

turn it backwards towards the scrotum and coccyx, when the under surface of the prostate gland, the neck and base of the bladder, the vesiculae seminales, and vasa deferentia will be exposed.

The *Prostate Gland* is placed immediately in front of the neck of the bladder, around the prostatic portion of the urethra, its base being turned backwards, and its under surface towards the rectum. It is retained in its position by the Levator prostatae and by the pubo-prostatic ligaments, and is invested by a dense fibrous covering, continuous with the posterior layer of the deep perinaeal fascia. The longest diameters of this gland are in the antero-posterior direction, and transversely at its base; and hence the greatest extent of incision that can be made in it without dividing its substance completely across, is obliquely outwards and backwards. This is the direction in which the incision is made through it in the operation of lithotomy, the extent of which should seldom exceed an inch in length. The relations of the prostate to the rectum should be noticed: by means of the finger introduced into this gut, the surgeon readily detects enlargement or other disease of this organ; he is enabled also, by the same means, to direct the point of a catheter when its introduction is attended with much difficulty, either from injury or disease of the membranous or prostatic portions of the urethra.

Behind the prostate is the posterior surface of the neck and base of the bladder; a small triangular portion of this organ is seen, bounded in front by the prostate gland, behind by the recto-vesical fold of the peritoneum, on either side by the vesiculae seminales and vasa deferentia, and separated from direct contact with the rectum by

361.—A View of the Position of the Viscera at the Outlet of the Pelvis.

the recto-vesical fascia. The relation of this portion of the bladder to the rectum is of extreme interest to the surgeon. In cases of retention of urine, this portion of the organ is found projecting into the rectum, between three and four inches from the margin of the anus, and may be easily perforated during life without injury to any important parts: this portion of the bladder is frequently selected for the performance of the operation of tapping the bladder. If the finger is introduced into the bowel, the surgeon may learn the position, as well as the size and weight, of a calculus in the bladder; and in the operation for its removal, if, as is not unfrequently the case, it should be lodged behind an enlarged prostate, it may be easily displaced from its position by pressing upwards the base of the bladder from the rectum.

Parts concerned in the Operation of Lithotomy. The triangular ligament must be replaced, the rectum drawn forwards so as to occupy its normal position, and the student should then consider the position of the various parts in reference to the lateral operation of lithotomy. This operation is usually performed on the left side of the perinaeum, as it is most convenient for the right hand of the operator. A staff having been introduced into the bladder, the first incision is commenced about an inch and a half in front of the anus, a little on the left side of the raphe, and carried obliquely backwards and outwards to midway between the anus and tuberosity of the ischium. This incision divides the integument and superficial fascia, the external haemorrhoidal vessels and nerves, and the superficial and transverse perinaeal vessels: if the fore-finger of the left hand is thrust upwards and forwards into the wound, pressing at the same time the rectum inwards and backwards, the staff may be felt in the membranous

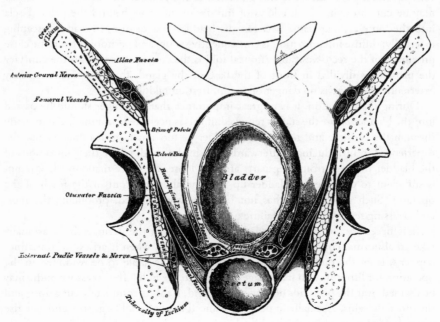

362.—A Transverse Section of the Pelvis; shewing the Pelvic Fascia.

portion of the urethra. The finger is fixed upon the staff, and the structures covering it are divided with the point of the knife, which must be directed along the groove towards the bladder, the edge of the knife being carried outwards and backwards, dividing in its course the membranous portion of the urethra and part of the left lobe of the prostate gland, to the extent of about an inch. The knife is then withdrawn, and the fore-finger of the left hand passed along the staff into the bladder; the staff having been withdrawn, and the position of the stone determined upon, the forceps are introduced over the finger into the bladder. If the stone is very large, the opposite side of the prostate should be notched before the forceps is introduced; the finger is now withdrawn, the blades of the forceps opened, and made to grasp the stone, which must be extracted by slow and cautious undulating movements.

Parts divided in the operation. The various structures divided in this operation are as follows; the integument, superficial fascia, external haemorrhoidal vessels and nerve, the posterior fibres of the Accelerator urinae, the Transversus perinaei muscle and artery, (and probably the superficial perinaeal vessels and nerves), the deep perinaeal fascia, the anterior fibres of the Levator ani, part of the Compressor urethrae, the membranous and prostatic portions of the urethra, and part of the prostate gland.

Parts to be avoided in the operation. In making the necessary incisions in the perinaeum for the extraction of a calculus, the following parts should be avoided. The primary incisions should not be made too near the middle line, for fear of wounding the bulb of the corpus spongiosum or rectum; nor too far externally, otherwise the pudic artery may be implicated as it ascends along the inner border of the pubic arch. If the incisions are carried too far forward, the artery of the bulb may be divided; if carried too far backwards, the entire breadth of the prostate and neck of the bladder may be cut through, which allows of infiltration of urine behind the pelvic fascia into the loose cellular tissue between the bladder and rectum, instead of escaping externally; diffuse inflammation is consequently set up, and peritonitis from the close proximity of the recto-vesical peritoneal fold is the consequence. If, on the contrary, the prostate is divided in front of the base of the gland, the urine makes its way externally, and there is no danger of any infiltration taking place.

During the operation, it is of great importance that the finger should be passed into the bladder *before* the staff is removed: if this is neglected, and the incision made through the prostate and neck of the bladder be too small, great difficulty may be experienced in introducing it afterwards; and in the child, where the connections of the bladder to the surrounding parts are very loose, the force made in the attempt is sufficient to displace the bladder up into the abdomen, out of the reach of the operator. Such a proceeding has not unfrequently occurred, producing the most embarrassing results, and total failure of the operation.

It is necessary to bear in mind that the arteries in the perinaeum occasionally take an abnormal course. Thus the artery of the bulb, when it arises, as sometimes happens, from the pudic, opposite the tuber ischii, is liable to be wounded in the operation for lithotomy, in its passage forwards to the bulb. The accessory pudic may be divided near the posterior border of the gland, if this is completely cut across: and the prostatic veins, especially in people advanced in life, are of large size, and give rise when divided to troublesome haemorrhage.

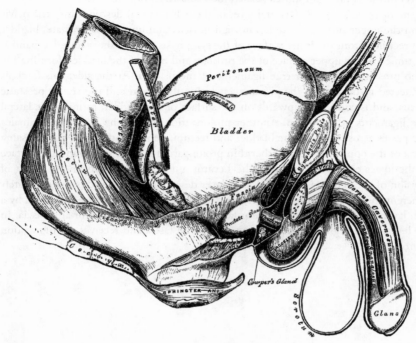

363.—Side View of the Pelvic Viscera of the Male Subject, showing the Pelvic and Perinæal Fasciæ.

Pelvic Fascia

The pelvic fascia is a thin membrane which lines the whole of the cavity of the pelvis, and is continuous with the transversalis and iliac fasciae. It is attached to the brim of the pelvis for a short distance at the side of this cavity, and to the inner surface of the bone around the attachment of the Obturator internus. At the posterior border of this muscle, it is continued backwards as a very thin membrane in front of the Pyriformis muscle and sacral nerves, behind the branches of the internal iliac artery and vein which perforate it, to the front of the sacrum. In front, it follows the attachment of the Obturator internus to the bone, arches beneath the obturator vessels, completing the orifice of the obturator canal, and at the front of the pelvis is attached to the lower part of the symphysis pubis; being continuous below the pubes with the fascia of the opposite side so as to close the front part of the outlet of the pelvis, blending with the posterior layer of the triangular ligament. At the level of a line extending from the lower part of the symphysis pubis to the spine of the ischium, is a thickened whitish band; this marks the attachment of the Levator ani muscle to the pelvic fascia, and corresponds to its point of division into two layers, the obturator and recto-vesical.

The *Obturator fascia* descends and covers the Obturator internus muscle. It is a direct continuation of the pelvic fascia below the white line above mentioned, and is attached to the pubic arch and to the margin of the great sacro-sciatic ligament. This fascia forms a canal for the pudic vessels and nerve in their passage forwards to the perinaeum, and is continuous with a thin membrane which covers the perinaeal aspect of the Levator ani muscle, called the *ischio-rectal* or *anal fascia*.

The *Recto-vesical fascia* (visceral layer of the pelvic fascia) descends into the pelvis upon the upper surface of the Levator ani muscle, and invests the prostate, bladder, and rectum. From the inner surface of the symphysis pubis a short rounded band is continued to the upper surface of the prostate and neck of the bladder, forming the pubo-prostatic or anterior true ligaments of the bladder. At the side, this fascia is connected to the side of the prostate, enclosing this gland and the vesical prostatic plexus, and is continued upwards on the surface of the bladder, forming the lateral true ligaments of the organ. Another prolongation invests the vesiculae seminales, and passes across between the bladder and rectum, being continuous with the same fascia of the opposite side. Another thin prolongation is reflected around the surface of the lower end of the rectum. The Levator ani muscle arises from the point of division of the pelvic fascia; the visceral layer descending upon and being intimately adherent to the upper surface of this muscle, while the under surface is covered by a thin layer derived from the obturator fascia, called the ischio-rectal or anal fascia. In the female, the vagina perforates the recto-vesical fascia and receives a prolongation from it.

INDEX

Abdomen, 692; apertures found in, 692; boundaries of, 692; lymphatics of, 494; muscles of, 261; regions of, 692; viscera of, 696

Abdominal aorta, 423, branches of, 425; surgical anatomy of, 423; muscles, 261; ring external, 263, 802, internal, 806; viscera, position of, 693

Abducens nerve, 555

Abductor minimi digiti muscle (hand), 303, (foot), 343

Abductor pollicis muscle (hand), 301, (foot), 342

Aberrant duct of testis, 783

Absorbent glands, 494

Absorbents, 493

Accelerator urinae muscle, 778

Accessory obturator nerve, 605; palatine canals, 52; pudic artery, 441

Acervulus cerebri, 542

Acetabulum, 91

Acromial end of clavicle, fracture of, 306

Acromial nerves, 584; region, muscles of, 281; thoracic artery, 399

Acromion process, 99; fracture of, 306

Actions of muscles. *See each group of Muscles*

Adductor brevis muscle, 322; longus muscle, 321; magnus muscle, 322; pollicis muscle (hand), 303, (foot), 346

Aggregate glands, 705

Air cells, 759

Air tubes, 749

Alae of vomer, 56

Alar ligaments of knee, 199; thoracic artery, 406

Alimentary canal, 675; subdivisions of, 675

Allantois, 769

Alveolar artery, 379; process, 48

Alveoli of lower jaw, 59; of upper jaw, 48; formation of, 682; of stomach, 701

Amphiarthrosis, 152

Ampullae of semicircular canals, 672

Ampullae of tubuli lactiferi, 798

Amygdalae, 686; of cerebellum, 546

Anal fascia, 832

Analysis of bone, how conducted, 1

Anastomosis of arteries, 351

Anastomotica magna of brachial, 412, of femoral, 451

Anatomy, descriptive, 1; of femoral hernia, 809; of inguinal hernia, 800; study of, 1; surgical, 1

Anconeus muscle, 297

Aneurisms of abdominal aorta, 423; of arch of aorta, 357; of thoracic aorta, 421

Angle of jaw, 59; of pubes, 90; of rib, 81

Angular artery, 370; movement, 156; process, external, 28; internal, 28; vein, 468

Animal constituent of bone, 1

Ankle joint, 201; arteries of, 203; bones of, 201; ligaments of, 201; relations of tendons and vessels, in front, 202, behind, 203

Annular ligament of radius and ulna, 184; of wrist, anterior, 300; posterior, 300; of ankle, anterior, 340; external, 340; internal, 340; of stapes, 666

Annulus ovalis, 730

Anomalus muscle, 221

Anterior annular ligament (wrist), 300, (ankle), 340; dental canal, 45; ethmoidal cells, 41; fontanelle, 26; fossa of skull, 62; nasal spine, 48; palatine canal, 52, 65; palatine fossa,

52, 65; region of skull, 70; triangle of neck, 379; crural nerve, 605
Antihelix, 658; fossa of, 658
Antitragicus muscle, 660
Antitragus, 659
Antrum of Highmore, 47
Anus, 820; muscles of, 822, 826
Aorta, 353; abdominal, 423; abdominal branches of, 425; abdominal, surgical anatomy of, 423; arch of, 353; ascending part of arch, 353; branches of, 358; descending part of arch, 356; peculiarities of arch, 357; peculiarities of branches of, 358; surgical anatomy of, 357; transverse portion of, 355; descending, 420; thoracic, 420; thoracic branches of, 421; surgical anatomy of, 421
Aortic opening of diaphragm, 273; of heart, 734; plexus, 626; semilunar valves, 732; sinuses, 732
Apertura iter chordae anterius, 662; posterius, 662; scalae vestibuli et cochleae, 669
Aperture of larynx, 745; of posterior nares, 73
Aponeurosis, 214; of deltoid, 281; of external oblique in inguinal region, 802; infra spinous, 283; of insertion, 211; of investment, 211; of occipito-frontalis, 214; subscapular, 282; supra-spinous, 283; vertebral, 253
Appendages of eye, 654; of skin, 632; of uterus, 795
Appendices epiploicae, 697
Appendix of right auricle, 729; of left auricle, 734; vermiformis, 706
Aqua labyrinthi, 672
Aquaeductus cochleae, 33, 671; vestibuli, 669; Fallopii, 663; Sylvii, 543
Aqueous chamber, epithelial lining of, 645; humour, 645; secreting membrane of, 651
Arachnoid of brain, 514; of cord, 513; structure of, 515

Arbor vitae uterimis, 793; vitae of cerebellum, 549
Arch of aorta, 353, peculiarities of, 357, branches of, 358, surgical anatomy, 357; of pubes, 94; of a vertebra, 6; supra-orbital, 27; of colon, 707; crural or femoral, 803; palmar superficial, 290; palmar deep, 293; plantar, 458; zygomatic, 65
Arciform fibres of medulla oblongata, 526
Areola of breast, 798
Arnold's ganglion, 619
Arteria centralis retinae, 651
Arytaeno-epiglottideus inferior muscle, 748; superior, 748
Arytaenoideus muscle, 749
Aryteno-epiglottidean folds, 743
Arytenoid cartilages, 743; glands, 749
Arm, back of muscles of, 287; front of muscles of, 285; arteries of, 407; bones of, 107; fascia of, 285; nerves of, 589; veins of, 476
Arnold's nerve, canal for, 34
Arteriae receptaculi, 385
ARTERIES, *General Anatomy of*, 351; anastomoses of, 351; coats of, 352, external, 352, internal, 352, middle, 352; contractile coat of, 352; distribution of, 351; epithehal lining of, 352; mode of division, 351; mode of origin of branches, 351; nerves of, 353; physical properties of, 353; sheath of, 353; structure of, 353; subdivision of, 351; systemic, 351; tortuosity of, 352; vessels of, 353
ARTERIES OR ARTERY, *Descriptive and Surgical Anatomy of*, 351
accessory pudic, 439; acromial thoracic, 406; alar thoracic, 406; alveolar, 379; anastomotica magna of brachial, 412, of femoral, 451; angular, 372; anterior cerebral, 388, choroid, 389, ciliary, 387, 654, communicating, 388, intercostal, 400, spinal, 396; *aorta*,

INDEX

353, abdominal, 423, arch of, 353, ascending part, 354, descending part, 356, 420, surgical anatomy of, 357, transverse portion, 355, thoracic, 420; articular, knee, superior internal, 454, external, 454, inferior internal, 454, external, 454; ascending cervical, 398, pharyngeal, 374; auditory, 397, 673; auricular anterior, 375, posterior, 374; axillary, 403; azygos of knee, 454

basilar, 397; brachial, 407, carpal anterior, 415; bronchial, 422; buccal, 378; of bulb, 826; bulbosi urethrae, 440

calcaneal internal, 462; carotid common, 361, external 366, internal 382; carpal ulnar anterior, 420, posterior, 420; carpal radial anterior, 418, posterior, 418; centralis retinae, 387; cerebellar anterior, 397, inferior, 397, superior, 397; cerebral anterior, 388, middle, 389, posterior, 389; cervical ascending, 398, superficial, 399, princeps, 401, profunda, 401; choroid anterior, 389, posterior, 398; ciliary, 387, anterior, 387, long, 387, short, 387; circle of Willis, 398; circumflex of arm anterior, 407, posterior, 407, of thigh external, 450, internal, 450, iliac, 445, superficial, 449; cochlear, 442; coccygeal, 442; colica dextra, 430, media, 430, sinistra, 431; coeliac axis, 425; comes nervi ischiadici, 442, phrenici, 400; common carotid, 361, iliac, 433, left, 434, right, 434; communicating anterior cerebri, 389, posterior cerebri, 389; communicating branch of ulnar, 420; coronary of heart, 359, of upper lip, 372, of lower lip, 372; cremasteric, 444; crico-thyroid, 368; cystic, 427

deep branch of ulnar, 420, cervical, 401, palmar arch, 412, temporal, 375; deferent, 432; dental inferior, 378, superior, 379; descending aorta, 420,

palatine, 379; digital plantar, 463; digital of ulnar, 420; dorsal of penis 441, of scapula, 399; dorsalis hallucis, 458, indicis, 416, linguae, 369, pedis, 457, pollicis, 416

epigastric, 444, superior, 410, superficial, 449; ethmoidal, 387; external carotid, 366, plantar, 462

facial, 370; femoral, 445, deep, 449; frontal, 387

gastric, 425, 428; gastro-duodenalis, 426, gastro-epiploica dextra, 426, gastro-epiploica sinistra, 428; gluteal, 442, inferior, 440

helicine, 779; haemorrhoidal external, 440, middle, 438, superior, 431; hepatic, 425; hyoid branch of lingual, 369, of superior thyroid, 368; hypogastric in foetus, 737, 740

ileo-colic, 429; iliac, common external, 443, internal, 436; ilio-lumbar, 442; inferior cerebellar, 397, dental, 378, labial, 372, laryngeal, 368, mesenteric, 431, profunda, 411, pyloric, 426, thyroid, 398; infra-orbital, 379; innominate, 360; intercostal, 422, anterior, 400, superior, 410; internal auditory, 673, carotid, 382, iliac, 436, mammary, 400, maxillary, 376, plantar, 462; inter-osseous ulnar, 418, of foot, 458, of hand, 418, anterior, 418, posterior, 419; intestini tenuis, 429

labial inferior, 372; lachrymal, 386; laryngeal, superior, 368; lateral sacral, 442, spinal, 396; lateralis nasi, 372; left common carotid, relations of, 365; lingual, 368; long ciliary, 387, 654, thoracic, 406; lumbar, 433

malleolar external, 457, internal, 457; mammary internal, 400; masseteric, 378; maxillary internal, 476; median, 419; mediastinal, 400, posterior, 422; meningeal anterior, 377, middle, 377, small, 378; from occipital, 373; from pharyngeal 374; from vertebral, 395;

mesenteric inferior, 431, superior, 428; metacarpal, 415; metatarsal, 458; middle cerebral, 389, sacral, 433; musculo-phrenic, 400; mylo-hyoid, 381
nasal, 379; of ophthalmic, 387, lateral of septum, 372; nutrient of humerus, 411, femur, 449, fibula, 461, radius, 419, tibia, 461, ulna, 418
obturator, 438; occipital, 373; oesophageal, 398; ophthalmic, 385; ovarian, 432
palatine, ascending, 370, descending, 379, posterior, 379, of pharyngeal, 374; palmar arch, superficial, 417, deep, 419; palmar interossei, 305; palpebral, 387; pancreatic, 428; pancreatico-duodenalis, 427, inferior, 428; perforating of hand, 416, of thigh, 450, intercostal, 400; plantar, 463; pericardiac, 400, 421; perinaeal superficial, 440, transverse, 440; peroneal, 460, anterior, 461; pharyngea ascendens, 374; phrenic, 432; popliteal, 451; posterior auricular, 374, cerebral, 397, communicating, 389, meningeal from vertebral, 396, palatine, 379; princeps cervicis, 373, pollicis, 416; profunda of arm, inferior, 411, superior, 411, cervicis, 373, femoris, 449; pterygoid, 378; pterygo-palatine, 379; pudic external deep, 449, internal, 439, superficial, 449; pulmonary, 463; pyloric inferior, 426, of hepatic, 425
radial, 412, relations of in fore-arm, 413, relations of in the hand 413, surgical anatomy of, 414; radialis indicis, 416; ranine, 369; recurrent interosseous posterior, 418, radial, 416, ulnar anterior, 418, posterior, 419, tibial, 455; renal, 431
sacral lateral, 442, middle, 443; scapular posterior, 399; sciatic, 491; short ciliary, 387, 654; sigmoid, 431; spermatic, 432; spheno-palatine, 379;

spinal anterior, 396, lateral, 396, posterior, 397, in neck, 396, in loins, 421, in thorax, 423, median, 397; splenic, 427; sterno-mastoid, 366; stylo-mastoid, 385; subclavian, 389, right, first part of, 390, left, first part of, 391, second portion of, 392, third portion of, 392, surgical anatomy of, 393; sublingual, 369; submaxillary, 371; submental, 371; subscapular, 406; superficial cervical, 399, circumflex iliac, 445, perinaeal, 440, palmar arch, 417; superficialis volae, 415; superior cerebellar, 397, epigastric, 401, haemorrhoidal, 431, intercostal, 401, laryngeal, 368, mesenteric, 428, profunda, 411, thoracic, 406, thyroid, 368; supra-orbital, 386, supra-renal, 431, supra-scapular, 398; sural, 454
tarsal, 458; temporal, 375, anterior, 375, deep, 378, middle, 375, posterior, 375; thoracic, acromial, 406, alar, 406, aorta, 420, long, 406, superior, 406: thyroid axis, 398; thyroid inferior, 398, superior, 368; tibial, anterior, 455, posterior, 459, recurrent, 457; tonsillitic, 371; transverse facial, 375; transverse of basilar, 397; transversalis colli, 399; tympanic, from internal carotid, 374, from internal maxillary, 377
ulnar, 416, relations at wrist, 417, relations in fore-arm, 416, relations in the hand, 417; recurrent, anterior, 418, posterior, 418, umbilical in foetus, 737, 739; uterine, 438
vaginal, 438; vasa, aberrantia of arm, 410, brevia, 428, intestini tenuis, 429; vertebral, 395; vesical, inferior, 438, middle, 438, superior, 437; vestibular 673; Vidian, 379

ARTICULATIONS in general, 150 acromio-clavicular, 184; ankle, 201; astragalo-calcaneal, 203, astragalo-scaphoid, 206; atlo-axoid, 160

INDEX

calcaneo-astragaloid, 203, calcaneo-cuboid, 205, calcaneo-scaphoid, 205 carpo-metacarpal, 190; carpal, 188 classification of, 152; coccygeal, 174 costo-clavicular, 177; costo-sternal, 169; costo-transverse, 168; costo-vertebral, 167
elbow, 182
femoro-tibial, 200, of foot, 207
hand, 208; hip, 193
immoveable, 152
knee, 195
larynx, 742; lower limb, 193
metacarpal 190; metacarpo-phalangeal, 192; metatarso-phalangeal, 208; metatarsal, 208; mixed, 153; moveable, 154; movements of, 156
occipito-atloid, 163; occipito-axoid, 164; of pelvis, 172; pelvis with spine, 174; phalanges, 192; pubic, 175
radio-carpal, 187; radio-ulnar, inferior, 186, middle, 185, superior, 184
sacro-coccygeal, 174; sacro-iliac, 172; sacro-sciatic, 174; sacro-vertebral, 156; scapulo-clavicular, 178; scapulo-humeral, 181; shoulder, 180; sterno-clavicular, 177; of sternum, 178
tarso-metatarsal, 207; tarsal, 203; temporo-maxillary, 165; tibio-fibular, inferior, 200, middle, 200, superior, 200; of tympanic bones, 667
upper extremity, 177; of upper limb, 177; of vertebral column, 156
wrist, 187
Arytenoid cartilages, 743
Ascending colon, 707
Astragalus, 141
Atlas, 7; development of, 13
Atlo-axoid articulation, 160
Atrabiliary capsules, 767
Attollens aurem muscle, 215
Attrahens aurem muscle, 215
Auditory artery, 673; canal, cartilaginous portion, 664; osseous portion, 664
Auditory meatus, external, 33, internal,

33; nerve, 554, 658; process, 35
Auricle of ear, 658; cartilage of, 659; ligaments of, 659; structure of, 659
Auricle of heart, left, 732; appendix of, 733; sinus of, 732; right, 729; openings in, 729; valves in, 730; sinus of, 729
Auricular artery posterior, 374, anterior, 375; fissure, 34; lymphatic glands, 497; nerve, of vagus, 576, posterior from facial, 559; veins, anterior, 469, posterior, 469; surface of sacrum, 14
Auricularis magnus nerve, 584
Auriculo-temporal nerve, 571
Auriculo-ventricular groove of heart, 728; opening, left, 733, 734; opening right, 730, 731
Axes of the pelvis, 94
Axilla, 401; dissection of, 275
Axillary artery, 403; first portion of, 403; second portion of, 404; third portion of, 404; peculiarities, 405; surgical anatomy of, 405; branches of, 406; lymphatic glands, 502; vein, 478
Axis, 7; cerebro-spinal, 509; coeliac, 425; thyroid, 398
Azygos artery, articular, 454; vein, left, lower, 482; vein, right, 481; uvulae muscle, 243

BACK, muscles of, first layer, 248; second layer, 251; third layer, 252; fourth layer, 254, fifth layer, 258
Bartholine, duct of, 689
Base of brain, 509; of skull, external surface, 62, internal surface, 62
Basilar artery, 397; process, 23; suture, 61
Basilic vein, 477, median, 477
Basio-glossus muscle, 238
Bauhin, valves of, 707
Beale, Dr., his researches on the liver, 716
Bend of elbow, 408
Berzelius, his analysis of bone, 1
Biceps muscle, 286, 329
Bicipital groove, 104

INDEX

Bicuspid teeth, 678
Biliary ducts, 716; glands of, 718; structure of, 718
Biventer cervicis muscle, 257
BLADDER, surgical anatomy of, 829; base of, 828; false ligaments of, 770; female, relations of, 790; interior of, 771; ligaments of, 770; lymphatics of, 772; nerves of, 772; shape, position, relations, 769; structure of, 771; summit, body, base, and neck of, 769; trigonum vesicae of, 771; true ligaments of, 770; uvula vesicae of, 772; vessels of, 772
Blood, circulation of, in adult, 728, in foetus, 738
Body of a tooth, 677; of a vertebra, 6
BONE, *General Anatomy of:* affected with rickets, analysis of, 2; animal constituent of, 1; cancellous tissue of, 2; chemical analysis of, 1; compact tissue of, 2; diploë of, 3; development of, 4; earthy constituent of, 1; general properties of, 1; growth of, 4; inorganic constituent of, 1; lymphatics of, 4; medullary canal of, 3; medullary membrane of, 3; nerves of, 4; organic constituent of, 1; ossification of, 4; spongy tissue of, 2; structure of the extremities, 3, of the shaft, 2; veins of, 4; vessels of, 3
Bones, forms of, 2; flat, 3; irregular, 3; long 2; mixed, 3; short, 3
BONES OR BONE, *Descriptive Anatomy of,* 1
  ankle, 201; astragalus, 141; atlas, 7; axis, 8
  calcaneum, 138; carpal, 115; carpus, 115; clavicle, 96; coccyx, 18; cranial, 21, articulations of, 55; cuboid, 140; cuneiform of carpus, 115; of tarsus, 138
  ear, 666; ethmoid, 41
  facial, 21, 43; femur, 124; fibula, 135; finger, 121; foot, 138; frontal, 27
  hand, 115; humerus, 102; hyoid, 76
  ilium, 86; incus, 665; inferior maxillary, 56, turbinated, 54; innominate, 85; ischium, 86
  lachrymal, 48, lesser lachrymal, 50, lingual 76
  magnum, 119; malar, 50, malleus, 665; maxillary, inferior, 56; metacarpus, 121; metatarsal, 145
  nasal, 44; navicular of carpus, 115, of tarsus, 138, number of, 5
  occipital, 21; orbicular, 666
  palate, 52; patella, 130; parietal, 25; pelvic, 85; phalanges, of foot, 147, of hand, 122; pisiform, 117; pubic, 90
  radius, 112; ribs, 80; ribs, peculiar, 83
  sacrum, 14; scaphoid of carpus, 115, of tarsus, 143; scapula, 97; semilunar, 116; sesamoid, 148; sphenoid, 36; sphenoidal spongy, 40; spongy, 54; stapes, 666; sternum, 77; superior maxillary, 44
  tarsal, 147; temporal, 30; tibia, 131; trapezium, 117; trapezoid, 117; triquetral, 43; turbinate, superior, 48, inferior, 47; tympanic, 35
  ulna, 107; unciform, 119; ungual, 123
  vertebra dentata, 8, prominens, 9; vertebra cervical, 6, dorsal, 9, lumbar, 11; vomer, 56
  Wormian, 43
Bostock's analysis of a rickety bone, 2
Bowman, on structure of kidney, 765
Brachia of optic lobes, 544
Brachial artery, 407, branches of, 411, peculiarities of, 409, surgical anatomy of, 410; lymphatic glands, 499; plexus, 587, branches above clavicle, 589, below clavicle, 589; region, posterior, muscles of, 296, 297, anterior 289
Brachialis anticus muscle, 287
Brain, 509; base of, 521; dura mater of, 513; interior of, 519; lateral ventricles of, 529; lobe of, anterior, 530, middle 530, posterior, 530; membranes of,

INDEX

dissection, 513; subdivision into parts, 530; upper surface of, 546; weight of, 545

Breasts, 798

Brim of pelvis, 93

Broad ligaments, formation of, 695, of uterus, 793

Bronchi, right and left, 751, structure of in lobules of lung, 758, in lung, 757

Bronchial arteries, 422, 759; lymphatic glands, 499; veins, 478; tubes, *See bronchi.*

Bronchus, mode of subdivision in lung, 751

Brunner's glands, 704

Bubonocele, 808

Buccal arteries, 378; glands, 677; lymphatic glands, 496; nerve, 560; veins, 468

Buccinator, muscle, 224

Bulb, artery of, 439, 826; of corpus cavernosum, 777; of corpus spongiosum, 778; olfactory, 552

Bulbi vestibuli, 790

Bulbous portion of urethra, 773

Bulbs of the fornix, 533

Bursae mucosae, 152

Bursal synovial membranes, 152

Caecum, 706

Calamus scriptorius, 525

Calcanean arteries, internal, 462

Calcaneo-astragaloid ligament, external, 203, posterior, 204

Calcaneo-cuboid ligament, internal, 205, superior, 205, long, 205, short, 205

Calcaneum, 138

Calices of kidney, 765

CANALS OR CANAL, accessory palatine, 52; alimentary, 675; anterior dental, 46, palatine, 48, 65; for Arnold's nerve, 34; auditory, 661; carotid, 33; central of modiolus, 670; for chorda tympani, 662; of cochlea, 669; crural, 813; dental posterior, 45; ethmoidal, anterior, 29, 59, posterior, 29, 59; femoral, 813; incisive, 49, 67; inferior dental, 58; infra-orbital, 45; inguinal, 779; for Jacobson's (tympanic) nerve, 34, 68; lachrymal, 48; malar, 50; nasal, 48; naso-palatine, 56; of Nuck, 787, 797; of Petit, 653; portal, 714; palatine posterior, 46, anterior, 48; pterygoid, 39; pterygo-palatine, 37; spermatic, 784; of spinal cord, 516; spiral of cochlea, 670; of modiolus, 671; semicircular, 668; for tensor tympani, 662; vertebral, 20; Vidian, 38; of Wirsung, 720

Canaliculi of eyelids, 657

Canalis spiralis modioli, 671

Cancellous tissue of bone, 2

Canine eminence, 45; fossa, 45; teeth, 678; of lower jaw, 678; of upper jaw, 678

Canthi of eyelids, 654

Capillaries, 725

Capitula laryngis, 743

Caput caecum coli, 706; gallinaginis, 772

Capsular ligament of hip, 194; of knee, 197; of shoulder, 181; of thumb, 189

Capsule of Glisson, 696; of the lens, 649; of lens, in foetus, 649; of Malpighian bodies of kidney, 765

Capsules, suprarenal, 767

Cardiac lymphatics, 507; nerves, 621, inferior, 622, middle, 622, superior, 621; nerves from pneumogastric, 576; plexus of nerves, deep, 585, superficial, 584; veins, 491, anterior, 491, great, 491, posterior, 491

Carotid arteries, common, 361; artery, external, 366, surgical anatomy, 367, internal, 382, cervical portion, 384, petrous portion, 384, cavernous portion, 384, cerebral portion, 384, surgical anatomy, 385; left, common, relations of, 360, peculiarities of, 364, surgical anatomy of, 365; canal, 33; ganglion, 619; plexus, 619, triangle

INDEX

inferior, 380, triangle suprior, 380
Carpal arteries, from radial, 415; from ulnar, 420; ligaments, 187
Carpo-metacarpal articulations, 191
Carpus, 115; development of, 123, articulations of, 188
CARTILAGE OR CARTILAGES, *General Anatomy of*, 150; articular, 150; arytenoid, 743; of auricle, 658; of bronchi, 759; costal, 84; cricoid, 742; cuneiform, 743; of ear, 659; ensiform, 77; of epiglottis, 743; fibro, 151, interarticular, 151, interosseous, 151, circumferential, 151, stratiform, 151; of larynx, 741, structure of, 744; of the nose, 639; of the pinna, 658; of Santorini, 743; semilunar of knee, 198; of septum of nose, 640; tarsal, 655; temporary, 4; thyroid, 741; of trachea, 749; of Wrisberg, 743; xiphoid, 77
Cartilago triticea, 744
Caruncula lachrymalis, 656; mammillaris, 552
Carnuculae myrtiformes, 790
Cauda equina, 514, 608
Cava inferior, 487, peculiarities, 488; superior, 481
Cavernous body, artery of, 441; groove, 37, 64; nerves of penis, 627; plexus, 620; sinus, 475
Cavities of reserve (teeth), 683
Cavity cotyloid, 91; glenoid, 100; of pelvis, 93; sigmoid, 109
Cells, ethmoidal, 42
Cement of teeth, 680; formation of, 682
Centres of ossification, 4
Centrum ovale majus, 534, minus, 534
Cephalic vein, 477
Cerato-glossus muscle, 238
Cerebellar arteries, anterior, 388, superior, 388, inferior 388; veins, 472
Cerebelli incisura, anterior, 546, posterior, 546
CEREBELLUM, 545; corpus dentatum of, 549; hemispheres of, 546; laminae of, 549; lobes of, 547; lobulus centralis of, 546; median lobe of, 546; peduncles of, 549; structure of, 549; under surface of, 546; upper surface of, 546; the valley of, 546; ventricle of, 548; weight of, 545
Cerebral arteries, 388; anterior, 388, middle, 389, posterior, 389; convolutions, 529; lymphatics, 497; veins, 472; ventricles, 537
CEREBRO-SPINAL AXIS, 509, fluid, 514, 521; nerves, 510
CEREBRUM, base of, 530; commissures of, 543; convolutions of, 529; crura of, 532; fibres of, 533; fissures, of Sylvius, 532, longitudinal, 527; general arrangements of its parts, 534; grey matter of, 529; hemispheres of, 528; interior of, 534; labia of, 534; lobes of, 530; peduncles of, 532; structure of, 545; sulci, 529; superior ganglia of, 537, 541; under surface, 530; upper surface, 528; ventricles of, 534
Cervical artery, ascending, 398, superficial, 399; ganglion inferior, 577, middle, 577, superior, 577; lymphatic glands, deep, 497, superficial, 496; nerves, 582, anterior branches of, 583, posterior branches of, 586, roots of, 582; plexus, 583, deep branches of, 585, posterior, 587, superficial branches of, 584; veins, transverse, 470; vertebrae, 6
Cervicalis ascendens muscle, 256
Cervico-facial nerve, 559
Cervix uteri, 794
Chambers of the eye, 651
Check ligaments, 164
Cheek, muscles of, 223
Cheeks, 676; structure of, 676
Chemical analysis of bone, 1; of cerebro-spinal fluid, 521; dentine or ivory, 680; enamel, 681; of nervous substance, 509; synovia, 152; thymus, 762; thyroid, 761

INDEX

Chest, muscles of front, 276, side, 280
Chiasma or optic commissure, 553
Chondro-glossus muscle, 238
Chorda tympani nerve, 558, 668
Chordae tendinese, of right ventricle, 732, of left, 734; vocales, 747; Willisii, 474
Choroid arteries, anterior, 389, posterior, 398; coat of eye, 646; plexus of lateral ventricle, 538; of fourth ventricle, 549; of third ventricle, 541; structure of, 646; veins of brain, 473
Chyli receptaculum, 495
Cilia or eyelashes, 656
Ciliary arteries, 387, anterior, 387, 654, long, 387, 654, short, 387, 654; ganglion, 565; ligament, 649; muscle, 649; nerves, long, 565, short, 565; processes of eye, 647; structure of, 648
Circle of Willis, 398
Circular sinus, 476
Circulation of blood in adult, 728; in foetus, 738
Circumduction, 156
Circumferential fibro-cartilage, 151
Circumflex artery of arm, anterior, 407, posterior, 407; of thigh, external, 438, internal, 438; iliac artery, 445, 800, superficial, 449; iliac vein, 486, superficial, 484; nerve, 590
Circumflexus palati muscle, 243
Clavicle, 95; acromial end, fracture of, 306; articulations of, 97; attachment of muscles to, 97; development of, 96; fracture of, 306; peculiarities, 96; sternal end, fracture of, 306; structure of, 96; tubercle of, 96
Clavicular nerves, 583
Clinoid processes, anterior, 39, 62, middle, 36, posterior, 36, 64
Clitoris, 789; fraenum of, 789; lymphatics of, 504; muscles of, 825; prepuce of, 789; structure of, 789
Coccygeal artery, 442; nerves, 608; nerve, anterior branch of, 609, posterior branch of, 608

Coccygeus muscle, 821
Coccyx, 18; articulations of, 19; attachment of muscles to, 18; cornua of, 18; development of, 18
Cochlea, 670; arteries of, 673; central axis of, 670; cupola of, 670; denticulate lamina of, 672; hamular process of, 671; infundibulum of, 670; lamina spiralis of, 670; membranous zone of, 671; nerves of, 674; osseous zone of, 671; scala tympani of, 671; scala vestibuli of, 671; scalae of, 671; spiral canal of, 670; veins of, 674
Cochlear artery, 673; nerve, 674
Cochlearis muscle, 671
Coeliac axis, 423; plexus, 624
Colica dextra artery, 430, media, 430, sinistra, 431
Colles fracture, 310
Colliculus bulbi urethrae, 778
Colon, 707
Columella cochleae, 670
Columnae carneaee of left ventricle, 734, of right ventricle, 732; papillares, 732, 734
Columns of abdominal ring, 802; of medulla oblongata, 517; of rectum, 710; of spinal cord, 517; of vagina, 792
Columnar layer of retina, 650
Comes nervi ischiadici artery, 442; phrenici artery, 400
Commissura simplex of cerebellum, 546
Commissure of flocculus, 547; optic, 554
Commissures, 509; of brain anterior, 543, middle or soft, 543, posterior, 543; of, spinal cord, grey, 518, white, 518
Communicans noni nerve, 585
Communicating artery of brain anterior, 389, posterior, 389
Compact tissue of bone, 2
Complexus muscle, 257
Compressor narium minor, 221; nasi, 221; sacculi laryngis, 746; urethrae, 826
Conarium, 543

Concha, 658
Condyles of bones. *See Bones*
Condyloid foramina, 23; fossae, 23; process, 59; veins posterior, 475
Congenital hernia, 808
Conglobate glands, 494
Coni vasculosi, 784
Conjoined tendon of internal oblique and transversalis, 265, 803
Conjunctiva, 645
Conoid ligament, 180
Constrictor inferior muscle, 240; medius, 240; superior, 241; isthmi faucium, 239; urethrae, 826
Conus arteriosus, 731
Convolution of corpus callosum, 530; of longitudinal fissure, 530; supra-orbitar, 530
Convolutions of cerebrum, structure of, 529; cortical substance of, 529; white matter of, 529
Coraco-acromial ligament, 180; coraco-brachialis muscle, 286; coraco-clavicular ligament, 179; coraco-humeral ligament, 181
Coracoid ligament, 180; process, 101; process, fracture of, 307
Cord spermatic, 780
Cordiform tendon, 273
Corium of skin, 630
Cornea, 644; arteries and nerves of, 645; elastic laminae of, 644; proper substance of, 644; structure of, 644
Cornicula laryngis, 744
Cornu Ammonis, 538
Cornua of the coccyx, 18; of hyoid bone, 76; of the sacrum, 15; of thyroid cartilage, 744
Corona glandis, 776
Coronal suture, 60
Coronary arteries of lip, 372; arteries of heart, left, 362; right, 362; peculiarities, 364; ligaments of liver, 711; plexus, anterior, 623; posterior, 623; sinus, 491, opening of, 730; valve, 730

Coronoid depression, 106; process, 59, of ulna, 109, fracture of, 309
Corpora albicantia, 533; Arantii, 732, 735; cavernosa penis, 777, crura of, 777, structure of, 777; cavernosa clitoridis, 789; geniculata, 544; mammillaria, 552; olivaria, 524; pyramidalia, 524; quadrigemina, 544; restiformia, 524; striata, 534, veins of, 473
Corpus callosum, 531, 534, convolution of, 530, genu of, 535, peduncles of, 532, ventricle of, 536; deutatum of cerebellum, 549; of olivary body, 525; fimbriatum, 538, 539; Highmorianum, 782; luteum, 797, structure of, 797; spongiosum, 778, arteries of, 779, colliculi bulbi, 778, structure of, 778
Corpuscles, Malpighian, of kidney, 764; of spleen, 723
Corrugator supercilii muscle, 217
Cortical substance of brain, 509; of cerebral convolutions, 529; of kidney, 764; of supra-renal capsules, 767
Costal cartilages, 84
Costo-clavicular ligaments, 178; costo-coracoid fascia, 278; costo-sternal articulation, 169; costo-transverse articulation, 168; costo-vertebral ligaments, 167; costo-xiphoid ligaments, 171
Cotyloid cavity, 91; ligament, 194; notch, 91
Coverings of direct inguinal hernia, 803, of femoral hernia, 817; of oblique, 802; of testis, tunica albuginea, 782, tunica vaginalis, 782, tunica vasculosa, 783
Cowper's glands, 776, 826
Cranial bones, 216, articulations of, 60;
CRANIAL NERVES, 551, classification of, 551, first pair, 551, second, 552, third 554, fourth, 555, fifth, 562, sixth, 555, seventh, soft portion, 553, hard portion, 557, eighth, glosso-

pharyngeal, 573; vagus, 576; spinal accessory, 576, ninth, 561
Cranial sutures, 60
Cremaster muscle, 780, formation of, 781
Cremasteric artery, 779; fascia, 780
Crest, frontal, 28; of ilium, 87; nasal, 44; occipital, 21, internal, 65; turbinated of palate, 52; of superior maxillary, 47, 48; of pubes, 90; of tibia, 133
Cribriform fascia, 811; plate of ethmoid, 41
Crico-arytaenoideus lateralis muscle, 747, posticus muscle, 747
Crico-thyroid artery, 368; membrane, 744; muscle, 747
Cricoid cartilage, 742
Crista galli, 41; ilii, 87; pubis, 90
Crown of a tooth, 678
Crucial ligaments of knee, 198, 199
Crura cerebri, 533, structure of, 533; of corpora cavernosa, 777; cerebelli, 549; of clitoris, 789; of diaphragm, 273; of fornix, 540
Cruraeus muscle, 318
Crural arch, 803, 813, deep, 814; canal, 815; nerve anterior, 605; ring, 815; sheath, 813
Crusta petrosa of teeth, 680
Crypts of Lieberkühn, 704
Crystalline lens, 652
Cuboid bone, 140
Cuneiform bone, hand, 116, foot, external, 144, internal, 143, middle, 144; cartilages, 743
Cupola of cochlea, 670
Curling, Mr., on the development of the testes, 786
Curvatures of the spine, 19
Cuspidate teeth, 678
Cutaneous branches of musculo-spiral, 597, of ulnar nerve, 595; nerve, of arm external, 591, internal, 592, lesser internal, 592, of buttock and thigh, 612, of inguinal region, 800, of ischio-rectal region, 821, from obturator, 606, from peroneal, 615, of thigh external, 603, of thigh internal, 606, of thigh middle, 606, of thorax lateral, 600; of plantar nerve, 613
Cuticle of skin, 631
Cutis vera, 629
Cystic artery, 427; duct, 718, valve of, 718; plexus of nerves, 625; veins, 491

Dartos, 780
Decussation of optic nerves, 553; of pyramids, 524
Deciduous teeth, 677
Deep crural arch, 814; palmar arch, 412; perinaeal fascia, 825
Deltoid aponeurosis, 281; muscle, 281
Dens sapientiae, 679
Dental artery, inferior, 372; canal, anterior, 46, inferior, 58, posterior, 45; follicles, 682; groove, 682; nerves anterior, 567, inferior, 572, posterior, 567; pulps, 684; sacs, 683, structure of, 684; tubuli, 680; vein, inferior, 469
Denticulate lamina of cochlea, 672
Dentine, 680, chemical composition of, 681; formation of, 684
Depressions for Pacchionian bodies, 62
Depressor alae nasi, 221; auguli oris, 223; epiglottidis muscle, 748; labii inferioris, 223
Derma, or true skin, 629
Descending aorta, 420; colon, 708
Descendens noni nerve, 561
Descent of testicle, 786
DEVELOPMENT of atlas, 13; axis, 13; bone, 4; carpus, 123; clavicle, 96; coccyx, 18; ethmoid, 42; femur, 129; fibula, 137; foot, 147; frontal bone, 30; hand, 123; humerus, 106; hyoid bone, 76; inferior turbinated bone, 55; lachrymal bone, 50; lower jaw, 59; lumbar vertebrae, 13; malar bone, 51; metacarpus, 123; metatarsus, 147; nasal bone, 44; occipital bone, 24; os innominatum, 92; palate bone, 54; parietal bone, 26;

INDEX

patella, 131; permanent teeth, 683; phalanges of hand, 123, of foot, 147; radius, 114; ribs, 84; sacrum, 17; scapula, 102; seventh cervical, 13; sphenoid, 39; sternum, 79; superior maxillary bone, 49; tarsus, 147, temporary teeth, 682; temporal bone, 35; tibia, 135 ulna, 112; vertebrae, 12; vomer, 56

Diameters of pelvis, 93

Diaphragm, 271; lymphatics of, 507

Diaphysis, 4

Diarthrosis, 152; rotatorius, 154

Digastric muscle, 235; fossa, 32; nerve, from facial, 559

Digestion, organs of, 675

Digital arteries from ulnar, 420, plantar, 463; cavity of lateral ventricle, 537; fossa, 126; nerves from median, 592, from ulnar, 596

Dilator naris, anterior, 221, posterior, 221

Diploë, 3; veins of, 471

Direct inguinal hernia, 803, comparative frequency of, 807, course of, 803, coverings of, 803, diagnosis of, 805, incomplete, 808

DISSECTION of abdominal muscles, 261; arch of aorta, 353; arm, 285; auricular region, 215, back, 328, epicranial region, 214; eye, 645; femoral hernia, 809; face, 216; foot, 340; fore-arm, 289; gluteal region, 323; hand, 300; head, 214; heart, left auricle, 732, left ventricle, 734, right auricle, 729, right ventricle, 730; hernia, 800, 809; inferior maxillary region, 223; infra-hyoid region, 233; inguinal hernia, 800; ischio-rectal region, 819; lingual region, 237; leg, 330; neck, 229; orbit, 218; palatal region, 242; palm of hand, 300; palpebral region, 216; pancreas, 719; pectoral region and axilla, 275; perinaeum, 819; pharynx, 239; pterygoid muscles, 227; radial region, 294; sole of foot, 340; spinal cord and membranes, 513; supra-hyoid region, 235; temporal muscle, 226; thigh, front of, 315, back of, 328, internal, 320

Dorsal artery of penis, 441; nerves, 598, anterior branches of, 599, peculiar, 600, posterior branches of, 598, roots of, 598; vertebrae, 9, peculiar, 10

Dorsales pollicis arteries, 416

Dorsalis hallucis artery, 458, indicis, 416, linguae, 369, pedis, branches of, 458, peculiarities of, 458, surgical anatomy of, 458

Dorsum of scapula, 98

DUCTS OR DUCT, of Bartholine, 689; biliary 716; of Cowper's glands, 773; cystic, 718; ejaculatory, 786; galactophorous, 798; hepatic, 718; of kidney, 765; lactiferous 798; of liver, 716; lymphatic, right, 496; nasal, 658; of pancreas, 720; parotid, 688; Stenonian, 688; thoracic, 495; Whartonian, 688

Ductless glands, 760, 767

Ductus arteriosus, 737; how obliterated in foetus, 739, communis choledocus, 718; pancreaticus minor, 720; Riviniani, 689; venosus, 739, how obliterated, 740

Duodenal glands, 704

Duodenum, 702; vessels and nerves of, 703

Dura mater of brain, 519, arteries of, 519, nerves of, 520, processes of, 520, structure of, 519, veins of, 519; of cord, 513, peculiarities of, 514

Ear, 658; arteries of, 374; auditory canal, 661; cochlea, 670; internal or labyrinth, 668; membranous labyrinth, 672; muscles of, 215; ossicula of, 665; pinna or auricle of, 658; semicircular canals, 669; tympanum, 667; vestibule, 668

Earthy constituent of bone, 1

INDEX

Ecker, on supra-renal capsules, 767
Eighth pair of nerves, 573
Ejaculatory ducts, 786, structure of, 786
Elbow, bend of, 408; joint, 182; vessels and nerves of, 184
Eminence of aqueductus Fallopii, 663; canine, 45; frontal, 27; ilio-pectineal, 89; jugular, 22; nasal, 29; parietal, 25; Eminentia articularis, 30; collateralis, 537, 538
Enamel of teeth, 681; chemical composition of, 681; formation of, 684; membrane, 684; organ, 684; rods, 681
Enarthrosis, 154
Endocardium, 735
Endolymph, 673
Ensiform appendix, 77, 79
Epidermis, 631, development of, 631; growth of, 631; structure of, 631
Epididymis, 781
Epigastric artery, 444, peculiarities, 444, relation to femoral ring, 815, with internal ring, 806, superficial, 449, 800; superior, 401; plexus, 624; region, 693; vein, 488, superficial, 485
Epiglottis, 743
Epiphyses, separation of, 4
Epithelium. *See various Organs.*
Erectile tissue, its structure, 778; of penis, 778; of vulva, 789
Erector clitoridis, 824; penis, 824; spinae, 254
Eruption of the teeth, 684
Ethmo-sphenoidal suture, 62
Ethmoid bone, 41, articulations of, 43, cribriform plate of, 41, development of, 43, lateral masses of, 42, perpendicular plate of, 41, os planum of, 42, unciform process, of, 43
Ethmoidal artery, 387; canal anterior, 29, 62, posterior, 29, 62; cells, 42; notch, 29; process of inferior turbinated, 55; spine, 36
Ethmoido-frontal suture, 62

Eustachian tube, 664, tympanic orifice of, 664; valve, 730, in foetal heart, 737
External abdominal ring, 802; annular ligament, 340; inguinal hernia, 800; orbitar foramina, 38; pteryoid plate, 39
Extensor brevis digitorum muscle, 341; carpi radialis brevior, 295, longior, 295; carpi ulnaris, 297; communis digitorum (hand), 296; indicis, 299; longus digitorum (foot), 332; minimi digiti, 297; ossis metacarpi pollicis, 298; primi internodii pollicis, 299; proprius pollicis, 332; secundi internodii pollicis, 299
EYE, 642; appendages of, 654; chambers of, 651; ciliary ligament, 649, muscle, 649, processes of, 647; humours of, 645, 651, aqueous, 651, crystalline lens, 652, vitreous, 652; membrana pupillaris, 648, choroid, 646, conjunctiva, 656, cornea, 644, hyaloid, 652, iris, 648, Jacob's, 650, retina, 649, sclerotic 643; pupil of, 648; tunics of, 642; uvea of, 648; vessels of globe of, 653
Eyeball, muscles of, 218, nerves of, 654, veins of, 654
Eyebrows, 654
Eyelashes, 656
Eyelids, 654, cartilages of, 655, Meibomian glands of, 655, muscles of, 216, structure of, 654, tarsal ligament of, 655
Eye teeth, 678

Face, arteries of, 372; bones of, 43, 69; lymphatics of 496; muscles of, 222; nerves of, 559; veins of, 468
Facial artery, 370, transverse, 375; bones, 21, 43; nerve, 557, branches of, 558, its communications, 557, course of, in temporal bone, 557; vein, 468
Falciform process of fascia lata, 811
Fallopian tubes, 795; fimbriated

INDEX

extremity of, 795; lymphatics of, 505; nerves of, 797; structure of, 795; vessels of, 797

False corpora lutea, 797; ligaments of bladder, 770; pelvis, 93; ribs, 80; vertebrae, 5, 13

Falx cerebelli, 520; cerebri, 520

Fangs of teeth, 679

FASCIAE OR FASCIA, 209, 210

anal, 832; aponeurotic, its structure, 211; of arm, 289

cervical, deep, 230, superficial, 229; costo-coracoid, 278; cremasteric, 780, 804; cribriform, 811

deep, 211; dentata, 539; dorsal, of foot, 341

fibro-areolar, its structure, 210; of foot, 340; of fore-arm, 289

of hand, 300

iliac, 312; infundibuliform, 805; intercolumnar, 263, 802; intercostal, 269; intermuscular of arm, 286, of thigh, 319, of foot, 341; ischio-rectal, 821;

lata, 317, 811, falciform process of, 811, iliac portion, 811, pubic portion, 811; of leg, 330; lumborum, 266

of mamma, 276

of neck, 228

obturator, 831

palmar, 300; pelvic, 831; perinaeal, deep, 825, superficial, 822; plantar, 340; propria, 780

recto-vesical, 832

spermatic, 802, superficial, 822, 211, of head, 212, of inguinal region, 800, of ischio-rectal region, 820, perinaeal, 822, of thigh, 315, of upper extremity, 274, 281

temporal, 226; of thorax, 269, 275; transversalis, 805; of trunk, 247

Fasciculi graciles, 524, teretes, 525, 545

Fasciculus unciformis, 532

Fauces, isthmus of, 686

FEMALE ORGANS OF GENERATION; bulbi vestibuli, 790; carunculae myrtiformes, 790; clitoris, 789; fossa navicularis, 789; fraenulum pudendi, 789; glands of Bartholine, 790; hymen, 789; labia majora, 788, minora, 789; nymphae, 789; praeputium clitoridis, 789; uterus, 792; vagina, 791; vestibule, 789

Femoral arch, 813, deep, 814; artery, 445; branches of, 448, peculiarities of, 447, surgical anatomy of, 447, deep femoral, 449; canal, variation in size of, according to position of limb, 817; hernia, complete, 817; coverings of, 817; cutaneous vessels and nerves of, 809; descent of, 817; dissection of, 809; incomplete, 817; seat of stricture, 818; superficial fascia, 809; surgical anatomy of, 809; varieties of, 817

ligament (Hey's), 811

region, muscles of anterior, 314, internal, 320, posterior, 328; ring, 815, position of surrounding parts, 815; sheath, 813

vein, 486; relation to femoral ring, 815

Femur, 124, articulations of, 129, attachment of muscles to, 130, development of, 129; fracture of above condyles, 348, below trochanters, 348; neck, fracture of, 347

Fenestra ovalis, 663; rotunda, 663

Ferrein, pyramids of, 765; tubes of, 765

Fibrae arciformes, 524, transversae, 548

Fibrine of muscle, 209

Fibro-cartilage, 151; circumferential, 151; inter-articular, 151; inter-osseous, 151; stratiform, 151

Fibro-cartilages, acromio-clavicular, 179; intervertebral, 157, of knee, 199; of lower jaw, 166; pubic, 176; radio-ulnar, 186; sacro-coccygean, 175, sterno-clavicular, 177

Fibula, 135; articulations of, 137; attachment of muscles to, 138; development of, 137; fracture of, with dislocation of the tibia, 349

846

INDEX

Fibular region, muscles of, 338
Fibrous rings of heart, 735
Fifth nerve, 562; ventricle of brain, 555
Fimbriae of Fallopian tube, 795
Fissura palpebrarum, 654
Fissure, auricular, 34, horizontal of cerebellum, 547; of ductus venosus, 713; for gall-bladder, 713; Glaserian, 31, 662; of liver, 712; longitudinal of liver, 712; longitudinal of cerebrum, 528, 530; of lung, 757; maxillary, 45; of medulla oblongata, 523; portal, 713; pterygo-maxillary, 70; spheno-maxillary, 69; sphenoidal, 38, 64; of spinal cord, anterior lateral, 517, median, 516, posterior lateral, 517, median, 516; Sylvian, 532; transverse, of liver, 713, of cerebrum, 539; umbilical, 712; for vena cava, 713
Flat bones, 3
Flexor accessorius muscle, 344; brevis digitorum, 343; brevis minimi digiti (hand), 304, (foot), 346; brevis pollicis (hand), 302, (foot), 345; carpi radialis, 290, ulnaris, 291; digitorum sublimis, 291; digitorum profundus, 292; longus digitorum, 337; longus pollicis (hand), 293, (foot), 336; ossis metacarpi pollicis, 301
Floating ribs, 80
Flocculus, 547
Foetus, circulation in, 738, Eustachian valve in, 737; foramen ovale in, 737; liver of, distribution of its vessels, 739; ovaries in, 797; relics in heart of, 729; vascular system in, peculiarities, 737
Folds, aryteno-epiglottidean, 743; recto-uterine, 695; recto-vesical, 695; vesico-uterine, 793
Follicle of hair, 633
Follicles sebaceous, 634
Follicular stage of development of teeth, 682
Fontanelle, anterior, 26; posterior, 24
Foot, bones of, 138; development of 147; dorsum, muscles of, 341; fascia of, 340; sole of, muscles of, 342, nerves of, 613, 614, vessels of, 457, 463

FORAMEN, caecum, 28, 62, of frontal bone, 28, of medulla oblongata, 523, of tongue, 635; carotid, 33; condyloid, 23; dental inferior, 58; incisive, 67; infra-orbital, 45; intervertebral, 20; lacerum anterius, 39, 64, medium, 64, posterius, 34; magnum, 22, 65; mastoid, 32; mental, 57; of Monro, 538; obturator, 86; optic, 36, 39; ovale of heart, 38, of sphenoid, 38, 64; palatine anterior, 48, posterior, 53; parietal, 25; pterygoid, 67; pterygo-palatine, 67; rotundum, 38, 64; sacro-sciatic, great, 87, small, 89; of Soemmering, 650; spheno-palatine, 54; spinosum, 38, 64; sternal, 77; stylo-mastoid, 34; supra-orbital, 28; thyroid, 91; vertebral, 6; Vesalii, 38, 64; of Winslow, 696

Foramina, sacral, 14; of diaphragm, 273; external orbitar, 38; malar, 53; olfactory, 41; Thebesii, 730

Fore-arm, fascia of, 289; muscles of, 289, vessels, 413, nerves of, 592, bones of, 107

Foreskin, 776
Form of bones, 2
Fornix, 540; bulbs of, 533; crura of, 540
Fossa, of antihelix, 658; canine, 45; condyloid, 23; cystis felleae, 713; digastric, 32; digital, 126; glenoid, 30; of helix, 658; iliac, 87; infra and supra spinous, 98; incisive, 45, 57; innominata, 658; ischio-rectal, 819; jugular, 68; lachrymal, 29; myrtiform, 45; navicular of urethra, 773, of vulva, 789; occipital inferior, 65; ovalis, 730; palatine anterior, 48, 65; pituitary, 36, 64; pterygoid of sphenoid, 39, of lower jaw, 59; scaphoid, 39; scaphoidea, 658; spheno-maxillary, 70; of skull, anterior, 62, middle, 62, posterior,

INDEX

64; subscapular, 97; sublingual, 57; submaxillary, 58; temporal, 69; trochanteric, 126; zygomatic, 69

Fossae, nasal, 73, 640

Fourchette, 789

Fourth nerve, 555; ventricle, 548

Fovea centrails retinae, 650; hemispherica, 669; hemi-elliptica, 669

Fracture of acromian process, 306 centre of clavicle, 306, acromial end of, 306; coracoid process, 307; coronoid process of ulna, 308; femur above condyles, 348, below trochanters, 348; fibula, with dislocation of tibia, 349; humerus, anatomical neck, 307, shaft of, 307, surgical neck, 307; neck of femur, 347; olecranon process, 309; patella, 349; radius, 309, lower end of, 310, neck of, 309, shaft of, 310, and ulna, 310; sternal end of clavicle, 306; tibia, shaft of, 349; ulna, shaft of, 310

Fraena of ileo-coecal valve, 707

Fraenulum cerebri, 544; pudendi, 789; of Vieussen's valve, 544

Fraenum clitoridis, 789; labii superioris et inferioris, 676; preputii, 776

Frontal artery, 387; bone, 27, articulations of, 30, attachment of muscles to, 30, development of, 30, structure of, 30; crest, 28; eminence, 27; nerve, 564; process of malar, 47; sinuses, 29; suture, 30, 60; vein, 468

Fronto-malar suture, 73

Fronto-sphenoidal suture, 62

Fundus of uterus, 698

Funiculi of nerve, 511

Funiculus, 511

Furrow, auriculo-ventricular, 736; interventricular, 736

Furrowed band of cerebellum, 546

Galactophorous ducts, 798

Gall bladder, 717; fissure for, 713; structure of, 717; valve of, 718

GANGLIA, *General Anatomy and Structure of,* 510; cardiac, 622; cephalic, 565; of fifth nerve, 565; lumbar, 626; lymphatic, 494; mesenteric, 626; renal, 625; sacral, 626; semilunar of abdomen, 624; solar, 624; of spinal nerves, 582; of sympathetic nerve, 617, branches from, 617; thoracic, 623

GANGLION, Arnold's 572; of Andersch, 574; carotid, 619; cervical, inferior, 620, middle, 620, superior, 619; ciliary, 565; on circumflex nerve, 591; diaphragmatic, 624; on facial nerve, 557; Gasserian, 563; glosso-pharyngeal, 574; impar, 617; inferior cervical, 621; inter-carotid, 620; on posterior interosseous nerve, 598; jugular, 574, 576, 577; lenticular, 565; lingual, 620; Meckel's, 565; middle cervical, 621; ophthalmic, 565; otic, 572; petrous, 574; pharyngeal, 620; pneumo-gastric, 578; of portio dura, 557; of Ribes, 617; root of vagus, 577; semilunar, 563; spheno-palatine, 567; submaxillary, 573; superior cervical, 619; supra-renal, 624; temporal, 620; thyroid, 620; trunk of vagus, 576; of Wrisberg, 622

Ganglionic branch of nasal nerve, 565

Gasserian ganglion, 563

Gastric arteries (vasa brevia), 428; artery, 425; follicles, 701; nerves from vagus, 576; plexus, 625; vein, 490

Gastro-colic omentum, 696

Gastro-duodenal artery, 426; plexus, 625

Gastro-epiploica dextra artery, 426, sinistra, 428

Gastro-epiploic plexus, 625; vein left, 490

Gastro-hepatic omentum, 695, 696

Gastro-phrenic ligament, 699

Gastro-splenic omentum, 697

Gastrocnemius muscle, 333

Gemellus inferior muscle, 327, superior, 327

GENERATIVE ORGANS, female, 788, male, 775

Genial tubercles, 57
Genio-hyo-glossus muscle, 237
Genio-hyoid muscle, 236
Genito-crural nerve, 603
Genu of corpus callosum, 535
Gimbernat's ligament, 262, 803, 813
Ginglymus, 154
Gladiolus, 77, 79
GLAND OR GLANDS, secreting, accessory of parotid, 688; aggregate, 705; arytenoid, 749; of Bartholine, 790; of biliary ducts, 716; Brunner's, 704; buccal, 676; ceruminous, 662; Cowper's 776, 826; duodenal, 704; ductless, 721; epiglottic, 749; gastric, 701; of Havers, 152; kidney, 763; labial, 676; lachrymal, 657; of larynx, 749; lingual, 637; of Littré, 773; liver, 710; lymphatic, 493; mammary, 798; Meibomian, 655; molar, 677; mucilaginous of Havers, 152; oesophageal, 692; of Pacchioni, 520; palatine, 686; pancreas, 719; parotid, 687; peptic, 701; Peyer's, 705; pharyngeal, 690; pineal, 543; pituitary, 532; prostate, 775; salivary, 687; sebaceous, 634; solitary, 705; sublingual, 688; submaxillary, 688; sudoriferous, 634; supra-renal, 767; thymus, 761; thyroid, 760; tracheal, 752; of Tyson, 777; uterine, 794; of vulva, 790
GLANDS DUCTLESS, 721; spleen, 721; suprarenal, 767; thymus, 761; thyroid, 760
GLANDS LYMPHATIC, *General Anatomy of*, 494, conglobate, 494; axillary, 499; bronchial, 507; cervical, 498; iliac, 502; inguinal, 501; intercostal, 506; lumbar, 503; mediastinal, 507; mesenteric, 506; oesophageal, 508; popliteal, 501; sacral, 502; thoracic, 506
Glandulae odoriferae, 777; Pacchioni, 520

Glans penis, 776; clitoridis, 789
Glaserian fissure, 31, 662
Glenoid cavity, 100; fossa, 30; ligament, 181
Gliding movement, 156
Glisson's capsule, 696, 716
Globus major of epididymis, 781, minor, 781
Glosso-epiglottidean ligaments, 743
Glosso-pharyngeal nerve, 574
Glottis, 745; rima of, 745
Gluteal artery, 442, inferior, 442; lymphatic glands, 502; nerve inferior, 612, superior, 610; region, lymphatics of, 504, muscles of, 523; veins, 486
Gluteus maximus muscle, 324, medius, 325, minimus, 325
Gomphosis, 153
Graafian vesicles, 796; membrana granulosa, of, 797; ovicapsule of, 797; structure of, 796
Gracilis muscle, 320
Granular layer of retina, 650
Granular lids, 656
Great omentum, 695, 696; cavity of, 695; sciatic nerve, 612
Greater wings of sphenoid, 37
Grey matter of cerebellum, 549; of fourth ventricle 549, of third, 543; of medulla oblongata, 526; of spinal cord, 510; of cerebrum, 509, chemical analysis 509
Groin, cribriform fascia of, 811; cutaneous vessels and nerves of, 809; region of, 810; superficial fascia of, 809; surgical anatomy of, 809
Groove, auriculo-ventricular, 728; bicipital, 104; cavernous, 37, 64; dental, 682; infra-orbital, 46; lachrymal, 48; mylo-hyoid, 58; nasal, 44; occipital, 32; optic 36, 64; subclavian, 96
Grooves in the radius, 114; ventricular, 728
Growth of bone, 4
Gubernaculum testis, 786

INDEX

Gums, 677
Gustatory nerve, 571
Gyri operti, 532
Gyrus fornicatus, 534

Haemorrhoidal artery, external, 440, middle, 438, superior, 431; nerve, inferior, 610; plexus of nerves, 626; veins, inferior, 486, middle, 486, superior, 486; venous plexus, 486
HAIRS, follicles of, 633, structure of, 633; medulla of, 634; root of, 633, sheath of, 633; shaft of, 634
Ham, region of the, 451
Hamstring tendons, surgical anatomy of, 330
Hamular process of sphenoid, 39; of cochlea, 671; of lachrymal, 50
Hand, bones of, 117; development of, 123; fascia of, 300; muscles of, 300, 301; nerves of, 594, 596, 597, 598; vessels of, 417, 420; ligaments of, 188
Hard palate, 685
Harmonia, 153
Havers glands of, 152
Head, muscles of, 212; veins of, 468
Head of scapula, 100; of ulna, 112
Heart, 727; annular fibres of auricles, 736; arteries of, 736; circular fibres of, 736; endocardium, 735; fibres of the auricles, 735, of the ventricles, 736; fibrous rings of, 735; foetal relics in, 730; left auricle, 732, ventricle, 734; looped fibres of auricles, 736; lymphatics of, 507, 737; muscular structure of, 735; nerves of, 621, 622, 736; openings in right ventricle, 731; position of, 727; right auricle, 729, ventricle of, 731; septum ventriculoiaim, 731; size and weight, 727; spiral fibres of, 736; structure of, 735; subdivision into cavities, 728; veins of, 736; vortex of, 736
Helicine arteries, 779
Helicis major muscle, 660, minor, 660

Helicotrema of cochlea, 671
Helix, 658; fossa of, 658; muscles of, 659; process of, 659
Hepatic artery, 425, 714, 716; cells, 716; duct, 714, 718; plexus, 625; veins, 489, 714, 716, 717
HERNIA congenital, 808; direct inguinal, 808; femoral, coverings of, 817, descent of, 817, dissection of, 809, varieties of, 817; infantile, 808; inguinal, 800, dissection of, 800; oblique inguinal, 807; scrotal, 808
Hesselbach's triangle, 808
Hey's ligament, 811
Hiatus Fallopii, 33, 64
Highmore, antrum of, 45
Hilton's muscle, 746
Hilus of kidney, 763; of spleen, 721
Hip joint, 193; muscles of, 323
Hippocampus major, 538, minor, 537
Horizontal plate of palate, 52
Humerus, 102; anatomical neck, fracture of, 308; articulations of, 107; attachment of muscles to, 107; development of, 106; head of, 102; neck of, 102; nutrient artery of, 411; shaft of fracture of, 307; surgical neck, fracture of, 307; tuberosities of, greater and lesser, 104
Humours of the eye, 651
Hyaloid membrane of eye, 652
Hymen, 789
Hyo-epiglottic ligament, 742
Hyo-glossal membrane, 637
Hyo-glossus muscle, 238
Hyoid artery of superior thyroid, 368; bone, 76, attachment of muscles to, 76, cornua of, 76, development of, 76; branch of lingual artery, 368; region, muscles of, infra, 233, supra, 235
Hypochondriac regions, 692
Hypogastric arteries in foetus, 740, how obliterated, 740; plexus, 627, inferior, 627; region, 693
Hypoglossal nerve, 561

INDEX

Ileo-coecal or ilio-colic valve, 706
Ileo-colic artery, 429
Ileum, 703
Iliac arteries, common, 433, left, 434, right, 434, peculiarities, 435, surgical anatomy of, 435; external, 443, surgical anatomy of, 444; internal, 436, peculiarity in the foetus, 436, at birth, 436, surgical anatomy of, 437; fascia, 312; fossa, 87; lymphatic glands, external, 502, internal, 502; region, muscles of, 312; vein, common, 487, peculiarities of, 487; external, 486, internal, 486
Iliacus muscle, 314
Ilio-femoral ligament, 194
Ilio-hypogastric nerve, 602
Ilio-inguinal nerve, 603
Ilio-lumbar artery, 442, vein, 487
Ilio-pectineal eminence, 90
Ilium, 86; crest of, 87; dorsum of, 87
Impressio colica, 714; renalis, 714
Incisive canal, 49, 67; foramina, 67; fossa, 45, 57
Incisor teeth of lower jaw, 678, of upper, 678
Incisura intertragica, 659
Incus, 665; ligament of, 666, suspensory, 666
Infantile hernia, 808
Inferior dental canal, 59; maxillary bone, 56; meatus of nose, 75; occipital fossa, 65; profunda artery, 411; turbinated bones; 54, articulations of, 55, development of, 55, ethmoidal process of, 54, lachrymal, process of, 55; vena cava, opening of, 729
Infra-costal muscles, 270
Infra-orbital canal, 45; foramen, 45; groove, 46
Infra-spinatus muscle, 283
Infra-spinous aponeurosis, 283
Infra-trochlear nerve, 565
Infundibula of kidney, 765
Infundibulifonn fascia, 780

Infundibulum of brain, 543; of ethmoid, 42; of cochlea, 670; of heart, 731
Inguinal canal, 805; glands, deep, 501, superficial, 501, 810; hernia, 807, direct, 807, incomplete, 808, oblique, 807, surgical anatomy of, 800; region, 693; dissection of, 800
Inlet of pelvis, 93
Innominate artery, 360, peculiarities of, 360, surgical anatomy of, 361; bone, 85, articulations of, 92, attachment of muscles to, 92, development of, 92; veins, 480, peculiarities of, 480
Inorganic constituent of bone, 1
Interarticular fibro-cartilage, 151, of scapulo-clavicular joint, 178; of jaw, 166, of pubes, 176, of radio-ulnar joint, 186, of sterno-clavicular joint, 178; ligament of ribs, 167
Intercarotid ganglion, 620
Interclavicular ligament, 178
Intercolumnar fascia, 263, 780, 802; fibres, 802
Intercondyloid notch, 128
Intercostal arteries, 422, anterior, 400, superior, 401; fasciae, 269; ligaments, 171; lymphatics, 507; lymphatic glands, 506; muscles, 270; nerves, 599, lower, 600, upper, 599; spaces, 81; veins, superior, 481
Intercosto-humeral nerve, 599
Interlobular, biliary plexus, 715; veins, 716
Intermaxillary suture, 66
Internal abdominal ring, 806; annular ligament, 340; carotid artery, 382; cutaneous nerve, 592; inguinal hernia, 807; oblique muscle, 803; occipital crest, 65; pterygoid plate, 39; sphincter, 821
Internasal suture, 70
Interossei muscles, dorsal of hand, 305, of foot, 346, palmar, 305; plantar, 347
Interosseous artery, 418, anterior, 418, posterior, 419; fibro-cartilage, 151;

851

INDEX

ligament of radio-ulnar joint, 185; nerve, anterior, 594, posterior, 598
Interpeduncular space of brain, 534
Interspinales muscles, 259
Intertransversales muscles, 259
Intervertebral notches, 6; substance, 157
Intestine, large, 705
Intestine, small, 702
Intralobular veins, 717
Intumescentia ganglio-formis, 557
Involuntary muscles, 209
Iris, 648; arteries of, 648; structure of, 648
Irregular bones, 3
Ischiatic lymphatic glands, 501
Ischio-rectal fascia, 832; fossa, 821, position of vessels and nerves in, 821; region, surgical anatomy of, 819
Ischium, 86; body of, 90; ramus of, 90; spine of, 90; tuberosity of, 90
Island of Reil, 532
Isthmus of the fauces, 686; of thyroid gland, 760
Iter ad infundibulum, 543; a tertio ad quartum ventriculum, 544
Ivory of tooth, 680

Jacob's membrane, 650
Jacobson's nerve, 668; canal for, 34
Jaw, lower, 56; development of, 59; articulations of, 60, ligaments of, 164, attachmeut of muscles to, 60, oblique line of, 57, pterygoid fossa of, 59; rami of, 58; symphysis of, 59
Jejunum, 703
Joint. *See Articulations.*
Jugular foramen, 65; fossa, 68; ganglion, 574, 576, 577; process, 22; vein, anterior, 470, external, 469, external posterior, 470, internal, 470, sinus or gulf of, 470

Kidney, 763; calyces of, 765; cortical substance of, 764; ducts of, 765; hilus of, 763; infundibula of, 765; lymphatics of, 505, 766; Malpighian bodies of, 765; mammillae of, 764; medullary substance, 764; nerves of, 766; papillae of, 764; pelvis of, 765; pyramids of Ferrein, 765; relations of, 763; renal artery, 766; sinus of, 764, 766; tubes of Ferrein, 765; tubuli uriniferi, 764; veins of, 766; weight and dimensions, 763
Knee-joint, 195
Kürschner, on structure of heart's valves, 731

Labia cerebri, 534; pudendi, majora, 788, minora, 789, lymphatics of, 501
Labial artery, 372; glands, 676; veins, superior, 468, inferior, 468
Labyrinth, 658; arteries of, 673; fibro-serous membrane of, 672; membranous, 672
Lachrymal apparatus, 657; artery, 386; bones, 49, articulations of, 50, attachment of muscles to, 50, development of, 50; canals, 49, 657; caruncula, 657; fossa, 29; gland, 657; groove, 48; nerve, 555; papilla, 657, 654; process of inferior turbinated bone, 52; puncta, 657; sac, 658; tubercle, 48
Lacteals, 493, 506
Lactiferous ducts, 798
Lacuna magna, 773
Lacus lachrymalis, 654
Lambdoid suture, 60
Lamella, horizontal of ethmoid, 41; perpendicular of ethmoid, 41
Lamina cinerea, 532; cribrosa of sclerotic, 643; fusca, 643; spiralis of cochlea, 670, membranacea, 671
Laminae of cornea, elastic, 643; of the vertebrae, 6
Laminated tubercle of cerebellum, 547
Large intestine, 705: cellular coat, 709; caecum, 706; colon, 707; ilio-caecal valve, 707; mucous coat, 709; muscular

coat, 709; rectum, 708; structure of, 709; serous coat of, 709

Laryngeal artery, inferior, 368, superior, 368; nerve, external, 578; internal, 578; recurrent, 579, superior, 578; from sympathetic, 620; pouch, 746; veins, 470

Laryngis sacculus, 746

Laryngo-tracheotomy, 752, 753

Laryngotomy, 752

Larynx, 741; actions of muscles of, 748; arteries of, 749; cartilages of, 741; cavity of, 745; glands of, 749; glottis, 745; interior of, 745; ligaments of, 744; lymphatics of, 749; mucous membrane of, 749; muscles of, 747; nerves of, 749; rima glottidis, 746; veins of, 749; ventricle of, 746; vocal cords of, false, 743, true, 743

Lateral ginglymus, 154; ligaments of liver, 711; masses of ethmoid, 41; region of skull, 68

Lateralis nasi artery, 372

Latissimus dorsi muscle, 250

Laxator tympani major muscle, 667; minor, 667

Leg, muscles of, 330: back of, 333; front of, 331; fascia of, 330, deep, 335

Lens, 653; changes produced in, by age, 653; structure of, 653; suspensory ligament of, 653

Lenticular ganglion, 565

Lesser lachrymal bone, 50; omentum, 695, 696; sciatic nerve, 612; wings of sphenoid, 38

Levator anguli oris, 222, scapulae, 252; ani, 827; glandulae thyroidae, 761; labii inferiori, 223, superioris alaeque nasi, 221; superioris proprius, 222; palati, 242; palpebrae, 217; prostatae, 827

Levatores costarum, 271

Lieberkühn, crypts of, 704

LIGAMENTS, structure of, 151
  acromio-clavicular, superior and inferior, 179; alar of knee, 199; of ankle, anterior, 201, lateral, 201; annular of radius, 184, of wrist, anterior, 300, posterior, 300, of ankle, 340, external, 340, internal, 340, of stapes, 666; anterior of knee, 195; arcuate, 272; aryteno-epiglottic, 743; astragalo-scaphoid, 206; atlo-axoid anterior, 160, posterior, 160
 of bladder, false, 770, true, 770 calcaneo-astragaloid external, 203, posterior, 160, interosseous, 204; calcaneo-cuboid, internal, 205, long, 205, short, 205, superior, 205; calcaneo-scaphoid, inferior, 206; superior, 205; capsular of hip, 194, of jaw, 165, of knee, 197, of shoulder, 181, of vertebrae, 159; carpo-metacarpal, dorsal, 190, interosseous, 191, palmar, 191; of carpus, dorsal, 188, 189, interosseous, 189, palmar, 188, 189; central of spinal cord, 515; ciliary of eye, 649; common vertebral anterior, 157, posterior, 157; conoid, 180; coraco-acromial, 180; coraco-clavicular, 179; coraco-humeral, 181; coracoid, 180; coronary of liver, 712; costo-clavicular, 178; costo-sternal, anterior, 169, posterior, 169; costo-transverse, anterior, 168, middle, 168, posterior, 169; costo-vertebral, or stellate, 167; costo-xiphoid, 171; cotyloid, 194; crico-arytenoid, 744; crico-thyroid, 747; crucial of knee, 198; cruciform, 160
 deltoid, 201; dorsal of carpus, 188, metacarpus, 190, tarsus, 203, metatarsus, 207, tarso-metatarsal, 207 of elbow, 182; anterior, 183, external lateral, 183, internal lateral, 183, posterior, 183
 falciform of liver, 711; femoral (Hey's), 811
 gastro-phrenic, 699; Gimbernat's, 262, 803, 813; glenoid, 181; glosso-epiglottidean, 743

INDEX

of hip, 193; hyo-epiglottic, 743 ilio-femoral, 194; ilio-lumbar, 172; of incus, 666; interarticular of ribs, 167; interclavicular, 178; intercostal, 171; interosseous, calcaneo-astragaloid, 204, calcaneo-cuboid internal, 205, carpal, 188, 189, carpo-metacarpal, 190, metacarpal, 191, metatarsal, 208, posterior, sacro-iliac, 173, radio-ulnar, 186, of ribs, 167, tarsal, 123, tarso-metatarsal, 207, tibio-fibular, inferior, 201; interspinous, 159; intertransverse, 160; intervertebral, 157

of jaw, 164, lateral external, 165, lateral internal, 165

of knee, 195

of larynx, 744; lateral of bladder, 770; of knee, 195, of liver, 711, of elbow, 183, of carpus, 188, of ankle, 201, of jaw, 165, longitudinal of liver, 711; long plantar, 205; lumbo-iliac, 172; lumbo-sacral, 172

metacarpo-phalangeal, 192; metacarpal, 191; metatarsal, 208; metatarso-phalangeal, 208; mucosum of knee, 199

nuchae, 250

oblique, 185; obturator, 176; occipito-atloid, anterior, 163, lateral, 163, posterior 163; occipito-axoid, 164; odontoid, 164; orbicular, 184; of ossicula, 666; of ovary, 797

palpebral or tarsal, 655; of patella, 195; of pelvis, 172; of the phalanges, hand, 192, foot, 208; of the pinna, 659; plantar, long, 205; posterior of knee, 196; posticum Winslowii, 196; Poupart's, 262, 803, 813; pterygo-maxillary, 225; pubic anterior, 175, posterior, 176, superior, 176; pubo-prostatic, 770

radio-carpal, 187; radio-ulnar joint, inferior, 186, middle, 185, superior, 184; recto-uterine, 793; rhomboid, 178; round of uterus, 797, of liver, 712, of radius and ulna, 185, of hip, 194

sacro-coccygeal, anterior, 175, posterior, 175; sacro-iliac, anterior, 173, oblique, 173, posterior, 173; sacro-sciatic, greater, 174, lesser, 174; sacro-vertebral, 172; of scapula, 178; of shoulder, 178; stellate, 167; sterno-clavicular, anterior and posterior, 177; of sternum, 171; stylo-maxillary, 165; sub-flavous, 159; sub-pubic, 176; supra-spinous, 159; suspensory of incus, 666, of lens, 655, of liver, 711, of malleus, 666, of mamma, 276, of penis, 776, of spleen, 721

tarsal of eyelids, 655; tarso-metatarsal, 207; of tarsus, 203; teres of hip, 193; thyro-arytenoid, inferior, 746, superior, 746; of thumb, 191; tibio-tarsal, 201; thyro-epiglottic, 743; thyro-hyoid, 744; tibio-fibular, 200; transverse of atlas, 160, of hip, 193, of knee, 195, of scapula, 180; trapezoid, 179; triangular of urethra, 791; of tympanic bones, 665, of incus, 666, of malleus, 666, of stapes, 666, of urethra, 826

of uterus, 793

of vertebrae, 156; vesico-uterine, 793

of wrist, anterior, 187, lateral external, 187, lateral internal, 187, posterior, 188

of ZINN, 219

Ligamenta subflava, 159

Ligamentum, arcuatum externum, 272, internum, 272; denticulatum, 515; latum pulmonis, 755; nuchae, 250; patellae, 195

Ligature of arteries. *See each Artery.*

Limbus luteus, 650

Linea alba, 268; aspera, 127; ilio-pectinea, 87, 89, 92; quadrati, 127; splendens, 515

Lineae semilunares, 269; transversae of abdomen, 269; transversae of fourth ventricle, 548

Lingual artery, 368; surgical anatomy of,

INDEX

369; bone, 76; ganglion, 621; nerve, 571; veins, 471
Lingualis muscle, 238
Lips, 675, structure of, 675; arteries of, 372
Liquor Cotunnii, 672; Morgagni, 652; Scarpae, 673; seminis, 786
Lithotomy, parts concerned in operation of, 821, avoided in operation, 821, divided, 821
LIVER, changes of position in, 711; distribution of vessels to in foetus, 738; ducts of, 714; fibrous coat of, 715; fissures of, 712; hepatic artery, 714, 716; hepatic cells, 716; hepatic duct, 714; hepatic veins. 714, 717; ligaments of, 711, lateral, 711, coronary, 712, round, 712, longitudinal, 711; lobes of, 713; lobules of, 715; lymphatics of, 505, 714; nerves of, 714; portal vein, 714; situation, size, and weight, 710; structure of, 714; its surfaces and borders, 710; vessels of, 714
Lobes of cerebrum, 530; of cerebellum, 547; of liver, 713, left, 714, lobus caudatus, 714, quadratus, 714, Spigelii, 714; of lung, 757; of prostate, 775; of testis, 783; of thyroid, 760; of thymus, 761
Lobular biliary plexus, 716
Lobule of the ear, 659
Lobules of liver, 715; of lung, 758
Lobuli testis, 783
Lobulus caudatus, 714; centralis of cerebellum, 546; quadratus, 714; Spigelii, 714
Locus coeruleus, 548; niger, 533; perforatus anterior, 532, posterior, 533
Long bones, 2
Long saphenous nerve, 607
Longissimus dorsi muscle, 256
Longitudinal fissure, of brain, 531; of liver, 712, ligament of liver, 711
Longus colli muscle, 246
Lower extremity, arteries of, 445; bones of, 124; lymphatics of, 501; ligaments of, 193; muscles of, 311; nerves of, 603; veins of, 484
LOWER, tubercle of, 729
Lumbar arteries, 433; fascia, 266; ganglia, 626; glands, 503; nerves, 600, anterior branches of, 601, posterior, 601, roots of, 600; plexus of nerves, 601, branches of, 601; region, 693; vein ascending, 488; veins, 488; vertebrae, 11, development of, 12
Lumbo-iliac ligament, 172
Lumbo-sacral ligament, 172
Lumbricales muscles, hand, 304, foot, 345
Lungs, 756; air-cells of, 759; bronchial arteries, 759, veins, 760; capillaries of, 759; in foetus, 739; lobes and fissures of, 757; lobules of, 758; lymphatics of, 507, 760; nerves of, 760; pulmonary artery, 759, veins, 760: root of, 757; structure of, 756; weight, colour, etc., 758
Lunulae of nails, 632
Lymphatic duct, right, 496
LYMPHATIC GLANDS, *General Anatomy of*, 493
Descriptive Anatomy:
    anterior mediastinal, 507; auricular posterior, 496; axillary, 499
    brachial, 499; bronchial, 507; buccal, 496
    in front of elbow, 499
    gluteal, 501
    iliac, external, 502, internal, 502; inguinal, deep, 501, superficial, 501; intercostal, 506; internal mammary, 506; ischiatic, 501
    of large intestine, 506; of lower extremity, 501; lumbar, 503
    of neck, 496
    occipital, 496
    parotid, 496; of pelvis, 502; popliteal, 501
    radial, 499

sacral, 502; of small intestine, 506;
of spleen, 506; of stomach, 505;
submaxillary, 496
of thorax, 506; tibial anterior, 501
ulnar, 499; of upper extremity, 499
zygomatic, 496

LYMPHATICS, *General Anatomy of*, 493,
coats of, 493; origin of, 493; plexus
of, 493; subdivision into deep and
superficial, 493; valves of, 494; vessels
and nerves of, 494; where found, 493

Descriptive Anatomy:
abdomen, 504; arm, 499
bladder, 504; bone, 4; broad ligaments, 504
cerebral, 497; cervical, superficial and deep, 498; of clitoris, 504; of cranium, 497
diaphragm, 507
face, superficial, 497, deep, 497; Fallopian tubes, 505
gluteal region, 504
head, superficial, 496; heart, 507
intercostal, 506; internal mammary, 506; intestines, 506
kidneys, 505
labia, 504; large intestine, 506; leg, 502; liver, 505; lower extremity, 501; lung, 507; lymphatic duct, 496
meningeal, 497; mesentery, 506; mouth, 497
neck, 498; nose, 497; nymphae, 504
oesophagus, 508; ovaries, 504
pancreas, 505; pelvis, 504; penis, 504; perinaeum, 504; pharynx, 497; pia mater, 497; prostate, 504
rectum, 504
scrotum, 504; small intestine, 506; spleen, 506; stomach, 505
testicle, 505; thoracic duct, 495; thorax, 507; thymic, 508; thyroid, 508
upper extremity, superficial, 500, deep, 501; uterus, 504
vagina, 504

Lyra of fornix, 541

Macula cribrosa, 669
Magnum of carpus, 115
Malar bone, 50, articulations of, 51, attachment of muscles to, 51, development of, 51, frontal process of, 50, maxillary process of, 51, orbital process of, 51; canals, 50; process, 47
Male urethra, 772
Malleolar arteries, external and internal, 457
Malleolus, external, 136, internal, 135
Malleus, 665; suspensory ligament of, 666
Malpighian bodies of kidney, 765; corpuscles of spleen, 723, relation with arteries, 724; veins, 724
Mamma, areola of, 798; lobules of, 798; nerves of, 799; nipple or mammilla of, 798; structure of, 798; vessels of, 799
Mammary artery, internal, 400; glands, 798; lymphatic glands, 506; veins, internal, 480
Mammilla of breast, 798; of kidney, 764
Manubrium of sternum, 77; of malleus, 665
Masseter muscle, 226
Masseteric arteries, 378; nerve, 570; veins, 468
Masto-occipital suture, 61
Masto-parietal suture, 60
Mastoid cells, openings of, 664; foramen, 32; portion of temporal bone, 32; process, 32; vein, 469
Matrix of nail, 632
Maxillary artery, internal, 376; bone, inferior, 56, superior, 44; fissure, 45; nerve, inferior, 570, superior, 566; process of inferior turbinated, 55, of malar bone, 51; sinus, 47; tuberosity, 45
Meatus auditorius externus, 33, 661, internus 33; of nose, inferior, 75, middle, 75, superior, 75; urinarius, male, 773, female, 789
Meatuses of the nose, 75

INDEX

Meckel's ganglion, 568
Median artery of fore-arm, 419; of spinal cord, 397; nerve, 592; vein, 477
Mediastinal arteries, from internal mammary, 400, posterior, from aorta, 422; lymphatic glands, 505
Mediastinum, anterior, 755; middle, 756; posterior, 756; testis, 783
MEDULLA OBLONGATA, 523; anterior pyramids of, 524, 525; back of, 525; fissures of, 523, grey matter of, 526; lateral tract, 524, 525; olivary body, 524, 525; posterior pyramids, 524; restiform bodies, 524, 526; septum of, 526; structure of, 525
Medulla spinalis, 516
Medullary canal of bone, 3, membrane, of bone, 3; substance of brain, 532, of kidney, 764, of supra-renal capsules, 767; velum posterior of cerebellum, 547
Meibomian glands, 655
Membrana fusca, 646; granulosa, 797; limitans, 651; nictitans, 656; pupillaris, 648; sacciformis, 187; secundarii, 663, structure of, 663; tympani, 671
Membrane, of aqueous chamber, 651; arachnoid, spinal, 514, cerebral, 521; choroid, 646; costo-coracoid, 278; crico-thyroid, 744; fenestrated, 352; hyaloid, 652; Jacob's, 650; limiting, 651; pituitary, 658; pupillary, 652; thyro-hyoid, 744; Schneiderian, 640
MEMBRANES OF SPINAL CORD, 513; of brain, 518
Membranous labyrinth, 672, structure of, 673; portion of urethra, 773; semicircular canals, 672
Meninges, cerebral, 519, spinal, 513
Meningeal artery, from ascending pharyngeal, 374, anterior, from internal carotid, 385, middle, from internal maxillary, 377, from occipital, 373, posterior from vertebral, 396, small, from internal maxillary, 378;

lymphatics, 497; veins, 469
Menisci, 151
Mental foramen, 57; process, 57
Mesenteries, 696, 697
Mesenteric artery, inferior, 431, superior, 428; glands, 506; plexus of nerves, inferior, 626, superior, 625; vein, inferior, 489, superior, 489
Mesentery, 697
Meso-caecum, 697, 706
Meso-colon, ascending, 697, descending, 697, transverse, 695, 697
Meso-rectum, 697
Mesorchium, 787
Metacarpal artery, 415; articulations, 191
Metacarpo-phalangeal articulations, 192
Metacarpus, 121; common characters of, 121; development of, 123; peculiar bones of, 121
Metatarsal articulations, 207; bones, 143
Metatarsea artery, 458
Metatarso-phalangeal articulations, 208
Metatarsus, 145, development of, 147
Middle clinoid processes, 36; ear, or tympanum, 658; fossa of skull, 62; meatus, 75
Milk teeth, 677
Mitral valve, 734
Mixed bones, 3
Modiolus of cochlea, 670
Molar glands, 677; teeth, 679; teeth, peculiar, 679
Mons Veneris, 788
Monticulus cerebelli, 546
Morsus diaboli, 795
Motor oculi nerve, 554
Mouth, 675; mucous membrane of, 675; muscles of, 223
Movement admitted in joints, 156
Mucilaginous glands, 152
Multicuspidate teeth, 679
Multifidus spinae muscle, 258
MUSCLES, *General Anatomy of,* 209; arrangement of fibres of, 209; bipenniform, 210; derivation of names,

210; form of, 209; fusiform, 210; involuntary, 209; meaning of the terms origin, 210, insertion, 210; penniform, 210; radiated, 210; sheath of, 209; size of, 210; structure of, 209; voluntary, 209

MUSCLES OR MUSCLE, *Descriptive Anatomy:*

of abdomen, 261; abductor minimi digiti (hand) 303, (foot) 343, indicis, 305, pollicis (hand), 301, (foot) 342; accelerator urinae, 823; accessorii orbicularis oris, 224; accessorius pedis, 344; accessorius ad sacro lumbalem, 256; of acromial region, 281; adductor brevis, 322, longus, 321, magnus, 322, pollicis (hand), 303, (foot) 346; anconeus, 297; anomalus, 221; antitragicus, 659; of anus, 820, 825, of arm, 285; aryteno-epiglottideus, inferior, 748, superior, 748; arytgenoideus, 747; attollens aurem, 215; attrahens aurem, 215; azygos uvulae, 243

of back, 246; basio-glossus, 238; biceps (arm) 286, (leg), 329; biventer cervicis, 257; brachial region, anterior, 289, posterior, 296, 297; brachialis anticus, 287; buccinator, 224

caninus, 222; cerato-glossus, 238; cervicalis ascendens, 256; of chest, 276; chondro-glossus, 238; ciliary of eye, 649; circumflexus palati, 243; coccygeus, 827; cochlearis, 671; complexus, 257; compressor narium minor, 221, nasi, 221, sacculi laryngis, 746, urethrae, 826; constrictor, isthmi faucium, 239, pharyngis inferior, 240, medius, 240, superior, 241, urethrae, 826; coraco-brachialis, 286; corrugator supercilii, 217; cremaster, 803; crico-arytsenoideus lateralis, 747, posticus, 747; crico-thyroid, 747; cruraeus, 319

deltoid, 281; depressor anguli oris, 223; depressor alae nasi, 221, epiglottitis, 748, labii inferioris, 223; diaphragm, 271; digastric, 235; dilator naris, anterior, 221, posterior, 221; dorsum of foot, 341

of external ear, 215; erector clitoridis, 824, penis, 824, spinae, 254; external sphincter, 820; extensor brevis digitorum, 341, carpi radialis brevior, 295, longior, 295, ulnaris, 297; coccygis, 259; digitorum communis, 296; indicis, 299; longus digitorum, 332; minimi digiti, 297; ossis metacarpi pollicis, 298; primi internodii pollicis, 299; proprius pollicis, 332; secundi internodii pollicis, 299; of eyelids, 216

of face, 212; femoral region, anterior, 314, internal, 320, posterior, 328; fibular region, 338; flexor accessorius, 344, brevis minimi digiti (hand), 304, (foot), 346, digitorum, 343, pollicis (hand), 293, (foot), 336, carpi radialis, 290, ulnaris, 291, digitorum sublimis, 291, longus digitorum, 337, pollicis (hand), 293, (foot), 336, ossis metacarpi pollicis, 301, profundus digitorum, 292; of foot, 339; of forearm, 289

gastrocnemius, 333; gemellus superior, 327, inferior, 327; genio-hyo-glossus, 237; genio-hyoid, 236; of gluteal region, 323; gluteus maximus, 324, medius, 325, minimus, 325; gracilis, 320

of hand, 301, 303; of head and face, 212; helicis, major, 660, minor, 660; Hilton's, 746; of hip, 323; humeral region, anterior, 285, posterior, 287; of hyoid bone and larynx, 233; hyo-glossus, 238

iliac region, 312; iliacus, 314, infra-costal, 270; infra-spinatus, 283; intercostal, 270; internal sphincter, 821; interossei of foot, 346, palmar,

305, dorsal, 305; interspinales, 259; intertransversales, 259

labial, 224; of larynx, 741; latissimus dorsi, 250; laxator tympani, major, 667, minor, 667; of leg, 320; levator anguli oris, 222, scapulae, 252, ani, 827, glandulae thyroidae, 761; labii inferioris, 223, superioris alaeque nasi, 221, proprius, 222, proprius alae nasi, posterior, 221, anterior, 221, menti, 223, palati, 242, palpebrae, 218, prostatae, 827; levatores costarum, 271; lingualis, 238; of lips, 223; longissimus dorsi, 256; longus colli, 246; lumbricales (hand), 304, (foot), 345

masseter, 226; of mouth, 225; multifidus spinae, 258; musculus accessorius ad sacro-lumbalem, 256; mylo-hyoid, 236; myrtiformis, 221

naso-labialis, 224; of neck, 228; of nose, 220

obliquus auris, 660, abdominis externus, 262, internus, 264; capitis, superior, 260, inferior, 260; oculi, inferior, 216, superior, 215; obturator, externus, 195, 328, internus, 326; occipito-frontalis, 214, omo-hyoid, 234; opponens minimi digiti, 304, pollicis, 301; orbicularis oris, 224, palpebrarum, 216; of orbit, 218

palate, 242; palato-glossus, 239, 243; palato-pharyngeus, 244; palmaris brevis, 303, longus, 290; pectineus, 321; pectoralis major, 276, minor, 278; of penis 823; of perinaeum, male, 824, female, 825; peroneus brevis, 339, longus, 338, tertius, 332; of pericranium, 215; of pharynx, 239; of the pinna, 659; plantaris, 335; platysma myoides, 229; popliteus, 336; pronator quadratus, 293, radii teres, 289; psoas magnus, 313, parvus, 314; pterygoid, internal, 227, external, 228; pyramidalis abdominis, 262, nasi, 221; pyriformis, 326

quadratus femoris, 327, lumborum, 268, menti, 223; quadriceps extensor cruris, 318

radial region, 294; rectus abdominis, 267, capitis anticus major, 245, minor, 245; posticus major, 260, minor, 260, femoris, 318, oculi, externus, superior, inferior, and internal, 219, lateralis, 246; retrahens aurem, 216; rhomboideus major, 252, minor, 252; risorius, 225; rotatores spinae, 259

sacro-lumbalis, 256; sartorius, 317; scalenus anticus, 246, medius, 247, posticus, 247; scapular region, anterior, 282, posterior, 283; semi-membranosus, 329; serratus posticus, superior, 253, inferior, 253; semi-spinalis dorsi, 258, colli, 258; semitendinosus, 329; serratus magnus, 280; sole of foot, 342, first layer, 342, second layer, 344, third layer, 345; soleus, 334; sphincter, external, 820, internal, 821, vaginae, 825; spinalis dorsi, 257, cervicis, 257; splenius, 253, capitis, 254, colli, 254; stapedius, 667; sterno-cleido-mastoid, 231; sterno-hyoid, 233; sterno-thyroid, 234; stylo-glossus, 239; stylo-hyoid, 236; stylo-pharyngeus, 241; subanconeus, 288; subclavius, 278; subcruraeus, 319; sub-scapularis, 282; supinator brevis, 297, longus, 294; supra-spinales, 259, supra-spinatus, 283

temporal, 226; tensor palati, 243; tarsi, 217; tympani, 662; vaginae femoris, 317; teres major, 284, minor, 284; thoracic region, anterior, 276, lateral, 280; of thorax, 269; of thigh, 312; thyro-arytaenoideus, 747; thyro-epiglottideus, 748; thyro-hyoid, 234; tibialis anticus, 331, posticus, 337; tibio-fibular region, anterior, 331; of tongue, 237; trachelo-mastoid, 257; tragicus, 660; transversalis abdominis, 265; colli, 256; trausversus auriculae, 660, pedis, 346, perinaei, 824, (female) 825; trapezius, 248; triangularis sterni, 270; triceps, extensor cubiti, 287, extensor cruris, 316; of trunk, 247; of tympanum, 662

upper extremity, surgical anatomy of, 306; of ureters, 766; of urethra, 823

vastus externus, 318, internus and cruraeus, 319; vertebral region, anterior, 245, lateral, 246

zygomaticus major, 222, minor, 222

Musculi papillares, left ventricle, 735, right, 732; pectinati in left auricle, 734, in right, 730

Musculo-cutaneous nerve of arm, 591, from peroneal, 615

Musculo-spiral nerve, 596

Musculo-phrenic artery, 400

Musculus accessorius ad sacro-lumbalem, 256

Mylo-hyoid groove, 58; muscle, 236; nerve, 572; ridge, 57

Myrtiform fossa, 45

Myrtiformis muscle, 221

Nails, chemical composition of, 632; lunula of, 632; matrix of, 632; root of, 632; structure of, 632

Nares, anterior, 73, posterior, 67; septum of, 74

Nasal angle, 44; artery, 379, 387; bones, 44, articulations of 44; cartilages, 639; crest, 44; duct, 658; eminences, 30; fossae, 73, 640, arteries of, 641, mucous membrane of, 640, nerves of, 642, veins of, 642; groove, 44; nerve, 564; nerves from Meckel's ganglion, 568; notch, 28; process, 47; spine, 29, anterior, 48, posterior, 51; venous arch, 468

Naso-maxillary suture, 70

Naso-palatine canal, 56; nerve, 569

Nates of brain, 544

Neck, muscles of, 228; triangle of, anterior, 380, posterior, 382, surgical anatomy of, 380; veins of, 468, 469

NERVES, *General Anatomy of,* 510; afferent or centripetal, 510; cerebro-spinal, 511, composition of, 510, junction of funiculi, 511, neurilemma of, 511, origin of, 512, apparent, 512, real, 512, plexus of, 511, sheath of, 511, structure, 510, 511, subdivisions of, 511, termination of, 512, vessels of, 511; compound, 551; efferent or centrifugal, 510; of motion, 551, 554; reflex or excito-motory, 510; spinal, roots of, 581; of special sense, 551

NERVES OR NERVE, *Descriptive Anatomy of:* abducens, 555; accessory obturator, 605, spinal, 581; acromiales, 585; anterior crural, 605; of arachnoid, 514; articular of elbow, 595, hip, 605, 612, knee, 605, 607, 613, 615, shoulder joint, 589, 590, wrist, 595; auditory, 553, 668; auricular, posterior, 559, of vagus, 576; of auricularis magnus, 584; of auriculo-temporal, 571, of second cervical, 583; of small occipital, 585; axillary, 583

of bone, 4; buccal, 570, of facial, 560 cardiac, 621, middle, 622, inferior, 622, superior, 621, of pneumogastric, 579; cavernous of penis, 627; cervical anterior, 583, posterior, 586, superficial, 584; cervico-facial,559; chorda tympani, 558, 668; ciliary, long, 565, short, 565; circumflex, 590; claviculares, 585; coccygeal, 608, anterior branch, 609, posterior, 608; cochlear, 674; communicans noni, 585, peronei, 613; of Cotuunius, 569; cranial, 551; crural anterior, 605; cutaneous, abdominal anterior, 600, lateral, 600, of arm, external,

595, small internal, 597, cervical, 587, of circumflex, 590, coccygeal, 608, dorsal, 598, of lesser sciatic, 612, lumbar, 600, of musculo-spiral, 596, of obturator, 605, palmar, 594, patellae, 607, peroneal, 615, plantar, 613, radial, 597, sacral, 608, of thigh, external, 603, middle, 606, internal, 607, thoracic anterior, 599, lateral, 599; ulnar, 595

deep temporal, 570; dental anterior, 567, inferior, 572, posterior, 567; descendens noni, 561; digastric from facial, 559; digital (foot), dorsal, 615, 616, plantar, 613, 614, (hand), dorsal, 595, 598, palmar, median, 592, ulnar, 595, radial, 597; dorsal of penis, 610; dorsi lumbar, 601, of dura mater, 520

eighth pair, 573; of eyeball, 654

facial, 557; of femoral artery, 605; fifth, 562; fourth, 555; frontal, 564

ganglionic branch of nasal, 565; gastric branches of vagus, 580; genito-crural, 603; glosso-pharyngeal, 574; gluteal, inferior, 612, superior, 610; great petrosal, 569, great splanchnic, 623; gustatory, 571

haemorrhoidal, inferior, 610; plexus superior, 625, inferior, 626, of heart, 621, 736, hepatic, 625, 714; hypoglossal, 561

ilio-hypogastric, 602; ilio-inguinal, 603; incisor, 572; inferior cardiac, 579; inferior maxillary, 570; infra-maxillary, 560, of facial, 559; infra-trochlear, 565; inter-costal, lower, 600, upper, 599; intercosto-humeral, 599; interosseous, anterior, 594, posterior, 598, of intestines, 625; ischiadic, great, 612, small, 612

Jacobson's, 668

of labyrinth, 672, labial, 568; lachrymal, 563; of Lancisi, 536; large cavernous, 627; laryngeal, external, 578, internal, 578, recurrent, 579, superior, 578; lesser splanchnic, 623; lingual, 561, of fifth, 571, of glossopharyngeal, 574; long ciliary, 565; lumbar, 600; lumbo-sacral, 601

malar branch of orbital nerve, 567, of facial, 559; masseteric, 570; maxillary, inferior, 570, superior, 566; median, 592; mental, 572; middle cardiac, 622; motor of eye, common, 554, external, 556; musculo-cutaneous of arm, 591; of leg, 615; musculo-spiral, 596; mylo-hyoid, 572

nasal, of ophthalmic, 564, from Meckel's ganglion, 569, from Vidian, 569; naso-palatine, 569; ninth, 561

obturator, 605; oesophageal, 580; occipital, great, 587, small, 584, of third cervical, 587, of facial, 558; olfactory, 551; ophthalmic, 563; optic, 552; orbital nerves, their relation, 556, in cavernous sinus, 556, in orbit, 556, in sphenoidal fissure, 556; orbital, 567

palatine, anterior or large, 568, external 569, posterior or small, 569; palmar-cutaneous, 594, of median, 594, ulnar, 596; palpebral, 567; pathetic, 555; perforans Gasserii, 591; perineal, 610, superficial, 610; peroneal, 615; petrosal, superficial external or large 558, 569, small, 573; pharyngeal of pneumogastric, 578, of glosso-pharyngeal, 575, of sympathetic, 620, of Meckel's ganglion, 570, from external laryngeal, 578: phrenic, 585; plantar, cutaneous, 613, external, 614, internal, 613; pneumogastric, 576; popliteal, external, 615, internal, 613; portio dura, 553, portio inter duram et mollem, 557; portio mollis, 553; posterior auricular, 559; pterygoid, 570; pterygo-palatine, 569; pudendal, inferior, 610; pudic, 610; pulmonary from vagus, 580 radial, 597; recurrent laryngeal, 579; renal splanchnic, 623; respiratory external, 589

sacral, 608; saphenous long or internal, 607, short or external, 613; sciatic, great, 612, small, 612; short ciliary, 565; sixth, 555; small cavernous, 627; spermatic, 625; spheno-palatine, 567; spinal, 581, accessory, 576; splanchnic, great, 623, small, 623, smallest, 624; splenic, 625; sternales, 584; stylo-hyoid of facial, 559; subclavian, 589; suboccipital, 583, posterior branch of, 587; subscapular, 590; superficialis colli, 584; superior cardiac, 621, maxillary, 566; supra-clavicular, 585; supra-orbital, 564; supra-scapular, 589; supra-trochlear, 564; sympathetic, 513, 617

temporal deep, 570, of facial, 571, of auriculo-temporal, 571, of orbital nerve, 567; temporo-facial, 559; temporo-malar or orbital, 567; third, or motor oculi, 554; thoracic posterior, 589; anterior, 590; thyro-hyoid, 561; tibial, anterior, 615, posterior, 613; of tongue, 638; tonsillar, 576, trifacial, 562; tympanic, of glosso-pharyngeal, 574, 668; of facial, 558

ulnar, 595; uterine, 627

vaginal, 627; vagus, 576; vestibular, 673; Vidian, 569

of Wrisberg, 590

Nervous substance, chemical analysis, 509

NERVOUS SYSTEM, *General Anatomy of,* 509; cerebro-spinal axis, 509; ganglia, 510; grey or cortical substance, 509; nerves, 510; subdivision of, 509; sympathetic, 512, composition of, 513, gelatinous fibres of, 513, structure of, 513, tubular fibres of, 513; white or medullary substance, 509

Nervi nervorum, 512

Nervus cardiacus magnus, 622, minor, 622; superficialis cordis, 621

Neurilemma, 511, of cord, 515

Ninth nerve, 561

Nipple, 798

Nodule of cerebellum, 547

Noduli Arantii, 732

Nose, 638; arteries of, 640; bones of, 44; cartilages of, 639; cartilage of septum of, 640; fossae of, 73; mucous membrane of, 640; muscles of, 220, 640; nerves of, 640; veins of, 468, 640

Notch, cotyloid, 91; ethmoidal, 29; inter-condyloid, 128; nasal, 28; pterygoid, 39; sacro-sciatic, greater, 87, 94, lesser, 87, 94; sigmoid, 59; spheno-palatine, 54; supra-orbital, 28; supra-scapular, 100

Nuck, canal of, 787, 797

Nummular layer of retina, 650

Nymphae 789; lymphatics of, 504

Oblique inguinal hernia, 807, course of, 807, coverings of, 807: ligament, 185; line of the clavicle, 96, lower jaw, 57, of radius, 113

Obliquus auris muscle, 660; externus abdominis, 262, internus, 264; inferior cervicis, 260, superior, 260; inferior oculi, 219, superior, 219

Obturator artery, 438, peculiarities of, 439, relation of to femoral ring, 816; externus muscle, 328, internus, 326; fascia, 831; foramen, 91; ligament or membrane, 176; nerve, 605, accessory, 605; veins, 486

Occipital artery, 373; bone, 21, articulations of, 25, attachment of muscles to, 25, development, 24, structure of, 24; crests, 21, protuberances, 21; fossae inferior, 65; groove, 32; lymphatic glands, 496; sinus, 496; triangle, 382; vein, 469

Occipitalis, major nerve, 587, minor, 584

Occipito-atloid articulation, 163;

INDEX

occipito-axoid articulation, 164; occipito-frontalis, muscle, 214

Occiput, arteries of, 373

Odontoid ligament, 164, tubercle for, 21; process of axis, 8

Oesophageal arteries, 422; branches of vagus nerve, 580; glands, 692

Oesophagus, 690, lymphatics of, 508, structure of, 691, surgical anatomy of, 691; relations of, in neck, 690, in thorax, 690

Oesterlen, on supra-renal capsules, 762

Olecranon process, 109; fracture of, 309

Olfactory bulb, 552; foramina, 41; nerve, 552, peculiarities of, 552

Olivary bodies of medulla oblongata, 524, 525; process, 36, 64

Omenta, 696

Omentum, gastro-colic, 696; gastro-hepatic, 696; gastro-splenic, 697, great, 695, 696, lesser, 695, 696; sac of, 694

Omo-hyoid muscle, 234

Opening, of aorta in left ventricle, 734; aortic in diaphragm, 273; caval in diaphragm, 273; of coronary sinus, 730; of inferior cava, 729; left auriculo-ventricular, 733, 734; oesophageal in diaphragm, 273; of pulmonary artery, 731, veins, 733; right auriculo-ventricular, 730; saphenous, 315, 810; of superior cava, 729

OPERATION for club foot, 339; of laryngotomy, 752; of laryngo-tracheotomy, 752; of lithotomy, 821; of oesophagotomy, 691; of staphyloraphy, 244; for strabismus, 220; tracheotomy, 752

ligature of the anterior tibial, 456, over instep, 456, in lower third of leg, 456, in upper part of leg, 456; axillary artery, 405

brachial artery, 410, 411

common carotid artery, 365, above omo-hyoid, 366, below omo-hyoid, 366; common iliac artery, 435

dorsalis pedis artery, 458

external carotid, 367, iliac artery, 441

femoral artery, 447

innominate artery, 361; internal iliac artery, 437

lingual artery, 369

popliteal artery, 453, in upper part of its course, 453, in lower part, 453; posterior tibial, 459, in middle of leg, 460, at lower third of leg, 460, at ankle, 460

radial artery, 414

subclavian artery, 393; superior thyroid artery, 368

ulnar artery, 418

Opercula of dental grooves, 682

Ophthalmic artery, 385; ganglion, 565; nerve, 563; vein, 475

Opponens minimi digiti muscle, 304; pollicis muscle, 301

Optic commissure, 532; foramen, 36, 39; groove, 36, 64; lobes, 544; nerve, 552, 553; thalami, or lobes, 541; tract, 532

Ora serrata, 649

Orbicular bone, 666; ligament, 184

Orbicularis oris muscle, 224; palpebrarum, 216

Orbit, 72; arteries of, 385; muscles of, 218; relation of nerves in, 555

Orbital artery, 379; nerve, 567; process of malar, 49, of palate, 53

ORGANS of circulation, 726; of deglutition, 689; of digestion, 675; of generation, female, 788, male, 795; of respiration, 741; of sense, 629; urinary, 763; of voice, 741

Organic constituent of bone, 1

Orifice, oesophageal of stomach, 698; of prostatic ducts, 773; pyloric of stomach, 698; of uterus, 771; of vagina, 789

Os calcis, 138; hyoides, 76; innominatum, 85; orbiculare, 666; planum, 42

Os uteri, 792

Ossa triquetra, 43
Ossicula, 666; ligaments of, 666
Ossification of bone, 4; of spine, progress in, 13
Osteo-dentine, 682
Osteology, 1
Ostium abdominale of Fallopian tube, 795, uterinum, 795; uteri internum, 795
Otic ganglion, 572
Otoliths, 673
Outlet of pelvis, 93
Ovarian arteries, 432: plexus of nerves, 625; veins, 488
Ovary, 796; corpus luteum of, 797; Graafian vesicles of, 796; ligament of, 797; lymphatics of, 504; nerves of, 797; ovisacs of, 796; shape, position, and dimensions, 796; situation in foetus, 797; stroma of, 796; tunica albuginea of, 796; vessels of, 797
Ovicapsule of Graafian vesicle, 797
Oviducts, 795
Ovisacs of ovary, 796
Ovula of Naboth, 794
Ovum, 797

Pacchionian depressions, 62
Palatal glands, 685
Palate, arches of, 686, hard, 685, soft, 685; bone, 52, articulation of, 54, attachment of muscles to, 54, development of, 54, turbinated crests of, 52, vertical plate of, 52; horizontal plate of, 52; muscles of, 242; orbital process of, 53; process, 48; sphenoidal process of, 53
Palatine artery, ascending, 370, descending, 379, posterior, 379; canal, anterior, 48, 65, accessory, 52; fossa, anterior, 48, 65; nerves, 568; process of superior maxillary, 45; veins, inferior, 468
Palato-glossus muscle, 239, 243; pharyngeus 244

Palmar arch, deep, 413, superficial, 416; cutaneous nerve, 594; fascia, 281; interossei arteries, 416; nerve, deep of ulnar, 596, superficial, 596
Palmaris brevis muscle, 303; longus muscle, 290
Palpebrae, 654
Palpebral arteries, 387; cartilages, 654; fissure, 654; folds of conjunctiva, 656; ligament, 654; muscles, 216; veins, inferior, 468, superior, 468
Pampiniform plexus of veins, 488
Pancreas, 719; structure of, 720; vessels and nerves of, 720
Pancreatic arteries, 428; duct, 720; plexus of nerves, 625; veins, 490
Pancreatica magna artery, 428
Pancreatico-duodenal artery, 428, inferior, 428; vein, 490; plexus of nerves, 625
Papilla lachrymalis, 657
Papillae of tooth, 682; conicae et filiformes, 637; conjunctival, 656; fungiformes (mediae), 636; of kidney, 764; maximae (circumvallate), 636; of skin, 634; of tongue, 636
Papillary stage of development of teeth, 682
Par vagum, 576
Parietal bones, 25, articulations of, 27, attachment of muscles to, 27, development of, 26; eminence, 25; foramen, 25
Parotid duct, 688; gland, 687, accessory portion of, 688, nerves of, 689, vessels of, 689; lymphatic glands, 496; veins, 469
Patella, 130, articulations of, 131, attachment of muscles to, 131, development of 131; fracture of, 349
Pectineus muscle, 321
Pectiniform septum, 789
Pectoral region, dissection of, 275
Pectoralis major, 276, minor, 278
Peculiar dorsal vertebrae, 10
Peculiarities of clavicle, 96; of sacrum, 16;

of arteries. *See each Artery.*
Pedicles of a vertebra, 6
Peduncles of cerebellum, 549; of cerebrum, 534; of corpus callosum, 532; of pineal gland, 543
Pelvic fascia, 831, parietal or obturator layer, 831; visceral layer, 832; plexus, 627
Pelvis, 85, 92; arteries of, 434; articulation of, 171; axes of, 94; boundaries of, 768; brim of, 93; cavity of, 93; diameters of, 94; false, 93; inlet of, 93; ligaments of, 171; lymphatics of, 504; male and female, differences of, 94; outlet of, 93; position of, 92; position of viscera at outlet of, 827; true, 93; of kidney, 765
Penis, 776; arteries of, 779; body of, 777; corpora cavernosa, 777; corpus spongiosum, 778; dorsal artery of, 441, nerve of, 610; ligament suspensory, 776; lymphatics of, 504, 779; muscles of, 823; nerves of, 779; prepuce of, 776; root of, 776
Penniform muscles, 210
Perforans Gasserii nerve, 591
Perforated space, anterior, 532, posterior, 533; perforating arteries, of hand, 416; from mammary artery, 400; from plantar, 463; from profunda, 450, inferior, 450, middle, 450, superior, 450
Pericardiac arteries, 400, 421
Pericardium, relations of, 726; structure, 727; fibrous layer of, 727; serous layer of, 727; vessels of, 727
Perilymph, 672
Perinaeal artery, superficial, 440, transverse, 440; fascia, deep, 825, superficial, 822; nerve, 610
Perinaeum, abnormal course of arteries in, 830; deep boundaries of, 821; lymphatics of, 504; muscles of, 823; surgical anatomy of, 821, 823
Periosteum, 3

Peritoneum, lesser cavity of, 695; ligaments of, 696; mesenteries of, 697; omenta of, 696; reflections traced, 694
Permanent teeth, 677
Peroneal artery, 460, anterior, 461, peculiarities of, 461; nerve, 615; veins, 485
Peroneus brevis muscle, 339; longus, 338; tertius, 332
Perpendicular plate of ethmoid, 41
Pes accessorius, 538; hippocampi, 538
Petit, canal of, 653
Petro-occipital suture, 61
Petro-sphenoidal suture, 61
Petrosal nerve, superficial large, from Vidian, 569, external, 557, small, 573; sinus, inferior, 476, superior, 476
Petrous ganglion, 574; portion of temporal bone, 30
Peyer's glands, 705
Phalanges (hand), 122, articulations of, 192, development of, 123; (foot), 147, articulations of, 147, development of, 147
Pharyngeal aponeurosis, 690; artery ascending, 374; ganglion, 620; glands, 690; nerve, from external laryngeal, 578, from glosso-pharyngeal, 574, from Meckel's ganglion, 570, from sympathetic, 620, from vagus, 577; plexus of nerves, 577, 620; spine, 23; veins, 471
Pharynx, 689; aponeurosis of, 690; arteries of, 374; mucous membrane of, 690; muscles of, 239; structure of, 689
Phlebolites, 487
Phrenic arteries, 432; nerve, 585; plexus of nerves, 624; veins, 489
Pia-mater of brain, 522; vessels and nerves of, 522; of cord, 515, structure of, 515; testis, 783
Pigmentary layer of choroid, 647
Pillars of external abdominal ring, 802; of diaphragm, 273; of fauces, 686; of fornix, 540

INDEX

Pineal gland, 543; peduncles of, 543
Pinna of ear, 658; ligaments of, 659; muscles of, 659; nerves of, 660; structure of, 658; vessels of, 660
Pisiform bone, 117
Pituitary body, 532; fossa, 36, 64; membrane, 640
Plantar artery, external, 462, internal, 462; fascia, 340; cutaneous nerve, 613; nerve, external, 614, internal, 613; veins, external, 485, internal, 485
Plantaris muscle, 335
Platysma myoides, 229
Pleura, cavity of, 754; costalis, 754; pulmonalis, 754; reflections of traced, 754; vessels and nerves of, 755
Pleurae, 754
PLEXUS OF NERVES, aortic, 626; brachial, 587; cardiac, deep, 622, superficial, 622; carotid, 619, external, 620, cavernous, 620; cerebral, 620; cervical, 583, posterior, 623; coeliac, 625; colic, left, 625, middle, 625, right, 625; coronary anterior, 623, posterior, 623; cystic, 625; diaphragmatic, 624; epigastric or solar, 624; facial, 621; gastric, 625; gastro-duodenal, 625; gastro-epiploic, 625, left, 625; great cardiac, 622; haemorrhoidal, superior, 626, inferior, 627; hepatic, 625; hypogastric, 627, inferior, 627; ileo-colic, 625; infra-orbital, 568; lumbar, 601; magnus profundus, 622; meningeal, 621; mesenteric, inferior, 626, superior, 625; oesophageal, 580; ovarian, 625; ophthalmic, 620; pancreatic, 625; pancreatico-duodenal, 625; patellar, 607; pharyngeal, 578, 620; phrenic, 624; prostatic, 627; pulmonaiy anterior, 579, posterior, 580; pyloric, 625; renal, 625; sacral, 609; sigmoid, 626; solar, 624; spermatic, 625; splenic, 625; superficial cardiac, 622; supra-renal, 624; tonsillar, 576; tympanic, 574,
668; vaginal, 627; vertebral, 626; vesical, 627
PLEXUS OF VEINS, choroid, 538; haemorrhoidal, 486; pampiniform, 488; prostatic, 486; pterygoid, 469; ovarian, 488; spermatic, 488; uterine, 487; vaginal, 487; vesico-prostatic, 486
Plica semilunaris, 656
Pneumogastric nerve, 576
Points of ossification, 4
Pomum Adami, 741
Pons hepatis, 712; Tarini, 533
Pons Varolii, 526; longitudinal fibres of, 527; septum of, 527; structure of, 526; transverse fibres of, 527
Popliteal artery, 452, branches of, 453, peculiarities of, 453; surgical anatomy of, 453; lymphatic glands, 501; nerve, external, 615, internal, 613; space, 451; vein, 486
Popliteus muscle, 336
Pores of the skin, 634
Portal canals, 716; fissure, 713: vein, 465, 491, 714, 716
Portio dura of seventh nerve, 553; mollis, 554; inter duram et mollem, 557
Porus opticus of sclerotic, 644
Position of pelvis, 94
Posterior annular ligament, 300; aperture of nares, 67; dental canals, 45; ethmoidal cells, 40; fontanelle, 24; fossa of skull, 64; nasal spine, 52; palatine canal, 46; triangle of neck, 380
Pott's fracture, 349
Pouches laryngeal, 746
Poupart's ligament, 262, 803, 813
Prepuce, 777
Preputium clitoridis, 789
Princeps cervicis artery, 373; pollicis artery, 416
PROCESSES OR PROCESS, acromian, 99; alveolar, 48; angular, external, 28, internal, 28, auditory, 33 basilar, 24

INDEX

clinoid, anterior, 39, 62, middle, 36, posterior, 36, 64; cochleariform, 664; condyloid of lower jaw, 59; coracoid, 101; coronoid of lower jaw, 59; of ulna, 109

ethmoidal of inferior turbinated, 55

frontal of malar, 50

hamular, 39; of helix, 671

of Ingrassias, 38

jugular, 22

lachrymal of inferior turbinated bone, 55

malar, 47; mastoid, 32; maxillary, 51, of inferior turbinated, 55, of malar bone, 51; mental, 57

nasal, 47

odontoid of axis, 9; olecranon, 106; olivary, 36, 64; orbital of frontal, 28; of malar, 51, of palate, 53, of superior maxillary, 46

palate, 48; palatine of superior maxillary, 46; pterygoid of palate bone, 52, of sphenoid, 39

sphenoidal of palate, 54; spinous of tibia, 131, of ilium, 87, of sphenoid, 38; styloid of temporal, 34; of ulna, 112; of radius, 113; superior vermiform of cerebellum, 546

unciform, 119, of ethmoid, 43

vaginal, 34

zygomatic, 30

Processes ciliary, 647

Processus ad medullam, 550; ad pontem, 550; brevis of malleus, 665; cochleariformis, 35, 664; e cerebello ad testes, 544; gracilis of malleus, 665

Profunda cervicis artery, 401; femoris artery, 449, vein, 486; inferior artery, 411, superior, 411

Progress of ossification in the spine, 13

Promontory of tympanum, 663; of sacrum, 14

Pronator quadratus muscle, 293; radii teres muscle, 289

Prostate gland, lobes of, 775; levator muscle of, 827; position of, 775; size and shape, 775; structure of, 775; surgical anatomy of, 829; vessels and nerves of, 776

Prostatic plexus of nerves, 627; of veins; 486; portion of urethra, 772; fluid, 776, sinus, 773

Protuberance, occipital, external, 21, internal, 23

Psoas magnus muscle, 313, parvus, 314

Pterygo-maxillary fissure, 70, ligament, 224

Pterygo-palatine artery, 379, canal, 37, nerve, 570

Pterygoid arteries, 378; muscles, external, 228, internal, 227; fossa of sphenoid, 39, of lower jaw, 59; nerves, 570; notch, 39; plexus of veins, 468; process of palate bone, 53; processes of sphenoid, 39; ridge, 38

Pubes, 90; angle of, 90; crest of, 90; spine of, 90; symphysis of, 90

Pubic arch, 94; articulations of, 176

Pubo-prostatic ligaments, 770

Pudendum, 788

Pudic artery, in male, 439, peculiarities of, 439, in female, 441, deep external, 449, superficial, 800, superficial external, 449; nerve, 610; vein, external, 485, internal, 486

Pulmonary artery, 463, 759, opening of in right ventricle, 731; capillaries, 759; nerves from vagus, 580; veins, 465, 492, 759, openings of in left auricle, 733

Puncta vasculosa, 534; lachrymalia, 657, 658

Pulp cavity of tooth, 679; of teeth, development of, 682

Pupil of eye, 647; membrane of, 648

Pyloric artery, 426, inferior, 426; plexus, 625

Pylorus, 699

Pyramid in vestibule, 669; of cerebellum, 546; of thyroid gland, 760

INDEX

Pyramidalis muscle, 268; nasi, 221
Pyramids, anterior, 664, decussation of, 524, posterior, 524; of Ferrein, 765; of Malpighi, 764; of the spine, 19
Pyriformis muscle, 326

Quadriceps extensor cruris muscle, 318
Quadratus femoris muscle, 327; lumborum, 268
Quadrigeminal bodies, 544

Radial artery, 412; branches of, 413; peculiarities of, 414; relations of in fore-arm, 413, in the hand, 413, at the wrist, 413; surgical anatomy of, 414; lymphatic glands, 499; nerve, 597; recurrent artery, 415; region, muscles of, 294; vein, 477
Radialis indicis artery, 416
Radiating fibres of retina, 648
Radio-ulnar articulations, inferior, 186, middle, 185, superior, 184
Radius, 112, articulations of, 115, development of, 114, fracture of, 309, of lower end, 310, of neck, 309, of shaft, 310; grooves in lower end of, 114; muscles attached to, 115; oblique line of, 113; sigmoid cavity of, 113; tuberosity of, 113; and ulna, fracture of, 310
Rami of the lower jaw, 58; of pubes, 92
Ramus of ischium, 92
Ranine artery, 369; vein, 468
Raphe of corpus callosum, 536; of palate, 685; of perinaeum, 821; of tongue, 635
Receptaculum chyli, 495
Recto-uterine ligaments, 793
Recto-vesical fascia, 832; fold, peritoneal, 695
Rectum, relations of, male, 708, female, 791; columns of, 710; lymphatics of, 504; sugical anatomy of, 827
Rectus abdominis, 267; capitis anticus major, 245, minor, 245; posticus major, 260, minor, 245; lateralis, 246; femoris muscle, 318; oculi, internus, superior, inferior, and external, 219
Recurrent artery, radial, 416, tibial, 457; ulnar, anterior, 418, posterior, 457; nerves to tentorium, 555
REGION, of abdomen, 261, 692; acromial, muscles of, 281, auricular, 215
back, muscles of, 248, brachial, anterior, 289, posterior, 296
cervical, superficial muscles of, 229
diaphragmatic, 271
epicranial, muscles of, 241; epigastric, 693
femoral, muscles of, anterior, 314, internal, 320, posterior, 328, fibular, 331; foot, dorsum of, 341, sole of, 340
gluteal, muscles of, 323; groin, 809
of hand, muscles of, 301 humeral, anterior, 285, posterior, 287; hypochondriac, 693; hypogastric, 693
iliac, muscles of, 312, infra-hyoid, 233; inguinal, 693, 800; inter-maxillary, muscles of, 223; ischio-rectal, 819
laryngo-tracheal, surgical anatomy of, 752; lingual, muscles of, 237; lumbar, 693
maxillary, muscles of, inferior, 223, superior, 223
nasal, muscles of, 220
orbital, muscles of, 218
palatal, muscles of, 242, palpebral, 216; perinaeum, 819, 821; pharyngeal, muscles of, 239; popliteal, 451; pterygo-maxillary, muscles of, 227
radial, muscles of, 294
scapular, muscles of, anterior, 282, posterior, 283; Scarpa's triangle, 446; supra-hyoid, muscles of, 235
temporo-maxillary, muscles of, 225, thoracic, 269, anterior, 276, lateral, 280, tibio-fibular, anterior, 331, posterior, 333

umbilical, 693
vertebral, muscles of, anterior, 245, lateral, 246
Renal artery, 431, 766; plexus, 625; veins, 489
Reservoir of thymus, 762
RESPIRATION, organs of, 741
Respiratory nerve of Bell, external, 589, internal, 585
Restiform bodies of medulla oblongata, 524, 526
Rete mucosum of skin, 631; testis, 784
Retina 649, arteria centralis of, 387, 651; fovea centralis of, 650; limbus luteus of, 650; membrana limitans of, 651; nervous layer of, 650; nummular layer of, 650; radiating fibres of, 651; structure of, 650
Retinacula of ileo-caecal valve, 707
Retrahens aurem muscle, 216
Rhomboid impression, 96; ligament, 178
Rhomboideus, 221, major, 252, minor, 252
Ribs, 80; angle of, 82; attachment of muscles to, 84; development of 84; false, 80; floating, 80; head of, 81; ligaments of, 167; neck of, 81; peculiar 83; tuberosity of, 82; vertebral, 80; vertebro-costal, 80; vertebro-sternal, 80
Rickets, 2
Ridge, internal occipital, 23; mylo-hyoidean, 57; pterygoid, 37; superciliary, 27; temporal, 25, 69
Rima glottidis, 745
Ring, abdominal, external, 262, 802, internal, 806; femoral or crural, 815; fibrous of heart, 735
Risorius muscle, 225
Root of lung, 757
Roots of spinal nerves, 581; of teeth, 678; of the zygomatic process, 30
Rostrum of sphenoid bone, 37; of corpus callosum, 535
Rotation, 156
Rotatores spinae muscles, 259

Round ligaments of uterus, 797; relation of, to femoral ring, 815; of liver, 712
Rugae of stomach, 701; of vagina, 792
Rupture of urethra, course taken by urine in, 802

Sac, lachrymal, 658; of omentum, 695
Saccule of vestibule, 672
Sacculus laryngis, 746
Sacra-media artery, 443
Sacral arteries, lateral, 442; canal, 16; cornua, 15; foramina, 14; ganglia, 608; lymphatic glands, 502; nerves, 608, anterior branches of, 608, posterior branches of, 608, roots of, 608; plexus, 609; vein lateral, 487, middle, 487
Sacro-coccygeal ligaments, 175
Sacro-iliac articulation, 174
Sacro-lumbalis muscle, 256
Sacro-sciatic foramen, greater, 174, lesser, 174; ligaments, 174; notch, greater, 87, 94, lesser, 89, 94
Sacro-vertebral angle, 14
Sacrum, 14, articulations of, 18, attachment of muscles to, 18, development of, 17, peculiarities of, 16, structure of, 14
Sacs, dental, 684
Sagittal suture, 60
Salivary glands, 687, structure of, 689
Santorini, cartilages of, 743
Saphenous nerve, long or internal, 607, short, 613; opening, 317, 811; vein, external or short, 485, internal or long, 484, 810
Sartorius muscle, 317
Scala tympani of cochlea, 671; vestibuli of cochlea, 671
Scalae of cochlea, 671
Scalenus anticus, 246; medius, 247; posticus, 247
Scaphoid bone, hand, 115, foot, 143; fossa of sphenoid, 39
Scapula, 97; articulations of, 102; attachment of muscles to, 102;

development of, 102; dorsum of, 98; glenoid cavity of, 100; head of, 100; ligaments of, 180; muscles of, 282; spine of, 99; venter of, 97

Scapular artery, posterior, 399, supra, 298; region, muscles of, anterior, 282, posterior, 283; veins, 470

Scapulo-clavicular articulation, 178

Scarfskin, 631

Scarpa's triangle, 446

Schindylesis, 153

Schneiderian membrane, 640

Schreger's analysis of bone, 2

Sclerotic, 643, structure of, 644, vessels and nerves of, 645

Sciatic artery, 441; nerve, greater, 612, lesser, 612; veins, 486

Scrotal hernia, 808

Scrotum, 779; dartos of, 780; lymphatics of, 504; nerves of, 781; septum of, 780; vessels of, 780

Sebaceous glands, 634

Sella turcica, 36, 64

Semen, 786; liquor seminis of, 786; seminal granules of, 786; spermatoza of, 786

Semicircular canals, 669, external, 670, posterior, 670, superior, 669; membranous, 672

Semilunar bone, 116, cartilages of knee, 198; ganglion of fifth nerve, 563, of abdomen, 624; valves aortic, 734, pulmonic, 732

Semimembranosus muscle, 329

Seminal granules, 786; ducts, 786; vesicles, 785

Seminiferous tubes, 783

Semispinalis dorsi, 258, colli, 258

Semitendinosus muscle, 329

Senac, on structure of heart's valves, 731

Senses, organs of the, 629

Separation of epiphyses, 4

Septum auricularum, 730, 734; crurale, 816; lucidum, 540; of medulla oblongata, 526; of nose, 74, cartilage of, 639; pectiniforme, 778; of pons Varolii, 527; scroti, 780; subarachnoid, 514; of tongue, 637; ventriculorum, 731

Serratus magnus, 280; posticus, inferior, 253, superior, 253

Sesamoid bones, 148

Shaft of a bone, its structure, 2

Sheath of arteries, 353; of muscles, 209; of nerves, 510; femoral or crural, 813; of rectus muscle, 268

Short bones, 3

Shoulder joint, 180; muscles of, 281; vessels and nerves of, 182

Sigmoid artery, 431; cavity, greater and lesser of ulna, 109, of radius, 114; flexure of colon, 708; meso-colon, 697; notch of lower jaw, 59

Simon, on supra-renal capsules, 762

Sinus, aortici, 735; cavernous, 475; circular, 476; circularis iridis, 649; coronary, 491, opening of in heart, 730; of jugular vein, 470; of kidney, 763, 766; lateral, 475; longitudinal inferior, 474, superior, 473; maxillary, 47; occipital, 475; petrosal inferior, 476, superior, 476; pocularis, 773; prostatic, 773; of right auricle, 729; of left, 732, straight, 474; transverse, 476

Sinuses, 466; confluence of the, 474; ethmoidal, 41; frontal, 29; maxillary, 47; sphenoidal, 37; of Valsalva, aortic, 735, pulmonary, 732

Sixth nerve, 555

Skeleton, 1, its number of pieces, 5

SKIN, *General Anatomy of,* 629; appendages of, 632; areolae of the, 630; colour of, its origin, 631; corium of, 630; cuticle of, 631; derma, or true skin, 629; epidermis of, 631; furrows of, 630, 631; hairs, 633; muscular fibres of, 630; nails, 632; nerves of, 632; papillary layer of, 630; rete mucosum of, 631; sebaceous glands of, 634; sudoriferous, or sweat-glands of,

INDEX

634; vessels of, 632

SKULL, 21, 61; anterior region, 70; base of external surface, 65, internal surface, 62; fossa of, anterior, 62 middle, 62, posterior, 64; lateral region of, 68; tables of, 3; vertex of, 61; vitreous table of, 3

Small intestine, cellular coat of, 703; mucous coat of, 703; muscular coat of, 703; serous coat of, 703; simple follicles, 704; structure of, 703; valvulae couniveutes, 704; villi of, 704

Small intestines, 702; duodenum, 702; ileum, 703; jejunum, 703

Socia parotidis, 688

Soft palate, 685; aponeurosis of, 686; arches or pillars of, 686; muscles of, 686; structure of, 685

Solar plexus, 624

Sole of foot, muscles of, first layer, 342, second layer, 344, third layer, 345

Soleus muscle, 334

Solitary glands, 705

Space, anterior perforated, 532; axillary, 401; intercostal, 81; popliteal, 451; posterior perforated, 533

Spermatic artery, 432, 781; canal, 805; cord arteries of, 781, course of, 781, lymphatics of, 781, nerves of, 781, relation to femoral ring, 815, relations of in inguinal canal, 781, veins of, 781; fascia, external, 802; plexus of nerves, 625, of veins, 488; veins, 488, 781

Spermatozoa, 786

Spheno-maxillary fissure, 69, fossa, 70

Spheno-palatine artery, 379; foramen, 54; ganglion, 568; nerves, 567; notch, 54

Spheno-parietal suture, 60

Sphenoid bone, 36, articulations of, 40, attachment of muscles to, 40, development of, 39, greater wings of, 37, lesser wings of, 38, pterygoid processes of, 39, spinous process of, 38, vaginal processes of, 37

Sphenoidal fissure, 38, 64; process of palate, 54; sinuses, 37; spongy bones, 40

Sphincter muscle of bladder, 771; of rectum, external, 820, internal, 821; of vagina, 825

SPINAL ARTERIES, anterior, 396, lateral, 396, posterior, 397, median, 397

SPINAL CORD, 516, arachnoid of, 514, arrangement of grey and white matter in, 518, central ligament of, 515, columns of, 517, dura mater of, 513, fissures of, 516, foetal peculiarity of, 518, grey commissure of, 518, grey matter of, 517, internal structure of, 517, ligamentum denticulatum of, 515, membranes of, 513, neurilemma of, 515, pia mater of, 515, sections of, 517, substantia cinerea gelatinosa, 517, white commissure of, 516, white matter of, 518

Spinal nerves, 581, arrangement into groups, 581, branches of, anterior, 582, posterior, 582, ganglia of, 582, origin of roots, anterior, 581, posterior, 581

Spinal veins, 482, longitudinal anterior, 483, posterior, 483

Spinal accessory nerve, 576

Spinalis cervicis muscle, 257; dorsi, 257

Spinal column, 5

Spines of bones, ethmoidal, 36; of ischium, 89; nasal, 29, anterior, 48, posterior, 52; pharyngeal, 23; of pubes, 90; of scapula, 98

Spinous process of sphenoid, 38, of tibia, 131, of vertebrae, 6

Spiral canal of cochlea, 671

Splanchnic nerve, greater, 623, lesser, 623, smallest or renal, 624

SPLEEN, 721; capillaries of, 725; fibrous elastic coat of, 722; lymphatics of, 506, 725; Malpighian corpuscles of, 723; nerves of, 725; proper substance of, 722; relations of, 721; serous coat of, 721, size and weight, 721; artery of, 724; structure of, 721; trabeculae of,

INDEX

722; veins of, 725
Splenic artery, 427, distribution of, 724; plexus, 625; vein, 490
Spongy portion of urethra, 773; tissue of bone, 2
Squamo-sphenoidal suture, 61; parietal suture, 60
Squamous portion of temporal bone, 30
Stapedius muscle, 667
Stapes, 666; annular ligament of, 666
Stellate ligament, 167
Steno's duct, 688
Sternal end of clavicle, fracture of, 306; foramen, 77; ligaments, 169; nerves, 585
Sterno-clavicular articulation, 177
Sterno-cleido mastoid muscle, 231
Sterno-hyoid muscle, 233
Sterno-thyroid muscle, 234
Sternum, 77, articulations of, 80, attachment of muscles to, 80, development of, 79, ligaments of, 171
STOMACH, 697; alteration in position of, 699; alveoli of, 701; cellular coat of, 700; curvatures of, 698; fundus of, 698; gastric follicles of, 701; ligaments of, 699, lymphatics of, 505; mucous glands of, 701, mucous membrane of, 701; muscular coat of, 700; orifices of, 698; pyloric end of, 698; pylorus, 699; serous coat of, 700; simple follicles of, 701; splenic end of, 698; structure of, 700; surfaces of, 699; vessels and nerves of, 701
Stratiform fibro-cartilage, 151
Striae longitudinales, 536, laterales, 536
Stricture, seat of in direct inguinal hernia, 809, in femoral, 818, in oblique, 807
Stroma of ovary, 796
Stylo-glossus muscle, 239
Stylo-hyoid muscle, 236; nerve from facial, 559
Stylo-mastoid artery, 374; foramen, 34; vein, 469
Stylo-maxillary ligament, 165

Stylo-pharyngeus muscle, 241
Styloid process of temporal bone, 34, of radius, 113, of ulna, 112
Subanconeus muscle, 288
Subarachnoid space of brain, 521; of cord 514; fluid, 521; septum, 515
Subclavian arteries, 389, branches of, 395, first part of, left, 391, right, 390, peculiarities of, 392, second portion of, 391, third, 392, surgical anatomy of, 393; groove, 96; nerve, 589; triangle, 382; vein, 479
Subclavius muscle, 278
Subcruraeus muscle, 319
Sublingual artery, 369; fossa, 57; gland, 688, vessels and nerves of, 689
Sublobular veins, 717
Submaxillary artery, 371; fossa, 58; ganglion, 573; gland, 688, nerves of, 688, vessels of, 688; lymphatic glands, 496; triangle, 370; vein, 468
Submental artery, 371; vein, 468
Sub-occipital nerve, 583, posterior branch of, 587
Sub-peduncular lobe of cerebellum, 547
Subscapular aponeurosis, 282; artery, 406; fossa, 97; nerves, 590
Subscapularis muscle, 282
Substantia cinerea gelatinosa, 517
Sudoriferous glands, 634
Sulci of cerebrum, 529; transversales, 24
Supercilia, 654
Superciliary ridge, 27
Superficial palmar arch, 417
Superficialis colli nerve, 584; volae artery, 415
Superior maxillary bone, 44, articulations of, 49, attachment of muscles to, 49, development of, 49; nerve, 566
Superior meatus, 75; profunda artery, 411; turbinated crest, 48, of palate, 53; vena cava, orifice of, 729
Supinator brevis muscle, 297; longus, 294
Supra-clavicular nerves, 585

Supra-orbital arch, 28; artery, 386; foramen, 28; nerve, 564; notch, 28
Supra-renal arteries, 431; capsules, 767, nerves of, 768, relations of, 767, structure of, 767, vessels of, 768; plexus, 624; veins, 489
Supra-scapular artery, 398; nerve, 589
Supra-spinales muscles, 259
Supra-spinatus muscle, 283
Supra-spinous aponeurosis, 283
Supra-trochlear nerve, 564
Sural arteries, 454; veins, 486
SURGICAL ANATOMY, 1; of abdominal aorta, 423; of arch of aorta, 357; of axilla 401; of axillary artery, 405
of base of bladder, 829; of brachial artery, 410
of common carotid artery, 365; of common iliac artery, 435
of external carotid, 367
of facial artery, 372; of femoral artery, 447, of femoral hernia, 809
of hamstring tendons, 330
of inguinal hernia, 800; of internal carotid, 385, of iliac arteries, 437; of ischio-rectal region, 819
of laryngo-tracheal region, 752; of lingual artery, 369
of muscles of eye, 220, of lower extremity, 347, of soft palate, 244, of upper extremity, 306
of oesophagus, 691
of perinaeum, 819, 821; of popliteal artery, 453; of prostate gland, 827
of radial artery, 414
of subclavian artery, 393; of superior thyroid, 368
of talipes, 339; of temporal artery, 375; of thoracic aorta, 421; of triangles of neck, 379
of ulnar artery, 418
Suspensory ligament of incus, 666; of lens, 653; of malleus, 666
Sustentaculum tali, 138
Sutura, 153; dentata, 153; harmonia, 153; limbosa, 153; notha, 153; serrata, 153; squamosa, 153; vera, 153
SUTURE basilar, 61; coronal, 60; cranial, 60; ethmo-sphenoidal, 62; ethmoido-frontal, 62; frontal, 29, 61; fronto-malar, 73, fronto-sphenoidal, 62; intermaxillary, 66; inter nasal, 70; lambdoid, 60; malo-maxillary, 73; masto-occipital, 61; masto-parietal, 60; naso-maxillary, 70; petro-occipital, 61; petro-sphenoidal, 61; sagittal, 60; spheno-parietal, 60; squamo-parietal, 60; squamo-sphenoidal, 61; temporal, 33; transverse, 61
Swallow's nest of cerebellum, 547
Sweat-glands, 634
SYMPATHETIC NERVE, 617; cervical portion, 619; cranial portion, 563; lumbar portion, 626; pelvic portion, 626; thoracic portion, 623
Symphysis of jaw, 56; pubis, 90
Synarthrosis, 152
Synchondrosis sacro-iliac, 172
Synovia, 151
SYNOVIAL MEMBRANE, 150, articular, 151, bursal, 152, vaginal, 152; ankle, 202; astragalo-scaphoid, 206; atlo-axoid, 162; calcaneo-astragaloid, 204; carpal, 191; chondro-sternal, 169; costo-transverse 168; costo-vertebral, 167; elbow, 183; hip, 194; intercostal, 169; interpubic, 176; knee, 199; metacarpal, 191; occipito-atloid, 164; phalanges, 192; radio-ulnar, inferior, 187, superior, 184; sacro-iliac, 173; scapulo-clavicular, 180; shoulder, 181; sterno-clavicular, 178; tarsal, 206; tarso-metatarsal, 207; temporo-maxillary, 166; thumb, 191; tibio-fibular, inferior, 201, superior, 200; wrist, 188
Systemic arteries, 351; veins, 465

Tables of the skull, 3
Taenia hippocampi, 538, 539; semi-circularis, 537; violacea, 548

INDEX

Tarsal bones, 138, ligaments, 204; cartilages of eyelid, 655; ligament of eyelid, 655

Tarsea artery, 458

Tarso-metatarsal articulations, 207

Tarsus, 138; development of, 147

TEETH, 677; bicuspid, 678; body of, 677; canine, 678; cement, of, 681; crown of, 678; crusta petrosa of, 681; cortical substance of, 681; cuspidate, 678; deciduous, 677; dental tubuli of, 680; dentine of, 680; development of, 682; enamel of, 681; eruption of, 684; eye, 678; false molars, 678; fang of, 678; general characters of, 677; growth of, 682; incisors, 678; intertubular substance of, 680; ivory of, 680; milk, 677; molar, 679; multicuspidate, 679; permanent, 678; pulp cavity of, 681; root of, 678; structure of, 680; temporary, 677, 6769; true or large molars, 679; wisdom, 679

Temporal artery, 375; anterior, 375; deep, 378; middle, 375; posterior, 375; surgical anatomy of, 375

Temporal bone, 30; articulations of, 35; attachment of muscles to, 36; development of, 35; mastoid portion, 32; petrous portion, 32; squamous portion, 30; structure of, 35

Temporal fascia, 226; fossae, 69; ganglion, 620; muscle, 226; nerves, deep, 570; ridge, 25, 69; suture, 33; vein, 468, middle, 469

Temporary cartilage, 4; teeth, 677, 679

Temporo-facial nerve, 559; maxillary articulation, 165, vein, 469

Tendo Achillis, 334; palpebrarum or oculi, 216

Tendon, central, or cordiform of diaphragm, 273; conjoined of internal oblique and transversalis, 803; structure of tendon, 210

Tensor palati muscle, 243; tarsi muscle, 217; tympani muscle, 667; vaginae femoris muscle, 317

Tentorium cerebelli, 520

Teres major muscle, 284; minor, 284

Testes, 781; coni vasculosi of, 784; coverings of, 779, 782; tunica albuginea, 782, vaginalis, 782, vasculosa, 783; gubernaculum testis, 786; lobules of, 783; lymphatics of, 504; mode of descent, 786; rete testis, 784; size and weight of, 782; structure of, 783; tubuli seminiferi of, 783; vas deferens of, 784; vasa efferentia of, 784; vasa recta, 784; vasculum aberrans of, 784

Theca vertebralis, 483

Thalami optici, 538

Thigh, back of muscles of, 328; deep fascia, or fascia lata, 315; fascia of, 315; muscles of front, 314

Third nerve, 554; ventricle of the brain, 542

Thoracic acromial artery, 406; aorta, 420, surgical anatomy of, 421; artery, alar, 406; artery, long, 406; duct, 495; ganglia of sympathetic, 580; nerves, anterior, 590, posterior, or long, 589; region, muscles of anterior, 276, lateral, 280

THORAX, *General Description of*, 726; base of, 726; bones of, 77; boundaries of, 726; cutaneous nerves of, lateral, 599; fasciae of, 275; muscles of, 269; parts passing through upper opening of, 726

Thumb, muscles of, 300

Thymus gland, 761; chemical composition, 762; lobes of, 761; lymphatics of, 508; structure of, 762; vessels and nerves of, 762

Thyro-arytenoid ligament, inferior, 746; superior, 746

Thyro-arytaenoideus muscle, 747

Thyro-epiglottic ligament, 743

Thyro-epiglottideus muscle, 748

Thyro-hyoid ligaments, 744; membrane, 744; muscle, 234; nerve, 561

INDEX

Thyroid artery, inferior, 398, superior, 368, surgical anatomy of, 368; axis, 398; branches of sympathetic, 621; cartilage, 741; foramen, 91; ganglion, 621; gland, 760, chemical composition, 761, isthmus of, 760, lymphatics of, 508, situation of, 760, structure of, 761, vessels and nerves of, 761; veins, inferior, 481, middle, 471, superior, 471

Tibia, 131; articulations of, 135; attachment of muscles to, 135; crest of, 133; development of, 135; fracture of shaft of, 349; spinous process of, 131; tubercle of, 132; tuberosities of, 132

Tibial artery, anterior, 455, branches of, 457, peculiarities of, 455, surgical anatomy of, 456; posterior, 459, branches of, 460, peculiarities of, 459, surgical anatomy of, 459; lymphatic glands, 501; nerve, anterior, 615, posterior, 613; recurrent artery, 457; veins, anterior, 485, posterior, 485

Tibialis anticus muscle, 331; posticus muscle, 337

Tibio-fibular articulations, superior, 200, middle, 200, inferior, 200; region, anterior, muscles of, 331

Tongue, 635; arteries of, 638; fibrous septum of, 637; follicles of, 637; mucous glands of, 637; mucous membrane of, 635; muscular fibres of, 638; muscles of, 237; nerves of, 638; papillae of, 636

Tonsillitic artery, 371

Tonsils, 686; nerves of, 686; vessels of, 686

Torcular Herophili, 24, 474

Trabeculae of corpus cavernosum, 777; of spleen, 722; of testis, 783

Trachea, cartilages of, 751; glands of, 752; relations of, 750; structure of, 751; surgical anatomy of, 752; vessels and nerves of, 752

Trachelo-mastoid muscle, 257

Tracheotomy, 752, 753

Tractus opticus, 553

Tragicus muscle, 660

Tragus, 659

Transversalis fascia, 805; muscle, 265, 805

Transversalis colli artery, 399; muscle, 256

Transverse arteries of basilar, 397; colon, 707; facial artery, 375, vein, 468; fissure of brain, 539, of liver, 695; ligament of hip, 194, of knee, 199, of scapula, 180; meso-colon, 695, 697; process of a vertebra, 6; sinus, 476; suture, 70

Transversus auriculae, 660; perinaei muscle, 824, perinaei (in female), 825

Trapezium bone, 117

Trapezius muscle, 248

Trapezoid bone, 117; ligament, 179

Triangle, inferior carotid, 380; of Hesselbach, 808; of neck, anterior, 380, posterior, 382, surgical anatomy of, 379; occipital, 382; Scarpa's, 446; subclavian, 382; submaxillary, 381; superior carotid, 381

Triangular ligament of urethra, 803, 825

Triangularis sterni muscle, 270

Triceps extensor cruris, 319; extensor cubiti, 288

Tricuspid valves, 731

Trifacial nerve, 562

Trigone of bladder, 771

Trochanteric fossa, 126

Trochanters, greater and lesser, 126

Trochlea of humerus, 106

Trochlear nerve, 555

True pelvis, 93; ligaments of bladder, 770; ribs, 80; vertebrae, 5

Trunk, muscles of, 247

Tube, Eustachian, 664; Fallopian, 795

Tuber cinereum, 532; ischii, 89

Tubercle of the clavicle, 96; of the femur, 126; lachrymal, 48; laminated of cerebellum, 547; of Lower, 729; for odontoid ligaments, 22; of scaphoid, 115; of the tibia, 132; of ulna, 109; of zygoma, 31

INDEX

Tubercles, genial, 57; of ribs, 84
Tubercula quadrigemina, 544
Tuberculo cinereo, 526
Tuberosities of humerus, greater and lesser, 104; of tibia, 131
Tuberosity of ischium, 86; maxillary, 45; of palate bone, 53; of radius, 113
Tubes, bronchial, 751, structure of in lung, 757
Tubuli of Ferrein, 765; lactiferi, 798; recti, 783; seminiferi, 783; uriniferi, 764
Tubulus centralis modioli, 671
Tuft, vascular, in Malpighian bodies of kidney, 765
Tunica albuginea, 782; of ovary, 796; Ruyschiana, 647; vaginalis, 782, 787; vaginalis propria, 782, reflexa, 782; vasculosa testis, 783
Turbinated bone, superior, 42, inferior, 54, middle, 42
Tutamina oculi, 654
Tympanic artery, from internal carotid, 385, from internal maxillary, 377; bone, 35; nerve, 558, 574, 668
Tympanum, 662; arteries of, 667; cavity of, 662; membrane of, 665; mucous membrane of, 667; muscles of, 666; nerves of, 667; ossicula of, 665; veins of, 667

Ulna, 107, articulations of, 112, development of, 112; fracture of coronoid process of, 308, of olecranon, 309, of shaft, 310; muscles attached to, 112; sigmoid cavities of, 109; styloid process of, 112; tubercle of, 109
Ulnar artery, 416, branches of, 418, peculiarities of, 417; relations in forearm, 416, in hand, 417, at wrist, 417, surgical anatomy of, 418; lymphatic glands, 499; nerve, 595; artery, recurrent, anterior, 418, posterior, 418; vein, anterior, 477, posterior, 477
Umbilical arteries in foetus, 737, 740, how obliterated, 740; fissure of liver, 712; region, contents of, 693; vein, 737, 740, how obliterated, 740
Umbilicus, 261
Unciform bone, 119
Ungual phalanges, 123
Upper extremity, arteries of, 389; bones of, 95; ligaments of, 177; lymphatics of, 499; muscles of, 274; nerves of, 590; veins of, 476
Urachus, 769
Ureters, nerves of, 767; relations of, 766; structure of, 720, 766; vessels of, 767
URETHRA, male, 772; bulbous portion of, 773; caput gallinaginis, 772; membranous portion, 773; prostatic portion, 772, sinus of, 773; rupture of, course taken by urine, 823; sinus pocularis of, 773; spongy portion of, 773; structure of, 774; verumontanum, 772; female, relations of, 791, structure of, 794
Urinary organs, 763
Uterine arteries, 438; nerves, 627; plexus of veins, 487
UTERUS, appendages of, 795; arbor vitae of, 793; broad ligaments of, 793; cavity of, 793; in foetus, 794; fundus, body, and cervix of, 792; ligaments of, 793; lymphatics of, 504; during menstruation, 795; in old age 795, after parturition, 795; during pregnancy, 795; at puberty, 794; round ligaments of, 797; shape, position, dimensions, 792; structure of, 793; vessels and nerves of, 794
Uterus masculinus, 773
Utricle of vestibule, 672
Uvea, 648
Uvula, 686; of cerebellum, 546; vesicae, 772

Vagina, columns of, 792; lymphatics of, 504; orifice of, 788; relations of, 792; situation, direction, shape, dimensions, 791; structure of, 792

Vaginal arteries, 438; plexus of nerves, 627; plexus of veins, 487; portal plexus, 716; process of temporal, 31; processes of sphenoid, 37; synovial membranes, 150

Vagus nerve, 576, ganglion of root of, 577, of trunk of, 577

Valve of Bauhin, 707; coronary, 730; of cystic duct, 718; Eustachian, 730; of gall-bladder, 718; ilio-caecal, 707; of Kerkring, 704; mitral, 734; of Vieussens, 544

Valves in right auricle, 729; semilunar aortic, 734; pulmonic, 732; tricuspid, 731

Valvulae conniventes, 704

Vas deferens, 784, structure of, 784; aberrans, 784

Vasa aberrantia of brachial artery, 410; afferentia of tymphatic glands, 494; brevia arteries, 428, veins, 490; efferentia of testis, 784, of lymphatic glands, 495; vasorum of arteries, 353, of veins, 467, intestini tenuis arteries, 429, recta, 784

Vascular system, changes in, at birth, 740, peculiarities in foetus, 737

Vasculum aberrans, 784

Vastus externus muscle, 319; internus and cruraeus, 319

VEINS, *General Anatomy of,* 465; anastomoses of, 465; coats of, 466; muscular tissue of, 466; plexuses of, 465; size, form, etc., 465; structure of, 466; valves of, 467; vessels and nerves of, 467

VEINS OR VEIN. *Descriptive Anatomy of,* 468
of alae nasi, 468; angular, 468; articular of knee, 485; auricular anterior, 469, posterior, 469; axillary, 478; azygos, left lower, 482, left upper, 482, right, 481
basilic, 477: basi-vertebral, 483; of bone, 4; brachial, 478; brachio-cephalic or innominate, 480; bronchial, 482; buccal, 469
cardiac, 491, anterior, 491, great, 491, posterior, 491; cava superior, 481, inferior, 487; cephalic, 477; cerebellar, 473; cerebral, 472; choroid of brain, 473; circumflex, iliac, 485, superficial, 484; comites brachial, 478, interosseous, 478, radial, 477, ulnar, 477; condyloid posterior, 475; of corpora cavernosa, 777; of corpus spongiosum, 778; of corporis striati, 473; cystic, 491
deep, or venae comites, 465: dental inferior, 469; diaphragmatic or phrenic, 489; of diploë, 471; dorsal spinal, 482, of penis, 487; dorsalis nasi, 468, pedis, 485
epigastric, 486, superficial, 484; of eyeball, 654
facial, 468; femoral, 486; frontal, 468 of Galen, 473; gastric, 490: gastro-epiploic left, 490; gluteal, 486
haemorrhoidal inferior, 486, middle, 486, superior, 486; of head, 468: hepatic, 489
iliac, common, 487, external, 486, internal, 486; ilio-lumbar, 487; inferior cava, 487; innominate, 480; intercostal superior, 481; interlobular, 716; interosseous of forearm, 478; intralobular, 717
jugular anterior, 470; external, 470; external posterior, 470; internal, 470 of kidney, 766
labial inferior, 468, superior, 468; laryngeal, 470; lateral sacral, 487; lingual, 471; of liver, 714; longitudinal inferior, 474; lumbar, 488, ascending, 488
mammary internal, 480; masseteric, 468; mastoid, 469; maxillary internal, 469; median cutaneous, 478, basilic, 477, cephalic, 477; medulli-spinal, 484; meningeal, 469; meningo-

rachidian, 484; mesenteric inferior, 489, superior, 489
nasal, 468; of neck, 469
oblique, 491; obturator, 486; occipital, 469; ophthalmic, 475; ovarian, 488
palatine inferior, 468; palmar deep, 478; palpebral inferior, 468, superior, 468; pancreatic, 490; pancreatico-duodenal, 490; parotid, 469; peroneal, 485; pharyngeal, 471; phrenic, 489; plantar external, 485, internal, 485; popliteal, 486; portal, 465, 491; profunda femoris, 486; pterygoid plexus, 468; pudic external, 485, internal, 486; pulmonary, 465, 492; radial, 477; ranine, 468; renal, 489
sacral, middle, 487; salvatella, 477; saphenous external, or short, 485, internal or long, 484; sciatic, 486; spermatic, 488; spinal, 482; splenic, 490; stylomastoid, 469; subclavian, 479; sub-lobular, 717; submaxillary, 468; sub-mental, 468; superficial, 465; supra-orbital, 468; supra-renal, 489; supra-scapular, 470; sural, 486; systemic, 465
temporal, 468, middle, 469; temporo-maxillary, 469; thyroid inferior, 481, middle, 471, superior, 471; tibial anterior, 485, posterior, 485; transverse cervical, 470, facial, 468
ulnar anterior, 477, posterior, 477; umbilical, 738, 740
vaginal of liver, 716; vasa brevia, 428; ventricular, 473; vertebral, 479
Veins, plexuses of, ovarian, 488, 797; pampiniform, 488, 781; pterygoid, 468; spermatic, 488, 781; uterine, 487, 794; vaginal, 487; vesico-prostatic, 486
Velum pendulum palati, 685; interpositum, 538, arteries and veins of, 541
Vena cava, superior, 481, inferior, 487
Venae comites, 485; Thebesii, 491; minimae cordis, 491; vorticosae, 646

Venter of ilium, 87; of scapula, 97
Ventricular veins, 473
Ventricles of brain, third, 542, grey matter of, 543; fourth, 548, lining membrane of, 548; fifth, 540; lateral, 537; of corpus callosum, 534; of heart, left, 734, right, openings in, 731: of larynx, 746
Vermiform process, of cerebellum, superior, 546, inferior, 546
Vertebra dentata, 7; prominens, 7
Vertebrae, cervical, 6; development of, 12; dorsal, 9; false, 5, 13; general characters, 6; ligaments of, 156; lumbar, 11; pedicles of, 6; sacral, 14; structure of, 12; true, 5
Vertebral aponeurosis, 253; artery, 395; column, 19; ossification of, 13; foramen, 6; ligaments, 156; region, muscles of, anterior, 245, lateral, 246; ribs, 80; vein, 479
Vertebro-costal ribs, 80
Vertebro-sternal ribs, 80
Vertex of skull, 61
Verumontanum, 773
Vesical artery, inferior, 438, middle, 438, superior, 437; plexus of nerves, 627
Vesico-prostatic plexus of veins, 486
Vesico-uterine ligaments, 793
Vesicles Graafian, 796
Vesicula prostatica, 773
Vesiculae seminales, form and size, 785, relations of, 786, structure of, 786, vessels and nerves of, 786
Vestibular artery, 673; nerve, 674
Vestibule, of ear, 668; aqueduct of, 33; of vulva, 789
Vidian artery, 379; canal, 38; nerve, 569
Vieussens, valve of, 544
Villi, 704
Viscera abdominal, position of in regions, 692; pelvic, position of at outlet of pelvis, 827
Vitreous humour of the eye, 652; table of the skull, 3

INDEX

Vocal cords, inferior or true, 746, superior or false, 746
Voice, organs of, 741
Voluntary muscles, 209
Vomer, 56; alae of, 56; articulations of, 56; development of, 56
Vortex of heart, 736
Vulva, 788

Wharton's duct, 571
White substance of brain, chemical analysis of, 510
Winslow, foramen of, 696
Wisdom tooth, 679
Womb. *See Uterus.*

Wormian bones, 43
Wrisberg, cartilages of, 743; nerve of, 592
Wrist joint, 187

Xiphoid appendix, 77, 79

Y-shaped centre of acetabulum, 92
Yellow spot of retina, 650

Zygoma, 30
Zygomatic arch, 65; fossae, 67; lymphatic glands, 496; process, 30
Zygomaticus major muscle, 222, minor, 222

REFER TO PAGE 354

187.—The Arch of the Aorta and its Branches.

REFER TO PAGE 363

189.—Surgical Anatomy of the Arteries of the Neck.   Right Side.

REFER TO PAGE 395

197.—Plan of the Branches of the Right Subclavian Artery.

REFER TO PAGE 407

200.—The Surgical Anatomy of the Brachial Artery.

REFER TO PAGE 402

199.—The Axillary Artery, and its Branches.

REFER TO PAGE 424

204.—The Abdominal Aorta and its Branches.

REFER TO PAGE 426

205.—The Cœliac Axis and its Branches, the Liver having been raised, and the Lesser Omentum removed.

REFER TO PAGE 427

206.—The Cæliac Axis and its Branches, the Stomach having been raised, and the Transverse Meso-Colon removed.

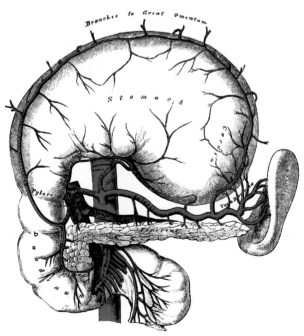

REFER TO PAGE 429

207.—The Superior Mesenteric Artery and its Branches.

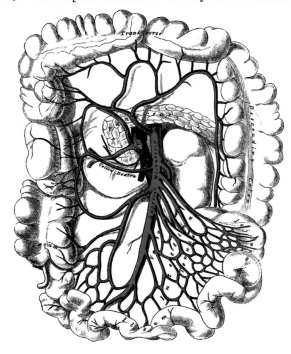

REFER TO PAGE 434

209.—Arteries of the Pelvis.

REFER TO PAGE 445

211.—Surgical Anatomy of the Femoral Artery.

REFER TO PAGE 452

212.—The Popliteal, Posterior Tibial, and Peroneal Arteries.

REFER TO PAGE 483

222.—Transverse Section of a Dorsal Vertebra, showing the Spinal Veins.

REFER TO PAGE 483

223.—Vertical Section of two Dorsal Vertebræ, showing the Spinal Veins.

REFER TO PAGE 495

227.—The Thoracic and Right Lymphatic Ducts.